Permanent Magnet Design and Application Handbook

Permanent Magnet Design and Application Handbook
Second Edition

Lester R. Moskowitz, Registered Professional
    Engineer: Pennsylvania, Wisconsin, California
Formerly Dean, Technology, Spring Garden
    College, Philadelphia, PA
President, L. R. Moskowitz & Associates,
    Consulting Engineers, Coopersburg, PA

KRIEGER PUBLISHING COMPANY
MALABAR, FLORIDA
1995

Original Edition 1976
Second Edition 1995

Printed and Published by
**KRIEGER PUBLISHING COMPANY
KRIEGER DRIVE
MALABAR, FLORIDA 32950**

Copyright © 1976 by Lester R. Moskowitz
Copyright © 1995 (new material) by Lester R. Moskowitz

All rights reserved. No part of this book may be reproduced in any form or by any means, electronic or mechanical, including information storage and retrieval systems without permission in writing from the publisher.
No liability is assumed with respect to the use of the information contained herein.
Printed in the United States of America.

> **FROM A DECLARATION OF PRINCIPLES JOINTLY ADOPTED BY A COMMITTEE OF THE AMERICAN BAR ASSOCIATION AND A COMMITTEE OF PUBLISHERS:**
> This publication is designed to provide accurate and authoritative information in regard to the subject matter covered. It is sold with the understanding that the publisher is not engaged in rendering legal, accounting, or other professional service. If legal advice or other expert assistance is required, the services of a competent professional person should be sought.

Library of Congress Cataloging-In-Publication Data

Moskowitz, Lester R., 1926-
    Permanent magnet design and application handbook / Lester R. Moskowitz. -- 2nd ed.
      p.   cm.
    Includes bibliographical references and index.
    ISBN 0-89464-768-7
    1. Permanent magnets -- Handbooks, manuals, etc. I. Title.
QC757.9.M67   1995
621.34--dc20
                                          95-8106
                                            CIP

10   9   8   7   6   5   4   3   2

Contents

    Preface to the Second Edition   vii

    Preface to the First Edition   viii

1. A New Approach to the Study of Permanent Magnets   1
2. A Brief History of the Permanent Magnet   4
3. Terms and Definitions   8
4. Classification of Permanent Magnets and Materials   11
5. Basic Manufacturing Process   22
6. Fundamentals of Magnetism   35
7. General Design Considerations   40
8. Leakage and Fringing   51
9. Circuit Effects   61
10. Exact Design Methods   94
11. Environmental Effects   108
12. Measurement and Testing   128
13. Magnetization, Demagnetization, Stabilization and Calibration   150
14. Mechanical Considerations   171
15. Standards, Specifications, and Purchasing   180
16. Typical Circuits and Applications   214

    Appendixes   271

1. Demagnetization Curves   273
2. Data Sheets—Magnetic & Physical Properties   287
3. International Index of Permanent Magnet Materials   789
4. Conversion Factors   919

    Bibliography   934

    Glossary   947

    Index   953

Preface to the Second Edition

Almost 20 years have passed since the first edition (1976) of this handbook. During that time there have been great changes in the available materials, properties, producers and applications of permanent magnets. The reprint with new addenda (1986) and this edition have not been changed as to the basic principles of permanent magnet design since these principles are still valid. Also, it was impractical to expand this edition to cover even a fraction of the new applications for permanent magnets, many of which were made technically and economically possible by the availability of new, very high energy materials. Most of the applications covered in this edition are still in use, albeit with more power, reduced size and/or reduced cost.

The first edition as is this edition is dedicated to C.D. Sloan, my primary mentor in magnetics and to my late wife, Dorothy Bay Moskowitz.

A special acknowledgment is made in this edition to my dedicated secretary, Mary Jane Wagner who was my right hand, half of my brain, caught my mistakes and without whom this edition would not have been possible.

Lester R. Moskowitz
Coopersburg, PA

Preface to the First Edition

No comprehensive book dealing with the very diverse subject of the design and application of permanent magnets has ever been published in the United States. The need for such a book has long been evident, and some attempt has been made to meet this need in the form of a few individual handbooks or company-sponsored manuals. The single, hardcover book in the field is disjointed, quite limited in scope, and assumes that the reader is almost as expert in the field as the authors. The only other published data in the field are articles in trade publications or technical journals, which, of course, deal primarily with individual aspects of permanent magnets and make no attempt to broadly cover the field.

This book is written from a practicing engineer's viewpoint. It will prove disappointing to the physicist or the metallurgist who is interested in the theory or the compositional details of permanent magnets. It will, however, prove eminently practical to engineers and inventors in all fields who need information on magnets in order to develop and design their products properly. This book will also be a useful reference to purchasing agents seeking information on sources, costs, and so on, to manufacturing engineers and quality control personnel in need of data on assembly techniques or testing, and to students who are interested in how magnets are designed and used. The book may even be of value to scientists who propose to use magnets in their own work, when their interest in magnetics is related more to how and not to why.

I have had the unique opportunity of working in virtually every phase of magnetics from the manufacturing of basic magnets to the development, design, and application of practically every type of assembly. This experience has spanned 20 years in the basic mills, in equipment manufacturing, and with magnet distributors. Being a "Johnny-come-lately" to the field has given me the advantage of recognizing the limitations of available published data and of being in a position, hopefully, to correct these limitations.

I am heavily indebted to the countless men whose work and publications form the basis of my own education in magnetics and whose inadvertent contributions to this book are immeasurable. So much of this material is included without conscious realization of the source that a detailed reference to each contributor is not possible. Such a reference list would, in all probability, be as large as the book itself. Where possible, particularly in relation to specific product designs or principles, I have incorporated the name of the individual (rather than his employer at the time) into the product or principle title.

I am particularly indebted to two people who, more than any others, made this book possible—C.D. Sloan and my late wife, Dorothy R. Moskowitz. The former taught me most of what I know about magnetics. The latter provided the encouragement and conditions that made this monumental task possible. They are, respectively, the godfather and godmother of this book.

I am also indebted to various manufacturers, who contributed illustrative materials, and to my children, who tolerated Daddy during the years of homework on his "book."

Lester R. Moskowitz
Philadelphia, Pennsylvania

A New Approach to the Study of Permanent Magnets

The use of permanent magnets in a wide variety of equipment has expanded far beyond the realization of those not intimately acquainted with this field. A.G. Clegg's estimate, published in 1962, placed worldwide usage of magnets at about 20,000 tons in the form of 800,000,000 magnets. Since 1962, heavy metal-base magnets (such as Alnico) have been increasingly displaced by the lighter weight ceramic magnets (ferrites). While this has resulted in little or no tonnage increase, the number of *individual* magnets in use has substantially increased to well over a billion units. The use of magnets in such heretofore economically impractical applications as automotive motors (windshield wipers, blowers, alternators, and so on) has resulted in a large increase in the quantity of magnets used. A better insight into the importance of permanent magnets can be gained by a consideration of the six basic functions that can be performed by a magnet (see Figure 1-1). Chapter 4 will provide a complete list of permanent magnet applications by industry and specific product. The varied forms of the basic magnets used in these applications are shown in Figure 1-2.

With so many magnets used in so many places, it is surprising that so little coordinated information about magnets is available to the potential user. Permanent magnets have, up to the present, been a "nether-nether land" in the technical fields. Permanent magnet design and application is not a part of any particular curriculum in any U.S. institution of higher learning [1] nor is

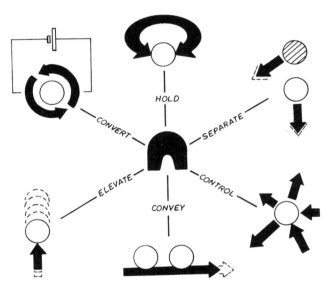

Figure 1-1. Six basic functions that can be performed by a permanent magnet.

such an elective offered in any of these institutions. In the course of most electrical engineering curriculums, permanent magnets may receive a casual mention—usually in the study of D'Arsonval meters, galvanometers, or specialty motors. Similarly, electrical engineering textbooks usually dismiss permanent magnets with a few general statements that acknowledge their existence but imply that they are unimportant in modern technology. Physicists and metallurgists devote more attention to magnetics than most other disciplines,

---

1. The exception to this is Spring Garden College, where C.D. Sloan and the author have developed and taught a major course in magnetics. This course provides substantial coverage of permanent magnets at the practical level and is offered as a junior or senior elective to both electronic/electrical engineering technology and mechanical engineering technology students.

Figure 1-2. Arnold Engineering Magnetron surrounded by the many shapes of cast and sintered Alnico magnets. (Courtesy of Arnold Engineering Co.)

but their approach is primarily directed toward the theoretical or compositional aspects. Most of those regularly involved in permanent magnet design and application are electrical engineers, mainly because permanent magnet design shares many similarities with electromagnet design and with electrical circuit design. Also, many applications of permanent magnets are components in or are related to electrical equipment. Evidence that there are exceptions to this rule is provided by the fact that one such "expert" is a chemist by training.

While most magnet experts do not share a common technical background, they do share other characteristics that may explain, in part, the lack of more and better information in this field. First, there are few broad-gage experts in magnet design and still fewer in magnet application engineering. It would be virtually impossible to find more than ten men in the United States who can properly design a permanent magnet for a wide variety of products in which the magnet is a critical operational element.

A characteristic most magnet experts have in common is that they became experts by the "mill" route; that is, they developed their expertise as employees of basic magnet *materials* (or magnet) producers rather than through experience with magnetic *equipment* producers. This mill background has produced experts who are so highly *magnet* oriented that they often fail to adequately consider the magnet as a part of a total *system* or product.

Another common characteristic of most magnet experts is an unwillingness or inability to communicate their knowledge to others. An occasional oral or written presentation by such experts is either so general that it is useless or so technical that it is unintelligible. Such experts speak and write solely for each other, and even then, on occasion they do not understand one another.

Still another common characteristic of most magnet experts is their insistence on "super" precision. In speaking or writing in the field, they are so determined to be 100 percent correct that they usually lose the nonexpert in a veritable swamp of detail. If they cannot show the nonexpert every exact, minute detail of how to design a magnet, they will show him nothing!

The approach of this book is to cover permanent magnets as:

1. an *interdisciplinary* subject covering magnetic electrical, mechanical, and economic factors.
2. a "rough" science that can be quickly learned and readily used to approximate designs and application solutions.
3. a refined science that can be learned by those who have the time, patience, and need to study the subject in great depth.
4. a continuing source of ready reference data for the engineer, purchasing agent, or quality control technician.

The interdisciplinary treatment will consist of relating the magnet to the total product system and to the environment in which it is used. Factors such as the primary manufacturing methods for the basic magnet itself will be described only to the extent that they have a bearing on how the magnet is designed, which material is selected, how it is incorporated into the product, and so on. This book will cover the economic considerations of magnets and the relationship that the magnets may have to fastening, structures, and so on, but it will not delve into the detailed chemistry or internal structure of the magnet except in the most rudimentary manner. Analogies and simplifications (even if imperfect) will be used for clarification when necessary.

Magnetics will also be covered as a "rough" science, and at times such coverage may seem to border on the artistic. No attempts will be made to "prove" equations, and no variables will be considered in demonstrating design-application methods unless these variables are of sufficient significance to have a serious effect on the whole method. Using this information, the student of permanent magnets will be able to design and apply permanent magnets on a "reasonable approximation" basis. As most magnet experts will admit (if properly cornered), the most refined design calculations are often limited and the final proof of the design is the actual construction of the unit and its subsequent test.

Magnetics will be covered as a refined science for those who are interested and who require higher precision calculation methods. However, every attempt will be made to avoid unexplained "fudge factors" related primarily to experienced designers and not readily attainable by the novice.

Finally, every attempt has been made to make this book a continuing reference source. Wherever possible, graphic illustrations rather than words have been used, and these illustrations have been specifically chosen to tell a *complete* story with a minimum of references to the text. Those tables and graphs most likely to find frequent use have been clearly identified, and the entire book has been exhaustively indexed.

Above all, every part of this book has been written for the man with a general technical background but no specific magnetic, metallurgical, mechanical, or electrical background.

2

A Brief History of the Permanent Magnet

Since their initial discovery around 600 B.C. in the form of the "natural magnet" (or lodestone), permanent magnets have been a source of unending fascination and interest both to the general public and to serious scientific investigators and innovators (engineers). Since the first important systematic studies were recorded by Sir William Gilbert in 1600, scientists have been interested in the basic *cause* of magnetism. Engineers, in turn, have been intrigued by the potentialities of a phenomenon that can do work *without physical contact* and, unlike gravity, *independently of direction.*

The most powerful magnet in Sir William's day was still the *lodestone.* He described how natural magnets could be improved by "arming" them with soft iron "caps" at each end. This is the first recorded use of what engineers today call "pole pieces." Some insight into Gilbert's investigations can be gained by observing that he noted that pole pieces could increase by fivefold the "holding power" of a magnet in contact with the mass being held, but that such pole pieces would not increase the attraction at a distance ("reaching power"). As will be evident later in this book, modern engineers can and do affect the pulling and the reaching power, not simply by adding pole pieces, but also by regulating their size and configuration.

In addition to investigating natural magnets, Gilbert was very interested in the requirements for making "artificial" permanent magnets. He clearly recognized that "hard" iron was far superior in retentivity and that "soft" iron was preferable for "caps." Other discoveries by Gilbert include methods for making a permanent magnet by proper contact (with a lodestone), by forging or drawing a steel bar while it faces north-south to the Earth's magnetic field, and by cooling a red-hot bar properly oriented with relation to the Earth's field. It is interesting to note that some 350 years later geophysical scientists are using this latter method to study the age of the Earth, to establish the field reversals that have taken place in the Earth, or to locate mineral deposits.

Robert Boyle's study of the destruction of magnetism by heat in the seventeenth century was significant later as it formed the basis of the heat-treatment and "aging" steps used to make modern metallic magnets. The first "compound" magnet was made by Servington Savery around 1730. Savery bound artificial magnets (made by contact with a lodestone) that were 1/20 inches in diameter by 2¾ inches in length with their "like" poles together and placed a fitted, common pole piece on each end. This same basic method is currently in use by at least one manufacturer of permanent magnet "rolls" for sheet steel conveying. Gowin Knight (1713–1772) was probably one of the first to produce magnets for commercial purposes and enjoyed a considerable reputation in Europe supplying magnets to scientific investigators and terrestrial navigators. Knight kept most of his methods secret, but the limited information available about his methods indicate that he was one of the most knowledgeable men in magnetics in his time. Other "fabricators" who built upon the work of Knight were the George Adams' (Sr. and Jr.), J. Fothergill, Benjamin Wilson, John Canton, and John Mitchell. Christie is credited with arranging a group of magnets made by Fothergill to form a *horseshoe* shaped unit. This horseshoe magnet was used by Michael Faraday in the famous experiment in which he generated electricity by cutting magnetic lines of flux with a copper conductor (disc). Most remarkable, Knight anticipated the modern "ferrite" permanent magnets by almost 200 years. He is known to have produced iron oxide magnets by grinding the oxide to fine particles in a water slurry, reducing the mixture to a thick paste by combining it with linseed oil, molding it into shape, baking it to a solid state in a moderate fire, and magnetizing it with another permanent magnet. The magnets Knight produced by this method were exceptionally powerful and were the ancestors of today's ferrite and fine particle (Lodex) materials.

In 1750, the first book on making steel magnets was published by John Mitchell in England. From the middle of the eighteenth century until the early 1900s, England was the primary font of knowledge and the primary source of improved magnetic materials. Afterwards, the role of English investigators and producers waned and important new developments came from Japan, Germany, Holland, and the United States.

The work of Ampère, Faraday, and Maxwell, which began with Oersted's discovery in 1820 of the relationships between electricity and magnetism, led to the rapid development of electromagnets which resulted in a decline in the interest in permanent magnets. With no industrial demand for permanent magnets, the state of the art advanced slowly for more than 100 years. Work by W. Arderon (in 1758), T. Cavallo (1786), and W. Sturgeon (1846), and information published in a German handbook in 1867 were important factors in the 1930s in forming the basis for producing permanent magnets from materials other than iron and steel and for alloying iron and steel with "non-magnetic" materials. In particular, Hensler's alloys of copper, manganese, and aluminum (first announced in 1901) were the godfathers of the modern Alnico grade magnet materials.

The first commercially available permanent magnets were made of quench-hardened steels. Among these, the most satisfactory material was developed by K. Honda in 1920. Honda's magnets were based on alloys of manganese, chrome, and tungsten, along with the primary constituents cobalt, carbon, and iron. Work on precipitation-hardened permanent magnet alloys of the iron-cobalt-molybdenum groups was done in the early 1930s by W. Koster in Germany and by B.A. Rogers and K.S. Seljesater in the United States.

Interest in oxide magnets (dormant since Knight's time) was revived by the work of Y. Kato and T. Takai in Japan in 1933. These men prepared mixtures of iron and cobalt oxides by compacting and sintering and used heat treatment in a magnetic field to improve their properties. This type of material (as improved by Neel and sold commercially for some years under the trade name Hardyne) was the first of the high-coercive materials [1] that increasingly dominate applications today.

In 1952, J.J. Went, G.W. Rathenan, E.W. Gorter, and G.W. Van Oosterhout, all of the Philips Company, announced a new class of magnet materials based on barium, strontium, or lead-iron oxides. This class of materials is commonly available today in the "ceramic" or ferrite magnets. Several rubber or vinyl-bonded varia-

1. High resistance to demagnetization.

tions of these ferrites were developed at low cost for low-power "consumer" applications such as refrigerator door gaskets, toys, and so forth.

In 1937, H. Neumann, A. Buchner, and H. Reinboth in Germany developed a series of ductile, copper-nickel-iron alloys that was commercially marketed in the United States under the trade name Cunife. W.H. Dannohl and N. Neumann found that replacing the iron with cobalt in the Cunife alloys improved the coercive force, but at the expense of ductility. The material based on this concept is known by the trade name Cunico.

In 1932, H. Potter substituted silver for copper in the Hensler alloys (Cu-Mn-Al) to produce a very high, intrinsic coercive [2] material that was commercially available as Silmanol. The related work of L. Graf and and A. Kussman in 1935 produced another high, intrinsic coercive alloy, platinum-iron. W. Jellinghaus extended this system by considerably improving its properties to produce commercial platinum-cobalt magnets.

In 1940, T.C. Kelsall and E.A. Nesbitt of the Bell Telephone Laboratories announced a ductile alloy group based on iron, cobalt, and vanadium. The material became available in several grades under the trade name of Vicalloy. A similar material developed for applications where high hysteresis characteristics are required is commercially available as P-6 Alloy.

Still another type of permanent magnet material was developed by C. Guilland based on manganese bismuthide (MnBi). This material, known as Bismanol, has unusually high coercive properties. In 1960, an alloy of manganese-aluminum was announced by A.J. Koch et al., but this material never achieved commercial availability.

The largest step in the improvement of permanent magnet materials since Gilbert's discoveries started with T. Mishima's development of an aluminum-nickel-iron alloy known as the Alni materials. Mishima's work led to wide variations in alloy composition, heat treatment, and forming techniques developed by other investigators. The addition of cobalt to the Mishima system by G.B. Jonas in 1939 produced the proliferation of alloys known today as Alnico(s).[3] Contributors to the development of improved (or variations of) Alnicos included D.G. Ebeling in the United States and M. McCraig in England (directional grain Alnico), and in the Netherlands A.J. Koch, M.G. Syag, and K.J. de Voss (isothermic full treatment), and H. Howe (sintering process).

A more recent magnet material is based on the work of F.E. Luborsky, L.I. Mendelsohn, and T.O. Paine on

2. See footnote 1.
3. Alnico was known by the trade name *Alni* before cobalt became a constituent. *Alnico* has always been a generic name, but it is commonly written "Alnico" rather than "alnico."

Figure 2-1. Solid cobalt, copper, iron, and a rare-earth element (either cerium or samarium) are placed in disk-shaped molds and fused by an arc furnace. When cooled to room temperature and magnetized, their strength will be comparable to that of any known magnet. (Photograph courtesy of Bell Telephone Laboratories)

elongated single-domain particles of iron or iron-cobalt in a lead matrix and is known commercially as Lodex.

Since the introduction of the Alni materials, followed shortly by the vastly improved Alnicos, the rate of development of new materials accelerated rapidly during the 1950s. Most of the material developments of the mid-1900s were made by metallurgists and metallurgical engineers who based their work on the theoretical pre-dictions which had evolved from the study of magnetism at the atomic level by physicists. Strategic war shortages of the critical metals comprising Alnico spurred the development of ferrites,[4] while the manufacturing limitations of both the Alnicos and the ferrites helped to spur the development of Lodex. These developments reached their peak in the decade from 1950 to 1960. New grades of Alnico, ferrite, and Lodex may still be introduced, but these will be variations of or improvements to existing materials rather than totally new systems. Many previously available materials have become obsolete (Cunico, Bismanol, and Vectolite are some examples). The Alnicos and the ferrites represent approximately 90 percent of the magnet materials made today.

The most recent advance in magnetic materials was the introduction of a whole new class of magnets based on the rare earths—most specifically, samarium and cerium. In 1968, the Raytheon Company began offering a samarium-cobalt magnet on a limited commercial basis. This material has a maximum energy capability approximately twice that of the best available grade of Alnico and six times that of the best available grade of ferrite. In 1969, the Sel-Rex Corporation—another new entry to the magnetic materials business—began offering a cerium-copper based material developed by the Bell Telephone Laboratories (see Figure 2-1). The rare-earth materials, while of great technical significance, have not achieved wide usage due to their high cost (80-100 times the cost of Alnico and 175-200 times that of ferrite!)

It may be of historical interest as well as useful for

---

4. *Ferrites* is also a generic name, but it is not usually capitalized unless used with a specific grade number.

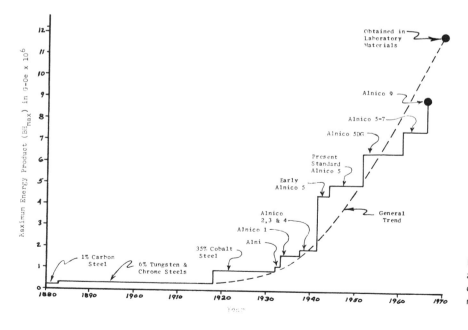

Note: By 1985 Magnet Materials with Maximum Energy Products of 35 MGO have become commercially available. See Addendum to Appendix 2

Figure 2-2. The progressive increase in available energy (BH_{max}) for commercially common permanent magnet materials.

future planning that an extrapolation in 1967 by the author of a plot of the historical advances in magnet material predicted that the "next" material would have the performance level that is actually achieved by samarium-cobalt. This plot is shown in Figure 2-2.

The role of the engineer in the history of permanent magnets has been primarily that of applying available magnet materials. However, he has also played an important role in developing commercial manufacturing processes for new materials.

Related Readings and References

1. E.N. da C. Andrade, "The Early History of the Permanent Magnet," *Endeavor*, Vol. XVII, No. 65 (January 1958). A detailed history. The author is a professor at the Imperial College of Science and Technology, University of London, England.

2. "Magnetic Materials in the Electrical Industry," (London: Macdonald & Company, 1955), p. 73. A table of applications.

3. Sir William Gilbert's original investigation, "De Magnete Magneticisque Corporibus et de Magno Magnete Tellure Physiologia," published in 1600.

4. J. Robinson, "A System of Mechanical Philosophy," Vol. IV, No. 209 (Edinburgh, 1822).

5. George Adams, "Lecture on Natural and Experimental Philosophy," Vol. IV, No. 449 (1794).

6. W. Scoresby. "Magnetical Investigations," Vol. I (London, 1844).

7. Patents (United States)

K. Honda	1,338,132	(cobalt steels)
K. Honda	1,338,133	(cobalt steels)
K. Honda	1,339,134	(cobalt steels)
Neel	2,463,413	(iron-cobalt oxides)
Neel	2,543,178	(iron-cobalt oxides)
Kelsall and Nesbitt	2,190,667	(Vicalloy)
Kelsall and Nesbitt	2,298,225	(Vicalloy)
Guillaud	2,576,679	(manganese bismuthide)
Mishima	2,027,994–9	(Alni)
Jonas	2,295,082	(Alnico)
Hansen	2,499,860–2	(effects of the addition of minor elements to Alnico)
Ebeling	2,578,407	(directional grain Alnico)
Howe	2,192,743	(sintered Alnico)
Neel	2,497,268	(fine-particle magnets)

8. R.M. Bozorth, *Ferromagnetism*, New York: Van Nostrand, 1955.

9. Y. Kato and T. Takai, *IEE Journal* (Japan), No. 53 (1933), 408–412.

10. NAVORD *Reports* 2440 and 2686 on manganese bismuthide.

11. Mendelsohn, Lubovsky, and Paine, *Journal of Applied Physics*, No. 26 (1955), 1274, and No. 28 (1957), 344, on Lodex.

Update:

As was predicted in the first edition (1976) and the 1986 edition, commercially available permanent magnet materials have attained the 35 million Gauss-Oersted level. Other important changes that have occurred over the last 10 years are:

- **The ownership of many U.S. mills and distributor/fabricators has changed, there have been new entries at the mill level along with the demise of other long established producers.**

- **There has been a world wide expansion of the number of mills that produce permanent magnet materials.**

- **There has been an increased production and use of all Rare Earth materials, particularly Neodymium-Iron-Boron, both in sintered and bonded form.**

- **There has been a major increase in the "customization" of materials with each mill producing many materials with varied properties and using separate (instead of generic) designations and individual trade names.**

- **A serious lag in the availability and/or publication of technical information in such respects as temperature effects, mechanical strength, etc.**

3

Terms and Definitions

The permanent magnet industry, like most high technology fields, has its own organizational structure and specialized terms (both technical and nontechnical) and a knowledge of these is necessary for effective design, application, and purchasing. Terms will be defined and their appropriate symbols will be shown as the need occurs in each chapter. In addition, a *complete* glossary of symbols, terms, and definitions is provided in the back of the book.

The magnet industry structurally consists of *mills, distributor-fabricators, and users. Mills* are the primary producers of permanent magnets from pure raw materials (iron, nickel, barium carbonate, and so forth) to substantially the final, formed shape. All mills, depending upon the materials they produce, are essentially large-scale, job-shop type ferrous foundries, ceramic plants, and/or metal rolling mills. Normally, mills also have facilities for cutting, surface grinding, testing, and magnetizing the magnets they produce. Their normal production includes bars, slabs, rings, horseshoes, and special shapes *in relatively large quantities.* They are also the producers of very large magnets in small quantities. Mill delivery times, depending upon tooling availability, are from six to fourteen weeks. A complete list of U.S. and foreign mills and the materials they produce appears in Chapter 15.

Distributor-fabricators purchase magnets in quantity from the mills. Usually these are "standard" magnets, or magnets for which proprietary tooling is already available. Distributor-fabricators serve two functions. First, they provide a stocking, small quantity source for standard magnets. Second, they can dimensionally modify "standards." Their operations essentially consist of cutting, grinding, and magnetizing facilities geared toward rapid delivery. Distributor-fabricator delivery times are normally twenty-four hours to two weeks.

Some distributor-fabricators distribute the magnets of only specific mills. Others do not indicate their mill sources, but obtain magnets from any available source that is economically advantageous. A complete list of distributor-fabricators also appears in Chapter 15.

Users are all those manufacturers in whose products permanent magnets are incorporated as a component. Users range from the manufacturers of such sophisticated precision products as seismographs to producers of children's toys and novelties.

The following technical and nontechnical terms must be defined now in order to understand the next few sections.

Ferromagnetic Materials: Those elements, alloys, or compounds whose inherent atomic structures (uncompensated electron spins and specific dimensional characteristics) create an inherent, substantial magnetic "moment" or force. Ferromagnetic materials are attracted to or may become magnets. Iron, nickel, cobalt, and barium ferrite are examples of ferromagnetic materials.

Nonmagnetic Materials: Those materials whose atomic structures produce little or no inherent magnetic moments. Nonmagnetic materials are not affected by a magnet and can never produce a magnetic field. Nonmagnetic materials include aluminum, copper, brass, zinc, plastics, and wood.

"Soft" Magnetic Material: A relative term used to describe a ferromagnetic material in which all moment alignment is substantially lost when the external (magnetizing) field is removed. The term "soft" is relative, since it depends on the product application for which the material is considered. In permanent magnet design and application, all grades of iron and mild steel (for example, C-1010, C-1018) are considered "soft."

"Hard" Magnetic Materials: Those ferromagnetic materials that retain a substantial degree of magnetic moment alignment after the external (magnetizing) field has been removed. "Hard magnetic material" is synonymous with "permanent magnet."

Permanent Magnet Materials: Any combination of elements that, when properly processed, are capable of retaining a magnetic field after exposure to an external field.

Properties: The *inherent* capabilities or characteristics unique to a particular material. These may be magnetic, physical, or chemical properties.

Orientation of a Magnet (magnetic material): The alignment of all or substantially all of the available "elemental magnets" (ferromagnetic domains) within a material. This alignment is accomplished in the basic manufacturing process. Magnets in which such alignment has been accomplished are also called *anisotropic* magnets; they have greatly enhanced magnetic properties in the direction of orientation and greatly reduced properties in all other directions.

Unoriented Magnets (materials): Magnets that are manufactured without any specific alignment of domains. These are also called *isotropic* magnets; have substantially the same magnetic properties in all directions.

Basic Permanent Magnets: For purposes of this book, we will define this as any physical entity *made entirely of a single permanent magnet material.*

Performance: The ability of a basic magnet or magnetic circuit to produce a specific field level in an air gap, to do work, and so forth, under specific conditions.

As-Cast Surface or *As-Pressed Surface:* The physical surface of a magnet as it emerges from the basic production process and before any finishing operation (except cleaning) is performed. The exact surface finish obtained will vary with the process used, the size of the magnet, a particular mill's capabilities, and other factors.

Ground Surfaces: Those surfaces of any magnet that have been subjected to a secondary finishing operation. Since the most widely used and highest performance magnet materials are *physically* too hard to be finished by any process except abrasive machining (grinding or cutting), the term "ground surface" is applied to any machined surface.

Standard Magnet: Either a basic magnet alone or a magnet assembly whose performance is precisely known and which is commonly available in any quantity without tooling charges from several commercial sources. The term may also refer to precision magnets usually available as calibrated *assemblies* and used as references for magnetic testing and measurement. Commonly available commercial *basic* magnets are also known as "stock" magnets, although both the terms "standard" and "stock" are misnomers.[1]

Special Magnet: Either a basic magnet made to a specific customer's order or a standard magnet magnetized, calibrated or stabilized in a specific unusual way. A special magnet may be unique in dimensional, magnetic, or mechanical properties.

Large Magnets: In general, those magnets weighing ten or more pounds and/or measuring at least $1 \times 5 \times 5$ inches in overall dimensions. Magnets weighing less than ten pounds or measuring less in overall dimensions are considered "small."

Small Quantities/Large Quantities: A function of magnet size and material type. In general, however, for small magnets, 1000 pieces or less are considered "small quantity"; for large magnets, quantities of 100 or less are considered small.

Magnet Assembly: A physical entity comprised of one or more basic magnets plus auxiliary parts to direct, contain, shunt, or control the magnetic field. A magnet assembly may also include any enclosures, covers, and parts necessary to secure assembly.

Magnetic Field: A natural phenomenon where, under certain conditions, energy or force transfer can occur through space. A magnetic field is similar to, but not the same as, a gravitational field in that it cannot be measured directly but only established by its effect.

Magnetic Flux (ϕ): A concept contrived by early magnetics investigators in an attempt to describe a magnetic field.

Magnetic Circuit: The sum of the total number of paths the magnetic flux may follow as it passes from and returns to its points of origin (or to and from some selected reference point). A magnetic circuit includes the magnetic flux source (basic permanent magnet) and any pole pieces or any ferromagnetic parts that are carrying some portion of the flux. Magnetic circuits may be either *open circuits* or *closed circuits*. Closed magnetic circuits are applicable only in certain testing procedures and in computer memory cores. Most work-

---

1. A number of basic sizes and shapes of round and rectangular bars, rings, horseshoes, etc., have achieved such widespread usage that many mills have proprietary tooling for these specific shapes and sizes. These mills commonly publish catalogs of *their* "standard" magnets. As a study of these "standard" magnet catalogs reveals, there is, in fact, no true standardization.

producing circuits are *open* in that some portion of the circuit consists of an air path (gap).

Air Gap: Any nonmagnetic discontinuity in a magnetic circuit.

Saturation: A condition where *all* of the available elementary magnetic moments (domains) in a ferromagnetic material are aligned in the same direction. Complete saturation is a condition that exists only while an external magnetizing force of sufficient magnitude is being applied to the magnetic material. This external magnetizing force may be either an electromagnet or a suitable permanent magnet. When the force is removed, some or all domains (depending on whether the material is hard or soft) will normally become misaligned in a random manner.

Reluctance: The relative opposition of a material, including air, to the passage of magnetic flux. Reluctance is essentially analogous to resistance in an electrical circuit.

Keeper: Any ferrous material of sufficient mass to carry all of the available flux produced by a magnet. It is placed across the air gap and completely closes the magnetic circuit. This term usually refers to a separate, mild steel part used with a magnet for storage and/or handling. Keepers are not intended to be part of the operating magnetic circuit.

Magnetization: The process by which a permanent magnet is initially "energized." This is accomplished by subjecting the magnet to a suitably high, properly shaped field from an electromagnet (or, for small magnets, from another considerably larger permanent magnet). Magnetization may be performed on the basic magnet at the mill level, by the distributor-fabricator, or by the user (on either the basic magnet or on the magnet after it is in the sub- or final assembly). Electromagnets or permanent magnets especially designed for this purpose are called *magnetizers;* they are available in a variety of operating types and designs.

Demagnetization: Either (1) the *deliberate* application of a decaying AC field or heat to eliminate the initial magnetization, or (2) the *unintentional* reduction in the level of initial magnetization through environmental effects or the "self-demagnetization" that occurs when a magnet is open circuited. Deliberate demagnetization is accomplished by specially designed electromagnets powered by alternating current (AC) and called *demagnetizers.*

Stabilization: The deliberate treatment of a magnet by magnetic, thermal, physical, or nuclear means, or by means of time, in order to preclude or minimize future changes in performance.

Calibration: The deliberate application of a controlled, external magnetic field to reduce the level of magnetization to a precise, predetermined level.

4
Classification of Permanent Magnets and Materials

Permanent magnet materials and/or magnets made of these materials can be categorized in various ways. These classifications are important because they may relate to technical properties, application potential, or availability—all factors in the proper design, application, specification, and purchasing of permanent magnets.

Magnet materials or magnets can be categorized by:

- Application
- Shape
- Material type
- Manufacturing process
- Magnetic properties
- Magnetization direction
- Physical properties
- Electrical properties
- Relative cost
- Commercial availability

Many of these classifications are interrelated. For example, the shape of a magnet is often determined or limited by the manufacturing process used. Similarly, electrical properties are inherently established by the composition of the material. Finally, relative cost is invariably determined by the extent of industrial use and commercial availability.

Applications of Permanent Magnets by Industry

- **Aircraft/Spacecraft
- **Advertising Displays
- ***Agricultural Equipment
- **Athletic/Sports Equipment
- **Automatic Controls/Systems
- ***Automobile, Truck, and Other Transportation Equipment
- **Basic/Applied Research Equipment
- **Biomedical Equipment
- **Burglar and Fire Alarm Systems
- *Cattle Raising
- *Ceramics Manufacturing
- *Chemical Manufacturing
- ***Electric Utility Equipment
- ***Electrical Appliances
- ***Electronic/Communications Equipment
- *Food Processing (all types)
- *Foundries
- **Furniture
- **Graphic Systems
- ***Iron Ore Beneficiation Equipment
- *Meat/Meat Byproduct Processing
- *Metal Finishing (plating and painting)
- **Mining Equipment (all types)
- **Navigational Systems
- **Oceanographic Equipment
- ***Office Equipment/Machines
- ***Office/Factory Structures
- *Oil/Gas Discovery/Drilling/Refining
- *Plastics Molding
- *Printing/Lithography/Typography
- *Shipbuilding/Repair
- *Steel Fabricating (structural)
- *Steel Fabricating, Stamping, Rolling, Forging
- *Steel Manufacturing (basic)
- *Synthetic Films and Fibers Manufacturing
- *Textiles (all levels of manufacturing)
- **Tools and Dies
- ***Toys/Novelties/Games

---

*Magnets are used primarily in the basic manufacturing process.
**Magnets are used primarily as components of the final products.
***Magnets are used both in manufacturing and as components.

Applications of Permanent Magnets by Products

Accelerometers
Advertising Displays
Aircraft Fuel Controls
Aircraft Motors
Alternators
Arc Quenching Devices
Artificial Eye Controls
Athletic Force Limiting Devices
Automatic Control Devices
Automobile and Truck Emergency Lights
Automobile and Truck Signs (removable)
Automobile Windshield Covers
Birth Control (Loop) Detectors
Blood Testers
Blood Vessel Damage Repair Methods
Blood Vessel Rupture Control
Burglar and Fire Alarms
Cabinet Latches (furniture)
Can Handling Systems
Can Openers
Cardiac Pacer Manipulators
Cardiac Pumps
Catheter Guidance Systems
Ceiling Panel Supports
Chip Handling Systems
Circuit Breakers
Clutches and Brakes
Coating Thickness Gauges
Coercive Liquid Pumps
Coin Sorters and Slug Detectors
Compasses
Contactors
Container Sorting, Testing, and Cleaning
Conveyors
Convoy Markers
Coolant Cleaners
DC Motors (all types and sizes)
Deflection Yokes
Dental Plate Retainers
Dial Indicator Bases
Drill Press Bases (portable)
Electric Welding Grounds
Electrocardiographs
Electron-Beam Switching Tubes
Electronic Component Assembly Fixtures
Electronic Keyboards
Electroplating Racks
Electroplating Tank Filters
Electroplating Tank Sweepers
File Card Systems
Floculators
Floor Sweepers
Flow Meters
Fluid Line Traps
Food Contamination Protection
Foundry Flask Clamping Systems
Frequency Meters
Funeral Procession Markers
Galvanometers
Gaskets, Seals, and Closures
Generators (industrial, military, aircraft)
Grate Separators
Gyroscopes
Home Appliance Motors
Hydroencyphalic (Brain) Pumps
Hysteresis Motors
Ion Pumps
Iron Ore Beneficiation Equipment
Jewelry Clasps
Laboratory and Industrial Mixers
Lamp Holders
Latching Switches
License Plate Holders
Liquid Level Indicators and Controls
Liquid Slurry Traps
Livestock Protectors (internal and external)
Load Isolators (microwave)
Machine Tool Chucks
Magnetoresistance Devices
Magnetos (small engine)
Magnetostrictive Devices
Magnetron Tubes
Microphones
Musical Instrument Pickups
Novelties
Oceanographic Equipment
Office Partitions (temporary)
Organs (air and electronic)
Paint Spray Masks
Painting Racks
Panel Meters
Particle Accelerators
Parts Sorters
Phonograph Pickups
Photograph Film Coating and Processing
Pipe Galvanizing
Pipe Testing
Pipe Welding
Plastics Molding
Plate Magnets
Plating Racks
Polarized Relays
Polarized Switches
Portable Appliance Motors
Positional Indicators
Pressure Vessel Liquid Agitators

Printing Plate Holders
Production Control Charts
Prosthetic Devices
Radar Arrays
Recording Tapes
Recovery Tools
Reed Switches
Refrigerator Door Latches and Seals
Relays
Road Marker Placement and Recovery Systems
Road Sweepers
Runout Table Motors (steel mill)
Scales and Balances
Seismographic Devices
Serving Trays and Glass Holders
Sheet Steel, Plate Handling, and Sorting Systems
Solenoids
Spectrographs
Speedometers
Splinter Extractors (human body)
Starters (engine)
Steel Drum Handling
Stepping Motors
Stepping Relays
Synchronous Motors
Tachometers
Tape Erasers
Telephone Bell Ringers
Telephone Receivers
Television Focusing Controls
Template Holders
Testing (magnetic materials)
Textile Bobbin Tension Controls
Textile Crushing Rolls (carding)
Textile Thread Tension Controls
Thermostats
Tool Racks
Torque Couplings
Toys and Games
Traveling Wave Tubes
Vacuum Pumps
Valves
Vibratory Feeders and Conveyors
Watthour Meters (bearings and dampers)
Wave Guides
Welding Clamps and Positioners
Welding Tables
Wire Forming
Wrenches and Screwdrivers

The preceding list is as comprehensive as possible, but new applications are constantly being developed. The specific circuit details and design-application factors for many of these applications will be provided in succeeding chapters (particularly in Chapter 16, "Typical Circuits and Applications").

Classification by Shape

The classification of basic permanent magnets by shape is of considerable importance in standardizing physical descriptions for nongraphically supported specifications. The descriptions "U-shape," "arc," or "horseshoe," for example, can be a matter of individual perception that varies from person to person. A rational system of classification by shape, developed by C.D. Sloan, Westinghouse Electric Corporation, is shown in Figure 4–1. This system conforms with the accepted terminology used by the mills and can serve to eliminate considerable confusion in verbal or written descriptions.

Classification by Material Type

Modern, commercially available, permanent magnet materials can be classified into four generic categories containing several subdivisions:

Metallic Materials

1. Cast, Unoriented (Example: Cast Alnico 3)
2. Cast, Heat-treat Oriented (Example: Cast Alnico 5STD)
3. Cast, Partial Crystal Oriented (Example: Cast Alnico 5DG)
4. Cast, Full Crystal Orientation (Example: Cast Alnico 5-7)
5. Sintered, Unoriented (Example: Sintered Alnico 2)
6. Sintered, Heat-treat Oriented (Example: Sintered Alnico 5)
7. Wrought, Unoriented (Example: Silmanal)
8. Wrought, Oriented in Forming (Example: Cunife)

All of the Alnicos, rare-earth materials, platinum-cobalt, and the cobalt, manganese, vanadium, tungsten, nickel, and chrome steels fall in the above categories.

Ferrite (or Ceramic) Materials

1. Dry Pressed, Unoriented (Example: Grade 1 Ferrites)
2. Dry Pressed, Oriented (Example: Grade 7 Ferrites)
3. Wet Pressed, Unoriented (Example: Grade 1 Ferrites)
4. Wet Pressed, Oriented (Example: Grade 5 Ferrites)

All barium, strontium, or lead ferrite materials (and mixes of these) fall in the above categories. They are alternately known as "ceramic" materials because their physical properties are similar to those of porcelain or china.

	Alnicos			Alnicos, continued	
A		Blocks	J		Side Pole Rotors
B		U-Shapes or Arcs less than 180°	K		Salient or Radial Pole Rotors
C		U-Shapes 180° or more	L		Internal Radial Pole Rings
D		C-Shapes	M		E-Shapes
E		Rods or Bars (Round, square, or rectangular)		**Ferrites**	
F		Chain Cast or Break-off Bars	N		Plates, Blocks, or Slabs
			O		Disks and Rings
G		Slugs or Plugs	P		Rectangular Bars
H		Rings	Q		Motor Stator Segments or Arcs
I		Bowls or Cups			

Figure 4-1. Basic magnet shapes (as classified by C.D. Sloan, Metals Division, Westinghouse Electric Corporation).

Elongated Single Domain (ESD) Magnets

1. Unoriented (Example: Lodex 42)
2. Oriented in Pressing (Example: Lodex 32)

While ESD materials are made of pure iron or iron-cobalt alloys, they are classified separately from metallic materials since they differ in both inherent properties and basic manufacturing method.

Composite (Bonded) Magnets

1. Alnico or Rare Earth Materials in Rubber Matrix, Unoriented (no available U.S. example)
2. Alnico or Rare Earth Materials in Plastic Matrix, Unoriented (no available U.S. example)
3. Alnico or Rare Earth Materials in Rubber Matrix, Oriented (no available U.S. example)
4. Alnico or Rare Earth Materials in Plastic Matrix, Oriented (no available U.S. example)
5. Ferrite in Rubber Matrix, Unoriented (Example: Plastiform, brand name of 3M Company material)
6. Ferrite in Plastic Matrix, Unoriented (Example: Koroseal, brand name of B.F. Goodrich Company material)
7. Ferrite in Rubber Matrix, Oriented (Example: Plastiform)
8. Ferrite in Plastic Matrix, Oriented (Example: Koroseal)

The plastic-bonded materials can be further subdivided by the specific bonding material used. This bonding material may be rigid, semirigid, or an elastomer. It may also be thermoplastic or thermosetting. Many of these variations relate only to custom-molded materials.

Note: By 1985 Composite Magnet Materials of the Rare Earth Materials in various matrixes have become commercially available. See Appendix 2

A complete list of every permanent magnet material made currently and in the past in the United States (and many foreign countries) is provided in the International Index of Permanent Magnet Materials (Lester R. Moskowitz, Permanent Magnet Users Association, 1969). This Index includes the magnetic properties of each material and the producer (where trade designations are used).

Classification by Manufacturing Process

Basic permanent magnets are currently manufactured by one or a combination of the following processes:

Shell or Baked Sand Molding: Conventional shell or baked sand molding methods are used in making most cast-metal magnets. Techniques for adding special cores, inserts, or "chills" (to produce partial or full crystal orientation) are widely used. All grades of cast Alnicos may be made by this method.

Sand Casting: "Green" sand, prepared sand, or special mold "washes" may be used in this casting process. Ordinary sand casting is seldom used except for unusually large magnets. The cast Alnicos are typical of the materials made by this process.

Investment Casting: Also known as the "lost wax process," this method is used only when very high "as-cast" precision is necessary or when the complexity of the magnet's shape does not permit shell molding. Not all grades of Alnicos can be properly cast by this method.

Sintering: Dry granules of elemental materials are compacted under pressure in metal dies and then sintered to form the final part. The process is widely used with Alnicos to give a lower-grade (magnetic properties), lower-cost magnet than the equivalent cast grade. When the sintered magnets are to be oriented, the orientation is achieved by heat treatment in a magnetic field *after* pressing and sintering. This method is only suitable for relatively small magnets produced in volume.

Dry Pressing: In this process, used to make certain grades of ferrite magnets, the prepared ferrite powder is formed to the proper shape under pressure and fused into a ceramic by high temperature firing. A high degree of orientation or partial orientation may be obtained by applying of a magnetic field during pressing.

Wet Pressing: This process is used to manufacture most oriented ferrite (ceramic) permanent magnets. A water-borne slurry of prepared ferrite powder is forced into a complex metal die mounted in a hydraulic press. Most of the water is removed by pressure. A Direct Current (DC) magnetic field of the proper intensity and direction may be applied during water removal and pressing. After pressing, the "wet" part is dried and fused into a ceramic solid by high temperature firing.

Rolling: Some magnet materials, notably the wrought materials such as Vicalloy, Remalloy, and cobalt steel, can be hot or cold worked into sheet, strip, or rod forms by conventional metal-rolling mill methods after casting to a billet. In many cases, such rolling produces oriented properties in the direction of elongation. Where such orientation is undesirable, it may be eliminated by annealing. Rolling is used to form the "stampable" metallic magnets and magnetic "tapes."

Electrodeposition: Elongated particles are prepared by electrodepositing iron or iron-cobalt into mercury. The eventual powder is pressed into the final magnet shape in metal dies with or without an aligning field. This process is used only to produce the ESD magnets.

Extrusion: This process is used primarily to produce the unoriented vinyl-bonded magnet materials of the type used in refrigerator door gaskets. Extrusion methods are not normally used to produce oriented magnets or magnetic strips. Instead, conventional extrusion techniques and equipment are used.

Colandering: Conventional rubber manufacturing technology is used to make either oriented or unoriented magnetic stock. This process is used to produce rubber-bonded ferrite materials in sheet form.

Molding: This process is primarily used to produce special, low-cost, high-volume, plastic-bonded ferrite magnets. Conventional injection or compression molding methods are used.

Since the detailed steps in the manufacturing process play a key role (along with the composition) in determining the ultimate capabilities of a magnet, the processes for the most common materials (Alnico and the ferrites) will be covered in Chapter 5.

Classification by Magnetic Properties

The classification of permanent magnet materials by magnetic properties is, of necessity, both selective and arbitrary, since it is dependent upon the particular property that is of most significance to the ultimate application. The factors (and their symbols) that may be of significance are:

Maximum Energy Product (BH_{max})

Residual Induction (B_r)

Normal Coercive Force (H_c)

Intrinsic Coercive Force (H_{ci})

Hysteresis Loss (E_h)
Temperature Stability

The first three of these factors relate to the direct performance of a magnet made from a material in a particular circuit under *static* conditions (when the "load" on the magnet does not change). The intrinsic coercive force H_{ci} is related to the ability of a magnet to withstand any type of *demagnetizing* force and is a function of the "flatness" of the "demagnetization curve" for the material. Hysteresis loss E_h represents the energy loss or gain in the material (magnet) under "dynamic" conditions and is a function of the area inside the "hysteresis loop" for the material. For purposes of classification, it is sufficient to know that, in some applications, a "flat" demagnetization curve (a high H_{ci}) may be desirable; in others, it may be a serious detriment. Similarly, in some applications, a material with the highest possible hysteresis loss (E_h) may be necessary, while in others, this factor is of small significance.

Temperature stability relates to both the temporary and permanent changes in any *magnetic* property with both elevated and reduced ambients. The special effects of temperature are covered where such information is available in Chapter 11.

Current materials classified by maximum energy product
Very High BH_{max}: Samarium Cobalt, Neodymium Iron Boron.
High BH_{max}: Alnico 5–7, cast; Alnico 9, cast; cerium-copper [1]; platinum-cobalt.
Medium BH_{max}: Alnico 5, cast; Alnico 5 DG, cast; Alnico 8l, cast; Alnico 8S, cast.
Low BH_{max}: Alnico, sintered (all); cerium-copper [1]; chrome steels (all); cobalt steels (all); Cunico; Cunife; ferrites, bonded (all); ferrites, sintered (all); Lodex (all); P-6 alloy; Vicalloy.
Note: High BH_{max} = 6.5 ×10⁶ gauss-oersteds (G-Oe) or more; Medium BH_{max} = 4.5–6.5 × 10⁶ G-Oe; Low BH_{max} = less than 4.5 × 10⁶ G-Oe.

Materials currently classified by residual induction:
Very High B_r: Samarium Cobalt; Neo-Iron-Boron.
High B_r: chrome steels (all); cobalt steels (all); Alnico 5, sintered; Alnico 5, cast; Alnico 5DG, cast; Alnico 5–7, cast; Alnico 6, cast; Alnico 9, cast; P-6 alloy, cast; samarium-cobalt; Vicalloy.
Medium B_r: Alnico 2, sintered; Alnico 8, sintered; Alnico 8l, cast; Alnico 8S, cast; cerium-copper [1]; Cunife; Lodex [1]; platinum-cobalt.
Low B_r: cerium-copper [1]; Cunico; ferrites, bonded (all); ferrites, sintered (all); Lodex [1].

1. Material is listed in both categories, since specific properties vary with chemical composition and process conditions.

Note: High B_r = 9000 gauss (G) or more; medium B_r = 5000–9000 G; Low B_r = less than 5000 G.

Materials classified according to normal coercive force:
High H_c: Alnico 8, sintered; Alnico 8l, cast; Alnico 8S, cast; Alnico 9, cast; cerium-copper; ferrites, sintered (all); platinum-cobalt; samarium-cobalt.
Medium H_c: Alnico 2, sintered; Alnico 5, sintered; Alnico 5, cast; Alnico 5DG, cast; Alnico 5-7, cast; Alnico 6, cast; Cunico; Cunife; ferrites, bonded (all); Lodex (all).
Low H_c: chrome steels (all); cobalt steels (all); P-6 alloy; Vicalloy.

Materials classified according to intrinsic coercive force:
Very high H_{ci} **(flat curve):** samarium-cobalt; Neodym-Iron-B.
High H_{ci} **(flat curve):** cerium-copper; ferrites, bonded (all); ferrite 1, sintered; ferrite 2, sintered; ferrite 3, sintered; ferrite 4, sintered; ferrite 6, sintered; platinum-cobalt.
Medium H_{ci} **(flat to slight knee in curve):** Alnico 8, sintered; Alnico 8l, cast; Alnico 8S, cast; ferrite 5, sintered; Lodex [2].
Low H_{ci} **(sharp knee in curve):** chrome steels (all); cobalt steels (all); Alnico 2, sintered; Alnico 5, sintered; Alnico 5, cast; Alnico 5DG, cast; Alnico 5–7, cast; Cunico; Cunife; Lodex [2]; P-6 alloy; Vicalloy.
Note: Very high H_{ci} = 9000 Oe or more; high H_{ci} = 3000–9000 Oe; medium H_{ci} =1000–3000 Oe; low H_{ci} = less than 1000 Oe.

Materials classified according to hysteresis loss:
High E_h: Alnico 2, sintered; Alnico 5, sintered; Alnico 5, cast; Alnico 5DG, cast; Alnico 6, cast; Alnico 8, sintered; Alnico 8l, cast; Alnico 8S, cast; Alnico 9, cast; cerium-copper; ferrites, bonded (all); ferrites, sintered (all); platinum-cobalt; samarium-cobalt.
Medium E_h: cobalt steel (17 percent); cobalt steel (36 percent); Cunico; Cunife; Lodex; Vicalloy.
Low E_h: chrome steel (5 percent); P-6 alloy.
Note: High E_h = .10 Joules per cubic centimeter per cycle (J/cm³/cycle) or more; medium E_h = .025–.10 J/cm³/cycle; Low E_h = less than .025 J/cm³/cycle.

Materials classified according to *reversible* temperature stability:[3]
Very high stability: Alnico 8, sintered; Alnico 8l, cast; Alnico 8S, cast; Alnico 9, cast; cerium-copper; samarium cobalt.

2. See footnote 1.
3. The bonded ferrites are not included in this category, since the rubber or vinyl matrix precludes exposure to appreciably high temperatures.

High stability: Alnico 5, sintered; Alnico 5, cast; Alnico 5DG, cast; Alnico 5-7, cast; platinum-cobalt.
Medium high stability: Alnico 2, sintered; Alnico 6, cast.
Medium stability: Cunife.
Low stability: none.
Very low stability: ferrites, sintered (all grades).

Materials classified according to *irreversible* temperature stability:
Very high stability: Alnico 5, sintered; Alnico 5, cast; Alnico 6, cast; Alnico 8, sintered; Alnico 8l, cast; Alnico 8S, cast; Alnico 9, cast; cerium-copper; ferrites, sintered; (all grades); samarium-cobalt.
High stability: Alnico 2, sintered; Alnico 5DG, cast; Alnico 5-7, cast; platinum-cobalt.
Medium stability: Cunife.

Note: Very high stability = less than 2 percent loss; high = 2-5 percent loss; Medium high = 5-10 percent loss; medium = 10-25 percent loss; Low = 25-50 percent loss; very low = 50 percent of total loss. Both reversible and irreversible temperatures range from 20°C to 250°C.

A study of these classifications reveals that with the exception of samarium-cobalt, no single material has high properties in every category. The selection of a material for a specific application will ultimately be based upon compromises between various properties and upon economics. A comparison of these materials and of all their respective properties can most effectively be made by studying Table 4-1 at the end of this chapter.

Classification by Direction of Magnetization

In addition to composition and manufacturing process, a key element in the performance of a magnet is the direction of magnetization. There are, for example, *31* possible two-pole magnetization arrangements for a basic rectangular-shaped magnet made of an unoriented (osotropic) material.[4] If multipole (4, 6, 8, and so on) arrangements are included, the number of possibilities becomes virtually infinite. The proper magnetization direction is even more critical if the magnet is inherently oriented (anisotropic) in the manufacturing process.

The fundamental magnetization method (classification) for rectangular and ring or slug magnets is shown in Figures 4-2 and 4-3. No further classification is necessary for the other shapes shown in Figure 4-1, since the basic geometry of such shapes as horseshoes and side pole rotors directly establishes the proper magnetization method.

4. Not all magnetization possibilities are used since some arrangements have no practical value.

Classification by Physical Properties

Permanent magnet materials may be classified by various physical properties, such as density and tensile strength (see the Physical Properties Tables in Appendix 3 for a complete listing). While certain properties may be of significance in selected applications, in general, only *density, ductility, machinability* and *corrosion resistance* are factors in *most* applications. Density itself is seldom a factor, but *energy per unit volume* or *energy per unit weight* (which are related to density) are of considerable significance. Figures 4-4 and 4-5 classify in bar graph form the commercially available materials on the basis of BH_{max} per unit volume and BH_{max} per unit weight.

On the basis of *ductility*, these materials can be classified as: *Very Hard and Brittle* [5]: ferrites, sintered (all grades). *Hard and Brittle:* Alnico 5, cast; Alnico 5DG, cast; Alnico 5-7, cast; Alnico 9, cast; samarium-cobalt. *Hard and Tough:* Alnicos, sintered (all grades); Alnico 6, cast; Alnico 8l, cast; Alnico 8S, cast; chrome steels (all); cobalt steels (all); Cunico; Cunife; P-6 alloy. *Tough:* cerium-copper; Vicalloy. *Weak and Crumbly:* Lodex (all grades). *Soft and Flexible:* ferrites, bonded (all grades). *Ductile:* platinum-cobalt.

Magnet materials may also be classified by machinability:
Grind with controlled atmosphere:
Grind [6] with aluminum oxide wheel in finished state: Alnicos, sintered (all grades); Alnico, cast (all grades); cerium-copper [7]; samarium-cobalt.
Grind with diamond wheels in finished state: ferrites, sintered (all grades).
Cut [8] or stamp in finished state: chrome steels (all); cobalt steels (all); Cunife; Lodex (all grades); Vicalloy; cerium-copper.
Cut [8] or stamp before final heat treatment (grind only after heat treatment): Cunico P-6 alloy; platinum-cobalt.
Classified by corrosion resistance: [9]
Exceptional: ferrites, sintered (all grades); platinum-cobalt.
Excellent: Cunico; Lodex (all grades); Vicalloy.
Good: Alnicos, sintered (all grades); Cunife.
Fair: Alnicos, cast (all grades); cobalt steel (36 percent); P-6 alloy.
Poor: chrome steels (all); cobalt steel (17 percent).

5. Not measurable on Rockwell C scale.
6. Refers to wet or dry abrasive grinding.
7. See footnote 1.
8. Refers to conventional chip-cut machining.
9. Materials not listed are omitted for lack of information.

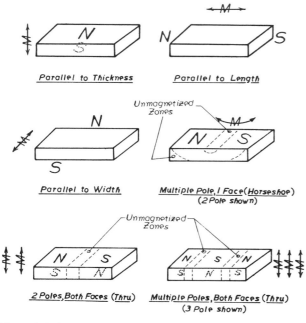

Figure 4-2. Standard magnetizing methods for block, bar, plate, and slab magnets.

Classification by Electrical Properties

In general, magnets based on alloys of pure metals are conductors and magnets based on metal oxides are nonconductors. Since some materials are composites or metals in a nonmetallic matrix, the simple categorical classifications of "conductor" or "insulator" are not always sufficient. More correctly, magnet materials classified by conductivity are:

Good conductors [10]: cerium-copper; chrome steels (all); cobalt steels (all); Cunico; Cunife; P-6 alloy; platinum-cobalt; samarium-cobalt; Vicalloy.

Fair conductors [11]: Alnicos, sintered (all grades); Alnicos, cast (all grades).

Poor conductors [12]: Lodex (all grades).

Insulators [13]: ferrite, bonded (all grades); ferrites sintered (all grades).

Classification by Relative Cost

A cost classification of *basic magnets* is virtually impossible, since the cost of such finished magnets is closely related to their shape, size, quantity, commercial availability, and many other factors. These comparisons are

  10. Resistivity of less than 40 micro-ohms/cm/cm^3 at 25°C.
  11. Resistivity of 40 to 100 micro-ohms/cm/cm^3 at 25°C.
  12. Resistivity of 100 to 200 micro-ohms/cm/cm^3 at 25°C.
  13. Resistivity of 10^4 micro-ohms/cm/cm^3 or higher at 25°C.

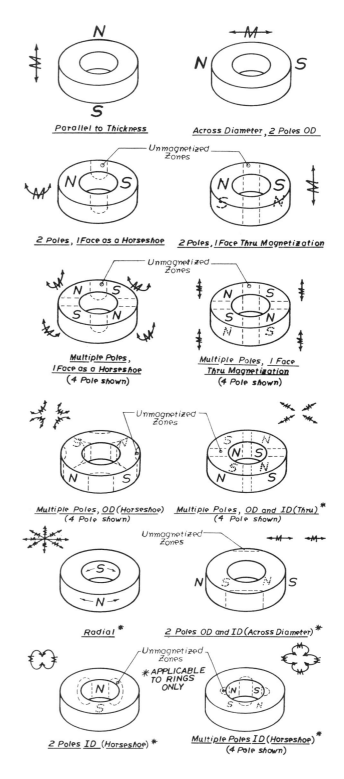

Figure 4-3. Standard magnetizing methods for ring, slug, or disk magnets (ring magnets shown).

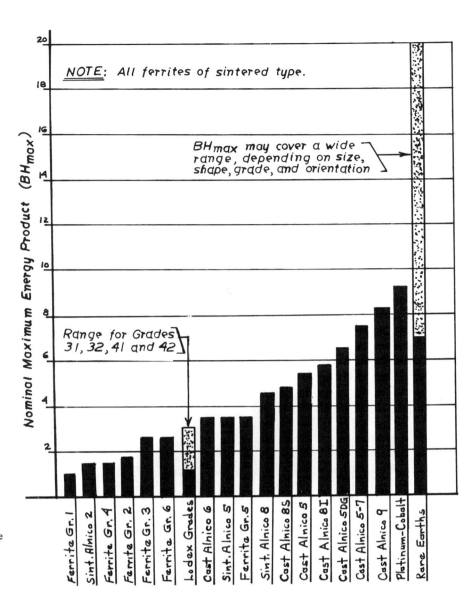

Figure 4-4. Comparison of magnet materials on an energy per unit volume basis.

further complicated by the fact that each type of material has limitations (or preferences) in shape, so that identical geometries of magnets cannot be formed from all materials.

A general classification by relative cost on a *material* basis is possible however, if shape, quantity, size, availability, and tooling costs (for nonstandard parts) are not considered. This classification is:

Very high cost (over $100/lb.): cerium-copper [14]; platinum-cobalt; samarium-cobalt [14].

High cost ($6 to $100/lb.): Alnico 5DG, cast; Alnico 5-7, cast; Alnico 8, sintered [14]; Alnico 8I, cast; Alnico 8S, cast; Alnico 9, cast; Vicalloy [14].

Moderate cost ($2 to $6/lb.): Alnico 2, sintered [14]; Alnico 5, sintered [14]; Alnico 5, cast; Alnico 6, cast; chrome steels (all); cobalt steels (all); Cunico; Cunife; Lodex (all grades) [14]; P-6 alloy [14].

Low cost $.50 to $2/lb.): ferrites, bonded (all grades); ferrite 1, sintered; ferrite 5, sintered; ferrite 2, sintered; ferrite 3, sintered; ferrite 4, sintered; ferrite 6, sintered.

Classification by Commercial Availability

A good criterion of commercial availability is the number of vendors who will supply a magnet on a normal

14. Available only in relatively small size basic magnets.

CLASSIFICATION OF PERMANENT MAGNETS AND MATERIALS 19

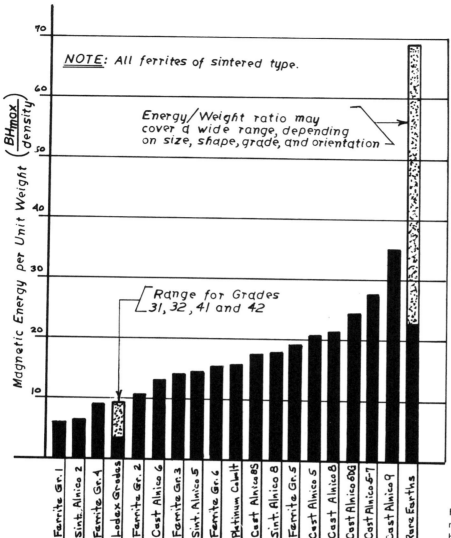

Figure 4-5. Comparison of magnet materials on an energy per unit weight basis.

basis to any potential customer. (Those producers who make magnets for in-house needs only are not included below.)

Arranged on the basis of decreasing availability, the materials are:

Ferrites, sintered: 11 sources.
Alnicos, cast (all grades): 9 sources.
Alnicos, sintered (all grades): 4 sources.
Ferrites, bonded (all grades): 4 sources.

Wrought Alloys: 3 sources
    Chrome Steels
    Cobalt Steels
    Cunico
    Cunife
    P-6 Alloy
    Vicalloy

Platinum-Cobalt: 2 sources

Rare-Earths: 2 sources
    Cerium-Copper
    Samarium-Cobalt

Lodex: 1 source

1985 AVAILABILITY UPDATE

Ferrites, Sintered	– 12 Sources
Ferrites, Bonded	– 7 Sources
Alnicos, Cast	– 7 Sources
Rare Earths, Sintered	– 7 Sources
Wrought Alloys	– 7 Sources
Alnicos, Sintered	– 4 Sources
Rare Earths, Bonded	– 3 Sources
Precious Metals (Pt.Co.)	– 2 Sources
ESD (Lodex)	– 1 Sources

Table 4-1

Material	Relative Properties*					Reversible Temperature Stability	Irreversible Temperature Stability
	BH_{max}	B_r	H_c	H_{ci}	E_h		
Alnico 2, Sintered	Low	Medium	Medium	Low	High	Medium High	High
Alnico 5, Sintered	Low	High	Medium	Low	High	High	Very High
Alnico 5, Cast	Medium	High	Medium	Low	High	High	Very High
Alnico 5DG, Cast	Medium	High	Medium	Low	High	High	High
Alnico 5-7, Cast	High	High	Medium	Low	High	High	High
Alnico 6, Cast	Low	High	Medium	Low	High	Medium High	Very High
Alnico 8, Sintered	Low	Medium	High	Medium	High	Very High	Very High
Alnico 8I, Cast	Medium	Medium	High	Medium	High	Very High	Very High
Alnico 8S, Cast	Medium	Medium	High	Medium	High	Very High	Very High
Alnico 9, Cast	High	High	High	Medium	High	Very High	Very High
Cerium-Copper, (all grades)	Low to High	Low to Medium	High	High	High	Very High	Very High
Chrome Steel (5 percent)	Low	High	Low	Low	Low	†	†
Cobalt Steel (17 percent)	Low	High	Low	Low	Medium	†	†
Cobalt Steel (36 percent)	Low	High	Low	Low	Medium	†	†
Cunico	Low	Low	Medium	Low	Medium	†	†
Cunife	Low	Medium	Medium	Low	Medium	Medium	Medium
Ferrites, Bonded (all grades)	Low	Low	Medium	High	High	Magnetic stability same as sintered ferrites. Physical stability depends on bonding material.	
Ferrite 1, Sintered	Low	Low	High	High	High	Very Low	Very High
Ferrite 2, Sintered	Low	Low	High	High	High	Very Low	Very High
Ferrite 3, Sintered	Low	Low	High	High	High	Very Low	Very High
Ferrite 4, Sintered	Low	Low	High	High	High	Very Low	Very High
Ferrite 5, Sintered	Low	Low	High	Medium	High	Very Low	Very High
Ferrite 6, Sintered	Low	Low	High	High	High	Very Low	Very High
Lodex (all grades)	Low	Low to Medium	Medium	Low to Medium	Medium	†	†
P-6 Alloy	Low	High	Low	Low	Low	High	High
Platinum-Cobalt	High	Medium	High	High	High	High	High
Samarium-Cobalt	High	High	High	Very High	High	Very High	Very High
Vicalloy	Low	High	Low	Low	Medium	†	†

*Based on claimed properties published by magnet material producer.
†No information available.

1985 Note: All rare earth materials have relative properties shown for Samarium Cobalt.

5 Basic Manufacturing Processes

There are 16 factors that determine the actual performance of a specific basic magnet in a particular circuit. These will be covered in detail in Chapter 7. One of the most important factors is the *material* from which the magnet is made. In turn, the magnetic and physical properties of the material are directly dependent on the following factors in the manufacturing process: chemical composition, crystal or particle size, crystal or particle shape, forming and/or fabrication method, and heat treatment. Some of the basic material properties change only with chemical composition and are independent of size, shape, or any other characteristic. Saturation induction, curie point, and magnetostriction are examples of such *structure-insensitive* properties. However, other properties, such as permeability, coercive force, and hysteresis loss, are very sensitive—not only to composition, but also to particle size, particle shape, fabrication methods, and heat treatment. These *structure-sensitive* properties are specifically affected by gross composition, impurities, strain, temperature, crystal structure, and crystal orientation. The effects of each of these factors are metallurgically complex and beyond the scope of this book.[1]

In most cases, the type of material to be produced will determine or at least limit the choice of the manufacturing process. For example, if the material is to be Alnico 5-7, the basic magnet can *only* be made by shell molding or by green sand casting. Furthermore, constraints may be imposed by the manufacturing process required, since it may control the shape and/or size of the basic magnet that can be produced by that process. For example, Alnico 5-7 cannot be produced in conventional horseshoe shapes or in very large sizes. Still another example of size limitation by process occurs in the manufacture of sintered Alnico magnets. Basic magnets of any grade of sintered Alnico can only be produced in sizes weighing a few *ounces*.

It is evident that the design engineer must have a sound understanding of each manufacturing process if he is to properly relate properties, shape, and size to the application. It is not unusual for engineers without this knowledge to select a material only on the basis of properties and to later learn that the constraints of geometry and size make it impossible to produce a magnet from this material.

The mill process for various magnet materials will be discussed now. The processes for the three most widely used materials—cast Alnicos, sintered Alnicos, and the sintered ferrites—will be described first and in the greatest detail. The processes for other available but less widely used materials will be described in less detail. The processes for the many obsolete materials has been omitted.

Mill Processes

Cast Alnicos are a group of aluminum-nickel-cobalt-iron [2] alloys among the most established and widely used permanent magnet materials. Twelve specific grades of this material in a cast form have been developed, but only seven are commonly available commercially at the present time. Within the historical grades, both unoriented (isotropic) materials, such as cast Alnico 2, and oriented (anisotropic) materials, such as cast Alnico 5, are possible. Furthermore, within the oriented grades, two "levels" of orientation are possible. The first level

---

1. A very complete but older textbook covering this aspect of all ferromagnetic materials is Richard M. Bozorth, *Ferromagnetism* (New York: Van Nostrand Reinhold Company, 1955). A more recent book dealing with metallurgical factors is F.N. Bradley, *Materials for Magnetic Functions* (New York: Hayden Book Company, 1971).

2. Small amounts of other elements such as copper, silicon, and titanium may be included for particular grades.

of orientation is developed by cooling the basic magnet after casting through its curie point while in a magnetic field of the proper shape and magnitude (heat-treat orientation). The second and higher level of orientation is attained by first producing elongated *crystal* growth in the material and then heat-treat orienting the completed casting. An example of a heat-treat oriented material is standard cast Alnico 5. An example of a material that is both heat-treat and crystal-oriented is Cast Alnico 5-7. The effect of heat-treat orientation alone is to substantially raise the residual induction (B_r) and the maximum energy product (BH_{max}), with some increase in normal coercive force (H_c). Similarly, the effect of both heat-treat and crystal orientation is to raise these properties further over materials that are only heat-treat oriented. It should be noted, however, that the metallurgical composition of materials intended for either type of orientation is not the same as that of unoriented materials. Further, crystal orientation is never successful when used alone, but only when combined with heat-treat orientation. Crystal orientation takes place during the actual casting process by adding "chill plates" in the molds. Heat-treat orientation is done after the casting has been made (and often after finish grinding of the required surfaces).

The unoriented grades of Alnico (1, 2, 3, 4, and 12) are not normally available in *cast* form. Certain of these grades and some oriented grades are, however, still widely used for small *sintered* magnets.

Figure 5-1 is a flow diagram of the processing steps involved in making all grades of cast Alnicos. In the case of an unoriented material, such as Alnico 3, step (7) of the flow diagram is omitted. This flow diagram does not show the step involved in producing crystal oriented materials, since this is part of the shell making process. The "chill plate" technique of drawing off heat rapidly from one plane of the material (in the orientation direction) is shown in Figure 5-2. Figure 5-3 is a photograph of a block of crystal oriented Alnico. (The block has been broken to clearly show the linear crystal growth pattern.)

It should be noted in Figure 5-2 that the degree of crystal orientation is shape and size limited. Crystals can only be grown to a limited depth (in block magnets) and cannot be grown around curves (as in a horseshoe magnet). Thus, in every case, Alnico 5-7 properties cannot necessarily be developed *throughout* the magnet. Frequently, large blocks or horseshoes will, in reality, have partially standard Alnico 5 properties and partially Alnico 5-7 properties. Such "mixed" cases exhibit properties between the two grades and are, by industry terminology, Alnico 5DG.

Since heat-treat orientation requires not only the proper *level* of magnetic field (as the formed basic magnet cools through the curie point) but also the proper *shape* of field, special "orienting" fixtures must be used. Typical orienting fixtures for the various basic magnet shapes are shown in Figures 5-4 to 5-7. Most mills that produce Alnicos have a wide variety of such fixtures to fit the various shapes and sizes normally oriented. In some cases, however, the construction or modification of a particular orienting fixture may be part of the tooling or setup charge for the order.

A potential source of variation in magnetic properties from mill to mill (or even within the same mill) is the orientation fixture used. For example, large channel horseshoe magnets may be oriented with a fixture like the one at the bottom of Figure 5-4 or the one shown at the top of Figure 5-7. Since the manual loading time for the fixture in Figure 5-7 is considerably shorter, the cost of channel horseshoes oriented by this method would be less. Note, however, that the fixture in Figure 5-7 produces a *through* field in the *entire* horseshoe. Thus, the "yoke" section will not contribute to the effective length of the final magnet. In fact, since the orientation at this yoke area is at right angles to the flux path, the yoke area will actually produce a reduction in the capability of the properly oriented "legs." Obviously, the better orientation method is the one shown in Figure 5-4.

Orientation fixtures should not be confused with *magnetizing* fixtures, which we will cover in later chapters. Magnetizing and orientation fixtures are similar in many ways, but the orientation fixture is invariably more costly and complex since it must withstand the heat of the cooling magnets. It must also *allow* the magnet to cool at the proper rate and must carry alternating current (AC) for induction heating and/or direct current (DC) for orientation *continuously* for an appreciable period of time. Water cooling of orientation fixtures is quite common, and design, construction, and useful life are serious problems for very small sizes and complex shapes of oriented magnets. Orientation fixtures and the necessary electrical and heating capabilities are not part of the facilities of distributor-fabricators but are normally available only at the mill level.

Finally, it should be noted that the magnets emerge from the orientation process (see step (10), Figure 5-1) in a *magnetized* state. It is standard procedure at the mill level, however, to demagnetize a magnet before further processing (to facilitate handling, and so on). Thus, an extra charge is made for magnets ordered in a magnetized state, since a remagnetization step is required before packing and shipping.

Sintered Alnicos are the form of the aluminum-nickel-cobalt-iron group most widely used for the manufacture of very small magnets (up to a few ounces). The process is advantageous because it produces magnets with a

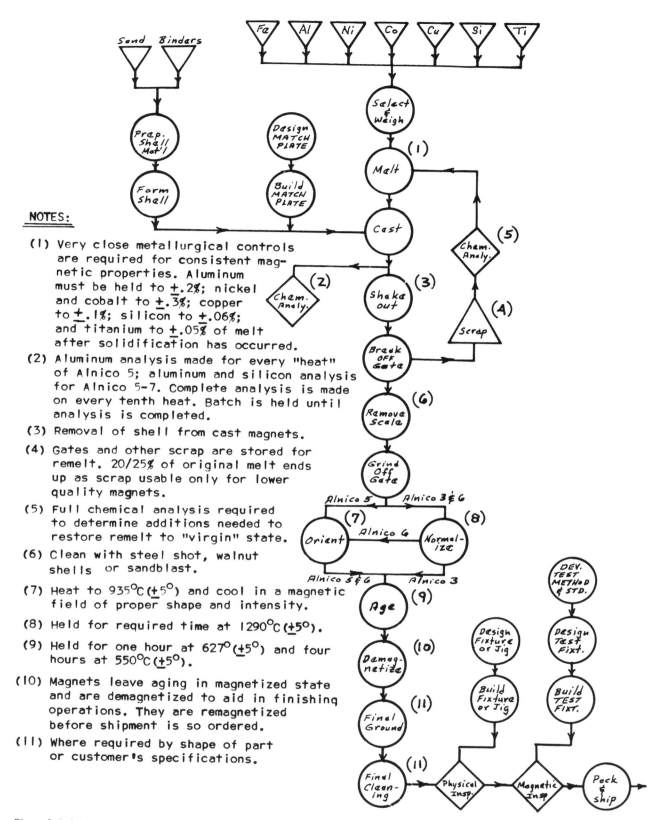

Figure 5-1. Basic manufacturing methods: process chart for cast Alnico magnets.

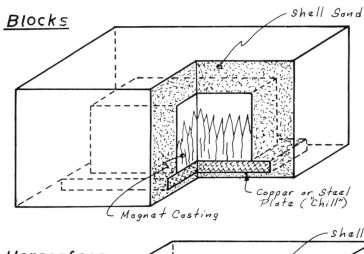
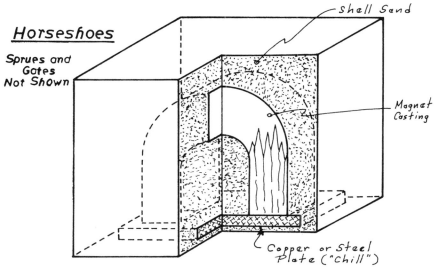

Figure 5-2. Basic manufacturing methods: producing crystal orientation in Alnico magnets.

Figure 5-3. Broken section of cast Alnico 5-7 block, showing crystal structure. (Photo courtesy of Arnold Engineering Co.)

BASIC MANUFACTURING PROCESSES 25

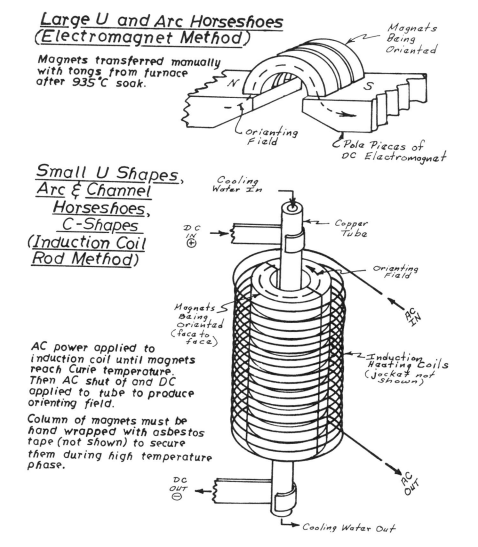

Large U and Arc Horseshoes (Electromagnet Method)

Magnets transferred manually with tongs from furnace after 935°C soak.

Small U Shapes, Arc & Channel Horseshoes, C-Shapes (Induction Coil Rod Method)

AC power applied to induction coil until magnets reach Curie temperature. Then AC shut off and DC applied to tube to produce orienting field.

Column of magnets must be hand wrapped with asbestos tape (not shown) to secure them during high temperature phase.

Figure 5-4. Basic manufacturing methods: orienting Alnico magnets. (For other shapes see Figures 5-5 to 5-7.)

better surface finish, greater homogeneity, higher tensile strength, and better dimensional control. However, compared to cast Alnicos, grade-for-grade the magnetic properties of sintered Alnicos are somewhat lower, particularly with regard to maximum energy product (BH_{max}).

The flow diagram for the manufacture of all grades of sintered Alnicos is shown in Figure 5-8. Currently, sintered Alnicos are commercially available in grade 2 (unoriented) and in grades 5 and 8 (both oriented). The primary orientation method for sintered Alnicos is by heat-treatment after forming. The orienting fixtures used are essentially the same as those for the cast Alnicos, although the small sizes of sintered magnets usually prohibit the use of more complex fixtures (see illustration at the bottom of Figure 5-5).

Sintered ferrites are a widely used group of materials commonly known as "ceramic" magnets and based on barium or strontium ferrites. Lead ferrites are theoretically usable, but no such material is currently in production. The majority of ferrite magnets are entirely of barium composition ($BaO\text{-}6Fe_2O_3$), although some mills may use a mixture of barium and strontium ferrites for some grades. Sintered ferrites are available in seven grades, ranging from the unoriented (isotropic) ferrite 1 to the highly oriented (anisotropic) grades 5, 6, and 7. Grades 2, 3 and 4 generally represent partially oriented materials that are limited to partial orientation by their specific sizes or shapes; radially oriented motor segments provide one example.

A flow diagram of the manufacturing process for sintered ferrites is shown in Figure 5-9. The high shrinkage inherent in the firing of a "wet" brick (formed in the pressing operation) and the normal "miniscus" effect preclude the production of basic ferrite magnets *directly* to close tolerances. After final forming, the

Rods, Bars, Slugs, Side Pole Rotors, 2 Pole Salient or Radial Pole Rotors, (Solenoid Method)

Radially Oriented Rings, Bowls, and Cups (Circular Electromagnet Method)

Magnets transferred from furnace manually.

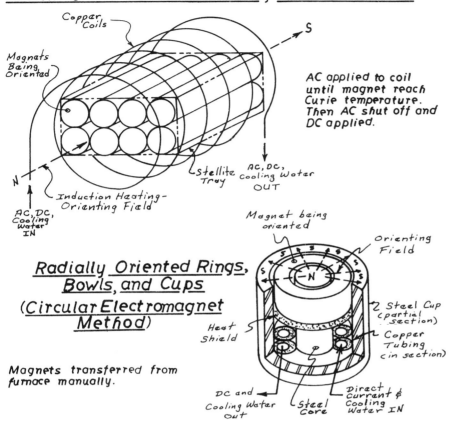

Figure 5-5. Basic manufacturing methods: orienting Alnico magnets (continued).

magnets are "wet" ground with diamond wheels on at least two surfaces where contact and/or dimensional control is critical. Chemical composition, particle size, particle shape, additives, pressing pressure, pressing rate, firing rate, magnet size, and configuration are all important factors in determining physical dimensions, finish, homogeneity, mechanical strength, and magnetic properties. In addition, the field strength and field shape applied during the pressing of the oriented grades materially affect both physical and magnetic properties.

Figure 5-10 shows the basic die structure and operating sequence used to press the basic magnet. It is evident from study of the process and pressing methods that the size, shape, and degree of orientation possible in ferrite magnets is highly limited. For example, wide cross sectional variations, because of either required holes or "steps," pose serious problems in controlling moisture removal during pressing and shrink rate during firing. Cracking and delamination are common quality control problems, although the effects of these defects on magnetic properties are often negligible. The final "yield" in ferrite production is closely related to part configuration and size, but seldom exceeds 95 percent. It is evident that in designing products requiring special shapes or sizes, the economics of the product can be considerably affected by the basic magnet design. The author knows of one case where the yield on a special ferrite magnet was less than 50 percent. Because of potential yield problems as well as both the complexity and the cost of adequate ferrite magnet dies, catalog "standard" magnets made from mill-owned dies should be used wherever possible. While the ferrite magnets can be "fabricated" with diamond cutting wheels (for straight cuts) and diamond core drills (for circular cuts and holes), this method of manufacture may be prohibitively expensive except for prototypes or small-volume requirements.

Another economic consideration for ferrite dies is the number of die cavities in relation to the pressing cycle (time). A small number of cavities in the die results in a high press-time cost per part. If the press-time cost is reduced by increasing the number of cavities, the die cost increases rapidly. A typical ferrite die costs upwards of $3000.

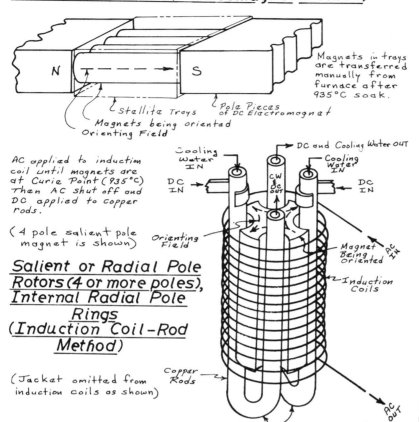

Figure 5-6. Basic manufacturing methods: orienting Alnico magnets (continued).

Bonded ferrites are made by the dispersion of barium or strontium ferrites (in powder form) in a rubber or vinyl matrix. Barium ferrite powder can be purchased from any mill producing sintered ferrites or from specialized powder producers. The bonded ferrite producer adds the matrix material and extrudes, colanders, presses, or molds the basic material stock by conventional rubber or plastics forming methods.

The magnetic properties of the bonded ferrites are similar to those of ferrite grade 1, but have a reduced maximum energy product (BH_{max}). Physical properties are determined largely by the matrix material, although the highly abrasive ferrite powder modifies some properties substantially.

Commercially available bonded ferrites are made in sheet stock form to various thicknesses or as continuous strips of a limited number of configurations and dimensions. Both the sheet and the strip may be of the oriented or unoriented type. Orientation is achieved by closely guarded techniques that are applied to the material in the *forming* stage and not to the production of the ferrite powder.

In most cases, the fabrication of the sheet stock or the strip into the basic magnet shape is done by the user or by subcontractors to the user or the mills.

Samarium-cobalt is a rare earth magnet material produced by first making a single phase alloy of $SmCo_5$ in a high-purity alumina crucible, using a radio-frequency induction furnace with helium as a protective atmosphere. The ingots cast from this melt are ground to less than 10 micron-size particles, using vibratory grinders with toluene as the fluid medium. The toluene is then removed by vacuum evaporation to produce a dry powder. This powder is pressed to the basic magnet shape in a very strong magnetic field and finally sintered in an inert atmosphere.

Other possible ways of producing basic magnets of this material are by hot pressing the powder or by extru-

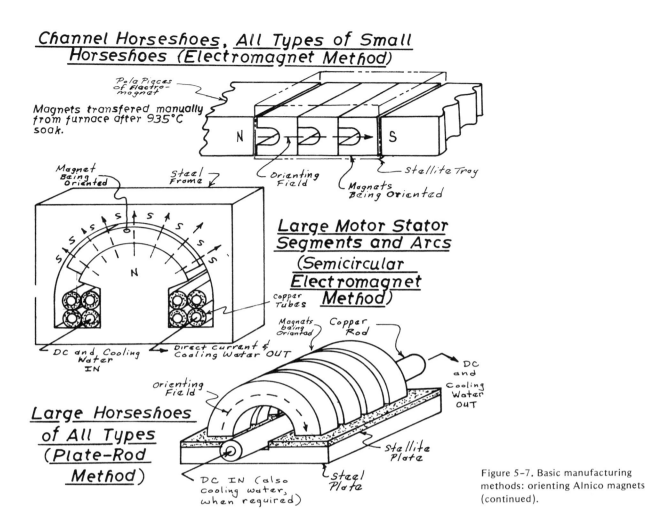

Figure 5-7. Basic manufacturing methods: orienting Alnico magnets (continued).

sion. Neither technique is currently believed to be in production, due to properties degradation and processing problems.

Particle size, partical alignment, density, and orientation field strength are important factors in producing the best magnetic properties. All samarium-cobalt magnets are of the oriented type and are limited in shape and size.

Cerium-copper is another form of rare earth magnet material. The cerium-copper manufacturing process closely resembles the manufacturing process for cast Alnicos, except that the former employs even more exacting metallurgical temperature and contamination controls. Metallic cerium-copper is cast by shell or investment techniques. The anisotropic form is heat-treat oriented in a suitably shaped, high-strength magnetic field. This form of the material is quite hard and requires abrasive machining to finish close dimension surfaces. The isotropic form of cerium-copper can be finished with conventional machine tools.

Lodex is the only one of the group of Elongated Single Domain (ESD) materials that is commercially available in the United States. ("Lodex" is a Hitachi Magnetics Corp. trade name).

This material is made by an electrolytic process where a large pool of mercury acts as a cathode for the electro-depositing of iron or iron-cobalt from an anode made of these materials. This basic "plating process" is carefully controlled to produce uniformity of composition and particle sizes of 100 to 200 angstroms (Å) in diameter that are greatly elongated axially. These elongated particles are removed from the mercury carrier by a magnetic separation process to form a heavy slurry. The particle shape is further affected and optimized by a heat treatment. Lead monoxide (PbO) is added to the slurry to surface coat the particles and to act as a binder (matrix) for the final magnet. The residual mercury content is reduced further by compacting the slurry into a solid ingot while under the influence of a strong magnetic field. The final traces of mercury are removed by a vacuum. The resulting mercury-free oriented ingot of iron or iron-cobalt in a PbO matrix is again powdered,

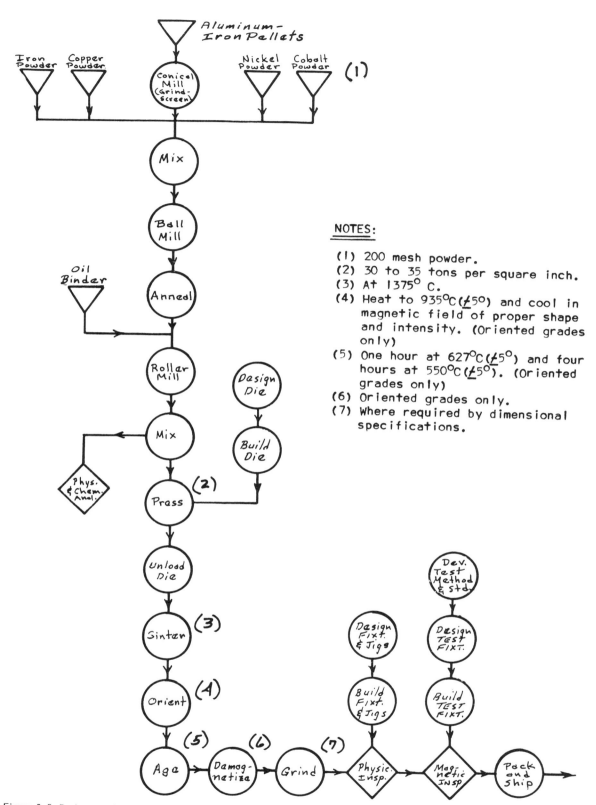

Figure 5-8. Basic manufacturing methods: process chart for sintered Alnicos.

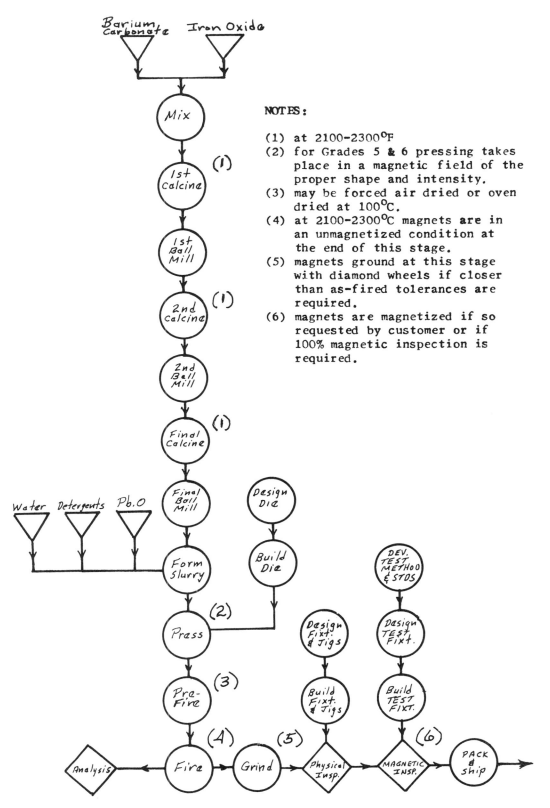

Figure 5-9. Basic manufacturing methods: process chart for sintered ferrites.

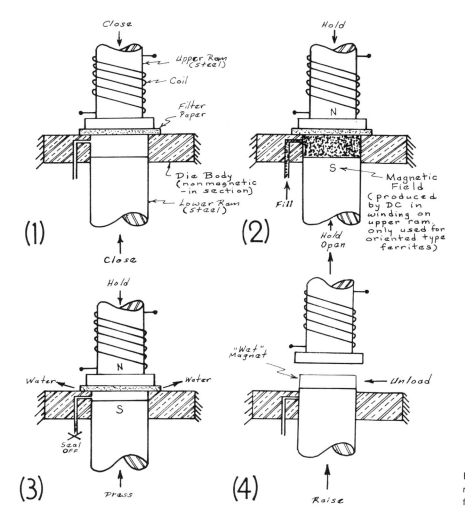

Figure 5-10. Basic manufacturing methods: pressing method for sintered ferrites.

and this powder pressed in metal dies to form the final magnet. The PbO content controls the "packing factor" in the slurry stage, and the use of (or absence of) an orienting magnetic field during actual part pressing controls the magnetic characteristics of the final magnet.

One advantage of Lodex is the great versatility possible in its properties.[3] By controlling the packing factor and the presence (or absence) of an orienting field in final pressing, a wide "family" of magnetic properties can be obtained. Currently, 11 specific grades of Lodex are offered commercially. The most important advantage of this material is the dimensional control possible in producing basic magnets without expensive final finishing operations. In precision metal dies "as pressed" the dimensions can be held to .0005 in. and intricate shapes can be produced without excessive costs. The relatively high die costs and limited sizes (small magnets only) in which Lodex can be made usually restrict the use of this material to complex, high-precision, large-volume applications. Further, the limited commercial availability (Lodex is made only by the Hitachi Magnetics Corp.) and the lack of stock sizes at mill or distributor levels present technical development problems in model making and in sample production runs.

For model or sample runs (if the basic material stock can be obtained), Lodex can be machined with conventional cutting tools. However, great care and expertise must be exercised to avoid "crumbling" of the material during machining.

Chrome steels and cobalt steels are manufactured by the same basic process used for general-purpose, high-alloy steels.

Conventional melting procedures are followed by casting or "hot" forming into bars, sheets, or plates at 900°C to produce the final magnet shape. A final heating to 800 to 900°C followed by a rapid oil or water quench is necessary. The composition and/or heat treatment requires close control to produce optimum magnetic properties. The final parts are machinable with conventional cutting tools even after heat treatment.

---

3. At times, this is also a disadvantage, particularly in maintaining quality.

The material has poor metallurgical stability at room temperature as its structure changes from the martensitic state. To improve temperature characteristics, the material is stabilized by temperature cycling up to its maximum expected operating limit. Both chrome and cobalt steels are available in limited quantities from magnetic materials mills specializing in wrought rather than in cast or ceramic materials.

Cunico and cunife are ductile or semiductile alloys based on groupings of copper-nickel-cobalt (Cunico) or copper-nickel-iron (Cunife). Both materials have been developed for those applications where ductility is mechanically or economically necessary, but at the expense of magnetic properties.

The constituent materials are normally first cast into ingot molds. These ingot molds must be of metal to provide a rapid "chilling" effect and to prevent separation of the precipitates. Next, the ingots are "cold worked" to an optimum size for final finishing and are given a homogenizing solution treatment at 1100°C.

Cunife materials are then quenched (from 1100°C) in oil and cold worked to final size. Cunico is cold worked without quenching to final shape, treated again with a homogenizing solution at 1100°C, and then oil quenched. Both alloys are given a final age at 600°C to develop their maximum magnetic properties. Both are metallurgically stable to 500°C.

The magnetic properties of each alloy are highly dependent upon the amount and type of cold working. The best magnetic properties for commercial Cunife are obtained in a round wire form .200 in. in diameter, which has been "cold reduced" and elongated in one direction. Cunife must be cold worked and reduced in area by 90 to 95 percent before reasonably good magnetic properties can be obtained. Where the final size of the material is more or less than the optimum .200 in. in diameter, the magnetic properties will be less than the best. Practical round-wire sizes range from a minimum of .020 in. to .250 in. in diameter. Wires with a rectangular cross section should not be thinner than .040 in. nor wider than .625 inches. Maximum practical length is approximately .12 inches. The material properties are highly anisotropic and the published properties are in the direction of elongation (rolling). Cunife 1 is the simplest copper-nickel-iron alloy. A modification of this alloy (identified as Cunife 2), obtained by adding 2.5 percent cobalt, produces a somewhat mechanically stronger alloy that is preferred for magnetic recording, since it has a higher residual induction and a lower coercive force (also a lower BH_{max}). The most readily available commercial alloy (simply called Cunife) has magnetic properties that approximate those of Cunife 1. Cunife, within the limits described above, compares favorably in magnetic properties with cast Alnico 3.

Cunife is highly machinable in all states (and after all heat treatment). Cunico is considerably harder and is easiest to machine before quenching and aging, although some machining can be done afterward. For best magnetic properties, the heat treatment should be applied to the final part.

Cunico I and Cunico II are substantially identical, except that the gross composition has been varied to provide a highly residual material in Cunico II and a highly coercive material in Cunico I. The magnetic properties of the Cunico alloys are lower than those of the Cunife alloys. Currently, Cunico is substantially obsolete. It has a decided advantage in some applications over Cunife, because Cunico is isotropic in its final form.

Vicalloy and P-6 alloy are alloys similar to the copper-nickel-iron/cobalt alloys (Cunife, Cunico) in physical properties (ductility) and in manufacturing process (cold reduction of ingots).

Vicalloy I is isotropic and Vicalloy II is anisotropic in the direction of cold reduction. Both grades can be freely machined in any condition and are normally used in wire or tape form for magnetic recordings in applications where the more common plastic-film, base-oxide, magnetic tapes are not suitable.

Magnetically, both alloys have considerably higher residual induction than Cunife or Cunico. The BH_{max} of Vicalloy I (isotropic) is approximately the same as that of Cunife 2 (anisotropic) or Cunico I. Vicalloy II (anisotropic), however, has a BH_{max} approximately 100 percent higher than the best grade of Cunife and 300 percent higher than the best grade of Cunico. The coercive force (H_c) of the Vicalloys is generally lower than the H_c for Cunife or Cunico.

Another alloy in this system, known commercially as P-6 Alloy, consists of the iron-cobalt-vanadium system with the addition of nickel. This alloy is normally made in cold-worked strip or wire form and is used where high hysteresis loss is desirable. P-6 Alloy represents a considerable improvement over 17 percent cobalt and 6 percent chrome steels originally used for this purpose. The material has the highest residual induction (B_r) of the Vicalloy-Cunife-Cunico group, with a very low H_c (under 100 Oe) and a low BH_{max} (less than $.5 \times 10^6$ G-Oe).

In the manufacture of Vicalloy, the elements are melted in an induction furnace and cast into sand or graphite molds. The ingots are hot swagged at 1000°C, either oil quenched or furnace cooled, and then given a six- to eight-hour anneal at 600°C. Vicalloy II is normally cold reduced 80 to 95 percent to develop its anisotropic properties.

P-6 Alloy is made by substantially the same methods, except that a final heat treatment of two hours at

575–600°C under a protective atmosphere is necessary after final punching, forming, or winding. The magnet must not be stressed after final heat treatment, since such stress will alter its magnetic properties.

Platinum-cobalt is a member of the precious metal magnet group. (Other versions, such as silver-manganese-aluminum, called Silmanol, are obsolete.) Platinum-cobalt is made by forming a cast ingot in a graphite mold or by sintering a mixture of platinum and cobalt powders in a reducing atmosphere. The alloy is then quenched very rapidly in a controlled atmosphere from a temperature of 1000 to 1100°C. Since the alloy is readily machinable and ductile at this stage, the required basic magnet can now be formed to size and shape. A final aging of from five to ten hours at 600°C is applied next. The aged alloy is quite hard and any machining after this stage must be done by abrasive methods. The quenching and aging rate is quite critical to develop maximum magnetic properties.

Fabricating Processes

In many cases, because of tooling costs, small quantities, or rapid availability, it becomes necessary to form the required magnet by modifying an existing available basic magnet. A common example is the forming of smaller, rectangular-shaped, sintered ferrite block magnets from much larger blocks or slabs. Such fabrication is most often performed by distributor-fabricators, although some mills are also equipped for this type of fabrication. Users generally avoid in-house fabrication because they lack the know-how, and cutting by conventional methods is difficult. Actually, many modern industrial plants with wet-type, abrasive cutting and surface machining equipment are basically capable of doing their own fabricating. With the proper equipment and a small amount of specialized knowledge, the user can effect important economic, flexibility, and delivery-time savings by doing their own fabricating. Most catalog standard magnets are readily available on a one-day, off-the-shelf delivery basis. This includes sintered ferrite slabs approximately 6 in. X 10 in. and rings up to 10 in. in diameter, both in various thicknesses.

Also, larger than standard sizes, shapes, or thicknesses can often be formed (with the specialized know-how) by soldering, adhesive bonding, encapsulation, or mechanical fastening.

These methods of fabricating will intentionally be covered later in Chapter 14, *after* the reader has achieved thorough understanding of other magnetic factors, such as circuit effects, environmental effects, and magnetization-demagnetization requirements. Without this understanding, fabrication may be a physical success and an operational (magnetic) failure.

1985 UPDATE—Rare Earth MATERIALS

The rare earth-cobalt alloys are produced either by melting the metal under an inert gas atmosphere or by reducing the rare earth oxides in the presence of cobalt and/or cobalt oxide. The alloys are crushed and milled to a fine powder of a few microns in size. The powder is then magnetically alligned during pressing to form a "green" compact. These compacts are sintered in an inert atmosphere and heat treated to increase the coercive force. The resulting magnet is hard and very brittle. Special grinding, slicing with diamond wheels, etc., is used to finish the pieces. The magnets are then magnetized in a very strong magnetic field and thermally stabilized if long-term stability is desired.

The Rare Earth Elements are: Lanthanum (La); Cerium (Ce); Praseodymium (Pr); Neodymium (Nd); Promethium (Pm); Samarium (Sm); Europium (Eu); Gadolinium (Gd); Terbium (Tb); Dysprosium (Dy); Holmium (Ho); Erbium (Er); Thulium (Tm); Ytterbium (Yb); Lutetium (Lu) and Yttrium (Y).

The most widely used rare earth element is Samarium in combination with Cobalt in an $SmCo_5$ composition.

6

Fundamentals of Magnetism

Any study of the theory of permanent magnets must begin with a *general* understanding of "what makes a magnet." While "what makes a magnet" has changed many times since the discovery of the lodestone hundreds of years ago, the current *atomic-domain theory* is adequately supported by basic research findings to a reasonable certainty of validity. Largely on the basis of the pioneering work of Pierre Weiss and its further development by many others (see Chapter 2), the current concept is that the *external* effects of a permanent magnet are the result of the internal structuring of atomic forces in the *ferromagnetic* class of materials.

In all ferromagnetic materials, particular structures can occur that are groups of approximately 10^{14} atoms. These structures, known as *domains*, represent the smallest unit of ferromagnetic material that exhibits the external properties of a basic magnet (a specific North and a specific South pole). These domains, hypothetically represented as tiny bar magnets within the larger magnet, are traditionally used to explain magnetism at the secondary-school level.

Current theory and its relationship to permanent magnets has been well illustrated semigraphically by A.F. Israelson, as shown in Figure 6-1. A careful study of this illustration will reveal that domain theory is, in a sense, a "unifying" theory, since it relates magnetism produced by permanent magnets to magnetism produced by electromagnets. According to this theory, *all* magnetism is produced by unbalanced spins or by electron flow. In a permanent magnet, the electron unbalance occurs at the domain level and is *internal* to the material. In an air-core electromagnet, the electron flow is in a conductor or wire. In the Earth, magnetism is produced by the electron flow in the molten iron core. The external field produced by adding a ferromagnetic core to any electromagnet is increased when the electromagnetic field aligns the domains always present in the material.

To relate theory to practice, it should be noted that, in industries associated with permanent magnets:

The proper structuring and control of domains is the purview of permanent-magnet *materials engineering* as applied at the mill level.

The design of specific magnets made of such materials to produce a desired end effect is the purview of permanent-magnet *design engineering*.

The design of a specific magnet and its relationship to the end function of the total product is the purview of permanent-magnet *application engineering*.

Figure 6-2 is both a graphic and a pictorial illustration of the relationship of domain theory to the curves that provide the necessary data for permanent magnet design and application engineering. Before proceeding with a study of this figure, magnetizing force (H) and induction (B) should be reviewed in the glossary of terms and definitions.

The central portion of Figure 6-2 shows what happens in *any* ferromagnetic material when a magnetizing force (H) is applied to a material that has never been subjected to a magnetizing force before. The effect, in the form of an induction (B), is plotted from $H = 0$: $B = 0$. Directly *below* the graphical plot of B and H in the figure is a pictorial representation using the bar magnet concept to show the step-by-step effects of an increase in the magnetization force. Directly *above* the plot of B and H is a more scientifically correct pictorial representation, showing the shift of domain "wall boundaries". This theory can best be understood by referring to the pictorial representation below and above the plot of B and H at each step. It will be noted that as H increases, at first there is only a small increase in B produced in the material. At this point, the magnetizing force is

36 CHAPTER 6

1

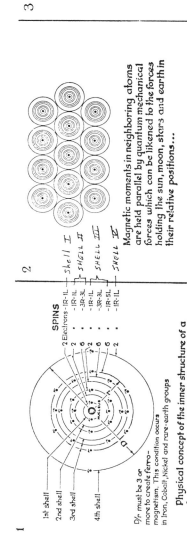

SPINS
- 2 Electrons — 1R·1L
- 2 · — 1R·1L
- 6 · — 3R·3L
- 2 · — 1R·1L
- 6 · — 3R·3L
- 6 · — 1R·5L
- 2 · — 1R·1L

D/r must be 3 or more to create ferromagnetism. This condition occurs in Iron, Cobalt, Nickel and rare-earth groups.

Physical concept of the inner structure of a ferromagnetic atom showing the electron arrangement necessary for the creation of magnetism...

The uncompensated, or off-balance, planetary spin of the electrons in the third incomplete quantum shell, together with specific dimensional characteristics creates a magnetic moment, or force....

2

Magnetic moments in neighboring atoms are held parallel by quantum mechanical forces which can be likened to the forces holding the sun, moon, stars and earth in their relative positions...

3

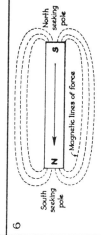

It is equally probable that magnetism will occur in any one of six directions.

DOMAIN AREA

The atoms possessing these magnetic characteristics are grouped into regions called domains... 6000 domains would occupy an area comparable in size to the head of a common pin.
A domain is composed of approximately one quadrillion (1,000,000,000,000,000) atoms...
If an atom were the size of a 1/2 inch ball, then a domain would contain enough of these balls to surround the earth with a band 30 miles wide...

4

In unmagnetized ferromagnetic materials the domains are randomly oriented and neutralize each other...
HOWEVER THE MAGNETIC FORCES ARE PRESENT!

5

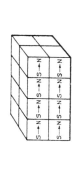

Application of an external magnetic field causes magnetism in the domains to be aligned so that their magnetic moments are added to each other and to that of the applied field.....
With soft magnetic materials such as iron, small external fields will cause great alignment, but because of the small restraining force only a little of the magnetism will be retained when the external field is removed...
With hard magnetic materials such as Alnico a greater external magnetic field must be applied to cause orientation of the domains, but most of the orientation will be retained when the field is removed, thus creating a larger permanent magnet, which will have one North and one South pole......

6

South seeking pole — North seeking pole
Magnetic lines of force

A freely suspended bar magnet will always tend to align itself with the North and South magnetic poles of the earth. — for example — the magnetic compass...
This occurs because unlike poles of a magnet are always attracted to each other by invisible lines of force whereas like poles repel each other. The earth, of course is the largest known permanent magnet...

7

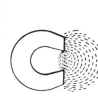

Permanent magnets can be scientifically designed and made in countless shapes and sizes to perform various tasks....
The horse-shoe shape is most commonly used in magnetic separators because its lines of force are more adaptable to the tasks which must be performed in the separation of ferrous from non-ferrous materials.
A piece of iron placed within the effective range of a magnet will, in turn become magnetized. It will have its own North and South poles which will be attracted to the parent, or larger magnet in proportion to its mass.

Data on this page compiled by
ARLO F. ISRAELSON — CHIEF ENGINEER
ERIEZ MANUFACTURING CO.
ERIE, PENNSYLVANIA, U.S.A.

Figure 6-1. Domain theory of magnetism.

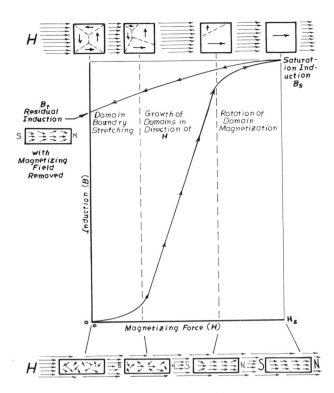

Figure 6-2. Orientation of magnetic domains.

"stretching" domain boundaries (top pictorial) or, considered another way, aligning relatively few elemental bar magnets (bottom pictorial).

As H increases further, a level is reached where small increases in H produce *large* increases in B. The rapid increase in B results from the growth of domains in the direction of H (top pictorial) or, in other words, the alignment of the majority of the elemental bar magnets (bottom pictorial).

With a still greater increase in H, a level is reached where further increases in H produce progressively *smaller* increases in B. Ultimately, a magnetizing force level is reached where further increases in H produce *no* increases in induction *in the material itself*. At this level, known as the *magnetizing force required to saturate* (H_s), all of the domains are aligned in the direction of H (pictorially shown at the top and bottom of Figure 6-2 on the far right-hand side). When this condition, known as the *saturation induction* (B_s), has been reached any further increases in H will have no effect on *domain* alignment in the material itself.

If, at this stage, the magnetizing force is discontinued and the material is a *soft* magnetic material, all induction in the material will also cease. In other words, there will be no residual alignment of domains, since the material will have returned to its original disoriented state. If, on the other hand, the material is a *hard* magnetic material, some but not all of the domains aligned during the application of the magnetizing force will *remain* aligned. The domains that remain aligned will continue to produce a *residual induction* (B_r) and will form the basis for the external field that can be produced by any permanent magnet.

Thus, it is evident that the degree of residual induction (B_r) that remains differentiates the hard magnetic materials (permanent magnet materials) from the soft magnetic materials ("mild" steels and others). Hard magnetic materials, such as Alnico 5, have a far lower saturation induction (B_s) than the soft magnetic materials but a far higher *retentivity* or residual induction (B_r). Conversely, grade C-1010 steel (one of the mild or "soft" steels) has a much higher saturation induction but virtually *no retentivity*.

The hard magnetic phenomenon described above is the basis for all permanent magnets that supply an external field. The soft magnetic phenomenon is the basis for ancillary parts used with permanent magnets to control, direct, or divert the magnetic flux (pole pieces, keepers, shields, shunts, and so on).

The S-shaped portion of the BH curve, from $H = 0 : B = 0$ to $B_s - H_s$, represents the *virgin magnetization curve* for any magnetic material. Unless the material is completely demagnetized by external means, the behavior reflected by this portion of the curve will never be repeated again.

In the preceding discussion of Figure 6-2, no reference was made to the method of *obtaining* the magnetizing force (H). This force is provided by a DC energized electromagnet (shown in the lower right-hand corner of Figure 6-3). According to Lenz' Law, an electric current flowing in a conductor produces a magnetic field. If this conductor is coiled around a ferromagnetic material or *core*, it will provide a magnetizing force to that core. In the circuit shown in Figure 6-3, a measure of the current (an ammeter) provides a direct relationship with H. If an instrument is also provided to measure the induction (B) produced in the ferromagnetic material, the relationship can be plotted between B and H. In the particular circuit shown in the figure, a Hall-effect gaussmeter element is inserted in an *infinitely small gap* in the ring-shaped ferromagnetic core. (Other methods can also be used, as we will see in Chapter 12). This particular method, while not the most accurate, is the easiest to explain to those not highly familiar with magnetic testing.

The central portion of Figure 6-3 is a plot of H vs. B. Note that the *solid line* confined to the first quadrant is identical in basic shape to the curve in Figure 6-2. Repeating the development of the first quadrant curve in Figure 6-3, zero current (I) produces zero magnetizing force (H) and zero induction (B). Note also that the

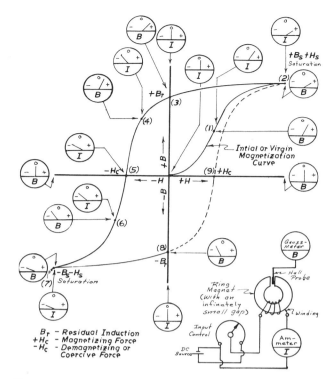

Figure 6-3. Analysis of a magnetic hysteresis loop. (The hysteresis curve shown is typical for Alnico 5.)

current is represented as flowing in one direction to produce $+H_c$. Point (1) along the initial or virgin magnetization curve represents the level of magnetizing force where the B is increasing very rapidly with small increases in H. Point (2) represents the saturation level for the material where further increases in H (or the current that produces it) result in no further increase in B. If, after saturation has been achieved at point (2), the current or magnetizing force is completely discontinued (or gradually reduced), a plot of B vs. H will produce the portion of the curve from point (2) to point (3). Point (3) represents the residual induction (B_r), which is one of the basic parameters that defines the type of material (hard or soft). The relative hardness of the material is defined by the distance from $H = 0$: $B = 0$ to point (3). With no current flow in the magnetizing coil and therefore no (H), flux will continue to flow in the magnetic material at the B_r level infinitely.

Next, consider the effect of *reversing the direction of current* in the coil and thereby reversing the direction of the magnetizing force. Since this is being applied to the same ring of ferromagnetic material that already has a residual induction (B_r), the reversed magnetizing force will, in effect, become a *demagnetizing force*, or a $-H$. The effect of this demagnetizing force is shown by the plot in the second quadrant of Figure 6-3. As the demagnetizing force ($-H$) is increased [to point (4), for example] the effect is *not* to produce an induction in the opposite direction ($-B$) but rather to *reduce the residual induction* ($+B$) in the material. Further, increasing $-H$ to point (5) will, at this stage, completely cancel the residual induction, *but only as long as this level of demagnetization is being applied by the electromagnet.*[1]

If the demagnetizing force is *not* discontinued but continues to increase in the $-H$ direction, flux will begin to flow in the ring in the *opposite* direction ($-B$) to its original flow. This reversal of flux at an intermediate point in the increase of $-H$ is shown at point (6) in the third quadrant of the figure. It should be noted that point (6) has *no* relationship whatever to point (1) in the first quadrant. Point (1) occurs on the initial or virgin magnetization curve for the ferromagnetic ring, but point (6) occurs only when the material is reversed from saturated state in one direction to saturation in the *other* direction.

As $-H$ continues to increase, $-B$ will continue to increase, as shown between points (6) and (7). Point (7) represents the level of $-H$ that has again produced saturation but now all of the domains are aligned in a direction *opposite* to their direction at point (2). If $-H$ is now, in turn, gradually reduced or discontinued, the plot of the results will be as is shown from point (7) to point (8). Point (8) represents the condition where there is no magnetizing or demagnetizing force ($I = 0$) and the residual induction in the material has reversed its direction ($-B_r$). Points (8) and (3) are identical in magnitude, except that the residual induction is opposite in direction. In other words, *the polarity of the flux has been reversed.*

The dashed line plotted between points (8), (9), and (2) in the second and fourth quadrants shows the effect of again increasing the current in the coil in a $+H$ direction. Note that the curve comprised of points (2), (3), (5), and (7) is identical to the curve comprised of points (7), (8), (9), and (2). The values of $+B_r$ and $-B_r$ are identical, and the values of $+H_c$ and $-H_c$ are identical. The parameter H_c, called the normal coercive force, is one of the basic properties of a permanent magnet material (its significance in design and application will be discussed later). The curve represented by points (2), (3), (5), (7), (8), (9), and (2) is known as the *hysteresis loop;* it completely describes the behavior of a ferromagnetic material under various conditions of magnetizing force. For purposes of permanent magnet design and application engineering, the complete hysteresis loop is not required and is seldom presented. Only the value of H_s, the *magnetizing force required to saturate the material*, and the second quadrant portion

---

1. One of the least understood facts about permanent magnets is that if the demagnetizing force is completely discontinued at point (5), the residual induction will *not* be zero. This is due to a "recoil" effect that will be explained in Chapter 11.

38 CHAPTER 6

of the curve are necessary. The second quadrant portion [points (3), (4), and (5) in the figure] is called the *demagnetization curve*. The properties of a permanent magnet material presented in tabular form, instead of as a demagnetization curve, normally state only the values of B_r and H_c and the *maximum energy product* (BH_{max}). BH_{max} is the *highest* value that can be obtained when multiplying values of B and H along the demagnetization curve. BH_{max} defines the greatest *capability* of the material, but whether that capability can be achieved in the actual magnet or magnetic circuit is determined by its ultimate design and application in a circuit. It must be stressed again that the properties that define a material, whether in tabular or in curve form, describe only the *capabilities* of that material. They do not *alone* establish the performance that will be obtained from a specific magnet made of that material.

One further important note with regard to the complete hysteresis loop. If the full value of H required to saturate the material is never applied, a hysteresis loop will be produced that falls *inside* the loop shown in Figure 6-3. Such loops within the "major loop" are known as *minor hysteresis loops*. If some level of magnetization less than saturation has been applied, then the maximum value of residual induction (B_r) of which the material is capable will never be attained. It should be noted that in a fully closed ring (zero air gap), while there can be a residual flux endlessly traversing the ring, *the circuit is incapable of doing mechanical work*. In order for mechanical work or energy transfer to take place, there must be some type of *air gap* in the circuit. The only exception to this is the closed ferromagnetic rings used for computer memory cores. (In this application, *information is stored;* work is not performed.) In other words, to do work or transfer energy, the magnetic material (magnet) must *produce a field external to itself and to do this there must be an air gap. However, the effect of any air gap is essentially the same as the demagnetizing effect produced by* -H *in a closed ring.*

Let's examine the demagnetizing effect of an air gap by again considering a closed ring of uniform cross section with a zero internal reluctance (or resistance) to the passage of flux.[2] With this condition, the *flux density* in the closed ring magnet (B_m) or, as used by some authors, B_d) will equal the residual induction (B_r) shown on the hysteresis loop or demagnetization curve. In symbolic form:

$$B_m \text{ or } B_d = B_r$$

2. This is not entirely true. However, as will become evident in later chapters, the internal reluctance of the material is usually so small compared to the air-gap reluctance that for most engineering purposes it can be neglected. Exceptions occur when the internal circuit length is very large and the air gap is very small.

WHEN $L_g = 0$ and $R_m = 0$

where L_g is the length of the air gap and R_m is the internal reluctance of the magnet.

Next, assume that a very thin saw blade is used to make a radial cut in the closed ring. The air gap so created represents an increase in the total reluctance in the circuit, since the relative ability of air to "conduct" flux is much lower than that of a ferromagnetic material. In an electrical analogy, by Ohm's Law, when the resistance (R) of a circuit increases, the flow of current (I) decreases for the same applied pressure (E). Similarly, in a magnetic circuit, with an increase in reluctance (R) caused by the introduction of an air gap, the total flux ($\emptyset$) will be reduced.[3] There is, in fact, an "Ohm's Law" for magnetic circuits:

Equation 1: $\quad \emptyset = \dfrac{F}{R}$

where $\emptyset$ is the total flux, F is the magnetomotive force, and R is the reluctance.

Obviously, if the total flux ($\emptyset$) has decreased and the cross sectional area through which it passes has not changed (the ring cross section), then the flux density (B_m) has decreased. Thus the flux density in the magnet (B_m) is no longer equal to B_r as it was for the closed ring; B_m has been reduced by some amount, depending on the *length* of the air gap (L_g) that has been introduced.

Referring again to the demagnetization curve portion of the hysteresis loop in Figure 6-3 (the second quadrant), it will be evident that the introduction of an air gap has had the same effect as applying an *external* -H would have to a *closed* ring. Stated another way, the introduction of an air gap produces a "self-demagnetizing" force which is directly related in magnitude to the length of this air gap (assuming again that the length of the magnet has remained constant). It is also evident that the greatest air gap possible will occur when the ring magnet is *straight;* in other words, when it is a *bar* magnet.

The demagnetization curves for the widely used magnetic materials appear in the Appendix. The B_r, H_c, and BH_{max}, as well as the H_s, are given in the Appendix Tables for these and many other materials.

The demagnetization curve is the fundamental tool used in designing a basic magnet to meet the specific performance requirements of a circuit. The properties tables in the Appendix are chiefly useful for comparing materials in preliminary selection.

Chapter 7 will begin (with a simplified design method) to show how demagnetization curves are used.

3. It is assumed that total length of the magnet (L_m) has not been reduced by the addition of the air gap.

7
General Design Considerations

It has often been the fervent wish of this author that permanent magnets were physically more complex. If each magnet contained a dozen semiconductors, a liquid crystal display, a gear train, cogs, and other parts, product design engineers would hesitate to simply "buy a magnet and try it." It has been the author's experience that a magnet selected in this way seldom performs the required function well, often does not perform it at all, and is not, in the end, economically feasible. As a result, many potentially sound applications of permanent magnets are abandoned at the development or design stages for other methods. Usually only in cases where there is no alternative to the unique capability of magnets *to provide a force, do work, or transfer energy without contact* does the product engineer persist until he has a functionally and economically "acceptable" magnetic system.[1] All too often, the system that results has cost a disproportionate amount of development time and does not perform as well (or cost as little) as it might have if the product manager had had the necessary expertise in magnetics. Again, it should be pointed out that there are few sources of practical, published information to provide the product engineer with such expertise. Mill-level technical assistance, once readily available, now is often not available at all or is not available quickly. Such assistance is seldom available to potentially low-volume users.

The operational complexities of permanent magnets are evident when the factors that can affect the performance of magnets are considered.[2]

---

1. Magnets are invariably a *key operational component* in any product where there is any reason at all to consider their use. Therefore, their proper function is vital to the performance (or the nonperformance) of the whole product.

2. These factors apply to the basic magnet alone. When all other factors that relate to a complete product are included, the number of factors to be considered may be gigantic!

Factors that Affect Performance

The performance of any permanent magnet is a function of the magnet itself and the total environment in which it operates. Specifically, the following factors determine and/or affect performance:

Magnet *material*
Magnet *size*
Magnet *shape*
Location of magnet *in circuit*
Level of magnetization
Location of magnetic *poles*
Magnetization before or after placement in final circuit
Material of which *auxiliary parts are made*
Shape of auxiliary parts (pole pieces, screws, and so on)
Temperature, radiation, shock, demagnetizing fields, and so on
Physical *handling during assembly*
Time
The *material* of which the part on which the magnet acts is made
The *size* of the part on which the magnets acts
The *shape* of the part on which the magnet acts
The *location* of the part on which the magnet acts

An analogy between a permanent magnet and an electric motor will further serve to illustrate the above factors. An electric motor has certain *inherent capabilities* that are determined by size, design, materials, construction, and quality. The actual work done by a motor (its specific performance under certain conditions) is a function of these capabilities, the characteristics of the power source *and load connected to the shaft*. In addition, other factors, such as ambient temperature and moisture, will affect both the inherent capabilities and the characteristics of the load.

Permanent magnets are similar to motors in that their capabilities are determined by factors that are part of their design. However, the "load" (the part on which the magnet acts) and the total environment (temperature, shock, and so on) determine the final performance within the limits of inherent capabilities. It is evident that the "systems" approach is necessary to effectively apply permanent magnets. This approach, however, has seldom been recognized by product engineers or adequately stressed by the mills. It invariably explains the common problems that occur in product engineering when permanent magnets are a component. The first step in the study of permanent magnet design methods is to become familiar with the format of the typical demagnetization curves published by the producers (mills).

Figures 7-1, 7-2, and 7-3 show one mill's typical

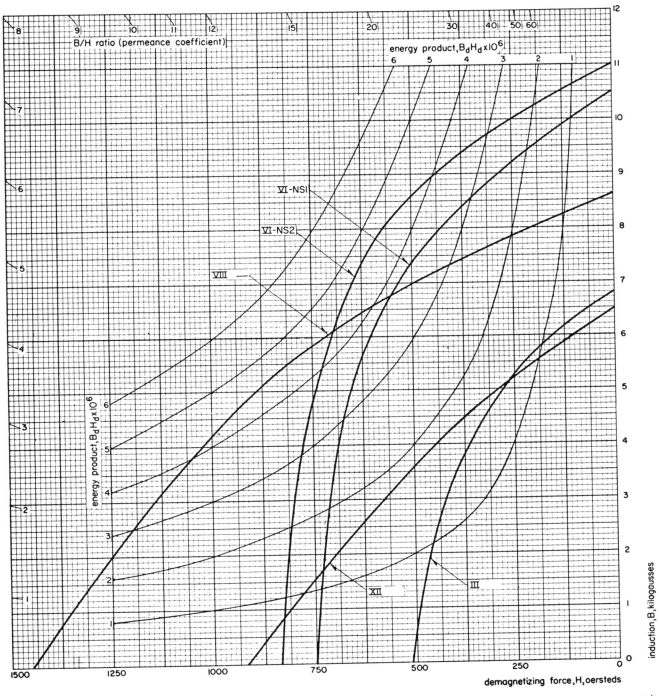

Figure 7-1. Typical demagnetization/energy product curves for cast Alnicos 3, 6, 8, and 12 (Westinghouse designations shown on curves).

GENERAL DESIGN CONSIDERATIONS 41

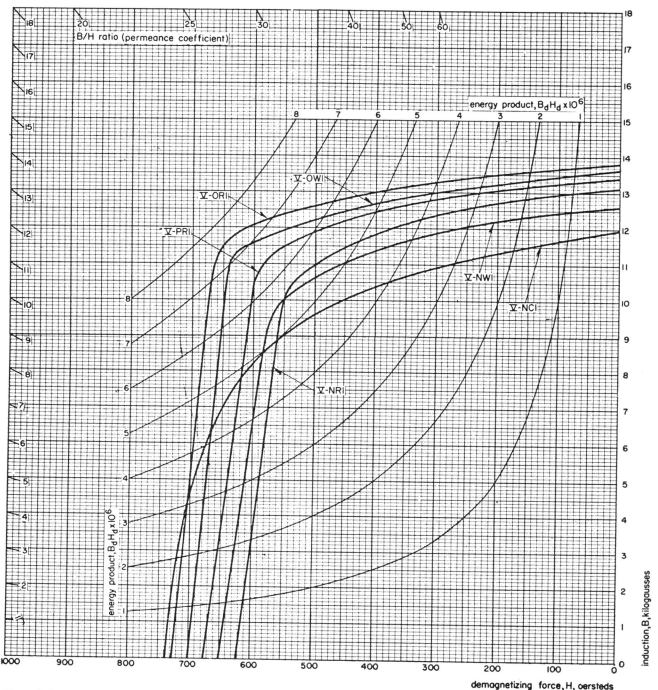

Figure 7-2. Typical demagnetization/energy product curves for variations of cast Alnico 5.

published demagnetization curves for various grades of cast Alnicos and one grade of sintered ferrite. These curves appear as *heavy* curved lines labeled with Roman numerals in the figures. We use Roman numerals here to distinguish between grade curves and scalar values (Arabic numerals are used for scales). It is commonplace for mills to use either numeral system to designate material grades.

Note that these figures are provided with four scales. The right-hand scale measures induction (B) in kilogausses (kG); the bottom scale, the demagnetizing force ($-H$) in oersteds (Oe). Two other important scales (whose use will become evident later) are also provided. The scale of *fine* solid curved lines extending from the upper-right to the lower-left corner of each graph is the energy product ($B_d H_d$) scale. The scale along the top

42 CHAPTER 7

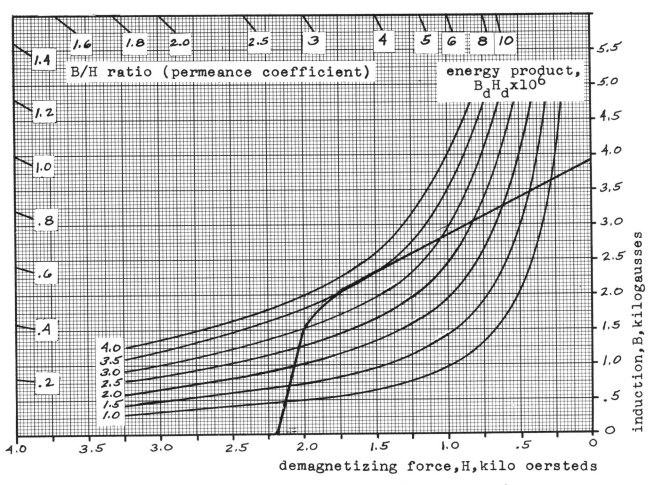

Figure 7-3. Typical demagnetization/energy product curve of grade 5 sintered ferrite (at room temperature).

and left-hand side of these curves is the *B/H* ratio or the permeance coefficient scale.

At this point, a brief review of some of the facts developed in Chapter 6 is advisable.

1. The demagnetization curves shown in Figures 7-1, 7-2, and 7-3 are *only* the second quadrant portions of the complete hysteresis loops for the materials.
2. The demagnetization curves do *not* show the saturation induction (B_s) or the magnetizing force required to saturate the material (H_s). These values can be obtained only from a properly scaled hysteresis loop or from tabular data provided by the mills themselves. However, the curves shown in these three figures assume that the material in question *was* saturated. Also, these demagnetization curves are *typical* or "average" for the particular grade of material indicated, and a normal and allowable variation (tolerance) of properties will occur. If the material has not been saturated, a "minor loop" curve within, parallel to, and of the same shape as each demagnetization curve will apply.
3. For any particular material, the point at which the demagnetization curve meets the *B* axis is the residual induction (B_r), and this is *the maximum flux density that can occur* in the magnet. B_r will occur only in a fully closed ring (where the air gap is equal to zero).
4. For any particular material, the point at which the demagnetization curve meets the *H* axis is the coercive force (H_c).
5. A magnet *with* an air gap has a "self-demagnetizing" effect that acts to *reduce* the flux density in the magnet, so that B_m (in the magnet) does not and can never equal B_r.

With these facts in mind, the purpose of the energy product scales and the *B/H* ratio scales can be better understood.

The energy-product scale is developed from multiplying specific values of *B* by the value of *H* that occurs on the *H* axis. Looking at Figure 7-1, for example,

GENERAL DESIGN CONSIDERATIONS 43

selecting a B value of 8 kG and projecting it to the H axis results in $H = 6.25$ Oe. Then, by multiplying, we obtain:

Energy Product = $(8 \times 10^3)(6.25) = 5000 \times 10^3$ G-Oe

$= 5000 \times 10^3 = 5 \times 10^6$ G-Oe

$= 5$ MGO (million gauss-oersteds)

With this energy product scale added to the demagnetization curve, the *maximum energy product* (BH_{max}) can be obtained graphically for any material by noting where the highest point of the *knee* of the material curve falls in relation to the energy product scalar plot. In Figure 7-2, for example, the highest point in the knee of the demagnetization curve for cast Alnico V-NC1 falls exactly on the 5×10^6 energy product line. This shows that the greatest energy *capability* [3] of this material per unit of volume is 5×10^6, or:

$BH_{max} = 5 \times 10^6$ G-Oe for cast Alnico V-NC1

It is an obvious goal of good magnet design to "operate" the magnet so that its maximum energy product capability is fully realized. There are, however, cases where operation at the BH_{max} point is deliberately avoided because such operation can conflict with other functional requirements (this will become evident in later chapters).

The B/H ratio is a relationship that describes the actual operating "point" of a particular magnet on the demagnetization curve for the material from which the magnet is made. Since a magnet with an air gap operates with an internal flux density (B_m) lower than B_r, this operating point along its demagnetization curve can be defined as the ratio of the B in the magnet to the H value projected on the H scale. The B/H ratio scale arrangement in Figures 7-1, 7-2, and 7-3 simplifies establishing the B/H ratios that may be desired to obtain BH_{max} or the B/H ratio for any other level of B_m. For example, to find the B/H ratio that will yield the highest BH_{max} for cast Alnico V-OR1 (see Figure 7-2), a straight line is drawn from $B = 0 : H = 0$ to the highest point of *its* demagnetization curve and a scaler reading of $B/H = 18$ is obtained.

Conversely, consider a particular magnet made of the same Alnico but operating with a B/H ratio of 13. Under these circumstances, the greatest energy product possible is 6×10^6 G-Oe. Also note that the magnet is operating below the knee of its demagnetization curve.[4]

Simplified Method of Design Calculation

From the study of magnetics to this point, it is evident that the B/H ratio is the key to how a basic magnet will perform in a circuit. It is, in fact, the key to selecting the most appropriate *material* from which to construct a magnet. The B/H ratio is also known as the *permeance coefficient* or the *load line*. This author prefers the designation B/H ratio, since this accurately describes what it *is*. Load line is a second preference, since it describes the functional aspect of the B/H ratio. Permeance coefficient is nice "jargon," but really describes nothing; furthermore, it can easily be confused with such terms as "permeate" and "permeability," which have entirely different meanings.

When applied to the demagnetization curve of a particular material, the B/H ratio or load line defines what a specific magnet will *do* in a specific circuit. The B/H ratio is a parameter that is *entirely determined by the* physical *relationships of the basic magnet and the* total *magnetic circuit in which it functions*. The B/H ratio is *independent of the material* from which the magnet is made.

This is most evident when a few basic magnet shapes are considered *without* any associated circuit (no pole pieces, no attractive object, and so on). Figures 7-4, 7-5, 7-6, 7-7, and 7-8 are plots of the B/H ratio for five simple magnet shapes. These plots are based on the equations which appear in each figure.[5] Note that all of the terms in these equations relate to length, diameter, thickness, width, or other *dimensional* terms. There is a small, "material-*type*" related difference (see Figure 7-4) that is significant only for round bar magnets, so that, in this case, separate curves are indicated. It should be noted that the curves in these illustrations apply only to the shapes *magnetized as shown*.

Using the dimensional ratio conversions of Figures 7-4, 7-5, 7-6, 7-7, and 7-8, the B/H ratio for these five magnet shapes can readily be determined. For example, a round bar magnet of length (L) 1.5 in. and diameter (D) 1/2 in. would have:

$$\frac{L}{D} = \frac{1.5}{.5} = 3$$

From Figure 7-4, if cast Alnico were to be used, the B/H ratio = 10 with $L/D = 3$. Drawing the load line from $B = 0 : H = 0$ to $B/H = 10$ on the demagnetization curves in Figures 7-1 and 7-2, it is evident that, with this B/H ratio, a magnet made of:

---

3. Again, a reminder that *capability* is not the same as *performance*. The actual energy product "delivered" will be determined by other factors than the capability of a material.

4. The significance of operating above or below the knee of the curve is related to the sensitivity of the magnet to external demagnetizing influences. This will be covered in detail in Chapter 11.

5. The equation for horseshoes (see Figure 7-8) is not given. This equation is quite long and complex, and presenting it here would serve no useful purpose.

Based on the equation:
$$\frac{B}{H} = \frac{K_1}{\left(\frac{D}{2}\right)^2}\sqrt{\frac{D}{2}\left(\frac{D}{2}+L\right)}$$

For all Alnicos $K_1 = .7L$
For all Ferrites $K_1 = L$

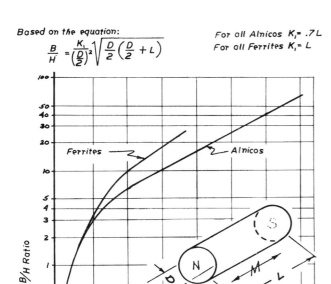

Figure 7-4. L/D ratio to B/H ratio conversion curves for axially oriented round bar magnets (applicable to simple bar magnets only).

Based on the equation:
$$\frac{B}{H} = \frac{4}{L}\sqrt{\frac{D}{2}\left(\frac{D}{2}+L\right)}$$

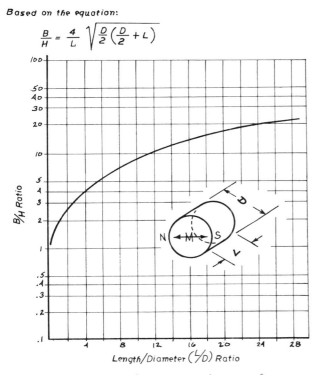

Figure 7-5. L/D ratio to B/H ratio conversion curves for diametrically oriented round bar magnets (applicable to simple Alnico or ferrite bar magnets only).

Based on the equation:
$$\frac{B}{H} = \frac{4K_3}{OD^2-ID^2}\sqrt{\frac{L}{2}(OD-ID) + \frac{OD^2+ID^2}{4}}$$

For all Alnicos $K_3 = .7L$
For all Ferrites $K_3 = L$

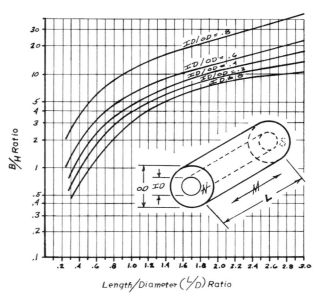

Figure 7-6. L/D ratio to B/H ratio conversion curves for axially oriented ring magnets (applicable to simple ring magnets only).

Based on the equation:
$$\frac{B}{H} = \frac{1.77K_2}{wt}\sqrt{L(w+t) + wt}$$

For all Alnicos $K_2 = .7L$
For all Ferrites $K_2 = L$

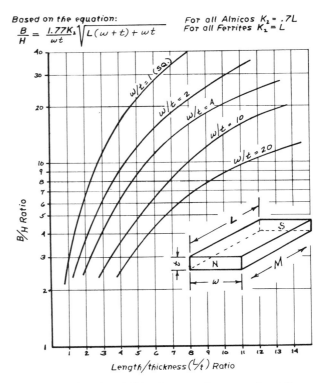

Figure 7-7. L/t ratio to B/H ratio conversion curves for rectangular bar magnets (applicable to simple bar magnets only).

GENERAL DESIGN CONSIDERATIONS 45

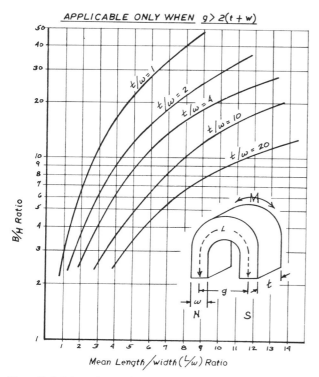

Figure 7-8. L/w ratio to B/H ratio conversion curves for horseshoe magnets.

Cast Alnico III would operate at a B/H_{max} of approximately 1.5 X 10^6 G-Oe and below the knee of the curve.

Cast Alnico V-NC1 would operate at a B/H_{max} of approximately 4.3 X 10^6 G-Oe and below the knee of the curve.

Cast Alnico V-NR1 would operate at a B/H_{max} of approximately 3.3 X 10^6 G-Oe and below the knee of the curve.

Cast Alnico V-NW1 would operate at a B/H_{max} of approximately 3.5 X 10^6 G-Oe and below the knee of the curve.

Cast Alnico V-PR1 would operate at a B/H_{max} of approximately 3.9 X 10^6 G-Oe and below the knee of the curve.

Cast Alnico V-OW1 would operate at a B/H_{max} of approximately 4.4 X 10^6 G-Oe and below the knee of the curve.

Cast Alnico V-OR1 would operate at a B/H_{max} of approximately 4.7 X 10^6 G-Oe and below the knee of the curve.

Cast Alnico VI-NS1 would operate at a B/H_{max} of approximately 3.75 X 10^6 G-Oe and below the knee of the curve.

Cast Alnico VI-NS2 would operate at a B/H_{max} of approximately 4.6 X 10^6 G-Oe and below the knee of the curve.

Cast Alnico VIII would operate at a B/H_{max} of approximately 4.2 X 10^6 G-Oe and above the knee of the curve.

Cast Alnico XII would operate at a B/H_{max} of approximately 1.8 X 10^6 G-Oe and above the knee of the curve.

On the basis of the above and considering *only* energy per unit volume, the best material for this particular basic magnet would be cast Alnico V-OR1. But if the magnet operates above or below the knee of the curve and, thus, its external demagnetization resistance is considered, the best choice would be cast *Alnico VIII*.

Having made the points that the *B/H* ratio is determined by dimensional factors and that material selection is, in turn, based upon the *B/H* ratio, it would be misleading not to mention that the conversion curves in Figures 7-4, 7-5, 7-6, 7-7, and 7-8 are of very limited practical use. The true *B/H* ratio is not just a function of magnet dimensions; it is a function of the dimensional factors *of the whole magnetic circuit*. Applications where a basic magnet is used alone with no auxiliary magnetic parts and where the object on which it works is negligible (in terms of the circuit) are few and far between. The only case that occurs to this author is a bar magnet used to trip a reed switch. The utility of these conversion curves is further limited by the fact that it is, at best, highly inaccurate (and sometimes impossible) to calculate the *flux output* at any point in space around a basic magnet alone. On the other hand, in the case of a basic magnet of any shape used *in a circuit*, the output in a working air gap *can* be calculated with reasonable precision.

Some appreciation for this problem can be obtained by studying Figures 7-9 and 7-10, which show the flux densities obtained for cast Alnico 5 round and square rods of various cross sections and lengths.[6] The solid, central curves show the maximum flux densities at the end *surface edge* (see the sketches in the upper right-hand corners of both figures); the dashed curve shows the percentage reduction of this flux as the probe is moved away from the surface. The table in the upper left-hand portion of Figure 7-10 shows the reduction in flux if the probe is located at the surface on the *center line* (*cl*) of the rod. The data shown are based on actual measurements and not on calculations, since there is no known way to calculate these values. It should be noted

---

6. These curves were developed to meet the specific needs of the author in one of his projects. They will, however, be useful to others who may have occasion to need the flux densities of these particular points or to know the optimum length of a basic magnet for various cross sections.

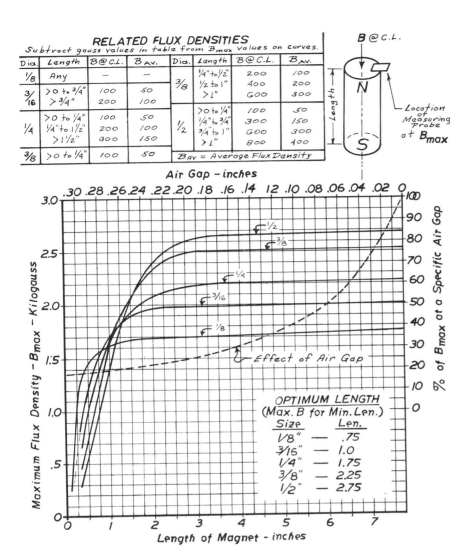

Figure 7-9. Flux densities obtained for Alnico 5 round bar magnets.

that there are literally an infinite number of points surrounding the magnet where similar measurements might be made.

The primary value of the conversion curves in Figures 7-4, 7-5, 7-6, 7-7, and 7-8 is for the rare basic magnet applications and to *quickly establish the lowest B/H ratio that can ever occur when used alone in a circuit.* Circuit parts added to the basic magnet (such as pole pieces) serve to *reduce* the air gap and therefore *increase* the B/H ratio. Thus, the engineer can use these conversion curves to give himself some idea of the *lowest* B/H ratio that can occur and then eliminate certain materials from further consideration with this knowledge. This becomes clear if the earlier example of B/H = 10 is reconsidered. Since B/H = 10 is for the maximum "open circuit," adding pole pieces to direct the flux to a specific gap would *increase the B/H ratio* to some value *greater* than 10. Since Alnico XII and ferrite 5 (in Figures 7-2 and 7-3) respectively, were already poor material choices, reducing their air gaps would make them even less acceptable. On the other hand, as the B/H ratio increases, the various grades of Alnico V become more appropriate.

Before undertaking a simple magnet design problem, the general equations that apply to magnetic circuits must be understood. These equations express the relationships of the factors in the magnetic circuit and enable the engineer to mathematically establish performance.

In a closed ring of a hard magnetic material that has been subjected to a magnetizing force, some total number of *lines of flux* (ϕ) will flow endlessly. Consider what will happen if an infinitely small air gap is introduced to the ring. The total flux that flows in the circuit will remain unchanged, so that all of the flux that flowed in the closed ring will also flow in the air gap. Therefore:

Equation 3: $\phi_g = \phi_m$

GENERAL DESIGN CONSIDERATIONS 47

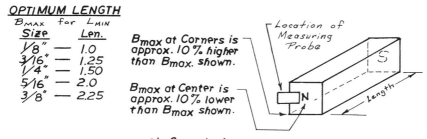

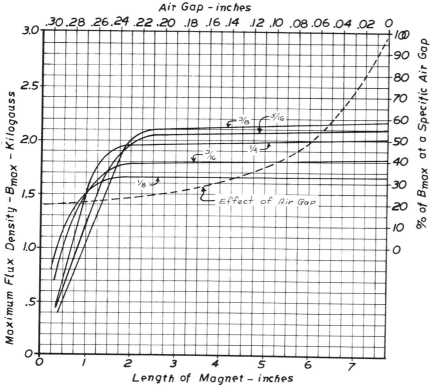

Figure 7-10. Flux densities obtained for Alnico 5 square bar magnets.

where ϕ_m is the total flux in the magnet (ring), and ϕ_g is the total flux in the air gap.

This relationship is valid provided that the ring *length* does not change and that no "leakage" of flux occurs in such a way as to bypass the air gap.[7] If the entire circuit has a uniform cross-sectional area, the flux density (*B*) is the total flux (ϕ) divided by the area (*A*) that it traverses, or:

Equation 4: $\quad B = \dfrac{\phi}{A}$

or

Equation 5: $\quad \phi = BA$

so that for the ring part of the circuit:

$$\phi_m = B_m A_m$$

7. This assumption of zero leakage is only true for infinitely small air gaps. Leakage *does* occur in most cases. How it can be included in these equations will be shown later.

and for the air gap:

$$\phi_g = B_g A_g$$

But since $\phi_g = \phi_m$, by substituting we obtain:

Equation 6: $\quad B_g A_g = B_m A_m$

It is evident that if there is flux in the circuit, some magnitude of magnetomotive force (*F*) must be producing it. If the total magnetomotive force is divided by the distance (*L*) over which it is applied, then:

Equation 7: $\quad \dfrac{F}{L} = H$

where *H* is the "per unit" *magnetizing force* (mmf) and *L* is the length.

But in a ring with an air gap, the total mmf is *also* applied across that gap, so that:

Equation 8: $\quad F_g = F_m$

and, for the ring portion of the circuit (from Equation 7):

$$F_m = H_m L_m$$

and for the air gap:

$$F_g = H_g L_g$$

Substituting in Equation 8, we obtain:

Equation 9: $\quad H_g L_g = H_m L_m$

Considering further what is happening in an air gap, the absolute permeability (μ) of air is established as unity.[8] This is based on the concept that the absolute permeability of a material is the ratio of the flux density produced to the magnetizing force causing it. Expressed mathematically:

Equation 10: $\quad \mu = \dfrac{B}{H}$

but if μ for air = 1, then for an *air gap:*

Equation 11: $\quad B_g = H_g$

On the basis of Equation 11, substituting B_g for H_g in Equation 9 gives us:

Equation 12: $\quad B_g L_g = H_m L_m$

or, solving for B_g, results in:

Equation 13: $\quad B_g = \dfrac{H_m L_m}{L_g}$

From Equation 13 it, is evident that the flux density produced in an air gap is directly proportional to the magnetizing force in the magnet and to the length of the magnet. The flux density is *inversely* proportional to the length of the air gap.

With these relationships established, all the equations needed for a simple, short-method calculation are available. Others will be introduced when necessary in discussing more accurate and complex designs.

DESIGN PROBLEM 1

When a design engineer uses a permanent magnet, one of his primary requirements is to establish the volume and configuration of magnet that most efficiently produces the required field. In other words, the selected magnet material should be so utilized that it produces the required flux while operating at its maximum energy product (BH_{max}). On the other hand, selecting the material requires knowing the *B/H* ratio that will be established by the configuration.

8. Actually, $\mu = 1$ is for a vacuum, but air is so close that it can be considered the same for most design purposes.

First, a particular magnet material that meets other requirements (cost, availability, temperature stability, and so on) is *tentatively* selected. Next, a trial calculation is made to determine the suitability of that material. Then the selection is reviewed, recalculated, and verified. Several such trial-and-error selections and calculations may be necessary.

Normally, the magnet designer has already established or has been given the air gap requirements. In this problem, assume that these requirements are:

Length of air gap (L_g) = 2.0 cm

Cross-sectional area of air gap (A_g) = 4.0 cm²

Required flux density in air gap (B_g) = 5000 G

Tentatively, the engineer selects cast Alnico V-NR1 (Figure 7-2). Then he must determine the *length* of the magnet (L_m) required. Since B_g and L_g are given, if H_m is established, Equation 12 can be used to solve for L_m. Referring to the demagnetization curve for Alnico V-NR1 in Figure 7-2, H_m at BH_{max} can be obtained [9] (H_m = 535 Oe).
Therefore, Equation 12

$$B_g L_g = H_m L_m$$

transposed for L_m is:

$$L_m = \dfrac{B_g L_g}{H_m} = \dfrac{(5000)(2)}{535} = \underline{18.7 \text{ cm}}$$

This calculation establishes the *minimum* length of a magnet measuring 18.7 cm from pole to pole. However, any circuit will have imperfect joints (small air gaps) between the magnet and the pole pieces, and there will be some slight reluctance in these pole pieces, as well as in other small imperfections in the circuit. All of these factors that are essentially *in series* with the magnet length (L_m) require some magnetomotive force (F) to overcome. Since the only source of the magnetization force (mmf) is the magnet itself, the magnet length will have to be *greater* than 18.7 cm to overcome the *series losses.* Determining the reluctance factor (f) to allow for these series losses is a matter of detailed calculations or experience in making a general estimate. *In most cases,* f *will be between* 1.1 *and* 1.5.

Assuming in this case that f = 1.2, the *actual* magnet length required is:

$$L_m f = (18.7)(1.2) = \underline{22.1 \text{ cm}}$$

Since B_g and A_g are given, if B_m is established, the area

9. Most properties tables, including those in the Glossary, give the H_m, B_m, and *B/H* ratio at maximum energy product (BH_{max}).

of the magnet (A_m) can be calculated by means of Equation 6. Again, referring to the demagnetization curve for Alnico V-NR1 in Figure 7-2, the B_m at BH_{max} = 10,500 G. Therefore, Equation 6

$$B_g A_g = B_m A_m$$

can be transposed to solve for A_m and the known values can be substituted, or:

$$A_m = \frac{B_g A_g}{B_m} = \frac{(5000)(4)}{10,500} = \underline{1.905 \text{ cm}^2}$$

Thus, the *minimum* cross-sectional area of the magnet must be 1.905 cm², but here again, the minimum must be increased to allow for series losses. In this case, the losses are *parallel losses* that occur because some of the flux produced by the magnet *will not reach the air gap*. Since flux lines will always follow the path of least reluctance, some flux will "jump across" the length of the magnet. This leakage flux must also be supplied by the magnet, and it requires an increase in area by some leakage factor (σ). Leakage factors can be calculated by tedious but effective methods that we will cover in Chapter 8. In practice, these factors range from as little as 1.1 to as much as 50, so that arbitrarily selecting them on the basis of experience can prove highly inaccurate.

For this problem, let us assume that the engineer has designed and tested similar circuits and found that σ = 15. Then the *actual* magnet area required will be:

$$A_m \sigma = (1.905)(15) = \underline{28.6 \text{ cm}^2}$$

Having made the above calculations for the magnet length and area, the engineer has a better idea of the space requirements for the magnet and can proceed to make a preliminary layout of the complete product. If this layout shows that either the length or the area of the magnet will not fit the available space, there are two possible courses of action:

1. Select another operating point along the demagnetization curve for the *same* material and repeat the calculations.
2. Work backward through the calculations from the available space limits and select a material that will provide the necessary output efficiently.

Several precautions should be pointed out before we proceed to the study of leakage and of more exact design methods in the next two chapters.

First, in using the short method illustrated in Design Problem 1, the reluctance factor (f) and the leakage factor (σ) can never be ignored in the final design. If there is no technical basis for knowledgeably establishing f or σ, an estimate on the high side is preferable. If the output is too high in the actual circuit, it can always be reduced by deliberate demagnetization-calibration or by using pieces of mild steel to shunt part of the flux from the air gap. On the other hand, too little output may be impossible to correct.

One of the best uses of this method is to determine *minimum* lengths and areas (by ignoring f and σ!) for *preliminary* design purposes. Obviously, if these minimums are calculated for a selected material and then do not fit, changes will have to be made in material, space, and so on.

The last caution—and one that applies to *all* magnetic calculations—is to exercise great care in keeping units consistent by using *one* of the three unit systems: the cgs, the mks, or the English system. For example, it is a common practice (and an erroneous one) to give the air gap flux density in gausses (cgs system) and the magnet length in inches (English system). In some equations, such as Equation 13, inconsistent units may be unimportant, depending upon what you are solving for, but it is safest *not to mix systems*. (The unit names and conversion factors for all three systems appear in Appendix 5.

8

Leakage and Fringing

It is evident from Chapter 7 that, in magnet design, serious consideration must be given to the flux that never reaches the air gap. With leakage factors that require magnet area increases of 10 to 500 percent (σ = 1.1-50), leakages must first be *minimized* and then be accommodated in the design. Figure 8-1 is a photograph of two magnets, a small horseshoe and a disc, that have been covered with iron filings. From this photograph, we can see that the flux travels many pathways and not just across the primary air gap.

In magnet design, the leakage fluxes can be reduced to:

1. The flux in the vicinity of the air gap that does not pass *directly* across the gap but runs parallel to it. This is commonly called the *fringing flux*.

2. The flux that radiates *between* the legs or across the back of all parts of the circuit. This alone is normally referred to as the *leakage flux*.

Both fringing and leakage flux are illustrated in Figure 8-2.

The leakage factor (σ) can be expressed as the ratio of the total flux (ϕ_T) to the flux across the air gap (ϕ_g), or:

Equation 15: $$\sigma = \frac{\phi_T}{\phi_g}$$

Since the total flux (ϕ_T) is difficult to calculate, σ is more readily expressed either on the basis of the reluctance (R) of the paths or as the *reciprocal* of R, which is termed the *permeance* (P). Since permeance is the

Figure 8-1. Iron filings showing field pattern of horseshoe and disk magnets. (Photograph courtesy of Bell Telephone Laboratories)

ϕ_g, R_g, or P_g = Air Gap
ϕ_f, R_f, or P_f = Fringing
ϕ_ℓ, R_ℓ, or P_ℓ = Leakage

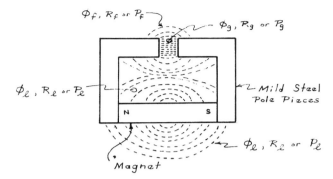

Figure 8-2

simplest basis for all magnetic path calculations, σ can most usefully be expressed as:

Equation 16: $\sigma = \dfrac{P_T}{P_g}$

The most effective way of understanding the paths that may be taken by flux, and therefore how leakage and fringing can be minimized, is to study the following conceptual relationships:

1. Flux lines, like electrical currents, will always follow the paths of least resistance. In magnetic terms, this means that *flux lines will follow the paths of greatest permeance* (lowest reluctance).
2. *Flux lines repel each other* if their direction of flow is the same.
3. As a corollary to condition (2), *flux lines can never cross* each other.
4. As a corollary to condition (1), flux lines *will always follow the shortest path* through any medium. They therefore can travel only in straight lines or curved paths, and they can never make true right-angle turns. Meeting the condition of condition (2), flux lines will normally *always move in curved paths*, although over short distances these paths may be considered straight for practical purposes.
5. *Flux lines always leave and enter the surfaces of ferromagnetic materials at right angles.*
6. All ferromagnetic materials have a "limited ability" to carry flux. When they reach this limit, they are "saturated" and behave as though they do not exist (like air, aluminum, and so on). *Below the level of saturation, a ferromagnetic material will substantially* contain *the flux lines passing through it*. As saturation is approached, because of conditions (1) and (2), the flux lines *may* travel as readily through air as through the material.
7. *Flux lines will always travel from the nearest north pole to the nearest south pole in a path that forms a closed loop.* They need *not* travel to their *own* opposite pole directly (although they do *ultimately*) if opposite poles of another magnet are closer and/or there is a lower path of reluctance (greater permeance) between them.
8. Magnetic poles, contrary to the convenient concept used by physicists, *are not* unit *poles*. In a magnetic circuit, any two points equidistant from the neutral axis *function as poles, so that flux will flow between them* (assuming that they meet the other conditions stated above).

Careful consideration of the above axioms will clearly show the causes of leakage and fringing flux.

The dichotomy between "path of least resistance" in condition (1) and "repulsion of lines" in condition (2) provides the explanation for fringing effects. Leakage effects are readily explained by conditions (1), (4), (7), and (8).

The often unexpected and unwanted increase in leakage and fringing effects resulting from the improper design of other circuit elements is explained by conditions (7) and (8). For example, pole pieces with inadequate cross sections, of unnecessary lengths, of improper shapes, or whatever, result in a substantial increase in "losses" at the expense of the working air gap flux.

Minimizing Leakage Fluxes

As the first step in studying how to minimize leakage flux (usually the larger of the two "losses" by a substantial amount), consider the effect of the location of the basic magnet in the circuit.

Figure 8-2 shows one commonplace method for locating the permanent magnet in a simple circuit. This method is frequently used because it minimizes the number of physical parts to be made and assembled (two pole pieces and one magnet). It is, however, magnetically the least effective way of designing the circuit.

Figure 8-3A is the same circuit as the one in Figure 8-2, except that the permanent magnet is shown in two sections of equal length. Figure 8-3B shows this circuit with the magnet sections redistributed closer to the air gap, and Figure 8-3C shows this circuit with the magnets moved as close as possible to the air gap. The effects of these relocations are indicated at the right of each illustration. If the flux density at the center of the air gap (B_g) of each circuit (A, B, or C) is used as the basis of comparison, we note that:

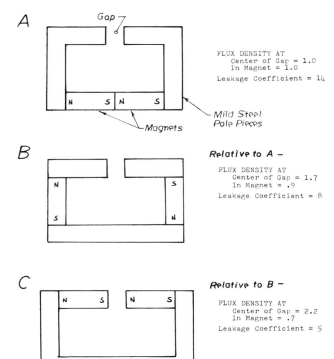

Figure 8-3. Effect of magnet location. (All magnets shown are identical cast Alnico 5 bars.)

B_{gB} is 70 percent greater than B_{gA}

B_{gC} is 220 percent greater than B_{gA} and

B_{gC} is 150 percent greater than B_{gB}

Also note that as the flux density in the air gap has been increased with each relocation, the flux density in the magnet (B_m) has been reduced (by 7 percent in circuit B and 26 percent in circuit C).

Since the same length, volume, and magnet material was used in each case, it is evident that relocating the magnet closer to the air gap increases the *efficiency* of the circuit. This is directly reflected in the reduction of the leakage coefficients (σ) noted in Figure 8-3. These coefficients drop from $\sigma_A = 14$, to $\sigma_B = 8$, to $\sigma_C = 5$. The reduction of leakage is readily explained by considering these circuits in relation to the preceding flux concepts. In Figure 8-3A there are large, magnetized surfaces with a magnetizing force (mmf) between them (the legs of the mild steel pole pieces). These pole pieces provide alternate paths to the air gap for flux and therefore represent *leakage* paths. *These leakage paths are* in addition to *the leakage that occurs from pole to pole in each magnet section.*

With the magnets relocated in Figure 8-3B, the flux paths (and hence the leakage) between the pole piece legs are eliminated, leaving only the pole-to-pole leakage in the magnet sections. Actually, this circuit is less efficient than it might be, since the pole-to-pole magnet leakage has increased compared to circuit A. This increase occurs because the top poles (that form the air gap) and the bottom "plate" are closer together and comprise a larger area than they did in circuit A.

The obvious improvement and best circuit is shown in Figure 8-3C. With this arrangement, "interleg" leakage is eliminated but the pole-to-pole magnet section is *not* greater than it is in circuit A. However, C is still not necessarily the best circuit possible because:

1. Problems of *uniformity* in air-gap flux density may occur, since one end surface of each permanent magnet is used as the gap surface. The composition of most permanent magnets is not sufficiently homogeneous to preclude possible variations in surface reluctance. Also, mechanically, it is poor practice in many applications to have the ends of a brittle material serve as the gap "faces." Usually, relatively thin, properly shaped pole pieces will assure uniform flux distribution without appreciably increasing leakage or fringing.
2. The magnets themselves are not shaped properly to minimize pole-to-pole leakage, as indicated by axiom (8).

Figure 8-4 illustrates the four methods of minimizing leakage flux, comparing poor arrangements with better arrangements in each case. The four methods of minimizing leakage are:

A. Optimizing the magnet shape.
B. Properly locating the magnet in the circuit.
C. Adding "blocking" poles to prevent the leakage flux from traversing a particular area.
D. Optimizing the inherent, heat-treat (or pressing) orientation of the basic magnet.

The principles on which A above and in Figure 8-4 are based are self-evident from the axioms covered previously. Method B has already been covered in detail in connection with Figure 8-3.

Blocking-pole principle C above and in Figure 8-4 is a relatively recent development that has been made possible by highly coercive, highly demagnetization-resistant materials such as the sintered ferrites. The illustration shows the blocking-pole principle applied to three common magnetic circuit configurations. In the top circuit, a major area of flux leakage occurs from the edges of the top pole piece to the bottom pole piece. *Extending the magnet beyond* all *of the edges of the pole pieces*,[1] permits the extended section, *which has*

1. In this figure, which is a side view of the assembly, only the back extension can be shown.

LEAKAGE AND FRINGING 53

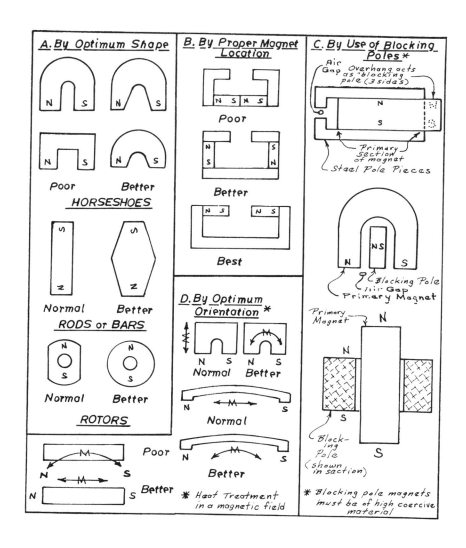

Figure 8-4. Methods of minimizing leakage flux losses in a magnetic circuit.

the same polarity as the primary section, to deliver a "counter mmf" (flux) that opposes and reduces the leakage flux that would otherwise occur. The blocking-pole section is indicated by the dashed lines and N and S pole markings in Figure 8-4C. A perceptive reader will notice that this illustration does not show *all* of the improvements that could be made by blocking-pole methods. As shown here in the figure, *substantial leakage will occur in the space between the back of the air gap and the side of the magnet.* This could be eliminated by having the magnet *fill* this space. Such filling would, in fact, eliminate the *fringing flux* that would normally occur at the back of the air gap.[2]

The blocking-pole arrangement for the horseshoe shape in Figure 8-4C eliminates interleg leakage by means of a "counter mmf." For this magnet configuration, leakage in this area is particularly large if the legs

2. The circuit shown is for a subminiature galvanometer magnet; the space behind the pole legs is open to accommodate for the mechanical parts required in the movement.

of the horseshoe are long and/or spaced closely together.

The ring-shaped blocking pole shown surrounding the rod magnet (again, in Figure 8-4C) provides a counter mmf to minimize the normally large pole-to-pole leakage associated with this magnet shape.

The inherent orientation of the magnet applied by heat treatment at the mill can have a material effect on the leakage flux. Figure 8-4D shows three common heat-treat arrangements. For the two general "horseshoe" shapes, heat treatment *straight through* will substantially increase leakage across the back (at the expense of the desired flux). In the case of the bar magnet, "curved" orientation will also increase leakage by effectively bringing the north and south poles closer together.

Determining which path flux will follow and how to minimize "lost" flux can be based either on experience or actual calculation. Needless to say, properly carried out calculations, while more tedious, are more accurate, consistent, and reliable.

Minimizing Fringing Flux

As can be noted in Figure 8-2, fringing starts at the very corners of the pole-piece tips and extends back some distance along the pole-piece surfaces. The diagram in Figure 8-5 shows how the fringing flux can be considered as two separate but parallel components. P_{SC} has the space configuration of a semicircular cylinder directly bridging (parallel to) the air gap. P_{HA} is "parallel" to both P_{SC} and the air gap (P_g) and has the space configuration of a half annulus. P_{SC} is normally considerably greater than P_{HA}.

Fringing fluxes can never be completely eliminated, since they are a direct result of the axiom that flux lines repel each other. However, since flux lines will also take the path of least reluctance and the shortest path, leakage fluxes can, at least, be minimized.

A study of the permeance equations given in Figure 8-5 clearly indicates how minimization can be effected in the case of P_{SC}. These equations are expressed in terms of permeance rather than reluctance for convenience in calculation. The relationship between reluctance (R) and permeance (P) is:

Equation 17: $R = \dfrac{1}{P}$

or

Equation 18: $P = \dfrac{1}{R}$

Permeance is obviously analogous to *conductance* in electrical circuits, so that, in these equations, the greater the permeance, the *easier* the flux path and the greater the flux.

Since the permeance of the air gap is:

$$P_g = \dfrac{A_g}{L_g}$$

increasing the air-gap area and/or *reducing the air-gap length* will increase the ease with which flux travels across this gap and thereby tend to reduce P_{SC} (if only by increasing the ratio of flux across the gap to the fringing flux). Care must be exercised, however, in increasing the air-gap area, since P_{SC} is also related to the gap dimensions and increases with these dimensions. *Thus, the most effective way of reducing the largest component of fringing flux (P_{SC}) is to make the air-gap length as small as possible.*

In the case of the second small component of fringing flux (P_{HA}), the most effective reductions can be made by tapering or beveling the pole pieces toward the air gap. Such tapering or beveling increases the length of the path for any flux in the P_{HA} area, thus increasing the reluctance, decreasing the permeance, and therefore decreasing this fringing flux loss.

However, in tapering or beveling pole pieces care must be exercised to be sure that the pole tips that form the final air gap *are not at or approaching saturation* for the grade of mild steel used. This precaution must be followed in determining pole-piece cross sectioning *anywhere* in the circuit, since one of the previously stated axioms is that fringing, leakage, and internal reluctance losses will increase as the material approaches saturation. (Acceptable levels of flux and calculations related to pole pieces will be covered in Chapter 9.)

Calculating Permeance

The primary requirement for magnetic circuit design, and therefore the design of the basic magnet used in the circuit, is to be able to calculate the relationships between flux paths. As pointed out in the previous section, the easiest approach to such calculations is by means of *permeances*.

Herbert C. Rotor [3] derived a series of simple equations for estimating permeances for various space configurations. Six of these equations and the space sections to which they apply appear in Figure 8-5. This author has found that in the vast majority of the circuits encountered, these six equations alone make every

---

3. Herbert C. Rotor, *Electromagnetic Devices* (New York: John Wiley & Sons, Inc., 1941) Chapter 5.

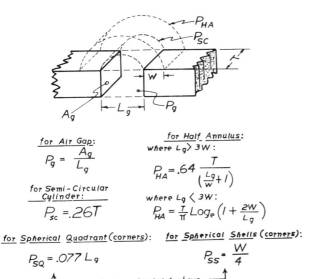

for Air Gap:
$$P_g = \dfrac{A_g}{L_g}$$

for Semi-Circular Cylinder:
$$P_{sc} = .26 T$$

for Half Annulus:
where $L_g > 3W$:
$$P_{HA} = .64 \dfrac{T}{\left(\dfrac{L_g}{W}+1\right)}$$

where $L_g < 3W$:
$$P_{HA} = \dfrac{T}{\pi} \log_e\left(1 + \dfrac{2W}{L_g}\right)$$

for Spherical Quadrant (corners):
$$P_{SQ} = .077 L_g$$

for Spherical Shells (corners):
$$P_{SS} = \dfrac{W}{4}$$

↑ — not shown in sketch above —↑

Note: All permeance calculations must be in three planes; all equations are for flux paths *in air*.

Figure 8-5. Basic permeance equations.

permeance calculation possible within reasonable accuracy. Virtually any circuit can be reduced to a series of spatial configurations to which these equations will apply. Then, the "component" permeances, when properly recognized as *series* or *parallel*, can be combined using an analogy of Kirchhoff's electrical laws to provide a complete analysis of the circuit in question. Sometimes reducing the magnetic circuit to elements by this method is somewhat tedious, but this provides two primary advantages:

1. Since the final calculations are the sum of a large series of smaller calculations, errors in the smaller calculations tend to average out.
2. The method lends itself to *computer* calculations once the total equation for a specific circuit has been developed.

The most effective way to illustrate the use and value of these equations is by means of a typical problem.

DESIGN PROBLEM 2

Determine if a working air-gap flux density (B_g) of 4500 G will be developed in the circuit shown in Figure 8-6 if the permanent magnet is made of cast Alnico III-NF1 and the dimensions are as follows:

L_m = 7.61 cm

$w_m = w_g$ = 3.18 cm

t_m = 1.9 cm

b = 2.54 cm

$c = t_g$ = .954 cm

L_g = .635 cm

Assume material properties are uniform throughout the magnet length.

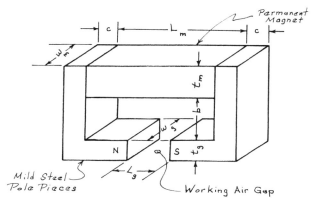

Figure 8-6

Solution:

The maximum gap flux will exist if there are no leakage or fringing flux losses. Assume, as a first approximation, that this is the case and determine if the stated flux level is possible.

Using Equation 14 from Figure 8-7:

$$\frac{B}{H} = P_T \frac{L_m}{A_m}$$

and substituting P_g for P_T, since fringing and leakage are assumed to be zero:

$$\frac{B}{H} = P_g \frac{7.61}{(3.18)(1.9)}$$

But the equation for P_g from Figure 8-5 is given as:

$$P_g = \frac{A_g}{L_g}$$

and in this circuit:

$$P_g = \frac{(.954)(3.18)}{.635}$$

Therefore, substituting in the *B/H* equation:

$$\frac{B}{H} = \frac{(.954)(3.18)}{.635} \times \frac{7.61}{(3.18)(1.9)}$$

gives us:

$$\frac{B}{H} = 6.02 \approx \underline{\underline{6.0}}$$

Thus, the *B/H* ratio or load line at which this magnet would operate is approximately $\underline{\underline{6.0}}$. Drawing a straight line from $B = 0 : H = 0$ to $B/H = \overline{\overline{6}}$ on Figure 7-1 (page 41) establishes this load line on the demagnetization curve for cast Alnico III-NF1. Projecting vertically from where the load line crosses the Alnico III curve gives H_m = 437.5 Oe. Solving Equation 13 for B_g gives us:

$$B_g = \frac{H_m L_m}{L_g}$$

Since H_m, L_m, and L_g are known:

$$B_g = \frac{(437.5)(7.61)}{.635} = \underline{\underline{5250 \text{ G}}}$$

From the foregoing calculation, it is evident that the required flux density in the air gap *is possible. Whether this gap flux is actually attained depends upon the losses produced by the leakage and fringing fluxes.* These leakage and fringing fluxes can be introduced to the calculation by considering the permeance of the whole circuit (not the gap alone) and recalculating the flux that will be produced in the air gap.

Using Rotor's method and equations in Figure 8-5,

Eq. 1: $\phi = \dfrac{F}{R}$

Eq. 2: $\phi = FP$

Eq. 3: $\phi_g = \phi_m$

Eq. 4: $B = \dfrac{\phi}{A}$

Eq. 5: $\phi = BA$

Eq. 6: $B_g A_g = B_m A_m$

Eq. 7: $H = \dfrac{F}{L}$

Eq. 8: $F_g = F_m$

Eq. 9: $H_g L_g = H_m L_m$

Eq. 10: $\mu = \dfrac{B}{H}$

Eq. 11: $B_g = H_g$ (in air, $\mu = 1$)

Eq. 12: $B_g L_g = H_m L_m$ (in air, $\mu = 1$)

Eq. 13: $B_g = \dfrac{H_m L_m}{L_g}$

Eq. 14: $\dfrac{B}{H} = P_T \dfrac{L_m}{A_m}$

Eq. 15: $\sigma = \dfrac{\phi_T}{\phi_g}$

Eq. 16: $\sigma = \dfrac{P_T}{P_g}$

Eq. 17: $R = \dfrac{1}{P}$

Eq. 18: $P = \dfrac{1}{R}$

Eq. 19: Work = ergs = $\dfrac{E_k}{8\pi}$

Eq. 20: Force = $\dfrac{B^2 A}{72 \times 10^6}$ (English units)

Eq. 21: Force = $\dfrac{B^2 A}{8\pi}$ (cgs units)

Eq. 22: $F = mmf = .4\pi NI = 1$ Gilbert

Eq. 23: $E_{max} = 8\pi BH_{max}$
(E_{max} in ergs/cm³ produced by a magnetic material in an EXTERNAL field.)

ALL UNITS MUST BE CONSISTENT

where:
- ϕ = Total Flux
- F = Magnetomotive Force
- R = Reluctance
- B = Flux Density
- A = Area
- H = Magnetizing Force or Coercive Force
- L = Length
- P = Permeance
- μ = Permeability
- σ = Leakage Factor
- E = Energy

Figure 8-7. General equations used in magnet design.

the circuit can be reduced to the group of permeance paths shown in Figure 8-8. In the three plan views in this figure, the permeance paths at the *corners* (Rotor's "spherical quadrants") have not been shown because of the limitations of a plan-view drawing.

Successively identifying and calculating *all* of the permeances in the circuit:

$$P_g = \dfrac{A_g}{L_g} = \dfrac{(.954)(3.18)}{.635} = \underline{\underline{4.77}}$$

as calculated in the first approximation above, and:

$P_1 = P_2 = .26T = (.26)(3.18) = \underline{.826}$

$P_3 = P_4 = .26T = (.26)(.954) = \underline{.248}$

$P_5 = P_6 = P_7 = P_8 = .077\, L_g = .077(.635) = \underline{.049}$

In order to calculate P_9, P_{10}, and P_{11}, the W at that end of the circuit must be determined and the appropriate Rotor's equation for P_{HA} must be selected:

$$W_1 = \dfrac{L_m + 2c - L_g}{2} = \dfrac{7.61 + 2(.954) - .635}{2}$$

$W_1 = 4.44$

Since $L_g < 3W_1$,

$$P_9 = P_{10} = \dfrac{T}{\pi} \log_e \left(1 + \dfrac{2W_1}{L_g}\right)$$

$P_9 = P_{10} = \dfrac{.954}{.314} \log_e \left(1 + \dfrac{2 \times 4.44}{.635}\right)$

$P_9 = P_{10} = .3 \log_e 15$

$P_9 = P_{10} = (.3)(1.708) = \underline{.512}$

LEAKAGE AND FRINGING

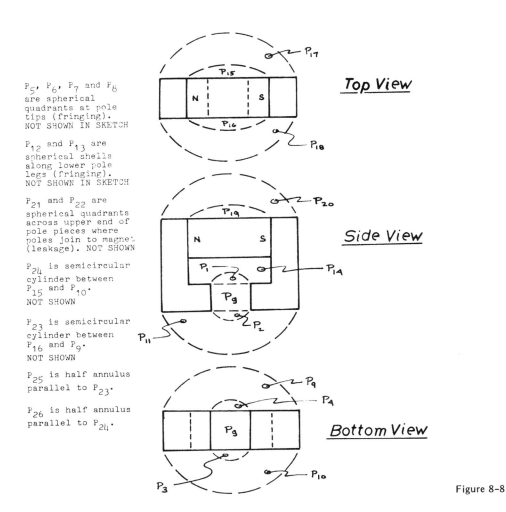

P_5, P_6, P_7 and P_8 are spherical quadrants at pole tips (fringing). NOT SHOWN IN SKETCH

P_{12} and P_{13} are spherical shells along lower pole legs (fringing). NOT SHOWN IN SKETCH

P_{21} and P_{22} are spherical quadrants across upper end of pole pieces where poles join to magnet (leakage). NOT SHOWN

P_{24} is semicircular cylinder between P_{15} and P_{10}. NOT SHOWN

P_{23} is semicircular cylinder between P_{16} and P_9. NOT SHOWN

P_{25} is half annulus parallel to P_{23}.

P_{26} is half annulus parallel to P_{24}.

Figure 8-8

$$P_{11} = \frac{T}{\pi} \log_e \left(1 + \frac{2W_1}{L_g}\right)$$

$$P_{11} = \frac{3.18}{3.14} \log_e \left(1 + \frac{2 \times 4.44}{.635}\right)$$

$$P_{11} = (1.015)(1.708) = \underline{1.73}$$

$$P_{12} = P_{13} = \frac{W_1}{4} = \frac{4.44}{4} = \underline{1.11}$$

$$P_{14} = \frac{A}{L} = \frac{W_g b}{L_m} = \frac{(3.18)(2.54)}{7.61} = \underline{1.06} \text{ (see footnote 4)}$$

$$P_{15} = P_{16} = .26T = (.26)(1.9) = \underline{.495}$$

In order to calculate P_{17}, P_{18}, and P_{19}, the W at that end of the circuit must be determined and the appropriate Rotor's equation for P_{HA} must be selected:

4. This space between the legs is treated as though it were an air gap permeance. P_1, which is part of this space and which was calculated earlier, is still treated separately.

$W_2 = c = .954$, since $L_g < 3W_2$

$$P_{17} = P_{18} = \frac{T}{\pi} \log_e \left(1 + \frac{2W}{L_g}\right)$$

$$P_{17} = P_{18} = \frac{1.9}{\pi} \log_e \left(1 + \frac{2 \times .954}{.635}\right)$$

$$P_{17} = P_{18} = .606 \log_e 4$$

$$P_{17} = P_{18} = (.606)(.8629) = \underline{.522}$$

$$P_{19} = .26T = (.26)(3.18) = \underline{.826}$$

$$P_{20} = \frac{T}{\pi} \log_e \left(1 + \frac{2W}{L_g}\right)$$

$$P_{20} = \frac{3.18}{3.14} \log_e 4$$

$$P_{20} = (1.015)(.8629) = \underline{.875}$$

$$P_{21} = P_{22} = \frac{W}{4} = \frac{.954}{4} = \underline{.238}$$

$$P_{23} = P_{24} = .26T = (.26)(2.54) = \underline{.66}$$

To find P_{25} and P_{26}, since the W is the same as it is for P_{17}, P_{18}, and P_{20} ($W_2 = .954$):

$$P_{25} = P_{26} = \frac{T}{\pi} \log_e \left(1 + \frac{2W_2}{L_g}\right)$$

$$P_{25} = P_{26} = \frac{2.54}{3.14} \log_e \left(1 + \frac{2 \times .954}{7.61}\right)$$

$$P_{25} = P_{26} = .81 \log_e (1.249)$$

$$P_{25} = P_{26} = (.81)(.2314) = \underline{.187}$$

The next step is to calculate the true *total* permeance (P_T) of the complete circuit. To do this, consideration must be given to whether the various fringing and leakage permeance paths are in *series* or in *parallel* with the air-gap permeance. One of the easiest ways to analyze this is to set up an *electrical circuit analog* of the magnetic circuit.

In this particular simple circuit, it is readily apparent that all of the fringing and leakage paths are in *parallel* with the air-gap path. However, electrical analogs are especially useful for the analysis of complex magnetic circuits. (Figure 8-9 illustrates such an analysis.)

The particular arrangement of the permeance symbols (—⌇⌇—) in Figure 8-9 should be noted. While all of the permeances are parallel to the air gap, the fringing flux permeances are arranged around the air gap permeance *where they occur*. The leakage permeances that occur in the pole-piece legs are arranged *below* the magnet (—┤├—) while those that occur at, around, or at the back of the magnet are arranged *above* the magnet.

In calculating the total permeance, since all permeances are in parallel, *they can be added arithmetically* (like electrical conductances):

$$P_T + P_g + P_1 + P_2 + \cdots + P_{26}$$

$$P_T = 4.77 + (2)(.826) + (2)(.248) + (4)(.049) + (2)(.512) + 1.73 + (2)(1.11) + 1.06 + (2)(.495) + (2)(.522) + .826 + .875 + (2)(.238) + (2)(.66) + (2)(.187)$$

$$P_T = \underline{13.356}$$

Now we return to the basic question: "Will there actually be 4500 G in the air gap?"

Again, using Equation 14 from Figure 8-7:

$$\frac{B}{H} = P_T \frac{L_m}{A_m} = 13.356 \frac{7.61}{(3.18)(1.9)}$$

$$\frac{B}{H} = \underline{16.8}$$

the *true B/H* ratio, with fringing and leakage considered.

Using this B/H ratio on the demagnetization curve for cast Alnico III-NF1 gives $H_m = 288$ Oe. Solving Equation 13 for B_g:

$$B_g = \frac{H_m L_m}{L_g}$$

Therefore:

$$B_g = \frac{(288)(7.61)}{.635} = \underline{3450 \text{ G}}$$

and the magnet and/or magnetic circuit is *not* acceptable.

The reason why only 3450 G is produced in the air gap instead of the maximum possible is evident if the leakage factor (σ) is determined from the permeance calculations:

$$\text{Leakage factor} = \sigma = \frac{P_T}{P_g} = \frac{13.356}{4.77} = 2.8$$

According to this factor, only about 35 percent of the flux reaches the air gap. The rest is lost in fringing and leakage!

Furthermore, using the permeance analysis and calculations, the percentage of flux lost in leakage alone can be calculated:

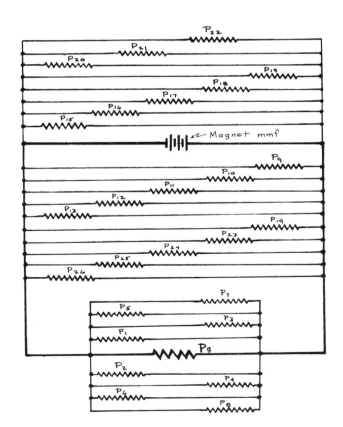

Figure 8-9

Percent of Leakage Loss

$$= \frac{P_T - P_f}{P_T} \times 100$$

$$= \frac{P_T - (P_1 + P_2 + \cdots + P_8)}{P_T}$$

$$= \frac{13.356 - (2)(.826) - (2)(.284) - (4)(.049)}{13.356} \times 100$$

$$= \frac{13.356 - 2.416}{13.356} \times 100$$

$$= \frac{10.940}{13.356} \times 100 = \underline{82 \text{ percent}}$$

and the percent of fringing loss = $\underline{12 \text{ percent}}$.

Next, consider what might be done to obtain the desired 4500 G. The engineer has the following alternatives:

1. *Increase the area of the magnet* without increasing the working air-gap area. This will lower the B/H ratio and increase H_m and B_g.
2. *Increase H_m by decreasing the permeance (P).* This can be achieved by a change in the shapes, spacing, etc., of the circuit elements.
3. *Use a different type of magnetic circuit*—one which will produce a more favorable (lower) B/H ratio, either through changes in dimensional parameters (L and A) or through a lower permeance (P).
4. *Use a different magnet material*—one that will produce a higher H_m (and, therefore, a higher B_g) for the same B/H ratio.

None of the alternatives, with the possible exception of changing the material, represents a certain solution to the problem. In each case, the entire analysis and calculation must be repeated, *taking into account every effect of changed conditions.* For example, an increase in the magnet area will also increase certain permeance factors and the generally expected results may not be realized. Changing permeance is not always a simple matter, and often the results of an attempted change in permeance produces insignificant results. In fact, it is not uncommon to find that a change expected to decrease permeance actually increases it.

To complete this problem, assume that the engineer is constrained by other considerations from changing anything *but* the material.

The most logical selection of a new material would be one that gives *the highest H_m* at a B/H ratio of 16.8 and that *operates as closely as possible to its BH_{max} point.* If operation at BH_{max} is not possible, the material should be *above* the knee of the demagnetization curve.

Applying a load line (B/H) of 16.6 to Figures 7-1, 7-2, and 7-3, the material that best meets the above criteria is cast Alnico V-NC1. For this material operating at this B/H, H_m = 550 Oe (see Figure 7-2, page 42).

Then, using this material:

$$B_g = \frac{H_m L_m}{L_g} = \frac{(550)(7.61)}{.635}$$

$$B_g = 6690 \text{ Oe}$$

The replacement material is, in fact, "too good," since only 4500 G is required. Working "backward," the H_m required of the magnet at a B/H ratio of 16.6 can be determined:

$$B_g = \frac{H_m L_m}{L_g}$$

$$4500 = \frac{H_m (7.61)}{.635}$$

$$H_m = \underline{375 \text{ Oe. required}}$$

Again, referring to Figures 7-1, 7-2, and 7-3, at a B/H ratio of 16.6, does any material produce exactly H_m = 375?

Actually, no material produces exactly this H_m. Figure 7-1 shows that cast Alnico VIII comes the closest, with H_m = 438. Next are Alnico VI-NS1 and VI-NS2, successively. However, all three of these materials are operating very "inefficiently" (well above their BH_{max} points). In addition, all three materials are less readily available and more costly than the cast Alnico V-NF1 that produced B_g = 6690 G!

The best material then *would* be cast Alnico V-NF1. If it produces a flux in the air gap that is *too* high, the engineer has two alternatives:

1. Reduce the length of the magnet (and recalculate).
2. Demagnetize and *calibrate* the magnet *after it is in the circuit* [5] by means of a suitable *external* demagnetizing force. Demagnetization-calibration would have the added advantage of *stabilizing* the magnet, so that it would be less subject to future environmental effects.

After reviewing the discussion that preceded this problem, it is evident that the reluctance factor (f) was never considered. This subject will be treated in detail in Chapter 9.

5. The reasons for this will be covered in subsequent chapters.

9

Circuit Effects

A brief review of the 16 factors that affect the performance of a magnet discussed in Chapter 7 shows that *every* part of the magnetic circuit is interrelated to every other part.

Pole Spacing

One of the least recognized factors in the ultimate performance of the magnet and/or the circuit *is the spacing between the poles.* This spacing is important, whether it is in reference to the poles of a basic magnet used alone (which is uncommon) or to the spacing of the soft iron pole pieces (that are supplied with flux by a basic magnet elsewhere in the circuit). Of course, in many applications such as galvanometer magnets, the spacing is determined by the other "hardware" that must fit in the air gap. In these cases, the air gap should be kept to a minimum consistent with this hardware. There are, however, countless other applications, such as tractive magnets (used for holding, separation, etc.), where the spacing between poles can be set at the discretion of the designer.

The effect of pole spacing is most simply illustrated by a study of the Pull Curves in Figure 9-1. This series of experimentally determined curves shows the holding force for various air gaps of 18 cast Alnico 5 horseshoe magnets of various sizes. The magnets range from .005 to 39.76 pounds; the space between their poles ranges from ¼ to 8¾ in. In each case, the magnet has been tested without pole pieces, after being fully saturated, on a very thick, cold-rolled steel plate that fully spanned the poles and could carry all the flux produced by the magnet. The particular significance of these curves is evident from two considerations. These are a comparison of:

1. The holding force (F) *per pound* (W) *of magnet* (rather than by the total force produced by each magnet).
2. The parallelism of the various curves.

Comparing the largest and the smallest magnet on a F/W basis *at zero air gap:*

Largest: $W = 39.8$ lbs.; $G = 8.75$ lbs.: $\dfrac{F_L}{W_L} = \dfrac{630}{39.8} = \underline{15.8}$

Smallest: $W = .005$ lbs.; $G = .25$ lbs.: $\dfrac{F_S}{W_S} = \dfrac{1.5}{.005} = \underline{300}$

On the basis of comparison, with a zero air gap between the magnet and the test piece, the small magnet produces almost 20 times the holding force per pound of the larger magnet. There obviously must be some reason for the smaller magnet's greater "efficiency" *in holding.*

Next, compare these same magnets in the same way at an air gap of .010 in.:

Largest: $\dfrac{F'_L}{W_L} = \dfrac{500}{39.8} = \underline{12.5}$

Smallest: $\dfrac{F'_S}{W_S} = \dfrac{.33}{.005} = \underline{66}$

In this case, the smaller magnet is only a little over *five* times more efficient in holding than the larger magnet.

A glance at Figure 9-1 shows that with any air gap over .020 in. the holding force produced by the small magnet is zero and, of course, must be considered to have zero efficiency for this purpose.

Finally, comparing the relative parallelism of the curves shows that in the first section of the curves (from $L_g = 0$ to $L_g = .010$ in.), the successive curves are almost parallel. In the next section (from $L_g = .010$ to L_g

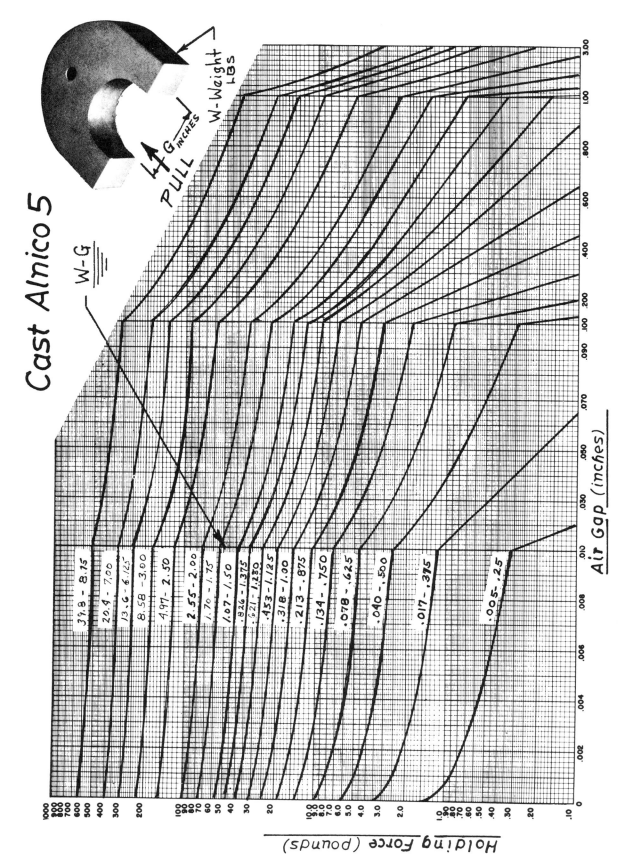

Figure 9–1. Pull curves for a family of horseshoe magnets. (Courtesy of Crucible Steel Company)

= .100 in.) the curves for the 13 largest magnets are still relatively parallel, but for the smallest three they diverge very widely. In the third section (from L_g = .100 to L_g = 1.00 in.), the curves for the eight largest magnets are essentially parallel, while the next ten smaller magnets diverge.

The inescapable conclusion to be drawn from a study of parallelism is that the larger the magnet, the greater the efficiency *in reaching force.*

The terms "holding power" and "reaching power" are widely employed by the manufacturers and users of magnetic separators and lifting magnets. The differences in operational characteristics (holding vs. reaching) cannot be explained on the basis of material, geometry, or size. The explanation lies in the concept of the magnetic *gradient.* Gradient can be defined as *the rate at which the flux density* decreases *with distance from the pole faces.*

In the smallest magnet in Figure 9-1, the pole faces are closer together (G = ¼ in.) and the *gradient* is higher, producing a high *holding power.* In the larger magnet, the pole faces are further apart (G = 8¾ in.), resulting in a low gradient which produces a high *reaching power.*

Experiments conducted with horseshoe magnets of identical material, weight, area, and mean magnetic length, but different pole spacing, have clearly established the effects of the magnetic gradient. Unfortunately, little has been done to relate specific gradients to performance for most magnetic products.[1] Recognition of *general* gradient effects has, however, led to new design concepts and performance improvements in some industrial magnetic products. For example, industrial magnets (really "circuits") intended as "holding" devices are designed with a small space between the poles, while those intended as "reaching" magnets have more widely spaced poles. An example of close pole spacing is the permanent magnet machine tool chuck (see Figure 9-2). The reaching power of such chucks is quite limited. The work piece must be brought into close proximity with the surface in order to be "captured" and held. At a distance of .100 in., most such chucks are completely ineffective.

Figure 9-3 shows the internal construction of two circuits used in basic "tube magnets." These tube-shaped assemblies are widely used to make a variety of tramp iron [2] separators, floor sweepers, and material handling devices.

The axial pole type (Figure 9-3A) is made with cast Alnico 5 rod magnets 1¾ to 2¾ in. long (magnetized

---

1. One exception known to this author is the attempt to relate performance and establish specifications for the gradient in iron ore (taconite) beneficiation separators.

2. Tramp iron is the term used to describe any ferrous material from grindings to nails that can contaminate nonmagnetic materials.

Figure 9-2. Field pattern of a permanent magnet machine tool chuck. (Photograph courtesy of O.S. Walker Company)

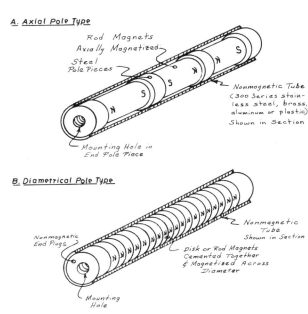

Figure 9-3. Construction of basic tube magnets used as part of various separator and material handling devices.

axially, of course). The magnets are separated by mild steel pole pieces and, like magnetic poles, face each other. The flux from the repelling magnets passes radially out of the pole pieces and travels axially to the opposite pole of the same magnet.

The diametrical pole type (Figure 9-3B) is made with sintered Ferrite 1 disc magnets (magnetized *across the diameter*). The discs are secured so they cannot rotate (and "cancel" each other out), since all north and south poles are in alignment.

The difference in field pattern, "reach," and performance of these tube assemblies is very substantial. The axial pole type has a deep field (low gradient) that extends around the entire circumference. This type is excellent for capturing relatively large objects, such as pieces of wire, nails, and so on. It is not as effective, however, at *holding* these objects once they are captured, and "wash-off" by subsequent material flow may occur.[3] In this design, as in the diametrical pole type, the flux is more concentrated at the pole ends, but since these pole ends are further apart and are fewer in number (for the same unit length), fine iron particles may completely escape "capture."

In the diametrical pole design, the magnetic field is kidney-shaped and the north and south poles are continuous on opposite sides for the full length of the assembly. This type has a relatively shallow magnetic field (higher gradient) and is best for capturing fine iron particles like ball mill grindings. Unless the material flow is carefully directed at single-tube units or the spacing between tubes for multiple-tube units [4] is kept smaller than it is for the axial type, the unit is far less effective for either large tramp iron or fine tramp iron. This type *does* have the advantage of holding all

---

3. However, the tube shape is deliberate, to provide a "sheltered zone" on the underside (opposite flow direction) into which the captured ferrous material can slide and be protected against wash-off.

4. For multiple-tube units, the poles of adjacent tubes must be arranged so that they are and stay in line. They are also arranged with unlike poles facing.

Figure 9-4. A variety of tramp iron clings to a permanent magnet pipeline trap. (Photograph courtesy of Eriez Magnetics, Erie, Pa., USA)

APPLICATION

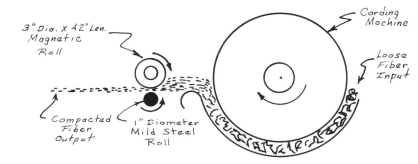

DESIGN

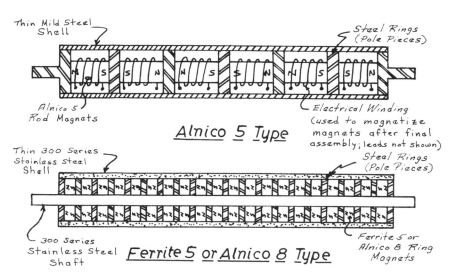

Figure 9-5. Crushing rolls used in textile fiber carding.

captured material better against wash-off (especially "fines," which can be quite hard to hold in a heavy material flow).

Figure 9-4 is a photograph of a multiple-element, tube magnet "pipeline trap" made with axial-type tube elements. Note the variety of tramp iron captured, both bridging from tube to tube and concentrating at the pole pieces along each tube.

Figure 9-5 illustrates another application where pole spacing has marked effects on gradient. In this application shown at the top of the figure, a 42 in.-long roll made with Alnico 5 magnets has seven poles while in the Ferrite 5/Alnico 8 type there are 20 poles. The Alnico 5 type has a considerably higher depth of field that makes it pull back the 1 in.-diameter (mild) steel lower roll if the compacted fiber becomes very thick or if it is thrust away by an unusually large, solid inclusion. The Ferrite 5/Alnico 8 type has a greater tractive force (on the 1 in. roll) at "normal" operating distances, but it has less reaching power for special operating conditions.

Figures 9-6, 9-7, and 9-8 are based on actual test results for typical rectangular and circular holding assemblies using grade 1 and grade 5 sintered Ferrites. The comparative holding power vs. the reaching power of each is quite evident from these curves.

Figure 9-9 shows pull curves for one rectangular and one circular holding assembly with two different grades of sintered ferrite magnets. Note that with a higher grade (a higher BH_{max}) material, the magnitude of the forces produced is higher, but the basic *shape* of the curves *has not substantially changed*.

Figure 9-10 compares the forces produced by a cast Alnico 5 magnet and a sintered ferrite 5 assembly of almost identical volume and weight. The Alnico 5 magnet is obviously a better "holding" magnet, while the Ferrite 5 assembly is a better "reaching" magnet. While the overall width of the Alnico unit is greater (4 in. vs. 3¾ in.), the *effective* distance between poles is greater in the Ferrite 5. Also (as will be shown in a later chapter), the force produced by any magnet is a function of

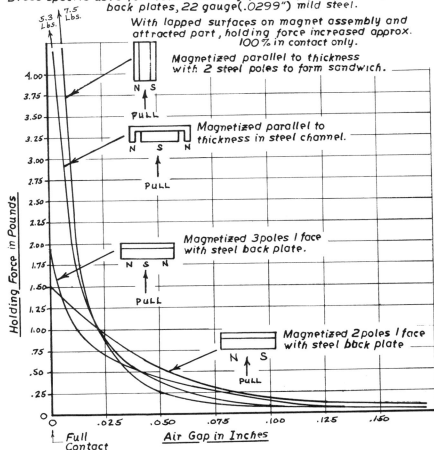

Figure 9-6. Force-air gap curves for typical small ferrite 1 rectangular holding assemblies.

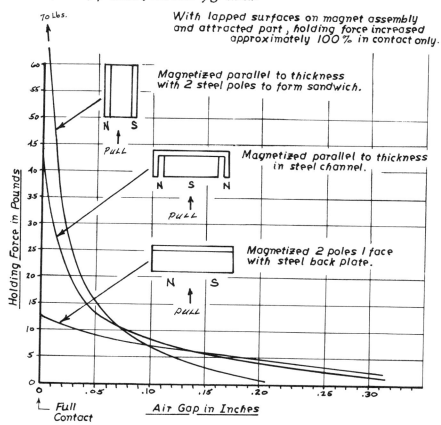

Figure 9-7. Force-air gap curves for typical small ferrite 5 rectangular holding assemblies.

the flux *density*. In contact (zero gap), the flux density produced by the Alnico magnet is greater (both because it has a higher flux operating level and a smaller pole area) than the flux density produced by the ferrite magnet.

The preceding discussion does not provide the engineer with concrete answers as to the best pole spacing or gradients to use in his circuit. Much experimental and analytical work remains to be done on this topic. But our discussion does give the engineer some insight into the matter of pole spacing and, hopefully, increases the possibility that this factor will be considered in future magnet designs and applications.

Use of Pole Pieces

As mentioned in previous chapters, basic permanent magnets are seldom used alone. They are usually combined with mild steel pole pieces that serve one or more of the following purposes:

1. *Direct the flux* from the basic magnet to the required area—often an absolute necessity to obtain the required path in anisotropic (oriented) materials.
2. *Change the flux* density level produced by the basic magnet to the level required in the "work" area. Pole pieces may serve to either concentrate or distribute this flux, depending upon the application requirements.
3. *Increase* the *uniformity* of the flux in the work area by distributing the normal variations that may occur in the basic magnet material itself.
4. *Eliminate* the need for *machining* costly, complex surfaces on the basic magnet pole faces.
5. *Protect the* brittle basic *magnet* from mechanical or environmental damage (heat, radiation, and so on).
6. *Provide adaptability* of stock basic magnets (or fewer "specials") to a wide variety of applications (products).
7. *Provide interchangeability* or replaceability, where required by the application.
8. *Complete the* magnetic *circuit* when the basic magnet

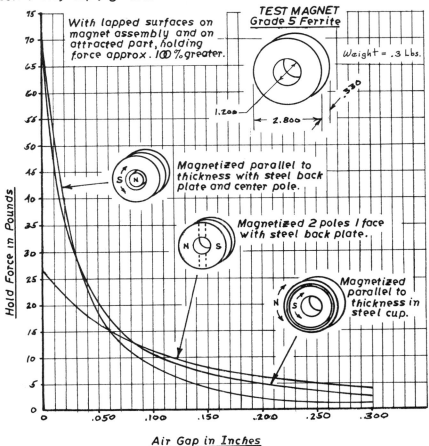

Figure 9-8. Force-air gap curves for typical small ferrite 5 circular holding assemblies.

length is less than the spatial requirements imposed by other hardware necessary to form the product.

The magnetic circuits used in loudspeakers or in motor-generator stators and rotors provide interesting examples of several of the above points. In all of the loudspeaker circuits shown in Figure 9-11, pole pieces serve the primary purposes of directing the flux to the air gap, concentrating the flux in this gap, eliminating the need for costly machining of the very hard basic magnet, and, in all cases except Figure 9-11D, completing the magnetic circuit. Secondary purposes of the pole pieces in these circuits may be to permit the use of highly oriented materials (such as cast Alnico 5DG and 5-7 or sintered Ferrite 5) to increase gap-flux uniformity, and other factors.

A similar application is shown in Figure 9-12 for a subminiature (gross volume = .375 cm^3) microphone.

In the typical motor-generator stator circuits (Figure 9-13) and rotor circuits (Figure 9-14) pole pieces are invariably used for a multiplicity of reasons. The basic magnet versions of the stators (Figures 9-13D and 9-13E) or of the rotors (Figures 9-14B and 9-14D) are the least used, except in small units where they can be made of an *unoriented* sintered Alnico or ferrite material. The photographs in Figures 9-15, 9-16, and 9-17 illustrate the diversity of the magnet shapes and sizes to which these circuits apply.

Figures 9-18, 9-19, and 9-20 show a simple, basic rectangular magnet magnetized in various ways, alone and also with typical pole pieces added. The comments under each illustration should be studied, and the fact that these circuits are *arranged in order of* increasing *holding force should be noted.*

Figures 9-21, 9-22, and 9-23 apply the principles in Figures 9-18, 9-19, and 9-20 to a simple, basic *ring* magnet. All six of these figures show how basic shapes can be magnetized for different effects [5] and how

---

5. That is, holding power vs. reaching power, as we saw in the preceding section of this chapter.

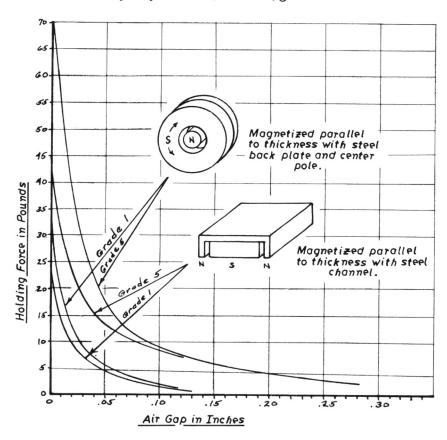

Figure 9-9. Comparison of pull curves for two grades of ferrite in two typical holding magnet assemblies.

pole pieces can be used to produce improvements in these effects.

Design of Pole Pieces

This section, while dealing primarily with pole pieces, applies equally well to other possible circuit, flux-carrying elements such as "back plates" and shunts. All pole pieces must meet three requirements:

1. They (it) must be of the proper size and shape to "collect" *all* of the flux available from the basic magnet (except the leakage flux related to the basic magnet itself).
2. They (it) must be of the proper size and shape to form the required working air gap.
3. They (it) must be capable of carrying the flux available from the magnet to the air gap with an "acceptable" level of internal (mmf) magnetomotive force loss.

Requirement (1) is determined by the design (area) of the basic magnet. Requirement (2) is determined by the function of the total product and/or the related hardware that must fit in the air gap. Requirement (3) is determined by the circuit design engineer. We will consider this requirement in detail first.

There are two kinds of mmf losses associated with the use of pole pieces in a circuit:

1. The losses due to the reluctance in the pole-piece material itself.
2. The losses due to the air gaps formed where the pole pieces join the magnet. (These air gaps are in *series* with the working air gap.)

Calculation of the mmf losses due to the reluctance of the material can be made by using the DC magnetization curves [6] in Figure 9-24 and the demagnetization curves for the material in the permanent magnet. The method involved is most easily explained by a sample

6. These curves are identical to the "virgin" or initial magnetization curves discussed in Chapter 6.

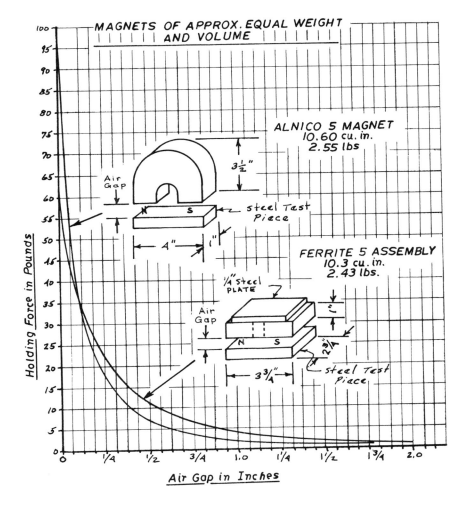

Figure 9-10. Comparison of pull curves for similar Alnico 5 and ferrite 5 horseshoe-type permanent magnets.

problem—a continuation of Design Problem 2 in Chapter 8.

DESIGN PROBLEM 2 (CONTINUATION A)

Assume that the final choice of material for Design Problem 2 was cast Alnico V-NF1 and that the *required* B_g = 6690 G (instead of the original 4500 G).

The calculations in this problem showed that the B/H ratio = 16.6, as established by the dimensional parameters of the circuit. Under these conditions, determine:

1. The mmf loss and magnet length increase needed to overcome this loss if the pole pieces are made of cast iron (as cast).
2. The "optimum" pole-piece material, the mmf loss for this material, and the increase in magnet length required to overcome this loss.

The first step in the solution of (1) and (2) is to determine the flux density in the pole pieces at the point where these pole pieces join the magnet. This can best be achieved by obtaining the flux density *in the magnet* (B_m) from the demagnetization curve for cast Alnico V-NC1 (see Figure 7-2, page 42). Since the magnet is operating at a B/H ratio of 16.6, projecting from the material curve at this B/H to the induction (B) scale gives B_m = 9100 G.

For the purposes of this calculation, we will assume that *all* of the flux in the magnet passes into the pole pieces at the joints, although this is not quite true (as can be seen from the permeance paths in Figure 8-8, page 58). In reality, part of the flux from the magnet along permeance paths P_{16}, P_{17}, and P_{19} (and a part of P_{14}) "leaks" out the sides of the magnet and never passes into the pole pieces. However, $B_p = B_m$ is a "safe" assumption, since the true B_p will be somewhat *lower* than the value used.[7]

---

7. If a design is very critical, B_p can be calculated quite accurately, since P_{14}, P_{16}, P_{17}, and P_{19} have already been calculated.

CIRCUIT EFFECTS 69

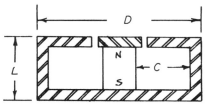

A. Center Slug Type

Used primarily with Alnico magnet materials. Seldom practical for Ferrites

B. Ring Type

Used with both Alnicos and Ferrites. Only practical design for Ferrites.

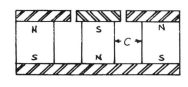

C. Slug & Ring Type

Used exclusively with Alnico as a means of reducing Length of Type A above.

D. W-Magnet Type

Used exclusively with Alnico to give shortest possible Length and minimum of parts. Not usable with Ferrites.

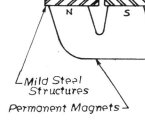

When Alnicos are used, magnet will be long with a small cross-section and coil will be small in diameter. When ferrites are used, magnet will be short with large area and coil will be large in diameter. With ferrites, C may be very small (1/16")

Figure 9-11. Basic magnetic circuits for loudspeakers and similar electromechanical transducers (all types shown in section; coils not shown in any of the designs).

To solve for (1) above, again refer to the DC magnetization curves in Figure 9-24, and locate the curve for "as-cast" cast iron (located in the right half of the figure). Projecting from the intrinsic-induction scale at $B = 9100$ to the cast-iron curve and then down from this curve to the magnetizing-force scale gives $H_p = 130$ Oe. Therefore, the mmf loss in the pole pieces is 130 Oe.

Since the only source of mmf in the circuit is the magnet, its length will have to be increased by the ratio:

$$\frac{H_p}{H_m} = \frac{130}{550} = .2361$$

The original length (from Design Problem 2) was $L_m = 7.61$, so the new length will have to be increased by 23.61 percent, or:

$$L_m' = L_m + .2361 \, L_m$$

$$L_m' = 7.61 + .2361 \, (7.61)$$

$$L_m' = \underline{9.41 \text{ cm}}$$

To solve for (2) above, we must decide on the appropriate criteria for the "optimum" pole-piece material. The two criteria that apply are:

1. The material should be consistent with the cost factors (both material acquisition and machining) of the product as a whole.
2. The material selected should operate at its point of maximum permeability (μ). This is the point at which the B/H ratio $(\mu = B/H)$ is greatest and, therefore, the greatest flux carrying ability is obtained with the least expenditure of mmf.

The cost factor is a decision that must be made by the design engineer. It is determined by product factors such as performance vs. cost. For example, a costly and difficult to machine material such as 4750 Alloy would hardly be appropriate for motors to be used in toys. (C1D18 steel or C1010 *would* be appropriate!)

The maximum permeability can be determined arithmetically with some effort, again from the DC magnetization curves in Figure 9-24. Maximum perme-

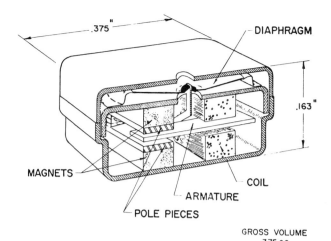

Figure 9-12. Cutaway view of Knowles Electronics subminiature microphone shows use of Alnico magnets as energy source. (Courtesy of Arnold Engineering Co.)

ability can also be estimated with reasonable accuracy by assuming that it will occur at the steepest slope of the DC magnetization curve.

For the purposes of this problem, assume that a "premium-price" material cannot be used.

At a B level of approximately 9000 G in Figure 9-24, cold-drawn carbon steel (annealed) will meet the above criteria. At a flux level of 9000 G, this material is operating at a high permeability with a required H of only approximately 4.3 Oe. The mmf loss in pole pieces made of this material would therefore be 4.3 Oe, and the magnet length increase necessary to overcome this force would be:

$$L_m'' = \frac{4.3}{550} \times 100 = \underline{\underline{.785 \text{ percent}}}$$

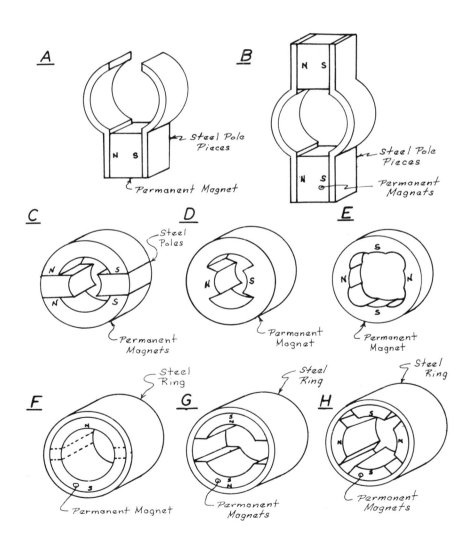

Figure 9-13. Permanent magnet stators.

CIRCUIT EFFECTS 71

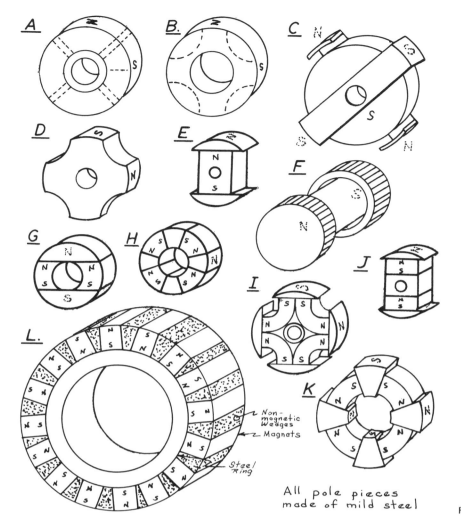

Figure 9-14. Permanent magnet rotors.

Figure 9-15. Core-type magnet for galvanometer. (Courtesy of Allevard-Ungine, and Magnetic Equipment Company)

Figure 9-16. Permanent magnet rotor for automobile alternator. (Courtesy of Sermag, and Magnetic Equipment Company)

Figure 9-17. Rotor of permanent magnet pilot generator for generating station. Output is 3 KVA at 75 rpm. (Courtesy of Alsthom, and Magnetic Equipment Company)

$$f = \frac{H_g + H_P + H_J}{H_g}$$

where H_g is the mmf required across the air gap, H_P is the mmf required to overcome losses in the pole pieces, and H_J is the mmf required to overcome losses in the joints where the magnet and pole pieces meet (series air gaps).

However, since $\mu = 1$ in air:

$$H_g = B_g$$

and the reluctance factor equation becomes:

$$f = \frac{B_g + H_P + H_J}{B_g}$$

Applying the reluctance factor to Equation 12 (Figure 8-7) gives us:

$$fB_gL_g = H_mL_m$$

or to Equation 13 gives us:

$$B_g = \frac{H_mL_m}{fL_g}$$

The reluctance factor equation above has introduced a new consideration to design calculation:—*joint reluctance* (R_J) and the mmf required to overcome it.

Joint reluctance occurs wherever two flux-carrying ferromagnetic parts (that is, magnet and pole pieces) meet *imperfectly*. As a practical matter, every joint is imperfect to *some* degree. Only two ground surfaces whose parallelism is well controlled can join so perfectly that the reluctance factor can be neglected. In the circuit in Design Problem 2, there are two joints—one at each end of the magnet. These joints introduce small but potentially significant air gaps that are in *series* with the working air gap. The problem in calculating the reluctance, permeance, or mmf loss in these gaps is in determining the *length* of the gaps. Obviously, their length will be a function of surface finish on adjoining parts, edge chipping, surface parallelism, and even surface contaminants (oil, for example). Unfortunately, there is no provable method of establishing these series air gaps, except by estimates or by measurements on previously made parts. Once the gap lengths *are* established (or guessed at!), only one method exists for calculating the H_J required to compensate for them.

This method incorporates the joint air gaps in the *initial B/H ratio* calculations. Referring back to Design Problem 2 in Chapter 8, it is evident that the joint air gaps can be considered as *series air gaps* and that their permeance can be calculated by the appropriate Rotor equation (see Figure 8-5, page 55). In this problem,

One final word on pole-piece design and material mmf losses. It will be evident to the discerning reader that the flux level at the pole *faces* will *not* be 9000 G, but will, in fact, be the *gap* flux density (6690 G). It would therefore be magnetically desirable to gradually reduce the cross-sectional area of the pole pieces from the magnet contact area to the gap. By *tapering* the pole pieces, all parts of the pole pieces can be made to operate at the same flux density, thus saving material and weight. In actual practice, except in the case of critical weight/space designs, tapering is seldom practical from an economic standpoint.

One important relationship in all magnet circuit design can be developed from the preceding reluctance calculations: the reluctance factor (f). In Design Problem 1 in Chapter 7, this reluctance factor was "assumed" to be between $f = 1.1$ and $f = 1.5$.

In reality, the reluctance factor is the ratio of the *total* mmf required in the circuit to the mmf needed for the air gap. Expressed as an equation:

CIRCUIT EFFECTS 73

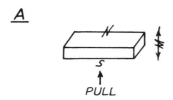

A

Basic Magnet - Magnetized Parallel to Thickness

Weakest type. Produces equal holding forces on both surfaces (N and S). Pull is relatively uniform over whole surface. Best with high coercive materials such as ferrites and Alnico 8.

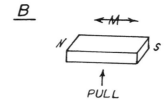

B

Basic Magnet - Magnetized Parallel to Length

Next strongest. May be used to hold on both sides with reduced pull on each. Force is greatest at poles and lowest in center. Object being held must span poles for best results. Used primarily with Alnicos.

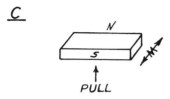

C

Basic Magnet - Magnetized Parallel to Width

Generally stronger than A, but strength compared to B depends on type of material and dimensional ratio. Generally similar to B. Any type of material may be used, but dimensional ratio must be properly selected to suit that material for best results.

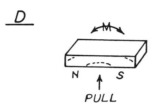

D

Basic Magnet - Magnetized as a Horseshoe

Generally stronger than C, but the performance is highly dependent on material used. Normally used with any type of unoriented material. Good depth of field for "reaching power".

← M → shows direction of magnetization

− − − shows unmagnetized zones

Figure 9-18. Simple rectangular holding magnet assemblies in order of increasing holding force for the same volume of material. (Continued in Figure 9-19 & 9-20.)

assuming an average air-gap length for each joint of .005 in., or .0127 cm, the permeance would be:

$$P_{g/1} = \frac{A_{g/1}}{L_{g/1}} = \frac{W_g \times t_m}{L_{g/1}} = \frac{(3.18)(1.9)}{.0127}$$

$$P_{g/1} = 476$$

But:

$$P_{g/2} = P_{g/1}$$

Therefore:

$$P_{g/2} = \underline{476}$$

The reader may be startled at the apparent magnitude of these joint permeances compared to the working air-gap permeance (P_g = 4.77). But recall that magnetic permeance is the equivalent of electrical conductance. Also recall that:

$$P = \frac{1}{R} \quad \text{or} \quad R = \frac{1}{P}$$

so that a *high* permeance represents *low* reluctance to the passage of flux.

The calculated joint permeance can be incorporated most correctly into the total permeance (P_T) calculations by first redrawing the electrical analog in Figure 8-9 (page 59). This redrawn analog is shown in Figure 9-25. Do not forget that permeances in series can be added arithmetically like electrical resistors in parallel, or:

$$\frac{1}{P_T} = \frac{1}{P_1} + \frac{1}{P_2} + \cdots + \frac{1}{P_x}$$

When the new P_T is calculated and used to determine the B/H ratio, the joint reluctance will be automatically included in the required magnet length calculations.

Concentration of Flux and Saturation

In the design of the magnetic elements of any circuit, caution must be exercised to avoid concentrations of

E
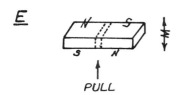

Basic Magnet - Magnetized 2 Poles - 1 Face (Thru)

Next strongest. Provides two working surfaces (N-S and S-N) of approximately equal strength. Pull is greater on each side when both sides are used at the same time. Fair depth of field. Used with all materials, but best with ferrites and Alnico 8.

F
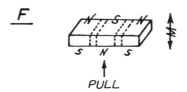

Basic Magnet - Magnetized 3 Poles - 1 Face (Thru)

Similar to E, but with greater pull in direct contact and lower depth of field (reduced reaching power). Ideally area of outer poles should be one-half of the center pole. Seldom used with Alnicos; best with ferrites.

G
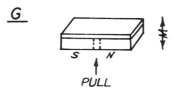

Magnet Assembly - 2 Poles 1 Face with Steel Back Plate

Identical to E, but with steel on one face always in place. Increases both in-contact pull and pull with air gap. Best with ferrites or with Alnico 8.

H
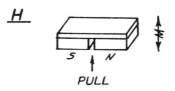

Magnet Assembly - 2 Magnets with Steel Back Plate

Similar to G, but with separate magnets used. Gives somewhat higher pull than G when used with an Alnico material.

I
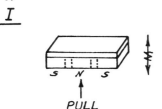

Magnet Assembly - Magnetized 3 Poles - 1 Face (Thru)

Similar to F, but with steel on one face always in place. Increases in-contact pull, but lowers pull with air gap. Seldom used with Alnicos; best with ferrites.

Figure 9-19. Simple rectangular holding magnets and magnet assemblies (continued in Figure 9-20).

flux at unexpected areas. Two common areas where such concentrations may occur are at the tips of the pole pieces and in the joint between the pole pieces and the basic magnet. We will now consider the causes and effects of such concentrations.

Recall one of the axioms from Chapter 8: "Flux lines repel each other." Since they do repel, flux lines will tend to be more concentrated toward the surface of a magnetic material than they are at its center. This repulsion-concentration effect accounts for the difficulty in obtaining a very high degree of flux uniformity in the working air gap and for the leakage-fringing effects already discussed.

Close examination of Figure 9-26A shows the effects of sharp corners on the edges of a pair of pole pieces that form a working air gap. The existence of such concentrations and the fact that they are not limited to soft magnetic materials can be supported experimentally by an examination of Figures 7-9 and 7-10 (pages 47 and 48). In the case of both round- and rectangular-shaped bar magnets, the flux density at the edges is higher than it is at the center of the magnet. For the round bar, the density around the circumference is uniformly higher than it is at the center. For the rectangular bar, the density is higher at the edges and higher *still* at the corners (although test data do not show corner measurements).

Figure 9-26B shows, in exaggerated form, the concentration of flux that can occur due to poor surface finishing at the joints. Since the permeability of the typical ferrous material is at least 10,000 times as great as air, flux will greatly "prefer" traveling through the iron and will avoid the air gap.

The effect of flux concentrations in any magnetic circuit is to increase the "losses." If the flux concentration occurs at pole tips, it will increase the mmf losses in the pole material and/or increase fringing-leakage losses. If it occurs at the joints, it will increase reluctance losses and it may also increase leakage or fringing (at the joints).

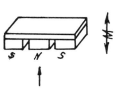 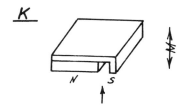

Magnet Assembly – 3 Magnets with Steel Back Plate

Similar to I, but with separate magnets. Gives somewhat higher pull than I when Alnicos are used.

Magnet Assembly – Parallel to Thickness with Steel Angle

Generally stronger for in-contact pull than previous types, but actual strength depends on material used and dimensional ratio. Large area pole (N) gives fair reaching power, while main gap (N-S) gives high holding power. Used with all types of materials.

 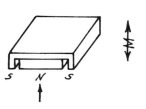 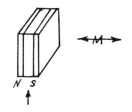

Magnet Assembly – Parallel to Thickness with Steel Channel

Fair reaching power from large area pole (N). Very high holding power from double gap (S-N and N-S). Used mostly with ferrites and Alnico 8.

Magnet Assembly – Parallel to Thickness with 2 Steel Plates

Highest holding and lowest reaching power of any type. Provides four holding surfaces. Seldom used with Alnicos, except Alnico 8; best with ferrites.

Figure 9-20. Simple rectangular holding magnets and magnet assemblies (continued).

In both the case of "edge" concentrations and the case of "surface irregularity" concentrations, as the flux density approaches the saturation level for that material, the losses will increase rapidly. A study of the DC magnetization curves in Figure 9-24 makes this quite clear. For example, in Design Problem 2 (continuation A) earlier in this chapter, the "optimum" material selected for the pole pieces was cold-drawn carbon steel (annealed). This material was selected on the basis of an internal flux density level of 9000 G, which resulted in an internal loss of only 4.3 Oe. If the flux level in the pole pieces as a whole was increased to 16,500 G, the internal loss would jump to 50 Oe. An 83 percent increase in flux would have produced almost a 1200 percent increase in losses! Similarly, an increase of flux density level to 18,000 G (100 percent) produces a loss of 125 Oe (2900 percent). When only a *part* of the pole pieces is approaching saturation, the losses will increase but not to this extent.

Throughout the preceding magnet and circuit calculations, the assumption has been made that the flux density in any one element of the circuit was uniform throughout the cross-section of that element. This assumption is "normal" in applied magnet design, since more exact methods become a mathematical horror. The assumption has practical validity, provided the design engineer avoids the identifiable flux concentrations.

Joint irregularity concentrations can only be avoided by the use of good quality machine finishing. Pole-tip and similar flux concentrations can be avoided by beveling the edges. This beveling may range from simply "breaking" the edge with a file to the substantial bevel illustrated in Figure 9-26C.

It should be very evident that the greatest "sin" in magnetic circuit design is designing all or any of a circuit's ferrous parts to operate near or at saturation.[8] The old magnetic designer's axiom that "there can't be too much iron in a circuit" is almost always true.[9]

8. There *are* times when a circuit is designed to saturation deliberately to achieve "special effects."
9. This is true, provided that the iron is not "misused." Ill-considered use of oversized parts can cause increases in leakage, fringing, *and* cost.

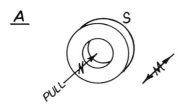

Basic Magnet-Magnetized Parallel to Thickness

Weakest type. Produces equal holding forces on both faces (N and S). Pull relatively uniform over whole surface. Used with all materials. For thin rings, use ferrites or Alnico 8; for thick rings, use Alnico 5.

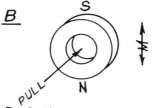

Basic Magnet-Magnetized Across the Diameter

Next strongest. May be used to hold on both faces, but pull is reduced when both surfaces are used. Force is greatest at both poles, but low in center. Object being held should span poles for best results. Used with all types of materials.

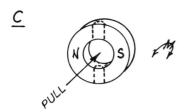

Basic Magnet-Magnetized as a Horseshoe

Stronger than A or B, with good depth of field. Normally used only with oriented materials of any type.

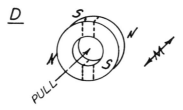

Basic Magnet-Magnetized 2 Poles-1 Face (Thru)

Strength compared to C depends on type of material and on dimensional ratio. Provides two working surfaces (N-S and S-N) with approximately equal pull. Pull is greater on each side when both sides are used at same time. Fair depth of field that improves when one side is close circuited. Best with ferrites and Alnico 8.

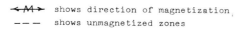

← M → shows direction of magnetization
– – – shows unmagnetized zones

Figure 9-21. Simple ring or slug holding magnets and magnet assemblies in order of increasing holding force for the same volume of material. (Continued in Figure 9-22 & 9-23.)

The Object in the Air Gap

In the last section, the "sin of saturation" was presented. This section will deal with the more common "sin" of not recognizing that *the object in the air gap is part of the total circuit.* This object *must be there* if the magnet is to do any work, and it must, of necessity, have *some* effect on the rest of the circuit. *It is not a passive element but an active one!*

Essentially, two effects may be produced by the object in the working air gap. The first of these is the very direct effect of distorting or decreasing the field in the gap. An example of a direct effect is produced by a closed circuited conductor [10] moving relative to the magnetic field. By Lenz' law, this conductor has a current flowing in it and, in turn, has a field of its own. The field around the conductor will interact with the magnetic field to cause distortion, to cause flux increases in some areas and flux decreases in other areas, or even

10. This may be (and, in fact, usually is) nonferrous.

to partially demagnetize the magnet (temporarily or permanently). Permanent magnet generators and motors, are products where this type of effect must be considered.

The second kind of effect is produced by a "passive" ferrous object that may be stationary or free to move. The presence of a ferrous object in the working air gap changes the reluctance (or permeance) of the total circuit and changes the *B/H* ratio (load line) of the magnet. If the object is free to move, the *B/H* ratio may not be a constant, but will change with the position of the object. This type of operation is common with products such as lifting magnets and stepping motors. It represents the so-called "dynamic" operation of a magnet. When the object in the gap is ferrous, the possibility that this object will reach *saturation* must also be considered (as in any other part of the circuit).

The interaction of magnetic fields in motors, generators, and similar products is adequately covered in many specialized books and need not be specifically considered here.

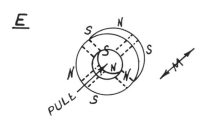

Basic Magnet – Magnetized 4 Poles – 1 Face (Thru)

Similar to D, but gives greater pull in contact and reduced pull with an air gap (lower reaching power). Seldom used with Alnicos; best with ferrites.

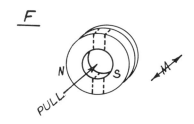

Magnet Assembly – 2 Poles 1 Face with Steel Back Plate

Identical to D, but with steel back plate on one face (cemented in place). Increased pull both in contact and with air gap. Best with Alnico 8 and ferrites.

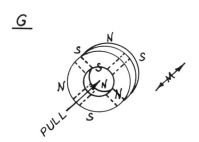

Magnet Assembly – 4 Poles 1 Face with Steel Back Plate

Identical to E, but with steel back plate on one face (cemented in place). Performance similar to E, but with greater holding power. Seldom used with Alnicos; best with ferrites.

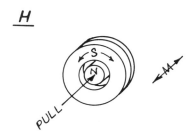

Magnet Assembly – Parallel to Thickness with Steel Back Plate & Center Pole

Highest holding with minimum diameter. Best pull in contact; low reaching power. Used with all types of materials.

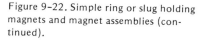

Figure 9-22. Simple ring or slug holding magnets and magnet assemblies (continued).

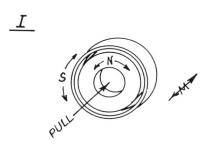

Magnet Assembly – Parallel to Thickness with Steel Cup

Highest holding power where center hole without obstruction is required. Very low reaching power; force falls off rapidly with air gap. Used with all materials, but thin designs require ferrites or Alnico.

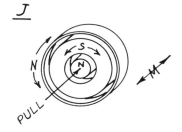

Magnet Assembly – Parallel to Thickness with Steel Cup and Center Pole

Best design for in contact holding power. Combination of H and I. Very low reaching power. Used with all materials, but best with Alnico 8 and ferrites where thin design is required.

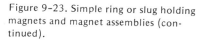

Figure 9-23. Simple ring or slug holding magnets and magnet assemblies (continued).

The dynamic operation of a magnet with a ferrous object in the air gap will be covered in detail in Chapter 10. However, at this point, it will be useful to consider the characteristics of this object that are relevant. These are *its magnetic properties, location in the gap, shape,* and *size*. Lifting or tractive magnets can serve as a single example to amplify each of the above points. The relevant magnetic property is the ability of the object being attracted to carry the available flux. Obviously, a part made of aluminum, brass, or wood will not be attracted at all, while pure iron and mild steels will be highly attracted. In between fall materials such as the

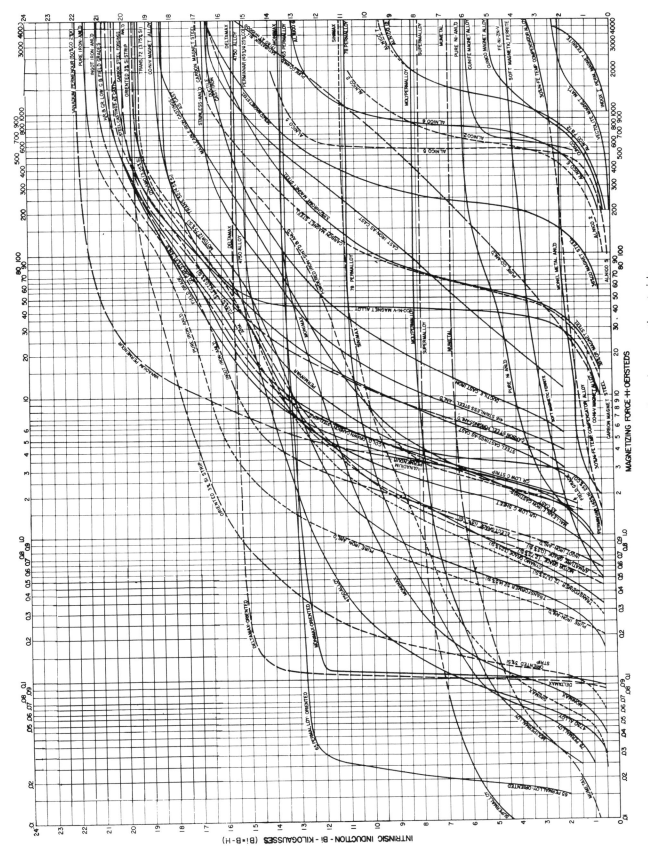

Figure 9-24. DC magnetization curves for various magnetic materials.

CIRCUIT EFFECTS 79

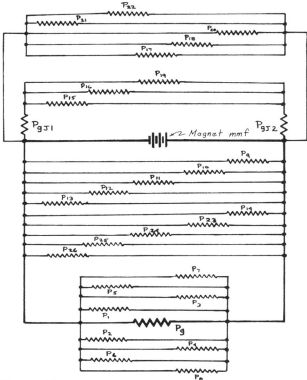

Figure 9-25

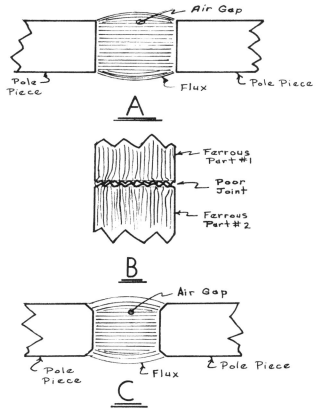

Figure 9-26

stainless steels: the 300 series is completely nonmagnetic;[11] the 400 series is weakly magnetic! The criteria of relative flux-carrying ability is the absolute permeability (μ). The μ of a vacuum or of air is 1. Nonmagnetic materials have a μ so close to 1 that, for practical purposes, they are considered the same as air. In contrast, for pure iron (annealed), μ = 10,000. Generally, the magnetic grades of stainless steel have permeabilities of between 3000 and 5000.

Assuming that the object being attracted *is* magnetic, the other three factors listed previously become relevant. The position of the object in the air gap, its size, and its shape all directly determine what the air gap really *is* and, thus, in turn, the *permeance*, *B/H* ratio, and flux output of the magnet.

The most often overlooked and misunderstood gap-related factor is the *shape* of the object in the gap. The tractive force of a specific permanent magnet on a one-pound steel *ball* will be quite different than the tractive force on a one-pound steel *bar*. Very little research that has practical application has been done on the "shape effect." Equations for calculating the permeance related to a cylindrical object have been derived, but even these are laborious to use and are of limited scope.

One practical area where the shape factor has long been recognized (but in which no real information has been developed) is the magnetic separator industry. Those involved with this industry discovered quickly that the performances of a separator on a nail, a piece of barbed wire, a hammer head, or a steel ball were all quite different—and seldom predictable! One manufacturer of such separators uses a 1 in.-diameter steel *ball* for its tests (and descriptive literature), while another uses a 1/8 in.-thick × 1 in.-wide × 3 in. steel *bar* for the same purpose. *Neither* test can be related to the performance of the separator on other shapes (including one test to the other) or to the actual gap flux densities obtained by more technical methods of measurement.

Many of the material, location, size, and shape effects are related to the general or localized saturation of the object being attracted. This can readily be observed on a simple shape that presents ideal conditions—the tractive force produced on a flat steel sheet or plate that fully spans the poles of the magnet. Figure 9-27 shows the actual test results of such an arrangement. Note that for both a zero air gap and a 1/16 in. air gap, the holding force increases at first as the sheet/plate thickness increases. It then reaches a maximum beyond which further thickness increases produce *no* force increase. *The region of rising force is the result of the ability of the magnet to deliver more flux than the sheet can carry.*

---

11. Interestingly enough, if 300 series stainless steel is cold-worked through forming or machining, it becomes weakly magnetic (because of internal metallurgical changes).

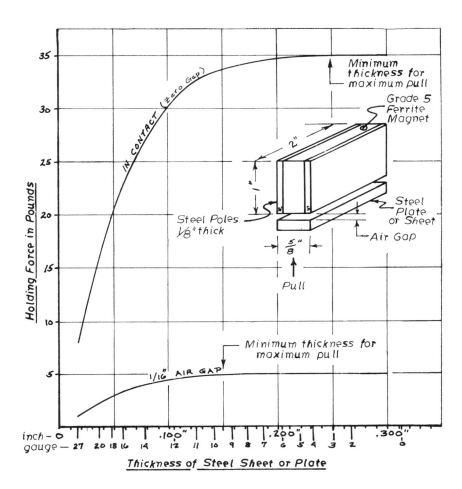

Figure 9-27. Effect of part thickness on the pull curve of a typical holding magnet assembly.

For example, a 27-gauge sheet cannot carry the flux delivered by the magnet at *either* a zero gap or a 1/16 in. gap. In both cases, the 27-gauge sheet is *saturated* and *the force-limiting factor in the system is not the magnet but the* object *in the gap!* With a 1/16 in. air gap, until 9-gauge steel is used, the magnet is always delivering more flux than the sheet can carry. At gauges thicker than 9, the force-limiting factor is no longer the object being attracted but *the magnet itself.* This condition does not occur with a *zero* air gap until the gauge thickness is 3, since the magnet can deliver more flux in contact than at a distance.

Another form of curve showing the saturation effect of the object in the air gap appears in Figure 9-28. This design is for a reaching-type magnet (Figure 9-27 is designed as a holding magnet). Note the difference in the pull force for various air gaps for the .0164 in.-thick sheet compared to the 3/8 in.-thick plate. Obviously, the 10-pound zero gap force (maximum) for the .0164 in.-sheet is limited by the saturation of this sheet and not by the "capacity" of the magnet. Not until the air gap is 1-1/8 in. are the flux output of the magnet and the flux-carrying ability of the sheet equal.

The evident saturation effects revealed by studying Figures 9-27 and 9-28 raise an interesting question of great practical importance. What *happens* to the "surplus" flux produced by the magnet if the object in the gap saturates before all of the flux is "used"? The answer: the flux goes elsewhere. In the case of sheet steel, for example, the flux goes *through the sheet.* In the case of either a holding or a reaching magnet, this surplus flux would attract and pick up other lower sheets in the same pile—as many as necessary to carry all the flux available. Recognition of this principle has important implications for the magnetic shielding [12] of other equipment and for the design of many products, such as material handling devices.

Sliding/Pull Ratio: Control of Sliding

Permanent magnets are so widely used in tractive applications that special consideration must be given to how they behave in "direct pull" vs. "lateral pull" condi-

12. Shielding is the *prevention* of the entry of magnetic fields.

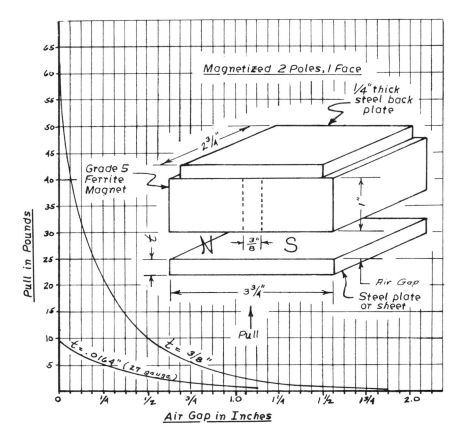

Figure 9-28. Pull characteristics of a typical plate-type magnet assembly.

tions. Direct pull is a force at right angles to the line formed by the pole pieces. For simplicity, these two types of force will be referred to as *pull* (for direct pull) and *sliding* (for lateral pull). The behavior of a magnet "in pull" can be determined from the approaches (calculation of permeance, B/H ratio, and so on) presented in Chapter 8. The method of actually calculating pull will be given in Chapter 10. In this section of this chapter, we will consider what can (or does) happen to a magnet in *sliding*. To simplify our explanation, a horseshoe magnet sliding on the surface of sheet of steel will be used as an example.

Sliding must be considered under two sets of conditions:

1. Sliding when the magnet poles are *in contact* with a ferromagnetic surface.
2. Sliding when the poles are in *no contact* with a ferromagnetic surface.

There must be a separate consideration of each of the above conditions when:

1. The ferrous surface is *continuous* under the poles of the sliding magnet.
2. The ferrous surface is *discontinuous* or *terminated* under the poles of the sliding magnet.

Consider first sliding *in contact*. A magnet is being attracted directly to the surface of the sheet. If a sliding force occurs, the resistance to such sliding will be determined only by the pull force and the coefficient of friction between the surfaces (the magnet and the sheet). As long as the magnet is moving over a *continuous* sheet (that is, there is always sheet steel fully spanning the poles), the performance of the magnet will not change, since the air gap is substantially constant, the B/H ratio is unchanged, and so on.

Next, consider a suspended magnet being moved at some distance from a steel sheet.[13] If the sheet is continous, the air gap is not changing, the frictional forces are practically nonexistent, and *there is no resistance to the magnet's motion nor any type of retarding force.*

In both of the above examples, however, if the magnet tries to move off the edge of sheet, there will no longer be a constant air gap but a *changing* gap. Thus, the operating point of the magnet will change, and it will, in effect, be operating under "dynamic" conditions. The physical effect can be anticipated by the fact that flux travels through steel with a lower reluctance than

13. If this idea is difficult to follow, imagine a magnet attached to the bottom of a wagon rolling along a steel sidewalk.

when it travels through air. *As any part of the magnet's pole moves off the sheet, a force will be produced that opposes the original sliding force.* This opposing force, unless it is exceeded by the sliding force, will try to keep both magnet poles in contact with or directly over the sheet!

If the sheet is *not* continuous but is made up of a series of parallel strips at right angles to the direction of the sliding (motion), the effect will be to produce an intermittent "drag" that opposes the sliding force. This author has, on one occasion, used this drag effect to produce a crude form of magnetic damping device that (by proper width and spacing of the steel strips) "locked in" at the maximum point of motion.

Only two other factors remain to be considered in relation to the sliding effects described here. These are *speed* or rate of sliding and *flux density in the material.* If the rate of sliding is high and/or the flux density is high, significant drag may be produced by eddy-current and hysteresis losses (in the sheet). The eddy currents are, of course, the result of relative motion between a magnetic field and a conductor (Lenz' law). The hysteresis loss is the result of successive magnetizations (and the energy required for such magnetizations) of different portions of the sheet as the magnet traverses its surface.

While in some circumstances sliding effects may be deliberately designed into a product, in most cases sliding is a serious problem. Lifting magnets provide an example where sliding effects are potentially a very serious *safety* problem.

There are three basic methods of minimizing or stopping *in-contact* sliding:

1. By making the magnet's tractive force great enough and/or the coefficient of friction between surfaces great enough to counterbalance the sliding forces.
2. By providing mechanical "stops" in the lateral direction.
3. By using ancillary friction "pads."

The first of these solutions to sliding is seldom the most economical. If the magnet is made larger than necessary for the required pulls, the cost increases rapidly. If the surface is made with a high frictional coefficient, the magnet air gap is usually increased (again requiring a larger magnet). Furthermore, the high frictional surface will be subject to wear, resulting in a gradual loss of the needed frictional coefficient.

It has been the experience of this author that *the typical ratio between the in-contact sliding force and the pull force of a magnet is* 10 : 1. This ratio has generally been found to be applicable for a magnet with normally ground surfaces that is in contact with the normal finish on unmachined cold-rolled flat steel.

Using this ratio, a magnet that has a pull force of 10 pounds will require only 1 pound to slide on a continuous sheet of steel. Or, conversely, it will take a magnet with a 100-pound pull force to provide a 10-pound resistance to sliding!

The second method of using mechanical "stops" is generally the most effective, most reliable, and most economical. In some cases, however, the use of such stops would be impractical and/or would interfere with some performance requirement of the product.

The use of ancillary friction pads is shown in Figure 9-29 (in this case, to stop the sliding of temporary office partitions). This figure illustrates two methods—one using sponge-rubber strips, the other using rubber tubing. Both methods require that the rubber be compressed (for sponge strip) or distorted (for tubing), and both therefore necessitate some increase in magnet pull (size) to overcome the oppositional force of the rubber. The tubing method has been found to be more effective. With this tubing projecting above the magnet surface by approximately one-sixth of its diameter in a groove *exactly equal in width to the tube diameter,* sliding can be stopped completely. The tubing (or sponge-rubber) *length* need *not* be more than 15 percent of the magnet axial length, so that the magnet size increase is relatively small. Surgical rubber tubing (pure gum), which is readily available in approximately 30 different diameters, is ideal for this friction pad.

Figure 9-30 shows an application of rubber-bonded ferrite magnets, where the inherently high frictional

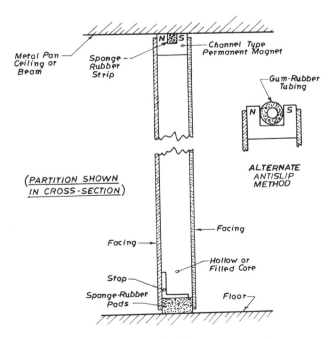

Figure 9-29. Magnetically held holding temporary walls.

Figure 9-30. Ceiling tiles held with permanent magnet holders. (Courtesy of Arelec, and Magnetic Equipment Company)

coefficient of this material is used to prevent sliding. In this case, the friction pad is essentially "built in" to the material.

Strength Control/Shutoff

Often, electromagnets are used in place of permanent magnets, because of the common misconception that the strength of a permanent magnet cannot be controlled (including complete "shutoff"). Fortunately, this is not the case, as shown in Figures 9-31, 9-32, 9-33, 9-34, and 9-35, which cover the following control methods: mechanical shunting, varying engagement, electrical shunting, magnetizing-demagnetizing, and mechanical "break-aways." While these illustrations are largely self-explanatory, a few words need to be added concerning the magnet materials that are appropriate to these systems.

The mechanical shunting methods shown in Figure 9-31 can use any magnet material effectively, although materials with a coercive force (H_c) of over 1000 Oe are generally preferable. If lower H_c materials are used and the shunts not properly designed, there may be some permanent demagnetization due to "contact demagnetization"[14] from the shunt. If lower coercive

14. Contact demagnetization will be covered in Chapter 11.

materials *are* used, the "shunt-side" pole pieces should project *beyond* the basic magnet, so that the engaged shunt can never come closer than within 1/8 in. of the magnet side. This projection of the poles increases leakage losses across the back of the magnet and either requires a larger magnet for the same force or reduces the force of the assembly. With a high coercive material, the shunt may contact the magnet without any permanent losses.

Varying engagement method A in Figure 9-32 *requires* the use of materials of $H_c > 1500$ Oe. If lower coercive materials are used, the basic magnet will partially demagnetize on the first "disengagement" and remain demagnetized to some less-than-maximum level (that is, it will be operating on a minor hysteresis loop). Varying engagement methods B, C, and D may be used with any magnet material effectively. The use of high coercive materials ($H_c > 1000$) does, however, make larger air gaps less critical.

The electrical shunting method A in Figure 9-33 requires the use of a material of $H_c < 700$ Oe. If a magnet with a higher coercive strength is used, the electrical power requirements of the control system become prohibitive. Electrical shunting method B requires the use of a material with an $H_c > 1500$ Oe to avoid partial demagnetization by the electrical control pulses.

The magnetize-demagnetize method in Figure 9-34 requires the use of a material with an $H_c < 700$ Oe. Higher coercive materials require prohibitively large power supplies for control.

The mechanical break-away methods illustrated in Figure 9-35 may be used with any material.

Mechanical shunting methods A, B, and C in Figure 9-31 are suitable for continuously variable control. Methods D and E are used primarily when simple high-low control is required. None of these methods provide for full "off" operation, but by adding a nonmagnetic cover to the working face, reductions of 90 percent of full "on" can be obtained.

Varying engagement methods A and C can be used for continuously variable operation, but 100 percent "off" is not practical. Method B is normally used only for simple "on-off" operation and *can* be operated so that it is 100 percent "off." Method D can provide continuous variability in very fine increments from full "on" to full "off." The dual magnet arrangement can be used to give a flux output at the working face that is the integration of two independent mechanical "inputs."

Electrical shunting method A can provide effective continuous control (including full "off"). Control can be from a remote point, using entirely electrical connections. The working face output can be made a function of a varying electrical input against the reference

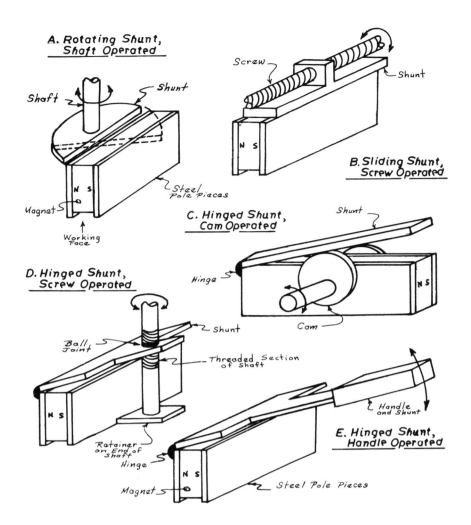

Figure 9-31. Mechanical shunting methods for controlling the output of a magnet. (For other control methods, see Figures 9-32 to 9-35.)

base provided by the permanent magnet. (For this type of application, materials with an $H_c > 1000$ should be used.) Method B is only effective as a simple "on-off" system; it does provide 100 percent "off" operation.

The magnetization-demagnetization method is ideal where polarity reversals may be required as well as "on-off" control. It is a poor but possible way of providing continuous variability. Continuous variability requires magnetization of the magnet or minor hysteresis loops, and the repeatability of this procedure is limited.

Both break-away methods are used primarily for simple "on-off" operation. Since the flux density varies inversely as the square of the distance from the pole faces, continuous variability is difficult and unreliable.

Multiple Magnet Circuits

Frequently, more than one permanent magnet may be used in a single circuit. Multiple magnet circuits may be of three kinds:

1. Two or more separate magnets connected in series or parallel to jointly contribute an mmf across a single air gap.
2. Two magnets arranged *in attraction* (with unlike poles opposite each other) in order to provide a *coupling force*.
3. Two magnets arranged *in repulsion* (with like poles opposite) in order to provide a particular kind of field pattern or a *levitating force*.

A simple example of multiple *series* magnets supplying one air gap can be seen in either Figure 8-3B or 8-3C (page 53). In multiple magnet circuits of this type, the performance and method of calculation are the same as they are for a single magnet circuit, except that care must be taken to correctly identify all leakage paths and properly locate each mmf in the calculations. The electrical analog method is particularly useful in avoiding "circuit errors" for even fairly simple multiple-magnet arrangements. Bear in mind that the magnets may be either in series or parallel, but that the flux will

CIRCUIT EFFECTS 85

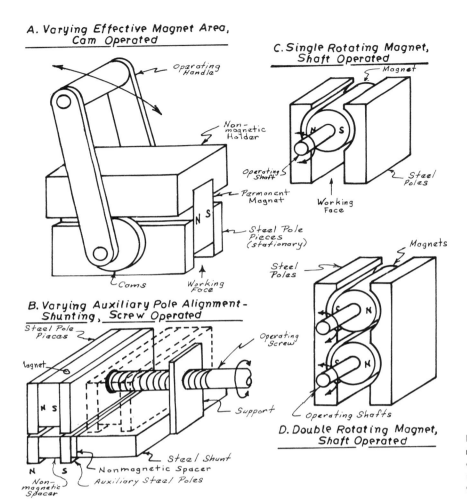

Figure 9-32. Varying engagement methods for controlling the output of a magnet. (For other control methods, see Figure 9-31 & Figures 9-33 to 9-35.)

always be in the same direction. Whether the magnets tend to demagnetize each other and whether they behave independently (each having its "own" B/H ratio) or as though they were a single magnet (with one B/H ratio applying to all) depends *upon how they are initially magnetized.* The importance of magnetizing methods will be covered in Chapter 13.

Figure 9-36 and 9-37 show the basic multiple magnet structure used in a wide range of magnetic separators and material handling equipment. These are simple examples of magnets in parallel, supplying the single longitudinal air gap between the steel pole pieces (strips). These pole pieces also serve to distribute the flux from each magnet more uniformly over the length of the gap.

Two examples of multiple magnets arranged in attraction (with unlike poles opposite) to provide a coupling or tractive force are shown in Figures 9-38 and 9-39.

In Figure 9-38, multiple magnets are used solely to increase the tractive force that can be obtained in a small space. If one roll were made of a permanent magnet and the other of mild steel, the force that could be produced in the one-in. diameter rolls (see the tube magnets in Figure 9-3) would be appreciably less than with both rolls made as magnets.[15] The calculation of magnet characteristics and performance for multiple magnets that perform a purely *tractive* function is both simple and accurate.

Where the multiple magnets are intended to perform a coupling or drive function, as shown in Figure 9-39, there is a reasonably simple, fairly accurate method of calculation for either wholly new designs or designs using available standard basic magnets. These methods, as we will see in Chapter 10, require a substantial num-

---

15. A precaution to the reader is in order here. It is not always true that a second magnet is "better" than mild steel. In the end, the possible increase in mmf must be weighed in relation to the lower permeability of permanent magnet materials. An example of when a second magnet is *not* better is when it is used to replace the iron core in a solenoid or a relay. Using a magnet may provide *polarized* operation of the solenoid or relay, but the *force* produced will generally be *lower.*

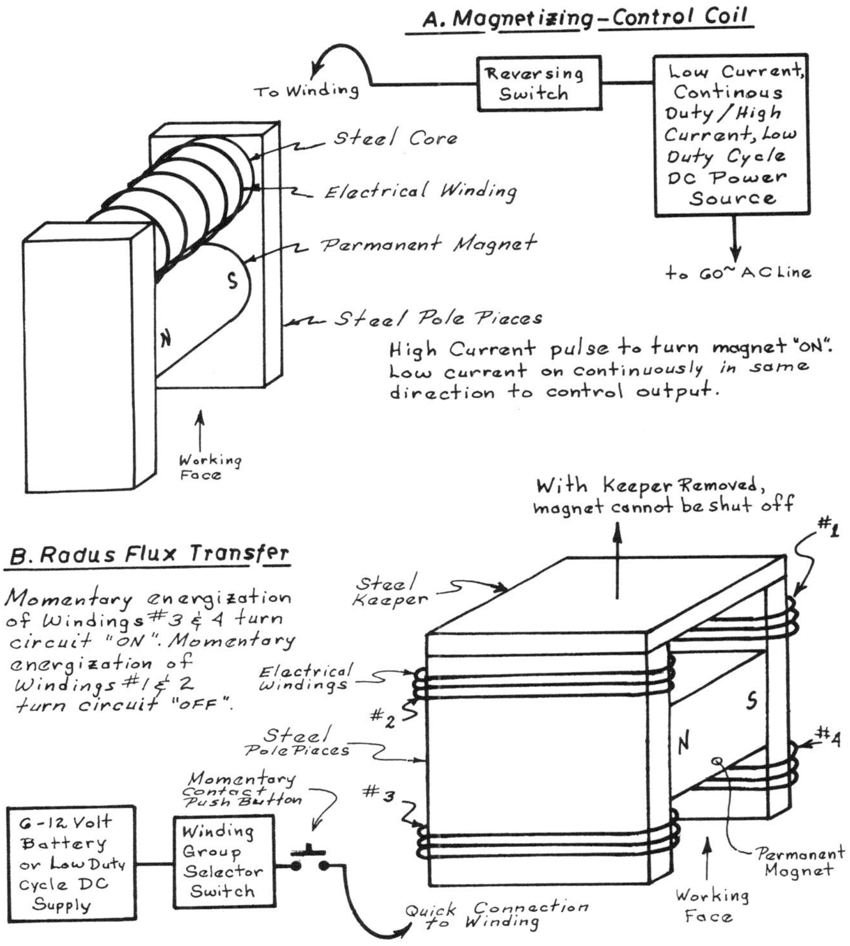

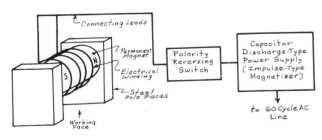

Figure 9-33. Electrical shunting methods for controlling the output of a magnet.

Figure 9-34. Magnetizing-demagnetizing method for controlling the output of a magnet.

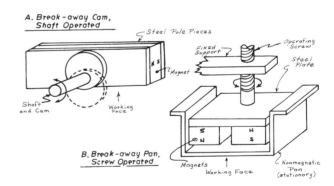

Figure 9-35. Break-away methods for controlling the output of a magnet.

ber of laborious steps, so that many coupling or drive designs are based on the actual test data from identical (or very similar) circuits.

The following considerations will be useful in relation to two-magnet, coupling-drive designs:

1. The magnetic forces act both in coupling (when in motion) and in axial attraction. Calculating this axial force is usually necessary in determining thrust bearing requirements and so forth.

2. The driven member and the drive member can *only operate synchronously*. This presents problems connected with initial starting or with synchronization loss due to overloads.

3. In operation, the angle between the attracting pole

CIRCUIT EFFECTS 87

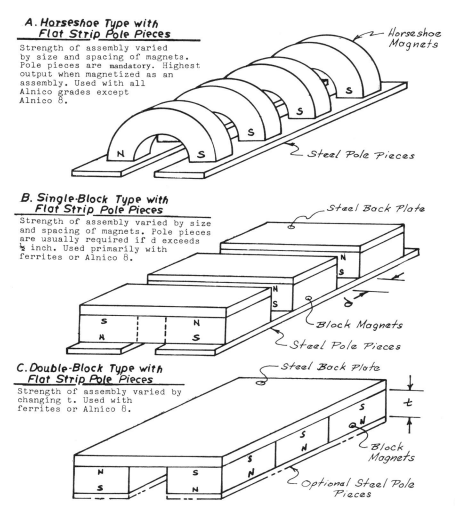

Figure 9-36. Basic plate magnet constructions.

centers is a function of *load on the driven member*, so that the circuit is one with a *variable* air gap. This coupling angle or the effects of synchronization loss may produce *cyclic* demagnetizing effects on either or both magnets. Of course, this can be most severe if the driven member "stalls." Further, the cyclic demagnetization effects, if they persist for any appreciable number of "cycles," can produce severe *heating* due to hysteresis and eddy-current effects. This potential demagnetization must be considered in the initial magnet design, or a magnet material must be used that has a high coercive force (H_c). Generally, materials with $H_c > 1000$ Oe are advisable.

4. The torque that a drive of this kind can transmit without loss of synchronization is a direct function of the *total flux produced across the air gap* (ϕ_g) by both magnets (in series). On the other hand, the undesirable axial tractive force is a function of air-gap flux *density* (B_g). Since:

$$\phi_g = B_g A_g$$

the torque capability can be maximized and the axial force can be minimized by designing the circuit with a low B_g and a large area.

5. The effect of leakage flux from the rotating magnetic field may produce eddy-current losses (and heating) in nearby or intervening *conductive* materials or both eddy-current and hysteresis losses in nearby *ferrous* materials. If any material is introduced between or near the magnets for environmental isolation, or other purpose, this material should be *nonmetallic*, if possible. It can *never* be a ferromagnetic material. If a conductive, nonmagnetic material is used, allowance for the eddy-current losses must be made in the input drive power. Since both eddy-current and hysteresis losses rise rapidly with speed, these effects can be minimized by using the lowest possible rotational rate.

Multiple magnets (usually pairs) may be arranged in repulsion for very specific effects. Figures 9–40 and 9–41 are two examples of magnets arranged to provide

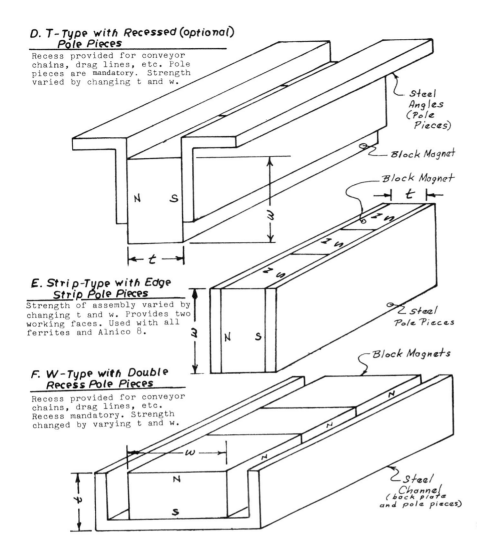

D. T-Type with Recessed (optional) Pole Pieces

Recess provided for conveyor chains, drag lines, etc. Pole pieces are mandatory. Strength varied by changing t and w.

E. Strip-Type with Edge Strip Pole Pieces

Strength of assembly varied by changing t and w. Provides two working faces. Used with all ferrites and Alnico 8.

F. W-Type with Double Recess Pole Pieces

Recess provided for conveyor chains, drag lines, etc. Recess mandatory. Strength changed by varying t and w.

Figure 9-37. Basic plate magnet constructions (continued).

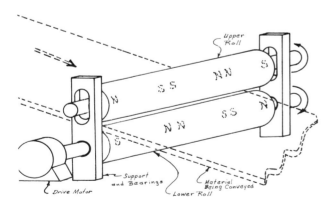

Figure 9-38. Light-duty pinch rolls for conveying nonmagnetic sheet stock. All rolls shown are axial-type tube magnet assemblies. Diametrical-type assemblies are not suitable for this application. For internal structure see Figure 9-3.

for the magnetic-field focusing of electronic beams. Figure 9-40 shows two large Alnico horseshoe magnets with soft magnetic pole pieces being used to direct the flux in a basically simple circuit. In Figure 9-41, the more complex "periodic" focusing arrangements are shown. Methods B, C, and D all provide periodic focusing, but only B and D use repelling magnets to obtain this effect.

In any focusing circuit, considerable care must be exercised in the selection of magnet materials, in the design of the magnets and ancillary parts, and in the assembly of the units. Nonuniform fields or demagnetization effects can occur from the other magnets in the circuit or from the magnetic field associated with the electron beam (flow). Highly coercive, high energy product-per-unit-volume materials *with low temperature sensitivity* are greatly preferred for these applications. The high coercivity greatly simplifies assembly and prevents demagnetization effects. High energy/unit

CIRCUIT EFFECTS 89

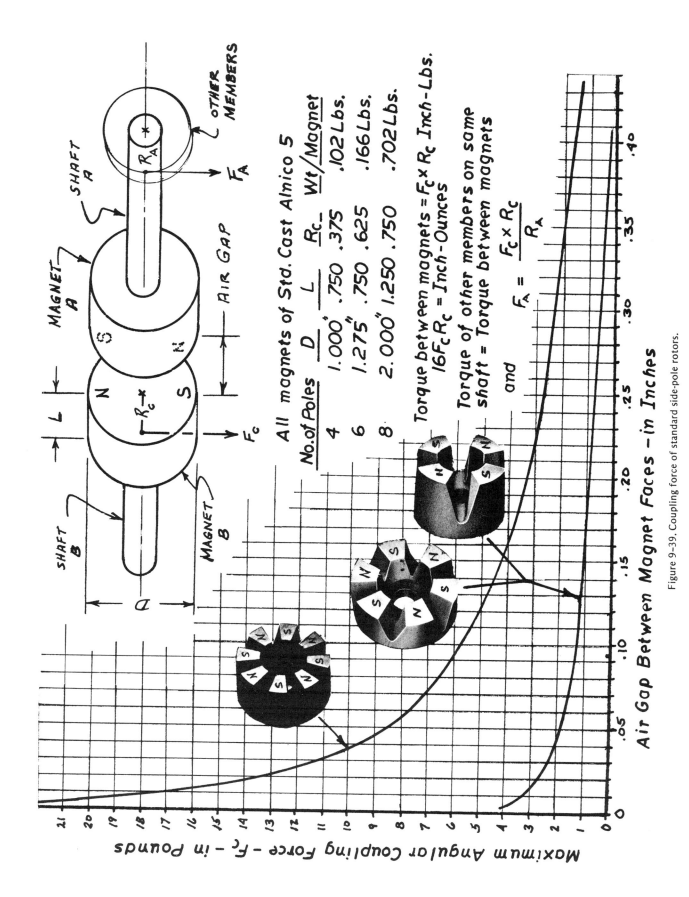

Figure 9-39. Coupling force of standard side-pole rotors.

Figure 9-40. Microwave tube focusing magnet assembly. (Courtesy of Varian Associates)

volume magnet materials provide the greatest focusing effects in the smallest possible tube design. Temperature stability is an important requirement, since most tubes of this type must operate over a wide range of ambients and have appreciable internal heating as well. The rare-earth magnet materials have largely replaced the platinum-cobalt materials in high-power traveling wave tube (TWT) applications. These difficult application requirements were largely instrumental in the commercial development of samarium-cobalt. In some older and/or larger designs, cast Alnicos are used, but temperature compensators are usually incorporated into the circuit. Sintered ferrites are *not* used because they are extremely temperature sensitive.

Designing TWT's and magnets for such circuits is a highly specialized area which is virtually independent of general magnetic product design.

Figure 9-42 shows four magnetic levitation circuits based on repulsion principles. While such "springs" have not been widely applied, they have significant and particular advantages. The rotating disc (measuring element) of the highest quality electrical watthour meters uses a magnet-levitating system as a bottom bearing, greatly reducing frictional losses and therefore greatly increasing measurement accuracy. (This will be discussed in more detail in Chapter 16.)

Experimental magnetic levitation systems have been developed for high-load applications, such as urban transportation vehicles, but economic and operational problems have precluded their commercial use.

The two primary factors to consider in levitation applications are the magnet material and the "centering" problems. Highly coercive materials are almost invariably required to prevent mutual demagnetization effects. The rare-earth materials and platinum-cobalt are ideal, if economy permits and space limitations require their use. In all repulsion systems, centering is invariably a problem, and "side thrusts" must be controlled by mechanical methods. This problem is directly related to the fact that the force produced by a magnet (or by groups of magnets) varies inversely with the square of the distance. Even the slightest "unbalance" in centering at one point results in large force changes that cause total lateral shifting.

The design of magnets in repulsion for levitating systems is very inexact. Since magnetic lines of flux repel and cannot cross each other, the magnetic field patterns are highly unpredictable. The known theoretical methods are complex and seldom worth using. Most of these circuits are designed empirically, with some use of scaling techniques to reduce or enlarge established designs. For preliminary design purposes, this author has found that calculating the force of *attraction* between the magnets and then reducing that force by 40 percent to give the "repulsion" force is as rapid and workable as any of the more elaborate methods. In this attraction calculation, the working air gap is assumed to be one-twentieth of the total magnet length (L_m). This method does not predict the "levitation distance" (air gap) that will actually be obtained and cannot be used directly to design for a specific distance.

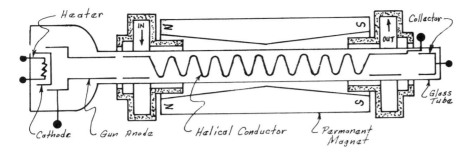
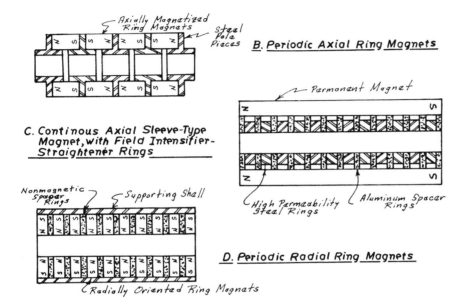

Figure 9-41. Traveling wave tube focusing magnets (all types shown in section).

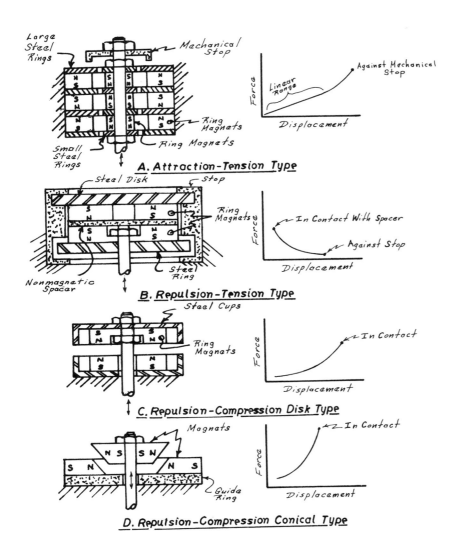

Figure 9-42. Typical magnetic spring systems and their force/displacement curves.

10

Exact Design Methods

In Chapters 7, 8, and 9, all the factors to be considered in permanent magnet design were developed and the basic equations used in magnet design were illustrated. The design problems presented in these chapters dealt with magnets in a "static gap" condition. In actual practice, relatively few true static gaps are encountered in magnet design. In most cases, the air gap or conditions within the gap are changing—indeed *must* change—if the circuit is to work. As examples, in a loudspeaker, there must be an interaction between the field of the magnet and the field of the moving coil; in a magnetic separator, the object being attracted moves toward the magnet, thus changing the air gap; in a motor, there are changes in both the air gap as the armature rotates and in the interaction of the permanent magnet with the electromagnetic winding. Thus, most magnetic circuits, if they are to be designed completely, must be considered under *dynamic conditions*.

This chapter will consist of two sections. The first will be devoted to the general methods for dynamic magnet design, using the demagnetization curves. The second section will cover typical design problems where a final, defined performance in terms of *work* is to be calculated.

Use of the Demagnetization Curve for Variable Gap Calculations

To understand what happens under variable gap conditions, consider the following sequence of circumstances:

1. A simple cast Alnico 5 horseshoe magnet is completely "keepered" (zero air gap). *While so keepered, the material is magnetized to saturation* (by one of the appropriate methods we will discuss later in Chapter 13).

2. The keeper is removed, fully opening the magnetic circuit.
3. A mild steel object to be attracted is brought within a distance of x cm of the magnet's pole face.
4. The magnet exerts a tractive force on the mild steel, pulling it to a distance y from the magnet face. At distance Y, the mild steel is restrained from further motion.
5. *While the mild steel is at a distance of Y,* the magnet is *remagnetized* to saturation.

The demagnetization curve for the cast Alnico 5 used in this magnet appears in Figure 10-1.

At the end of step (1) above, after the saturating magnetizing force (H_s) has been removed, the flux that will be flowing in the "closed ring" formed by the magnet and the keeper will be B_r.[1] In this case, B_r = 12,600 G. Since there is no air gap, there will be no external field and no gap mmf (H_m). In this state, the circuit is incapable of doing any work.

Assume that when the magnet is fully open-circuited in step (2), the B/H ratio is determined to be B/H_1 = 10. If this load line O–B/H_1 is drawn on Figure 10-1, it crosses the demagnetization curve for the material at point A. Then line A–B_{m1} gives a flux density in the magnet of B_{m1} = 6000 G, and line A–H_{m1} gives a magnetizing force in the magnet of H_{m1} = 600 Oe. According to Equation 13 from Figure 8-7:

$$B_{g1} = \frac{H_{m1} L_m}{L_{g1}}$$

where L_m is constant. H_m will be applied across the somewhat indeterminate air gap (L_{g1}) to produce a gap flux density of B_{g1}.

---

1. For the purposes of this explanation, ideal conditions of no reluctance in the materials and no leakage fluxes are assumed.

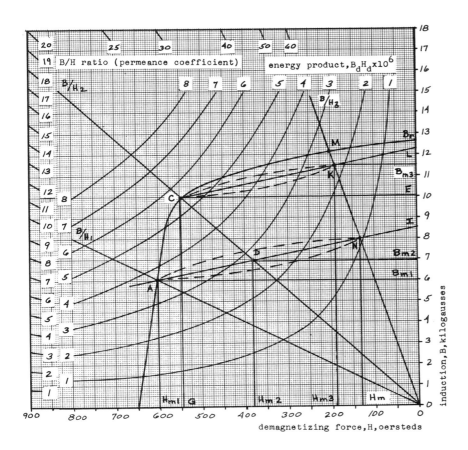

Figure 10-1. Demagnetization/energy product curve for cast Alnico 5.

If, in step (3), with the mild steel object at a distance of x cm, the new B/H ratio is determined to be B/H_2 = 18 (line O-B/H_2), we might *expect* B_m to be line C-E (10,000 G) and H_m to be line C-G (550 Oe). But *this will not be the case*. The B/H ratio could be *expected* to increase from 10 to 18 as the circuit became increasingly closed-circuited, since:

$$\frac{B}{H} = \frac{H_m L_m}{L_{g2}}$$

where L_m is a constant, since the magnet has not been physically altered, and L_{g2} is $x/2$, since there are two gaps in series. But there is *no* similar reason to expect the operating point of the material to go from point A to point C along the demagnetization curve. In fact, recalling the theoretical considerations, it is evident that once the magnet has been exposed to the level of self-demagnetization represented by B/H = 10 (point A), it will henceforth be operating on a *minor* hysteresis loop. To determine the true operating point of the magnet under these conditions at distance x, a straight line must be drawn through point A that is approximately parallel to the demagnetization curve from points C to B_r. This line, identified as A-I, crosses the load line O-C at point D. The horizontal projection D-B_{m2} gives the true B_m in the magnet (6900 G) at a load line of B/H = 18, and the vertical projection D-H_{m2} gives the true H_{m2} (380 Oe).

It is evident that "exposure" of the magnet to the larger air gap (in this case, the open-circuited gap) resulted in a substantial demagnetizing loss (G-H_{m2} = 550 − 380 = <u>170 Oe</u>) that will be directly reflected in the air gap flux, since:

$$B_{g2} = \frac{H_{m2} L_m}{L_{g2}}$$

where L_m is constant and $L_{g2} = y/2$. If, as is indicated in step (5), the magnet is remagnetized to saturation while the mild steel is *at distance x*, the *first air gap that the magnet is exposed to* when H_s is removed *will be the one with a B/H = 18*. Thus, the greatest self-demagnetization exposure will only be to this level, and the operating point of the magnet at distance x will actually be at point C. Then H_{m2} = 550 Oe and:

$$B_{g2} = \frac{H_{m2} L_{m2}}{g2}$$

Now consider the effect of allowing the mild steel object to move still closer to the magnet poles—to a distance of y. If, at this distance, B/H_3 = 60, then line O-B/H_3

represents the load line and point (K) represents the actual operating point of the magnet. Under these conditions, B_{m3} = 11,500 G and H_{m3} = 190 Oe. The magnet is *not* operating at point C, since it has not been resaturated.

If the mild steel is again moved *away* from the magnet face, from distance y to distance x *or any point in between*, the magnet's operating point *will always be along the C-K-L line*. The recoil line is really the joined end points of the minor hysteresis loop, shown by the dashed lines in Figure 10-1. The curved lines of this minor loop are so close that using the recoil line instead of the minor loop lines introduces a negligible error and calculations are greatly simplified.

Going back still another step, if the mild steel bar is moved further from the pole faces until the magnet is again *fully open-circuited* (operating at point A and then toward the face to distance Y), the magnet will "recoil" along the A-D-N-I line to operate at point N where B_m = 8000 G and H_m = 130 Oe.

Note, too, the variation in the available energy product in the above sequence of operating conditions. When the magnet is operating at:

Point A the product of $B \times H$ = 3.6 G-Oe $\times 10^6$
Point C the product of $B \times H$ = 5.5 G-Oe $\times 10^6$
Point M the product of $B \times H$ = 2.5 G-Oe $\times 10^6$
Point D the product of $B \times H$ = 2.7 G-Oe $\times 10^6$
Point K the product of $B \times H$ = 2.25 G-Oe $\times 10^6$
Point N the product of $B \times H$ = 1.05 G-Oe $\times 10^6$

Obviously, under either static or dynamic conditions, the energy *capability* of the magnet is not determined by the material alone but by the *load effect* of the object on which it acts. Further, this capability is determined by the "worst" condition to which the magnet is *initially* "exposed." Finally, the material's characteristics plus the "worst" condition *will establish a recoil line (line A-D-N-I or line C-K-L) along which the magnet will always operate under dynamic conditions.*

Special notice should be taken of the effect of remagnetizing the magnet after open-circuiting. If the final dynamic operating circuit is to be from distance x to distance y, the magnet should be initially magnetized with the mild steel at distance y. Stated more generally, *A magnet should be magnetized in its final operating circuit.* If the magnet is first magnetized, then open-circuited, and finally placed in the circuit, as shown in this example, the recoil line will be A-D-N-I and the energy levels available will range from 1.05 to 2.7 G-Oe $\times 10^6$. If, however, the magnet is magnetized *in the circuit*, the recoil line will be C-K-L and the energy levels available will be 100 percent higher, ranging from 2.25 to 5.5 G-Oe $\times 10^6$

If the magnet is magnetized in the circuit, but with the *smallest* air gap to which it will be exposed (distance y), then the magnet will "lose" strength on its first exposure to distance x and will remain at a lower level for all subsequent dynamic "operations."[2]

Use of the Demagnetization Curve for External Field Calculations

The preceding discussion related to dynamic operation represented by a varying air gap. Next, we will consider dynamic operation represented by an external [3] demagnetizing field like the one that occurs in a motor, a generator, or a similar electromechanical transducer.

Assume that the circumstances described previously for the varying air gap represent a *stepping motor. Since no external pulsing field has ever been applied,* the operating point at *minimum air gap* is point K. With *maximum air gap* (during rotation), the operating point is C. The recoil line is C-K-L.

Assume that during pulsing of the stepping motor, the electromagnet applies a demagnetizing field (F_{d1}) of 400 Oe to the permanent magnet. The effect of this demagnetizing force will be developed in Figure 10-2.

During the time that the demagnetizing force (F_{d2}) of 400 Oe is applied to the permanent magnet by the electromagnet, the flux level in the permanent magnet will be reduced by some level. Since the B/H_2 ratio (18) of the magnet cannot, in reality, be changed except by a *physical* change in the circuit, the B/H ratio *slope* must remain the same. Then the effect of F_{d1} = 400 Oe will be to "depress" the magnet along a line *parallel to* B/H = 18 (as shown by line F_{d1}-F'_{d2} in Figure 10-2), *but only while the demagnetizing force is applied*. At that moment, the permanent magnet will effectively be operating at point T.

Upon removal (discontinuance) of F_{d1} from the electromagnet, the magnet will "recover" along recoil line T-W. This recoil line will be parallel to line C-L. With the magnet now in the position of maximum gap (B/H = 18), the operating point will *no longer* be point C; instead, the magnet will operate at point U on line T-W. Similarly, when the magnet is at minimum gap (B/H_3 = 60), it will operate at point V instead of point K. The permanent effect of F_{d1} will have been to reduce B_m from 10,000 G to 5500 G and H_m from 550 to 305 G *at maximum gap*. At minimum gap, B_m will have been reduced from 11,500 to 6500 G and H_m from 193 to 109 G.

2. On initial magnetization at distance y, the operating point is M. On exposure of distance x to the gap, the operating point becomes C. Subsequent operations will be along line C-K.
3. External to the permanent magnet itself.

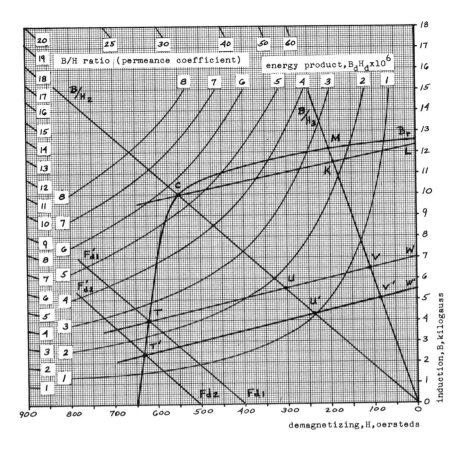

Figure 10-2. Demagnetization/energy product curve for cast Alnico 5.

Thus the *latching force* of the permanent magnet when the electromagnet is off will have been reduced, since the *gap flux density* will have been reduced by the ratio of 109 : 193. Similarly, the "pull-in" torque from the maximum air gap (after pulsing by the electromagnet) will have been reduced by the ratio of 305 : 550.

Now we will consider how this permanent demagnetizing effect of the electromagnet might be *avoided*. Referring to the situations developed in Figure 10-1, would fully *open-circuiting* the magnet result in enough *self*-demagnetization to prevent further demagnetization by the electromagnet? A negative answer is evident from a comparison of Figures 10-1 and 10-2. When the magnet was completely open-circuited (Figure 10-1), the lowest operating point (A) corresponded to a true B/H ratio of 10 and a recoil line of A-D-N-I. On the other hand, while the electromagnet exerted a demagnetizing force (Figure 10-2), it operated at point T and the recoil line was T-U-V-W. Since point T and the recoil line from it is *lower* than point A, full open-circuiting *alone* will *not* stabilize the magnet against the electromagnetic field. Since the magnet gap cannot be made larger (except by physically modifying the magnet), the only stabilization methods are:

1. Apply the primary electromagnetic field to reduce the operating level to recoil line T-U-V-W.
2. Momentarily apply a new *external* AC electromagnetic field whose peak-to-peak value produces at least 400 Oe.

In actual industrial practice, if an AC stabilizing field were used, this field would be 10 to 25 percent *greater* than the maximum "internal" demagnetizing field expected in the future. This would assure the designer that the internal demagnetizing would not change the performance of the stepping motor to even a minor degree.

Referring again to Figure 10-2, consider the effect of applying an external AC stabilizing field (F_{d2}) of 500 Oe (400 Oe + 25 percent = 500 Oe).

During the time this field (F_{d2}) is applied, operation of the magnet is depressed to point T'. When the field is removed, the magnet will "recover" along recoil line T'-W' to U' and V' in succession (maximum and minimum air gaps respectively). Judging from where points U' and V' fall on the energy product curves, it is evident that, under these conditions, the Alnico 5 is not being used efficiently. While this material is *capable* of an energy product of 5.5 *G-Oe* × 10^6 (at BH_{max}), here, *at*

EXACT DESIGN METHODS 97

best, the material is delivering a little over 1 *G-Oe* × 10⁶ and a maximum *H* of 240 Oe in the gap (at operating point *U'*). Under the circumstances, another material should be considered—one that *will* be used more efficiently and, if possible *that has little or no* knee *in the demagnetization curve*. A quick glance at the constructions of Figures 10-1 and 10-2 shows that *the severe effect of all demagnetizations* (whether self-demagnetization or demagnetization from the electromagnet) *is related to the sharp knee in a demagnetization curve*. The "flatter" the curve (the less knee), the closer the operating points and recoil lines.

Assuming that an *identical* magnet is made of Alnico 8, whose demagnetization curve appears in Figure 10-3, the constructions in Figure 10-2 will be repeated. In Figure 10-3, line 0-B/H_2 represents operation at a B/H of 18. Point *C* represents the operating point of the magnet under maximum gap conditions. Line 0-B/H_3 represents operation with the minimum gap ($B/H = 60$). If the magnet is *first* subjected to operating point *C*, the material will recoil along a line approximately parallel to the upper part of the demagnetization curve. This recoil line will be C-B_r, but the minimum gap operating point (*K*) is *also along this same line*. Thus, *there will be no self-demagnetization loss due to exposure to the maximum gap.*

Next, the construction for applying a "safe," external, AC stabilizing force of 500 Oe will be shown. This $F_{d2} = 500$ is represented by line F_{d2}-F'_{d2} while it is applied. On removal of F_{d2}, the recoil line will be line T'-W', the operating point at maximum gap will be point U', and the operating point at minimum gap will be point V'. *Note that the reduction in H_m is only* 18 Oe *between points C and U' and 12 Oe between points*

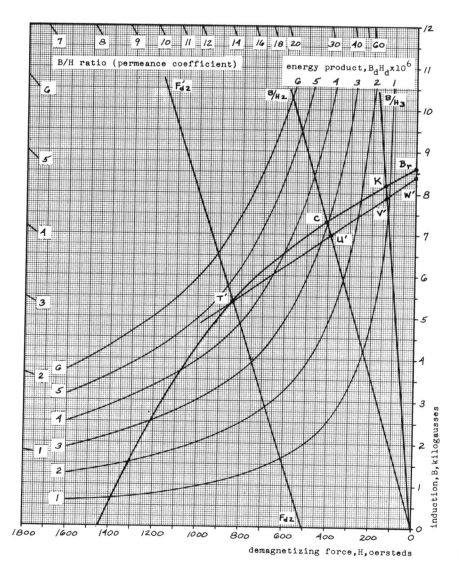

Figure 10-3. Demagnetization/energy product curve for cast Alnico 8.

K and V'. In other words, only a very small demagnetization has occurred as a result of stabilization! The Alnico 8 magnet has an energy product of approximately 2.8 $G\text{-}Oe \times 10^6$ at point U' and 1.1 $G\text{-}Oe \times 10^6$ at point V', so that this material is not operating with great efficiency either (BH_{max} = 4.6 $G\text{-}Oe \times 10^6$). However, compare H_m in Figures 10-2 and 10-3, and remember that the flux density in the air gap is:

$$B_g = \frac{H_m L_m}{L_g}$$

In Figure 10-2 (stabilized Alnico 5), H_m at U' = 240 Oe, while H_m at V' = 82 Oe. But in Figure 10-3 (stabilized Alnico 8), H_m at U' = 395 Oe and H_m at V' = 125 Oe.

If the 240 Oe and the 82 Oe of the stabilized Alnico 5 were sufficient for satisfactory operation of the stepping motor, then according to the equation above, the length of the magnet (L_m) could be *reduced* [4] by using Alnico 8 to produce the same B_g.

Now assume that the circuit has been *redesigned* to use Alnico 8 efficiently, with a minimum gap B/H of 10 and a maximum gap B/H of 6. These constructions are shown in Figure 10-4, including the desired 500 Oe stabilizing force. Compare this figure to Figure 10-5 for sintered Ferrite 5, using the same B/H ratios. Note that the completely "flat" demagnetization curve for Ferrite 5 produces an *identical* recoil line and material curve, resulting in zero demagnetization (either external or self).

4. Of course, there is more to it than just reducing L_m. If L_m is reduced, the B/H ratio of the circuit will be affected, which will result, in turn, in a new H_m. While this means a recalculation of the B/H ratio using the permeance method in Chapter 8, it is also an opportunity to design for the BH_{max} of Alnico 8.

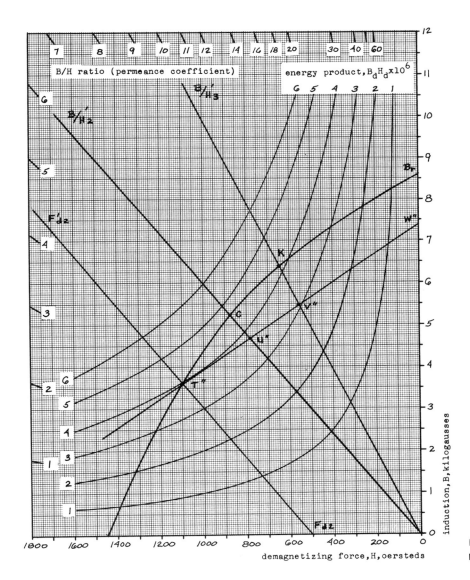

Figure 10-4. Demagnetization/energy product curve for cast Alnico 8.

EXACT DESIGN METHODS 99

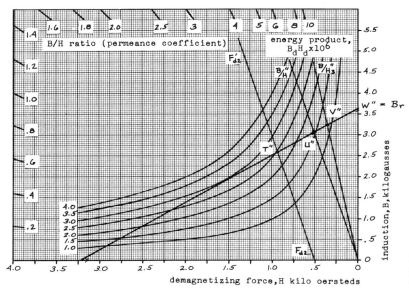

Figure 10-5. Demagnetization/energy product curve (at room temperature) for grade 5 sintered ferrite.

It might appear that Ferrite 5 is the "perfect" material for this circuit (despite the fact that it is also not operating efficiently and that it has poor temperature stability characteristics). But this may not be the case, since one *other* factor has not been considered throughout this sequence of material analyses: the flux density in the *magnet* (B_m). While H_m determines the flux *density* in the air gap, *all* of the total lines of flux (ϕ) originate in and are supplied to the gap by the *magnet*. Therefore, the total flux in the magnet (ϕ_m) must equal the total flux in the air gap (ϕ_g) plus the flux losses due to leakage and fringing. Mathematically:

$$\phi_m = \phi_g + \phi_1$$

But

$$\phi = BA$$

and

$$B_m A_m = \phi_g + \phi_1$$

So if the gap and leakage/fringing area dimensions are fixed, as is the flux density required in the air gap (B_g), then:

$$B_m A_m = K_1 + K_2$$

where K is the magnetic susceptibility, and the flux density in the magnet in relation to its area must be adequate to *supply* the gap flux plus lost fluxes. Thus, in Figure 10-2:

B_m at U' = 4300 G

B_m at V' = 5000 G

while in Figure 10-3:

B_m at U' = 7000 G

B_m at V' = 7900 G

and in Figure 10-4:

B_m at U'' = 4650 G

B_m at V'' = 5500 G

and in Figure 10-5:

B_m at U'' = 3000 G

B_m at V'' = 3300 G

Reviewing the above B_m values and assuming that Figure 10-2 represents a satisfactory *gap flux*, switching to Alnico 8 with *no* magnet *dimension* modifications (Figure 10-3) will produce far more flux than is needed. Switching to Alnico 8 and *reducing the magnet length* (L_m) will produce a small excess of flux, and the magnet area can also be reduced slightly. Switching to Ferrite 5 will *not* produce sufficient flux unless the area is *increased by over 56 percent*.[5]

The preceding method of determining the effects of external demagnetization is not entirely accurate in that it is based on the "normal" demagnetization curve for the material (*B* plotted against the normal coercive force). A really accurate method requires the use of *intrinsic* curves (*B* plotted against the intrinsic coercive force). Unfortunately, intrinsic curves or the value of H_{ci} on which to base such accurate calculations are very

---

5. Again, the reminder that changing lengths or areas will require recalculation of *B/H* ratios!

seldom available. For most purposes, the method presented is adequate, and it is certainly preferable to no method at all!

A number of general inferences can be drawn from the preceding study of operating points, demagnetization effects, and materials:

1. The length to area relationship for the BH_{max} of high-residual, low-coercive materials like cast Alnico 5 is quite different than that of the low-residual, high-coercive materials like sintered Ferrite 5. The high B_r, low H_c materials are most efficient at *high B/H* ratios (usually B/H = 8 or up). The low B_r, high H_c materials operate most efficiently at *low B/H* ratios ($B/H \approx 1$). Therefore, high B_r, low H_c materials generally require a long magnetic length (L_m) and a small area (A_m); low B_r, high H_c materials, a short magnetic length (L_m) and a large area (A_m).
2. High B_r, low H_c materials have demagnetization curves with a definite knee, while low B_r, high H_c materials have little or no knee in the curve. The rare-earth materials are the one exception to this rule. The rare earths have a high B_r *and* a high H_c, with a "flat" curve.
3. For materials with a knee in the curve, the effects of self- or external demagnetization can only be minimized or avoided by operating the magnet *above* the knee at its *maximum* open-circuit. In the case of external demagnetizing influences, the momentarily reduced load line must not fall below the knee.
4. Materials with a "flat" demagnetization curve are ideal for use in those circuits that are impractical to magnetize after assembly. With this type of material, the basic magnet can be saturated outside the final circuit and inserted afterwards, without loss of performance due to self-demagnetization.
5. It is seldom practical (without substantial losses) to saturate the basic magnet before assembly if the material used has any appreciable knee in the demagnetization curve.
6. Materials with a "flat" demagnetization curve are *not* practical when it is necessary or desirable to deliberately calibrate and/or stabilize the magnet. With no knee in the curve, the recoil line will always coincide with the material curve.
7. Materials with a "flat" demagnetization curve are not desirable when *total* demagnetization might be required at any time during manufacture of the product or in the future. Materials with a perfectly "flat" or nearly "flat" curve cannot be demagnetized by an external field, since any recoil line always follows the material curve until $-H_s$ is reached (and then the polarity of the magnet reverses!). This effect will be covered in detail in Chapter 13.

Before proceeding to the energy/work relationships in a magnet, one other fact concerning the demagnetization curve must be established. This is the relationship of the *published* demagnetization curves to the curve of any specific magnet made of that material. The published demagnetization curves are *typical* curves and may only approximate the true curve of a material in a specific magnet. This deviation from the typical curve (as we mentioned earlier) is due to the normal nonuniformity in each magnet itself. As with all other processes, the manufacture of magnet is imperfect and a normal "tolerance" on magnetic as well as physical properties can be expected. Reputable mills normally [6] maintain B_r and H_c at ±5 percent and BH_{max} at ±10 percent of the published typical values. Referring to Figure 10-6, it is evident that these variations can make a substantial difference in the results calculations based on the demagnetization curve. The experienced designer will allow for variations in material properties in the magnet and/or the product design.

Use of the Demagnetization Curve for Energy-Work Calculations

Using the demagnetization curve for a selected material combined with the parameters of a particular magnet made of that material, energy relationships can be determined directly. These energy relationships, in turn, can be used to calculate the work done by the magnet.

Assume that Figure 10-7 represents the material-magnet parameters for a particular circuit, with a minimum air gap of B/H = 30 and a maximum air gap of B/H = 14.

Theoretically, expressed in terms of B and H, the *total potential energy* in the *material* on a unit volume basis is *the area under the demagnetization curve*. Of course, not all of the energy available in the material can be "delivered." For this to occur, the magnet in the circuit would have to operate at a minimum air gap of B/H = 0 and a maximum air gap of B/H = ∞.

In a "real" circuit, the potential energy *in the magnet* will be the area 0-A-B_k in Figure 10-7. Of this energy, the area B_k-A-B_{m1} is the potential energy stored in the magnet when it is operating at point A, while area B_{m1}-A-0 is the potential energy stored in the field under these same maximum-gap conditions.

As the operating point changes from point A to point K, the magnet is providing some force (F) acting through some distance (d) or, in other words, doing work. Since doing work requires some expenditure of

---

6. Unless the user is advised otherwise.

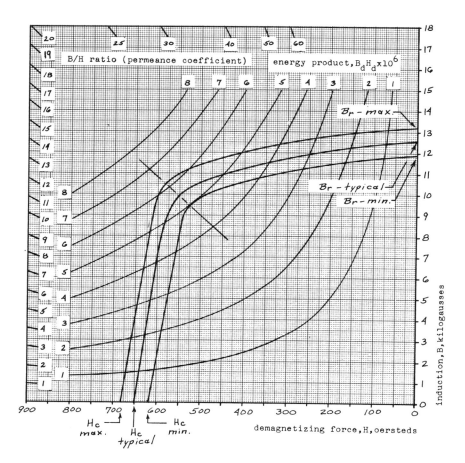

Figure 10-6. Normal variation in demagnetization/energy product curve for cast Alnico 5.

energy, and the only energy source is the magnet itself, some of the potential energy in the magnet will be converted to kinetic energy. The energy delivered by the magnet will therefore be represented by area 0–A–K. The useful energy (E_k) expressed in terms of this area is:

$$E_k = \frac{B_k (H_{m1} - H_{m2})}{2}$$

in G-Oe/cm³ *of magnet material*, or

$$E_k = .65 \, B_k (H_{m1} - H_{m2})$$

in ergs/in.³ of magnet material. *When the greatest volumetric efficiency* [7] *of a magnet material is desired*, it is necessary to establish the operating points A and K in the design so that:

$$AK = KB_k$$

It then follows that:

$$AD = D0$$

It can be shown graphically that only when the volu-

7. Volumetric efficiency is the greatest energy delivered for the smallest volume of material.

metric efficiency is maximum, can E_k be expressed in terms of area 0–A–B_k by:

$$E_k = \frac{B_k H_{m1}}{4}$$

Having determined the energy delivered by the magnet, this energy can be changed from B/H units to ergs (dyne-centimeters) by:

Equation 19   Work = ergs – $\dfrac{E_k}{8\pi}$ = dyn-cm³ of mechanical work

(Where desirable, dyn-cm can, in turn, be converted to lbs.-in. and/or cm³-in.³)

Where it is desirable to develop a force/distance (F/d) characteristic curve for a circuit, the E_k at *several* air gaps at and in between the maximum (point A) and the minimum (point K) can be calculated. The average force (F_w) at each energy level can be determined by:

$$F_w = \frac{E_k}{L_g}$$

When the force/distance curve is for a circuit where a

102 CHAPTER 10

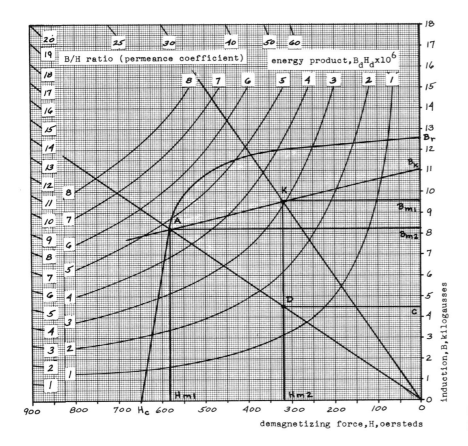

Figure 10-7. Demagnetization/energy product curve for cast Alnico 5.

zero air gap condition occurs, the *contact force* (F_c) can be obtained from:

Equation 20 $\quad F_c = \dfrac{B^2 A}{K}$

where $K = 72 \times 10^6$ for F_c in pounds, and $K = 8\pi$ for F_c in dynes. If the air gap has been effectively reduced to zero (in very good contact with the poles of the magnet), B in the above equation is B_r. If the air gap is imperfect but relatively small, B_g should be calculated by methods previously described and used for B in Equation 20. In all cases, A is the area of contact (or the air gap) between *one* magnet pole face and the part being attracted.

There is one important precaution and/or limitation to the development of force/distance curves as described above. *These equations and the resulting curves assume that there is full, matching, surface engagement between the magnet faces and the part being attracted.* In other words, this method does *not* work if a *flat* pole-faced magnet (circuit) is attracting a ferrous *sphere*. It *will* work when flat poles attract a flat bar that fully spans and covers the poles. Similarly, the method is usable when curved pole faces attract an identically curved "follower," as in some types of torque couplers.

One final note concerning the method for obtaining the maximum useful energy output from a magnet. In addition to maximizing area 0-A-K or area B_{m2}-K-D-C by the choice of operating points, it is obvious that the slope of recoil line A-K-B_k will have significant bearing on the size of these areas. The slope of this recoil line, called the *recoil permeability* (μ_r), can be expressed as:

$$\mu_r = \dfrac{B_{m2} - B_{m1}}{H_{m1} - H_{m2}}$$

In the example shown in Figure 10-7:

$$\mu_r = \dfrac{9600 - 8300}{580 - 320} = \dfrac{1300}{260} = \underline{5.0}$$

Recoil permeability (μ_r) is one of the magnetic parameters normally provided in the properties tables for permanent magnet materials. *In general, the lower the recoil permeability, the greater the capability of the material for producing high levels of useful energy.*

Before we study some typical design problems, consider the general question of why a permanent magnet does not lose its strength (magnetism) as it does work. This question is regularly asked by perceptive laymen and engineers alike who do not recognize the obvious answer. As we saw in the preceding development of the energy/work relationships, a magnet in a sense *does* "lose strength" when it does work. The energy level in the magnet is decreased in the process of doing work, but *this energy is returned* (restored) *to the magnet when the object it has attracted is removed by an external force*. If the object is not removed, the magnet is incapable of doing further work [8] on another object. In other words, the magnet does work by providing a force (F) over some distance (d). To remove an object on which work has been done, some external source of energy must apply an *equal* force across the same distance! The internal potential energy level of the magnet is thereby restored to its original level and the conservation of energy law is not violated.

An excellent analogy can be made to a ball rolling off a table. When the ball is *on* the table, it has some level of potential energy but no energy is being delivered. When the ball rolls off the table, the potential energy is converted to kinetic energy when it hits the floor. Resting on the floor, the ball has expended its maximum energy for that condition. The *potential* energy can only be restored if it is picked up and placed on the table again.

One last question arises: what about "losses"? Logically, we would expect the magnet's energy level to be reduced by internal losses each time it is "cycled." There is little doubt that such losses exist; in fact, they comprise the area inside the minor hysteresis loop (see Figure 10-1). This hysteresis loss does not reduce the potential energy level in the magnet, because *it is supplied for each "cycle" by the external* (restoring) *energy source*. Thus, the external energy supplied must be *greater* than the energy the magnet originally supplied. In most cases, the energy required to supply this loss is quite small. In typical mechanical force producing systems, it is small enough to be negligible. In some applications, such as hysteresis motors, the loss is deliberately made substantial enough to provide the particular type of force desired (torque, for example).

DESIGN PROBLEM 3

An elementary torque coupler like the one shown in Figure 10-8 is designed so that the calculated $B/H = 17$ at the minimum rotational gap (full engagement) and $B/H = 7$ at the maximum rotational gap (90° from full engagement). The axial air gap between the drive member (magnet assembly) and the driven member is .015.

---
8. If the air gap is not at zero, the magnet *may* do further work if it can move the same object closer.

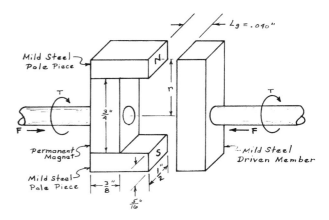

Figure 10-8

Other dimensions are as shown in the figure. The material selected is cast Alnico VIII.

1. Calculate the axial traction (F) on the shaft mounts and bearings.
2. Calculate the *approximate* average torque for a 1° angular displacement between the drive and the driven members.
3. Accurately calculate both the average and the maximum torque that can be maintained without loss of synchronism.

The demagnetization curve for the particular material indicated for this design (cast Alnico VIII) appears in Figure 10-9. The calculated B/H ratios (load lines) shown in the figure establish point A as the maximum-gap operating point and point K as the minimum operating point (after operating stabilization).

To solve for the axial traction (F), the flux density in the air gap must be determined first. This density will be greatest at minimum-gap operating point K. From Figure 10-9, $H_{k1} = 400$ Oe at point K, while from Figure 10-8, $L_m = .75$ in. $= (.75)(2.54) = 1.9$ cm, and $L_g = .040$ in. $= (.040)(2.54) = .102$ cm. Note, however, that the magnet must supply mmf to *two* air gaps *in series*, each .102 cm long. Using Equation 13 from Figure 8-7:

$$B_g = \frac{H_m L_m}{L_g} = \frac{(400)(1.9)}{.102} = \underline{\underline{7450 \text{ G}}}$$

Next, calculating the area per pole over which this flux acts, from Figure 10-10:

$$A_g = (.312)(.500) = .156 \text{ in.}^2$$
$$A_g = (.156)(6.45) = 1.005 \text{ cm}^2$$

and, using Equation 21 from Figure 8-7:

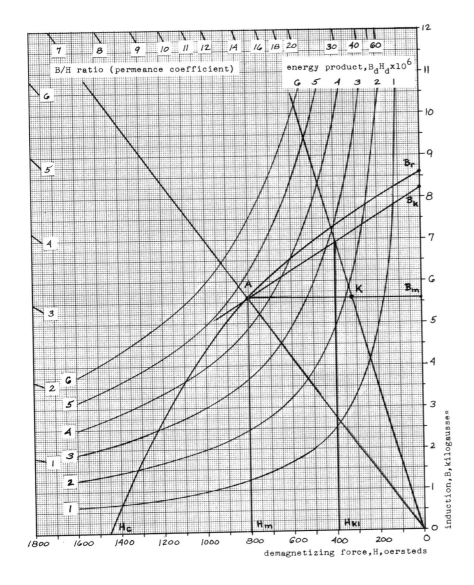

Figure 10-9. Demagnetization/energy product curve for cast Alnico 8.

$$F = \frac{B^2 A}{8\pi} = \frac{(7450)^2(1.005)}{8\pi} = \underline{2.22 \times 10^6 \text{ dyn}}$$

or

$$F = (2.22 \times 10^6)(2.25 \times 10^{-6}) = \underline{5 \text{ lbs}}$$

Since F is the force *per pole*, the *total* axial load that must be carried by the shaft mounts and bearings is 4.44×10^6 dyn or 10 lbs.

In solving for the approximate average torque (T_a) *for small angular displacements,* the following equation is adequate for a first approximation:

$$T_a = \frac{B_g^2 A_g}{10^6} \, r \, N \sin \theta$$

where

B_g = the air-gap flux density (in G)

A_g = the air-gap area in in.^{2}

r = mean radius from center of shaft to center of pole area in in.

N = number of poles in assembly

θ = angle of angular displacement between drive and driven poles

From the preceding calculations, B_g = 7450 G and A_g = .156 in.^{2}. From Figure 10-8:

EXACT DESIGN METHODS 105

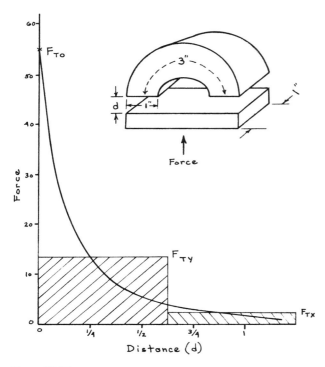

Figure 10-10

$$r = \frac{.75 + .312}{2} = .531 \text{ in.}$$

Substituting the known quantities in the torque equation gives us:

$$T_{av} = \frac{(7450)^2(.156)(.531)(2)(\sin 1°)}{10^6}$$

$$T_{av} = .161 \text{ in.-lbs TOTAL}$$

To obtain the average and maximum torque at *any* displacement angle (short of synchronization loss), the energy expended in shifting the magnet between operating points A and K must be determined first. Referring to Figure 10-9, as established in the preceding section, the energy expended is:

$$E_k = \text{area } 0\text{-}A\text{-}K$$

or

$$E_k = \frac{B_K(H_m - H_{K1})}{2} = \text{G-Oe/cm}^3 \text{ of magnet material}$$

or

$$E_k = .65 \, B_k \, (H_m - H_{k1}) = \text{ergs/in.}^3 \text{ of material}$$

But converting this equation into all English units gives us:

$$E_k = 5.77 \times 10^{-7} \, B_k \, (H_m - H_{k1}) \text{ in in.-lbs/in.}^3 \text{ of magnet material}$$

Since the change in energy occurs during rotation through an angle equal to ½ the pole spacing:

$$\text{Average Torque} = \frac{\text{Potential Energy}}{\theta \text{ in Radians}}$$

$$T_{av} = \frac{5.77 \times 10^{-7} \, B_k(H_m - H_{k1})}{\theta \text{ in Radians}}$$

Substituting the values obtained from Figures 10-8 and 10-9 and using $\theta = 90°$:

$$T_{av} = \frac{(5.77 \times 10^{-7})(8250)(800 - 400)}{1.57}$$

$$T_{av} = 1.21 \text{ in.-lbs/in.}^3 \text{ of material}$$

But the volume of magnet material (Figure 10-8) is:

$$V_m = (.75)(.375)(.5) = .1405 \text{ in.}^3$$

and the total average torque is:

$$T_{av} = (.1405)(1.21) = .17 \text{ in.-lbs}$$

The *maximum torque* (T_{max}) will approach the 1 : 63 ratio for a sinusoid, and:

$$T_{max} = \frac{(9.15 \times 10^{-7}) B_K (H_m - H_{K1})}{\theta \text{ in Radians}}$$

and

$$T_{max} = \frac{9.15 \times 10^{-7}(8250)(800 - 400)}{1.57}$$

$$T_{max} = 1.92 \text{ in.-lbs/in.}^3 \text{ of material}$$

Or the total maximum torque will be:

$$T_{max} = (1.92)(.1405) = .269 \text{ in.-lbs}$$

One important precaution: *The approximate method using B_g to calculate the approximate average torque is only accurate at a one-degree angular displacement.* At any other angles, the accuracy becomes progressively less. At angles of less than 1°, the error is toward lower than true torque values, while at angles of above 1°, the error is on the high side. At 1°, as shown in this example, $T_a = .170$ in.-lbs by the "short method," while the true average $T_{av} = .161$ in.-lbs by the "energy" method (an error of a little over 5 percent). A recalculation at other angles will show how rapidly accuracy decreases with angles other than 1°. For example, at 2°, $T_a = .321$ in.-lbs, which is 20 percent higher than the *maximum* torque obtained by the more accurate method ($T_{max} = .269$).

DESIGN PROBLEM 4

At the beginning of this chapter, the relationships of various operating conditions for a cast Alnico 5 horseshoe magnet were described, using Figure 10-1. For this magnet, calculate the following forces and develop the "pull vs. distance" curve for the magnet:

1. The force (F_x) that is produced *after* the magnet has been fully open-circuited and with the mild steel bar at distance x.
2. The force (F_y) produced on the bar at distance y.
3. The force (F_0) produced on the bar at zero air gap ($L_g = 0$).

Assume that the magnet volume is 3 in.^{3}, that the area of each pole is 1 in.^{2}, that distance $x = 1$ in., and that distance $y = 1/4$ in.

In the open-circuit condition (see Figure 10-1), the magnet is operating at point A. When the steel bar is attracted to distance x, the operating point is now point D and the area 0–A–D is the energy expended in moving the bar to this distance. Therefore:

$$E_{Kx} = \frac{B_K(H_{m1} - H_{m2})}{2}$$

But in this illustration, $B_k = 1 = 8500$ Oe, and from Figure 10-1, $H_{m1} = 600$ Oe and $H_{m2} = 380$ Oe. Substituting in the above equation and converting to English units gives us:

$$E_K = (5.77 \times 10^{-7})(8500)(600 - 380)$$

$$E_{Kx} = 1.08 \text{ in.-lbs/in.}^3$$

The magnet volume is given as 3 in.^{3}, and the total energy produced is:

$$E_{Tx} = 3(1.08) = 3.24 \text{ in.-lbs.}$$

The *average* force at distance x is:

$$F_x = \frac{E_{Tx}}{L_x} = \frac{3.24}{1.00} = 3.24 \text{ lbs}$$

When the magnet moves to distance y, the magnet operates at point N and the energy expended covers area 0–D–N.

$$E_{Ky} = \frac{B_K(H_{m2} - H_{m3})}{2} \text{ G-Oe per cm}^3$$

or

$$E_{Ky} = (5.77 \times 10^{-7})(8500)(380 - 130)$$

$$E_{Ky} = 1.225 \text{ in.-lbs/in.}^3$$

the magnet volume is 3 in.^{3}:

$$E_{Ty} = 3(1.225) = 3.68 \text{ in.-lbs}$$

and the average F_y is:

$$F_y = \frac{3.68}{.25} = 14.72 \text{ lbs}$$

The values of F_{Tx}, F_{Ty}, and F_{T0} calculated above can be plotted as shown in Figure 10-10 and the "pull curve" drawn accordingly. It is evident from this design problem that a product designer requiring a magnet with a particular force/distance characteristic could work backward from a plot of this characteristic to the basic magnet dimensions. While such a calculation is somewhat lengthy by manual methods, once the method has been set up with the appropriate circuit constraints, computer calculations can be used.

11

Environmental Effects

Note: On new materials listed in the Addendum to Appendix 2, Magnetic Properties Tables, contact producing mill for environmental effects data. See Addendum to International Index, Appendix 4 for names of mills producing material.

As is the case with any component or product, the performance of a permanent magnet is dependent on the ambient conditions in which it operates. These conditions may directly affect the magnet material at the domain level by:

1. Changing the energy level of the atoms that comprise each domain.
2. Changing the size, shape, or direction of the orientation of the domains.
3. Changing the permeability between adjacent domains.

These are basically *metallurgical* changes that cause a permanent change in the material; they cannot normally be recovered by simple remagnetization of the magnet. The older, quench-hardened magnet materials were relatively unstable, and metallurgical changes were not uncommon. Fortunately, modern Alnico and ferrite materials are metallurgically stabilized by *aging* at the mills and are generally very stable. In most applications, unless the magnet is or will be subjected to extremely high temperatures or to work harding, metallurgical changes will not occur.

Types of Change

All changes in the magnet that produce changes in the magnetic field can be classified into three categories: reversible, remagnetizable,[1] and nonremagnetizable or irreversible.

A *reversible change* is one that exists only as long as the condition producing it persists. For example, if a magnet is subjected to a rise in temperature that reduces its field strength but the field strength fully recovers

---

1. Some authors include this type of change in the "irreversible" category.

with return to "normal" ambient conditions, the change would be classified as "reversible."

A *remagnetizable* change does not "automatically" disappear with removal of the causative condition but *can* be fully reversed by remagnetization of the magnet. An example of this kind of change occurs when a basic magnet is removed from and then replaced in its circuit. The losses in strength due to open-circuiting can be restored by remagnetization of the circuit.

A *nonremagnetizable* or *irreversible* change is a permanent change in the material that remains even if the causative condition is removed; it cannot be restored by remagnetization or any other procedure normally available to magnet users. An example of this type of change is the annealing that can occur in metallic magnet materials [2] when they have been heated to the curie point and allowed to cool without the application of a suitable level and shape of external field.

In any particular environmental situation, *all three types of changes may occur simultaneously*. Most experimental work and published data relate to each type of change separately. For design purposes, it is important to know which changes are reversible, which are remagnetizable, and which are nonremagnetizable, as well as the extent of each. Table 11-1 provides a summary of the environmental factors, the types of changes, their effects, and the usual means for stabilization.

Environmental Factors

Changes in environment that may cause a change in magnet performance are:

Time
Temperature

---

2. Annealing has the effect of changing both the alignment and the permeability between domains.

Table 11-1
Summary of Environmental Factors

Environmental Factor	Type of Change	Magnetic Change in Material	Percent of Effect	Method of Stabilization
Time	Reversible	—	None	—
	Remagnetizable	Magnetic reorientation of some domains. CAUSE: Localized energy fluctuations within material.	See Figure 11-2 and Table 11-2.	Partial demagnetization (or) magnetic aging.
	Nonremagnetizable	Structural change. CAUSE: Metallurgical instability.	Less than ½ percent per 100 years in Alnicos and ferrites.	Metallurgical aging.
Temperature	Reversible	Energy-level change within atoms.	See Figure 11-3.	Cannot be stabilized; provide temperature compensation.
	Remagnetizable	Magnetic reorientation of some domains. CAUSE: Permeability increase at low temperatures; thermal agitation at elevated temperatures.	See Figures 11-3, 11-4, 11-5, 11-16.	Partial demagnetization (or) magnetic aging at temperatures slightly higher than operating condition.
	Nonremagnetizable	Structural change. CAUSE: Metallurgical instability.	See Figure 11-3 and Table 11-3.	Special metallurgical techniques. NOTE: If structural change occurs, some materials can be reworked by mills to recover initial structure.
Vibration, Shock, and Mechanical Stress	Reversible	Energy-level change within atoms.	No measured effect in Alnicos or ferrites.	Cannot be stabilized; provide isolation.
	Remagnetizable	Magnetic reorientation of some domains. CAUSE: Energy fluctuation within material.	See Figures 11-14, 11-15, 11-16.	Partial demagnetization will provide almost complete stability.
	Nonremagnetizable	Structural change. CAUSE: Cold-working of material, fracture, etc.	Fracture of magnet usually occurs before nonremagnetizable loss.	Magnet must be protected against fracture.
Nuclear Radiation	Reversible	Information is too incomplete to categorize.	No information available.	No information available.
	Remagnetizable		See Figure 11-19.	
	Nonremagnetizable		No information available.	
Spurious Contacts	Reversible	Energy-level change within atoms.	No information available.	Cannot be stabilized; prevent spurious contact by encapsulation.
	Remagnetizable	Magnetic reorientation of some domains. CAUSE: Concentration of magnetic-field energy in vicinity of contact.	Almost no effect in ferrites; see Figure 11-18 for Alnico 5.	Partial demagnetization with external field (or) controlled contact with ferromagnetic object (or) encapsulation to prevent contact.
	Nonremagnetizable	—	None	—
External Magnetic Fields	Reversible	Energy-level change within atoms.	Determine from demagnetization curve.	Cannot be stabilized; prevent effect of external fields by magnetic shielding.
	Remagnetizable	Magnetic reorientation of some domains. CAUSE: Change in energy of magnetic circuit; energy loss in form of hysteresis heating.	Determine from demagnetization curve; see Figures 11-20 and 11-21.	Partial demagnetization (or) magnetic shielding.
	Nonremagnetizable	—	None	—
Working-Gap Variation	Reversible	Energy-level change within atoms.	Determine from demagnetization curve.	Cannot be stabilized; effect can be reduced by proper circuit design to avoid large variations in B/H variation.
	Remagnetizable	Magnetic reorientation of some domains. CAUSE: Change in energy of magnetic circuit; energy loss in form of hysteresis heating.	Determine from demagnetization curve.	Partial demagnetization (or) cycling the magnet through normal working-gap limits. Loss can be reduced by proper circuit design to reduce B/H variation.
	Nonremagnetizable	—	None	—

Vibration, shock, and mechanical stress
Nuclear radiation
Spurious contacts
External magnetic fields
Working air-gap (permeance) changes

Before considering each of the above factors, a few notes of explanation and caution are necessary.

First, surprisingly little widely applicable, specific information on environmental factors is available. The greatest impetus (now largely expired) for research on environmental factors has been sponsored by the federal government. This research has been very limited and is now quite dated.

Second, virtually all of the environmental information available is on changes in *remanence* (B_m or B_d) for a particular material, rather than on B_r, H_c, and BH_{max}. Since the data is related to remanence and remanence, in turn, is a function of the dimensional ratio (L/D), the validity of the data is limited to the exact L/D ratio and to the particular material. It cannot be broadly applied to substantially different L/D ratios or to any other material.

Time Effects: Both remagnetizable and nonremagnetizable time-related changes can occur in any magnetic material. Remagnetizable changes occur when some of the less stable domains are reoriented by localized fluctuations in thermal and/or magnetic energy. This condition can occur even when the primary mass of the magnet is thermally stable. The demagnetizing effect decreases rapidly as the number of unstable or marginally stable domains decreases. This phenomenon is known as "after effect," "magnetic creep," "normal time change," "relaxation effect," or "magnetic viscosity."

Various studies have shown that magnetic viscosity is greater in soft magnetic materials than in permanent magnet materials and most rapid immediately following magnetization. Depending upon the material, no further changes occur 10 to 25 hours after initial magnetization.

Figure 11-1 shows the change in magnetic viscosity in cast Alnico 5 as a function of time. Most of the change in remanence occurs within the first five hours, and virtually complete stability is achieved after 25 hours.

The work done under the auspices of the U.S. Office of Technical Services by K.J. Kronenberg and M.A. Bohlmann [3] covered magnetic viscosity for a number of Alnico type materials. From this work, summarized in Figure 11-2, the following generalizations can be made:

(a) The higher the coercive force (H_c), the more stable the remanence (B_m or B_d).
(b) The higher the dimensional ratio (L/D), the more stable the remanence.

3. WAOC Technical Report 58-535, 1958.

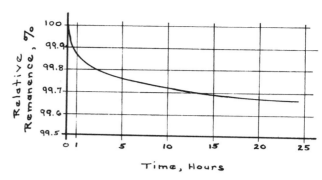

Figure 11-1. Remagnetizable time change for cast Alnico 5 at 24°C and L/D = 4.3.

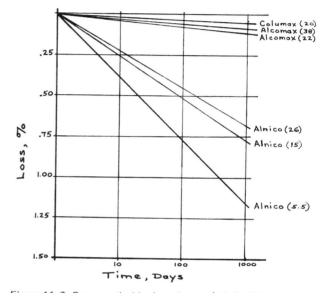

Figure 11-2. Remagnetizable time change. (All English materials; numbers in brackets are B/H ratios of test samples.)

(c) The remanence decreases linearly with the log of time.
(d) The magnet can be stabilized to avoid magnetic viscosity changes.

Table 11-2 presents the remagnetizable, time-related change that can be expected for both common Alnico and sintered ferrite materials. Note the high stability of the sintered ferrites and the marked effect on the Alnicos of the L/D ratio. While the change is relatively small and insignificant for most applications, in those cases where exceptionally high stability is required, remagnetizable time changes can be avoided or minimized by:

1. Operating the magnet with a large L/D ratio (high on the demagnetization curve).
2. Stabilizing the magnet by momentarily applying a 5-10 percent (of H_c) external demagnetizing AC field.

The nonremagnetizable time-related changes that can also occur in a magnet are caused by structural

Table 11-2
Remagnetizable Time Change

Material	L/D Ratio	Remanence (B_d) (kG)	Stability: Relative Remanence at 24° C, 5 log cycles (10,000 hrs) After Magnetization (percent)	Measuring Accuracy (percent)
sintered Ferrite 1	0.9	1.4	100.0	±0.1
sintered Ferrite 5	0.8	2.5	99.6	±0.1
cast Alnico 3	3.5	4.5	98.10	±0.04
	2.2	3.2	97.04	±0.05
cast Alnico 7	3.5	4.9	99.32	±0.04
	2.2	3.9	98.96	±0.06
cast Alnico 5	8.0+	12.3	99.95	±0.01
(long)	6.7	12.1	99.89	±0.02
	5.8	11.9	99.81	±0.02
(medium)	4.3	10.4	99.23	±0.02
(short)	3.5	8.2	98.84	±0.04
	3.3	7.6	98.97*	
	2.9	6.7	98.50	±0.05
	2.2	4.8	98.3	±0.07
	2.1	4.1	97.6*	
	1.4	2.6	98.2	±0.1

*Extrapolated 1 to 2 log cycles beyond last measurement.

Table 11-3
Critical Temperatures of Various Permanent Magnet Materials

Alloy or Grade	Maximum Temperature*		Curie Temperature		Alloy or Grade	Maximum Temperature*		Curie Temperature	
	°C	°F	°C	°F		°C	°F	°C	°F
Alnico 1	450	842	780	1436	Alnico 8	550	1022	860	1580
Alnico 2	450	842	815	1499	Ferrites	982	1800	450	842
Alnico 3	450	842	760	1400	Cunico	500	932	860	1580
Alnico 4	450	842	800	1472	Cunife	400	752	450	842
Alnico 5	550	1022	890	1634	Platinum-Cobalt	400	752	490	914
Alnico 6	550	1022	875	1607	Remalloy	500	932	900	1652
Alnico 7	550	1022	850	1562	Vicalloy	500	932	850	1562

*No structural change takes place up to this temperature.

changes in the material itself. These changes can be a problem in quench-hardened or carbide-hardened steels, but such changes are virtually unmeasurable in properly mill-aged Alnicos.

Temperature Effects: The effects of temperature must be considered in two ways: *operating* temperature and *exposure* temperature. Figure 11-3 shows the general effects of elevated operating temperatures on all grades of sintered ferrites and on cast Alnicos 5 and 8. For ferrites, the curve shows *reversible* changes up to 450° C. From 450° C to 982° C, the changes are *remagnetizable*. Above 982° C (1800° F), there is a major structural change in the material and the changes become nonremagnetizable or irreversible.

For the Alnicos, changes up to 550° C are substantially *reversible*, although a small part of the total change may be *remagnetizable*. From 550° C to the curie points (860° C and 890° C), the changes are largely structural and are *not* reversible or fully remagnetizable.

Note that from the standpoint of operating temperature, ferrites are totally useless above 450° C, since they have become totally nonmagnetic. The highly linear reduction in magnetization from 100 percent at 0° C to 30 percent at 400° C can present serious design problems, unless allowances are made for the changes or unless such changes are a *requirement* for the operation of the specific product (that is, a temperature-sensitive magnet desired to attain a particular kind of performance). Generally, ferrites are not used in products subjected to operating temperatures of over 50° C.

Exposure temperatures may, of course, produce any of the three types of changes. The "critical" temperature for any material is usually the temperature that will produce changes in structure, or nonremagnetizable changes. Table 11-3 shows the critical temperatures associated with 14 commercial materials. This table should be used as a primary design guide in temperature-material selection. Figure 11-4 shows the actual change

ENVIRONMENTAL EFFECTS

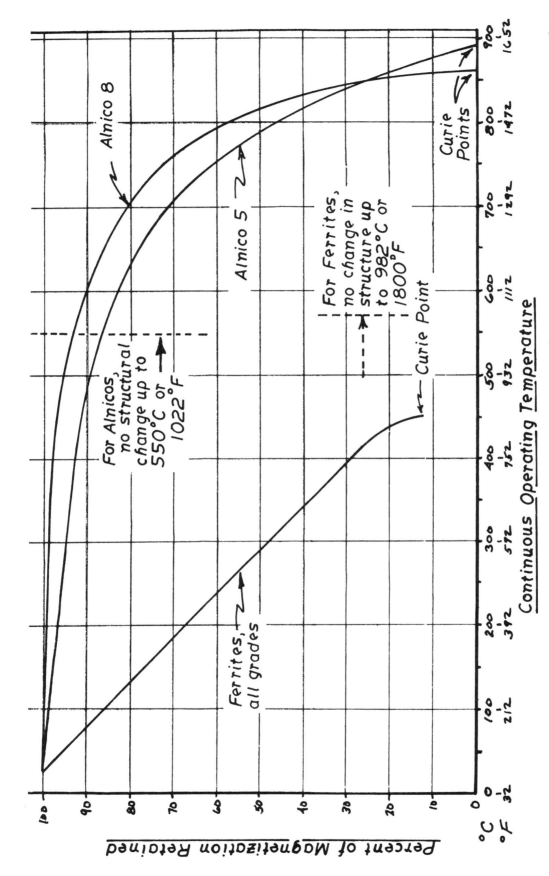

Figure 11-3. General effects of elevated operating temperatures on Alnico 5, Alnico 8, and all ferrites.

If Curie temperature is reached, magnet will be completely demagnetized. If exposure temperature has not produced chemical or structural change, it may be remagnetized to original strength. Curves based on magnet operating at B/H Ratio above the knee of the demagnetization curve.

in the demagnetization curve of cast Alnico 5DG at elevated temperatures. Note that as the temperature increases, the knee in the curve becomes flatter with a reduction in BH_{max}, reflecting changes in both B and H. However, also note that B and H do not necessarily change at the same rate for the same temperature.

Figures 11-5, 11-6, 11-7, 11-8, 11-9, and 11-10 provide specific percentage remanence change values for both reversible and irreversible changes in various grades of metallic magnet materials. Figures 11-5, 11-6, 11-7, and 11-8 are for both reduced and elevated temperatures, but Figures 11-9 and 11-10 are for elevated temperatures only. It should be noted from Figures 11-4 to 11-8 that there is a *reversible increase in remanence for reduced temperatures* (except for cast Alnicos 2 and 3) simultaneously with an *irreversible reduction in remanence*. Most reduced-temperature irreversible losses *are* remagnetizable.

Figure 11-11 compares sintered ferrites and Alnicos at elevated and reduced temperatures. Note that the ferrites show a much greater sensitivity to reduced temperatures (as they do to elevated temperatures) than the Alnicos. While this low temperature sensitivity has not greatly inhibited the use of ferrites for such potentially low-temperature applications as taconite-ore separators and automotive motors, the effects must invariably be considered in the primary product design.

Figure 11-12 shows the specific effects of reduced temperatures on three grades of cast Alnicos for various B/H ratios (operating load lines).

Figures 11-13 and 11-14 show the direct change in the demagnetization curves for two grades of sintered ferrites at various elevated and reduced temperatures. *Note that an irreversible loss will occur if the magnet operates at a B/H ratio below the knee of the curve at the lowest temperature to which the magnet may be exposed.* This loss will be in proportion to the change in H. A study of Figure 11-13 for Ferrite 5 indicates that in general as the temperature of ferrites *rises* above +20° C, B_r is reduced and H_c increases for temperatures up to +100° C. At +200° C, B_r is reduced, but H_c is again the same as it was at +20° C. Conversely, as the temperature *drops* below +20° C, B_r increases while H_c is always reduced. The demagnetization curves for Ferrite 6 in Figure 11-14 follow a similar pattern, but with different B_r/H_c values and temperatures.

Figure 11-15 provides a useful design tool for establishing the minimum B/H ratio at which grade 5 and 6 Ferrites may be operated without nonremagnetizable (irreversible) changes.

Vibration, Shock, and Mechanical Stress: The effect of single, cyclic, or continuous mechanical loads may cause some of the less stable domains in a magnet to lose their orientation. The effect of such mechanical loads is similar to the effects of thermal fluctuations or magnetic viscosity, where improperly aligned neighboring domains have a demagnetizing effect on each other at the expense of the external output capability. The degree of magnetization loss due to mechanical effects is dependent upon the type of magnet material and the operating point (B/H ratio or L/D ratio) on the demagnetization curve.

Relatively little consistent research has been conducted on mechanical effects, and most of this is virtually impossible to relate to practical operating conditions. However, the available information does indicate that the effect of mechanical loads on modern magnet materials is both small and remagnetizable, so it can be neglected in most applications.

Figures 11-16, 11-17, and 11-18 and Tables 11-4, 11-5, and 11-6 present the available information on the effects of shock, stress, and vibration for some of the common metallic magnet materials. Note that changes in the more modern, highly coercive grades of Alnicos are relatively small. No similar data are available on the ferrites, but their very high coercivity suggests that any changes would be appreciably smaller than they would be for the highly coercive Alnicos. Since the sintered ferrites are very brittle (as are all typical, ceramic-type materials), the major effects of shock, vibration, or stress would occur through the mechanical disintegration of the basic magnet. The edges of sintered ferrite are, in particular, highly subject to fracture. If mechanical failure occurs in a ferrite but the magnet is so contained (by enclosures, empotting, or whatever) that it

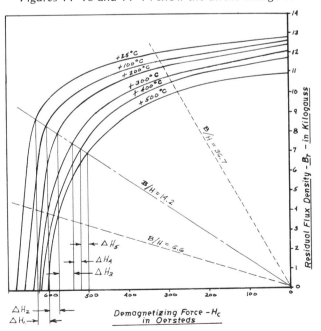

Figure 11-4. Reversible change in the demagnetization curve for Alnico 5DG at various temperatures.

ENVIRONMENTAL EFFECTS 113

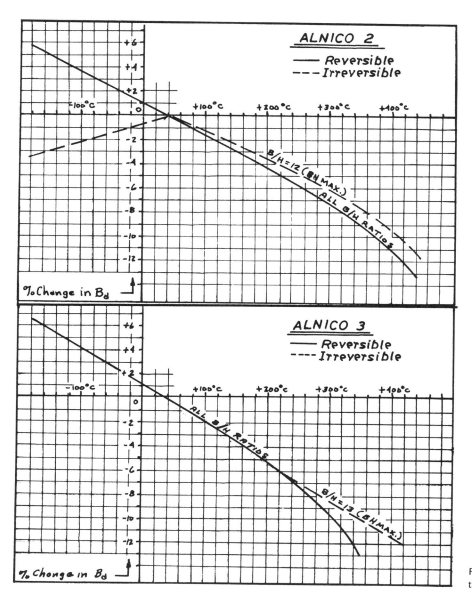

Figure 11-5. Effects of temperature on the residual induction of Alnicos 2 and 3.

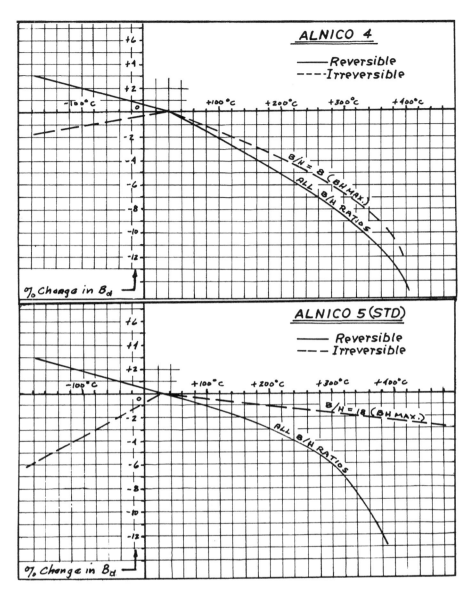

Figure 11-6. Effects of temperature on the residual induction of Alnicos 4 and 5.

is *fully* held together, the effect on its magnetic properties is small enough to be negligible. However, since individual fragments or parts have the *same* direction of magnetization, if these parts are not well held together, the inherent repelling action can result in separation (or realignment) and very substantial reductions in magnetic output.

AC field stabilization is most effective against changes due to mechanical loads, but it is only applicable to metallic magnet materials. Such stabilization may not completely eliminate shock, stress, or vibration changes, but it will materially reduce their levels. The only effect or necessary "preventive" measure against mechanically produced changes in sintered ferrites is the aforementioned mechanical *containment*.

Spurious Contacts: This type of change, also called *contact demagnetization,* is one of the least recognized and potentially most troublesome that can occur in permanent magnet materials. Contact demagnetization occurs when certain types of permanent magnets come into direct physical contact with *any* ferromagnetic object. This ferromagnetic object may be a tool, workbench, chasis, or shield, or another permanent magnet. *Contact demagnetization occurs only when the spurious contact is on some surface of the magnet other than the pole faces.* As a result of localized domain realignment, *instantaneous* changes occur in the performance of the magnet because the instant contact provides a "shorter flux path" for the domain. These changes are both partially reversible and remagnetizable.

ENVIRONMENTAL EFFECTS 115

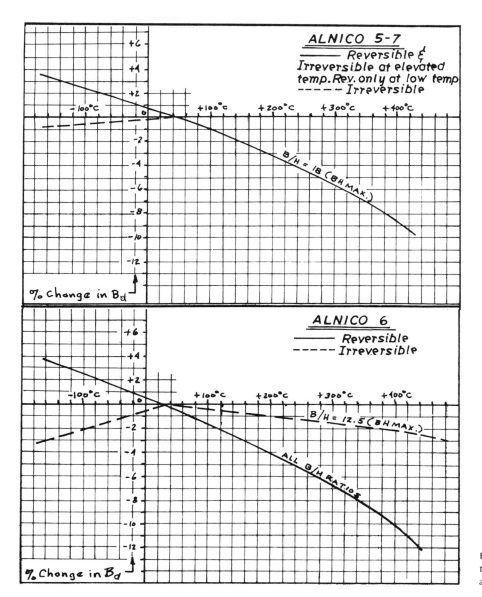

Figure 11-7. Effects of temperature on the residual induction of Alnicos 5-7 and 6.

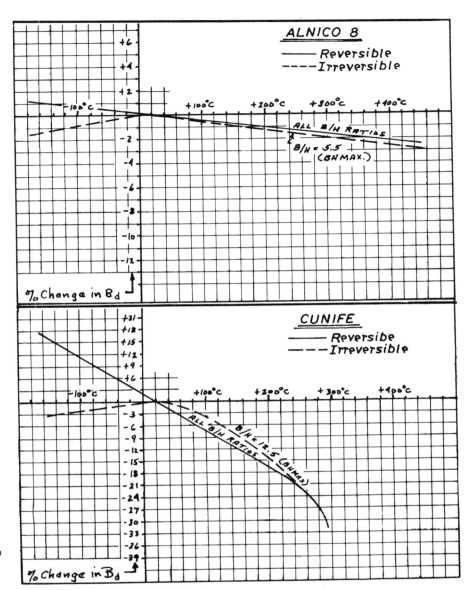

Figure 11-8. Effects of temperature on the residual induction of Alnico 8 and cunife.

Figure 11-19 presents the only available, empirically established information on the extent of spurious contacts. This information cannot, unfortunately, be applied to any other specific magnets except the two (one round rod and one horseshoe) used in the test.

Note in Figure 11-19 that with each contact or series of contacts, the working gap flux (B_g) is reduced. However, also note that if, after such contacts, the working gap is *closed* (for the horseshoe) or the primary poles are touched to a ferrous object (for the rod magnet), *there will be some "recovery" of B_g*. This clearly indicates that part of the initial loss is *reversible*. The remagnetizability of this loss can be substantiated by the fact that B_g is restored to the original level by simple resaturation of the magnet.

Two facts are of particular significance to the design engineer in anticipating the effects of contact demagnetization. It is evident from Figure 11-19 that *a condition of maximum demagnetization is eventually reached* (after approximately 40 contacts) and that this is *equivalent to an externally applied* AC *field of 300 Oe*. In this particular case, since the magnets were made of cast Alnico 5, the contact demagnetization effect is almost equal to 50 percent of the normal coercive force (H_c) for this material.

Two methods may be used to avoid changes in magnet output due to contact demagnetization:

1. Prevent the contact of any "foreign" ferrous object by using suitable enclosures.
2. Stabilize the magnet to the necessary level.

ENVIRONMENTAL EFFECTS 117

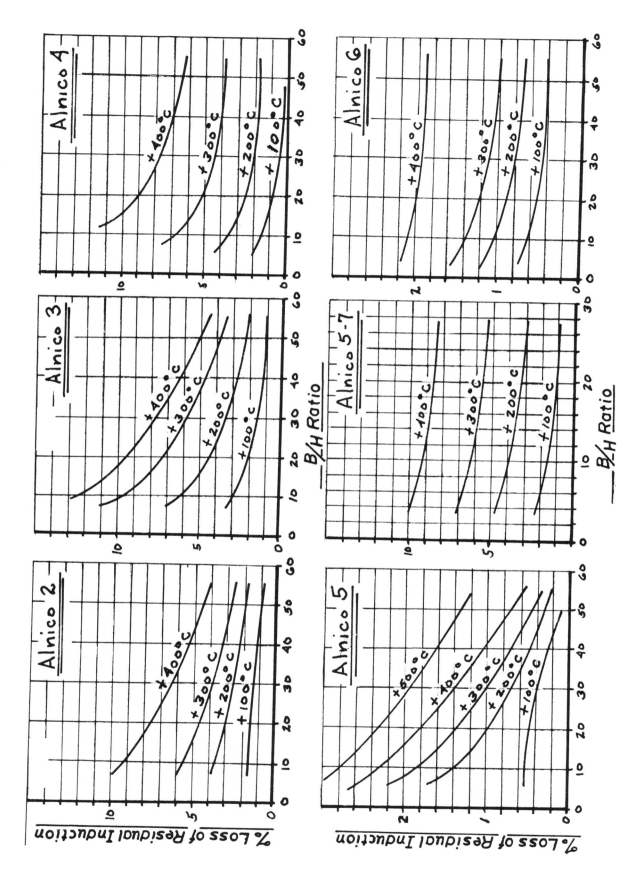

Figure 11-9. Remagnetizable loss of six grades of Alnico at various B/H ratios (at elevated temperatures).

NOTES

Losses shown will occur in magnet either with continuous operation at that temperature or if it once reaches that temperature and then returns to normal.

For TOTAL loss under CONTINUOUS elevated temperature ADD Reversible and Irreversible Losses at that temperature.

To determine effect of Residual Induction changes on actual magnet performance, establish effect on H_d from B-H Curve for that material.

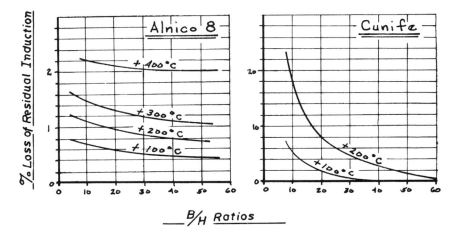

Figure 11-10. Irreversible loss in Alnico 8 and cunife at various B/H ratios at elevated temperatures.

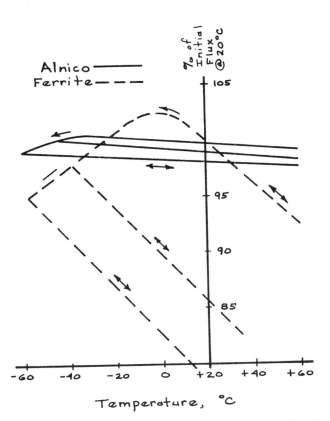

Figure 11-11. Reversible and remagnetizable temperature change of Alnico and ferrite at B/H_{max}.

Enclosures (including encapsulation) are the most effective way to prevent contact demagnetization, since the maximum capability of the magnet continues to be available. Enclosures *must not be of a ferromagnetic material*, unless they are located far enough from any surface of the magnet to prevent shunting losses as a result of the enclosure itself. Experience has shown that any enclosure or covering that keeps foreign ferrous objects even a *short* distance from direct contact with a magnet will effectively prevent contact demagnetization. *For small magnets, as little as 1/16 in. of protection is sufficient. For larger magnets, 1/8 in. is advisable.* However, enclosures are intended only to prevent this one type of loss. Even with an enclosure and at the spacing recommended above, there will be a loss in output if the foreign ferrous object is permitted to *remain* near the "nonworking" surfaces of the magnet.

Stabilization by using a controlled AC field is completely effective only if the level of stabilization exceeds the maximum effect that can occur due to future possible spurious contacts. In the examples in Figure 11-19, the "safe" stabilization of the rod magnet and the horseshoe, respectively, would be to 40 and 45 percent of the original maximum B_g. This represents a very substantial reduction in the efficiency of the material and should only be used where protective enclosures are not practical.

ENVIRONMENTAL EFFECTS 119

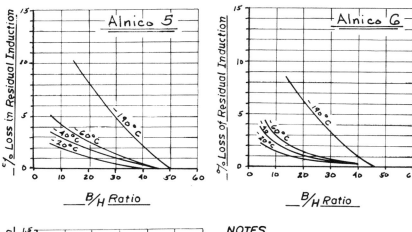
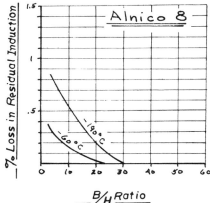

NOTES

Losses shown will occur in magnet either with continuous operation or if it once reaches that temperature and then returns to normal.

For TOTAL loss under CONTINUOUS elevated temperature ADD Reversible and Irreversible losses at that temperature.

To determine effect of Residual Induction changes on actual magnet performance, establish effect on H_d from B-H Curve for that material.

Figure 11-12. Irreversible loss in Alnicos 5, 6, and 8 at various B/H ratios at reduced temperatures.

Contact demagnetization is a phenomenon *exclusively* associated with the lower coercive magnet materials, including most Alnicos. It does not occur to *any* degree in any form of ferrite, platinum-cobalt, or rare-earth types. In the case of Alnicos, the more highly oriented types are more subject to contact demagnetization than the nonoriented or less-oriented types are. Thus, cast Alnico 5 is more sensitive than cast Alnico 2, and cast Alnico 5-7 is the most sensitive of all. The high coercivity of cast Alnico 8 makes it far less sensitive than all other Alnicos. However, even Alnico 8 is affected to some degree by spurious contacts.

Nuclear Radiation: Basic permanent magnets (and assemblies) are occasionally used in products that are subjected to nuclear radiation. The effects of nuclear radiation on the magnetic and physical properties of magnet materials has been investigated to a very limited extent. In most of these investigations, the percentage of change in magnetic properties that was measured was less than the estimated accuracy of the measuring instrument itself.

T.D. Owen et al. checked the feasibility of using one type of Alnico material for coolant flowmeters in an atomic reactor by exposing magnets to fast neutrons together with gamma radiation for periods up to 290 days. The instrumentation was estimated to have a possible error of 5 percent. No changes in magnetic field were detected in four samples exposed for periods up to 13 days. Decreases of 2 and 3 percent were found in samples irradiated for 144 and 290 days.

R.S. Sery, D.L. Gordon, and R.H. Lundsten investigated radiation effects by exposing most of the available permanent magnet materials to 3×10^{17} fast neutrons per cubic centimeter. On the basis of demagnetization curves and open-circuit measurements, they concluded that any changes attributable to radiation were obtained from radiation tests with various materials.

The information available at present is not complete enough to categorize radiation changes into reversible, remagnetizable, and nonremagnetizable changes. It is likely that all three forms of change can occur but that the percentages of change are very small. Table 11-7 summarizes the information currently available.

There is some evidence to indicate that the *physical* damage to magnets (and the possible subsequent magnetic change) of nuclear radiation is substantially the same as that observed for other *metallic* materials (such as those used in control rods). The physical effects on ferrites have not, to the knowledge of this author, ever been investigated.

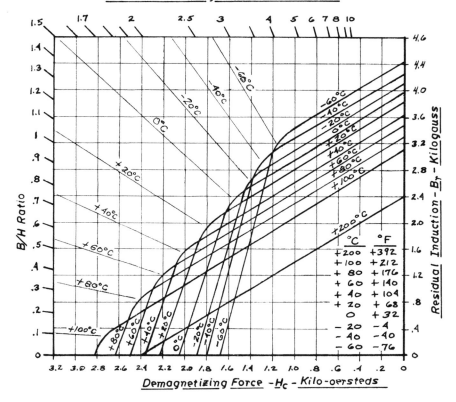

Figure 11-13. Demagnetization curves for grade 5 ferrites at various temperatures.

External Field Effects: External magnetic fields may cause reversible and remagnetizable changes in a permanent magnet. They do not cause nonremagnetizable (irreversible) changes directly, but a high-frequency external field could produce sufficient heating due to hysteresis and eddy currents to effect a structural change. As we saw in detail in Chapter 10, an external magnetic field can change the effective operating point of a magnet on the demagnetization curve for the material from which the magnet is made. Actually *three* factors determine the extent of both the reversible and the remagnetizable changes:

1. The operating load line (B/H ratio) of the magnet as determined by its physical parameters.
2. The peak magnitude of the applied external field.
3. The *intrinsic* demagnetization curve for the material.

In the development of external field effects in Chapter 10, the *normal* demagnetization curve was used. Technically, the use of the normal demagnetization curve is *not* correct. In the case of magnet materials with an H_c of less than 1000 Oe, the normal demagnetization curve and the intrinsic demagnetization curve are so nearly identical that the error introduced by using the normal curve is very small. In the case of highly coercive materials such as the ferrites, platinum-cobalts, and rare earths, the intrinsic demagnetization curve is quite different than the normal curve, so the error would be quite large. This difference between the normal and intrinsic curves is quite evident from examining the curves for sintered Ferrite 5 shown in Figure 11-20. Unfortunately, the intrinsic curves or even the intrinsic coercive force (H_{ci}) are not usually published or available from the mills for most of the highly coercive magnet materials, and the design engineer has no choice except to use the normal curve. To use the normal demagnetization curve, the following "correction" can be made for the highly coercive (or, really, for *any*) material.

$$H_e = H_f \left(1 + \frac{1}{P}\right)$$

where H_e is the *effective* demagnetizing external field in Oe, H_f is the actual demagnetizing external field in Oe, and P is the B/H ratio of the specific magnet affected by the field.

The effective demagnetizing field (H_e) calculated above is then used as it was in Chapter 9 to determine

ENVIRONMENTAL EFFECTS 121

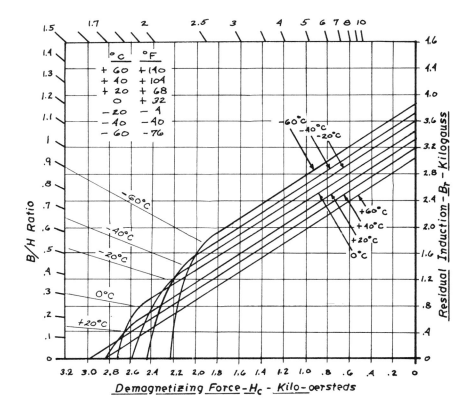

Figure 11-14. Demagnetization curves for grade 6 ferrites at various temperatures.

122 CHAPTER 11

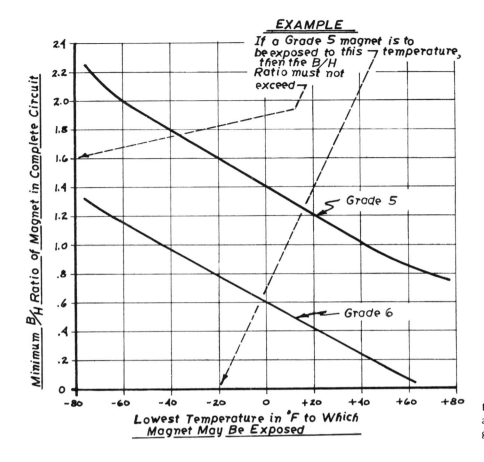

Figure 11-15. Minimum B/H ratio to avoid remagnetizable loss of strength in grade 5 and 6 ferrites.

the true operating condition of the magnet after exposure to the field.

In measuring or calculating the actual field (H_f), be sure to use the *peak* value if AC or pulsating DC currents produce the external field.

Remagnetizable losses occur only during the first cycle or, at most, during the first few cycles of demagnetization exposure. The only method of preventing external, field-produced changes in the magnet is to stabilize it by *deliberate* exposure to an equivalent, controllable AC field.

The effect on the remanence (B_m) *only* of three grades of cast Alnicos *operating at* BH_{max} is shown in Figure 11-21. Note the substantially greater "resistance to demagnetization" of Alnico 6 compared to Alnico 2 or 5. Until the development of Alnico 8 and the ferrites, Alnico 6 was the most widely used material where better stability against external fields was required, but now Alnico 6 is substantially an obsolete material. Unfortunately, an experimentally developed curve for Alnico 8 similar to the curves in Figure 11-21 is not available.

Working Air-Gap (Permeance) Changes: The possible demagnetization effect of increasing the air gap beyond the maximum, normal, working air gap was adequately explained and illustrated in Chapter 10 and needs no further amplification here. Instead the effects established in Chapter 10 can be summarized as follows:

1. As the air gap in a permanent magnet increases from $L_g = 0$, a self-demagnetizing effect will always occur, so that B_m will be less than B_r.
2. If the demagnetization curve of the material is *completely flat* (with no knee), the self-demagnetization change is *reversible* in that reducing or closing the air gap will result in B_m approaching or equaling B_r.
3. If the curve is *not* flat, the self-demagnetization effect will be *remagnetizable*.
4. The higher the curve and/or the more abrupt the knee in the demagnetization curve, the greater the self-demagnetization effect as the working air gap increases.
5. Self-demagnetization changes due to varying air gaps can be minimized by:
   a. Selecting a material with a relatively flat demagnetization curve.

ENVIRONMENTAL EFFECTS 123

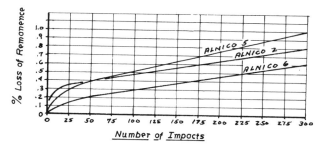

Figure 11-16. Effects of shock or impact on Alnicos based on magnet operating at B/H_{max}.

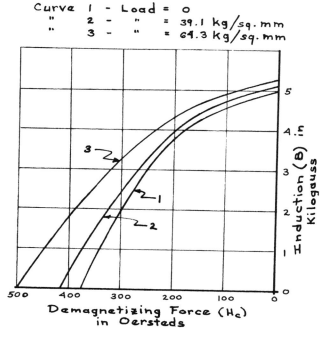

Figure 11-17. Reversible stress change in cunife.

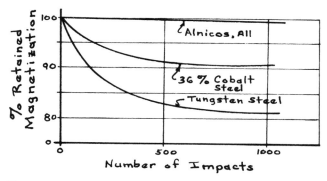

Figure 11-18. Remagnetizable change due to impact for all Alnicos, 36% cobalt steel and tungsten steel.

Table 11-4
Remagnetizable Loss Due to Impact
(percent remanence retained)

Number of Impacts**	Alnico 1	Alnico 2	Alnico 3	Alnico 4	Alnico 5
20	100	100	100	99.7	100*
50	100	100	100	99.7	99.9
100	99.2	100	100	99.7	99.8
200	99.2	100	100	99.7	99.8*
300	99.2	100*	99.8	99.7	99.6
500	99.2*	100	99.6	99.7	99.5
750	99.2	99.2*	99.6	99.7	99.5
1000	99.0*	99.2	99.6	99.7*	99.5*

*Indicates region where bars cracked or broke.
**Data based on bars 10 inches long and one-half inch square in cross section; one impact equivalent to dropping the magnet a distance of one meter onto a wooden platform with bar hitting on end.

Table 11-5
Remagnetizable Loss Due to Vibration
(free motion test*)

Material	Remanence Loss (percent)
Cobalt Steel (36 percent)	3.20
Chrome Steel (3.5 percent)	.00
Alnico 2	.33
Alnico 5	.32

*Data obtained after specimen vibrated with a total displacement of .050 inches at 10-50 cycles per second for 40 minutes.

Table 11-6
Remagnetizable Loss Due to Vibration
(clamped end test*)

Material	Remanence Loss (percent)
Cobalt Steel (36 percent)	.77
Chrome Steel (3.5 percent)	1.48
Alnico 2	.81

*Tested with one end of magnet securely clamped and a displacement of the free end .25mm at frequencies up to 120 cycles per second for 1.5 hours.

Table 11-7
Effects of Radiation on Permanent Magnet and Soft Magnetic Materials

Material	Radiation Only		
	Temperature °C	Threshold Dose/cm²	Effect
Alnico 2, 5, 12 Cunico	60–325	$>10^{20}$ Neutrons (>.5 eV)	Lower Remanence and Induction
Cunife Chrome Steel Cobalt Steel Barium Ferrites Silmanol Platinum-Cobalt Lodex	275	$>10^{19}$ neutrons (>.5 eV)	
Soft Ferrites	100	$\approx 10^{12}$ neutrons (>1 MeV)	Lower remanence and permeability: higher H_c.
Soft Metallic Alloys	<100	$>10^{17}$ neutrons (>.5 eV)	Lower Remanence Lower Permeability Higher H_c
Mumetal 4-79 Moly. Permalloy 5-79 Moly. Permalloy Binary Nickel Iron	<100	$>10^{15}$ neutrons (>.5 eV)	
Pure Iron	100	$\geqslant 10^{17}$ electrons (2 MeV)	
4-79 Moly. Permalloy 5-80 Moly. Permalloy Mumetal	60–80	$\approx 2 \times 10^{16}$ electrons (2 MeV)	
Pure Iron 5-80 Moly. Permalloy	100	$\approx 10^{14}$ protons (1.5–4 MeV)	Lower Remanence and permeability; no change in H_c
65 Nickel Iron	100	$>10^{17}$ electrons (2 MeV)	Lower remanence and permeability: higher H_c.

Material	In Presence of a Magnetic Field – Radiomagnetic Treatment	
	Radiation/cm²	Effect
5-80 Moly. Permalloy	$\approx 8 \times 10^{17}$ electrons (2 MeV)	Higher remanence: improved squareness; higher maximum permeability; no change in H_c.
65 Nickel-Iron	$\approx 7 \times 10^{17}$ electrons (2 MeV)	No change except small increase in H_c.
4-79 Moly. Permalloy Mumetal	3×10^{17} neutrons (>.5 eV)	Small increase in remanence with large increase in H_c.

  b. Operating the magnet as high as possible on the demagnetization curve (with a high B/H ratio).
6. Self-demagnetization changes with varying air gaps can be *eliminated* by:
  a. Using a material with a completely flat demagnetization curve (with no knee).
  b. Using a controlled AC field to deliberately demagnetize (stabilize) the magnet to the greatest level of self-demagnetization that can occur.

The most difficult problem in determining the effects of external fields or in establishing the magnitude of controlled AC needed for stabilization is the *measurement* of the external field. Some of the methods of measurement and testing which we will cover in Chapter 12 are applicable to this problem. However, as a practical matter, in many cases the external field demagnetization effects are often determined by empirical methods in which an "equivalent" DC field is often used where the real external field is produced by AC. DC fields are more easily generated, controlled, and measured by readily available equipment.

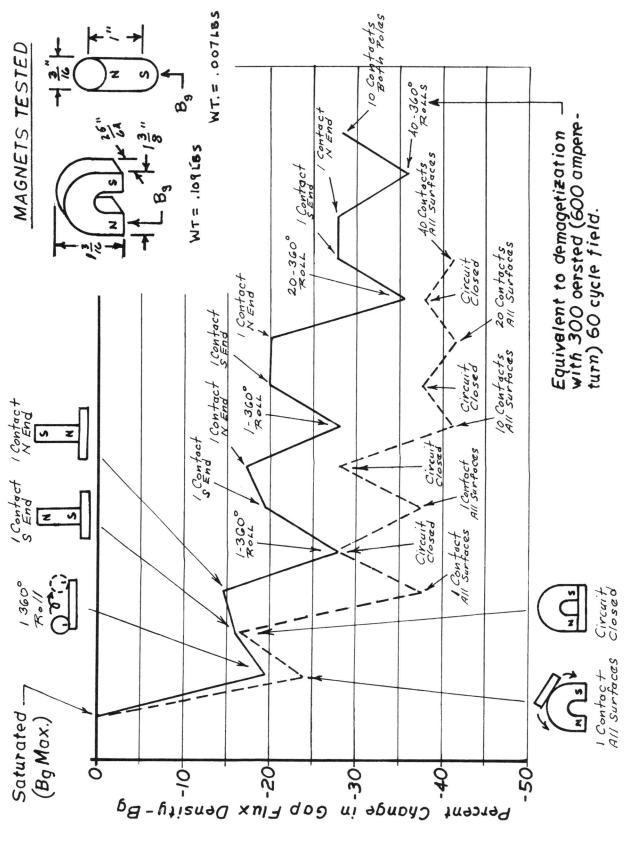

Figure 11-19. Examples of contact demagnetization of Alnico 5 magnets.

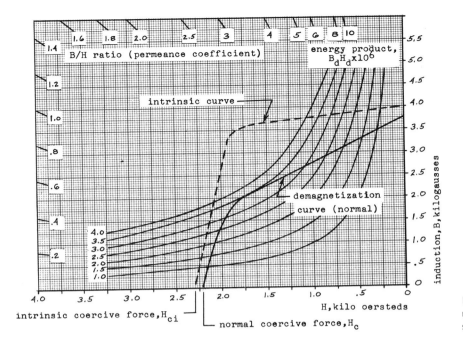

Figure 11-20. Comparison of demagnetization and intrinsic curves for sintered ferrite 5.

*2.02 Oersteds = 1 Ampere-Turn/Inch of magnet length

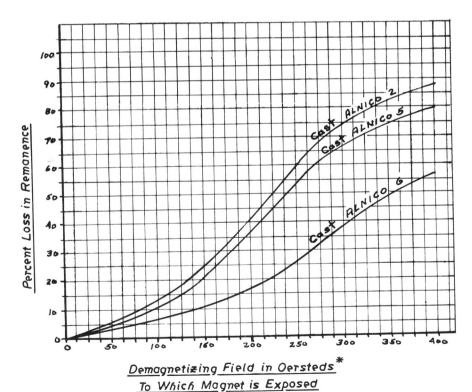

Figure 11-21. Effect of external fields on Alnicos based on magnet operating at B/H_{max}.

ENVIRONMENTAL EFFECTS 127

12

Measurement and Testing

Magnetic measurement and testing is one of the most confused and poorly understood technical areas of modern technology—probably ranking third in this respect to biological and to nuclear testing. Much of this confusion and misunderstanding stems from the inherent difficulty of measuring a phenomenon that cannot be measured directly but only *by its effects*. The confusion and misunderstanding is further compounded by the great differences in techniques and terms used by materials researchers (physicists) and product designers (engineers). Lastly, product designers are not always fully aware of the *objective* of their testing or the meaning of the results they obtain.

Magnetic testing of *any* kind has one of three basic purposes:

1. Materials testing: the measurement of the *inherent capabilities* of the *material*.
2. Magnet-performance testing: the measurement of the *capabilities* of a particular *specific basic magnet*.
3. Circuit-performance testing: the measurement of what a specific magnet made of a particular material is doing *under actual operating conditions*.

Equipment for magnetic testing can also be categorized by whether it is suitable for routine, incoming or in-process quality control testing, product development-design testing, or applied research testing.

Materials Testing

This is the most basic type of magnet testing conducted at the magnet mill or at user levels. The purpose of such testing is to determine the inherent capabilities of the material by measuring the complete hysteresis loop. The basic device used to perform such tests is called a permeameter or hysteresisgraph.

Permeameters are not generally suitable for routine incoming or in-process testing, except on a small, statistical sampling basis. All tests must be performed on precision-ground specimens cut from representative sections of the magnet to be tested.[1] Usually, this type of testing is therefore "destructive" testing. Testing time per specimen will normally range from one minute for permeameters equipped with integrating and recording devices to 30 minutes for the less automatic instruments. Permeameters are somewhat delicate instrument systems that require well controlled ambient conditions and skilled operators. They provide limited (indirect) information about the performance of the magnet as a whole and still less information about how the magnet performs in the actual circuit.

However, permeameters are invaluable laboratory devices that can provide the basic correlative information needed to establish other, more rapid test methods. Permeameters are most commonly used in the material development and quality control laboratories of the mills. Only the larger, more sophisticated magnets users find such facilities economically or technically feasible.

Standard commercial permeameters are available from relatively few sources. Many are custom-designed and custom-made when needed, often by the laboratory itself. Figures 12-1 to 12-12 show the various permeameter designs that have been developed for materials measurement purposes. The most popular types of permeameters are the Ewing Isthmus (Figure 12-1) and the Fahy Super H (Figure 12-2). The various permeameters shown were devised to overcome the successively realized shortcomings of earlier permeameters or in response to the need to measure improved magnet materials. Currently, much developmental work

---

1. If the basic magnet is a simple bar or rod of uniform cross section with precision ground ends (flatness and parallelism), the entire magnet may be suitable for testing.

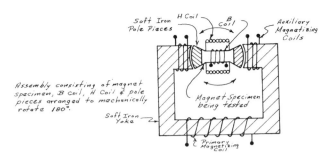

Figure 12-1. Measurement of basic material properties: Ewing Isthmus Permeameter.

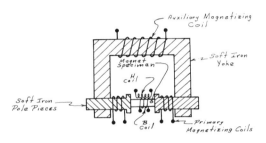

Figure 12-2. Measurement of basic material properties: Fahy Super H Permeameter.

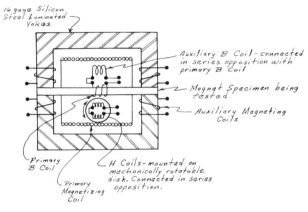

Figure 12-3. Measurement of basic material properties: Sanford-Winter M-H Permeameter.

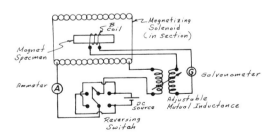

Figure 12-4. Measurement of basic material properties: solenoid-type low-permeability tester.

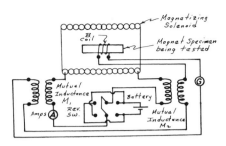

Figure 12-5. Measurement of basic material properties: ASTM null-method (low-permeability) solenoid test.

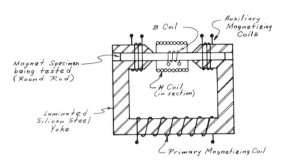

Figure 12-6. Measurement of basic material properties: Modified Ewing Permeameter.

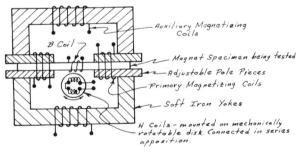

Figure 12-7. Measurement of basic material properties: Sanford-Bennett High H-type Permeameter.

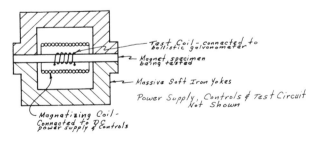

Figure 12-8. Measurement of basic material properties: Hopkinson Permeameter.

MEASUREMENT AND TESTING 129

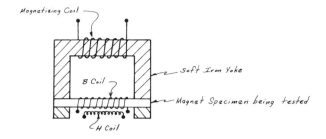

Figure 12-9. Measurement of basic material properties: Fahy Simplex Permeameter.

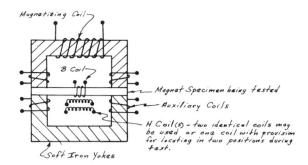

Figure 12-11. Measurement of basic material properties: Sanford-Bennett Permeameter.

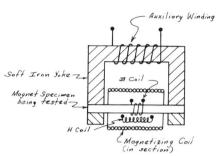

Figure 12-10. Measurement of basic material properties: Babbit Permeameter.

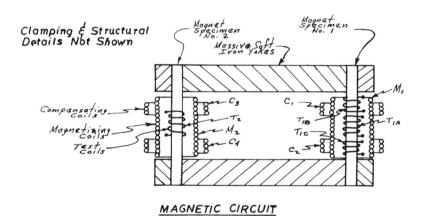

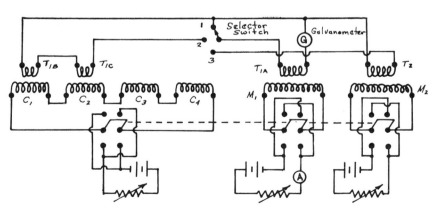

Figure 12-12. Measurement of basic material properties: Burrows Permeameter.

is in progress to devise improved permeameters to test the rare-earth magnet materials. Their very high intrinsic coercive force (H_{ci} = up to 25,000 Oe.) has made testing by presently available equipment unworkable.

Most permeameters have certain basic elements in common:

1. A very low-loss, soft-iron closing circuit to provide the test specimen with as close to a zero air-gap condition as possible. (The only exceptions are the solenoid types shown in Figures 12–4 and 12–5).
2. A primary source of a controllable (and reversible) DC field. This consists of a suitable winding (coil) on the closing circuit and a highly stable, *pure* DC power source.
3. A winding (coil) or Hall-effect element to measure B and H separately *directly* in the specimen being tested.

In addition, most permeameters will have one or more auxiliary magnetizing windings so located and controlled that their fields compensate for the almost inevitable air gap between the specimen and the yoke or for mmf losses in the yoke itself.

Great caution, time, and effort are required to design and build a reasonably accurate permeameter that will provide *absolute* B and H values. Where the permeameter is to be used solely for *comparative* values, the problems of design, construction, and operation are considerably lessened. However, even in comparative testing, a large data base is necessary to obtain valid results. In most instances, especially if absolute data are required, most users find it more economical and reliable to purchase complete permeameters from those manufacturers specializing in this type of equipment.

Figure 12–13 is a photograph of a typical, purchased permeameter designed for small specimens (up to 1 in. in length). In this case, the unit is basically the Burrows type shown in Figure 12–12, with the yoke magnetizing coils on the right side of the work area. The DC controls and automatic demagnetization curve plotter are on the left, and the visual instruments for measurement and control are in the center. The DC power source (not shown) is located in a separate panel rack next to the test station.

Figure 12–14 is a photograph of a custom-designed and custom-constructed permeameter which automatically provides a complete hysteresis loop on large basic magnets. It can be used on a production-line basis employing only semiskilled operators. Figure 12–15 is a photograph of the actual basic magnets that can be tested with this particular permeameter.

Figure 12–16 is a photograph of a complete, typical, small-magnet test facility of the type found at the mill level and sometimes at the magnet user's operational level. It includes a permeameter for material-properties testing, basic magnet performance testing, and circuit testing. An important part of the facility is equipment for magnetizing and controllably demagnetizing the magnets being tested.

Before considering magnet- and circuit-performance testing, it is necessary to consider the basic methods available for *absolute* measurement of the magnitude and direction of magnetic fields. The methods used to

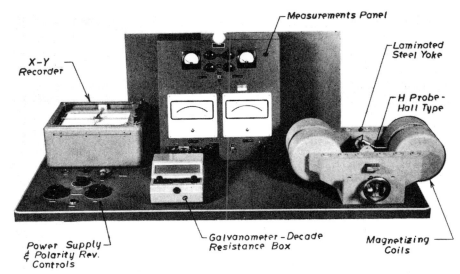

Figure 12–13. Measurement of basic material properties: typical laboratory permeameter for small specimens.

MEASUREMENT AND TESTING 131

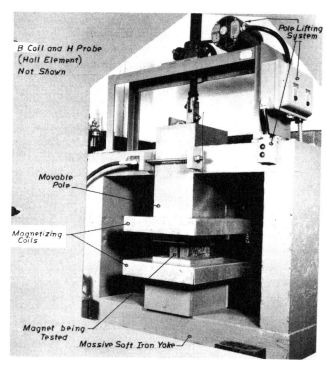

Figure 12-14. Measurement of basic material properties: magnetic circuit of automatic permeameter, which provides complete B/H curves on large magnets.

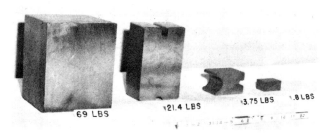

Figure 12-15. Cast Alnico 5 blocks that can be tested on the automatic permeameter shown in Figure 12-14. (Courtesy of Westinghouse Electric Corp.)

measure B and H in a permeameter [2] also form the basis of testing for other purposes.

Three basic types of absolute flux measuring devices are in current use:

> Fluxmeters [3]
> Gaussmeters
> Magnetometers

Fluxmeters work on the principle that voltage will be produced in a coil of wire if there is *relative* motion between the coil and a magnetic field (Lenz' Law). The basic principle and associated circuitry are shown in Figure 12-17.

Commercial fluxmeters are essentially galvanometers of the ballistic-, spotlight- or taut-band type that are designed with high internal damping. These special galvanometers combined with appropriate decade resistance boxes and "search coils" *provided by the user* form the complete measurement system.

Recently, commercial units became available that include electronic integrating circuits that provide a direct scaler or digital reading of *total* flux (or flux density, if the search coil area is programmed into the fluxmeter). In all cases, however, since the fluxmeter may be used on an infinite variety of magnet shapes and size, the construction of an appropriate search coil is the responsibility of the user.

In operation, the appropriately shaped search coil of one or more turns is placed in or around the area in which the flux is to be measured (see Figure 12-17). The search coil is then removed (or, alternately, the magnet can be removed) and the galvanometer deflection is noted. The *total flux* (ϕ) is then obtained from the equation:

$$\phi = \frac{100\, MID_s}{nD_c}$$

where M is the value of the standard mutual inductance in millihenries (mH), I is the value of the calibrating current in milliamperes (mA), D_s is the search-coil deflection, D_c is the calibration deflection ($\approx$ to D_s), and n is the number of turns on the search coil.

The average flux density over the area of the search coil can be determined next by dividing ϕ by the search-coil area (A_c) in square centimeters, or:

$$B_a = \frac{100\, MID_s}{A_c n D_c}$$

It is evident from a study of the principles and equations for a fluxmeter that it measures *total flux* ϕ directly and that when ϕ is reduced to B, the flux density value obtained will be for the *average B over the area of the search coil*. The fact that fluxmeters *do* measure ϕ directly rather than measure B is one of their major advantages. The other methods of basic measurement measure B directly and in a very limited, small area. Localized flux-density concentrations can be very misleading in general, and more so if they are used as the basis for calculating ϕ for any substantial area.

Fluxmeters may be used for all types of testing

---

2. The "B coil" surrounds the test specimen *closely* and measures the flux density *inside* it. The "H coil" is in the air *beside* but very close to the specimen and measures the B in the air (but since $B = H$ in air, an H reading is obtained).

3. The term "fluxmeter" is often applied (incorrectly) very broadly to cover *any* instrument that measures magnetic lines of flux or flux density. The term is often used to describe Hall-effect gaussmeters or even magnetometers.

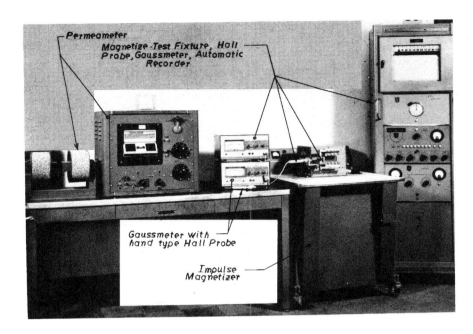

Figure 12-16. Typical small-magnet test facility.

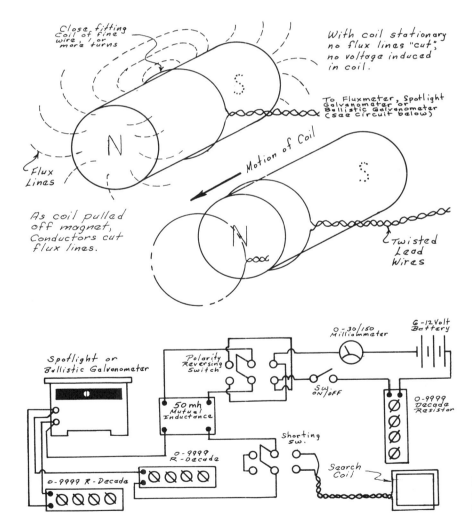

Figure 12-17. Basic measurement-test methods and instruments: search coil and galvanometer (fluxmeter).

MEASUREMENT AND TESTING 133

(material, magnet, or circuit), but they require considerable knowledge in both setting up and evaluating the measurements. Furthermore, some training and skill is required for accuracy and repeatability. Finally, depending on the set-up, the test may be relatively slow and not ideal for production-line testing. Fluxmeter arrangements can provide excellent *comparative* testers for one particular size and shape of magnet if the physical arrangements for the test (magnet position control, coil protection, and so on) are carefully set up and maintained.

Gaussmeters are measuring instruments (systems) based on the change in the electrical properties of certain crystals or metals when they are exposed to a magnetic field. The system currently used in all commercial gaussmeters is based on the Hall-effect phenomenon associated with indium arsenide crystals. Figure 12-18 shows the principles and circuitry for the Hall-effect gaussmeter. Another type of gaussmeter based on the element bismuth is shown in Figure 12-19. This system is not currently used in commercial instruments.

All gaussmeters consist of two basic parts: a flux-sensing element usually mounted in some type of probe and a separate electronic circuit (unit) to measure the change in the electrical properties of the sensing element. Figure 12-20 is a photograph of two models of commercial Hall-effect gaussmeters. Note the sensing probes in the center of the photograph and the tip-protection caps (on table behind probes). The actual Hall element is approximately .010 X 1/8 X 1/4 in. and is located in the extreme tip of each probe assembly. The high sensitivity to mechanical damage of these indium-arsenide wafers and their leads requires that they be handled with considerable care and that they be protected when not in use. Replacement Hall elements (wafers and leads) are normally available, but replacement should be accompanied by recalibration of the probe as a whole. In Figure 12-20, also note the

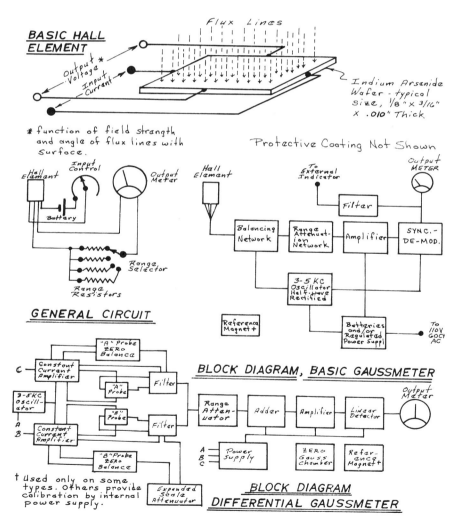

Figure 12-18. Basic measurement-test methods and instruments: Hall-effect gaussmeters.

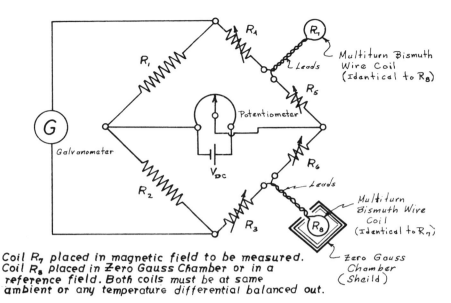

Coil R_7 placed in magnetic field to be measured.
Coil R_8 placed in Zero Gauss Chamber or in a
reference field. Both coils must be at same
ambient or any temperature differential balanced out.

PROPORTIONAL VARIATION OF RESISTANCE OF BISMUTH AT VARIOUS FIELD INTENSITIES & TEMPERATURES

H	-192°C	-135°C	-100°C	-37°C	0°C	+18°C	+60°C	+100°C	+183°C
0	.40	.60	.70	.88	1.00	1.08	1.25	1.42	1.79
2000	1.16	.87	.86	.96	1.08	1.11	1.26	1.43	1.80
4000	2.32	1.35	1.20	1.10	1.18	1.21	1.31	1.46	1.82
6000	4.00	2.06	1.60	1.29	1.30	1.32	1.39	1.51	1.85
8000	5.90	2.88	2.00	1.50	1.43	1.42	1.46	1.57	1.87
10,000	8.60	3.80	2.43	1.72	1.57	1.54	1.54	1.62	1.89
12,000	10.80	4.76	2.93	1.94	1.71	1.67	1.62	1.67	1.92
14,000	12.90	5.82	3.50	2.16	1.87	1.80	1.70	1.73	1.94
16,000	15.20	6.95	4.11	2.38	2.02	1.93	1.79	1.80	1.96
18,000	17.50	8.15	4.76	2.60	2.18	2.06	1.88	1.87	1.99
20,000	19.80	9.50	5.10	2.81	2.33	2.20	1.97	1.95	2.30
25,000	25.50	13.3	7.30	3.50	2.73	2.52	2.22	2.10	2.09
30,000	30.70	18.20	9.80	4.20	3.17	2.86	2.46	2.28	2.17
35,000	35.50	20.35	12.20	4.95	3.62	3.25	2.69	2.45	2.25

Figure 12-19. Basic measurement-test methods and instruments: Bismuth coil-resistance bridge gaussmeter.

Figure 12-20. Two models of Hall-effect gaussmeters. Note probe and reference magnet in center. (Courtesy of Magnetic Equipment Co.)

"standard magnet" (the cylindrical shape near the center of the photograph with a hole for the probe tip) that may be used to calibrate and/or check the proper operation of the Hall instrument as a whole. In most Hall-effect gaussmeters, an internal *electronic* calibration system is provided, but a standard magnet is still advisable as an absolute check on accuracy. Standard magnets available from the manufacturers of Hall-effect gaussmeters consist of a highly stabilized, precision-measured, permanent magnet assembly enclosed in a heavily shielded protective case (providing magnetic and mechanical protection).

Hall-effect gaussmeters are widely used in all types of magnetic testing because they are simple, reliable, moderately priced, and easy to use. These gaussmeters are increasingly used as the *H*-sensing element in permeameters.

Standard commercial Hall-effect gaussmeters are available from a number of reliable sources. They are especially adaptable for use in incoming and in-process quality control testing of both the basic magnet and magnetic circuits. Test fixtures can readily be designed and built in which the inexpensive Hall elements ($15 to $80 each) are permanently secured and connected to the electronics system as needed. Hall-effect gaussmeters can provide simple production-line, go-no go tests as well as a continuous recording of part-by-part performance.

MEASUREMENT AND TESTING 135

A primary limitation of gaussmeters is that they directly measure *flux density* (B) and do so for a very small area (the area of the indium arsenid wafer). Thus, localized flux concentrations can give high readings of B that can be misleading either in relation to the flux density elsewhere or in determining total flux (ϕ) over a larger area.

Another characteristic of Hall-effect gaussmeters is their sensitivity to the *angle* of the flux lines in relation to the sensor surface. The electrical change produced in the sensor is a direct function of the flux angle, with the maximum reading occurring when the flux lines are at right angles to the sensor surface. This angular sensitivity is both a very useful feature and a disadvantage. It is useful when knowing the angle of the flux lines is desired (which is quite often the case, especially when tracing leakage fluxes). On the other hand, it may not always be possible to position the sensor properly in order to obtain the true B reading; too often, the angular sensitivity is simply overlooked.

A *magnetometer* is not one specific type of device but a broad term applied to a wide variety of radically different devices. These may range from simple taut-band, mounted permanent magnets (like the one shown in Figure 12-21A) to elaborate, high-cost, high-precision instruments for measuring planetary magnetic fields. One type of magnetometer widely used by physicists is based on a mechanically vibrated coil, which (by Lenz' Law) results in an output voltage proportional to the field strength. In reality, these magnetometers are fluxmeters, but with a constant vibrating coil area and a fixed rate of flux "cutting."

In general, all instruments properly classified as magnetometers are based on a system where a permanent magnet or a current-carrying coil is attached to a pointer or an indicator which is caused to deflect in proportion to the magnetic field to be measured. Figure 12-21B is another simple form of magnetometer intended for the production testing of basic bar- or rod-shaped magnets.

There are currently no commercially available, "universal" magnetometers. This type of device is normally designed and built for a specific magnet by either the mill or the user.

Magnet-Performance Testing

Magnet-performance testing serves the primary purpose of shipping-receiving or in-process quality control at both the mill and the user levels. Ideally, the type of

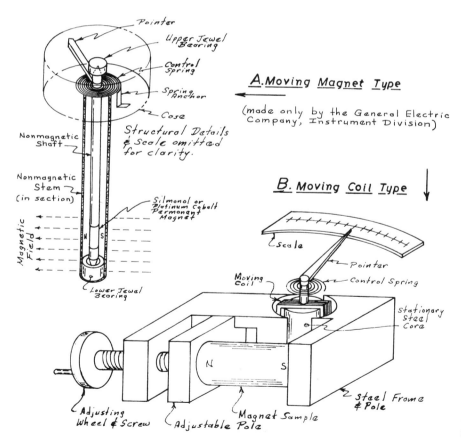

Figure 12-21. Basic measurement-test methods and instruments: moving magnet and moving coil magnetometers.

testing performed should directly reflect how the magnet will operate in the final circuit. Unfortunately, not many test methods can be correlated to final performance. The development of truly representative test methods requires considerable effort.

Magnet performance tests can be categorized as: field-pattern tests, tractive/repulsion force tests, potentiometer tests, total-flux tests, and flux-density tests.

One general consideration for *all* types of magnet-performance testing which should be noted here is that *all such tests must be conducted with fully magnetized magnets.* This can introduce the problem of insuring *proper* magnetization with no possibility of unintentional partial demagnetization (from spurious contacts). In addition, it may be necessary to provide for total demagnetization after testing in order to facilitate further processing or assembly. The equipment for magnetization and demagnetization are normally separate units, but complete "magnetize-test-demagnetize" testers are available from several commercial sources.

Field-pattern tests range from the elementary "iron-filings" and "compass" tests shown in Figure 12-22 to more elaborate (and accurate) methods using a Hall-effect gaussmeter. The simple iron-filings test is limited in utility to tracing leakage-fringing effects on new designs. Obviously, it is in no way relatable to magnet performance in the final product or, for that matter, to comparing the performance of one magnet to another.

Hall-effect gaussmeters are widely used to accurately determine primary and leakage field patterns by measuring the *level* of flux density and flux *angle* at various points surrounding the magnet. In a properly conducted test, a nonmagnetic template is prepared and positioned on the magnet. This template is marked with a large number of test points at which the gaussmeter readings (magnitude *and* angle) are to be taken. Based on these readings, an accurate "flux plot" can be made by connecting points of equal potential. Here again, this type of test is not practical for more than a few magnets and the test results are not directly relatable to final performance.

Tractive-repulsion force tests can be made with an infinite variety of simple devices. These range from the "spring-scale" method (shown in Figure 12-23), which is used in testing separators or separator magnets, to the somewhat more elaborate arrangements shown in Figures 12-24 and 12-25.

Tractive-repulsion force test methods are simple, inexpensive, and reasonably repeatable (if the equipment is properly made). They are extensively used in incoming quality control by users, particularly where large-sample or 100-percent inspection is required. Test equipment and methods are invariably devised by the user. No "standardized" commercial test equipment of this kind is currently available.

The greatest limitation of this type of testing is that it seldom provides any correlation between different magnet sizes and shapes. Also, except in carefully developed cases, the correlation between these tests and the final product performance can be relatively poor.

Potentiometer tests are particular tests occasionally used on simple bar magnets. In these tests, the *sole* objective is to measure the mmf (H_m) produced by a specific size and shape of magnet. This method does not measure B_r, B_m, or BH.

The simplest and most widely used type of magnetic potentiometer is shown in Figure 12-26. Commercial units of this type are not currently available and must be developed and built by the user as needed.

Potentiometer tests are simple, reliable (for what they measure), and repeatable. When properly set up, they are readily usable for incoming quality control and provide a reasonable check on the consistency of H, a key material property. The relationship of this test method to ultimate circuit performance or to other magnets of identical shape but different size can be good if adequate and reliable equipment and data bases have been established.

Total-flux tests can be used for magnet performance

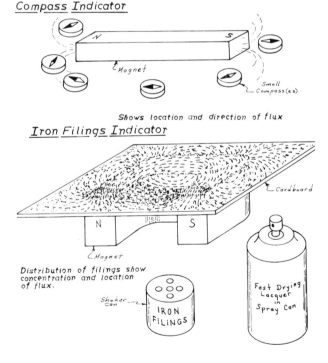

Figure 12-22. Basic measurement-test methods and instruments: simple iron filings and compass tests.

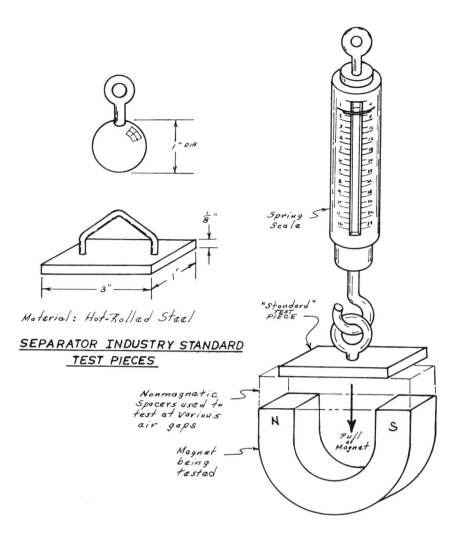

Figure 12-23. Basic measurement-test methods and instruments: mechanical-type tractive force tester.

testing of incoming or in-process parts if a fluxmeter is properly "fixtured" with a stable search-coil assembly. For reliable, repeatable results with minimum time expenditure by semiskilled personnel, the fixture assembly should include an accurate mechanical positioning holder for the magnet and a well protected search coil positively secured in the desired position in relation to the magnet.

One common form of fixture suitable for the production testing of rod-shaped magnets is a "drop-through" arrangement. As shown in Figure 12-27 a vertical, thin-walled, *nonmetallic* tube that is slightly larger on the inside diameter (ID) than the size of the magnets to be tested is wound at a selected spot along its length with a search coil. The search coil is permanently fastened to the outside diameter (OD) of the tube. A slot is cut in the tube so that a retractable mechanical stop will hold the magnet in exactly the right position where the search coil surrounds the midpoint (axially) of the magnet.[4]

4. The "neutral axis" of the magnet, midway between the north and south poles.

The operator drops the magnetized magnet into the top end of the tube, and from there it drops by gravity to a position against the mechanical stop. After the magnet is at rest against the stop, the operator retracts the stop, allowing the magnet to drop down the rest of the tube with the magnet's flux lines *cutting* the search-coil turns. The operator reads the resultant induced voltage on the fluxmeter connected to the search coil. While the fluxmeter is measuring total flux (ϕ) in this arrangement, since the search coil area is fixed, the output reading can be calibrated in terms of flux density in the magnet (B_m). With the "normal" B_m established by engineering from the demagnetization curve and the B/H ratio (or from empirical tests), the quality of individual magnets can be established. Care must be exercised in this type of test arrangement to use a tube with a very thin wall so that the search coil is as close to the perimeter of the magnet as possible. If the fit is not close, the flux measured will not be the approximate ϕ_m (or B_m) but will be a measure of the *external* field of the magnet. Measuring the external field introduces a number of

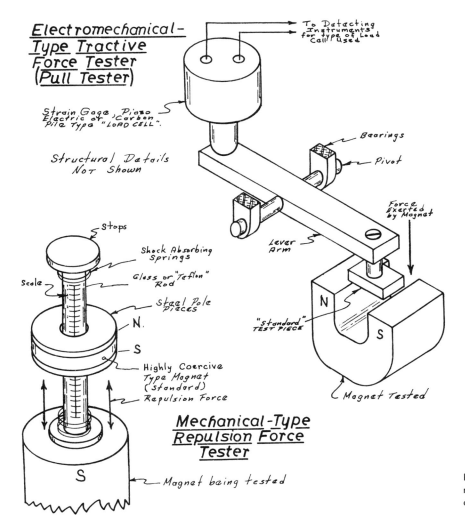

Figure 12-24. Basic measurement-test methods and instruments: other types of force testers.

other possible variables and the correlation of the test results becomes poor. The retractable mechanical stop arrangement is necessary; a simple drop-through system is unworkable because it provides two self-canceling, unreadable, opposing deflections of the fluxmeter—first as the north pole of the magnet passes through the search coil and then as the south pole passes the same point.

Figures 12-28 and 12-29 show two other total-flux type testers that can be used for production testing. With any type of fluxmeter/search-coil arrangement used for production testing, a *digital*-readout type of fluxmeter is highly desirable.

Total-flux testing methods have the advantage of measuring the true *internal total flux* of the magnet rather than the *flux density* at some arbitrary point *outside* the magnet. When properly set up, total-flux tests can give the most accurate, reliable indication of magnet quality in relation to the material demagnetization curve.

The test method described above is a so-called "single-point" test in that it measures the magnet's performance in relation to *one* particular B/H point on the demagnetization curve as determined by the physical parameters of that magnet. To more accurately reflect the possible variations in material properties (demagnetization curve variations), the magnet must be tested with at least *two* different B/H ratios. In total-flux test methods, "multiple-point" tests are possible, but they greatly complicate the design, construction, and reliable operation of the facility. Multiple-point testing is more easily applied to flux-density tests and will therefore be described in the next section. The principles of multiple-point testing can, of course, be applied to total-flux tests when necessary. The major limitation in total-flux tests is the necessity for the user to design and build most of the test system himself. While commercial magnetizers, fluxmeters, and demagnetizers are available, the search-coil/fixture design system integration and test procedure development is not standardized and is

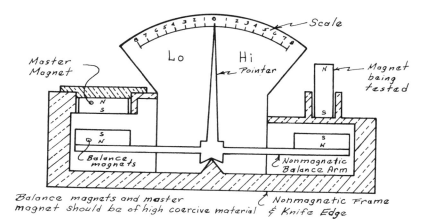

Figure 12-25. Measurement of magnet performance: repulsion balance.

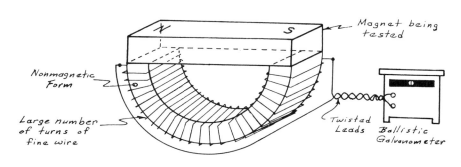

Figure 12-26. Measurement of magnet performance: Chattock potentiometer.

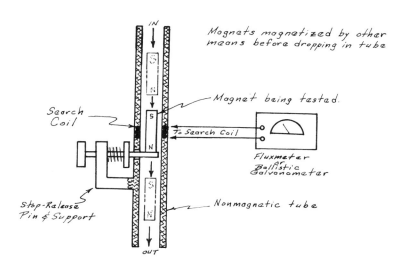

Figure 12-27. Measurement of manget performance: "drop-thru," search coil type total flux tests.

not normally available from commercial sources. Other problems are the wearing (with use) of fixtures, the training of operators, and the essentially "manual"[5] method of operation.

Flux-density tests are basically the measurement by means of a gaussmeter of the external field (in terms of flux density) produced by a magnet at one specifically defined and controlled *physical* location. One of the

---
5. Automated test systems are possible, but they are quite complicated and costly to develop.

---

following points of measurement is appropriate for this type of testing:

1. At the neutral axis of magnet.
2. At the center of the pole face.
3. At the edge of the pole face where B_{max} is expected.
4. At the center of the air gap.

The test location selected depends upon the shape of the basic magnet and upon empirically determined correlativity with final performance in the circuit.

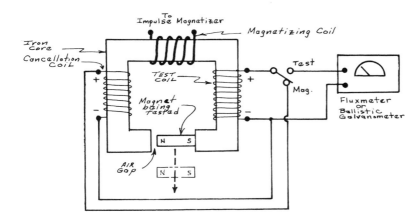

Figure 12-28. Measurements of magnet performance: pull out, search coil type total flux tests.

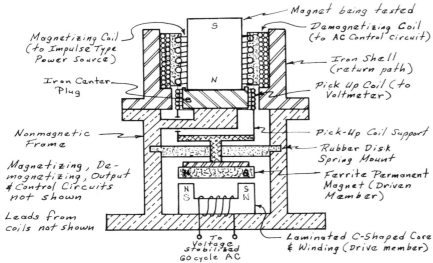

Figure 12-29. Measurement of magnet performance: vibrating-coil type total flux tests.

Figure 12-30 shows the flux-measuring locations and physical-fixturing concepts for a rod- or a bar-shaped magnet. For these magnet shapes, any of the first three test locations above may be used. The center of the air gap is not appropriate because there is no specifically defined working air gap. Figure 12-31 shows the testing arrangements for a basic horseshoe magnet. In this case, only the pole-center location (point 2 above) or the air gap center (point 4) are appropriate. A measurement at the neutral axis, which is at the *back* of the horseshoe, provides poor correlation with output in the working air gap. The edge of the pole face also represents a poor test location, since there is a very substantial difference between the flux density and the flux angles at the inner and outer edges of the pole. If the outer edge is used, the gaussmeter is really measuring an area of *leakage* field. If the inner edge is chosen, the flux is highly concentrated, and since it is at a substantial angle to the flux lines, the flux density cannot be measured with accuracy or consistency (minor variations in magnetization will produce misleadingly large changes in flux readings).

Experience has shown that, in general, where the magnet has a definable working air gap, the most effective measurements are those made at the center of this gap. Where the magnet does not *have a definable air gap, measurement at the neutral axis is generally most effective.* As used here, the term "effective" refers to how accurately the flux-density test reflects the true inherent properties of the magnet material and therefore whether it will perform to expectations in the final circuit.

A single test of the magnet after it has been fully saturated is *not* as effective as testing the magnet under *two or more* different operating conditions. Where a single-point test might indicate that a number of magnets are identical, a multiple-point test could show very substantial material variations. This is the case *regardless* of which test location has been selected.

The greater effectiveness of a multiple-point test can

MEASUREMENT AND TESTING 141

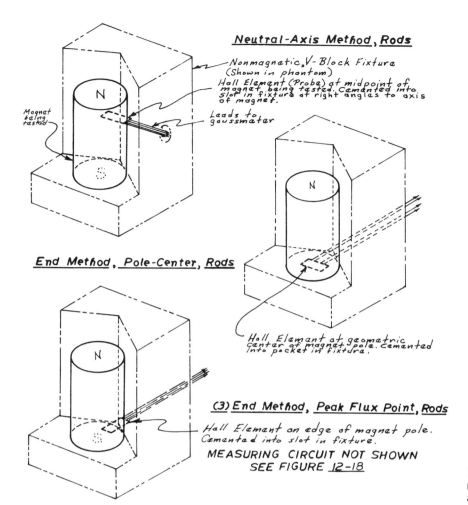

Figure 12-30. Measurement of magnet performance: Hall-effect gaussmeter arrangements for single-point tests.

most readily be understood by referring to the demagnetization curves shown in Figure 12-32. This figure shows the possible curves for two physically identical but magnetically different magnets. If the open-circuit B/H ratio is line 0–A in Figure 12-32 for both magnets,[6] then the values of B_m and H_m would be identical as shown. If a single-point, flux-density test were made under open-circuit conditions, the *inevitable* erroneous conclusion would be that the materials of which they are made are identical. If, on the other hand, the B/H ratios of *both* magnets were changed to line 0–B in the figure, the difference in material properties would clearly show in the flux-density test, since

$$B_{m_{B1}} \neq B_{m_{B2}} \text{ and } H_{m_{B1}} \neq H_{m_{B2}}.$$

In production testing, it is not practical to change the B/H ratio *physically* by cutting each magnet down to a shorter length. The same *effect* however can be

6. Since the B/H ratio is determined by physical parameters and the magnets *are* physically identical, their open-circuit B/H ratios would be the same.

achieved by first testing the whole magnet after saturation *and then applying a specific AC demagnetizing force (field) and repeating the flux-density test*. With this procedure, if two magnets test the same under both conditions, the magnets are, without any doubt, made of identical materials.

Obviously, it is possible to apply a three-, four-, or more step procedure by employing more than one "knockdown" step. In most circumstances, a two-point test (saturation plus one knockdown) is sufficient, particularly if the saturated test is for an operating point at or above the knee of the demagnetization curve and the knockdown applied is sufficient to effectively bring the operating point to well below the knee of the curve.

Figure 12-33 shows how the single-point test arrangements in Figures 12-30 and 12-31 can be modified by adding a demagnetizing coil and a control for multiple-point testing.

In designing the "fixturing" for flux-density type tests, every effort should be made to provide a fixture that closely duplicates the final operating circuit of the

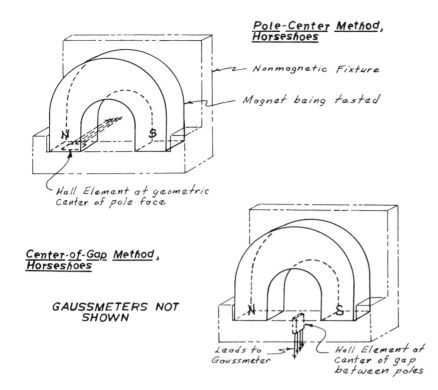

Figure 12-31. Measurement of magnet performance: Hall-effect gaussmeter arrangements for single-point tests (continued).

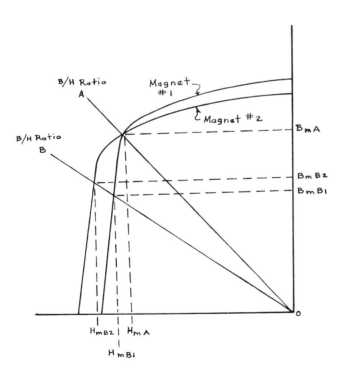

Figure 12-32. Possible demagnetization curves for two physically identical but magnetically similar magnets.

magnet. Ideally, the circuit is identical to the final circuit, and the magnet is saturated, tested, and knocked down (for a two-point test) *in this circuit*. Some compromises in duplicating the final circuit may be necessary, since the fixture must allow for rapid interchangeability of the magnets being tested.

It is also highly desirable that when a new fixture test system is developed, one or more "master" or "reference" magnets be established. In most cases, reference magnets that represent the minimum, maximum, and median values should be selected. These reference magnets and their test data should be kept under the careful control of the engineering or quality control department and treated like any other "standard."

Two flux-density test systems that use specially selected magnets as a reference base for a "comparitor" system are shown in Figures 12-34 and 12-35. The reference magnet may be selected so that it represents the mean value or the minimum acceptable value of the magnets to be tested.

A flux-density test system based on D'Arsonval-type electromechanical movement is shown in Figure 12-36. This illustration also shows another system that is generally similar to a permeameter, except that it measures B in an air gap using a Hall-effect gaussmeter.

MEASUREMENT AND TESTING 143

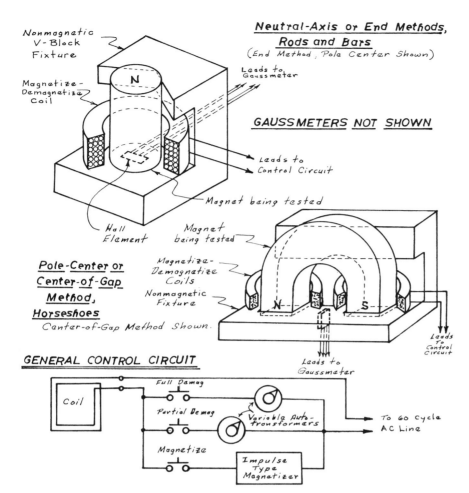

Figure 12-33. Measurement of magnet performance: Hall-effect gaussmeter arrangements for multiple-point tests.

A unique feature of this permeameter-type tester is the fact that the air gap can be changed (by changing the nonmagnetic spacer) to give an "equivalent-circuit" operating point to the magnet (B/H ratio same as in final circuit). At the same time, multiple-point testing is made possible by means of the magnetizing-demagnetizing coil.

Flux-density testing of the type described here is widely used at all levels of magnet manufacture and usage. It is simpler to use in developing fixtures and it is more reliable for production use with unskilled workers. The primary problems that may be encountered are properly locating the Hall-effect element, protecting it from mechanical damage, and correlating the test data with final performance standards.

Circuit-Performance Testing

Magnetic circuit testing always employs a fluxmeter (Figure 12-19) or a gaussmeter (Figure 12-20). When measurements are to be made in the working air gap, either a fluxmeter or gaussmeter may be used. In these cases, the gaussmeter is used more frequently because problems may be encountered in positioning and properly removing the search coil needed with any fluxmeter. Hall-effect gaussmeters have the advantage of being very small, fitting into spaces as small as .010 × 1/8 × 1/8 in.[7] The most serious potential problem encountered with gaussmeters is related to the fact that they measure flux *density*. Since this measurement is made for a small area (the size of the indium arsenide wafer), the readings cannot necessarily be extrapolated for a larger, *total* area. Thus, the small size of the magnetic sensor can be both an advantage and a disadvantage.

Properly used, Hall-effect gaussmeters can be particularly useful in determining the *angle* of flux lines when "tracing" leakage fluxes or in flux mapping. It is difficult if not impossible to use fluxmeters for this purpose.

Fluxmeters are used to measure the total flux over

7. Hall probes are available in a number of sizes, including "miniature" probes of this size.

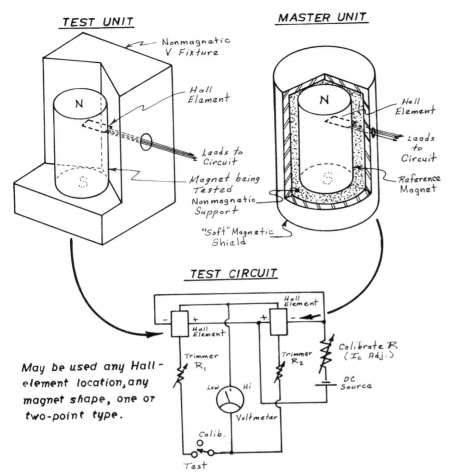

Figure 12-34. Measurement of magnet performance: Hall-effect gaussmeter arrangements for comparative tests.

an accessible area or *inside* a magnet, pole piece, or whatever. Such measurements *do*, however, require that the search coil be slipped *off* the magnet to obtain a reading. In many cases, the measurement can be made in two steps: first, the magnetized magnet with the search coil wrapped around its neutral axis is pulled out of the circuit and the galvanometer deflection is noted; then the search coil is slipped off the magnet and the

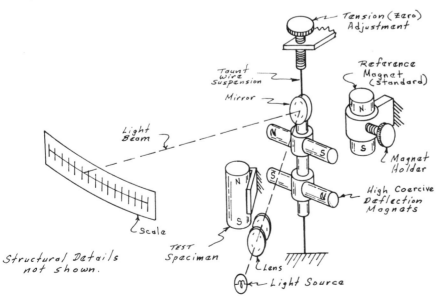

Figure 12-35. Measurement of magnet performance: comparison magnetometer.

MEASUREMENT AND TESTING 145

D'Arsonval Magnetometer

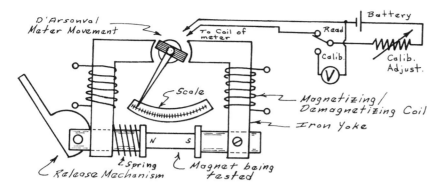

Production Permeameter

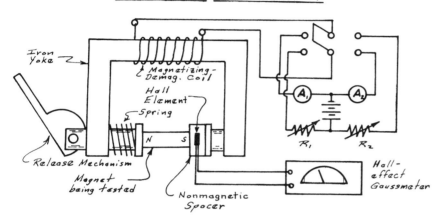

Figure 12-36. Measurement of magnet performance.

second deflection is noted. The sum of the two deflections is used in the flux calculation (see the preceding total-flux method section) and represents the total flux *in the magnet*.

Another case where the fluxmeter is useful is in measuring the *change* in flux level in the magnet (or in another part of the circuit) under variable (dynamic) operating conditions. A search coil connected to a fluxmeter will show a deflection produced for each *change* in flux level (but only *during* the period of change).

Obviously, circuit-performance testing is invariably different for each specific product. Aside from the preceding discussion on the equipment available, no other rules or principles can be developed that apply to all cases.

Methods for Correlating the Tests

Since the purpose of any user-level material or basic-magnet test is ultimately to assure final product performance, the proper correlation with this product performance is critical. The following guidelines will prove useful in establishing this correlation. The best method of assuring accurate correlation is the complete testing and tracing of a controlled lot of magnets. In this method, a group of basic magnets are tested at incoming inspection by a fluxmeter or a gaussmeter. As they are tested, the magnets are *indelibly* identified and the test results are recorded. They are then successively incorporated into a *single unit* [8] of the final product, and this product's performance is evaluated by the appropriate method. The final product's performance is then correlated with the basic-magnet tests and minimum, mean (and, if necessary, maximum) test values are established for the magnet. If the product's construction (standard or with minor modifications) is such that the foregoing steps are physically practical, good correlation can be obtained if as few as *ten* basic magnets are tested. The ten magnets selected should be representative of the full range of magnets that will

---

8. The unit selected for this test should be one that has been established as "representative," with all other possible variables controlled.

normally be supplied by the mill or the distributor-fabricator. The magnets can be selected from a larger, statistically valid lot (if such a lot is available) using either the fluxmeter or the gaussmeter test. If a sufficiently large lot is not available or if there is any doubt as to either the lot's "representativeness" or the limits of variation that *can* be allowed, an "artificial" range of samples (in the lot of ten) can be created by reducing the level of some magnets by deliberate steps of external field (AC) demagnetization.

If the product design and construction is such that it is impractical to interchange magnets in a single unit, then both the magnet lot size and product lot size must be increased considerably. In this second, less reliable and less effective method, the test lot of basic magnets is identified, tested, and then followed through the normal production process to final-product performance testing. The lot size chosen must provide a statistically valid correlation allowing for all the possible variables that may affect the final performance of the product. It has been the author's general experience in using this method that a lot size of less than 100 units is not adequate for good correlation, except on the simplest of products.

A third method of correlation can be used when bars or slabs of material are obtained for in-house fabrication into basic magnets for a variety of products. In this method, the true demagnetization curve is obtained [9] for *each* bar or slab in a representative, small lot. Using the representative sample lot with known properties, a *multiple-point*, fluxmeter- or gaussmeter-type test can be established and test limits can be set to assure that future bars or slabs will fall within the allowable limits of the material properties.

Production Testing for Quality Control

The statistical sampling plan used by reputable mills for final quality control tests is given in Table 12-1. This plan, based on MIL-STD-105C, Normal Level II, is recommended for incoming and in-process inspection by the user of the typical industrial products that incorporate magnets as a component. Where unusually severe levels of quality must be met in the final product, more stringent quality levels can be set (at a higher cost and with fewer available vendors).

The normal tolerances met by reputable mills for the material properties in relation to published (typical)

---

9. The demagnetization curve can be obtained by placing a special request to the supplier (mill or distributor-fabricator), through tests by an independent laboratory, or, if available, from an in-house laboratory.

properties follow. These tolerances are not published by any mill, but are limits that can be met with *normal* processing:

B_r ± 5 percent

H_c ± 5 percent

BH_{max} ± 10 percent

No particular tolerances are followed on the B/H ratio at which BH_{max} will occur.

The Magnetic Materials Producers Association (MMPA) Standard Specifications table No. 0100-72 (see Chapter 15) gives *nominal* values only for B_r, H_c, and BH_{max} that define the various grades of materials. This MMPA standard also gives the specific dimensional tolerances that can be met and the general characteristics for surface conditions, chips, burrs, and so on. No standards or tolerances for mechanical properties (tensile strength or whatever) have ever been established by the MMPA, since such testing is not a normal part of mill-level quality control. Where such factors are critical to the user, individual specifications should be set, using the normal, attainable properties given in the physical properties tables in the Appendix as a guide.

Costs of Commercial Test Equipment

The wide range [10] of specialized equipment available for magnetic testing precludes unit comparisons. However, for the major categories of test equipment, the following general cost guides may be useful:

Permeameters, hysteresisgraphs, loop tracers, B-H tracers: $8000–$18,000 for complete units.

Fluxmeters without search coils or fixturing: $400–$1500, depending on ranges, on whether integrating or nonintegrating, and on whether a scaler or a digital readout is provided.

Hall-effect gaussmeters, complete with one standard probe: $200–$2000, depending on range, portability, sensitivity, calibration provisions, and readout system.

"Pocket" testers: $20–$100, depending on principle of operation.

As a general rule, purchasing the lowest-cost magnetic testers is seldom sound economics. A saving of a few hundred dollars on a test unit can be offset many times by uncertain, unrepeatable, slow, or unreliable test results.

---

10. While a wide range of equipment is available, the number of sources is relatively small.

Table 12-1
Typical Statistical Sampling Plan for Magnet Quality Control. (Taken from MIL-STD-105C, Normal Level II)

AQL .25 percent

Lot Size		50 or less	51 to 800	801 to 3200	3201 to 8000	8001 to 22,000	22,001 to 100,000
Sample		100%	50	150	225	300	450
Accept		—	0	1	2	3	4
Reject		—	1	2	3	4	5

AQL .40 percent

Lot Size	35 or less	36 to 500	501 to 1300	1301 to 3200	3201 to 8000	8001 to 22,000	22,001 to 110,000
Sample	100%	35	110	150	225	300	450
Accept	—	0	1	2	3	4	5
Reject	—	1	2	3	4	5	6

AQL .65 percent

Lot Size	25 or less	26 to 300	301 to 800	801 to 1300	1301 to 3200	3201 to 8000	8001 to 22,000	22,001 to 110,000
Sample	100%	25	75	110	150	225	300	450
Accept	—	0	1	2	3	4	5	6
Reject	—	1	2	3	4	5	6	7

AQL 1.00 percent

Lot Size	15 or less	16 to 180	181 to 500	501 to 800	801 to 1300	1301 to 3200	3201 to 8000	8001 to 22,000	22,001 to 110,000
Sample	100%	15	50	75	110	150	225	300	450
Accept	—	0	1	2	3	4	5	7	10
Reject	—	1	2	3	4	5	6	8	11

AQL 1.50 percent

Lot Size	10 or less	11 to 110	111 to 300	301 to 500	501 to 800	801 to 1300	1301 to 3200	3201 to 8000	8001 to 22,000	22,001 to 110,000
Sample	100%	10	35	50	75	110	150	225	300	450
Accept	—	0	1	2	3	4	5	8	10	14
Reject	—	1	2	3	4	5	6	9	11	15

Table 12-1 (continued)

				AQL 2.50 percent							
Lot Size	7 or less	8 to 65	66 to 180	181 to 300	301 to 500	501 to 800	801 to 1300	1301 to 3200	3201 to 8000	8001 to 22,000	22,001 to 110,000
Sample	100%	7	25	35	50	75	110	150	225	300	450
Accept	—	0	1	2	3	4	6	8	11	14	20
Reject	—	1	2	3	4	5	7	9	12	15	21

				AQL 4.00 percent							
Lot Size	5 or less	6 to 40	41 to 100	111 to 180	181 to 300	301 to 500	501 to 800	801 to 1300	1301 to 3200	3201 to 8000	8001 to 22,000	22,001 to 110,000
Sample	100%	5	15	25	35	50	75	110	150	225	300	450
Accept	—	0	1	2	3	4	6	8	11	17	20	29
Reject	—	1	2	3	4	5	7	9	12	18	21	30

13

Magnetization, Demagnetization, Stabilization and Calibration

A permanent magnet or a circuit containing a magnet is simply an "inert" piece of metal or ceramic until it has been "energized" by an appropriate external field. This magnet may be a less than ideal source of a magnetic field in some applications, unless it has been stabilized and/or calibrated to required levels. Furthermore, it may be difficult to process in manufacturing or an actual safety hazard in some applications, unless it can be substantially "de-energized" or demagnetized. Specific equipment has been developed to magnetize, demagnetize, stabilize, and calibrate all types of permanent magnets. This equipment, whether "homemade" or commercially obtained, must be understood and used properly, since improper use can directly affect the performance of the magnet and the ultimate product of which the magnet is an operational part.

Magnetization

The basic purpose of any magnetizer is to supply a *unidirectional* magnetic field of the proper *magnitude, shape,* and *time duration* to fully saturate the permanent magnet to which it is applied. Magnetizers may be used on individual magnets, on groups of basic magnets, or on complete assemblies containing basic magnets.

All magnetizers fall into one of the following categories: permanent magnet type, impulse or capacitor-discharge types, semicontinuous DC electromagnetic type, half-cycle type, DC-bias/AC-inrush type, and special purpose types.

All of these electrical types contain the following subelements: power supply, control circuits, and pole pieces and/or windings and/or fixturing.

The primary determinant of the *field level* that can be developed by a particular magnetizer is the capability of the power supply. A secondary determinant is the design and construction of the iron circuit (where used), pole pieces, winding, and/or fixturing. The iron-circuit and pole-piece designs establish the internal leakage that will occur in the magnetizer at the expense of the field delivered to the magnet to be energized. The winding and/or fixturing designs establish the number of turns and the electrical resistance that will limit the power supply's current-delivering capacity.

The only determinant of the *shape* of the field produced by a magnetizer is the design of the pole pieces, winding, and/or fixturing.

The *time duration* of the field delivered by the magnetizer is determined by the power supply and by control-circuit characteristics, as well as by the *impedance* of the iron circuit and winding.

In a properly designed and functioning electrically operated magnetizer, a magnetic field equal to the required H_s is produced to saturate that particular material *for that magnet length*. The pole pieces, winding, and/or fixturing are so designed that the field produced has the *same shape* as the magnet to be magnetized. The field is "prolonged" for a sufficient length of time so that any transient electrical "skin effects" do not diminish the true field to which the magnet is exposed. Finally, when the magnetizing field is discontinued, *there are no inductively produced field reversals* (oscillations) that might partially demagnetize the magnet.

Permanent-magnet magnetizers consist only of a large, high-energy-per-unit-volume basic magnet (usually cast Alnico 5), with adjustable mild steel pole pieces so that the air gap can be varied to fit the magnets being energized. This type of magnetizer is only suitable for energizing relatively small basic magnets that have fairly simple shapes and where the manual labor required to remove the energized magnet is not objectionable. Typical commercially available PM (permanent-magnet type) magnetizers are generally suitable for energizing simple bars or slugs of a maximum mean magnetic length as follows:

Alnicos 2 and 5, cast or sintered	1/2 in.
Alnico 8, cast or sintered	3/8 in.
Ferrites, all types and grades	1/8 in.

The maximum length of magnet that can be placed within the pole pieces is generally one inch, and at that gap, a field of about 3000 Oe is obtained. Short-gap adjustments produce correspondingly higher magnetic fields.

Impulse magnetizers are also called "capacitor-discharge magnetizers," since large electrolytic condensers are used to store the energy over a period of several seconds. This energy is then rapidly discharged in several milliseconds (by a control circuit) into a winding or magnetizing fixture and from there into the magnet. The general and schematic circuit for a 100-joule impulse magnetizer is given in Figure 13-1. In this medium-power unit, gas-tube rectifiers provide DC to charge a 1000 mfd capacitor bank. When the capacitors

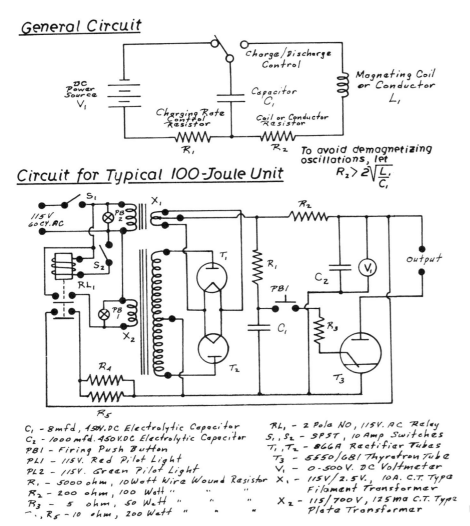

Figure 13-1. Impulse magnetizer circuits.

are charged as indicated by the voltmeter, a Thyratron tube is used to discharge the stored energy into the winding or fixture connected to the output terminals. This circuit is typical of the type used for impulse magnetizers, although the use of solid-state rectifiers and controls are being increased. In large-capacity models, for electrical and personnel safety, the use of an output transformer with a high-voltage primary and a high-current (low-voltage) secondary is common.

Impulse magnetizers are rated by their capacity to deliver energy in joules (watt-seconds). The time over which this energy is delivered is determined by the impedance of the conductor (winding) in relation to the output impedance of the magnetizer itself. Ideally for maximum energy transfer, the magnetizer's output impedance and the input to the conductor, winding, or fixture should match. Since the input impedance is not really controlled (it varies with each specific fixture), achieving impedance matching is seldom practical. For a given time constant, the actual current that will be delivered is determined by the conductor, winding, or fixture *resistance*. Typical commercial units in the 10,000 to 200,000 ampere turn capability are available. Power-line requirements vary, ranging from 1 to 10 amperes (A) at 115-230 volts (V) 60-cycle AC.

In most cases, the capacity of magnetizer required is determined by empirical methods and/or by guidance from the experience of the magnetizer manufacturer.

Proper fixturing is as much a key to the correct application of an impulse magnetizer as the magnetizer is itself. This fixturing *always* consists of a single or multiple conductor (winding) through which the current from the magnetizer will flow and around which the magnetizing field will be produced. In addition, the fixture *may* consist of soft iron members to reduce air gaps and to direct or concentrate the field; nonmagnetic parts to physically locate and hold the magnet in the proper position; and in the case of high-current or high-repetition usage, water cooling coils or air cooling fins.

Figures 13-2, 13-3, and 13-4 illustrate typical *single-*

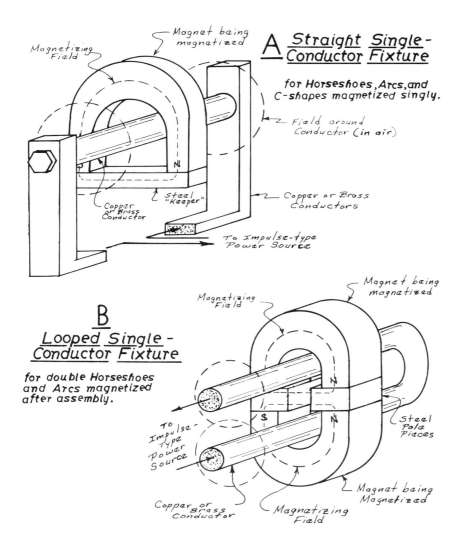

Figure 13-2. Magnetizing methods and fixtures: typical single-conductor magnetizing fixtures.

conductor magnetizing fixtures for common magnet shapes and desired pole arrangements. The following equation [1] can be used as a guide to the size of magnet that can be magnetized on single-conductor fixtures:

$$r = \frac{I}{5H}$$

where r is the radius of the magnet in inches (from the center to the *maximum* OD), I is the current through the conductor (in amperes), and H is the magnetizing force required (in Oe) for the material from which the magnet is made.

Figures 13-5 and 13-6 illustrate typical "wound" fixtures. The arrangements shown also illustrate the versatility and application flexibility of impulse magnetizers. A one-shot or small-run magnetizing fixture can be made in a few minutes by simply winding a few turns of heavy insulated wire on the magnet itself or onto a "shorting bar" of mild steel (see Figure 13-5). The fixtures shown in Figure 13-6 provide a single, "universal" magnetizing arrangement that can be used with a wide range of sizes of a particular basic magnet configuration.

Another outstanding feature of impulse magnetizers is obvious from the many magnet *shapes* being magnetized in Figures 13-2 to 13-6. Impulse magnetizers can be fixtured to energize virtually any shape of magnet: horseshoes, bars, disks, or multipolar slabs or rotors. Other types of magnetizers lack this degree of flexibility, particularly for multipolar magnetization.[2] Multipolar magnets require *simultaneous* magnetization of *all poles*. If the poles are magnetized in pairs (two poles at a

1. This equation does not take into account the volume or weight of magnet being magnetized. Volume and weight *do* affect magnetizing capability, but this effect must be determined empirically or must be based on the magnetizer manufacturer's recommendations.

2. Multipolar magnets are those magnets that have more than *two* poles.

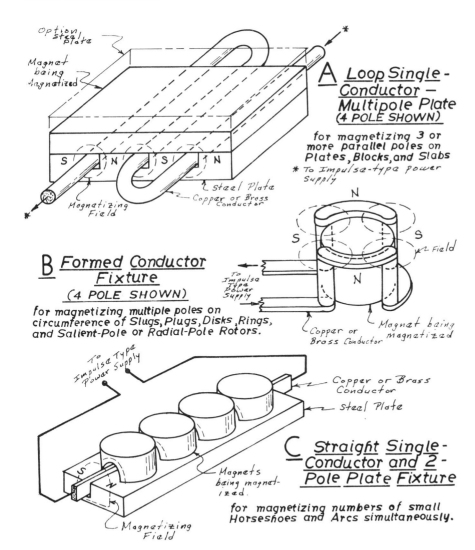

Figure 13-3. Magnetizing methods and fixtures (continued).

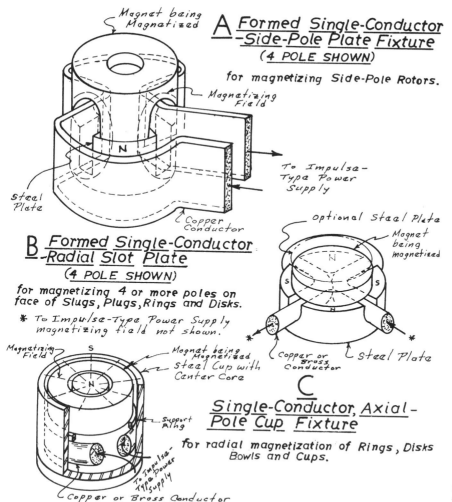

Figure 13-4. Magnetizing methods and fixtures (continued).

time), uneven field distribution and/or partial demagnetization of previously magnetized poles will usually occur. Typical multipolar magnetizing fixtures for impulse magnetizers are shown in Figures 13-3A, 13-3B, 13-4A, 13-4B, and 13-6B.

The most serious limitations of impulse magnetizers are in the selection of the proper capacity magnetizer and in the design of the proper fixture. The manufacturers of commercially available magnetizers can be consulted on proper capacity selection, and most of these manufacturers offer a design, build, and test service for fixtures to fit any basic magnet or complete magnet assembly.

Semicontinuous DC magnetizers are basically iron-core electromagnets energized by well-filtered direct current.[3] Figure 13-7 is a sketch of a typical vertical-coil, C-core design of an electromagnetic magnetizer. Figure 13-8 is a photograph of a commercially available horizontal coil, C-core unit. The power supplies are not shown in either figure.

Figures 13-9 and 13-10 and Tables 13-1 and 13-2 give the construction, circuitry, physical dimensions, winding requirements, DC power specifications, and magnetizing capabilities for three types of medium-power electromagnetic magnetizers. Sufficient information is provided for the construction of any of these units. In general, however, most users will find it more economical to purchase these magnetizers from commercial producers.

DC electromagnetizers are equipped with adjustable and interchangeable pole pieces to increase utility. Units with capacities of 300,000 ampere-turns (NI) with air-gap adjustments of up to 24 in. are commercially available. The magnetizing capacity will, of course, vary with the NI rating and the air-gap opening. Most commercial units can be ordered with a gap opening versus the field-strength curve to serve as a guide [4] to capability.

---

3. The coils and power supplies of this type of magnetizer are designed for intermittent duty (typical duty cycles of 30 to 50 percent). Hence, the designation "semicontinuous."

4. Since this field strength is measured in *air*, and a magnet

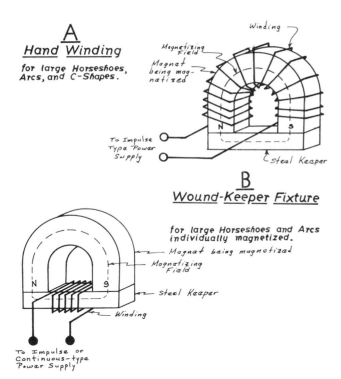

Figure 13-5. Magnetizing methods and fixtures: typical round fixtures.

This type of magnetizer is substantially limited in use to straight bar magnets, slabs, blocks, or moderately deep horseshoes. DC magnetizers are also almost invariably limited to simple bipolar magnetization.

The high inductance of DC electromagnetic magnetizers requires a time interval of 2 to 5 seconds (depending upon magnet size) for the magnetic field to reach full magnitude. Conversely, a substantial time interval is required for the field to collapse. Premature removal of the magnetized magnet is not only physically difficult but may result in partial demagnetization (or distorted initial magnetization). The relatively slow buildup and decay of this type of equipment is both a disadvantage and an advantage. It is a disadvantage, since it reduces the duty cycle (by causing coil heating) and the "production rate" of the equipment. It is a substantial advantage in magnetizing very large magnets or any magnet encased in an electrically conductive sheathing. In these cases, the rapid field buildup and decay of impulse magnetizers may produce eddy currents, which prevent proper flux penetration to the inner areas of the magnet. Also, physically removing larger magnets from the unit after they are magnetized can be a problem. Since the pole pieces and the iron core provide a substantial closed circuit for the magnet, a tractive force

---

to be magnetized closes the air gap and reduces the reluctance, the actual capability may (depending on the magnet material) be higher than is indicated. This will be discussed in greater detail later.

must be overcome to remove it. It is not uncommon to add mechanical or hydraulic break-away mechanisms to ease such physical removal.

Half-cycle magnetizers operate by using a *single* half cycle of an AC power line to provide a high, unidirectional current in a very low resistance conductor (fixture). The field produced around the conductor is, of course, also unidirectional. The half wave is "picked off" the power line, with the amplitude and direction controlled by a Thyratron. A typical half-wave magnetizer circuit is shown in Figure 13-11.

When connected to a 220-600 V 60-hertz (Hz) AC power line, half-cycle magnetizers that draw primary currents of 10,000 A are not uncommon. Since the current is drawn for such a short period of time (1/120th of a second) for each operation of the unit, typical industrial plant systems can be used without disruption. It is essential however, that the half-cycle unit be located close to the plant substation or the main bus to prevent distribution-line reluctance and voltage drops from affecting the magnetizer.

Where currents of 10,000 A through a single conductor (10,000 NI) are not adequate, to avoid exceeding the capacity of existing Thyratrons, a transformer may be used to supply energy to the magnetizing fixture.

Half-cycle magnetizers are used primarily to magnetize very large magnets (up to several hundred pounds of Alnico 5 *at once*). These units have the inherent advantages of impulse magnetizers in low-conductor (fixture) heating and a low, steady-state, power-line demand. Half-cycle magnetizers usually have a higher possible production rate (if water cooling is used for the ignitrons) than impulse units, since no time is required to "charge" capacitors.

Generally, the pulse duration of half-cycle equipment is sufficiently long to minimize the effects of eddy currents on flux penetration.

The need for critical location in the plant, the instantaneous power-system demand, the potential problems of power-line regulation, the reduced commercial availability, and the disproportionately high cost of small- and medium-power units have made half-cycle magnetizers less widely used than impulse or electromagnetic types.

DC-bias/AC-inrush magnetizers are very special magnetizers that have largely been ignored by commercial equipment producers and have therefore been unknown to magnet users. The construction and circuitry for a small magnetizer based on this principle are sketched in Figure 13-12. This particular unit will magnetize cast Alnico 5 magnets up to 2 in. in length and any ferrite magnet up to ½ in. in length. The cross-sectional area capability of this unit is approximately 1¼ square in.

In this type of magnetizer, a small DC-produced

MAGNETIZATION, DEMAGNETIZATION, STABILIZATION AND CALIBRATION 155

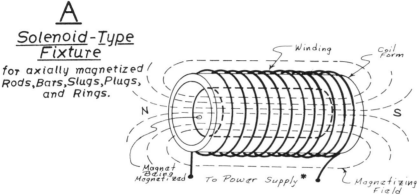

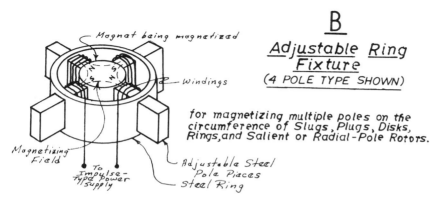

Figure 13-6. Magnetizing methods and fixtures (continued).

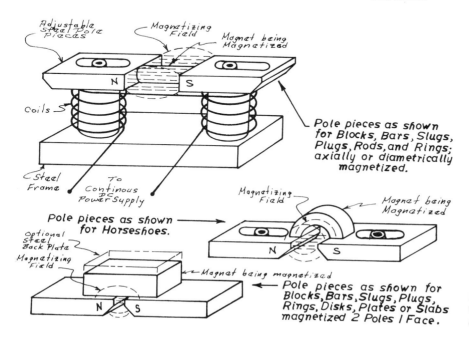

Figure 13-7. Magnetizing methods and fixtures: typical vertical-coil, C-core design electromagnetic magnetizers.

Figure 13-8. A commercially available, horizontal-coil, C-core electromagnetic magnetizer. (Courtesy of Magnetic Equipment Co.)

field is first applied to the coil. Then AC directly from the power line produces an alternating field which is superimposed on the steady-state, DC field. The initial AC current is quite high and is limited only by the internal ohmic resistance of the AC coil. After the first surge, the AC is reduced by the inductive reactance of the AC coil and the increased internal resistance of the four 300-watt incandescent lamps in series with the 115 V 60-Hz power source. While the AC is still applied, the AC coil is moved over the length of the magnet to produce full saturation.

An interesting feature of this magnetizer is that it also serves as a *demagnetizer* (but *only* for Alnico materials). If the "biasing" DC winding is not energized but the AC coil *is* (and is pulled over the magnet in the gap), full demagnetization will be affected.

The most serious *technical* limitation of the DC-bias/AC-inrush magnetizer is its suitability for only the small-length, small-area, bipolar magnetization of relatively small, cross-sectional area bars or rods. Further, the AC coil must surround the magnet, thus limiting the versatility of the unit.

The most serious *general* limitation is the unavail-

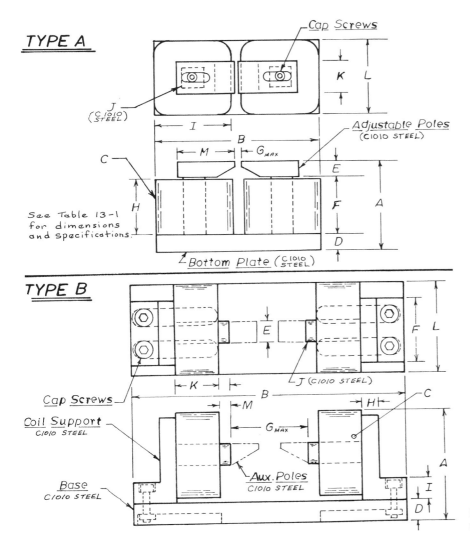

Figure 13-9. Electromagnetic magnetizer construction.

MAGNETIZATION, DEMAGNETIZATION, STABILIZATION AND CALIBRATION 157

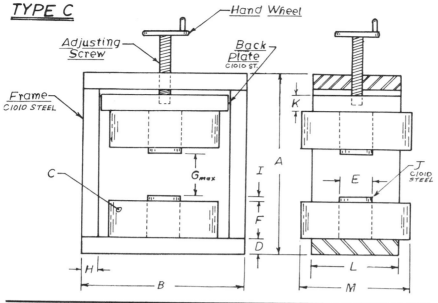

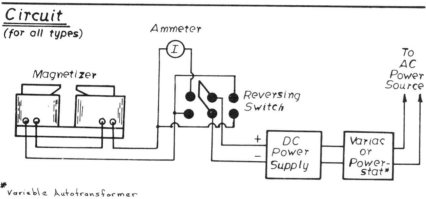

Figure 13-10. Electromagnetic magnetizer construction (continued).

ability of the unit from normal commercial equipment sources. However, units of this kind are relatively easily made by any magnet user who has metal-machining capabilities.

Special purpose magnetizers are custom-designed, custom-made magnetizers that are sometimes necessary for:

1. Assemblies that cannot be magnetized by basic "fixturing."
2. Volume-production magnetization of basic magnets and/or simple assemblies.

In Chapter 10, the importance of magnetizing a basic magnet in its final circuit was clearly established. For most products, this can be accomplished by designing suitable pole pieces for use with electromagnetic magnetizers or suitable fixtures for use with impulse magnetizers. Figure 13-13 shows a "special" magnetizer designed for one of these cases—the axial-type, tube-magnet assembly described in Chapter 9. These assemblies are made in lengths of 4" to 60", and for ease of assembly as well as magnetic reasons, they must be magnetized *after* assembly. The "pass-through" magnetizer in Figure 13-13 successively magnetizes each basic magnet *inside* the stainless steel or aluminum tube. Demagnetization of the preceding basic magnet is prevented by one "bucking coil," and partial, improper magnetization (wrong polarity) of the succeeding basic magnet is prevented by another bucking coil. The main (magnetizing) coil is energized by either an impulse magnetizer or a continuous, full-wave, DC power supply.

Figures 13-14 and 13-15 show two highly automatic systems that magnetize, test, and calibrate basic magnets in a volume-production operation. The system in Figure 13-14 also sorts the magnets into three lots. The system in Figure 13-15 delivers consistent magnets directly to the final shipping package.

Time Requirements: Internal to the material, reorientation of domains occurs instantaneously as soon as the magnetizing field has reached the required level. In that

Table 13-1
Dimensions and Specifications for Three Types of Medium-Power Electromagnetic Magnetizers (all linear dimensions in inches)

Type A—Mechanical

Size (NI = ampere turns)	A	B (min)	C (per coil)	D	E	F	G_{max}	H	I	J	K	L	M
10,000 NI	6 1/4	9	6000 turns #24DG wire	1	1	4 1/4	2 1/2	4	4	1 5/8 (sq.)	2 1/2	4	3 1/2
50,000 NI	10 1/2	15	3750 turns #15DG wire	2	2	6 1/2	4 1/2	6 1/4	7	2 5/8 (sq.)	3 1/2	7	5 1/2
125,000 NI	14 1/2	22	2650 turns #9DG wire	2 1/2	2 1/2	9 1/2	7 1/2	9 1/4	10 1/2	3 5/8 (sq.)	4 1/2	10 1/2	8

Type B—Mechanical

Size	A	B (min)	C (per coil)	D	E	F	G_{max}	H	I	J	K	L	M
10,000 NI	6	15	6000 turns #24DG wire	1	1 7/8	2 1/2	1 1/2	1	1	1 7/8 (dia.)	4	4 1/2	1/2
50,000 NI	10 1/2	28	3750 turns #15DG wire	2	2 7/8	5	4	2	2	2 7/8 (dia.)	6 1/4	7 1/2	3/4
125,000 NI	14 1/2	41	2650 turns #9DG wire	2 1/2	3 7/8	7	8	2 1/2	2 1/2	3 7/8 (dia.)	9 1/4	11	1

Type C—Mechanical

Size	A	B (min)	C (per coil)	D	E	F	G_{max}	H	I	J	K	L	M
10,000 NI	14 1/2	7 1/2	6000 turns #24DG wire	1	1 7/8	4	2 1/2	1	1/2	1 7/8 (dia.)	1	2 1/2	4 1/2
50,000 NI	24 1/2	13 1/2	3750 turns #15DG wire	2	2 7/8	6 1/4	4 1/2	2	3/4	2 7/8 (dia.)	2	5	7 1/2
125,000 NI	37	18	2650 turns #9DG wire	2 1/2	3 7/8	9 1/4	9	2 1/2	1	3 7/8 (dia.)	2 1/2	7	11

Electrical*

Size	Amps @ 115 Volts 60 Cycle (coils in parallel)	Amps @ 230 Volts 60 Cycle (coils in series)
10,000 NI	2.0	1.0
50,000 NI	14.0	7.0
125,000 NI	48.0	24.0

*AC current to rectifier; current input to coil is approximately 15 percent lower.

Table 13-2
Magnetizing Capacities for Three Types of Medium-Power Electromagnetic Magnetizers

Magnet	10,000 NI	50,000 NI	125,000 NI
Type A			
Alnico 5 Bars	1 1/2 in. long	4 in. long	8 in. long
Alnico 5 Horseshoes	1 1/4 in. mean length	3 1/2 in. mean length	7 in. mean length
Ferrite Blocks	3/16 in. thick	3/4 in. thick	1 1/2 in. thick
Type B			
Alnico 5 Bars	1 5/8 in. long	5 in. long	10 in. long
Alnico 5 Horseshoes	1 in. mean length	3 in. mean length	9 in. mean length
Ferrite Blocks	1/4 in. thick	1 in. thick	2 in. thick
Type C			
Alnico 5 Bars	1 5/8 in. long	5 in. long	10 in. long
Alnico 5 Horseshoes	⟶	not recommended	⟵
Ferrite Blocks	1/4 in. thick	1 in. thick	2 in. thick

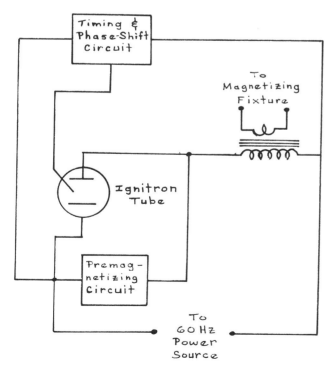

Figure 13-11. Half-cycle magnetizer circuit.

sense, there is no requirement for the magnetizing field to persist for any particular length of time. There is, however, a time interval required for:

1. The magnetizing field to reach full magnitude as determined by the time constant (reactance) of the magnetizer/winding circuit.
2. The transient effects of eddy currents produced by the field buildup (changing). These eddy currents oppose and/or distort the field to which the magnet is being exposed, temporarily [5] preventing full flux penetration.

The time constant of a magnetizer system is determined by the *type* and the *size* of magnetizer, and the *load connected to it* (conductor, winding, or fixture impedance). This time constant is significant because it:

1. Directly affects the production rate of the magnetizer.
2. Determines whether eddy currents will be produced.
3. *Presents possible quality variations when the magnetizer operator may not allow enough "on" time or "charge" time for the field to reach its peak.*

In general, size for size, half-cycle magnetizers have the shortest time constant, impulse units have the next shortest, and DC electromagnets have the appreciably longest buildup time. As a result, the half-cycle magnetizer has the highest production rate, the impulse unit has the next highest, and the DC electromagnet is the slowest. Conversely, impulse magnetizers potentially present the greatest eddy-current problems, while such problems are seldom encountered with DC electromagnets. Half-cycle types seldom present eddy-current problems because of the relatively long time interval (1/120th of a second) and the slow "rate of change" (*sinusoidal*) of the 60-Hz AC industrial power systems.

Quality variations from "premature" shutoff are most often encountered with DC electromagnetic types, especially in very large capacity units. Automatic interval timers (and visual signals) are highly advisable in these units to prevent this problem. Premature "firing" of an impulse magnetizer can occur if it is not equipped with at least a voltmeter that indicates charge level, or better still, a minimum-voltage, firing-control, lockout relay and visual indicator.

Improving and Determining Magnetizer Capacity Requirements: As was implied several times in preceding sections, determining the size of unit needed for a particular basic magnet or magnet assembly is partially technology, partially "art," and heavily *experience*. While commercial magnetizer manufacturers are the best source of assistance, the user may of necessity have to solve his own problems (particularly when they

5. Eddy currents disappear once the magnetizing field has reached a steady-state condition (once it has "peaked").

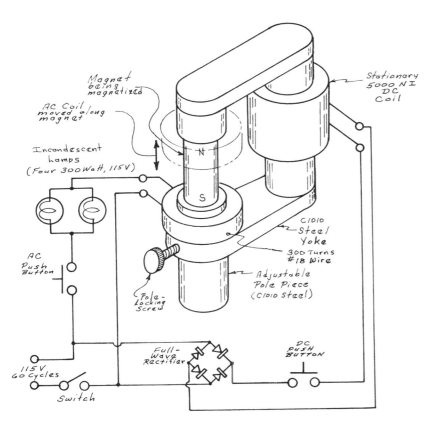

Figure 13-12. Becker DC bias/AC inrush magnetizer-demagnetizer.

relate to fixturing or to new uses of a magnetizer already on hand). The following guidelines and rules may be of assistance, both in this regard and in working with commercial sources of units:

1. Wherever possible, *soft iron* (mild steel) *should be added to the circuit* of impulse or half-cycle magnetizers, and if possible, this iron should be arranged so that it produces a completely closed circuit. The closer the circuit is to being closed, the greater the efficiency of the magnetizer will be.[6] The solenoid arrangement in Figure 13-6A is an example of an inefficient but highly flexible system. Figure 13-2A is an example of iron used to close-circuit a magnet, and Figure 13-2B is an example of two magnets magnetized simultaneously and acting as closed circuits for each other at the same time.

There are two *disadvantages* to adding iron to a circuit: one is the increase in impedance (and the time constant); the other is the physical problem of removing the magnetized magnet from the iron, to which it is attracted.

2. Wherever possible, *the basic magnet should be magnetized in its final circuit*, or better yet, as far along in final product assembly as possible. In addition to

6. The highest magnetizing field is produced with the minimum power (current) input.

improving the efficiency of the magnet (as we have previously discussed), this will usually improve the efficiency of the magnetizer (since most complete magnetic circuits contain soft magnetic pole pieces that partially close the basic magnet's gap). Other advantages to "assembled-circuit" magnetization are that the magnet is less likely to have spurious contacts that can result in contact demagnetization, less likely to pick up tramp iron (filings, chips, and so on), easier to modify in processing, and easier to handle and assemble.

3. Spurious electrical oscillations may occur in any type of magnetizer system, depending on the relationship of magnetizer impedance to "load" impedance. These oscillations produce field-direction reversals in the magnetizing conductor or winding that can partially demagnetize the magnet. This can be avoided by designing the fixture so that

$$R > 2\sqrt{\frac{L}{C}}$$

where R, L, and C are the ohmic resistance, inductance, and capacitance, respectively, of the fixture.[7]

7. Except in simple cases, L and C are difficult to calculate and are most easily determined by impedance bridge measurements of the actual fixture. Resistance can be added in series where necessary to meet the conditions of this equation.

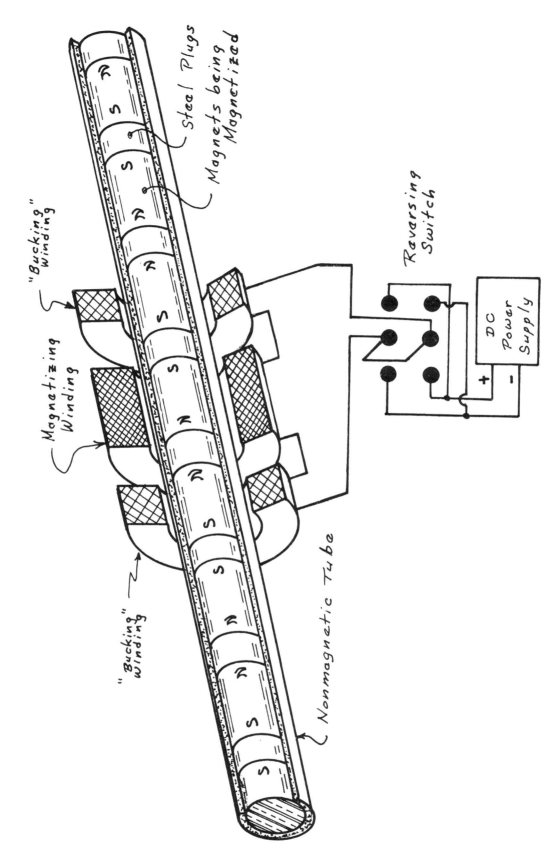

Figure 13-13. Special magnetizer for axial-type, tube magnet assemblies.

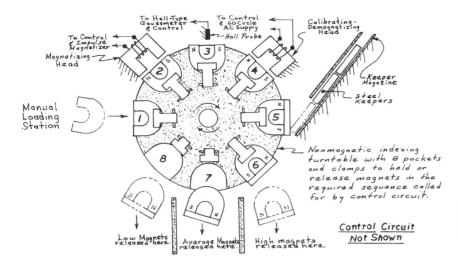

Figure 13-14. Custom semi-automatic magnetizer-calibrator-sorter for horseshoe magnets. (Courtesy of Magnetics Engineers, Philadelphia, Pa.)

4. Most engineers erroneously assume that if "enough" field to saturate the magnet is good, then "more is better." As a *generalization*, this is correct, but in some cases increasing the field strength beyond the necessary level (H_s plus some "safety factor") produces an unexpected problem. The magnetic field becomes "distorted" and it no longer "fits" the shape of the magnet being magnetized. This problem is most evident when a *shallow* horseshoe magnet is energized on an electromagnetic unit. Increasing the electromagnetic field beyond the level necessary to saturate this shallow horseshoe can result in a field of such depth (and strength) that the magnetic field does not fit the curved section of the horseshoe properly.

5. Occasionally, a situation arises where both an impulse magnetizer and an electromagnetic magnetizer are available in the same operation (plant), but neither alone has sufficient capacity to fully saturate a particularly large magnet. In such cases, the two types of units can be combined to magnetize a single magnet (see Figure 13-16). Preliminary tests must be made to establish the polarity of the field of each magnetizer, so that they may be operated "additively."

6. A common problem that occurs in the use of all types of magnetizers is determining whether the magnet is *really* fully saturated. This can be done by first calculating the B_m (or B_d) to be expected for the particular B/H ratio and material after exposure to full saturation. Several magnets can then be energized in the magnetizing system and tested by the fluxmeter method to determine if the proper level of B_m has been obtained. This method is slow, laborious, requires a good fluxmeter, and may still be in error, depending on the accuracy of calculation and testing.

A "quick" and effective method that is generally reliable uses the "polarity-reversal" method, in which one or a few magnets are first energized (presumably to saturation), on the magnetizing system being used. The energized magnets are then tested by *any* method available and a record is made of the test results. After testing, each magnet is re-energized (*without* any deliberate demagnetization) *with the polarity reversed*. Each magnet is then *retested* by the same method with its reversed polarity, and *the test results are compared. If the test results are substantially* [8] *identical, then it can be assumed that the magnetizer is fully saturating the magnet. Tests by other methods show that if the magnetizer is not saturating the magnet, it will be evident from the polarity reversal method by a variation of* over 10 percent *in the comparative data.*

7. The obvious effect of not fully saturating a magnet is that its performance is reduced. This inevitably leads to the temptation of using partial magnetization to deliberately reduce the output of a magnet, either for stabilization or calibration purposes. *This method of stabilization or calibration is not recommended.* As a method of calibration, partial magnetization is difficult to control with consistency. As a method of stabilization, partial magnetization tends to decrease rather than to increase stability.

8. The allowable variation in the two sets of data will depend upon the test method used. For the fluxmeter or gaussmeter methods, 2 percent is ideal but 5 percent is "acceptable."

MAGNETIZATION, DEMAGNETIZATION, STABILIZATION AND CALIBRATION 163

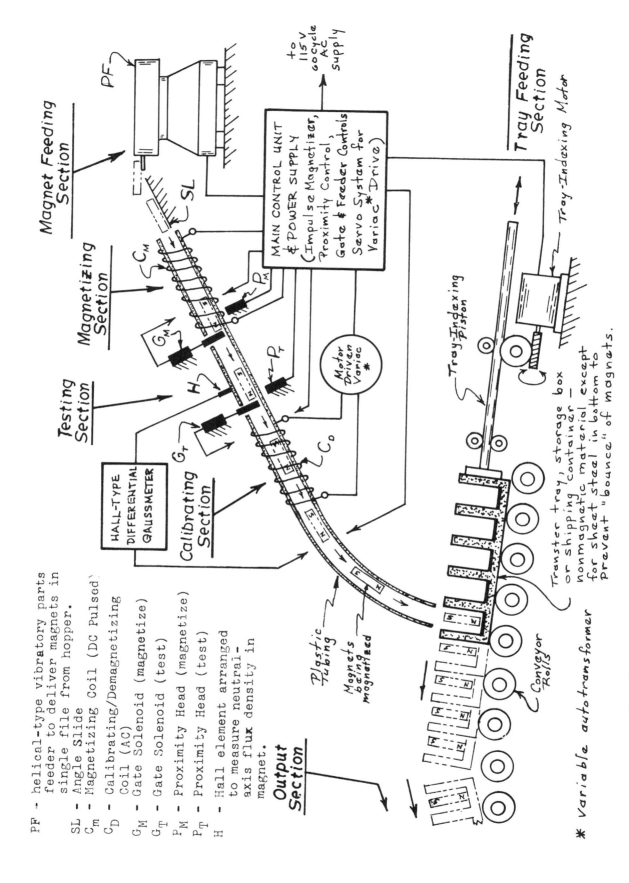

Figure 13-15. Custom automatic magnetizing-calibrating system. (Courtesy of Magnetics Engineers, Philadelphia, Pa.)

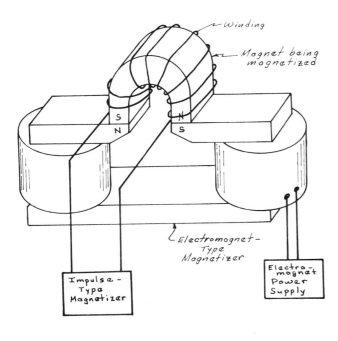

IMPULSE MAGNETIZER POLARITY AND WINDING DIRECTION ARRANGED TO PRODUCE A FIELD IN THE SAME DIRECTION (POLARITY) AS THE ELECTROMAGNET UNIT.

Figure 13-16. Method of magnetizing when mean length of magnet exceeds capability of a single magnetizer.

The only effective method of stabilization or calibration is to *fully* saturate the magnet *first* and then to use an appropriate means to reduce the output to the desired level.

8. Particular caution should be exercised with anisotropic (oriented) materials to be sure that the magnetizing field is really *fully parallel* to the inherent orientation of the material. If the field does not match the orientation direction, the performance of the material will be adversely affected. Problems are seldom encountered when the magnet shape clearly establishes the direction of magnetization (as in a horseshoe) for the oriented material. However, in the case of a bar or slab, particularly after it has been cut into smaller pieces of substantially the same dimensions in all planes (that is into *cubes*), identification of the "preferred direction" of magnetization (orientation) is easily lost. Magnetization of one of these cubes in the *wrong* direction can produce very misleading results. One result can be that the cube will perform as though the material is *isotropic* (unoriented).

The correct magnetization of a perfect, cube-shaped, oriented magnet can be identified by magnetizing it successively in each of the three possible directions and then measuring the magnetic output with a fluxmeter or a gaussmeter. Obviously, when magnetized in the preferred direction, the output will be substantially greater. If, however, the magnet is a general rectangular solid with substantially different dimensions in each plane, "successive-plane" magnetization can fail completely on the basis of output comparisons. For example, if an unknown magnet is ½ × 2 × 3 in. and its orientation direction is across the ½-inch dimension, magnetization across the 2- or 3-inch dimension might still produce a higher output because of the greater magnetic *length* (in spite of the lower internal capability). In such cases, either calculating what the output *should* be or cutting the magnet into a perfect cube and testing it are the only ways to identify orientation direction.

9. It is common practice to determine the capability of an electromagnetic magnetizer by measuring the field strength (with a gaussmeter or a fluxmeter) in the required air gap without the magnet actually being in the air gap. For example, a B measurement of 2000 G [9] would normally be considered inadequate to saturate a cast Alnico 5 magnet, since such materials require 3000 Oe. In actual fact, this may not be correct, since it may not have taken into account the reduction in reluctance of the *magnetizer* circuit when a *ferrous* material instead of air is in the gap. It could well be that with the magnetizer gap substantially "filled" with the magnets to be magnetized, the reluctance would be reduced enough for the magnetizer to deliver the needed 3000 Oe. Of course, there are only two ways to be sure: one is to *try* it and then to check for saturation by the polarity-reversal method (see preceding section); the other is to "play it safe" and increase the gap flux *in air* to 3000 Oe.

One important limitation: *the reluctance change effect only applies to iron-based magnet materials* (Alnico, Cunife, Lodex, and so on). It does *not* apply to the ferrites. The permeability (μ) of the ferrites is so close to 1 that an unmagnetized ferrite magnet in the gap does *not* substantially affect the reluctance.

10. The magnetizing force required to saturate (H_s) a permanent magnet is normally a function of the characteristics of the material. The magnetic properties tables in the Appendix give the H_s required for a wide variety of materials. For the three most common materials:

H_s for Alnico 5 = 3000 Oe

H_s for Alnico 8 = 6500 Oe

9. *Reminder:* In air, $B = H$, so that the magnetizing force measured is "also" 2000 Oe.

MAGNETIZATION, DEMAGNETIZATION, STABILIZATION AND CALIBRATION 165

H_s for all ferrites = 10,000 Oe

In English units, the relationship between Oe and NI is:

1 Oe = 2.02 NI/inch

so that

H_s for Alnico 5 ≈ 6000 NI/inch of magnet length

H_s for Alnico 8 ≈ 13,000 NI/inch of magnet length

H_s for all ferrites ≈ 20,000 NI/inch of magnet length

An example of how magnetizing requirements are used in determining magnetizer capacity follows:

A cast Alnico 5 magnet with a mean magnetic length of 4 inches is to be saturated with an available electromagnetic magnetizer. The magnetic properties tables show that this material requires an H_s = 3000 Oe.

On this basis, the *total* capacity of the (NI_t) magnetizer is:

$NI_t = 2.02 H_s L_m$

$NI_t = 2.02(3000)(4)$

$NI_t = 24,200$ NI *minimum*

Allowing for reluctance losses between the magnet pole pieces of the magnetizer, the magnetizer should be rated at 30,000 NI to be suitable for this application. It must also be mechanically adjustable to a 4-in. air gap.

One important qualification must be made regarding the level of H_s, both above and in the properties tables. *The H_s values given are those required to saturate the magnet in a substantially* closed-*circuit condition* (by keepering or through the magnetizer core). If the magnet is to be magnetized in an open-circuit condition (such as the solenoid fixture in Figure 13-6), *the H_s must be increased to $2H_s$ or as much as $3H_s$* to "compensate" for the large, self-demagnetizing effect of the magnet itself (since it opposes the applied magnetizing force).

Special consideration must be given to the magnetization of the rare-earth magnet materials. Their very high magnetic properties present difficult problems in obtaining initial saturation. Magnetizing forces from 20,000 Oe to as high as 60,000 Oe (40,000–120,000 NI/inch of magnet length) are necessary, depending on the specific composition of the material. Virtually no previously available magnetizer of any type can produce magnetizing fields of this magnitude. In fact, in at least one case, a superconducting electromagnet is being used to produce the necessary field of 60,000 Oe. The commercial suppliers of magnetizers (particularly of the impulse type) have developed or are developing magnetizers to meet this need.

11. Personnel safety must be considered in the selection, operating procedure, maintenance, and fixturing of any type of magnetizer. The potential safety hazards are: risk of electric shock; risk of eye damage from light emission, due to arcing at connections or due to short-circuiting; burns due to heat generated by arcing and/or short circuits; and finger and hand injuries due to highly tractive forces developed during initial magnetization or to "snap back" during magnet removal *after* magnetization. The risk of electrical shock is usually only a factor for impulse units. The risk of eye damage or burns may be a factor for impulse or for half-cycle units. The potential for limb injury is most prevalent for DC electromagnetic units. Limb injuries may occur to a lesser degree with half-cycle units, but seldom occur for impulse units.

The means of minimizing these risks is to use (and maintain) clean, high-pressure electrical connections, adequate insulation on lead wires and windings, two-hand "triggering" buttons, mechanical guards, and well-designed mechanical break-away mechanisms to assist in magnet removal.

12. Magnetization can be achieved at three levels in the magnet supply cycle. Magnets can be furnished in a magnetized (and stabilized/calibrated) condition by the mills or by the distributor-fabricator. They may also be initially magnetized by the user. As we pointed out in (2) above, there are important technical advantages to magnetization by the user. But, there are important operational and economic reasons for in-house magnetization. Prevention of spurious contact demagnetization can be controlled more directly,[10] packaging costs for incoming magnets can be minimized, and the mill or distributor's magnetizing charges (labor, equipment amortization, overhead, and profit) can be kept "in the house." Where the magnets *are* to be purchased in a magnetized state, the packaging method should be included in the specifications for the magnet. (We will discuss this in Chapter 15.)

Relative Costs: Since magnetizers have specific applications, and limitations, and vary in level of quality from source to source, their costs must be presented with

---

10. If proper in-house handling procedures are established *and enforced.*

some caution. In addition, the level of technical support provided by different commercial suppliers will directly affect prices. As a general guide, the following relative [11] costs of generally comparable "capacity," commercially available magnetizers will be useful:

Semicontinuous DC electromagnetic type	1.0
Impulse (capacitor-discharge) type	3.0
Half-cycle type	3.0
Permanent-magnet type [12]	.3

These comparisons are *exclusive* of the fixturing for magnetizing specific magnets, but they *do* include power supplies and controls. Fixturing for the DC electromagnetic magnetizer is generally the lowest in cost and also the least adaptable for a variety of magnet shapes. Fixturing for permanent-magnet types is seldom a cost factor because normally only a pair of "universal," adjustable pole pieces are needed. Fixturing costs for impulse and half-cycle units of the same capacity are usually comparable, except that the costs are relatedly higher for half-cycle units, since they are most often used for very large magnets.

The actual cost of commercially supplied impulse *fixtures* (not just a simple, single conductor) normally ranges from $200 to $1000, depending upon the complexity of the magnet's shape, duty cycle, and so forth. This range includes "development," design, and construction costs.

Demagnetization

The most basic type of demagnetizer is intended to totally eliminate the magnetization of the permanent magnet. For this application, the type of demagnetizer long used with magnetic machine tool chucks (to eliminate the residual magnetism of machined parts) and with magnetic tapes is *not* suitable. Such magnetizers seldom have the necessary field capability (or proper field configuration), since they are essentially intended for relatively "soft" magnetic materials.

There are five basic ways to totally demagnetize a magnet:

1. By elevated temperature.
2. By a decaying AC-produced field.
3. By a reversed DC field of sufficient magnitude to cause "recoil" to $B = 0 : H = 0$ on the demagnetization curve.
4. By deliberately introducing oscillatory currents in a conventional magnetizer.
5. By mechanical rotation and withdrawal from a

11. The base of comparison is the DC electromagnetic type.
12. Available in small-capacity units only.

steady-state DC field (produced by a DC electromagnetic magnetizer).

Demagnetization by heat may be used with virtually any magnet material, except the bonded ferrites. It is, however, a very "risky" way of demagnetizing any metallic type of permanent magnet. Unless the temperature is quite carefully controlled, there is a substantial risk of "overheating" the magnet and causing irreversible structural (metallurgical) changes in the material.

Thermal demagnetization is widely used for the sintered ferrites. It is, in fact, the *only* effective way of producing total loss of remanence. There are risks connected with using heat. The magnet must be raised to the required 450° C or more temperature *slowly* [13] and reduced to normal temperatures even more slowly to avoid thermal shocks that can mechanically break or crack it. Also, if the ferrite magnet is part of an assembly, either it must be removed from the assembly or all parts of the assembly must be able to withstand the required temperature. If properly demagnetized by this method, no dimensional changes, warpage, or mechanical damage will occur.

Demagnetization by a decaying AC field is *totally ineffective* for all ferrite or rare-earth magnets. Otherwise, it is the most widely used demagnetizing method for all types of metallic magnets. For permanent magnet applications there are three ways to obtain a "decaying" AC field. One method is to slowly draw the magnet through a stationary coil (winding) supplied by a constant-voltage AC source; the mechanical motion effectively "decays" the field to which the magnet is subjected. The second method is for the magnet to remain stationary while the AC energized coil is passed over (around) it. The third method is to use a special AC power supply to energize the coil with an electrically controlled decaying current, while both the magnet and the coil remain stationary.

Figure 13-17 is a photograph of a fixed-coil, variable input demagnetizer, and Figure 13-18 is a sketch of a conveyorized, continuous, pass-through unit.[14] Figure 13-19 presents the complete dimensions, electrical requirements, and capabilities for the "do-it-yourself" construction of four designs of simple pass-through demagnetizers.

The electrically controlled, decaying-field demagnetizer is vastly superior to these and other pass-through types. Its higher cost is usually more than offset by a higher production rate, reduced manual labor, and better total or *partial* demagnetization control (for stabilizing and calibrating).

13. For magnets of any substantial size, a "soaking" period will be required so that all parts of the material reach 450° C.
14. The upper belt on this demagnetizer is necessary to pull the magnet out of the end of the coil.

Figure 13-17. A fixed-coil, variable-input demagnetizer/magnetizer. (Courtesy of O.S. Walker Company)

Commercially made demagnetizers are available from a number of suppliers, although the electrically controlled decay types are available primarily only from those equipment manufacturers that specialize in producing magnetizers and test equipment.

A few considerations concerning AC demagnetizers are in order at this point:

1. With either type of pass-through unit, caution must be exercised not to move the magnet or the coil too slowly. The 60-Hz AC normally used produces eddy current and hysteresis loss in the magnet that can cause fairly rapid heating. While this heating may not reach a level that will cause metallurgical changes in the magnet, overheated magnets can cause personnel injury.

2. If a fairly high coercive material is to be demagnetized, and/or the capacity of the magnetizer is not quite adequate, the efficiency of the demagnetizer can be improved (without limiting the pass-through motion) by holding a "soft" magnetic part on both ends of the magnet as it passes through the coil. This concept is illustrated in Figure 13-20. A soft ferrite slug is preferred for this purpose, since ferrite does not become overheated from eddy currents and hysteresis losses.

3. In demagnetizing magnets of a general horseshoe shape, multiple passes may be necessary, and the position of the magnet must be changed with each pass. Such magnets should *not* be passed-through in a keepered state, since a keepered (closed-circuited) magnet is more "self-contained" and tends to be more stable (resists demagnetization).

4. Using the AC method, magnets can usually be demagnetized (and calibrated or stabilized) when they are in their final assembly (see Figure 13-21). The only limitation here is that AC does not affect other parts of the product. In these cases in particular, the electrically controlled, decay demagnetizer is preferred,

* for demagnetizing operation, Power Input = 60 cycle AC.
for magnetizing operation, Power Input = DC.

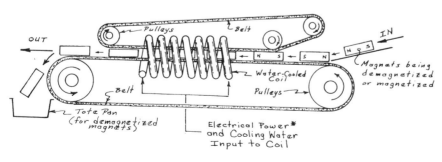

Figure 13-18. Continuous-belt demagnetizer/magnetizer.

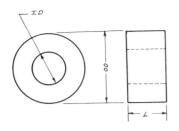
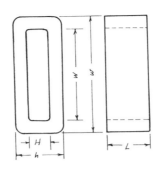

Circular Type   Rectangular Type

Type	ID or H & W	Field Stren. oer.	Duty	Cooling	OD or h & W	L	Winding	AC Volts	AC Amps
Circular	3"	1700	Intermittent	Air	4 3/4"	4"	400 Turns #12 DG Wire	110	30
Circular	4"	1700	Intermittent	Air	6 3/8	5	150 Turns #10 DG Wire	220	40
Rectang.	2 1/4 x 5 3/4	1500	Intermittent	Air	6 1/2 x 10 1/4	6 1/2	350 Turns #8 DG Wire	220	75
Circular	6"	1500	Continous	Water	5"	5	160 Turns 1/4" Copper Tubing	220	150

Figure 13-19. 60-cycle AC solenoid-type demagnetizers.

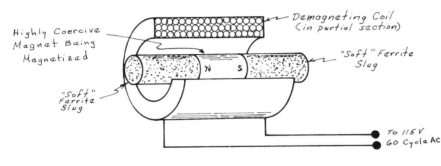

Figure 13-20. Method of improving the performance of a solenoid-type demagnetizer on highly coercive magnet materials.

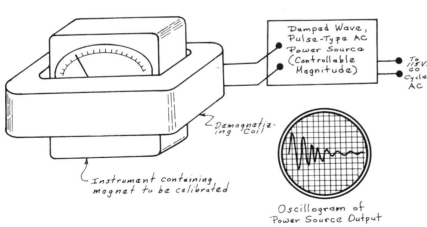

Figure 13-21. Magnet-calibrating demagnetizer.

so that the magnet can be demagnetized or changed with a *minimum* time exposure to the AC field.

5. Rare-earth magnets present difficult if not impossible problems if AC demagnetization is desired. With intrinsic coercive forces (H_{ci}) up to 25,000 Oe the demagnetizing capability required is impractical.

Demagnetization by recoil is accomplished by applying a steady-state DC field of *opposite* polarity to the original magnetizing field ($+H_s$). This $-H$ must be of sufficient magnitude so that when it is discontinued, the magnetic material recoils along its minor loop to where B_m and $H_m = 0$. The $-H$ must not equal $+H_s$ in

MAGNETIZATION, DEMAGNETIZATION, STABILIZATION AND CALIBRATION   169

magnitude, or the magnetic polarity will simply be *reversed* (since the magnet will be saturated in the opposite direction).

When a material has a very flat demagnetization curve (such as sintered Ferrite 1), the recoil line substantially coincides with the demagnetization curve itself and it is difficult if not impossible to control $-H$ to the precise magnitude necessary. If $-H$ is too low, no demagnetization takes place. If it is too high, reverse saturation occurs.

This method of demagnetization is relatively slow and requires an experienced operator. In general, it is only used for magnets too large to fit the available magnetizers or if no other demagnetization equipment is readily available.

Demagnetization by oscillatory matching is the design and adjustment of an impulse or DC electromagnetic system so that oscillations *deliberately* occur when the electrical input is discontinued. As we learned earlier, oscillations can be avoided in a magnetizer system if

$$R > 2\sqrt{\frac{L}{C}}$$

Obviously, if

$$R \approx 2\sqrt{\frac{L}{C}}$$

then *oscillations will be produced*. With some effort, it is possible to develop a "dual" system in which the resistance can be controlled to *avoid* oscillations when the unit is used as a magnetizer and to *create* oscillations when the unit is used as a demagnetizer.

The obvious disadvantages of this method are that the system can only be used for one purpose at a time and that the adjustments for demagnetization may be tricky, variable, and sometimes uncertain.

Demagnetization by rotational removal employs a DC electromagnetic magnetizer as a demagnetizer. In this technique, the magnet to be demagnetized is not allowed to come into full contact with the magnetizer pole pieces (this is accomplished by using nonmagnetic spacers or by extending the air gap). The magnetizer is continuously energized, and the magnet is slowly withdrawn manually from the field region while it is "rotated" and/or "tumbled."

This method is practical only for relatively small magnets, since the mechanical force required demands substantial manual labor and presents some personnel safety hazards.

Stabilization and Calibration

The purpose of stabilizing magnets has been discussed in preceding chapters. Calibration modifies the various characteristics of a group of magnets to a specific, lower, and less variant level. There are five methods of stabilizing and/or calibrating. These are by deliberate: thermal exposure (heat or cold), lapse of time, intentional mechanical shocks, contact with ferrous parts, and exposure to external magnetic fields (AC or DC). Technically (but not necessarily practically), all of the above methods can be used for purposes of stabilization. For calibration purposes, only the "contact" method or the "field" method can be used.

Stabilization by means of thermal exposure, lapse of time, or mechanical shock is slow, uncertain, uncontrollable, and often unpredictable. Thermal exposure, particularly at reduced temperatures, is occasionally the only practical way of stabilizing some types of equipment. One example is the deliberate exposure of large iron-ore beneficiation separators to low temperatures before installation. Since these separators are often operated in unheated buildings in cold climatic regions, the units are deliberately stabilized by exposure to the lowest anticipated ambient temperatures.

Stabilization and calibration by deliberate contact (spurious contact demagnetization) is fairly common practice when more "scientific" methods are not usable because of lack of equipment or equipment size limitations. This method is, of course, slower and more difficult to use where high consistency and precision are required.

Technically and operationally, the best and also the most widely used stabilization/calibration method is by deliberate exposure to a controlled AC field. In some circumstances (particularly for very large magnets), a DC-field system may be used, since AC fields can present problems of flux penetration. Normally, AC systems are most effective if the magnet is stationary and the input current is controllably decayed to zero.

Stabilization/calibration methods are largely restricted to the metallic permanent magnets. Ferrite and rare-earth materials cannot be calibrated by any method except thermal exposure, and even with this method, the results obtained are highly variable.

Costs of Demagnetizers and Stabilizers/Calibrators

The large variety of sizes and operating types of demagnetizer/stabilizer/calibrator (DSC) equipment precludes meaningful comparisons. Simple single-coil, bench-type, pass-through demagnetizers may cost as little as $25, while a complete, semiautomatic, combination DSC unit may range from $1000–$6000. Sophisticated "magnetreaters" are available from a number of commercial sources.

14
Mechanical Considerations

In many ways, a permanent magnet is a "custom" component—particularly in a physical sense. While there are defined magnet material types with specific magnetic and inherent physical *properties*, a magnet's geometry and dimensions are custom-made to the user's requirements in a high proportion of applications. This custom-forming may be accomplished in the basic magnet manufacturing process (by casting, pressing, etc.) or by "fabricating" (cutting, machining, etc.) from standard or stock materials. Obviously, the geometric and dimensional requirements established in the product design will have a substantial effect on the ultimate cost of the magnet.

This chapter deals with factors over and above the basic process that affect costs, determine material selection, and relate to overall design and assembly.

Tooling Costs and Custom Designs

The relative costs of *production* tooling (not "samples") to make basic magnets of various types and grades of material are:

Cast Alnicos (all grades except 5DG, 5-7, and 9)	- 1.0 (base)
Cast Alnico 5DG	- 1.2
Cast Alnico 5-7 and 9	- 1.3
Ferrite 1	- 3.5
Ferrite 5 (and all oriented grades)	- 3.6
Wrought Alloys (all)	- .5
Sintered Alnicos (all)	- .9
Bonded Ferrites (formed from sheet stock)	- .3
Bonded Ferrites (extruded)	- .6

In addition to ultimate production tooling, relative tooling costs for *samples* may be useful. Sample tooling for cast Alnicos can be quite modest ($150–$400). For ferrites, sample tooling is not appreciably less than production tooling. The cost of designing the ferrite die is substantially the same for single- or multiple-cavity dies. The only variation in sample versus production tooling cost is related to the construction labor and the materials. *Sample tooling for ferrites can often be minimized or avoided entirely by fabricating the sample parts from standard blocks or rings.*

Size, Shape, and Orientation

In general, the larger the magnet, the lower the cost per pound. Conversely, the smaller the magnet, the higher the cost per pound. Similarly, a complex shape will be more costly than a simpler shape. Oriented material is more costly than unoriented material. The more complex the orientation requirement the higher the cost.

On this basis, the most costly type of magnet would be very small and complex in shape and would require multipolar orientation (a multipole radial rotor, for example). The least costly magnet would be large, with a simple configuration, and would be made of an unoriented material.

Physical Imperfections

Partial cracks, edge chipping, burrs, voids, blow holes, and inclusions are common to all types of magnets, but particularly to the cast Alnicos and the sintered ferrites. *These imperfections are not necessarily indicative of poor magnetic quality.* It is not unusual for a magnet with a great degree of imperfections to have high magnetic properties. Of course, it does not follow that such a magnet will necessarily be of the highest *quality*. The relationship between imperfections and magnetic quality is largely a matter of degree. Generally, if imperfections are widespread in a substantial lot of magnets, the magnetic quality will also be poor, since the prevalence of the imperfections represents poor process control by the mill.

Unrealistic requirements for a "flaw-free" magnet, unless absolutely necessary for aesthetic reasons, increase the cost of a magnet substantially. Sintered Alnicos and the wrought-magnet materials are generally free of all but the most minor imperfections.

Inserts and Holes

Requiring inserts or holes in any type of magnet materially affects costs. Inserts add cost through both the value of the insert itself and the reduced yield in the production cycle. Inserts *can* be cast in place for Alnico magnets, but the production yield is usually substantially reduced due to shrinkage fractures that occur during solidification. The mold design and mold making are also complicated.

It is *not* practical to add inserts (during pressing or before firing) directly to the sintered ferrites. For these materials, inserts can only be added by drilling a hole in the "fired" piece and then cementing (*never* pressing) the insert in place.

Holes *can* be cast into Alnicos or pressed into sintered ferrites, but unless properly designed, the yield (and therefore the cost) will be adversely affected.

Sectional Variations

Thin sections or wide variations in section add cost by reducing production yield in the casting of Alnicos and in the pressing and firing of sintered ferrites. The differences in cooling-rate shrinkage provide a common source of cracking and reject magnets. The sintered ferrites are more critical in this regard than the cast Alnicos.

Figures 14-1 and 14-2 comprise a configuration-dimension design guide that is applicable to sintered ferrites. The situations designated as "undesirable" will materially increase costs and may completely preclude the manufacture of that particular magnet.

Vendors

In many cases, a particular magnet (standard or modified standard) may be purchased from either a mill or a distributor-fabricator. Generally, the distributor-fabricator may be the *only* source of *small* quantities of magnets, since mills are not organized or equipped to handle small-quantity orders. When either source is willing and able to supply a particular magnet, the costs will be comparable. When large quantities are involved, the mills are the primary source of low-cost magnets. Distributor-fabricators do not have in-house casting, pressing, or rolling facilities, although they usually have excellent machining capabilities.

Machining

Conventional chip-cutting machining cannot be used with most permanent magnet materials. Both Alnico and ferrite magnets are extremely hard, exceeding the levels at which even carbide tools are usable. All cutting, surface removal, and similar operations must be performed on these materials by *grinding*. Many operations, including counterbores and the tapping of holes, are impossible or, at best, difficult and costly.

Grinding and cutting equipment suitable for magnets can be obtained from normal, commercial machine-tool sources. In purchasing machine tools for this purpose, it should be noted that Alnico and ferrite grinding dust are highly abrasive. Machines used in grinding these materials must therefore be designed for minimal contamination of working parts *and for separation of grinding dust from the coolant system.*

Both Alnico and ferrite magnets should be machined or finished in an unmagnetized state, if at all possible.

The techniques used in finishing Alnicos and ferrites are generally similar. However, there are sufficient differences in details to make separate consideration necessary.

Alnicos
Cutting: Abrasive cut-off machines with aluminum oxide wheels will cut Alnico readily. To achieve an optimum cutting rate, the grit and wheel thickness should vary somewhat with the grade of Alnico. *For satisfactory results, a high-volume water flow must be directed into the cut.* This flow should also hit both sides of the wheel. Ordinary tap water with a rust inhibitor is satisfactory, and the water may be recirculated if a good coolant cleaning system is used. (Magnetic coolant cleaners or separators are only moderately efficient with Alnico grindings; a combination magnetic and paper filter is preferable.) To make precision cuts or to minimize edge chipping, mechanical clamping and controlled power feed of the cutting head are recommended.

Surface Removal: Surface-removal operations may be performed with any type of horizontal, vertical, center-

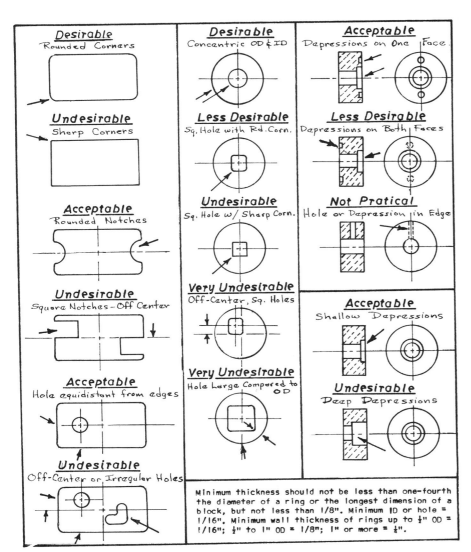

Figure 14-1. Configuration-dimension guide for sintered ferrite magnets.

less, double disk, or oscillating grinder. As with the cutting operations described above, aluminum oxide wheels are best, with a high-volume water flow onto the wheel and grinding area. A high degree of coolant filtering (for recirculating coolants) is a necessity for high-precision work (tolerances of .001 or more).

Magnetic chucks can be used to hold the material when Alnicos are ground, but *auxiliary blocking with ordinary steel bars is absolutely necessary.* [1] Where a magnetic chuck is used, small magnets or magnets made of low-coercive materials may require demagnetization between operations.

Holes/Counterbores: These operations can most readily be performed on Alnicos by *electrical discharge machining* (EDM or Elox). Copper-alloy electrodes are usually better than carbon electrodes. Electrical discharge machining of Alnicos is slower than it is for

1. Alnico is only 50 percent iron and cannot be securely held by the chuck.

common tool steels by 2 : 4. EDM operations have no effect on magnetic properties.

Deburring/Polishing: Alnico magnets can be deburred or polished either by hand methods or by tumbling. When deburring by hand, the small, abrasive wheels on hand-power tools are effective. For tumble deburring, commercially available polishing stones are used. When both deburring and polishing are required, tumbling for 24 to 36 hours with *walnut* shells will produce good results with minimum mechanical damage to the magnets.

Both sand blasting and shot blasting are most satisfactory in removing mill scale, but they are generally *not* suitable for deburring and polishing.

Ferrites

Cutting: Both the equipment and the technique employed in cutting ferrites are substantially the same as those used in cutting Alnicos, except that *diamond*

MECHANICAL CONSIDERATIONS 173

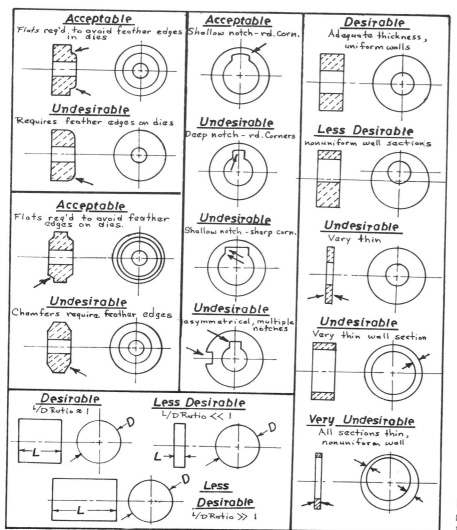

Figure 14-2. Configuration-dimension guide for sintered ferrite magnets (continued).

abrasive wheels are required. Multiple-head (parallel) cutting is practical, using modified horizontal milling machines with a suitable coolant system. It should be noted, however, that the coolant filtering system for ferrites *cannot be magnetic.* Unmagnetized ferrite particles are almost totally unaffected by common, commercial magnetic coolant filters.

Surface Removal: All surface removal operations require diamond wheels. The equipment and techniques used are the same as those previously described for Alnicos. As with Alnicos, magnetic chucks cannot be used directly with ferrites. The typical magnetic chuck will produce almost no holding force on a ferrite, so these chucks can only be used if mild-steel "blocking bars" secure the ferrite magnet against the lateral grinding forces.

Holes/Counterbores: These operations can only be performed on ferrites by using diamond-core drills mounted in a hollow-spindle drill press or in a vertical mill. *A high volume of water must be continuously pumped through the spindle during cutting.*

Since ferrites are electrical insulators, electrical-discharge machining is not usable.

Counterbores can be produced by core drilling to the required depth and then "breaking out" the remaining core section manually.

Deburring/Polishing: Ferrite magnets normally require no deburring or polishing operations. If a beveled edge is required, diamond-wheel grinding may be used. Tumbling, sand blasting, and shot blasting are all ineffective and invariably cause serious edge chipping.

Machining Cost Guide

In general, the two most significant factors influencing the cost of machining a magnet are: location and extent of surfaces to be machined (the pole face only, overall grind, etc.), and surface roughness (microinch finish) required. Other factors influencing machining costs include *type of material* (Alnico, ferrite, etc.); *grade* (Alnico 5, Alnico 8, etc.); *manufacturing process* (cast, sintered, etc.); *size* (large, medium, small); *shape* (horseshoe, rod, etc.); *precision required* (in dimensions, parallelism, flatness); and *special tooling required* (to hold parts during machining). Since an infinite number of combinations of these factors may exist, no meaningful relative costs can be stated. For the more common magnet materials (Alnicos and ferrites), however, the following generalizations may be useful:

1. The least expensive magnet will be the one with the fewest machined surfaces.
2. The cost of machining the pole faces of a magnet is generally less than the cost of machining secondary surfaces. (Most mills have a broad range of toolings for the pole-face grinding of common shapes.)
3. The machining of flat surfaces is less costly than the machining of curved surfaces. (A possible exception is where centerless grinding can be used.)
4. The machining cost is lower for sintered Alnico magnets than it is for cast Alnico magnets. (The surface quality of sintered Alnico is initially better.)
5. The machining cost of *any* Alnico is lower than that of any ferrite. (Ferrites require diamond wheels and slower feed rates.)
6. The machining cost of cast Alnico 5 is lower than that of cast Alnico 8. (Alnico 8 is "harder," more brittle—a tougher material.)
7. Grinding fillets and rounds is very costly (since these are largely "hand" operations.)
8. ID machining is more costly than OD machining.
9. The machining of special surfaces (spherical sections, for example) is very costly. (Such machining often requires special tooling, special setups, and usually some experimentation.)
10. Machining to tolerances tighter than those shown in the MMPA Standard Specifications No. 0100-72 table (Chapter 15) will add appreciable cost.
11. For Alnico, machining to surface roughnesses of less than 32 microinches will usually add cost. ("Normal" machining for grinding magnets ranges from 63 to 16 microinches). For ferrite magnets, surface roughnesses of 16 microinches are "normal" and roughnesses of 4 microinches are attainable (but at an additional cost).

Supplemental Finishing

Frequently, supplemental finishes are required for basic permanent magnets or magnet assemblies. The most common supplemental finishes are: plating, painting and coating, polishing, and encapsulation (metals and nonmetals). None of the mills or distributor-fabricators maintain in-house capabilities for applying these supplemental finishes. When required (particularly for large-quantity orders) they will supply magnets with such finishes, but the actual operations are performed by outside vendors.

For users who elect to purchase the basic magnets from mills or distributor-fabricators and apply their own supplemental finish (or use their own vendors), the following discussion will be of assistance.

Plating: This is one of the most common supplemental finishes. It is applied to magnets both to prevent corrosion and to improve appearance.

Cast Alnico magnets can be plated with most of the common plating materials (copper, cadmium, zinc, nickel, tin, and others). Standard plating solutions and equipment can be used. Alnico magnets are more difficult to plate than most wrought or cast materials because of their relatively poor adherence qualities; minute pits and cracks are inherent in Alnico. Although these minor imperfections seldom affect the magnetic properties, they tend to "spot out" or bleed after plating. This effect can be minimized with proper plating procedures, as we will see in Chapter 15.

Sintered Alnico is very difficult to plate because of its inherent porosity. It is, however, possible to "fill" the pores of the material with a suitable plastic, so that the magnet will plate satisfactorily. The plastic fillers used are those commonly applied to sintered ferrous parts.[2]

Flame spraying and similar techniques of metal coating are not recommended for direct application to Alnico magnets. Mechanical or magnetic "damage" to the magnet may occur if localized temperatures reach 980° F.

Since ferrites are electrical nonconductors (insulators), they cannot be electroplated in a conventional manner. The plating methods applied to plastic parts are usually satisfactory from a decorative standpoint, but mechanical properties are poor. Ferrites are seldom plated in any way, since their inherent corrosion resistance is very good and plating is seldom functionally necessary.

Ordinary plating methods can be applied to most wrought alloys. Alloys such as Lodex require procedures similar to Alnico, but because of special property limitations, the recommendations of the particular material manufacturer should be followed.

2. See *Finishing and Plating of Metal Powder Parts*, General Session on Powder Metallurgy of the 11th Annual Meeting Proceedings of the Metal Powder Association, Vol. 1, p. 6. New York City, 1969.

Painting and Coating: Paints may be applied to Alnico magnets to prevent rusting and/or to improve appearance. The Alnicos may be painted with any material that can be applied with equipment normally used with any ferrous part. A good rust-inhibiting primer is mandatory. Fluidized bed or PVC coating [3] methods are applicable to Alnico, since the temperature to which the magnet must be subjected in the process is well below the "damage" or "irreversible" change points.

Most commercial paints made for nonporous surfaces adhere well to sintered ferrites, if the surface is thoroughly cleaned and is free of all oil, grease, dirt, and loose ferrite particles. No primer is normally required. Epoxy-base paints provide particularly good adhesion qualities. PVC or fluidized bed coatings are not normally used with ferrite magnets.

Polishing: The primary purpose of polishing is to improve appearance. Alnico magnets without any supplemental finish can be polished by a variety of common metal-polishing techniques. The most readily used hand method employs a buffing wheel and a polishing range. Mass polishing can be accomplished by the prolonged tumbling of the magnets in rotary tumblers, using walnut shells as the buffing agent.[4] Needless to say, the level of polish that can be obtained is limited by the initial surface roughness and by the length of the polishing operation.

Sintered ferrite magnets are normally not polished, since standard surface machining (grinding) with diamond wheels produces the highest polish this material can achieve.

Encapsulation: Alnico magnets may be encased in molten metal or in plastics for one or several of the following purposes: prevent contact demagnetization; secure multiple magnets into a single assembly; provide mounting lugs, holes, etc., for the magnet or its associated parts; improve appearance; prevent mechanical damage; prevent corrosion.

Standard, commercial, die-casting metals and equipment are used with no special problems. Examples of magnets in a variety of die-cast configurations are shown in Figure 14-3. For plastics, encapsulation or empotting by injection molding, by compression molding, or with flexible molds are used with any commercial polymeric material. Thorough cleaning of the magnets is the only special requirement.

Sintered ferrite magnets may be encapsulated in polymeric materials for most of the same reasons given for the Alnicos. Encapsulation to prevent contact demagnetization or to prevent corrosion is not necessary

Figure 14-3. Alnico magnets in a variety of die-cast configurations. (Courtesy of Arnold Engineering Co.)

for ferrites, however. Also, direct [5] molten-metal encapsulation is not used for ferrites.

Fastening

Magnets may be secured into the product assembly by any of the following means: "through" bolting, welding, cementing, riveting, soldering or brazing, pressure clamping, or press fits.

The fastening technique most appropriate to a particular product varies with its characteristics and environmental requirements, the magnet material, cost, and available equipment. *Drilled and tapped holes are not generally practical for Alnico magnets and are* not technically possible *for sintered ferrite magnets.*

All of the above techniques are potentially applicable to metallic magnet materials. Only through-bolting, cementing, riveting, and pressure clamping are potentially applicable to ferrite magnets.

Obviously, when *through-bolting* is used, the magnet must be provided with integral "lugs," "ears," or through-holes. Such provisions may add considerable cost to the magnet, as well as increase its "fragility" to mechanical loads.

Welding can be used to directly fasten metallic magnets into an assembly, but care must be taken to avoid thermal damage (structural magnetic changes) at the weld area. Practically, *some* damage is almost unavoidable, particularly with gas-welding methods. Depending upon the size, shape, and location of the weld, damage from electric-arc welding may be so minor that the magnet is still usable. Spot welding of the Alnicos is not

---

3. On small magnets, a heavy PVC coating can also be used to prevent spurious contact demagnetization, as we saw in Chapter 11.

4. Using this technique, some breakage and chipping of the magnets can be expected.

5. More commonly, the ferrite magnet is cemented into a previously die-cast shell.

usually effective. Other metallic magnet materials can be spot welded with more success. In all cases, the welding technique for magnet materials is *not* the same as it is for ordinary ferrous materials: welding cannot be used with any form of ferrite.

Cementing is being used increasingly in production to fasten magnets into assemblies. The ferrites, in particular, cement quite well to one another and to other parts. The metallic materials are somewhat more difficult to cement, but they can be effectively fastened by this method if the proper materials and surface preparation are used. It is common practice for mills, distributor-fabricators, and users to cement ferrite rings, bars, slabs, and blocks into stacks to form units larger than those that could be formed by direct pressing.

The requirements for *riveting* are the same as those for through-bolting (a lug or hole). In addition, close control of the riveting pressure (force) is necessary (particularly with sintered ferrites) to prevent "loose" or broken magnets. Depending upon the location of the rivet (in relation to the magnet poles) and its size, rivets may be either ferrous or nonferrous.

As a fastening technique, *soldering or brazing* is preferred to welding, since the potentialities of thermal damage are greatly reduced. The technique is, of course, only applicable to metallic magnets, and these magnets should be copper-plated for a good bond.

Pressure clamping is one of the most widely used fastening methods for both metallic and ferrite magnets. It is generally one of the least expensive and flexible methods, particularly when post-assembly removal of the magnets might be required. Generally, the clamps must be of a nonferrous material to avoid "shorting" a portion of the magnet's field. Pressure clamps should, where possible, be designed with a small degree of "spring" action to prevent undue stress on the magnet.

Press fits are occasionally used with metallic magnet materials, but this fastening method tends to be a costly and a "touchy" one. The magnet must be machined to a high precision, and the risks—even with the "tougher" metallic magnets—of cracking, crumbling, or otherwise succumbing to stress are relatively great. Press fits are completely impractical for the sintered ferrites because of their extremely brittle properties.

Mechanical Properties

No permanent magnet should be designed in such a way that it carries structural loads when it is used in a product. While some materials have reasonably good tensile strengths, the homogeneity is not sufficiently well controlled in the manufacturing process to maintain strength values reliably.

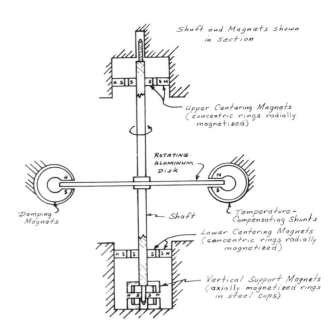

Figure 14-4. Magnets used in watthour meters.

The very high compressive strength, hardness, and abrasion resistance of the sintered ferrites may, on occasion, be utilized directly in the product to serve some purpose other than producing a magnetic field. In such cases, it is generally advisable to provide some way of "containing" or supporting the periphery of the magnet to prevent edge cracking and chipping. Also, full "back" support should be used to prevent bending or shear loads. *Shock loads that are directly applied to any surface of the magnet must be avoided.*

Thermal Properties

The metallic magnet materials are fairly good conductors of heat, while the ferrites are excellent thermal insulators. These properties can be used to advantage in a product, provided temperatures do not reach "critical" levels, possible reversible magnetic losses are anticipated (or "compensated" for), and thermal *shock* is avoided.

It is established practice, particularly for metallic magnets, to incorporate temperature-compensating elements [6] where high magnetic stability against *reversible* changes is required. An outstanding and long established use of such compensators is in electrical utility, watthour-meter damping magnets. As shown in Figure 14-4, a temperature-sensitive compensating "shunt" is formed *around the back surfaces* of the C-shaped magnets. The basic theory of this method is that

---

6. There are two methods of temperature compensation—the shunting method described here, and a second method which employs a bimetal strip to vary the air gap. The bimetal strip method is seldom used because of serious operational problems.

certain nickel-iron alloys change their magnetic permeability with temperature changes. At lower temperatures, the shunt permeability increases, diverting more flux from the air gap through the shunt. At higher temperatures, the permeability decreases, reducing the flux diverted from the working air gap. It should be noted, however, that *any* compensator is designed to divert flux from the air gap (to varying degrees). Therefore, allowance ("over-design") must be made for such diversion in the initial design of the magnet, with some inherent sacrifice in magnetic efficiency.

Special temperature-sensitive alloys are used for compensators. The most widely employed are approximately 30 percent nickel-iron alloys (also known as "Curie" alloys). Figures 14-5 and 14-6 are the temperature-permeability curves for Carpenter Nos. 30 and 32—two typical, commercially available compensating alloys.

Temperature compensation is only practical over a limited range between -60° C and 180° C. The curves in Figures 14-5 and 14-6 can be used to calculate the dimensions and the effects of compensators, using the equation

$$\phi = 46 \mu A$$

where ϕ is the total flux shunted at temperature t, μ is the permeability of shunt material from Figure 14-5 or Figure 14-6, A is the cross-sectional area of the shunt in cm², and 46 is the constant field strength H for the curves in Figures 14-5 and 14-6.

Typical products which may require temperature compensation are watthour meters, speedometers, traveling-wave tubes, and aircraft sensors. Typically, simple compensators are "linear," although they may be designed for "under-compensation," "over-compensation," or "flat" (uniform gap flux) compensation. Using "double" shunts, nonlinear compensation can be achieved.

Electrical Properties

The metallic magnets are basically electrical conductors (to varying degrees), while all ferrites are excellent electrical as well as thermal insulators. Using the conductive properties of metallic magnets for supplemental functions is undesirable. The conductivity of metallic magnets is poor compared to copper, and the field generated by the current passing through the magnet may directly affect the primary field.

The insulating properties of ferrites may be used in any appropriate situation where the magnetic and electrical insulating functions can be combined.

Chemical Properties

The chemical "reactivity" of *metallic* magnets is directly related to their chemical composition (as is reflected in the Appendix properties tables). The general resistance

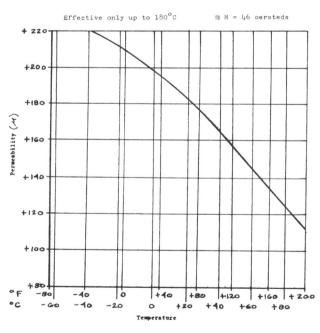

Figure 14-5. Typical temperature/permeability curve for Type 1 nickel-iron alloy.

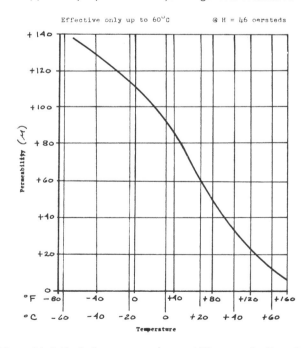

Figure 14-6. Typical temperature/permeability curve for Type 2 nickel/iron alloy.

of metallic magnets to corrosion (with the exception of platinum-cobalt) is substantially the same as the resistance of high-alloy steels.

Sintered ferrites share typical "ceramic" properties in that they are highly resistant to corrosion. Sintered ferrites are also substantially resistant to water, oils, and most chemicals. This property is particularly useful in severe environmental situations, but only when temperature, mechanical stress, etc., are not factors.

Polarity Identification

In some applications, it is necessary to identify (with suitable markings) the north and south poles of permanent magnets. This identification may be required for initial assembly or for field servicing of the particular product. The poles are commonly marked with paint spots, etched letters, or "cast-in" letters applied by the mill, the distributor-fabricator, or the user. A constant source of confusion and misunderstanding when the identification is added by the mill or by the distributor-fabricator is: "What is a north (or south) pole?" There is no *absolute* definition of polarity—only a *relative* definition. In a relative sense, polarity is "in the eye of the beholder."

The National Bureau of Standards' (NBS) definition is based on the concept that if a magnet is suspended so that it can move freely, the "end" of the magnet that points to the *geographically* identified North Pole is a *"north-seeking"* pole. This pole should be *marked on that magnet simply as a* north *pole*. Following this system, a magnet with "north" marked on a particular end is, in reality, a "South" since it is only "north seeking."

No type of commercially available fluxmeter, gaussmeter, or reference magnet has an established procedure for relating the meter to the NBS system. Such a procedure is not practical with these instruments or magnets, since the direction of the search-coil winding, the direction of "pull off," the direction of the connections, the position of the Hall probe, can all "reverse" the polarity measurements. One inexpensive pocket device is available primarily as an NBS system polarity indicator.[7]

Physical Testing

The normal quality control procedures of the mill or the distributor-fabricator contain no type of physical testing. Mill in-process controls cover only magnetic tests and chemical analysis. Where required by the user, the following tests can be performed directly by the mills (or by mill-connected vendors): penetrant tests for cracks, etc.; x-ray tests for voids and inclusions; centrifugal-vibration tests; tensile-strength tests; compressive-strength tests; shear tests; and electrical-resistivity tests. Magneflux testing, while possible on metallic magnets, is difficult to apply and is not recommended.

Requiring any of the above tests should be avoided unless absolutely necessary. Such tests increase costs, slow down deliveries, and more often than not, prove very little. Requiring certified chemical analysis should also be avoided. The basic interest the user has in a properly applied magnet is in its *magnetic performance*. Defining test methods with performance results and requiring certification of these results is appropriate and meaningful.

7. Available under the trade name of "Magtester" from the Sel-Rex Corporation, Nutley, New Jersey.

15

Standards, Specifications, and Purchasing

The last step in permanent magnet design and application is establishing standards specifications. For the design engineer, this often includes identifying sources.[1] As a "job-shop, custom-made" type of industry, the burden of establishing permanent magnet standards and preparing meaningful specifications has largely been left to the user. Even locating specific, *appropriate* supply sources for magnets, magnetizers, testers, and other equipment has often required extensive efforts on the part of the user. Directories such as *Thomas' Register* and *McCraes' Blue Book* contain a substantial number of categories related to permanent magnets, but the majority of source listings in each category really offer permanent magnet *assemblies* (tramp-iron separators, lifting magnets, etc.) Few of the source listings are actually *suppliers* of basic magnets (mills or distributor-fabricators). For example, there are currently only 23 mills and 16 distributor-fabricators in the United States, yet the leading purchasing directory lists over 200 different vendors in categories that indicate they are suppliers of "permanent magnets." This situation exists because there has never been any clear distinction between basic magnets as fundamental *components* and magnet *assemblies*.

Standards

The earliest "standard" widely applied voluntarily by both mills and users (and required for federal purchasing) was QQ-M-60, *Federal Specification, Magnet Materials, Permanent,* approved June 20, 1950. As will be noted from reviewing the copy of QQ-M-60 at the end of this chapter, while the specification is quite thorough, it is basically a *material* specification and not a *performance* specification. Futhermore, QQ-M-60 does not cover many of the current materials (cast Alnico 5DG, Alnico 5–7, Alnico 8, any of the ferrites, etc.), but it does cover obsolete materials such as Vecto-lite and Silmanal. QQ-M-60 is still useful as a guide for user-established standards, particularly with regard to surface finishes and *material-properties* testing methods.

The current "standard" sponsored by the Magnetic Materials Producers Association (MMPA Standard No. 0100–90)[2] was issued in August 1964 and revised January 1, 1972 and 1990. This standard is "advisory" and "voluntary" for the members of the MMPA and is intended for the "guidance" of governmental and industrial purchasers. As the foreword of MMPA Standard 0100–90 states in part: "It is hoped that the data in this publication will serve as a guide to governmental and industrial purchasers so that they may be assured of uniform quality manufactured to commercial standards. These standard specifications were developed under the auspices of the Magnetic Materials Producers Association and were voluntarily established by mutual consent of those concerned. These specifications are advisory only and their use or adaptation is entirely voluntary. The users of the specification are wholly responsible for protecting themselves against all liabilities for patent infringement."

As with QQ-M-60, the MMPA standards do not define performance, only *nominal* properties. Since no tolerances on these nominal values are specified in MMPA Standard 0100-66, individual mills have unlim-

---

1. While source identification is normally a purchasing function, in the case of highly specialized components, the engineer may need to know the available sources to obtain information for his own purposes and to assist in purchasing.

2. Adopted for Department of Defense use, September 19, 1966.
3. Emphasis added.

ited leeway in varying these values. *Reputable magnet mills maintain the following normal tolerances:*

B_r ± 5 percent of nominal

H_c ± 5 percent of nominal

BH_{max} ± 10 percent of nominal

These tolerances can be maintained for typical magnet shapes and sizes with normal manufacturing process controls. Closer tolerances can be maintained, but at a rapidly increasing cost. Technical problems may require meeting these tolerances by 100-percent testing and sorting of normal tolerance lots.

The information contained in MMPA Standard 0100–90 is provided at the end of this chapter. Particular attention should be given to the dimensional tolerances, since they represent normal process capabilities and "tighter" requirements can substantially increase costs. While the magnetic properties and tolerances suggested above can be used as a *guide*, examination of the properties tables (or curves) of individual mills will show considerable variation. It should also be noted that not all mills are members of MMPA and that nonmembers may not follow the MMPA's voluntary standards.

The industry-wide method for the cadmium plating of cast Alnico magnets appears below. This method has been developed from years of experience and is established practice for the most common supplemental finishing process applied to cast Alnicos.

Standard Method for Electroplating Cast Alnico Permanent Magnets

1. Demagnetize magnets as completely as possible.
2. Abrasive blast with aluminum-oxide grit or other nonferrous grit. Be sure *all* scale is removed.
3. Clean thoroughly with a hot alkaline cleaner.
4. Thoroughly rinse with cold running water.
5. Clean with suitable reverse-current cleaner.
6. Rinse in cold running water.
7. Dip in muriatic acid solution (30 to 40 percent acid) for 3 to 5 minutes.
8. Rinse in cold running water.
9. Dip in nickel chloride solution (3 oz./gal) for 5 minutes.
10. Rinse in cold running water.
11. Copper strike 3 to 5 minutes ($\approx$ .0005 in. copper deposit). Heavier deposit should be used if magnets tend to be abnormally porous.
12. Rinse in cold running water.
13. If heavy copper plate is used, buff to close cracks and pores.
14. Cadmium plate (.0005 in. minimum cadmium deposit).
15. Rinse in cold running water.
16. Dip in chromic acid (Irridite).
17. Rinse in cold running water.
18. Rinse in hot water.
19. Dry thoroughly.

Specifications

As we mentioned earlier, the "standardization" of magnets is voluntary and not necessarily industry-wide. Therefore, the specifications set by the *user* are the first step and the basic method for controlling the incoming component.

Any complete specification should contain the following details (or as many as are appropriate):

Configuration in the form of a suitable drawing and/or description, as given in Figure 4-1 (page 14).

Physical dimensions and tolerances in accordance with the user's requirements or MMPA Standard 0100–66, whichever is more restrictive.

Material type and grade in accordance with standard industry designations (cast Alnico 5, sintered Ferrite 1, etc.).

Magnetic properties: Nominal B_r, H_c, and BH_{max}, *plus* allowable variations on these properties.

Orientation and/or magnetization: State if material is to be oriented or unoriented and whether the magnets are to be shipped in an unmagnetized or magnetized condition. In either case, drawing markings (showing orientation and magnetization direction) should be in accordance with the nomenclature and methods shown in Figures 4-2, 4-3, and 4-4 (pages 17, 19, 20).[4] *When there is to be an unmagnetized or unoriented zone(s) on the magnet, state or show the width of this zone* (see Figures 4-2, 4-3, and 4-4). Unless otherwise specified, magnets are normally shipped in an unmagnetized condition. To preclude errors, it is advisable to specifically state whether they are "magnetized"[5] or "unmagnetized."

Machined finish: State machine finish (on required surfaces only) in accordance with user requirements, Par. 3.7 of QQ-M-60 or MIL-STD-10A, whichever is more restrictive. *Include percentage of machined surface over which this finish level must be met.* If machine

---

4. For more complex shapes, the same general methods should be used.
5. *Reminder:* If "magnetized" is stated, details of the magnetization method must be furnished.

finishes and percentage areas of MMPA Standard 0100-72 are sufficient, *so state.*

Supplemental finishes: State in full detail any supplemental finishes required (plating, PVC coatings, etc.), where these finishes are to be furnished by the mill or the distributor-fabricator. For plating, include a copy of the standard method for electroplating permanent magnets provided earlier in this chapter as a specification. For other supplemental finishes (fluidized bed coatings, die casting, etc.), obtain specifications from the appropriate material supplier.

Polarity markings: Where magnetized magnets are to be furnished and polarities are to be indicated, state that markings shall be in accordance with the National Bureau of Standards methods. Specify the *method* of marking (cast-in, paint-type, spot-size, color, location, etc.) required by the user.

Defects and imperfections: Where necessary, the allowable extent of cracks, chips, burrs, and voids should be specified by the user or the MMPA-0100-66 standards should be used. Meaningful specifications for such defects are extremely difficult, and "tight" specifications should be avoided unless they are absolutely critical to the function of the product.

Packaging and keepering specifications should be general and the primary responsibility for the detailed method left to the mill or the distributor-fabricator. The following are examples of adequate but not excessive specifications:

"Unmagnetized magnets shall be packaged in such manner that no more than ____ percent of any lot shall be fractured in shipment."

"Magnetized magnets shall be so handled and packaged that no demagnetization shall occur from spurious contact with any ferromagnetic material during shipment."

"Magnetized magnets shall be fully closed-circuited during or immediately after saturation by means of a suitable soft-steel keeper."

Where specific, special packaging is advantageous to the user to facilitate storage or assembly operations,[6] this packaging should be specified in detail.

Testing: Full details of the test method to be applied by the user should be furnished to the mill or the distributor-fabricator. Where possible, this should include drawings of the test fixture (or a duplicate of the fixture itself), the instrumentation to be used, expected test values, AQL levels, and so on. Ideally, a "standardized," complete product assembly or subassembly should be provided. Penetrant, X-ray, or other required physical testing procedures should be specified as clearly and as minimally as possible. *Mechanical test specifications* (like tensile strength) *should be avoided unless they are absolutely necessary to the application.* Many of these specification details may be either incorporated directly on drawings (where furnished) or covered in separate specification sheets. In some cases, adequate specifications can be provided on the purchase order, without drawings or detailed specification sheets.

Purchasing

Materials, General Comparisons: Table 15-1 lists the relative advantages and disadvantages of all the general types of magnet materials currently available.

Materials, Cost Comparisons: It is virtually impossible to establish absolute costs, since various magnet materials are seldom directly interchangeable (for example, a ferrite magnet cannot be replaced by an Alnico magnet with the same size, shape, and other characteristics). However, one general fact is particularly useful: The cost of all magnet materials, regardless of type, is approximately the same *when compared on the basis of "energy per dollar."*

Within a single *category* of materials, more specific comparisons are possible. For the common grades of Alnicos and ferrites, the following *relative costs* apply:

Cast Alnicos: Alnico 2, .8; Alnico 4, .8; Alnico 5, 1.0[a]; Alnico 5DG, 1.3; Alnico 5-7, 1.6; Alnico 6, 1.1; Alnico 8, 1.3; Alnico 9, 3.0.[b]

Sintered Alnicos: Alnico 2, .8; Alnico 5, 1.0.[a]

Ferrites: Grade 1, 1.0[a]; Grade 5, 1.3.

Magnets, Mill Sources: Table 15-2 lists all of the U.S. mills that currently furnish basic magnets. The particular materials made by each mill are indicated. Table 15-3 shows the foreign ("free-world" only) mills.

Magnets, Distributor-Fabricators: Table 15-4 lists the U.S. distributor-fabricators who supply basic magnets, modifications of basic magnets, and some assemblies. Most distributor-fabricators offer a variety of material types from various mill sources and have appropriate catalogs available.

---

6. As example, on one occasion, the author had reason to "package" small rod magnets in soda straws. On another occasion, small rods were packaged "cartridge-belt" style, using a plastic strip and masking tape.

[a] Indicates comparison base.
[b] Cast Alnico 9 is in very limited production and is therefore available from only a few sources. Furthermore, it is available in only relatively small-size magnets so it has a high unit cost.

Table 15-1
General Comparison of Magnet Materials

Material	Advantages	Disadvantages
Cast Alnicos	Readily available from many sources. Ideal where space requirements indicate long, thin magnet. Moderate tooling costs. High energy-per-unit weight (Alnico 5, 5DG, 5-7, 8, and 9). High energy-per-unit volume (Alnico 5, 5DG, 5-7, 8, and 9). Excellent temperature stability. Economical in small- to moderate-size runs. Most common grade. Can be made in virtually any size and shape, including very large blocks. Available in many grades to fit different applications. Standard grades can be modified readily for special applications.	Cannot be machined except by grinding (must be cast close to final shape). Close tolerances cannot be held in casting process. High raw-material cost. Uses "critical" materials. Must be heat treated in magnetic field to obtain best properties (Alnicos 5 through 9). Some grades can be cast in limited sizes and shapes (Alnico 5DG, 5-7, 8, and 9 only). Low corrosion resistance.
Sintered Alnicos	Better finish and closer tolerance possible than in cast grades. Costs less than cast grades. Mechanically stronger than cast grades. Ideal for large-volume runs of very small magnets. Good energy-per-unit weight. Good energy-per-unit volume. Excellent temperature stability.	Process suitable for only small magnets. Available from only a few sources. Available as standard in only a few grades. Small parts difficult to make in high-performance grades (Alnicos 5 and 8). Not suitable for small runs. Can be made only in limited shapes. Cannot be machined, except by grinding. High raw-material cost. Uses "critical" materials. Moderate corrosion resistance.
Sintered Ferrites	Low part cost in large-quantity runs. Electrical insulator. Very high resistance to demagnetization (external and self). Ideal where space requirements indicate thin, large, arc magnet. Readily available from many sources. Low raw-material cost. Uses no "critical" materials. High corrosion resistance.	High tooling costs. Very brittle with low tensile strength. Cannot be machined, except by grinding with diamond wheels. Low to medium energy-per-unit weight and volume. Close tolerances cannot be held in forming process. Limited in shape and size to simple blocks and rings. Limited grades available. Very high temperature sensitivity.
Bonded Ferrites	Lowest part costs. Can be formed to final part with steel rule or other cutting dies. Can be extruded to basic shape. Flexible. Corrosion proof. Good dimensional control. Low raw-material cost. Good where large magnetic area is required. Uses no "critical" materials. Low tooling cost.	Low energy-per-unit volume and medium energy-per-unit weight. Limited availability in sizes, shapes, and magnetic properties. Basic stock limited to sheets and strips. Poor temeprature stability.
Wrought Alloys (Steels)	Most grades can be formed into final shapes by conventional metal-cutting, forming, and stamping methods. Low tooling cost. Excellent temperature stability. Close tolerances can readily be attained in final forming. Excellent finish. Low part cost for large-quantity runs. High mechanical strength and toughness.	Low energy-per-unit weight and volume. Limited availability of grades. Limited sources. Usually economical in large-volume runs only. Suitable only for small parts. High raw-material cost. Uses "critical" materials. Poor corrosion resistance.
ESD Grades	Final part can be formed with good finish and close dimensional control (± .005) without further machining. Good temperature stability. Can be made economically in complex shapes. Available in a variety of grades to fit different applications. Magnetic properties are a good "compromise" between Alnico 5 and Ferrite 1. Ideal for large-volume runs.	Only one source (General Electric Co.) Medium energy-per-unit weight and volume. High tooling cost. Limited to small parts. Some grades use "critical" material (cobalt). Low mechanical strength.

(continued)

Table 15-1 (continued)

Material	Advantages	Disadvantages
Precious Metal Alloys	Very high energy-per-unit volume (platinum-cobalt). Very high resistance to demagnetization (platinum-cobalt). Excellent temperature stability. Excellent conductor of electricity. High corrosion resistance. High mechanical strength and toughness. Good finish and dimensional control in final-part forming. Low tooling cost.	Very high cost. Very limited availability. Suitable for only small parts. Available in few grade types. Uses precious metals.
Rare Earths	Highest available energy-per-unit volume. Highest available energy-per-unit weight. Highest available resistance to demagnetization. Cerium-cobalt-copper types are machinable with ordinary machine tools. Excellent temperature stability. Very thin magnets can be made.	High cost. Limited availability. Some grades are difficult to machine, except by grinding.

Table 15-2 U.S. Permanent Magnet Mills and Material Types Produced*

Mill & Location	Cast Alnico	Sintered Alnico	Sintered Ferrites	Wrought Alloys	Bonded FE/RE**	Rare Earths
Arnold Engineering Co., 300 N. West St., Marengo, IL 60152	Y	Y	Y	N	Y	Y
Ceramic Magnets, Inc., 16 Law St., Fairfield, NJ 07004	N	N	Y	N	N	N
Crucible Magnetics, 101 Magnet Dr., Elizabethtown, KY 42701	N	N	Y	N	N	Y
Electrodyne Co. Inc., 4192 Taylor Rd., Cincinnati, OH 45209	N	N	N	N	Y	N
Electron Energy Corp., 924 Links Ave., Landisville, PA 17538	N	N	N	N	N	Y
Flexmag Industries, Inc., 4480 Lake Forest Dr., Cincinnati, OH 45242	N	N	N	N	Y	N
GenCorp, 605 West Eichel Ave., Evansville, IN 47710	N	N	N	N	Y	N
General Magnetic Co., 5252 Investment Dr., Dallas, TX 75236	N	N	Y	N	N	N
Hitachi Magnetics Corp., 7800 Neff Rd., Edmore, MI 48829	Y	Y	Y	N	Y	Y
Hoskins Mfg. Co., 10776 Hall Rd., Hamburg, MI 48139	N	N	N	Y	N	N
Magnequench/Delco-Remy Div. General Motors 6435 S. Scatterfield Rd., Anderson, IN 46013	N	N	N	N	N	Y
Magnetic Specialty, Inc., 707 Gilman St., Marietta, OH 45750	N	N	N	N	Y	N
Magno-Ceram Co., 2612 S. Clinton Ave., South Plainfield, NJ 07080	N	N	Y	N	N	N
National Magnetics Group, Inc., 210 Win Dr., Bethlehem, PA 18107	N	N	Y	N	N	N
Permanent Magnet Co., 44th & Bragdon Sts., Indianapolis, IN 46226	Y	Y	Y	Y	N	Y
Poly-Mag, Inc., 685 Station Rd., Bellport, NY	N	N	N	N	Y	N
RJF International Corp., 11310 Rockfield Ct., Cincinnati, OH 45241	N	N	N	N	Y	N
RECOMA, Inc., 2 Stewart Place, Fairfield, NJ 07006	N	N	Y	N	Y	Y
Stackpole Corp., 700 Elk Ave., Kane, PA 16735	N	N	Y	N	Y	Y
TDK Corp. of America, 1600 Feehanville Dr., Mount Prospect, IL 60056	N	N	Y	N	Y	Y
Tengam Engineering, Inc., 545 Washington St., Otsego, MI 49078	N	N	N	N	Y	N
Thomas & Skinner, 1120 E. 23rd St., Indianapolis, IN 46205	Y	Y	N	N	N	N
3M Electrical Specialties, 6801 Riverplace Rd. Bldg. 130-3N-56, 3M Austin Center, Austin, TX 78726-2963	N	N	N	N	Y	N
UGIMAG, Inc., 405 Elms St., Valparaiso, IN 46383	Y	N	Y	N	Y	Y

* No known U.S. Mill currently produces Precious Metal (Pt-Co), ESD or Cerium-Cobalt-Copper materials.

** FE - Barium or Strontium Ferrite: RE - Samarium-Cobalt or Neodymium-Iron-Boron

Not all mills produce both Plastic Bonded Ferrites and Plastic Bonded Rare Earths. Plastic Bonded Alnicos offered in U.S. only by UGIMag. Most mills that offer plastic bonded materials can supply Plastic Bonded Alnicos on special order. Plastic bonding methods may be calendered, injection molded or compression molded rubber or thermoplastic compounds depending on producer.

Table 15-3
Foreign Producers of Permanent Magnets

This table list the 23 mills in 6 countries from which information could be obtained
See Data Sheets, Appendix 1 for Magnetic and Physical Properties

England:——— Cookson Group, PLC (Arlec/Magnet Applications, Ltd.)
　　　　　　　130 Wood St. London EC2V 6EQ
　　　　　　Magnet Applications, Ltd. (Cookson/Arlec)
　　　　　　　Northbridge Rd. Berkhamster Herts HP4 1EH
　　　　　　SG Magnets, Ltd. Tesla House 85 Ferry Lane
　　　　　　　Rainham Essex RM 13 9YHF

France:——— Aimants UGIMag, SA BPNo2, 38830 St. Pierre-D'Allevard
　　　　　　Arlec, Pau (see Magnet Applications, Ltd.)

Germany:——— Baermann, Max GMBH Postbox 100158, Wulfshoff
　　　　　　　D-5060 Bergisch Gladbach 1
　　　　　　Krupp Widia, GMBH, Postfach 10216
　　　　　　　D4300, Essen
　　　　　　Magnetfabrik,-Bonn, Postfach 2005
　　　　　　　Dortheenstrasse 215, Bonn 1
　　　　　　Magnetfabrik Schramberg, D-7230 Schramberg-Sulgen
　　　　　　Thyssen Magnettechnik GMBH, Postfach 44271
　　　　　　　Osterkirchstrasse 177, D-44271, Dorthmund
　　　　　　Vacuumschmeltz GMBH, Max Planck Strasse 15
　　　　　　　Guner Weg 37, D-6450 Hanau

Japan:——— Hitachi Metals Ltd, Marunouchi 2 Chrome Chi ku Tokyo
　　　　　　TDK Corporation, 13-1 Nihonbashi 1 Chrome, Chuo, Tokyo

Netherlands:——— Philips Components
　　　　　　　P.O.Box 218, 5600 MD, Eindhoven

Switzerland:——— UGIMag Recoma, AG, Industriestrasse 297 CH- 5242, Lupfig

Brazil:——— Eriez, LTDA Supergauss
China:——— Baotou Rare Earth Institute
　　　　　Changshu Magnet Material Factory
　　　　　China Neodymium Magnets Group
　　　　　Chuanxi Machinery Plant
　　　　　Chung Cheung Trading Co.
　　　　　General Rare Earth Factory
　　　　　Hongguan Magnet Steel Plant
　　　　　Ningbo Konit Industry
　　　　　Norinco
　　　　　San Huan Corporation
　　　　　Shanghai Yue Long Chemical Plant
　　　　　Sino-Asia
　　　　　West China Trading Co.
　　　　　Yokohama Kagaku Kinzoku Co.
Czech Republic:——— Paramet
England (UK):——— Boxmag
　　　　　Cermag
　　　　　Johnson Matthey Co.
　　　　　Magnet Developement Ltd.
　　　　　Magnet Material Group (OUTOK)
　　　　　Neosid Ltd.
　　　　　Rare Earth Products
　　　　　Roberts Magnetics, Ltd.
　　　　　Telcon Metals, Ltd.
France:——— Francosid
　　　　　Giffey-Pretre
　　　　　Rhohe Poulene
Finland:——— Outokumpu
Germany:——— BT Magnet Technology GMBH
　　　　　Bec Brezniker
　　　　　Bosch GMBH
　　　　　Frank und Schulte
　　　　　Goldschmidt AG
　　　　　Hupro Magnet
　　　　　I BS Magnet
　　　　　Ker. Werke Hermsd. (Tridelta)
　　　　　Magnetkeramik Lueneburg
　　　　　Rhein Magnet
　　　　　Spoerle Electronic
　　　　　Tridelta

Hong Kong:——— Atlas Corporation
　　　　　Better Electric Co. Ltd.
　　　　　Eastern Mall Ltd.
　　　　　Millennium Magnets Ltd.
　　　　　Trustwell Industrial Co.
　　　　　WAT LAP, Ltd.
　　　　　Wide Ocean Enterprises
Hungary:——— Electrimpex
India:——— G.P.Electronics
　　　　　Magneti, Ltd.
　　　　　PML
Israel:——— Magma Magnets Mfg. (M & E GMBH)
Japan:——— Cosmo Tokoyo Magnet Engineering
　　　　　Dia Ichi Chemical
　　　　　Daido Specialty Steel
　　　　　Diado Tokushko
　　　　　Daini Seikosha
　　　　　Dia Rare Earth Co.
　　　　　Fuji Denke
　　　　　Fuji Electrochemical
　　　　　Hokko Denshi
　　　　　Japan Industrial Material Co.
　　　　　Japan Specialty Steel
　　　　　Kanafuchi Chemicals
　　　　　Kanafuchi Kagaku Kogyo
　　　　　MG Company Ltd.
　　　　　Mag X Company
　　　　　Matsushita Electr.Components
　　　　　Meito Company
　　　　　Mitsubishi Metals Corp.
　　　　　Namiki Precision
　　　　　Nippon Ferrite Ind. Co.
　　　　　Nippon Miniature Bearing
　　　　　Nippon Steel
　　　　　Polymagnet
　　　　　Seiko Electronic
　　　　　Seiko Epson
　　　　　Shin-Etsu Chemical
　　　　　Showa Denko
　　　　　Sumitomo Special Metals
　　　　　Sumit. Metal Min. (Mitsui, Germ.)
　　　　　Sumitomo Bakelite
　　　　　Sumitomo Metal Mining
　　　　　Taiyo Uden
　　　　　Tohoku Metal Ind. (Tokin)
　　　　　Tokyo Magnet Eng. (TME)
　　　　　Tokyo Denki Kagaku
　　　　　Tokyo Ferrite
　　　　　Tokyo Magnet Chemical
　　　　　Toshiba
Korea:——— Han Young Magnet Co.
　　　　　Korea Ferrite
　　　　　Pacific Metal Tongkook
Mexico:——— Delco Products (Delredo/AB-TDK
Netherlands:——— Bakker Madava
Russia:——— Paramet
Spain:——— Durifer
　　　　　Hansa (ISSA)
　　　　　Imanes Sinterzados
　　　　　Pihar
Sweden:——— Sura Magnets AB
Switzerland:——— Andemars
　　　　　Bohli Magnettechnik AG
　　　　　Comadur SA
　　　　　Von Roll AG
Thailand:——— O.T.G.
Taiwan:——— General Magnetic Ltd.
　　　　　PMI
　　　　　Superrite
　　　　　Superior Prec. Mold Co
　　　　　Taigene
Yugoslavia:——— Electronska Industrija

*For over 2 years, this author has repeatedly corresponded with, researched and even telephoned the Commercial Attaches of every country that might make magnets. To date, these efforts have (disappointingly) produced only the information shown in this table.

Table 15-4
U.S. Distributor-Fabricators of Permanent Magnets

Adams Magnetic Products Co. 2081 N. 15th Ave. Melrose Park IL 60160
    Adams Southeast: P.O.Box 933, Apopka FL
    Adams Southwest: 2714 National Circle, Garland, TX 75041
    Adams West: 155 Mata Way, San Marcos, CA 92069
    Adams Northeast: 34 Industrial Way, Eatontown, NJ 07724
ALL Magnetics, Inc. P.O.Box 6495, Anaheim CA 92670-0495
Armtek, 2955 E. Hillcrest, Suite 104, Thousand Oaks, CA 91362
Armtek, 440 Constance Drive, P.O.Box 2694, Warminster, PA 18974
Bunting Magnetics, Box 468 Newton, KS, 67114-0468
Cookson Magnet Sales, 214 Ramins Bldg. 272 Titus Ave. Warrington, PA 18976-2483
Dexter/Permag, 1050 Morse Ave. Elk Grove Village, IL 60007
    Atlantic Division: 400 Karin La. Hicksville, NY 11801
    Midwest Division: 2960 South Ave. Toledo, OH 43609
    Minnesota Division: 14956 Martin Dr. Eden Praire, MN 55344
    Northeast Division: 10 Fortune Dr. Billercia, MA 01865
    Pacific Division: 0631 Humboldt St. Los Alamitos, CA 90720
    Southeast Division: 6730 Jones Mill Ct. Norcross, GA 30092
    Southwest Division: 1111 Commerce Dr. Richardson, TX 75081
Douglas International, Inc. 312 West State St. Geneva, IL 60134
Dura Magnetics, Inc. 5510 Schultz Drive, Sylania, OH 43560
Hennaco Industrial Enterprises, Inc. 5 Highview Ct. Montville, NJ 07045
Industrial Magnetics, Inc.1240 M-75S, P.O. Drawer 80 Boyne City, MI 49712
Jobmaster, P.O. Drawer 207, 9006 Liberty Rd. Randallstown, MD 21133-0207
Magnet Sales & Manufacturing, Inc. 11248 Playa Court, Culver City, CA 90230
Magnet Technology, 11356 Deerfield Rd. Cincinnati, OH 45242
Magnetool, Inc.505 Elmwood, Troy, MI 48083
Magnum Magnetics, Rt. 3, Box 347, Marietta, OH 45750
Master Magnetics, Inc. 607 Gilbert, Castle Rock, CO 80104
Miami Magnet Co. 6073 N.W. 167 St Miami, FL 33015
Polymag, 685 Station Rd. Bellport, NY 11713
Rochester Magnet Co. 1115 E. Main St. Rochester, NY 14609
Storch Products Co. 11827 Globe Rd. Livonia, MI 48151
TMC Magnetics, Inc. P.O. Box 614, Pine Brook, NJ 07058
U.S. Magnet & Alloy Corp. 85 N.Main St. Yardley, PA 19067

Table 15-5
U.S. Manufactures of Magnetizers, Demagnetizers & Test Equipment

Company	Address
A-L-L Magnetics, Inc.	P.O.Box 6495 Anaheim, CA 92806
Alpha Magnetics, Inc.	1868 National Ave. Bldg. A, Howard CA 94545
Annis, R.B. Co. Inc.	1100 N. Delaware St. Indianapolis, IN 46202
Applied Magnetics Laboratory Inc,	1404 Bare Hills Rd. Baltimore, MD 21209
Bell, F.W. Inc.	6120 Hanging Moss Rd. Orlando, FL 32807
Dowling Miner Magnetics Corp.	P.O.Box 1829, Sonoma, CA 95476
Electro-Matic Products Co.	2235-37 N. Knox Ave. Chicago, IL 60639
Electro-Technic Products, Inc.	4644 N. Ravenswood Ave. Chicago, IL 60640
Industrial Magnetics, Inc.	1240 M-75S, Boyne City, MI 49712
LDJ Electronics, Inc.	2202 Stephenson Hy. Troy, MI 48083
Lake Shore Measurement & Control Technologies	64 E. Walnut St. Westerville, OH 43083
Magnetic Instrumentation, Inc.	8350 East 48th St. Indianapolis, IN 46226
Mag-Netron/ Omnitech, Inc.	Rt.2 Box 66, Hockley TX 77447
Magnos, Inc.	9 Bonazzoli Ave. Unit 17, Hudson, MA 01749
Magnet Sales & Manufacturing	11248 Playa Ct. Culver City, CA. 9023
Magnet Source Master Magnetics, Inc.	607 S. Gilbert St. Castle Rock, CO 80104
Miami Magnet Co.	6073 N.W. 167th St. Miami, FL 33015
Ullman Devices Corp.	P.O.Box 398, Ridgefield CT 06877
Universal Magnetics	5555 Amy School Rd. Howard City, MI 49329
Walker Scientific Inc.	Rockdale St. Worcester, MA 01606

**MMPA STANDARD
No. 0100–90**

STANDARD SPECIFICATIONS FOR PERMANENT MAGNETIC MATERIALS

FOREWORD

This publication represents standard practices in the United States relating to permanent magnet materials. This standard is a revision of "MMPA Standard 0100—Standard Specifications for Permanent Magnet Materials" which was originally published in 1964, plus important information from the latest documents prepared by the International Electrotechnical Commission (IEC) Technical Committee 68.

IEC is the oldest continuously functioning standards organization in the world. In 1906, the IEC was given the responsibility of securing the cooperation of technical societies to consider the question of international electrical standardization. The membership of IEC consists of 41 national committees, one for each country. These committees represent the electrical interests of producers, users, government, educators and professional societies of each country. The MMPA is represented on the United States National Committee of IEC/TC68.

It is hoped that the data in this publication will serve as a guide to governmental and industrial purchasers so that they may be assured of uniform quality manufactured to commercial standards. These standard specifications were developed under the auspices of the Magnetic Materials Producers Association and were voluntarily established by mutual consent of those concerned. These specifications are advisory only and their use or adaptation is entirely voluntary. The users of the specification are wholly responsible for protecting themselves against all liabilities for patent infringement.

TABLE OF CONTENTS

Section I	General Information	3
Section II	Alnico Magnets	6
Section III	Ceramic Magnets	11
Section IV	Rare Earth Magnets	17
Section V	Iron–Chromium–Cobalt Magnets	22
Section A	Permanent Magnet Materials Not Covered in Product Sections (See Section I Scope)	25
Section B	Glossary of Terms	26
Section C	Magnetic Quantities (Symbols, Units and Conversion Factors)	28

COPYRIGHT MAGNETIC MATERIALS PRODUCERS ASSOCIATION

STANDARD SPECIFICATIONS FOR PERMANENT MAGNET MATERIALS

SECTION I

1.0 SCOPE & OBJECTIVE

1.1 Scope: This standard defines magnetic, thermal, physical and mechanical characteristics and properties of commercially available permanent magnet materials as listed in Table 1.

There are a large number of permanent magnet materials in use which are not described in this document. These materials generally fall into one of the following categories:

(a) Older materials that have been largely replaced by new materials.

(b) Materials made by only one company with a specialized and limited use.

(c) Materials evolving from development status to production which at this time are not mature from a commercial viewpoint.

For reference purposes, the principal magnetic properties of the materials in the above categories are listed in Appendix A.

1.2 Objective: The objective of this standard is to establish criteria by which users of permanent magnet materials may be assured of magnets manufactured to present commercial standards.

2.0 DEFINITIONS & TERMS

2.1 Definitions: The following definitions characterize materials covered in this standard:

2.1.1 Permanent Magnet (Magnetically Hard) Material: A permanent magnet material, also designated as a magnetically hard material, has a coercive force generally greater than 120 Oe.

2.1.2 Individual Magnet: The term *individual magnet* denotes a magnet purchased in a size and shape to be ready for direct incorporation into a magnetic circuit.

2.1.3 Bulk Magnet Material: The term *bulk magnet material* designates bar, rod, slab, strip, sheet, etc., from which the purchaser cuts, stamps or forms individual magnets.

2.1.4 Polarity of a Magnetized Magnet: The North Pole of a magnet is that pole which is attracted to the geographic North Pole. Therefore, the North Pole of a magnet will repel the north seeking pole of a magnetic compass.

2.1.5 Demagnetized Magnet: For the purposes of this standard, a magnet shall be considered demagnetized if, when any of its poles is dipped in soft iron powder (of -5, $+10$ mesh), not more than 3 particles of powder adhere to it anywhere upon withdrawal.

2.2 Terms

2.2.1 A glossary of terms commonly used with permanent magnetic materials is given in Appendix B.

3.0 CONDITION

Unless otherwise specified, bulk magnet materials shall be furnished in the non heat-treated condition, as rolled, as forged, or as-cast condition as applicable. Individual magnets shall be furnished in a fully heat-treated and demagnetized condition.

4.0 CLASSIFICATION & DESIGNATION

4.1 Classification: The classification of permanent magnet materials covered by this standard is given in Table 1. Section numbers for the material classes covered in this standard as well as reference to the International Electrochemical Commission (IEC) material code numbers are also given in the table.

TABLE 1
MATERIAL CLASSIFICATION

Material	MMPA Section	IEC Code
Alnico	II	R1
Ceramic	III	S1
Rare Earth	IV	R5
Iron-Chromium-Cobalt	V	R6

4.2 Designation: Permanent magnet materials in this specification will be divided into separate sections by the MMPA Class. Each standard section will address the relevant properties, characteristics and specifications of each class of materials and the established subgrades.

In general, reference will be made to historically recognized subgrade descriptions (such as Alnico 1, 2, etc., or Ceramic 5, 8, etc.) and to a system, referred to as the Brief Designation, that classifies each subgrade by typical normal energy product and typical intrinsic coercive force. In this system, for example, a material having maximum normal energy product of 5.0 megagauss-oersteds (MGO) and an intrinsic coercive force of 2000 oersteds (2.0 kOe) would be assigned a Brief Designation of 5.0/2.0. When similar grades exist the nearest IEC Grade Code Number will also be listed for cross reference.

5.0 MAGNETIC PROPERTIES, THERMAL PROPERTIES & OTHER CHARACTERISTICS

The magnetic, thermal, surface and internal structure, and other physical characteristics are set forth in tables in each section for the different classes of magnetic materials. The figures in these tables are intended to be descriptions of each of the materials. The properties of the materials produced by individual manufacturers may differ somewhat from those shown. For information concerning properties of actual grades produced, refer to individual manufacturer's literature. The properties shown in the tables with each class shall not be used as inspection criteria of either individual magnets or bulk magnet materials.

5.1 Principal Magnetic Properties: Permanent magnet materials are identified by the following principal magnetic properties:

Maximum value of energy product	$(BH)_{max}$	MGO
Residual induction	B_r	gauss
Coercive force	H_c	oersteds
Intrinsic coercive force	H_{ci}	oersteds

The measurements of the principal magnetic properties are made in a closed magnetic circuit permeameter by commonly accepted procedures such as given in IEC Standard Publication 404-5 *"Methods of Measurement of Magnetic Properties of Magnetically Hard* (Permanent Magnet) *Materials"* or the *"MMPA Guidelines on Measuring Unit Properties of Permanent Magnets."* They are accurate only for magnets having a straight magnet axis and produced with a constant cross section along the axis of magnetization. The minimum magnet volume of a sample used to measure these magnetic properties shall be one cubic centimeter and the smallest dimension shall be at least 5mm.

The performance of a permanent magnet circuit is dependent on the dimensions of all components and the properties of the other components of the circuit, as well as the properties of the permanent magnet. It is recommended not to use unit properties of a material as the specification. These are generally recommended to be *only* used on prints or drawings to show a subgrade within a material group. Section 8.0 of the specification describes the proper means of specifying the acceptable properties of a permanent magnet component part.

5.2 Thermal Properties: Predicting magnet performance as a function of the magnet's temperature requires knowledge of the following thermal properties:

Reversible temperature coefficient of the residual induction	TC (B_r)	%/°C
Reversible temperature coefficient of the intrinsic coercive force	TC (H_{ci})	%/°C
Curie temperature	T_c	°C
Maximum service temperature	T_{max}	°C

The values listed for each class of materials for thermal properties are typical values intended as design guidelines only and are not to be used as a basis for acceptance or rejection.

Values for irreversible temperature characteristics are not listed because they depend on the magnet material, geometry and circuit in which the magnet is used.

5.3 Surface and Internal Structure Characteristics: Permanent magnet materials have been developed primarily for their magnetic properties. The magnetic properties of some materials are produced using manufacturing techniques which are not consistent with producing perfect physical specimens. Minor physical imperfections rarely impair the magnetic capabilities of a magnet or compromise its stability or ability to resist demagnetization.

Imperfections commonly found in permanent magnet materials shall be judged acceptable if the following conditions are met:

(1) The magnet meets the magnetic performance criteria agreed upon between the magnet manufacturer and customer.

(2) The imperfections do not create loose particles that would interfere with proper assembly or functioning in the end use device.

Unless otherwise agreed, visual imperfection guidelines listed in the individual material sections apply.

5.4 Other Physical Properties: Typical values for other physical properties important to a magnet user are listed in the tables in the sections for each class of permanent magnet material and are intended to be descriptions of the material, not criteria for acceptance or rejection.

6.0 MECHANICAL CHARACTERISTICS

Most permanent magnet materials lack ductility and are inherently brittle. Such materials should not be utilized as structural components in a circuit. Measurement of properties such as hardness and tensile strength are not feasible on commercial materials with these inherent characteristics. Therefore, specifications of these properties are not acceptable. Measurements of mechanical properties shown in the tables were performed under very carefully controlled laboratory conditions. The values are shown only for reference and comparison to other classes of materials.

7.0 DIMENSIONS AND TOLERANCES

Dimensions and tolerances shall be as specified on the magnet drawing and must be agreed upon between the magnet manufacturer and user before an order is accepted. Normally the magnet user furnishes a drawing to the manufacturer showing all dimensions and tolerances. When no drawing is available from the user, the manufacturer may furnish a drawing to the user for his approval before manufacturing parts. The standard for drawing, drawing notation and tolerancing is that established in ANSI Y14.5. Although individual manufacturers will each have their own capability to hold a given tolerance, standard tolerance tables applying to specific classes of these materials are listed in the individual sections.

8.0 INSPECTION & TESTING

Unless otherwise agreed upon by the manufacturer and the user, all bulk magnet materials and individual lots of identical individual magnets will be inspected for all specified characteristics by the use of a statistical sampling plan derived from MIL-STD-105D. The specific recommended sampling plans from this MIL-Standard are indicated in the sections covering the individual classes of materials.

8.1 Performance Testing Approach—Magnetic Characteristics: The principal characteristics—B_r, H_c, H_{ci} and $(BH)_{max}$—of a magnetic material are used to identify a specific subgrade within a material class. Generally, individual manufacturers can hold unit magnetic property tolerances of $\pm 5\%$ for residual flux density, B_r and $\pm 8\%$ for coercive force, H_c. The range for the energy product, (BH) max is $\pm 10\%$. Intrinsic coercive force, H_{ci}, is generally specified as minimum value only.

The size and/or shape of the actual magnet to be produced may cause magnets to have properties considerably different from these characteristics. Therefore, use of these characteristics in specifying acceptable properties for a given magnet shape is not recommended. The recommended means is to specify the minimum magnetic lines of flux at one or more load lines on the major or minor hysteresis loop. A magnet producer can assist in magnetic circuit analysis which will determine this actual operating flux. From the analysis, a method of test shall be chosen which will cause the magnet being tested to operate at levels which duplicate the performance in the final circuit. The magnet user and supplier shall agree upon a reference magnet to be used to calibrate the test equipment. The acceptance limits shall be agreeable to both manufacturer and user. The acceptability of a magnet shall be judged solely by a comparison with the reference magnet tested in an identical manner.

Functional test methods typical for the different magnet materials and applications are described in the MMPA publication *Testing and Measurement of Permanent Magnets.*

8.2 Visual Characteristics: The recommended procedure for establishing acceptable levels for visual characteristics is for manufacturer and user to prepare a mutually agreed upon set of go/no-go standards or sample boards. In the absence of such a set of standards or other descriptions of acceptable criteria, the guidelines set forth in each individual section apply.

SECTION II
ALNICO MAGNETS

1.0 CHEMICAL COMPOSITION

Alnico alloys basically consist of aluminum, nickel, cobalt, copper, iron and titanium. In some grades cobalt and/or titanium are omitted. Also these alloys may contain additions of silicon, columbium, zirconium or other elements which enhance heat treatment response of one of the magnetic characteristics.

2.0 MANUFACTURING METHODS

The Alnico alloys are formed by casting or powder metallurgical processes. The magnetic performance of most grades can be increased in a preferred direction by applying a magnetic field during heat treatment thus producing magnetic anisotropy. These alloy systems are hard and brittle and do not lend themselves to conventional machining. The best properties of cast Alnico magnets are achieved with columnar or single crystal structure with the direction of magnetization parallel to the columnar grain axis.

3.0 MAGNETIC PROPERTIES

Typical magnetic properties and chemical compositions of the various commercial grades of Alnico are given in Table II-1.

4.0 DIMENSIONS AND TOLERANCES

Allowable tolerances for cast and sintered Alnico are given in Tables II-2 and II-3.

5.0 MECHANICAL CHARACTERISTICS

The following general specifications are for mechanical characteristics and visual imperfections.

5.1 Surface Conditions

5.1.1 All magnet surfaces shall be free of foreign materials which would tend to hold or collect extraneous particles on the magnet surface in the unmagnetized condition.

TABLE II-1
TYPICAL MAGNETIC PROPERTIES AND CHEMICAL COMPOSITION OF ALNICO MATERIALS

MMPA Brief Designation	Original MMPA Class	IEC Code Reference	Chemical Composition*					Magnetic Properties (nominal)			
			Al	Ni	Co	Cu	Ti	Max. Energy Product $(BH)_{max}$ (MGO)	Residual Induction B_r (kilogauss)	Coercive Force H_c (oersteds)	Intrinsic Coercive Force H_{ci} (oersteds)
ISOTROPIC CAST ALNICO											
1.4/0.48	Alnico 1	R1-0-1	12	21	5	3	—	1.4	7.2	470	480
1.7/0.58	Alnico 2	R1-0-4	10	19	13	3	—	1.7	7.5	560	580
1.35/0.50	Alnico 3	R1-0-2	12	25	—	3	—	1.35	7.0	480	500
ANISOTROPIC CAST ALNICO											
5.5/0.64	Alnico 5	R1-1-1	8	14	24	3	—	5.5	12.8	640	640
6.5/0.67	Alnico 5DG	R1-1-2	8	14	24	3	—	6.5	13.3	670	670
7.5/0.74	Alnico 5-7	R1-1-3	8	14	24	3	—	7.5	13.5	740	740
3.9/0.80	Alnico 6	R1-1-4	8	16	24	3	1	3.9	10.5	780	800
5.3/1.9	Alnico 8	R1-1-5	7	15	35	4	5	5.3	8.2	1650	1860
5.0/2.2	Alnico 8HC	R1-1-7	8	14	38	3	8	5.0	7.2	1900	2170
9.0/1.5	Alnico 9	R1-1-6	7	15	35	4	5	9.0	10.6	1500	1500
ISOTROPIC SINTERED ALNICO											
1.5/0.57	Alnico 2	R1-0-4	10	19	13	3	—	1.5	7.1	550	570
ANISOTROPIC SINTERED ALNICO											
3.9/0.63	Alnico 5	R1-1-20	8	14	24	3	—	3.9	10.9	620	630
2.9/0.82	Alnico 6	R1-1-21	8	15	24	3	1	2.9	9.4	790	820
4.0/1.7	Alnico 8	R1-1-22	7	15	35	4	5	4.0	7.4	1500	1690
4.5/2.0	Alnico 8HC	R1-1-23	7	14	38	3	8	4.5	6.7	1800	2020

✱Note: Balance iron for all alloys

TABLE II-2
TOLERANCES, CAST ALNICO MAGNETS

	Dimension (inches)	Tolerances (inches)
Size:		
Unfinished surfaces: (including draft)	0–1	±.016
	1–2	±.031
	2–3	±.031
	3–4	±.047
	4–5	±.047
	5–6	±.062
	6–7	±.062
	7–8	±.078
	8–9	±.078
	9–10	±.094
	10–12	±.094
Finished surfaces: any plane ground dimension		±.005
Center or centerless ground: 0 to 1.5 OD		±.004
Over 1.5 OD		±.008
Parallelism:		
Finished surfaces		$1/2$ total tolerance between surfaces
Angles, including squareness:		
Between two unfinished surfaces		± $1\frac{1}{2}°$
Between one finished & one unfinished surface		± $1\frac{1}{2}°$
Between two finished surfaces		±.005 from true angle as measured on the shorter of the two surfaces in question or ± $1/2°$, whichever is greater
Concentricity between inside and outside surfaces:		
Unfinished surfaces		
Hole diameter > its length		.032 TIR
Hole diameter < its length		1.5 x Total OD TOL. TIR
Finished surfaces		.007 TIR
Surface roughness:		
Unfinished surfaces		No surface roughness specification
Finished surfaces		63 microinches over at least 95% of surface

TABLE II-3
TOLERANCES, SINTERED ALNICO MAGNETS

	Dimension (inches)	Tolerances
Size:		
Unfinished surfaces: (including draft)	0 to .125	±.005 inch
	over .125 to .625	±.010 inch
	over .625 to 1.250	±.015 inch
Finished surfaces:		
Plane ground	any	±.005 inch
Center ground or centerless ground	0 to 1.5	±.002 inch
	over 1.5	±.005 inch
Parallelism:		
Finished parallel surfaces		½ total tolerance between surfaces
Angles including squareness:		
Between two unfinished surfaces		± 1°*
Between one finished, one unfinished surface		± 1°*
Between two finished surfaces:		±.005 from true angle as measured on the shorter of the two surfaces in question or ± ½°, whichever is greater
Concentricity: (between inside and outside surfaces):		
Unfinished surfaces:	0 to .5 OD	.005 inch TIR
	over .5 to 1 OD	.010 inch TIR
	over 1 to 1.5 OD	.015 inch TIR
Finished surfaces:	any	.003 inch TIR
Surface Roughness:		
Unfinished surface		No surface roughness specification
Finished surface:		63 microinches over at least 95% of the ground surface

*Tolerances may be greater for special shapes

5.2 Chips and Burrs

5.2.1 Magnets shall be free of loose chips and burrs. They shall be free of imperfections which will result in loose chips or particles under normal conditions of handling, shipping, assembly and service.

5.2.2 A chipped edge or surface shall be acceptable if no more than 10 percent of the surface is removed, provided no loose particles remain and further provided the magnet under examination meets the agreed upon magnetic specification.

5.3 Other Physical Imperfections

5.3.1 Imperfections such as cracks, porosity, voids, cold flow, shrinkage, pipe and others, all of the type commonly found in cast or sintered Alnico magnets, shall be judged acceptable if the following conditions are met:

5.3.1.1 The magnet meets the minimum magnetic performance criteria agreed upon.

5.3.1.2 The imperfections do not create loose particles or other conditions which will interfere with proper functioning of the end use device.

5.3.1.3 These visual imperfections do not extend more than 50% through any cross-section. However, this does not apply to the columnar materials (Alnico 5-7 and Alnico 9) which are particularly crack-prone due to their columnar grain. Magnets made of these materials shall be judged acceptable if they maintain their physical integrity satisfactorily for the application.

5.4 Other Conditions

5.4.1 Inspection methods such as the use of penetrants, microscopic inspection, magnetic particle analysis, spin tests, ultrasonics, or x-ray shall not be acceptable methods for judging the quality of cast or sintered Alnico magnets except as provided in 5.4.2 below.

5.4.2 In cases where the magnet is expected to withstand abnormal conditions or stresses, such conditions must be previously specified and a mutually acceptable service test devised to assure that the magnet shall not fail under the specified service conditions. Such tests should duplicate service conditions with appropriate safety factors.

6.0 PHYSICAL PROPERTIES

Typical physical properties for Alnico magnets are given in Table II-4.

7.0 THERMAL PROPERTIES

Typical thermal properties for Alnico magnets are listed in Table II-5.

8.0 INSPECTION SAMPLING PLANS

Unless otherwise specified, the sampling plan to be used by the manufacturer and the receiving inspection department of the user shall be in accordance with MIL-STD-105D. All attributes will be inspected using a single sampling plan with the Inspection Level and AQL as specified in Tables II-6 and II-7 which show sample sizes and rejection numbers for given lot sizes.

TABLE II-4
PHYSICAL PROPERTIES OF ALNICO MATERIALS

Brief Designation	Original MMPA Class	IEC Code Reference	Density g/cm³	Density lbs/in³	Tensile Strength psi	Transverse Modulus of Rupture psi	Hardness (Rockwell C)	Coefficient of Thermal Expansion Inches x 10⁻⁶ per °C	Electrical Resistivity Ohm-cm x 10⁻⁶ (at 20°C)
ISOTROPIC CAST ALNICO									
1.4/0.48	Alnico 1	R1-0-1	6.9	.249	4,000	14,000	45	12.6	75
1.7/0.58	Alnico 2	R1-0-4	7.1	.256	3,000	7,000	45	12.4	65
1.35/0.50	Alnico 3	R1-0-2	6.9	.249	12,000	23,000	45	13.0	60
ANISOTROPIC CAST ALNICO									
5.5/0.64	Alnico 5	R1-1-1	7.3	.264	5,400	10,500	50	11.4	47
6.5/0.67	Alnico 5DG	R1-1-2	7.3	.264	5,200	9,000	50	11.4	47
7.5/0.74	Alnico 5-7	R1-1-3	7.3	.264	5,000	8,000	50	11.4	47
3.9/0.80	Alnico 6	R1-1-4	7.3	.265	23,000	45,000	50	11.4	50
5.3/1.9	Alnico 8	R1-1-5	7.3	.262	10,000	30,000	55	11.0	53
5.0/2.2	Alnico 8HC	R1-1-7	7.3	.262	10,000	30,000	55	11.0	54
9.0/1.5	Alnico 9	R1-1-6	7.3	.262	7,000	8,000	55	11.0	53
ISOTROPIC SINTERED ALNICO									
1.5/0.57	Alnico 2	R1-0-4	6.8	.246	65,000	70,000	45	12.4	68
ANTISOTROPIC SINTERED ALNICO									
3.9/0.63	Alnico 5	R1-1-1	6.9	.250	50,000	55,000	45	11.3	50
2.9/0.82	Alnico 6	R1-1-4	6.9	.250	55,000	100,000	45	11.4	54
4.0/1.7	Alnico 8	R1-1-5	7.0	.252	50,000	55,000	45	11.0	54
4.5/2.0	Alnico 8HC	R1-1-7	7.0	.252		55,000	45	11.0	54

NOTE: Alnico permanent magnet materials lack ductility, and are inherently extremely brittle. They should not be designed for use as structural components. Measurement of properties such as hardness and tensile strength is not appropriate nor feasible on commercial materials but values are shown above for comparison. This data, determined experimentally under controlled laboratory conditions, is a composite of information available from industry and research sources.

TABLE II-5
THERMAL PROPERTIES OF ALNICO MATERIALS

Brief Designation	Original MMPA Class	IEC Code Reference	Reversible Temperature Coefficient % Change per °C			Curie Temperature		Max. Service Temperature	
			Near B_r	Near Max. Energy Prod.	Near H_c	°C	°F	°C	°F
1.5/0.57	Alnico 2	R1-0-4	−0.03	−0.02	−0.02	810	1490	450	840
5.5/0.64	Alnico 5	R1-1-1	−0.02	−0.015	+0.01	860	1580	525	975
3.9/0.80	Alnico 6	R1-1-4	−0.02	−0.015	+0.03	860	1580	525	975
5.3/1.9	Alnico 8	R1-1-5	−0.025	−0.01	+0.01	860	1580	550	1020
5.0/2.2	Alnico 8HC	R1-1-7	−0.025	−0.01	+0.01	860	1580	550	1020
9.0/1.5	Alnico 9	R1-1-6	−0.025	−0.01	+0.01	860	1580	550	1020

NOTE: The above data is a composite of information available from industry and research sources.

TABLE II-6
SINGLE SAMPLING, NORMAL INSPECTION

Lot Size	Level II	AQL 1.0	Level I	AQL 4.0	Level S-4	AQL 4.0
	Sample Size	Rejection Number	Sample Size	Rejection Number	Sample Size	Rejection Number
2 – 8	2	1	2	1	2	1
9 – 15	3	1	2	1	2	1
16 – 25	5	1	3	1	3	1
26 – 50	8	1	5	1	5	1
51 – 90	13	1	5	1	5	1
91 – 150	20	1	8	2	8	2
151 – 280	32	2	13	2	13	2
281 – 500	50	2	20	3	13	2
501 – 1,200	80	3	32	4	20	3
1,201 – 3,200	125	4	50	6	32	4
3,201 – 10,000	200	6	80	8	32	4
10,001 – 35,000	315	8	125	11	50	6
35,001 – 150,000	500	11	200	15	80	8
150,001 – 500,000	800	15	315	22	80	8
500,001 & over	1,250	22	500	22	125	11

TABLE II-7
ACCEPTABLE QUALITY LEVELS (AQL) FOR ALNICO MAGNETS

Characteristic	Inspection Level	AQL
Individual Magnetic parameters	II	1.0
Individual dimensional and appearance parameters:		
Categorized as major	I	4.0
Categorized as minor	S-4	4.0
Visual parameters	S-4	4.0

SECTION III

CERAMIC MAGNETS

1.0 CHEMICAL COMPOSITION

The general formula $MO \cdot 6 Fe_2O_3$ describes the chemical composition of ferrite (ceramic) permanent magnets, where M generally represents barium or strontium or any combination of the two.

2.0 MANUFACTURING METHOD

Ceramic magnets are generally formed by a compression or extrusion molding technique which is then followed by sintering. Finish grinding or shaping, when necessary for better control of dimensions, is normally done by using diamond grinding wheels.

The material to be molded can be in either a dry powder or wet slurry form. Magnetic performance can be increased in a preferred direction by applying a magnetic field in that direction during the molding process.

3.0 MAGNETIC PROPERTIES

The magnetic properties of the various commercial grades of ceramic permanent magnet materials are given in Table III-1. This data is compiled from information submitted by ceramic magnet manufacturers. Characteristics of each grade obtained from individual manufacturers may vary from the standard listing.

4.0 OTHER CHARACTERISTICS

4.1 Mechanical Characteristics: Ceramic magnets are used for their magnetic capability, not for their mechanical properties. It is recommended that they not be used for structural purposes since they are low in tensile and flexural strength.

4.2 Visual Characteristics: Imperfections such as cracks, porosity, voids, surface finish, etc., all of the type commonly found in sintered ceramic magnets, shall not constitute reason for rejection. They shall be judged acceptable if the following conditions are met:

1. The magnet meets the minimum magnetic performance criteria agreed upon.
2. The imperfections do not create loose particles or other conditions which will interfere with proper assembly or mechanical functioning of the end use device.

Mutually agreed upon acceptance criteria in the form of written descriptions, pictorial drawings or actual part sample boards can be especially useful for judging visual acceptability. In the absence of an agreed upon standard, the following visual inspection guidelines apply:

1. Magnets shall be free from loose chips and surface residue which will interfere with proper assembly.

TABLE III-1
TYPICAL MAGNETIC PROPERTIES AND CHEMICAL COMPOSITION OF CERAMIC MAGNET MATERIALS

MMPA Brief Designation	Original MMPA Class	IEC Code Reference	Chemical Composition	Magnetic Properties (nominal)			
				Max. Energy Product $(BH)_{max}$ (MGO)	Residual Induction B_r (gauss)	Coercive Force H_c (oersteds)	Intrinsic Coercive Force H_{ci} (oersteds)
1.0/3.3	Ceramic 1	S1-0-1	$MO \cdot 6Fe_2O_3$	1.05	2300	1860	3250
3.4/2.5	Ceramic 5	S1-1-6	$MO \cdot 6Fe_2O_3$	3.40	3800	2400	2500
2.7/4.0	Ceramic 7	S1-1-2	$MO \cdot 6Fe_2O_3$	2.75	3400	3250	4000
3.5/3.1	Ceramic 8	S1-1-5	$MO \cdot 6Fe_2O_3$	3.50	3850	2950	3050
3.4/3.9	—	—	$MO \cdot 6Fe_2O_3$	3.40	3800	3400	3900
4.0/2.9	—	—	$MO \cdot 6Fe_2O_3$	4.00	4100	2800	2900

(M represents Barium, Strontium or combination of the two)

NOTE FOR ALL MATERIALS:
Recoil Permeability Range—1.05 to 1.2
Recommended Magnetizing Force—10,000 oersted minimum

2. Chips shall be acceptable if no more than 5% of any surface identified as a magnetic pole surface is removed.
3. Cracks shall be acceptable provided they do not extend across more than 50% of any surface identified as a magnetic pole surface.

In cases where the magnet is expected to withstand abnormal conditions such as chemical corrosion, thermal shock or mechanical stresses, such conditions must be previously specified. A mutually acceptable service test should be devised to evaluate the acceptability of the magnets. Such tests should duplicate service conditions with appropriate safety factors.

Inspection methods such as the use of penetrants, magnetic particle analysis, ultrasonics, or x-ray shall not be acceptable methods for judging quality of sintered ceramic magnets.

4.3 Dimensions and Tolerances: Recommended tolerances for ceramic magnets are given in Table III-2. Functional gaging and dimensioning of ground arc shape segments is given in Table III-3.

5.0 PHYSICAL PROPERTIES

Typical physical properties for ceramic magnets are listed in Table III-4.

6.0 THERMAL PROPERTIES

Typical parameters that relate to temperature changes for ceramic magnets are listed in Table III-5.

7.0 INSPECTION SAMPLING PLAN

The standard means of inspection for magnetic, visual and dimensional characteristics for shipment by a manufacturer or receiving inspection by a user will be in accordance with MIL-STD-105D General Inspection Level I. The sampling plan will be in accordance with the MIL-STD Table III-A—Double sampling plans for normal inspection. The sampling plan shown in Table III-6 has been compiled from MIL-STD-105D to reflect this inspection criteria.

The specific Acceptable Quality Level (AQL) used for judging acceptability of a lot of ceramic magnets for various magnet characteristics is given in Table III-7. Unless otherwise agreed, the acceptance criteria will be in accordance with this section.

TABLE III-2
TOLERANCES, SINTERED CERAMIC MAGNETS
(See Table III-3 for arc segment dimensions)

	Dimensions (Inches)	Tolerance
Size:		
Unfinished surfaces:		
Dimensions perpendicular to pressing: (die formed)	any	± 2% or ± .025 inch, whichever is greater
Dimensions parallel to pressing (1): (punch formed)	0 to 0.4 0.4 to 0.8 over 0.8	± .016 inch ± .025 inch ± 3%
Cut dimensions:	any	± 3% or ± .025 inch, whichever is greater
Finished surfaces:		
Plane ground:	any	± .005 inch
Center or centerless ground:	0 to 1.0 over 1.0	± .003 inch ± .005 inch
Squareness:		
Between two unfinished surfaces:		90° ± 1°
Between one finished, one unfinished surface:		90° ± 1°
Between two finished surfaces:		90° ± 30′
Parallelism:		
Finished parallel surfaces:		½ total tolerance between surfaces
Surface Roughness:		
Unfinished surface:		Within standard dimensional tolerance
Finished surface:		125 microinches over at least 80% of the ground surface
Warp:		
Maximum allowable warp:		.011 inches per inch
Rings and Rounds:		
Out of round:		Within standard dimensional tolerances
Wall thickness:		± 1½% of nominal wall thickness or ± .010 inch whichever is greater.

(1) Wet compacted ceramic magnets are ground on the magnetic pole surfaces in most cases.

TABLE III-3
FUNCTIONAL GAGING AND DIMENSIONING OF GROUND ARC SEGMENTS
(Inches)

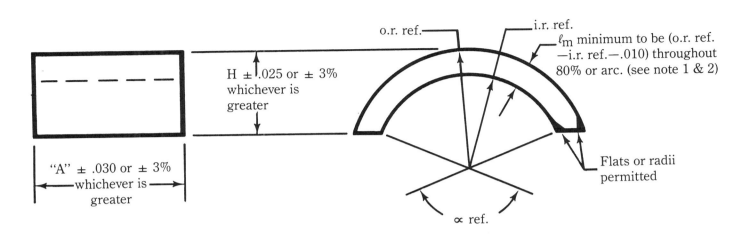

Gage Note:
Part must pass through a gage having an outside radius of o.r. ref. + .006 and an inside radius of i.r. ref. − .006. Minimum axial length of gage to be "A".

Notes:
1) When dimension "A" exceeds two inches, subtract an additional .003 from ℓ_m minimum for each additional one half inch.

2) Remaining 20% of arc to be no less than 0.9 x ℓ_m minimum.

TABLE III-4
PHYSICAL PROPERTIES OF CERAMIC MAGNETS

Property	Typical Value*
Density	4.9 g/cm³
Coefficient of thermal expansion (25°C to 450°C)	
—Perpendicular to orientation	10 x 10⁻⁶ cm/cm/°C
—Parallel to orientation	14 x 10⁻⁶ cm/cm/°C
Thermal conductivity	.007 cal/cm-sec°C
Electrical resistivity	10⁶ ohm-cm
Porosity	5%
Modulus of elasticity	2.6 x 10⁷ psi
Poisson ratio	0.28
Compressive strength	130,000 psi
Tensile strength	5000 psi
Flexural strength	9000 psi
Hardness	7 (Mohs)

*NOTE: The above data is a composite of information available from industry and research sources.

TABLE III-5
THERMAL PROPERTIES OF CERAMIC MAGNETS

Property	Typical Value
Reversible temperature coefficient of residual induction	−0.2%/°C
Reversible temperature coefficient of intrinsic coercive force	0.2 to 0.5%/°C
Curie temperature	450°C
Maximum service temperature* (Without metallurgical change)	800°C

*NOTE: Temperatures greater than 450°C will require remagnetization.

TABLE III-6
DOUBLE SAMPLING PLAN FOR NORMAL INSPECTION LEVEL I
FROM MIL-STD-105D

Lot Size	Sample	Sample Size	Cumulative Sample Size	AQL 1.5 Ac	AQL 1.5 Re	AQL 4.0 Ac	AQL 4.0 Re
Less than 501	*	—	—	—	—	—	—
		—	—	—	—	—	—
501 to 1200	First	20	20	0	2	1	4
	Second	20	40	1	2	4	5
1201 to 3200	First	32	32	0	3	2	5
	Second	32	64	3	4	6	7
3201 to 10,000	First	50	50	1	4	3	7
	Second	50	100	4	5	8	9
10,001 to 35,000	First	80	80	2	5	5	9
	Second	80	160	6	7	12	13
35,001 to 150,000	First	125	125	3	7	7	11
	Second	125	250	8	9	18	19
150,001 to 500,000	First	200	200	5	9	11	16
	Second	200	400	12	13	26	27
500,001 and over	First	315	315	7	11	11	16
	Second	315	630	18	19	26	27

*For lot sizes less than 501 pieces refer to MIL-STD-105D.

AQL — Acceptable Quality Level
Ac — Acceptance Number
Re — Rejection Number

The number of sample units inspected shall be equal to the first sample size given by the plan. If the number of defectives found in the first sample is equal to or less than the first acceptance number, the lot or batch shall be considered acceptable. If the number of defectives found in the first sample is equal to or greater than the first rejection number, the lot or batch shall be rejected. If the number of defectives found in the first sample is between the first acceptance and rejection numbers, a second sample of the size given by the plan shall be inspected. The number of defectives found in the first and second samples shall be accumulated. If the cumulative number of defectives is equal to or less than the second acceptance number, the lot or batch shall be considered acceptable. If the cumulative number of defectives is equal to or greater than the second rejection number, the lot or batch shall be rejected.

TABLE III-7
ACCEPTABLE QUALITY LEVELS (AQL) FOR CERAMIC MAGNETS

Characteristics	Inspection Level	AQL
Individual Magnetic parameters	I	1.5
Individual dimensional parameters:		
Categorized as major	I	1.5
Categorized as minor	I	4.0
Visual parameters	I	4.0

SECTION IV

RARE EARTH MAGNETS

1.0 CHEMICAL COMPOSITION

Rare earth magnet materials currently fall into two families of materials. They are rare-earth cobalt 5 and the rare-earth 2 transition metal 17 group.

1.1 1-5 Alloys (Rare-Earth Cobalt 5): These alloys are usually binary or ternary alloys with the approximate atomic ratio of one rare earth atom to five cobalt atoms. The rare earth element is most commonly samarium but can also be other light rare earth such as, but not limited to, praseodymium, cerium, neodymium or a combination, or a mixture known as misch metal. Heavy rare earths such as gadolinium, dysprosium and erbium can substitute for the light rare earth elements to give the magnetic material a lower temperature coefficient of remanence. The rare earth elements typically are 34-39 weight percent of the alloy.

1.2 2-17 Alloys (Rare-Earth 2 Transition Element 17): These alloys are an age hardening type with a composition ratio of 2 rare earth atoms to 13-17 atoms of transition metals. The rare earth atoms can be any of those found in the 1-5 alloys. The transition metal (TM) content is a cobalt rich combination of cobalt, iron and copper. Small amounts of zirconium, hafnium or other elements are added to enhance the heat treatment response. The rare earth content of 2-17 materials is typically 23-28 weight percent of the alloy.

1.3 Rare Earth Iron Alloys: These alloys have a composition of two rare earth atoms to 14 iron atoms with one boron atom. There may be a substitution of other rare earth and/or minor additions of other elements. The rare earth content of RE-Fe magnet alloys is typically 30 to 35 weight percent.

2.0 MANUFACTURING METHODS

The rare earth magnet alloys are usually formed by powder metallurgical processes. The magnetic performance of all grades is optimized by applying a magnetic field during the pressing operation, thus producing a preferred direction of magnetization. Pressing and aligning techniques can substantially vary the degree of orientation and the residual induction (B_r) of the finished magnet. The direction of the magnetic field during die pressing can be either parallel or perpendicular to the pressing direction. Magnets can also be formed by isostatic pressing.

After pressing, the magnets are sintered, heat treated and ground to the final dimensions. Rare earth magnets are inherently brittle and cannot be machined with conventional metal cutting processes such as drilling, turning or milling. The magnets can be readily ground with abrasive wheels if liberal amounts of coolant are used. The coolant serves to minimize heat cracking, chipping and also eliminates the risk of fires caused by sparks contacting the easily oxidized grinding dust.

3.0 MAGNETIC PROPERTIES

The magnetic properties and chemical compositions of the commercial grades of rare earth magnet materials are given in Table IV-1. Since many combinations of elements and orientations are possible, many additional grades are available from various producers.

4.0 DIMENSIONS AND TOLERANCES

Allowable tolerances for sintered rare earth magnets are given in Table IV-2.

5.0 MECHANICAL CHARACTERISTICS

The following general specifications are for mechanical characteristics and visual imperfections.

5.1 Surface Conditions

5.1.1 All magnet surfaces shall be free of foreign materials which would tend to hold or collect extraneous particles on the magnet surface in the unmagnetized condition.

5.2 Chips

5.2.1 Magnets shall be free of loose chips. They shall be free of imperfections which will result in loose chips or particles under normal conditions of handling, shipping, assembly and service.

5.2.2 A chipped edge or surface shall be acceptable if no more than 10 percent of the surface is removed, provided that no loose particles remain at the edge or surface, and further provided the magnet under examination meets the magnetic specification agreed upon between the producer and user.

5.3 Other Physical Imperfections

5.3.1 Imperfections such as minor hairline cracks, porosity, voids, and others, all of the type commonly found in sintered metallic magnets, shall be judged acceptable if the following conditions are met:

5.3.1.1 The magnet meets the minimum magnetic performance criteria agreed upon.

5.3.1.2 The imperfections do not create loose particles or other conditions which will interfere with proper functioning of the end device.

5.3.1.3 Cracks shall be acceptable provided they do not extend across more than 50 percent of any pole surface.

5.4 Other Conditions

5.4.1 Non-destructive inspection methods such as the use of penetrants, microscopy, magnetic particle analysis, ultrasonic inspection, or x-ray shall not be accept-

TABLE IV-1
MAGNETIC PROPERTIES—CHEMICAL COMPOSITION OF RARE EARTH MAGNETS*

MMPA Brief Designation	IEC Code Reference	Chemical Composition		Magnetic Properties (nominal)**			
		Alloys	Possible Elements	Maximum Energy Product $(BH)_{max}$ MGO	Residual Flux Density B_r (gauss)	Coercive Force H_c (oersteds)	Intrinsic Coercive Force H_{ci} (oersteds)
5/16	R5-1	RE Co$_5$	RE = Sm, Nd, MM	5	4700	4500	16000
14/14	R5-1	RE Co$_5$	RE = Sm, MM	14	7500	7000	14000
16/18	R5-1	RE Co$_5$	RE = Sm, Nd	16	8300	7500	18000
18/20	R5-1	RE Co$_5$	RE = Sm, Pr, Nd	18	8700	8000	20000
20/15	R5-1	RE Co$_5$	RE = Sm, Pr, Nd	20	9000	8500	15000
22/15	R5-1	RE Co$_5$	RE = Sm, Pr, Nd	22	9500	9000	15000
22/12	R5-2	RE$_2$TM$_{17}$	RE = Sm, Ce; TM = Fe, Cu, Co, Zr, Hf	22	9600	8400	12000
24/7	R5-2	RE$_2$TM$_{17}$	RE = Sm, Ce; TM = Fe, Cu, Co, Zr, Hf	24	10000	6000	7000
24/18	R5-2	RE$_2$TM$_{17}$	RE = Sm; TM = Fe, Cu, Co, Zr, Hf	24	10200	9200	18000
26/11	R5-2	RE$_2$TM$_{17}$	RE = Sm; TM = Fe, Cu, Co, Zr, Hf	26	10500	9000	11000
28/7	R5-2	RE$_2$TM$_{17}$	RE = Sm; TM = Fe, Cu, Co, Zr, Hf	28	10900	6500	7000
26/20	R7-3	RE$_2$TM$_{14}$B	RE = Nd, Pr, Dy, Tb; TM = Fe, Co	26	10400	9900	20000
27/11	R7-3	RE$_2$TM$_{14}$B	RE = Nd, Pr, Dy, Tb; TM = Fe, Co	27	10800	9300	11000
30/18	R7-3	RE$_2$TM$_{14}$B	RE = Nd, Pr, Dy, Tb; TM = Fe, Co	30	11000	10000	18000
33/11	R7-3	RE$_2$TM$_{14}$B	RE = Nd, Pr, Dy, Tb; TM = Fe, Co	33	11800	10800	11000

* Temperature compensated materials and materials with maximum energy products of 40 MGO are available from various manufacturers.
** To achieve the properties shown in this table, care must be taken to magnetize to technical saturation.

able methods for judging the quality of sintered rare earth magnets except as provided in Section 5.4.2.

5.4.2 In cases where the magnet is expected to withstand abnormal conditions or stresses, such conditions must be previously specified and a mutually acceptable service test devised to assure that the magnet shall not fail under the specified service conditions. Such tests should duplicate service conditions with appropriate safety factors.

6.0 PHYSICAL AND THERMAL PROPERTIES

Typical physical and thermal properties for rare earth magnets are given in Table IV-3.

7.0 INSPECTION SAMPLING PLANS

Unless otherwise specified, the sampling plan to be used by the manufacturer and the receiving inspection department of the user shall be in accordance with MIL-STD-105D. All attributes will be inspected using a single sampling plan with the inspection level and AQL as specified in Table IV-4.

TABLE IV-2
TOLERANCES
SINTERED RARE EARTH MAGNETS

	Dimension (inches)	Tolerances (inches)
Size: Unfinished surfaces: Dimensions perpendicular To pressing (die formed)	0 to .125 .126 to .625 .626 to .875 .876 and up	±.006 ±.012 ±.018 ±2.5%
Dimensions parallel to pressing (punch formed)	any	±2.5% or ±.02 whichever is greater
Finished surfaces: Plane ground	any	±.005
Center or Centerless ground	0 to 1.5 over 1.5	±.004 ±.008
Parallelism: Between finished surfaces		½ total tolerance between surfaces
Angles including squareness: Between two unfinished surfaces Between one finished, one unfinished Between two finished surfaces		± 1° ± 1° ± .005 from true angles as measured on the shorter of two surfaces in question or ± ½° whichever is greater
Concentricity: Between inside and outside surfaces: Unfinished surfaces	0 to .5 .5 to 1 1 to 1.5	±.010 TIR ±.020 TIR ±.030 TIR
Finished surfaces	any	±.007 TIR
Surface Roughness: Finished surfaces	any	63 microinches over at least 95% of the ground surface

TABLE IV-3
TYPICAL PHYSICAL AND THERMAL PROPERTIES
SINTERED RARE EARTH COBALT MAGNETS

	1-5 Alloys	2-17 Alloys	Nd-Fe-B
Mechanical Properties:			
Modulus elasticity	23×10^6 psi	17×10^6 psi	22×10^6 psi
Ultimate tensile strength	6×10^3 psi	5×10^3 psi	12×10^3 psi
Physical Properties:			
Density	8.2 g/cc	8.4 g/cc	7.4 g/cc
Coefficient of thermal expansion			
Perpendicular to orientation	13×10^{-6}/°C	11×10^{-6}/°C	-4.8×10^{-6}/°C
Parallel to orientation	6×10^{-6}/°C	8×10^{-6}/°C	3.4×10^{-6}/°C
Electrical resistivity	53μ ohm cm.	86μ ohm cm.	160μ ohm cm.
Magnetic Properties:			
Curie temperature	750°C	825°C	310°C
Reversible temperature coefficient of residual induction (−100°C to +100°C)	−.043%/°C	−.03%/°C	−.09 to −.13%/°C
Recoil permeability	1.05	1.05	1.05
Max. service temperature*	250°C	300°C	150°C

The mechanical properties listed are typical engineering properties. These materials are inherently brittle, lack ductility and should not be used as structural components in a design. Measurements of mechanical properties are not feasible except under controlled laboratory conditions and specifications of these properties are not acceptable to the manufacturer.

*See T_{max} definition in Appendix B, Glossary.

TABLE IV-4
SINGLE SAMPLING, NORMAL INSPECTION
NORMAL INSPECTION
SINTERED RARE EARTH COBALT MAGNETS
FROM MIL-STD-105-D TABLE I AND TABLE II-A

Lot Size	Level II, AQL 1.5		Level I, AQL 4.0		Level S-4, AQL 4.0	
	Sample Size	Rejection Number	Sample Size	Rejection Number	Sample Size	Rejection Number
2 – 8	2	1	2	1	2	1
9 – 15	2	1	2	1	2	1
16 – 25	3	1	3	1	3	1
26 – 50	5	1	5	1	5	1
51 – 90	5	1	5	1	5	1
91 – 150	8	1	8	2	8	2
151 – 280	13	1	13	2	13	2
281 – 500	20	2	20	3	13	2
501 – 1,200	32	2	32	4	20	3
1,201 – 3,200	50	2	50	6	32	4
3,201 – 10,000	80	3	80	8	32	4
10,001 – 35,000	125	4	125	11	50	6
35,001 – 150,000	200	6	200	15	80	8
150,001 – 500,000	315	9	315	22	80	8
500,001 & over	500	13	500	22	125	11

TABLE IV-5
ACCEPTANCE QUALITY LEVELS (AQL) FOR RARE EARTH MAGNETS

Characteristic	Inspection Level	AQL
Individual Magnetic parameters	II	1.5
Individual dimensional parameters:		
Categorized as major	I	4.0
Categorized as minor	S-4	4.0
Visual parameters	S-4	4.0

SECTION V

IRON-CHROMIUM-COBALT MAGNETS

1.0 CHEMICAL COMPOSITION

These alloys are primarily of the Iron-Chromium-Cobalt composition. Some grades may also contain additions of vanadium, silicon, titanium, zirconium, manganese, molybdenum or aluminum.

2.0 MANUFACTURING METHODS

The Iron-Chromium-Cobalt alloys are formed either by casting to size or by casting in the form of an ingot which is then rolled and/or drawn either to final shape or to form which can then be cut.

Heat treatment is essential to develop the magnetic properties. The magnetic properties can be increased in a preferred direction by applying a magnetic field during heat treatment. Although this alloy is hard and brittle in its fully heat treated condition, it is sufficiently ductile to be rolled, drawn, machined, turned or threaded prior to its final heat treatment.

3.0 MAGNETIC PROPERTIES

The magnetic properties of the commercial grades of Iron-Chromium-Cobalt alloys are listed in Table V-1.

4.0 DIMENSIONS AND TOLERANCES

Standard tolerances for cast, rolled and drawn Iron-Chromium-Cobalt magnets are listed in Table V-2.

5.0 MECHANICAL CHARACTERISTICS

The following general specifications are for mechanical characteristics and visual conditions.

5.1 Surface Conditions

There should be no cracks, chips, burrs or other conditions that would interfere with the proper functioning of the end use device.

5.2 Chips

5.2.1 Magnets shall be free of loose chips or imperfections which will result in loose chips under normal handling and shipping conditions.

5.2.2 A chipped edge or surface shall be acceptable if not more than 10% of the edge or 5% of the surface is removed.

5.3 Visual Standards The use of mutually agreed upon visual standards is recommended in cases where such properties are critical.

5.4 Inspection Methods Inspection methods such as the use of penetrants, magnetic particle analysis, ultrasonics or x-ray shall not be acceptable methods for judging the quality of cast, rolled or drawn Iron-Chromium-Cobalt magnets except in cases where the magnet is expected to withstand abnormal conditions. In those cases the service conditions must be specified and a mutually acceptable service test devised.

TABLE V-1
TYPICAL MAGNETIC PROPERTIES AND CHEMICAL COMPOSITIONS OF IRON-CHROMIUM-COBALT MAGNET MATERIALS

Brief Designation	MMPA Class*	IEC Code Reference	Magnetic Properties		(Nominal) BH$_{max}$ (MGO)
			B$_r$ (kilogauss)	H$_c$ (oersteds)	
ISOTROPIC					
1.6/0.46	Fe Cr Co 1	R6	8800	460	1.6
1.6/0.35	Fe Cr Co 2		9900	350	1.6
1.0/0.20	Fe Cr Co		10500	200	1.0
ANISOTROPIC					
5.2/0.61	Fe Cr Co 5	R6	13500	600	5.25
2.0/0.25	Fe Cr Co 250	R6	14000	250	2.0

*Composition is 15 to 35 weight percent chromium, 5 to 20 cobalt, balance iron with minor amounts of other elements present.

TABLE V-2
TOLERANCES
IRON–CHROMIUM–COBALT MAGNETS

	Dimensions (Inches)	Tolerances
ROLLED BARS Rounds or Squares	0 to .312	± .010
	.312 to .625	± .015
	.625 & over	± .020
ROLLED FLATS Thickness	.002 to .014	± 5%
	.014 to .095	± 5%
	0 to .125	± .010
	.125 to .250	± .015
	.250 to .500	± .020
	.250 to .500	± .020
Width	.125 to 8.000	± .005
	1.000 to 8.000	± .010
	0 to .500	± .015
	0 to .750	± .020
	.750 to 1.000	± .025
CENTERLESS GROUND BARS Diameter	0 to 1	± .002
DRAWN BARS Diameter	0 to .190	± .003

6.0 PHYSICAL PROPERTIES

Typical values for density, thermal expansion, thermal conductivity and electrical resistivity are listed in Table V-3.

7.0 THERMAL PROPERTIES

Typical values for the reversible temperature coefficients of residual induction and intrinsic coercive force, Curie temperature and maximum service temperature are listed in Table V-4.

8.0 INSPECTION AND TESTING

8.1 Unless otherwise agreed upon by the manufacturer and the user, all lots of bulk magnet material and individual magnets will be inspected by the use of a statistical sampling plan in accordance with MIL-STD-105D. All specified attributes will generally be inspected with the inspection level and AQL as specified in Table V-5. Reduced sampling may be used in cases where the confidence in the repeatability of a specific attribute is such that the user can be assured of an acceptable level of quality.

8.2 A method of magnetic tests shall be chosen which causes the magnet to operate at one or more points on its major and/or minor hysteresis loops as indicated by the operating points of the magnet in its final magnetic circuit. The acceptance limits chosen shall be mutually agreeable to both the manufacturer and user. A magnet shall be chosen to serve as a calibration reference. Acceptability of a magnet will then be judged solely by comparison to the reference magnet, tested under identical conditions.

TABLE V-3
PHYSICAL PROPERTIES OF IRON-CHROMIUM-COBALT MAGNETS

Brief Designation	MMPA Class	IEC Ref	Density		Electrical Resistivity ohm–cm x 10^{-6} (at 20°C)	Thermal Conductivity Cal/cm sec °C	Coefficient of Thermal Expansion $\times 10^{-6}$/°C
			lb/in³	g/cm³			
1.6/0.35	FeCrCo 2	R6	.278	7.7	70	.05	10
5.2/0.61	FeCrCo 5	R6	.278	7.7	70	.05	10
2.0/0.25	FeCrCo 250	R6	.278	7.7	70	.05	10

TABLE V-4
THERMAL PROPERTIES OF IRON-CHROMIUM-COBALT MAGNETS

Brief Designation	MMPA Class	IEC Ref	Temperature Coefficient B_r Reversible –50 to +200°C	Curie Temp °C	Max Service Temp °C
1.6/0.35	FeCrCo 2	R6	.036	640	500
5.2/0.61	FeCrCo 5	R6	.02	640	500
2.0/0.25	FeCrCo 250	R6	.03	640	500

TABLE V-5
SINGLE SAMPLING, NORMAL INSPECTION

Lot Size	Level II	AQL 1-5
	Sample Size	Rejection Number
1–15	100%	1
16–25	5	1
26–50	8	1
51–90	13	1
91–150	20	1
151–280	32	2
281–500	50	3
501–1,200	80	4
1,201–3,200	125	6
3,201–10,000	200	8
10,001–35,000	315	11
35,001–150,000	500	15

APPENDIX A
PERMANENT MAGNET MATERIALS
NOT COVERED IN PRODUCT SECTIONS
(See Section I Scope)

Magnet Material	$(BH)_{max}$ (MGO)	Magnetic Properties	
		B_r (gauss)	H_c/H_{ci} (oersteds)
3½% Cr Steel	.13	10300	60
3% Co Steel	.38	9700	80
17% Co Steel	.69	10700	160
38% Co Steel	.98	10400	230
Ceramic 2	1.8	2900	2400/3000
Ceramic 6	2.45	3200	2820/3300
Bonded Ceramic	1.5	2500	2100/2200
Alnico 4	1.35	5600	720
PtCo	9.0	6450	4000
Vicalloy 1	.80	7500	250
Remalloy	1.0	9700	250
Cunife 1	1.4	5500	530
ESD 31	2.3	5000	1000
ESD 32	3.0	6800	960
ESD 41	1.1	3600	970
ESD 42	1.25	4800	830
MnA1C	5.0	5450	2550/3150
Bismanol	5.3	4800	3650
Vectolite	.5	1600	900

APPENDIX B
GLOSSARY OF TERMS

A_g **Area of the air gap,** or the cross sectional area of the air gap perpendicular to the flux path, is the average cross sectional area of that portion of the air gap within which the application interaction occurs. Area is measured in sq. cm. in a plane normal to the central flux line of the air gap.

A_m **Area of the magnet,** is the cross sectional area of the magnet perpendicular to the central flux line, measured in sq. cm. at any point along its length. In design, A_m is usually considered the area at the neutral section of the magnet.

B **Magnetic induction,** is the magnetic field induced by a field strength, H, at a given point. It is the vector sum, at each point within the substance, of the magnetic field strength and resultant intrinsic induction. Magnetic induction is the flux per unit area normal to the direction of the magnetic path.

B_d **Remanent induction,** is any magnetic induction that remains in a magnetic material after removal of an applied saturating magnetic field, H_s. (B_d is the magnetic induction at any point on the demagnetization curve; measured in gauss.)

B_d/H_d **Slope of the operating line,** is the ratio of the remanent induction, B_d, to a demagnetizing force, H_d. It is also referred to as the permeance coefficient, shear line, load line and unit permeance.

$B_d H_d$ **Energy product,** indicates the energy that a magnetic material can supply to an external magnetic circuit when operating at any point on its demagnetization curve; measured in megagauss-oersteds.

$(BH)_{max}$ **Maximum energy product,** is the maximum product of $(B_d H_d)$ which can be obtained on the demagnetization curve.

B_{is} **(or J) Saturation intrinsic induction,** is the maximum intrinsic induction possible in a material.

B_g **Magnetic induction in the air gap,** is the average value of magnetic induction over the area of the air gap, A_g; or it is the magnetic induction measured at a specific point within the air gap; measured in gauss.

B_i **(or J) Intrinsic induction,** is the contribution of the magnetic material to the total magnetic induction, B. It is the vector difference between the magnetic induction in the material and the magnetic induction that would exist in a vacuum under the same field strength, H. This relation is expressed by the equation:

$$B_i = B - H$$

where: B_i = intrinsic induction in gauss; B = magnetic induction in gauss; H = field strength in oersteds.

B_m **Recoil induction,** is the magnetic induction that remains in a magnetic material after magnetizing and conditioning for final use; measured in gauss.

B_o **Magnetic induction,** at the point of the maximum energy product $(BH)_{max}$; measured in gauss.

B_r **Residual induction** (or flux density), is the magnetic induction corresponding to zero magnetizing force in a magnetic material after saturation in a closed circuit; measured in gauss.

f **Reluctance factor,** accounts for the apparent magnetic circuit reluctance. This factor is required due to the treatment of H_m and H_g as constants.

F **Leakage factor,** accounts for flux leakage from the magnetic circuit. It is the ratio between the magnetic flux at the magnet neutral section and the average flux present in the air gap. $F = (B_m A_m)/(B_g A_g)$.

F **Magnetomotive force, (magnetic potential difference),** is the line integral of the field strength, H, between any two points, p_1 and p_2.

$$F = \int_{p_1}^{p_2} H\, dl$$

F = magnetomotive force in gilberts
H = field strength in oersteds
dl = an element of length between the two points, in centimeters.

H **Magnetic field strength,** (magnetizing or demagnetizing force), is the measure of the vector magnetic quantity that determines the ability of an electric current, or a magnetic body, to induce a magnetic field at a given point; measured in oersteds.

H_c **Coercive force** of a material, is equal to the demagnetizing force required to reduce residual induction, B_r, to zero in a magnetic field after magnetizing to saturation; measured in oersteds.

H_{ci} **Intrinsic coercive force** of a material indicates its resistance to demagnetization. It is equal to the demagnetizing force which reduces the intrinsic induction, B_i, in the material to zero after magnetizing to saturation; measured in oersteds.

H_d is that value of H corresponding to the remanent induction, B_d; measured in oersteds.

H_m is that value of H corresponding to the recoil induction, B_m; measured in oersteds.

H_o is the magnetic field strength at the point of the maximum energy product $(BH)_{max}$; measured in oersteds.

H_s **Net effective magnetizing force,** is the magnetizing force required in the material, to magnetize to saturation measured in oersteds.

J, see B_i Intrinsic induction.

J_s, see B_{is} Saturation intrinsic induction.

ℓ_g **Length of the air gap,** is the length of the path of the central flux line of the air gap; measured in centimeters.

ℓ_m **Length of the magnet,** is the total length of magnet material traversed in one complete revolution of the center-line of the magnetic circuit; measured in centimeters.

ℓ_m/D **Dimension ratio,** is the ratio of the length of a magnet to its diameter, or the diameter of a circle of equivalent cross-sectional area. For simple geometries, such as bars and rods, the dimension ratio is related to the slope of the operating line of the magnet, B_d/H_d.

P **Permeance,** is the reciprocal of the reluctance, R, measured in maxwells per gilbert.

R **Reluctance,** is somewhat analogous to electrical resistance. It is the quantity that determines the magnetic flux, ϕ, resulting from a given magnetomotive force, F.
$$R = F/\phi$$
where:
R = reluctance, in gilberts per maxwell
F = magnetomotive force, in gilberts
ϕ = flux, in maxwells

T_c **Curie temperature,** is the transition temperature above which a material loses its magnet properties.

APPENDIX B

GLOSSARY OF TERMS (continued)

T_{max} **Maximum service temperature,** is the maximum temperature to which the magnet may be exposed with no significant long range instability or structural changes.

V_g **Air gap volume,** is the useful volume of air or non-magnetic material between magnetic poles; measured in cubic centimeters.

μ **permeability,** is the general term used to express various relationships between magnetic induction, B, and the field strength, H.

μ_{re} **recoil permeability,** is the average slope of the recoil hysteresis loop. Also known as a minor loop.

ϕ **magnetic flux,** is a contrived but measurable concept that has evolved in an attempt to describe the "flow" of a magnetic field. Mathematically, it is the surface integral of the normal component of the magnetic induction, B, over an area, A.

$$\phi = \int \int B \cdot dA$$

where:
ϕ = magnetic flux, in maxwells
B = magnetic induction, in gauss
dA = an element of area, in square centimeters
When the magnetic induction, B, is uniformly distributed and is normal to the area, A, the flux, ϕ = BA.

A closed circuit condition exists when the external flux path of a permanent magnet is confined with high permeability material.

The demagnetization curve is the second (or fourth) quadrant of a major hysteresis loop. Points on this curve are designated by the coordinates B_d and H_d.

A fluxmeter is an instrument that measures the change of flux linkage with a search coil.

The gauss is the unit of magnetic induction, B, in the cgs electromagnetic system. One gauss is equal to one maxwell per square centimeter.

A gaussmeter is an instrument that measures the instantaneous value of magnetic induction, B. Its principle of operation is usually based on one of the following: the Hall-effect, nuclear magnetic resonance (NMR), or the rotating coil principle.

The gilbert is the unit of magnetomotive force, F, in the cgs electromagnetic system.

A hysteresis loop is a closed curve obtained for a material by plotting (usually to rectangular coordinates) corresponding values of magnetic induction, B, for ordinates and magnetizing force, H, for abscissa when the material is passing through a complete cycle between definite limits of either magnetizing force, H, or magnetic induction, B.

Irreversible losses are defined as partial demagnetization of the magnet, caused by exposure to high or low temperatures external fields or other factors. These losses are recoverable by remagnetization.
Magnets can be stabilized against irreversible losses by partial demagnetization induced by temperature cycles or by external magnetic fields

A keeper is a piece (or pieces) of soft iron that is placed on or between the pole faces of a permanent magnet to decrease the reluctance of the air gap and thereby reduce the flux leakage from the magnet. It also makes the magnet less susceptible to demagnetizing influences.

Leakage flux is flux, ϕ, whose path is outside the useful or intended magnetic circuit; measured in maxwells.

The major hysteresis loop of a material is the closed loop obtained when the material is cycled between positive and negative saturation.

The maxwell is the unit of magnetic flux in the cgs electromagnetic system. One maxwell is one line of magnetic flux.

The neutral section of a permanent magnet is defined by a plane passing through the magnet perpendicular to its central flux line at the point of maximum flux.

The oersted is the unit of magnetic field strength, H, in the cgs electromagnetic system. One oersted equals a magnetomotive force of one gilbert per centimeter of flux path.

An open circuit condition exists when a magnetized magnet is by itself with no external flux path of high permeability material.

The operating line for a given permanent magnet circuit is a straight line passing through the origin of the demagnetization curve with a slope of negative B_d/H_d. (Also known as permeance coefficient line.)

The operating point of a permanent magnet is that point on a demagnetization curve defined by the coordinates ($B_d H_d$) or that point within the demagnetization curve defined by the coordinates ($B_m H_m$).

An oriented (anisotropic) material is one that has better magnetic properties in a given direction.

A permeameter is an instrument that can measure, and often record, the magnetic characteristics of a specimen.

Reversible temperature coefficients are changes in flux which occur with temperature change. These are spontaneously regained when the temperature is returned to its original point.

Magnetic saturation of a material exists when an increase in magnetizing force, H, does not cause an increase in the intrinsic magnetic induction, B, of the material.

A search coil is a coiled conductor, usually of known area and number of turns, that is used with a fluxmeter to measure the change of flux linkage with the coil.

The temperature coefficient is a factor which describes the reversible change in a magnetic property with a change in temperature. The magnetic property spontaneously returns when the temperature is cycled to its original point. It usually is expressed as the percentage change per unit of temperature.

An unoriented (isotropic) material has equal magnetic properties in all directions.

APPENDIX C
MAGNETIC QUANTITIES
SYMBOLS, UNITS AND CONVERSION FACTORS

Quantity	Symbol	CGS Unit	Conversion* Factors	SI Unit
MAGNETIC FLUX	ϕ	maxwell	10^{-8}	weber
MAGNETIC INDUCTION (magnetic flux density)	B	gauss	10^{-4}	tesla
MAGNETOMOTIVE FORCE (magnetic potential difference)	F	gilbert (oersted-cm)	$\dfrac{10}{4\pi}$	ampere-turn
MAGNETIC FIELD STRENGTH (magnetizing or demagnetizing force)	H	oersted	$\dfrac{10^3}{4\pi}$	ampere/meter
ENERGY PRODUCT (magnetic energy density)	$B_d H_d$	megagauss-oersted	$\dfrac{10^5}{4\pi}$	joule/meter^3

*Multiply quantity in CGS units by the conversion factor to obtain quantity in SI units

LIST OF CONTRIBUTORS

ALLEN-BRADLEY–TDK MAGNETICS
5900 N. Harrison Street
Shawnee, Oklahoma 74801

ARNOLD ENGINEERING COMPANY
300 West Street
Marengo, Illinois 60152

CRUCIBLE MAGNETICS
101 Magnet Drive
Elizabethtown, Kentucky 42701

ELECTRON ENERGY CORPORATION
924 Links Avenue
Landisville, Pennsylvania 17538

GENERAL MAGNETIC COMPANY
5252 Investment Drive
Dallas, Texas 75236

HITACHI MAGNETICS CORPORATION
7800 Neff Road
Edmore, Michigan 48829

I G TECHNOLOGIES, INC.
405 Elm Street
Valparaiso, Indiana 46383

NATIONAL MAGNETICS GROUP, INC.
250 South Street
Newark, New Jersey 07114

THE PERMANENT MAGNET COMPANY, INC.
4437 Bragdon Street
Indianapolis, Indiana 46226

RECOMA, INC.
400 Myrtle Avenue
Boonton, New Jersey 07006

SEMICON ASSOCIATES
1801 Old Frankfort Pike
Lexington, Kentucky 40504

THE STACKPOLE COMPANY
700 Elk Avenue
Kane, Pennsylvania 16735

D.M. STEWARD MANUFACTURING COMPANY
P.O. Box 510
Chattanooga, Tennessee 37401

THOMAS & SKINNER, INC.
1120 E. 23rd Street
Indianapolis, Indiana 46206

VACUUMSCHMELZE
186 Wood Avenue South
Iselin, New Jersey 08830

16

Typical Circuits and Applications

In preceding chapters, adequate emphasis was placed on the uses (Figure 1-1) and widespread applications (Chapter 4) of modern permanent magnets. There is hardly an area of modern science, technology, or industrial process in which permanent magnets are not now used or do not have a potential use when imaginatively and properly applied. Some of these current applications and their detailed circuits have been dispersed throughout preceding chapters, where they were appropriate to illustrate particular principles. This chapter is intended to present engineers with a wide range of effective applications as a stimulus to develop new or improved products. For example, a circuit developed for an iron-ore separator, a traveling-wave tube, or a pump might well be applied to an artificial heart, a nuclear reactor control, or an automotive transmission system. This chapter will also help engineers working with new products to become familiar with how magnets are or should be applied in these areas of technology. Finally, it may even help the "old dogs" in particular product areas to "learn a few new tricks."

This chapter cannot, of course, begin to present *all* of the present magnet applications. The ones included are those most widely used and, naturally, those closest to the experience of the author.

One word of caution: Many specific magnetic circuits are covered by current patents, so that they cannot be indiscriminately applied to other products without legal complications. To indicate all of the products covered by such patents and/or to list patent numbers would not only be beyond the scope of this book but would also be beyond the capabilities of the author.

There are many competent patent attorneys [1] who can conduct patent searches. However, an engineer can benefit considerably from proper use of the facilities of the U.S. Patent Office, alone [2] or in cooperation with a patent attorney.

Before beginning a study of the applications that follow, it is advisable to consider the advantages and limitations of magnetically "operated" products in general as well as the advantages and limitations of permanent magnets compared to DC electromagnets in particular.

Potential Advantages of All
Magnetically Operated Products
  May produce force without contact.
  Simplicity of construction.
  High forces can be developed in small spaces.
  Highly uniform forces can be obtained.
  Maximum force produced can be limited.
  High degree of control mechanically and/or electrically.
  High adaptability.
  Simple means for release of forces.
  Wide latitude of operation.

Potential Limitations of All
Magnetically Operated Products
  Application usually limited to ferromagnetic materials.
  Maximum force that can be produced is limited.
  Distance through which force can act is limited.
  Costly for some operations (functions).
  Lacks selectivity between ferromagnetic objects.
  Residual magnetism may result in problems.

---

1. While it is hardly common practice in this type of book to "endorse" specific individuals, the highly specialized nature of permanent magnet products has prompted the author to suggest that Patent Attorney Charles Lovercheck of Erie, Pennsylvania, is one of the most knowledgeable attorneys in this field.

2. "How to Use Patents" by L.R. Moskowitz in the April 1963 issues of *Product Engineering Magazine* will be of considerable assistance to engineers who have never had the training, occasion, or opportunity to directly use patents and the facilities of the U.S. Patient Office.

Stray (leakage) fields may produce problems.
Limited release rate of magnetically created forces.
Interpart forces (attraction or adhesion) can occur between objects subjected to the same magnetic force.

*Potential Advantages of
Permanent Magnets*
No source of electrical energy required.
May be operated in high-temperature areas.
Magnetic force not affected by power-system failures.
No electrical connections required.
No aging effects (insulation, thermal, moisture, etc.).
No maintenance required.
Less sensitive to shock, vibration, and other mechanical stresses.
No operating cost.
No elaborate control systems required.
High per-unit-volume forces produced.
Low cost of small units.
Can be used in hazardous locations (explosive vapors, dust, etc.).

*Potential Limitations of
Permanent Magnets*
Field-strength control limited.
Design is highly specialized technologically.
Polarity reversals (when necessary) are more difficult.
High cost of very large units.

*Potential Advantages of
DC Electromagnets*
Zero to maximum field strength readily controlled.
Design technology less specialized.
Polarity reversal simple.
Low cost of very large units.

*Potential Limitations of
DC Electromagnets*
Requires source of rectified power.
Limited by temperature conditions.
Electrical failures may produce hazards.
Power connections (cables, slip rings, etc.) required.
Repairs and burnout occur due to aging.
More sensitive to mechanical shocks, vibration, etc.
Costly designs required for hazardous atmospheres.
High cost of small units.

Switch/Relay/Solenoid Applications

One of the earliest, simplest, and still current commercial permanent magnet applications is to provide a *"toggle" action in switch mechanisms.* Two basic systems (as shown in Figure 16-1) are used in devices such as household thermostats. They operate on the principle that the magnetic force is nonlinear (that it varies inversely with the square of the distance), so that "snap" action is inherently created. Most magnets used for this application are small sintered Alnicos or, more recently, small sintered ferrite bars.

Another "high-power" application is *arc quenching for large power switches and circuit breakers.* A permanent magnet is so arranged that the magnetic field formed by the current of the arc interacts with the permanent magnet field, "stretching" the arc rapidly to form an arc "blow-out" system. The availability of sintered ferrite magnets, with their very high resistance to demagnetization, has substantially increased the practicality and use of magnets for this purpose.

One of the most important applications of permanent magnets both electrically and electromechanically has been the development of *reed switches.* Such switches are commonly incorporated into relays, snap switches, position sensors, synchronization detectors, flow gauges, high-frequency vibrators, weighing devices, alarm systems, office-machine and computer keyboards, and even cardiac pacemakers. Figures 16-2 and 16-3 show how reed switches are used in a few of these

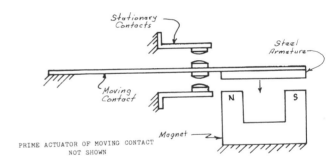

A. Simple Tractive Type

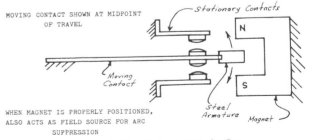

B. Gap Center-Instability Type

Figure 16-1. Magnets used to produce snap action of electrical contacts.

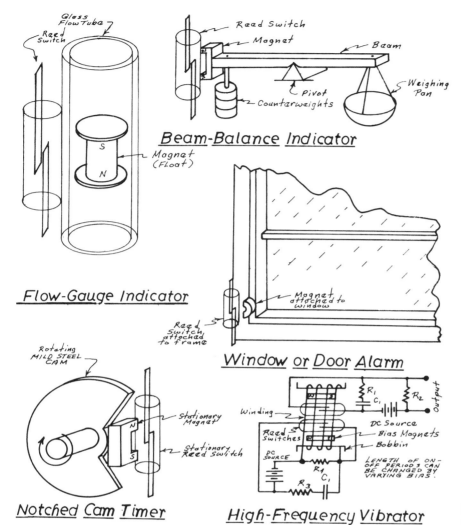

Figure 16-2. Typical reed-switch magnet applications.

applications. In spite of their widespread use, the available application information on permanent or electromechanical reed-switch magnet combinations has been surprisingly limited. The first substantive information [3] became available in 1968 and is covered herewith.

Reed-switch manufacturers have traditionally been concerned with reed/electromagnet relationships. This is evidenced by the rating system used: pull-in and drop-out ampere-turns. Ratings are derived by surrounding the switch with an electrically energized solenoid. Limited attention has been given to the reed/permanent magnet relationship by either switch or magnet manufacturers. Some switch manufacturers offer a few simple bar magnets as accessories. Others suggest that potential users contact magnet manufacturers. The magnet manufacturers, on the other hand, offer no information

3. L.R. Moskowitz, "Selecting Magnets for Reed-Switch Actuation," *Automation* (October 1968).

on this subject and refer the user to the switch manufacturer. As a result, most users are forced to develop empirical switch-magnet combinations for themselves. With a limited or often erroneous understanding of the magnetic principle involved, and with literally thousands of available magnets, this procedure is time consuming, costly, and often produces less than satisfactory results.

The basic concepts involved in the magnetic operation of reed switches will be presented here. The first step in properly applying or using permanent-magnet reed switches is understanding their construction and basic operation. The two reed elements are made of a nickel-iron alloy and are enclosed in a hermetically sealed glass envelope. If a magnetic field occurs in proximity to these nickel-iron reeds, localized poles will be created in the reeds by magnetic *induction*. If the field that creates these poles is produced by a permanent magnet (as shown in Figure 16-4) or a surrounding solenoid, unlike poles will be created,

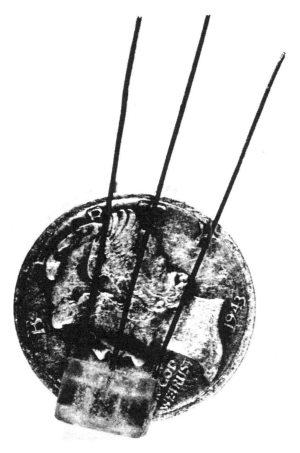

Figure 16-3. Reed switch for changing rate of dual-rate cardiac pacemaker. Implanted in body with pacemaker.

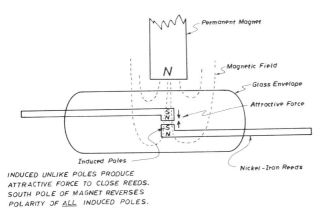

INDUCED UNLIKE POLES PRODUCE
ATTRACTIVE FORCE TO CLOSE REEDS.
SOUTH POLE OF MAGNET REVERSES
POLARITY OF ALL INDUCED POLES.

Figure 16-4. Operating principle of a reed switch.

attraction will occur, and closure will take place. *However, the permanent magnet does not attract the reeds directly.* Rather, operation is determined by the induction the magnet is capable of producing at a given distance. Thus, a large magnet designed to produce a "shallow" field may be less effective than a much smaller magnet that produces a "deep" field. Also, the direction of the inducing field must be such that *unlike poles* must be created in the contact area of the reeds. If the field direction produces *like* poles, the contacts may actually be repelled and closure will be impossible. This concept explains why "double trips" occur in some reed-magnet systems. In fact, recognition of these phenomena may be used to "lock out" a reed or to produce very rapid off-on operation.

When an electrically energized solenoid surrounds the reed switch, the field direction inherently produces unlike poles. The six general reed-switch/permanent-magnet/electromagnet systems shown in Figure 16-5 form the basis for any application.

The relative motion between a permanent magnet and a reed switch causes actuation of the switch. The actuation can be varied by any one of several factors, such as by selection of size and type of magnet, by relative spacing between switch and magnet, or by introducing a movable shield (in the event that motion is not possible between switch or magnet). Figures 16-6 and 16-7 show various types of permanent magnets used with reed switches and relative points of actuation. Figure 16-8 illustrates the seven electromagnetic methods for reed-switch actuation.

Table 16-1 is a practical engineering table that can be used to design reed-magnet systems and/or to minimize cut-and-try methods of system development. If the PI/DO (pull-in/drop-out) ampere-turns of any switch are known (given in the manufacturer's literature), the design engineer can establish the operating distances with any of ten standard permanent magnets by using this table. The magnets in the table are simple bar shapes that are normally available as stock items from most magnet manufacturers or distributors. Conversely, with established operating distances, the design engineer may select the magnet and switch combination that will suit these conditions.

This table is not intended to be comprehensive, particularly as it relates to standard magnets. All the magnets shown are simple shapes and are used for operations in which the direction of motion is at right angles to the major axis of the switch. Note that the lower section of this table gives PI/DO values for ten specific designs of *electrical solenoids*. This information is provided for users who elect to build or modify reed relays for special applications.

Table 16-2 lists the ten standard permanent magnets that can be used with Table 16-1. Table 16-3 provides all the necessary winding data to make electromagnetic actuators in conformance with the operating tables.

Finally, it is not uncommon for reed-switch users to require the addition of inverse operation, latching, or biasing to the system after initial switch selection and

TYPICAL CIRCUITS AND APPLICATIONS 217

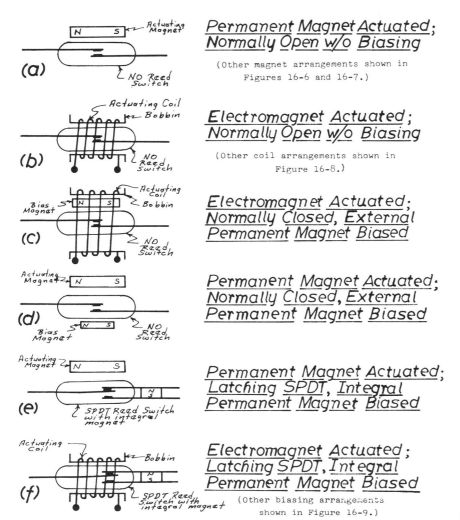

Figure 16-5. Basic reed-switch magnet systems.

installation. Since the available switches with these features are limited, Figure 16-9 presents nine ways in which an ordinary, normally open switch may be modified by the user for special operations. Figure 16-10 is a photograph of a reed switch with a small bonded ferrite bar magnet between two of the reeds to provide this type of action.

Permanent magnets are used in other relay (nonreed-switch type) applications, primarily to create an electrically directional effect (polarizing) or to provide a locking (latching) action when the relay coil is not energized. For polarized relays, the metallic materials (usually Alnico) are used rather than the ferrites, since the magnet is usually designed as part of the electromagnetic (coil) circuit. With this arrangement, the low permeability of the ferrites would greatly increase the reluctance of the electromagnetic circuit and therefore decrease its efficiency. The higher coercive grades of Alnico are preferred (although all grades have been used at one time or another) to avoid limiting the electromagnet's demagnetizing effects. In those cases where ferrites *are* used, they are arranged so that they simply polarize a soft-iron armature rather than serve as the armature itself.

For latching functions, either Alnicos or ferrites may be used, since the magnet can be located so that it is neither part of the electromagnetic circuit nor demagnetized by it.

Magnets are used in solenoids in place of the customary soft-iron core (or as a small section of such a core) *solely* to provide electrical direction-sensitive operation (a "push-pull" effect). The metallic materials are invariably used, because the low permeability of ferrites would ruin the efficiency of the solenoid. Even when Alnicos are employed, since they are only approximately 50 percent iron, the force-input characteristics of such solenoids will be substantially lower than the equivalent soft-iron core types.

Table 16-1
Permanent-Magnet and Electromagnetic Actuators for Reed Switches

Approximate Amp-Turn Rating of Reed Switch		Permanent Magnet				Approximate Amp-Turn Rating of Reed Switch		Electromagnet		
Pull In	Drop Out	Max. Air Gap to Pull In	Min. Air Gap to Drop Out	Magnet Number (See Table 16-2)	Min. Air Gap to Drop Out	Max. Air Gap to Pull In	Drop Out	Pull In	Coil Number (See Table 16-3)	Minimum Current (MA)	Maximum Applied Voltage
55	25	3/32	1/16	1	1/8	3/32	50	55	1	18.2	1.3
		3/16	3/8	2	1/4	3/16			2	15.1	1.5
		9/16	7/8	3	5/8	9/16			3	12.5	2.0
		9/32	5/8	4	5/16	9/32			4	11.0	2.5
		3/4	1 1/8	5	13/16	3/4			5	9.2	3.2
		9/32	5/8	6	5/16	9/32			6	7.9	3.8
		5/16	11/16	7	11/32	5/16			7	6.1	5.2
		3/4	1 1/8	8	13/16	3/4			8	4.2	6.5
		1/2	13/16	9	9/16	1/2			9	3.7	9.8
		21/32	1 in.	10	23/32	21/32			10	2.2	15.0
75	35	1/16	3/32	1	5/64	1/16	65	75	1	25.0	1.8
		1/8	5/16	2	5/32	1/8			2	20.0	2.1
		7/16	15/16	3	15/32	7/16			3	17.0	2.7
		1/4	9/16	4	9/32	1/4			4	15.0	3.4
		5/8	15/16	5	21/32	5/8			5	12.5	4.3
		1/4	9/16	6	9/32	1/4			6	10.7	5.2
		9/32	11/8	7	21/32	9/32			7	8.4	7.2
		5/8	15/16	8	21/32	5/8			8	5.8	9.0
		3/8	13/16	9	13/32	3/8			9	5.0	13.2
		1/2	1 1/8	10	9/16	1/2			10	3.0	20.5
95	35	1/32	1/8	1	1/16	1/32	80	95	1	31.4	2.2
		1/16	1/4	2	1/8	1/16			2	27.5	2.8
		7/16	11/16	3	1/2	5/16			3	21.5	3.1
		1/32	1/4	4	1/4	3/16			4	19.0	4.3
		1/2	15/16	5	17/32	1/2			5	15.9	5.5
		1/16	5/16	6	1/4	3/16			6	13.6	6.6
		1/8	7/16	7	9/32	7/32			7	10.6	9.0
		1/2	15/16	8	17/32	1/2			8	7.3	10.0
		3/16	9/16	9	13/32	7/32			9	6.3	16.6
		11/16	1 3/8	10	1/2	13/32			10	3.8	26.0

TYPICAL CIRCUITS AND APPLICATIONS

Table 16-1 (continued)

Approximate Amp-Turn Rating of Reed Switch		Permanent Magnet				Approximate Amp-Turn Rating of Reed Switch		Electromagnet		
Pull In	Drop Out	Max. Air Gap to Pull In	Min. Air Gap to Drop Out	Magnet Number (See Table 16-2)	Min. Air Gap to Drop Out	Max. Air Gap to Pull In	Drop Out	Pull In	Coil Number (See Table 16-3)	Minimum Current (MA)	Maximum Applied Voltage
110	45	—	—	1	—	—	85	110	1	36.6	2.6
		—	—	2	—	—			2	30.5	3.1
		1/4	5/8	3	7/16	1/4			3	25.0	4.0
		5/32	1/2	4	1/4	5/32			4	22.0	5.0
		7/16	7/8	5	7/16	7/16			5	18.4	6.3
		5/32	1/2	6	1/4	5/32			6	15.7	7.6
		3/16	9/16	7	9/32	3/16			7	12.3	10.2
		7/16	7/8	8	9/16	7/16			8	8.5	13.0
		5/32	9/16	9	9/32	5/32			9	7.3	19.2
		11/32	3/4	10	3/8	11/32			10	4.4	30.0
125	50	—	—	1	—	—		125	1	42.0	3.0
		1/4	5/8	2					2	34.8	3.5
		3/32	3/8	3					3	28.5	4.6
		13/32	13/16	4					4	25.4	5.8
		3/32	3/8	5					5	21.2	7.2
		5/32	15/32	6					6	17.8	8.6
		13/32	13/16	7					7	13.9	11.8
		1/16	7/16	8					8	9.6	14.7
		1/4	3/4	9					9	8.4	22.2
				10					10	5.0	34.0
140	95	—	—	1				140	1	46.6	3.3
		3/16	5/16	2					2	39.0	3.9
		1/8	3/16	3					3	32.0	5.1
		3/8	1/2	4					4	28.0	6.3
		1/8	3/16	5					5	23.4	8.0
		5/32	7/32	6					6	20.0	9.7
		3/8	1/2	7					7	15.6	13.2
		5/32	7/32	8					8	10.8	16.7
		9/32	11/32	9					9	9.3	24.6
				10					10	5.6	38.0

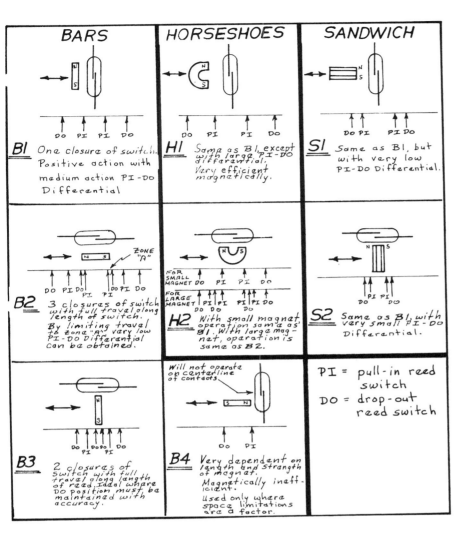

Figure 16-6. Permanent magnet actuation methods for reed switches.

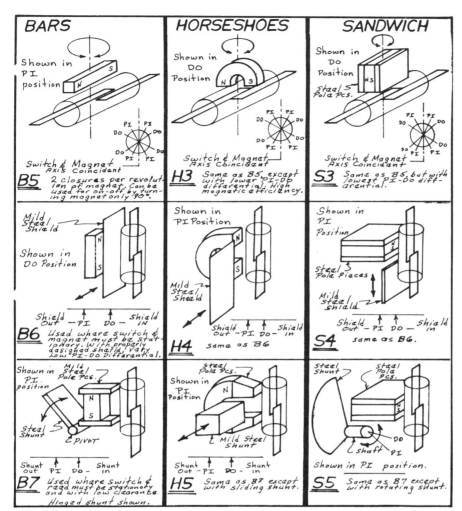

Figure 16-7. Permanent magnet actuation methods for reed switches (continued).

Table 16-2
Standard Permanent Magnets for Actuation of Reed Switches

Magnet Number	Grade of Material	Shape	Size
1	Sintered Alnico 5	Square	1/16 x 1/16 x 1/2 in.
2	Sintered Alnico 5	Square	1/8 x 1/8 x 3/4 in.
3	Cast Alnico 5	Square	3/16 x 3/16 x 1 in.
4	Sintered Alnico 2	Square	.205 x .205 x .680 in.
5	Cast Alnico 5	Square	1/4 x 1/4 x 1 in.
6	Sintered Alnico 2	Rectangle	.187 x .260 x .885 in.
7	Sintered Alnico 2	Rectangle	.160 x .350 x .885 in.
8	Cast Alnico 5	Rectangle	.125 x .500 x 1 in.
9	Cast Alnico 5	Round	.1870 dia. x 1 in.
10	Cast Alnico 5	Round	.250 dia. x 1.250 in.

Table 16-3
Electromagnets for Actuation of Reed Switches

Coil Number	Number of Turns[a]	AWG Wire Size	Maximum Resistance (ohms)
1	3,000	33	70.5
2	3,600	34	100.5
3	4,400	35	160.0
4	5,000	36	225.0
5	6,000	37	340.0
6	7,000	38	485.0
7	9,000	39	850.0
8	13,000	40	1540.0
9	15,000	42	2640.0
10	25,000	44	6820.0

[a] Uniformly wound coil on nylon forms .25 in. dia. x 2.0 in. long. Power dissipation = 1 watts maximum.

If permanent magnet is used for bias, minimum amp-turns of coil to operate = $2.02 \times H_m \times$ mean length of magnet, where:

for Sintered Alnico 2, $H_m = 520$
Sintered Alnico 5, $H_m = 590$
Cast Alnico 5, $H_m = 640$
Ferrites, Grade 1, $H_m = 1825$

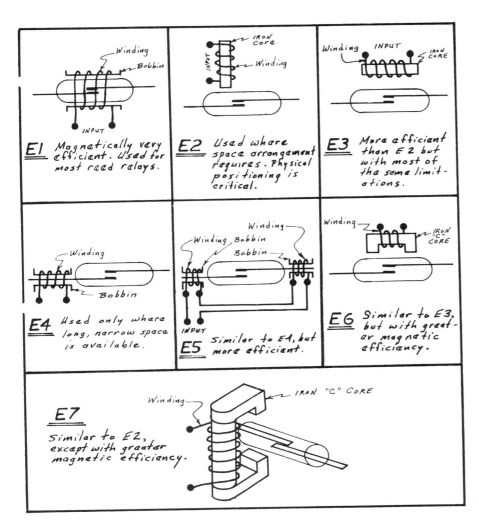

Figure 16-8. Electromagnetic actuation methods for reed switches.

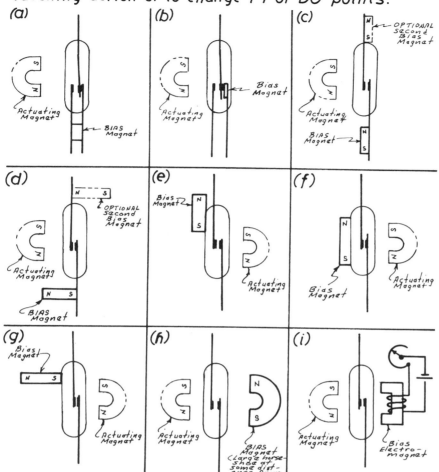

Figure 16-9. Biasing methods for reed switches.

Electronic Communications Applications

One of the earliest, large-volume uses of the then "high-power" cast Alnico 5 magnets were the magnetron tubes in World War II radar systems. Since that time, the electronic applications have expanded to klystrons, beam-switching tubes, traveling-wave tubes, cathode-ray tube focusing systems, and load isolators. Figures 16-11 and 16-12 show typical magnetron/klystron applications. Figure 9-41 shows the "circuits" used for traveling-wave tubes (TWTs), and Figure 16-13 is a photograph of a typical TWT magnet assembly. Figures 16-14, 16-15, and 16-16 show the remaining applications mentioned above. An indicator of how far magnet application in electronics has progressed is that approximately *100* cast Alnico 5 rods (2 in. in diameter × 30 in. long) are used in *each array* developed for an early-warning defense radar system.

On a less sophisticated scale, magnets are used in every telephone bell ringer, and in many types of receivers, microphones, phonograph pickups, and loudspeakers (see Figures 9-11 and 9-12 and 16-17).

The magnet materials used in electronic applications range from wrought alloys (for telephone receiver disks) to rare earths for traveling-wave tubes.

Instrument-Metering Indicator Applications

In addition to the electrical utility, watthour-meter "drag" application (Figure 14-4), permanent magnets are a basic operational element in all D'Arsonval meter-galvanometer movements. The four basic "systems" employed in such meters are shown in Figure 16-18. Alnico materials are invariably used in external magnet designs, while internal (core) magnet design is one of the more important uses for ESD-type Lodex materials. Of course, Alnico can be and is used for internal magnet design as well. Ferrites are not used for meter magnets, since temperature stability and maximum energy-per-unit volume are critical requirements for this type of product.

Other applications in this category include heat-

Figure 16-10. Small bonded ferrite magnet used to bias a reed switch. (Courtesy of 3M Company)

treatment temperature gauges and coating-thickness gauges (see Figure 16-19). Simple rod magnets or salient pole, rotor-type magnets are commonly combined with reed switches or electrical windings to provide fluid-flow detector-alarms, flow-measuring devices, pressure gauges, and rotary or linear motion tachometers. Miniature salient pole magnets are also widely used in all types of aircraft indicators to show flap positions, etc. Of course magnets are still widely used in aircraft and marine applications in high-quality navigational compasses and gyroscopes.

Two unique developments combine permanent magnets with Hall-effect gaussmeters: (1) a Hall-effect navigational compass, and (2) a continuous, spring-steel quality sensor marker. The compass, by converting the magnet's physical position into electrical signals through several Hall elements, provides very high accuracy with a remote "readout" that can be in either analog or digital form.

The spring-steel, quality-control device is used to *continuously* measure the metallurgical properties of the steel as it is unreeled from coils and enters the spring-forming machine. The system is based on the concept that the metal's properties (hardness, tensile strength, and so on) can be correlated with magnetic properties. As the steel passes a highly stable permanent magnet, a Hall element senses the change in reluctance

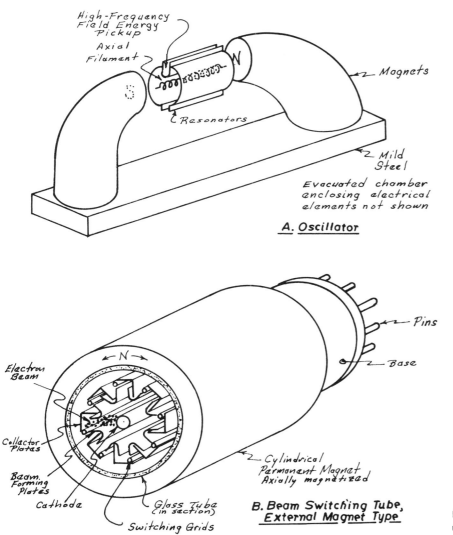

Figure 16-11. Magnets used in magnetron-type tubes.

TYPICAL CIRCUITS AND APPLICATIONS 225

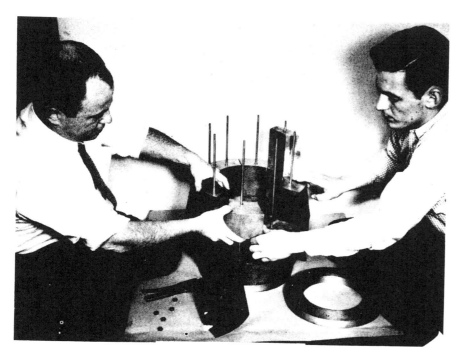

Figure 16-12. Assembling a barrel-type magnet for a Klystron tube. (Courtesy of General Magnetics Co.)

Figure 16-13. Sintered ferrite magnets used in focusing arrays for tubes. (Courtesy of D.M. Steward Mfg. Co.)

(or permeability). When deviations exceed preset limits, a small jet sprays quick-drying paint onto the portions of the steel that are outside these limits. Finally, the steel passes through a small demagnetizing head to eliminate any residual magnetism that might affect the forming operation or ultimate spring performance.

Another unique development combines two permanent magnets and an iron-core electromagnet to provide either a differentiator or an integrator (depending upon the connections) that can combine *any* number of electrical or mechanical signals (or combinations of both). The system has no electrically moving parts and can be used to provide a simple visual output or a usable mechanical torque.

Generators and Motors

Permanent magnets have been widely applied where mechanical-to-electrical or electrical-to-mechanical energy conversion is required. They have been used in such small, commercial, "generator" applications as tachometers virtually since the permanent-magnet principle was first recognized. A major impetus to the wider use of magnets in this application was the minimum weight and size to the high-power output requirements on military equipment (communications, land vehicles, and aircraft) that began with World War II and has been subsequently maintained by other military or commercial aircraft needs.

Permanent magnet generators are essentially of two types: (1) Engine-ignition system generators (magnetos), and (2) AC or DC power and signal system generators. The two most common ignition-system magneto arrangements are shown in Figure 16-20. The heavy duty, high-power types and/or high-quality, light-duty types use the high-energy Alnico materials, while the "consumer" product units more commonly use sintered ferrites. Obviously, where a wide range of temperature conditions must be met (such as in military applications), Alnicos are used in all cases for temperature stability.

The units used for AC/DC power [4] and signal

---

4. All generators really produce AC internally. For a DC unit, commutation or semiconductors are included for final AC-to-DC conversion.

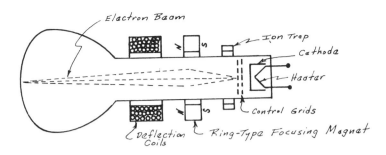

A. External Permanent Magnet

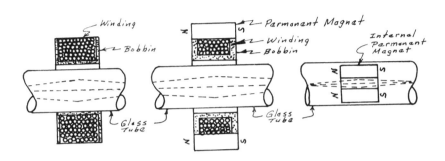

B. DC Electromagnet   C. External Electro-magnet & Permanent Magnet   D. Internal Permanent Magnet

Figure 16-14. Cathode ray tube focusing magnets (all types shown in section).

Figure 16-15. Two bonded ferrite magnets (arrows) used in television tube focusing assembly. (Courtesy of 3M Company)

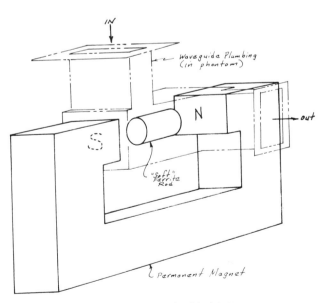

Figure 16-16. Typical microwave load isolator.

TYPICAL CIRCUITS AND APPLICATIONS   227

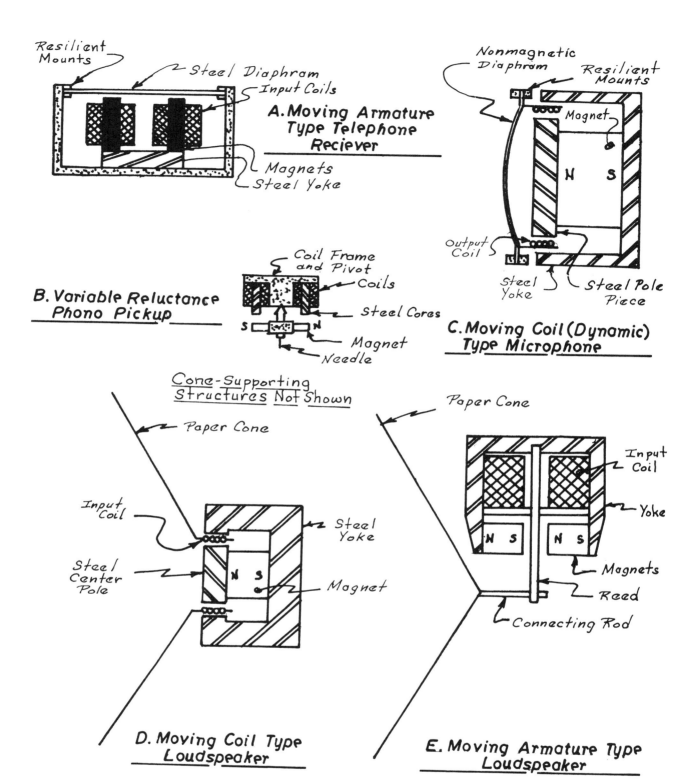

Figure 16-17. Typical electromechanical transducers (all shown in section).

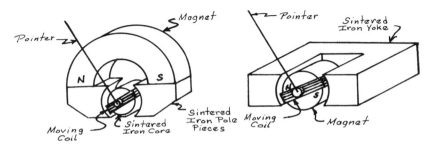
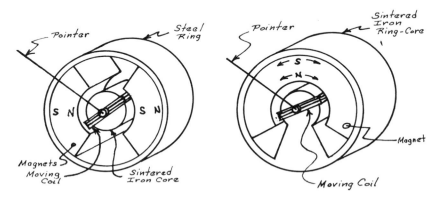

Figure 16-18. Magnet circuits used in D'Arsonval meters (bearings, springs, scale, etc., not shown).

TYPICAL CIRCUITS AND APPLICATIONS 229

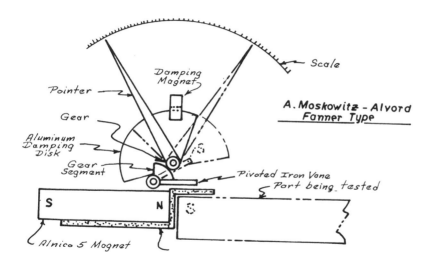
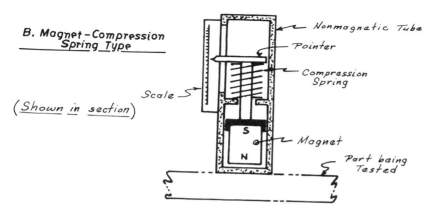

Figure 16-19. Coating thickness gauges.

generator systems are normally designed with the permanent magnet as the rotor, for self-evident technical and economic reasons. The configuration may vary widely with the application from a simple two-pole circuit to any number of poles on one rotor (Figure 9-14 on page 72 shows typical rotors). Among the most recent applications that employ a substantial volume of magnets, is the alternator used in automobiles and trucks. The long-established size, weight, and output advantages of permanent magnets developed for military and aircraft requirements were adaptable to commercial-consumer applications due to improvements in the energy-per-unit volume capabilities of Alnicos and the low cost of sintered ferrites.

The use of permanent magnets in both DC and AC motors, while more recent than their use in generators, developed more rapidly and extensively. Electric-clock motors using high-hysteresis wrought, metallic-magnet alloys became widely used in the mid-1930s. Since that time, magnets have been successfully used in such heavy-duty applications as steel-mill runout table motors of more than 40 horsepower.

Permanent-magnet motors, depending upon operating principle and circuit details, basically fall into the following types:

1. Rotational
    A. Discrete pole
        (1) General drive
        (2) Servo
        (3) Printed circuit
    B. Hysteresis-Reluctance
        (1) Conventional
        (2) Pyromagnetic
2. Stepping
3. Linear
4. Vibratory-oscillatory

Permanent-magnet motors may be designed with the magnet as either the stator or the rotor, depending upon the operational principle. Some types may be designed both ways. Discrete-pole motors are a case in point, although designs where the magnet is the stator are far more common. Linear or vibratory-oscillatory motors may also be designed with either a moving or a stationary magnet. For both hysteresis-reluctance and stepping types, the magnet must always be the rotor.

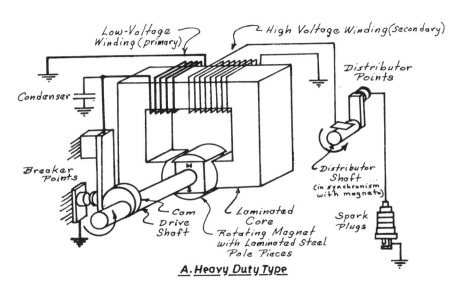

A. Heavy Duty Type

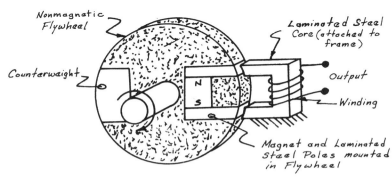

B. Small-Engine Type

Figure 16-20. Magneto ignition systems.

TYPICAL CIRCUITS AND APPLICATIONS 231

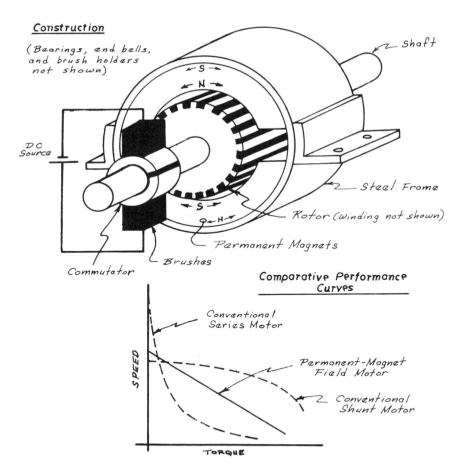

Figure 16-21. Typical permanent magnet field DC motor.

The most widely used type of motor is the discrete-pole, general drive, (torque) motor shown in Figure 16-21. Basically, as used in this product, the motor is similar to a conventional (all electromagnetic), externally excited, shunt-type DC motor. Figure 16-21 also shows the comparative speed–torque characteristics for a conventional shunt (self-excited), a conventional series, and a typical discrete-pole, permanent-magnet motor. Note that the performance of the permanent magnet motor is what would be expected of a conventional, separately-excited shunt motor that is "under excited." This type of motor is widely used in portable power tools, automotive windshield wipers, blowers and starters, and countless other consumer and industrial applications. Small permanent-magnet motors may employ rubber-bonded magnets as the stator poles. Alnicos and sintered ferrites are the primary materials used for larger, higher torque motors. Sintered ferrites are used in fractional horsepower in the low- to middle-quality range. For high-quality fractional horsepower and for all integral horsepower sizes, cast Alnicos 5 or 8 are used.

Discrete-pole servo motors and printed-circuit motors are basically the same as high-quality, general-drive motors, except for design details and configurational differences. These motors generally require the use of highly temperature-stable, high energy-per-unit-volume materials, such as cast Alnicos 5, 5DG, 5-7, 8, and 9 (and possibly the rare earths). The many potential rotor and stator configurations for this type of motor appear in Figures 9-13 and 9-14 (pages 71 and 72).

Cobalt-steel or Alnico hysteresis motors are widely used in gyroscopes, tape drives, and other applications where high starting torque and quiet operation at a very constant (synchronous) speed are desired. The elements and structure of this type of motor are shown in Figure 16-22A. The rotating magnetic field produced by the polyphase [5] stator winding induces a magnetic field in the permanent-magnet rotor. The flux in the rotor will lag the magnetizing current (field) in the stator and produce a torque. This torque is expressed by the equation:

5. The multiple phases may be available directly from the power source or they may be "created" by the same methods conventionally used for single-phase induction motors.

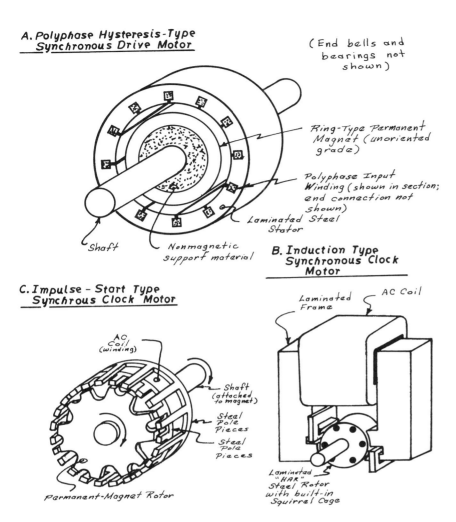

Figure 16-22. Typical permanent magnet AC motors.

$$T = KfVE_h$$

Where K is a constant depending on the units used, f is the frequency of applied AC, V is the volume of magnetic material in the rotor, and E_h is the energy-loss cycle characteristic of rotor material on a unit-volume basis.

From this equation, it is evident that the maximum torque will be determined by the hysteresis loss (E_h) in the rotor material. Of course, the stator electromagnet must be capable of providing the necessary H to realize the maximum E_h. Some designs use very thin Alnico rings backed by high-permeability steel to obtain high volumetric efficiency, but a high input to the winding is also required. Others (like the one shown in Figure 16-22A) employ a larger volume of a magnet material with a lower E_h, and the magnet is supported by a nonmagnetic core. The cobalt steels are used in such cases, reducing input power requirements.

Figures 16-22B and 16-22C show two designs of synchronous, self-starting AC motors widely used in consumer and industrial timers. Both of these motors differ from the motor in Figure 16-22A in that the rotors are magnetized *before* becoming a part of the assembly. In the impulse-starting type motor, the rotor-pole pitch is unequal and nonsymmetrical to the stator poles. When AC is applied to the winding to produce a stator field, the magnetized rotor will oscillate until the magnitude of this oscillation exceeds the pole pitch. At that point, the rotor "locks in," synchronizing with the rotating field of the stator and running at a constant speed which is a function of the applied frequency and the number of poles in the motor. To assure consistent starting in *one* direction, a mechanical pawl and cam arrangement (not shown in the illustration) is included in the design.

The synchronous induction-type motor "starts" on the same principle as a conventional squirrel-cage induction motor, but it "locks in" to synchronous speed when the lag angle permits. Since these motors are normally single-phase devices, the rotating magnetic field on the stator is produced by the shaded pole

TYPICAL CIRCUITS AND APPLICATIONS 233

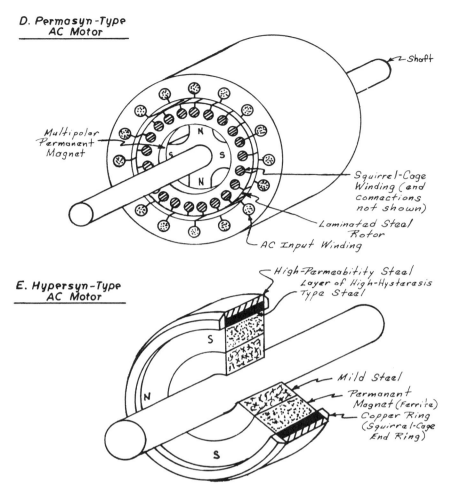

Figure 16-23. Typical permanent magnet AC motors (continued).

arrangement shown in Figure 16-22B. Relatively weak (wrought) permanent-magnet materials must be used in this type of motor to prevent the rotor from remaining "locked" at a standstill, in spite of the starting torque provided by the squirrel-cage winding.

Figure 16-23 shows two other types of synchronous AC permanent-magnet motors: the Permasyn type and the Hypersyn type. In the Permasyn design, a permanent magnet is used *inside* the rotor of what would otherwise be a conventional induction motor. The result is a "hybrid" motor that can start and run either as an induction motor or as a synchronous motor, depending upon the load. By this method, the advantages of both types of motor can be obtained, although at a considerable increase in cost. The Permasyn motor differs considerably from the motors shown in Figure 16-22 in both concept and performance. In the Permasyn design, the magnet is essentially "keepered" by the steel of the squirrel-cage rotor, but radial slots are added to force the flux into the stator. A highly coercive magnet material is necessary because of the demagnetizing influence during starting. Alternatively, the demagnetizing effect must be considered in the design.

The Hypersyn motor is unique in that it combines four different materials which give it particular characteristics. A conventional squirrel cage provides the best starting characteristics, a high-hysteresis element gives smooth torque characteristics, and a permanent-magnet rotor obtains "running" efficiency. The diffusion layer (outer ring) is made of a high-permeability material to improve the uniformity of the stator flux. The Hypersyn motor can accelerate heavy loads rapidly and then operate highly efficiently at a synchronous speed, with little or no "flutter" or "wow."

One of the most unique motors operates by converting radiant energy to mechanical motion. In this "pyromagnetic" motor, an incandescent lamp heats a *localized* section of a ferromagnetic rotor that is within the field of a permanent-magnet stator. Reduction of rotor permeability in the heated area reduces the force of magnetic attraction between the exposed area and the stator magnet. The unheated areas retain their higher permeability and are therefore more strongly attracted to the stator magnet. The result is a finite rotational force. The pyromagnetic motor is intended for use in advertising signs, lamps, or novelties; it produces a relatively high torque at low speeds.

Stepping motors are being increasingly used in a variety of product applications for electrical switching, mechanical transfer, or display purposes. There is some question as to whether these devices should be classified as motors or as stepping relays. Since they are normally designed to provide considerable torque rather than simply to perform a switching function, classification as a motor seems more appropriate. Stepping motors are designed with the permanent magnet as the rotor and varying numbers of coils (electromagnets) on the stator. The number of "steps" is determined by the number of permanent-magnet and electromagnetic poles. The holding force (locking) during nonactivated periods is determined by the strength of the field produced by the magnet in relation to the number of poles.

Stepping-motor magnets may range from the wrought materials to both the Alnicos and sintered ferrites. Materials with a high energy-per-unit volume reduce size and increase force-sensitivity characteristics, while ferrites permit high electrical pulse inputs without risk of rotor demagnetization.

A permanent-magnet type linear motor is essentially a directional solenoid with an infinite stroke. While specific products and applications have been largely limited to a few special actuators, the potential for this type of motor is quite high. Current developmental work on its applications ranges from material-handling drives to propulsion systems for high-speed transportation vehicles.

In linear motors, the magnet may be fixed or moving, but it must be highly coercive, and it must have high energy-per-unit weight and volume.

Permanent-magnet motors that produce reciprocating motion (linear or rotary) have been developed for both low-power and industrial applications. One such design, used to provide an oscillatory motion for an electronic clock escapement, employs six tiny magnets mounted on two Y-shaped yokes (three magnets per yoke). Three electromagnets are mounted between the magnets on each yoke. As the magnets of the escapement (which replaces the traditional clock balance wheel) move, they generate a drive signal to a transistor. The transistor in turn switches current through the drive coil to provide a constant reciprocating motion.

A well-established industrial application of a linear-type reciprocating permanent-magnet motor is shown in Figure 16-24. This photograph of five sizes of

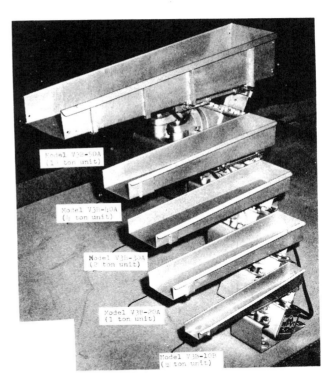

Figure 16-24. Typical AC-PM type (electromagnetically operated) vibratory feeders. (Courtesy of Eriez Magnetics, Erie, Pa., U.S.A.)

linear-type vibratory feeders [6] utilizes the permanent-magnet drive system shown in Figure 16-25. A fully electromagnetic form of this drive is shown in Figure 16-25A, and another permanent-magnet version appears in Figure 16-25C. The Type B drive utilizes either Alnico or ferrite materials successfully, while Type C is limited to the use of highly coercive ferrites. Vibratory feeders (and conveyors) using this drive are available with capacity ratings of 750 tons of dry sand per hour. This type of drive motor is also applied to storage bin "shakers," railroad car unloaders, and paper joggers.

Vibratory drive motors are limited by practical considerations to short "strokes" (up to three-eighths in.) and relatively high frequencies.[7]

The use of permanent magnets for existing motors or the invention of new types of motors is far from at an end. One recent development (for which no specific application has yet been found) is a dual-shaft motor that provides unique operating characteristics. The two shafts, located on opposite ends of the motor and *mechanically unconnected*, rotate in *opposite*

---

6. Used for industrial bulk materials or parts handling.
7. For DC power system operation, the normal operating frequency is 3600 vibrations per minute. Using special frequency generating equipment, any frequency is technically possible.

TYPICAL CIRCUITS AND APPLICATIONS

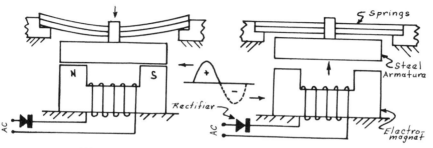
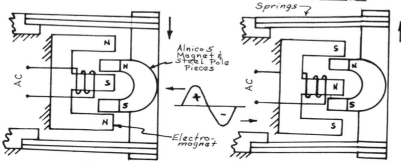
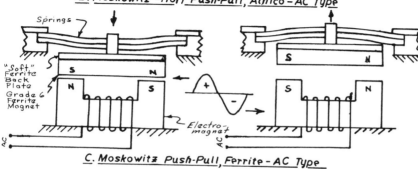

A. Weyant Pull-Release, Half-Wave Electromagnet Type

B. Moskowitz-Hoff Push-Pull, Alnico-AC Type

C. Moskowitz Push-Pull, Ferrite-AC Type

Figure 16-25. Vibratory motors.

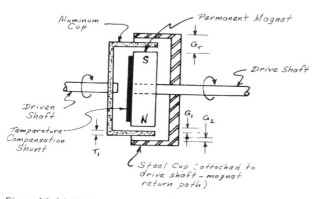

Figure 16-26. Eddy-current type automobile speedometer.

directions. Under no load or with the *same* load on both shafts, the speed of both shafts is a constant. If the load on one shaft increases, the speed of this shaft is reduced, with a proportionate increase in the speed of the *opposite* shaft. Thus, at all times and under all loads, where S_1 and S_2 are speeds of respective shafts and where K = constant:

$$S_1 + S_2 = K$$

Drives, Couplings, and Pumps

Two well-established applications of permanent magnets for these purposes are shown in Figures 16-26 and 16-27. The operating principle and advantages of magnets used in speedometers and tension controls are self-evident from a study of these illustrations.

Figure 10-10 and Figures 16-28 and 16-29 show the basic configuration of magnetic circuits that can be applied to drives or couplings, as well as to fluid-handling pumps. The obvious advantage of using magnets for these purposes is that torque can be transmitted without contact, *through* other materials, and without any

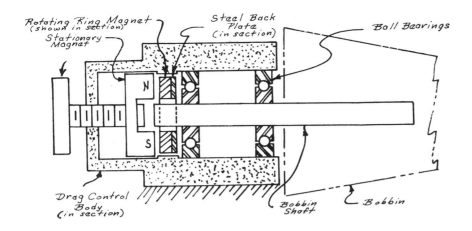

Figure 16-27. Hysteresis-type tension control for textile-machine bobbins.

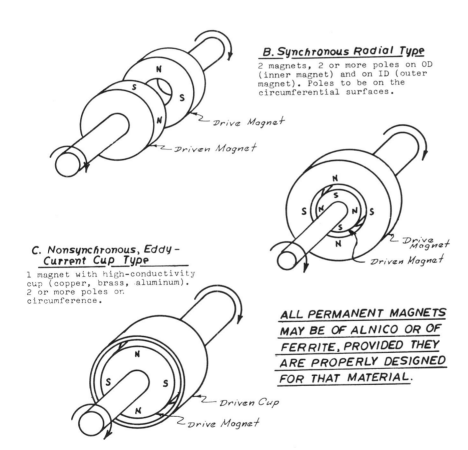

Figure 16-28. Magnetic drives and couplings. (Continued in Figure 16-29.)

TYPICAL CIRCUITS AND APPLICATIONS 237

D. Nonsynchronous, Eddy-Current Disk Type

1 magnet, 2 or more poles, 1 face. High-conductivity facing disk with mild-steel backing disk.

E. Nonsynchronous, Hysteresis-Cup Type

1 magnet, 2 or more poles on the circumference. Cup of high hysteresis alloy such as 17% cobalt steel.

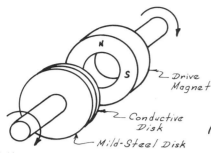
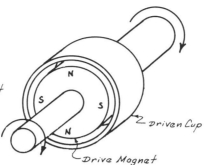

F. Nonsynchronous, Hysteresis Disk Type

1 magnet, 2 or more poles, 1 face. Disk of high-hysteresis alloy on face and mild-steel back.

G. Synchronous Linear Motion Type

Any number of "inner" magnets on common shaft with same number of "outer" magnets connected together. All magnets magnetized parallel to thickness. Usually all magnets are abutting, with mild steel rings (pole pieces) between adjacent magnets. NOTE POLARITY ARRANGEMENT

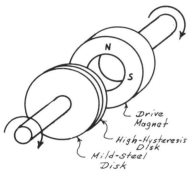
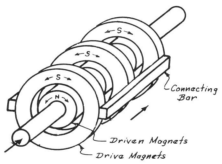

Figure 16-29. Magnetic drives and couplings (continued).

mechanical damage if the inherent torque limits are exceeded.

For pumps, using the high chemical stability of sintered ferrites, the "driven member" can operate directly in highly corrosive fluids. Drives and couplings based on high-energy Alnico magnets have been developed with ratings as high as 50 horsepower. Direct-acting pumps of the type shown in Figure 16-30A are commonplace in a variety of industries.

A unique pumping system, where the fluid being pumped also acts as part of the magnetic system (and thereby gets "pumped"), is shown in Figure 16-30B. This system is used to move conducting fluids in pipelines; it has the advantage of *no moving parts*. One version of this design contains almost one ton of cast Alnico 5 and produces over 1000 G in a 16½ in. air gap.

Many of the rotor-stator configurations shown in Figures 9-13 and 9-14 can be used in specific drive, coupling, and pumping applications.

Separation, Material Handling, and Automation Equipment

The earliest important magnetic separator application (albeit *electro*magnets) was in the beneficiation of iron ore. Before the turn of the century, electromagnetic "drums" were being used to remove large tonnages of nonferrous materials like rock from iron ore to form taconite. As permanent-magnet materials improved, electromagnetic separators were gradually displaced entirely by permanent-magnet types. Figure 16-31 shows the essential separator "system" for the magnetic drums used in iron mining and in countless other separation applications. Figures 16-32 and 16-33 illustrate the early electromagnetic circuits and the seven basic permanent-magnet circuits that are the primary element in drum separators. The magnetic circuits in these illustrations are self-evident, but should be studied for potential application to other products. Particular attention should be given to the circuit in Figure 16-33G.

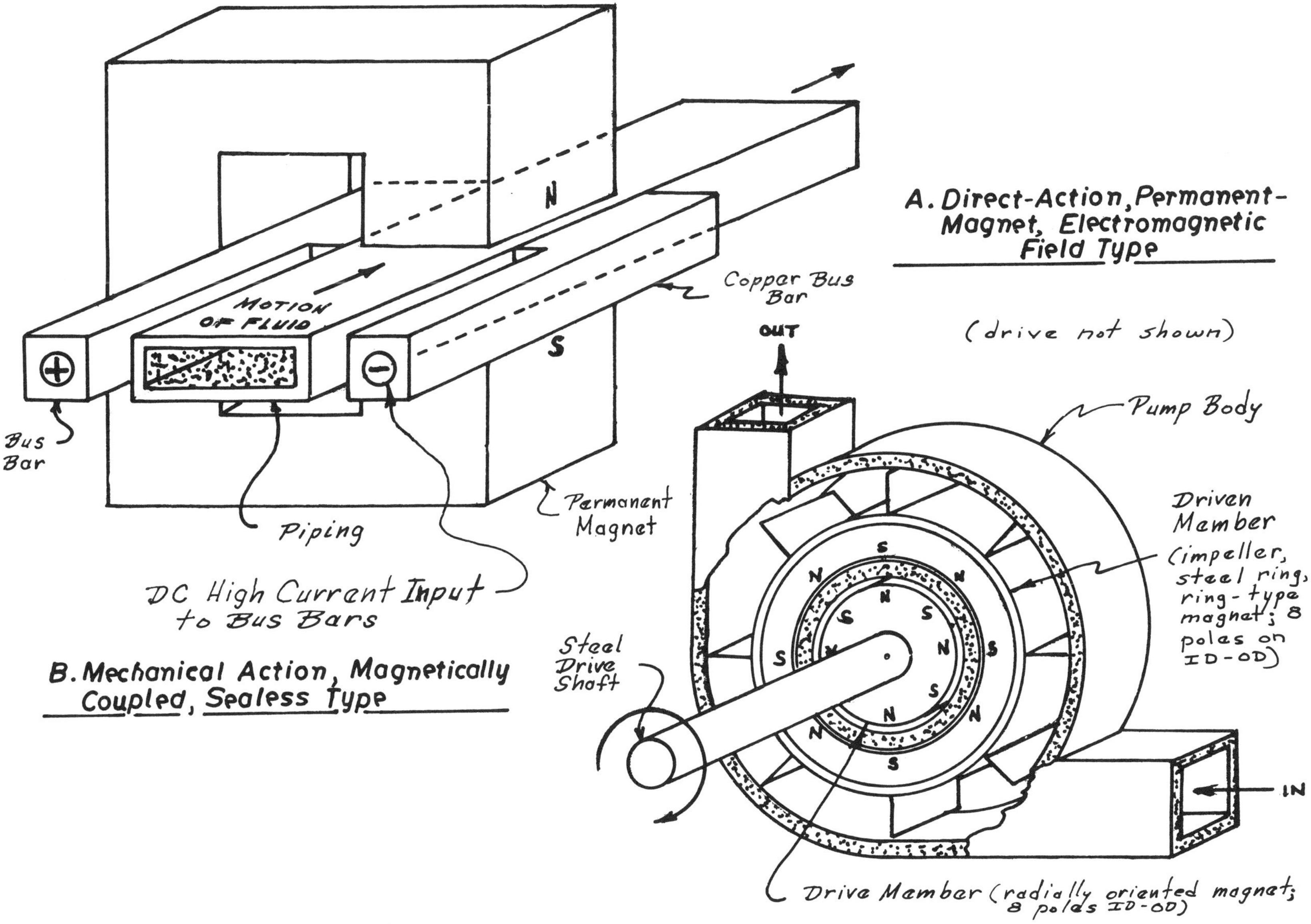

A. Direct-Action, Permanent-Magnet, Electromagnetic Field Type

(drive not shown)

Pump Body

Driven Member (impeller, steel ring, ring-type magnet; 8 poles on ID-OD)

Drive Member (radially oriented magnet; 8 poles as ID-OD)

Steel Drive Shaft

B. Mechanical Action, Magnetically Coupled, Sealess Type

Copper Bus Bar

Permanent Magnet

DC High Current Input to Bus Bars

Piping

Motion of Fluid

Bus Bar

Figure 16-30. Two types of magnetic pumps.

TYPICAL CIRCUITS AND APPLICATIONS 239

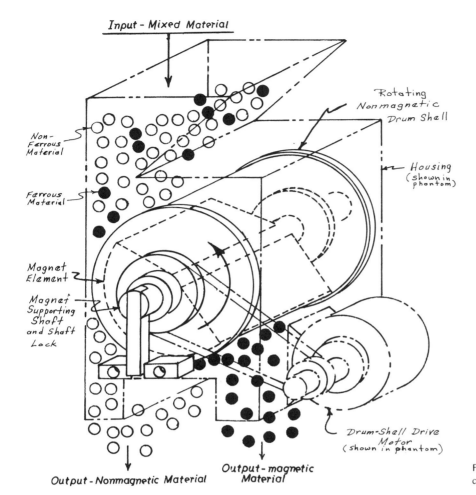

Figure 16-31. Typical drum-separator construction.

In this unique circuit, virtually all leakage flux has been eliminated to produce exceptional magnetic efficiency. The circuit in Figure 16-33F is also intended to minimize leakage (and *does*), while the configuration is adapted to the large magnet area and short-magnet length requirements of sintered ferrite materials.

Another form of heavy-duty, continuous-type magnetic separator appears in Figure 16-34 in two forms: the cross-belt type and the in-line type. The main operating section of this form of separator consists of three elements: an elongated "plate" magnet, two regular conveyor pulleys, and a conveyor belt. The only fundamental difference between the cross-belt and the in-line separator is the relative discharge position. The in-line type does have the advantage of providing the "mixed" materials with a longer "exposure" to the magnetic field.

Another form of continuous, "self-cleaning" unit appears in the photograph in Figure 16-35. In this system, a magnet *drum* is combined with a regular tail pulley and belt to provide a single-type, in-line separator. Various single drum-type, self-cleaning separators are widely used as coolant cleaners for machine tools that work ferrous materials.

As will be evident from the following examples, permanent-magnet separators range from large, complex, self-cleaning units to very simple assemblies that require periodic manual cleaning. One of the simplest of these special-purpose units is shown in Figure 16-36. This assembly of ring-shaped basic permanent magnets and mild-steel disks is used in plastic-molding machine infeeds and food processing lines to remove any fine iron contaminants before they can reach the molding or packaging operations. This type of separator is periodically lifted from the hopper, blown free of captured contaminants by an air hose, and replaced in the process lines.

Another simple assembly that is widely used for a variety of operational units is the plate magnet shown in Figures 9-36 and 9-37. Plate magnets are shown in more detail in the lower right-hand section of Figure 16-37. These magnets can be constructed of Alnico or sintered ferrite basic magnets mounted with common pole pieces (strips of mild steel). They are available

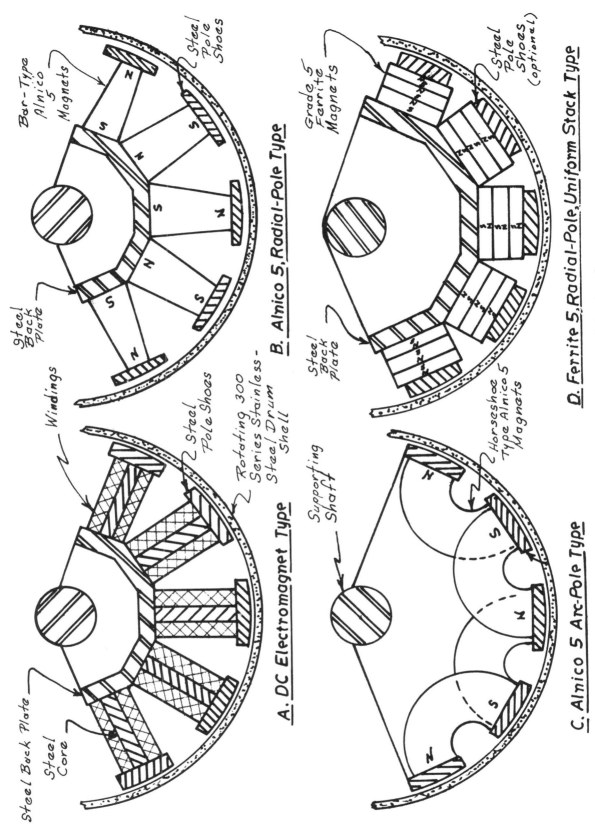

Figure 16-32. Typical drum separator elements (5-pole designs shown).

TYPICAL CIRCUITS AND APPLICATIONS 241

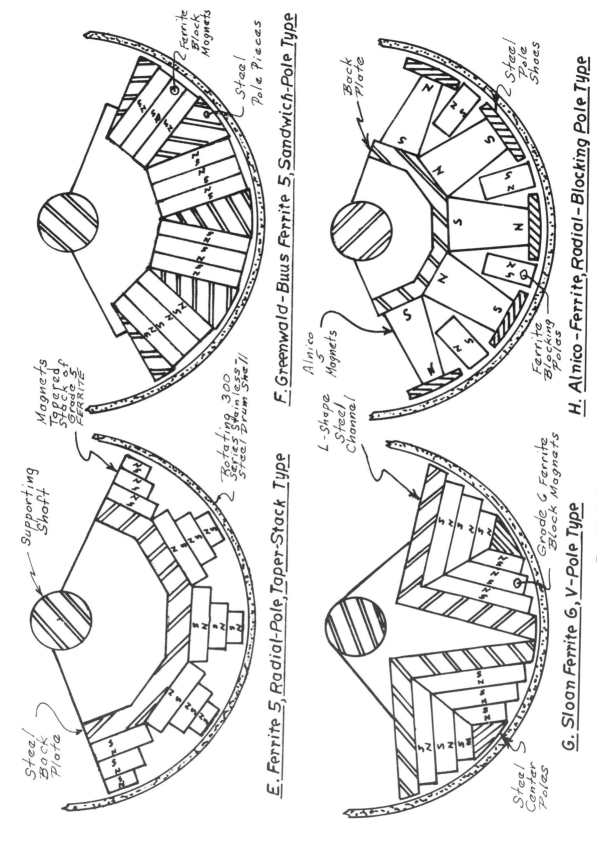

Figure 16-33. Typical drum-separator elements (continued).

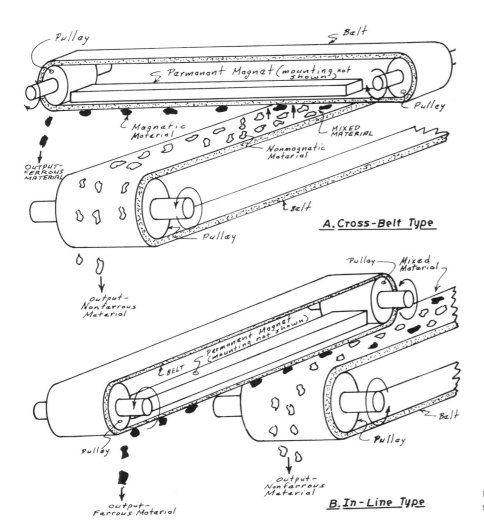

Figure 16-34. Continuous-belt type separators.

TYPICAL CIRCUITS AND APPLICATIONS 243

Figure 16-35. Self-cleaning suspension magnet assembly. (Courtesy of Bux-Shrader Co.)

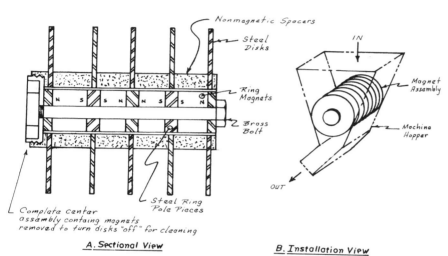

Figure 16-36. Simple disk-type hopper separator.

in a wide range of lengths, magnetic strengths, and "multiple-row" (parallel) units. The air gap is enclosed with a nonmagnetic material (aluminum, brass, stainless steel, or plastic) to prevent the captured magnetic materials from being drawn between the pole legs and shorting the magnets. Pole-face details vary from "recessed"-gap types to "stepped-pole" types or "flat-face" types as required for particular applications. For example, the recessed-gap type of plate magnet is ideal where fine iron contaminants are captured and may be washed off by subsequent material flow. The recessed gap provides a sheltered zone to minimize such wash-off.

Figure 16-37 also shows three typical applications of plate magnets in open-gravity chutes, combined with regular belt conveyors or in closed ducts or chutes called magnetic "humps". Figure 16-38 is a photograph of a typical hump separator, with one plate magnet opened to show the captured ferrous "tramp" materials. As we mentioned previously, plate magnets are also an essential element in most self-cleaning, continuous separators (like the one in Figure 16-34). As we will see later, the basic plate magnet configuration is also widely applicable to material-handling/automation applications.

The floor sweepers in Figures 16-39 and 16-40 are one "semi"-separation application of plate magnets. This type of sweeper, in either this simple form or in a more complex, self-cleaning form, is used for cleaning shop areas, airport runways, and so on. Hand-pushed, vehicle-towed, and fork-truck attachment types are available from a variety of manufacturers in various lengths and in assorted magnetic strengths.

The tube-magnet assemblies in Figure 9-3 are applied to many separator applications, as well as to many material-handling applications. Figures 16-41 through 16-45 show a variety of these separator applications, both used with dry-bulk materials and as liquid-line "traps."

The specific structures for magnetic separators are so varied that the preceding only provides a small sample of those used in actual practice. All separators are,

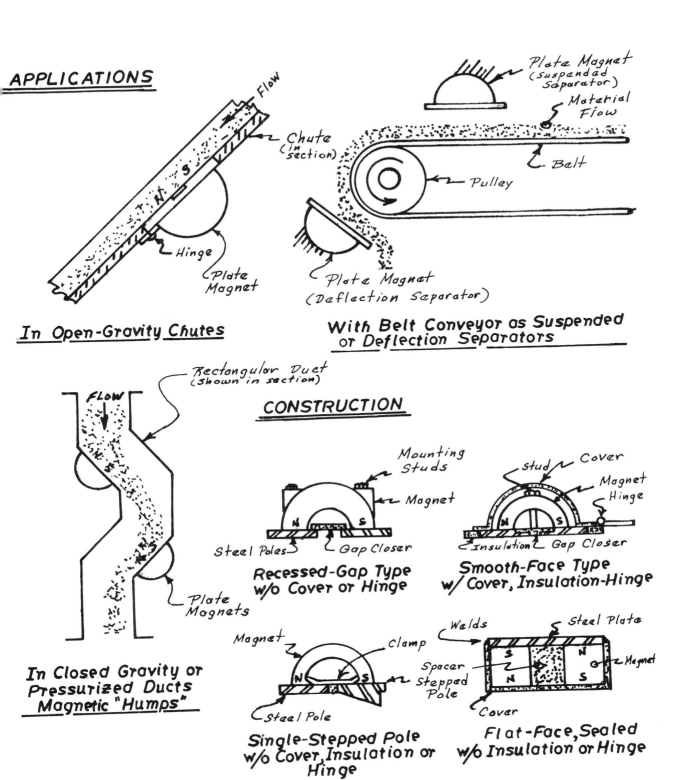

Figure 16-37. Simple plate-magnet separators. (Alnico horseshoe and double-block ferrite types shown; for other types see Figures 9-36 and 9-37.)

TYPICAL CIRCUITS AND APPLICATIONS 245

Figure 16-38. Magnetic hump separator consisting of two permanent plate magnets and special sheet-metal duct. Magnets shown swung open for cleaning. (Courtesy of Eriez Magnetics)

however, basically a plate-, tube-, pulley-, roll-, drum-, or rectangular-shaped magnet assembly, or some combination of these, with nonmagnetic ancillary elements (belts, chutes, etc.).

Categorizing magnetic equipment into separators versus material-handling devices (automation equipment) is often difficult and may be irrelevant. Many of the same assemblies (for example, plate magnets) may be used either as separators or material-handling units. This is particularly well illustrated by Figure 16-46, which shows a typical magnetic elevating conveyor for cans, steel boxes, machining chips, or general ferrous parts. Note that the system is essentially a conventional belt-conveyor system, with both straight and curved sections of plate magnets comprising the magnetic "holding" element. For more abrupt turns—or, in some cases, at the input (pick-up) point—magnetic pulleys or rolls [8] of the type shown in Figures 16-65, 16-66 and 16-67 may be substituted for the curved rail section or for the tail pulley.

The versatility of magnetic handling devices is illustrated in the can-conveying application in Figure 16-47, the can-washing application in Figure 16-48, and the can testing-sorting applications in Figures 16-49

8. Pulleys and rolls are essentially identical, except that rolls are normally "uncrowned."

and 16-50. Other unique applications are the bakery-pan handling system shown in Figure 16-51, the wooden barrel lifting system in Figure 16-52, and the foundry-flask clamping systems in Figure 16-53.

Control systems using magnets are shown in the can-routing system in Figure 16-54 and the bottle-cap "drag" control arrangement in Figure 16-55.

Sheet-steel handling applications range from multiple magnets mounted on a supporting grid to lift 4 × 8 feet sheets in warehouses to the sheet-"fanner" principle illustrated in Figures 16-56, 16-57, and 16-58 and 16-59. Steel tote-box applications range from the inter-floor elevator (Figure 16-60) to the shock-absorber, coupling-, slippage-control, and slide-control systems in Figures 16-61 to 16-64.

Figure 16-64 illustrates an interesting type of the "induced" magnetic roll, which differs from the "internally" powered magnetic rolls in Figures 16-65 to 16-67. Self-powered magnetic rolls are available in standard sizes from 3 ½ to 48 in. and in virtually any length. In addition to generally flat or contoured sheet-steel conveying, control for coating, slitting, etc., magnetic rolls or pulleys may be used in continuous separators. Rolls available with special contours are used for handling pipe in galvanizing, pressure testing and welding. A typical contoured-roll application for pipe coating is shown in Figure 16-68. Nonmagnetic materials like plastics and aluminum can often be handled by a magnetic "pinch-roll" system of the type shown in Figure 9-38.

Other forms of magnet assemblies used with chain conveyors for holding any type of steel part in painting, plating, degreasing, and coating processes appear in Figure 16-69. In addition to their applicability to a wide variety of part shapes and sizes, magnetic racks of these types have the advantage of being able to hold the parts without holes, hooks, etc. They can even hold small sheet-steel blanks by the edge alone.

Other material-handling applications (not illustrated) include magnetic trays for semiconductor manufacturing, circuit-holding boards, resistor lead-wire spacing controls (for packaging), and paint-spray masking systems.

Figure 16-70 shows the variety of commercially available standard holding assemblies that can be used for conveyor elements, as "die-set" units in stamping-pressing operations, and in general purpose holding applications. These types of units are available in hundreds of sizes, holding capabilities, and materials (both magnet and enclosing elements).

Permanent-magnet assemblies, both standard and custom-made, have substantial applications in the fabrication of steel assemblies by arc or gas welding. A more widespread use of magnets for this purpose has

APPLICATION

MANUAL OR VEHICLE TOWED

Without wheels and handle, may be hung on forks or on back of lift trucks.

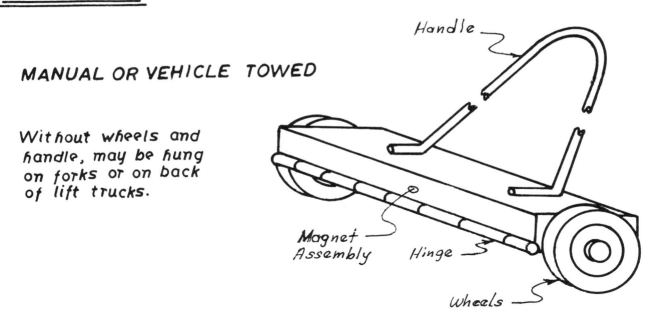

CONSTRUCTION

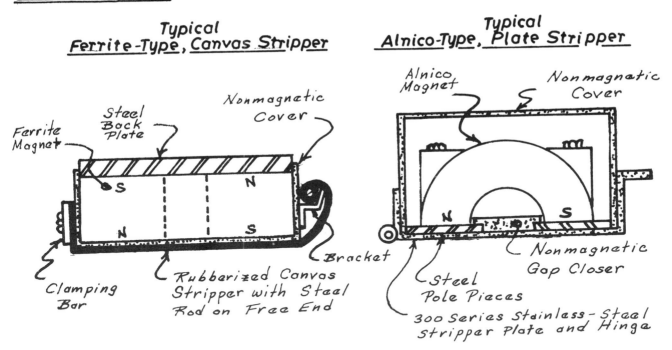

Figure 16-39. Heavy duty road, yard or floor sweeper (for other constructions see Figure 9-36 and 9-37).

TYPICAL CIRCUITS AND APPLICATIONS

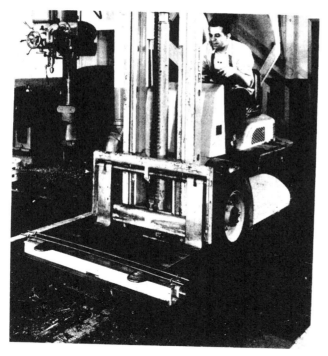

Figure 16-40. Permanent magnet floor sweeper on fork truck used to clean shop areas.

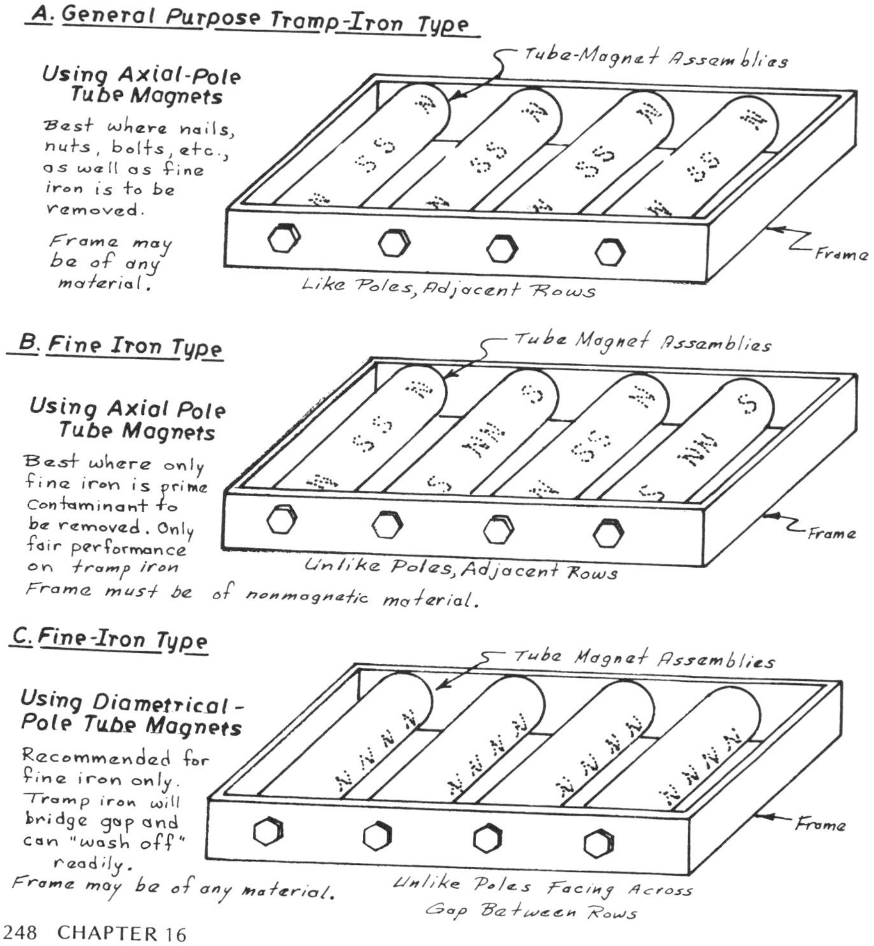

A. General Purpose Tramp-Iron Type
Using Axial-Pole Tube Magnets
Best where nails, nuts, bolts, etc., as well as fine iron is to be removed.
Frame may be of any material.
Tube-Magnet Assemblies
Frame
Like Poles, Adjacent Rows
B. Fine Iron Type
Using Axial Pole Tube Magnets
Best where only fine iron is prime contaminant to be removed. Only fair performance on tramp iron
Frame must be of nonmagnetic material.
Tube Magnet Assemblies
Frame
Unlike Poles, Adjacent Rows
C. Fine-Iron Type
Using Diametrical-Pole Tube Magnets
Recommended for fine iron only. Tramp iron will bridge gap and can "wash off" readily.
Frame may be of any material.
Tube Magnet Assemblies
Frame
Unlike Poles Facing Across Gap Between Rows

Figure 16-41. Typical single bank flat grate separators, housings not shown (For internal structure of tube-magnet assemblies see Figure 9-3.)

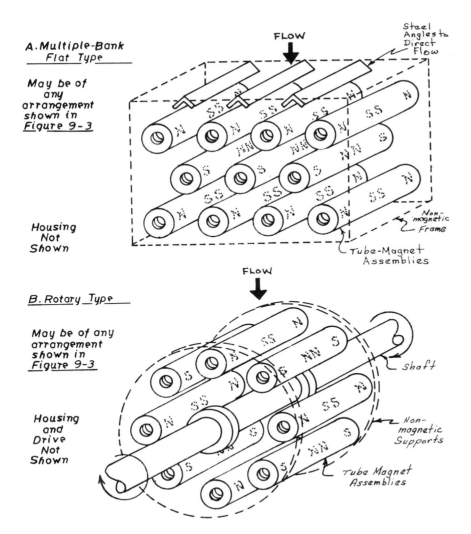

Figure 16-42. Other typical grate separators.

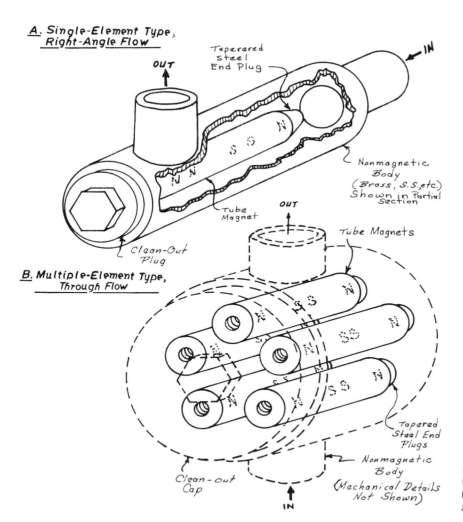

Figure 16-43. Liquid line filters or traps. (Axial-type tube magnets shown for both types; for internal structure of tube assemblies see Figure 9-3.)

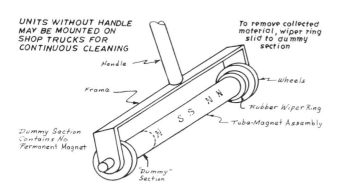

Figure 16-44. Light duty floor and tank sweepers. (Axial-type tube magnet shown; diametrical type not suitable; see Figure 9-3 for internal structure.)

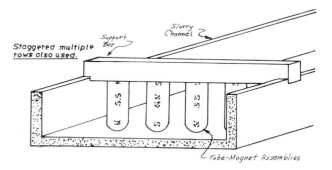

Figure 16-45. Open channel filters. (Axial-type tube magnets shown; for internal structures see Figure 9-3.)

APPLICATION

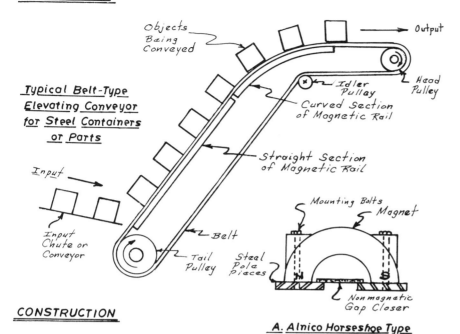

CONSTRUCTION

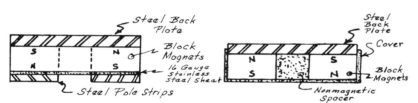

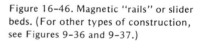

Figure 16-46. Magnetic "rails" or slider beds. (For other types of construction, see Figures 9-36 and 9-37.)

Figure 16-47. Permanent magnet rails deliver filled 8-ounce cans of tomato sauce to warehouse area. (Courtesy of Eriez Magnetics, Erie, Pa., U.S.A.)

TYPICAL CIRCUITS AND APPLICATIONS 251

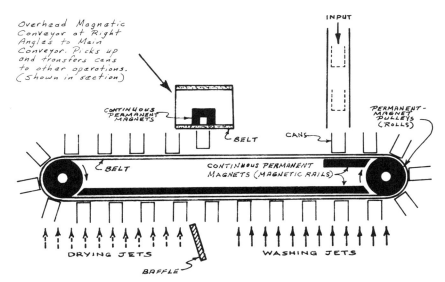

Figure 16-48. Continuous can washing/drying/inverting system.

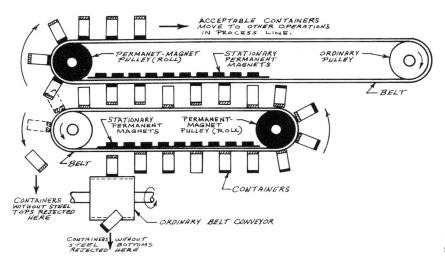

Figure 16-49. Automatic inspection and sorting of steel ended fiber containers.

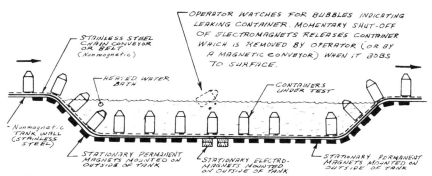

Figure 16-50. Inspection-sorting system for aerosol cans.

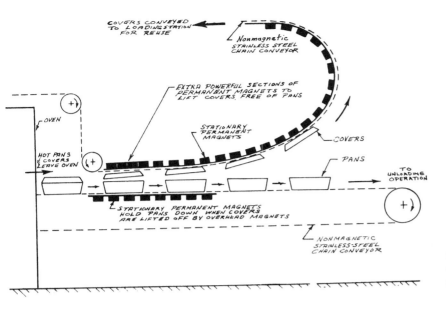

Figure 16-51. Automatic system for separating and recycling of hot baking pans and covers.

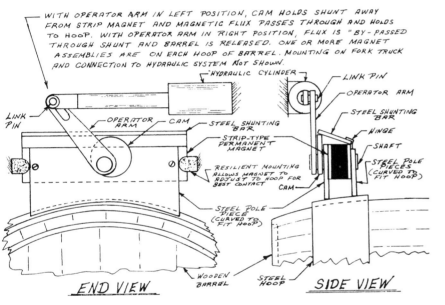

Figure 16-52. Fork truck attachment for handling wooden barrels.

TYPICAL CIRCUITS AND APPLICATIONS 253

A. DeCleene-Rankin Type for Wide-Flange Flasks

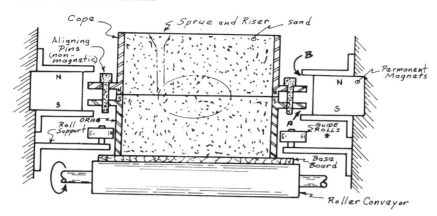

B. Moskowitz Type for Narrow-Flange Flasks

* Splash shields not shown

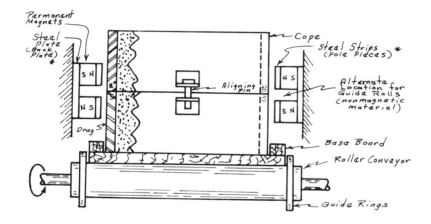

Figure 16-53. Continuous clamping of foundry flasks.

been inhibited by improper applications that have resulted in poor-quality arc welding or damage to the magnet in gas welding. Figures 16-71 and 16-72 show the correct methods for applying magnets in arc- and gas-welding operations. Figure 16-73 illustrates a specially designed, "releasable" welding *table* typical of those used in metal fabricating shops.

Miscellaneous Applications

Magnets are used in so many ways in products not covered in the preceding sections that they defy categorization. Magnetic applications range from toys and gadgets that have long been commercially available to sophisticated applications that are still in the earliest stages of development.

The *automotive industry* is a typical case in point. There magnets were initially used in speedometers (Figure 16-26) and have more recently been used in windshield-wiper motors, blower motors, alternators, and, most recently, starter motors. Automotive accessory items range from the old stand-by, the car compass, to dashboard clips, religious statues, medallions, banners, signs, and lights (Figures 16-74 and 16-75). Currently, considerable attention is being paid to devising new automotive drives for an "electric" automobile. It is a safe prediction that if and when an electric automobile is developed, it will use permanent-magnet motors (as some golf carts and industrial-plant personnel carriers do now).

Recently, a study was conducted in the *transportation field*, to prove the technical feasibility of using permanent magnets to increase the wheel-to-rail traction of lightweight railroad cars (for high-speed, intercity transportation systems). Other developmental work is currently being directed toward automated freight-car sorting systems and toward decreasing wear on the "third-rail shoe" used for electric trains. Office products and computer applications include permanent-magnet

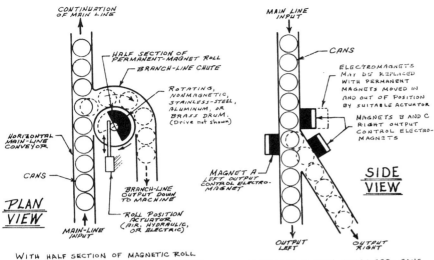

Figure 16-54. Horizontal or vertical can "routing" system.

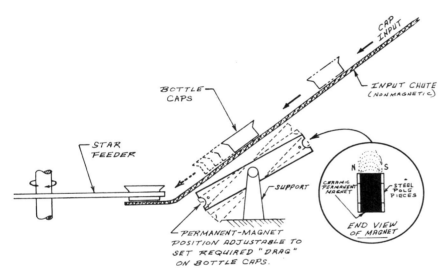

Figure 16-55. Magnetic "drag" system to control bottle caps.

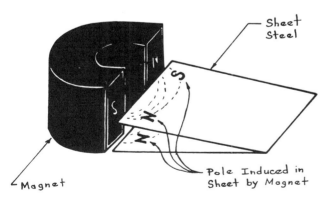

Figure 16-56. Principle of sheet fanning (separation).

TYPICAL CIRCUITS AND APPLICATIONS 255

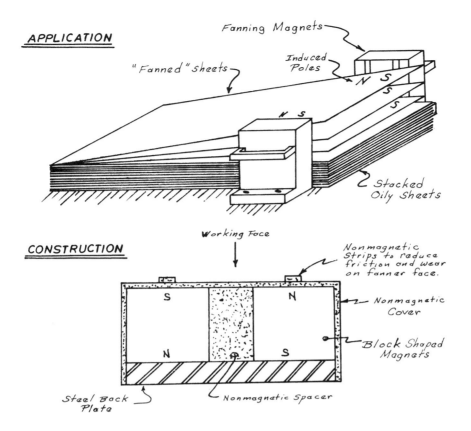

USED FOR SEPARATING OILY SHEET STEEL TO FACILITATE OR PERMIT MANUAL OR AUTOMATIC HANDLING

Figure 16-57. Sheet fanners or floaters. (Top-sectional view is the typical construction; for other constructions see Figures 9-36 and 9-37.)

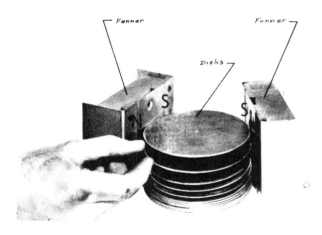

Figure 16-58. Sheet fanners separating oiled steel disks. (Courtesy of Magnetics Engineers, Philadelphia, Pennsylvania)

Figure 16-59. Single-sheet fanner separating perforated metal sheets. (Courtesy of The Permanent Magnet Company)

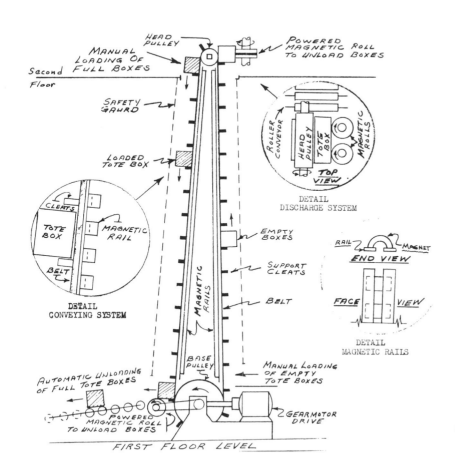

Figure 16-60. Vertical interfloor handling system for steel tote boxes (main structural details not shown).

TYPICAL CIRCUITS AND APPLICATIONS 257

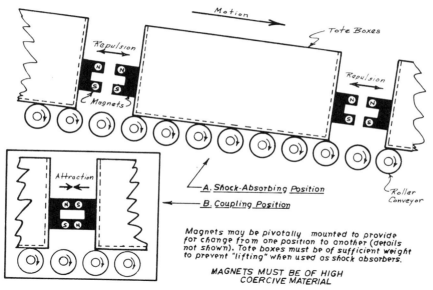

Figure 16-61. Magnets used as shock absorbers or "train" couplings.

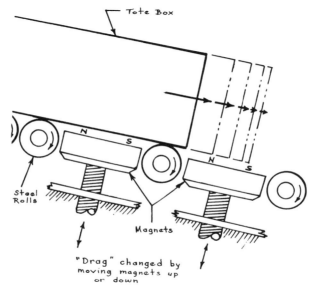

Figure 16-62. Controlling downward speed of steel tote boxes on inclined conveyors (structural details not shown).

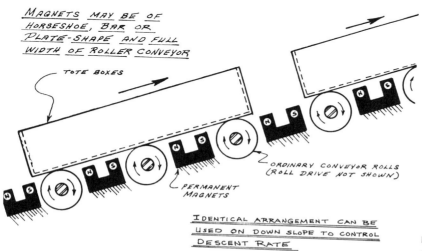

Figure 16-63. Eliminating tote-box slippage.

258 CHAPTER 16

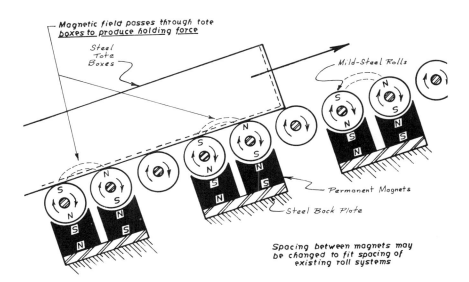

Figure 16-64. Induced magnetic rolls used to control tote boxes.

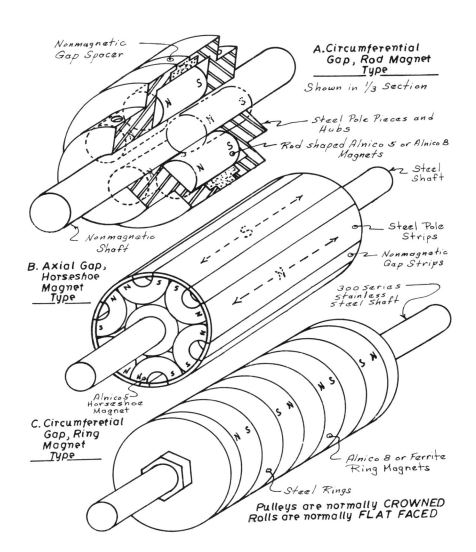

Figure 16-65. Separator or conveyor pulleys and rolls.

TYPICAL CIRCUITS AND APPLICATIONS 259

Figure 16-66. Small single- and multiple-conveyor rolls. (Courtesy of The Permanent Magnet Co.)

Figure 16-67. Large magnetic pulley (note crown). (Courtesy of Bux-Shrader Co.)

Figure 16-68. Using magnetic rolls to feed steel pipe through coating operation. (Courtesy of Eriez Magnetics, Erie, Pa., U.S.A.)

motors for tape drives, displays, printers, typewriters, adding machines, and blowers, among others. Reed-switch/magnet combinations are used for keyboards, selective panel switches, and safety interlocks. Among their more novel applications, is one (shown in Figure 16-76) that serves to improve the visibility of card files or folders. This system works on the sheet-fanner principle illustrated previously in Figure 16-56. A very popular use of both small Alnico magnets and bonded ferrites is for graphic displays such as production scheduling boards.

Textile machinery applications include the carding-machine crushing roll (Figure 9-5), bobbin tension controls (Figure 16-27), many forms of plate magnets (Figures 9-36 and 9-37), and several forms of rolls based on axial-pole tube magnets (Figure 9-3).

In the *construction industry*, permanent magnets are used for welding jigs (Figures 16-71 and 16-72) and for quick-removal ceilings (Figure 9-30). One preformed concrete products manufacturer of steps, beams, etc., has devised a quick-change system using high-power, permanent-magnet holders which act as clamps to secure the form sections until the poured concrete is set. Using this method for cast steps, the span, risers, and treads can be adjusted quickly to meet any specified dimensions.

Another "quick-change" application is used in the European *printing industry* to hold both rotary and flat printing mats. The mats are provided with a thin, flexible steel back, and permanent magnets are built into the printing roll (or platen) to secure the mat.

Two widely different and unusual *safety applications* are for *security locks* and for *sports*, specifically skiing. In the lock application, tiny magnets are embedded in the edge of what appears to be a conventional key. Similar magnets are built into the lock's tumbler system. The combination of the conventional pin-tumbler and the "coding" (polarity combinations) of the magnets must match to permit authorized operation of the lock. In the skiing application, rod-shaped Alnico 5 magnets are permanently attached to the skier's boots. High sensitivity Hall Detectors or similar magnetic sensing devices can then be used by search parties to locate lost skiers.

In *power generation* technology, permanent magnets are used as "through-the-wall" agitators in high-pressure systems, as safety devices for nuclear fuel rods, and even in high-powdered, fixed-station generators (Figure 9-17) A system currently under development uses magnetic fluids to convert thermal energy directly to electrical energy.

Environmental protection is provided by a unique process that cleans up oil spills. Fine iron is applied to, and combines with, the spilled oil. The iron, in turn,

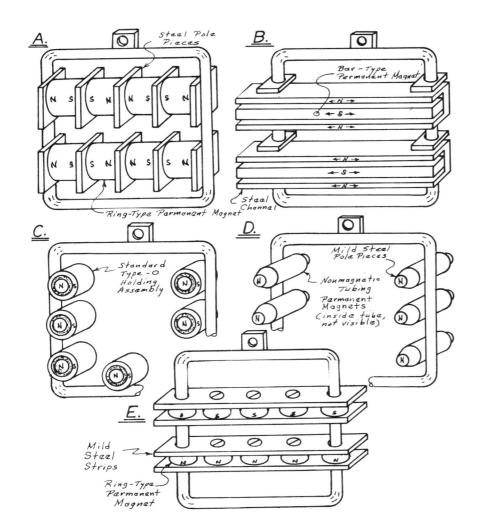

Figure 16-69. Typical painting, plating and degreasing racks for small parts and sheets.

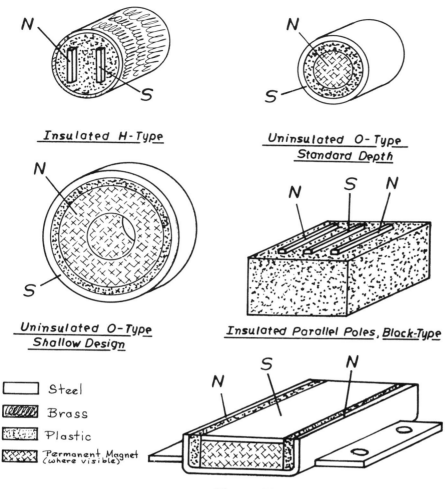

Figure 16-70. Typical commercially available standard holding assemblies.

can be recovered by conventional, continuous magnetic separation equipment.

The biomedical applications of permanent magnets range from very simple, well-established applications to new and unbelievably sophisticated neurosurgical systems. The most obvious (and oldest) biomedical application is a "fishing tool" to remove inadvertently swallowed foreign (and, hopefully, ferrous) objects from the stomach or lungs. Another simple (and economically important) application is the "cow magnet". In this application, a ½ in. diameter X 3 in. long Alnico magnet with rounded ends is placed in the cows first stomach. The heavy magnet remains in this tough stomach capturing ingested nails, barbed wire pieces, etc. before it can transfer to the cow's second, very delicate stomach. It prevents "tramp-iron disease" (also sometimes called "hardware disease").

Still another simple application is used in one European country (Sweden) as part of a human female contraceptive safety system. In this case, a small permanent magnet is molded into and becomes a part of a plastic intrauterine device (IUD). Part of the "kit" is a small pocket compass that permits checking to make sure that the IUD has not been expelled by the body.

A more significant biomedical application involving denture control is shown in Figure 16-77. Several patents have been issued covering denture control methods that use permanent magnets.

A well-documented but controversial application of magnets is to remove inoperable ferrous objects from the human body. The gradual, steady pull of a deep-field type magnet (attached for an extended period of time at the proper place on the patient's body) combines with the body's peristaltic action to gradually move the foreign object to an "operable" location. If the magnet is properly designed (for field gradient) and properly located, it is possible (with some good luck), for the foreign object to relocate to a surgically accessible position.

A. Using Parallel-Pole Magnets or Assemblies

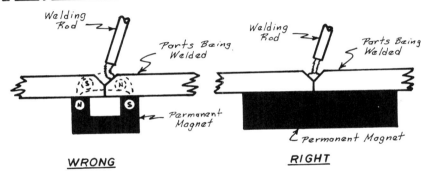

B. Using Rod-Magnet, Adjustable-Arm Assemblies

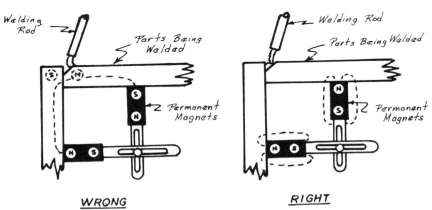

When improperly used, magnetic field across joint draws arc to one side. Properly used, magnet is "shorted" or localized away from weld.

Figure 16-71. Proper use of magnets in arc welding.

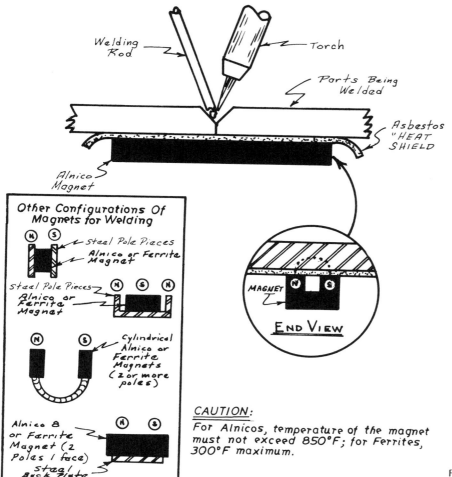

Figure 16-72. Proper use of magnets in gas welding.

Other biomedical applications (not illustrated) include means for making an artificial eye "follow" the remaining normal eye, eye lid control where nerve or muscle damage prevents the eye remaining open, a patient-proof blood shunt coupling for an artificial kidney machine, and telemetering of organ position and blood flow in animals. One of the most established of these applications combines a permanent magnet (external to the body) with a reed switch built into a dual rate heart pacemaker (implanted in vivo) to enable the patient to change the pacemaker rate from the "activity" mode to the rest mode.

Within the last ten years, a miniature implantable (in vivo) pump has been developed for the treatment of hydrocephalus ("water on the brain"). The pump element (shown in Figure 16-78) is surgically implanted permanently in the skull (see Figure 16-79). This pump consists of an impeller, a multipole permanent magnet, bearings, and a nonmagnetic case. The total pump is approximately 1 ¼ in. in diameter and 3/16 in. thick. The external drive unit (Figure 16-80) may be battery-powered for portable use or 115 V AC-powered for "home" use. The external unit consists of a "matching", multipole magnet driven by a small permanent-magnet motor. When the need for "pumping" is indicated, the external unit is placed against the head of the patient at the implant point of the pump. Magnetically coupled, the pump forces the unwanted fluid down an implanted drain tube and into the stomach.

Implications that the successful hydrocephalic pump could provide the power system for an artificial heart are obvious. While proposals for such a heart have already been made and actual developmental models have been built, no major effort is known to be in progress at this time.

Some of the most exciting new applications of magnets in biomedicine are occurring in neurology. The method shown in Figure 16-81 for repairing

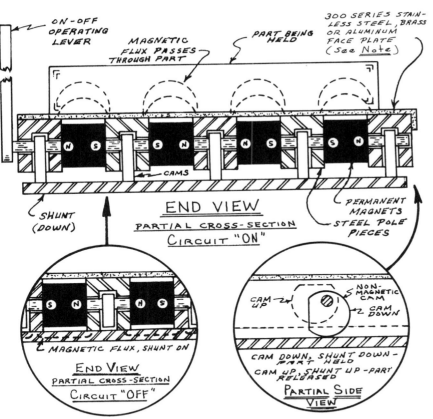

Note: Nonmagnetic face only required if total shut-off is necessary. As alternate, face may be omitted and double acting cam used to push part off. Magnets may be Alnicos or ferrites.

Figure 16-73. Releasable magnetic welding table.

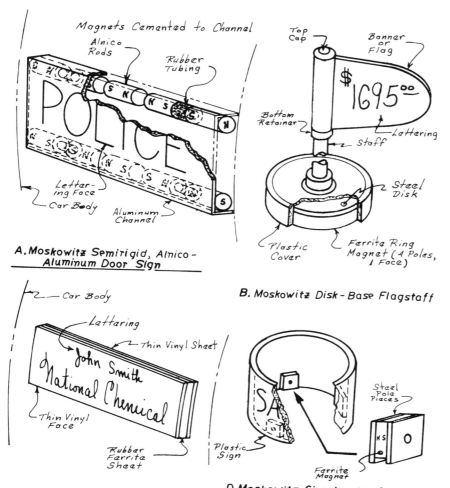

Figure 16-74. Removable vehicle signs, flags and lights.

aneurysms (blood-vessel "blowouts") is the forerunner of more complex and more effective systems. One such system uses a .6-.8 millimeter silastic catheter (tube) with one or more tiny permanent magnets on the tip. The catheter is introduced intravenously at the proper location and "guided" to the damage site by an external magnet. Progress of the catheter tip is followed fluoroscopically. A supplemental oscillatory magnetic field enables the catheter to move more freely without damage to the blood vessels, thus increasing maneuverability. This system has been successfully used to repair aneurysms (by filling them through the catheter with a nonreactive setting fluid) and to break embolisms (blockages).

Other biomedical applications in research or in very early developmental phases are subminiature blood-flow meters, blood-control valves, injected finely divided ferrites used in conjunction with external magnets for intestinal disfunction diagnosis, and magnetic-field measurement or the control of the electrical currents in the brain.

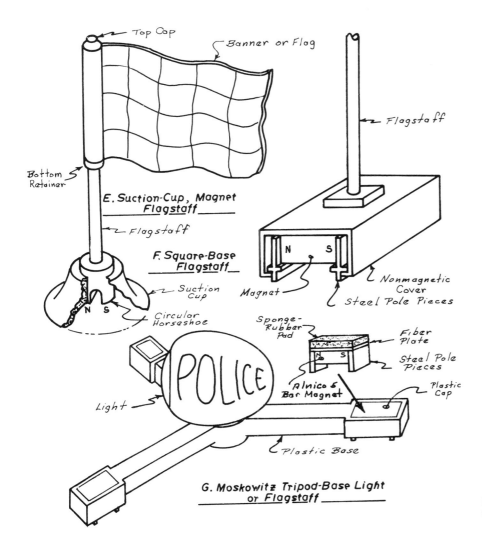

Figure 16-75. Removable vehicle signs, flags and lights (continued).

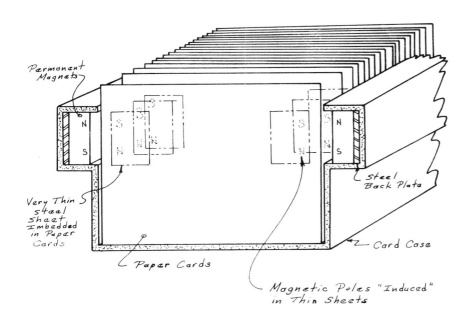

Figure 16-76. Magnetic sheet-fanning principle used to improve the visibility of filed cards.

TYPICAL CIRCUITS AND APPLICATIONS 267

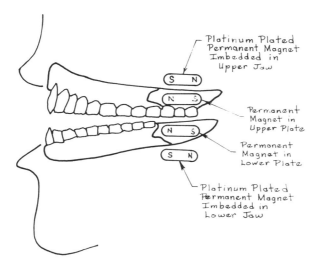

Figure 16-77. Magnets used to improve dental prosthetic device.

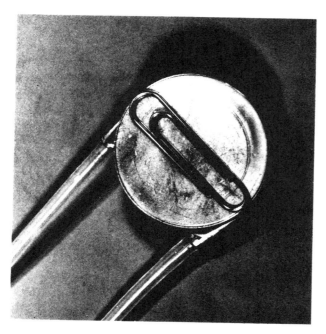

Figure 16-78. Implantable pump for encephalitics—pump unit. (Courtesy of Fairchild Hiller and Plenum Publishing Corp.)

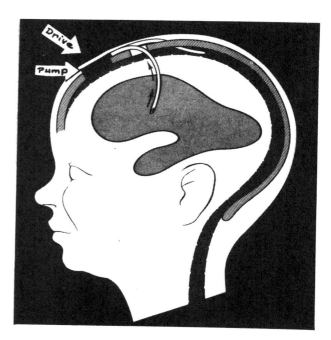

Figure 16-79. Implantable pump for encephalitics—location of pump unit. (Courtesy of Fairchild Hiller and Plenum Publishing Corp.)

Figure 16-80. Implantable pump for encephalitics—battery operated drive unit. (Courtesy of Fairchild Hiller and Plenum Publishing Corp.)

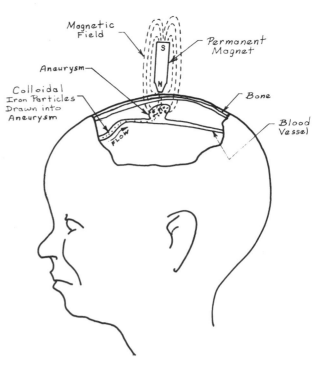

Figure 16-81. Repairing aneurysms using a magnet and colloidal iron particles.

TYPICAL CIRCUITS AND APPLICATIONS 269

Appendixes

Appendix 1

Demagnetization Curves

Note: Demagnetization Curves are not included for most of the new materials shown in the Data Sheets, Appendix 2. In many cases such curves were not published. Contact producing mill for such curves.

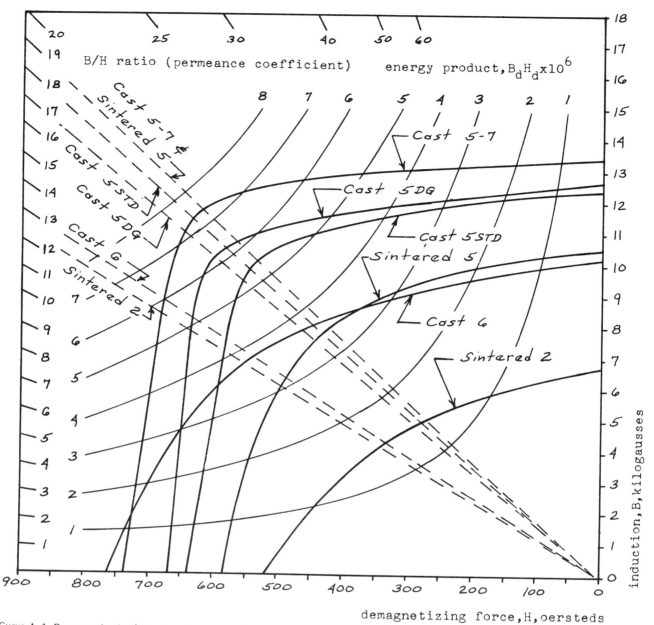

Curve A-1. Demagnetization/energy product curves for cast Alnicos 5, 5DG, 5-7, and 6 and for sintered Alnicos 2 and 5.

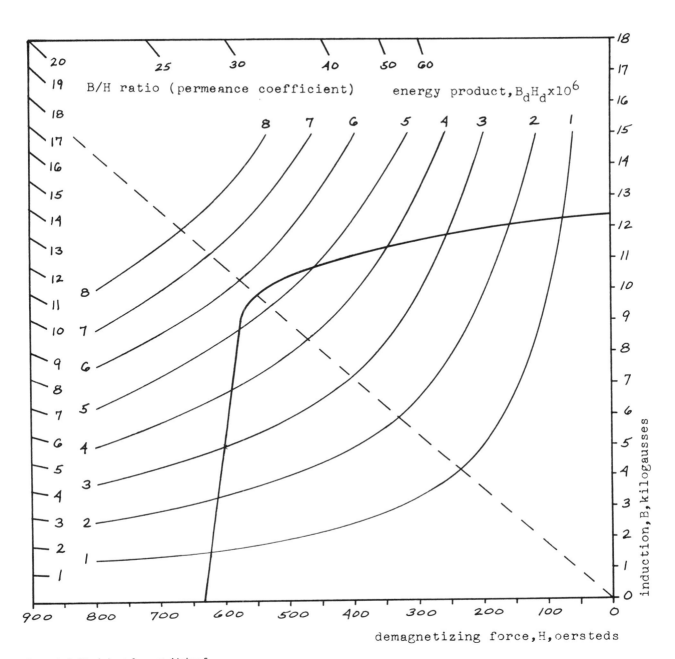

Curve A-2. Worksheet for cast Alnico 5.

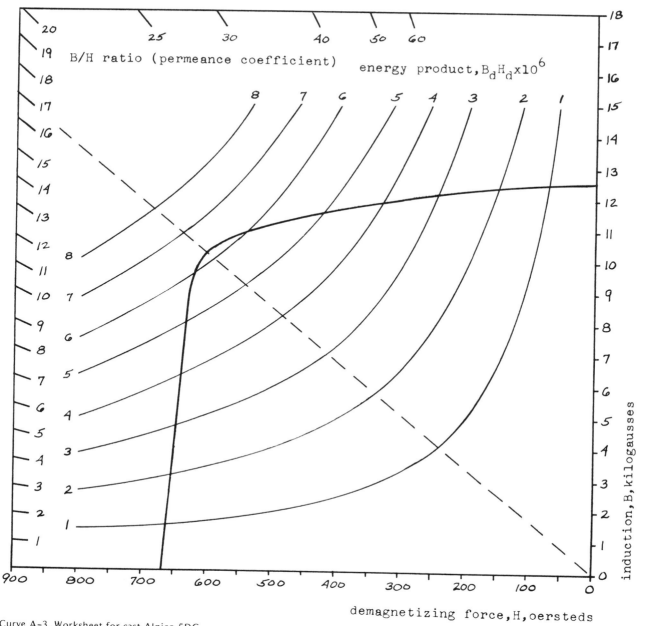

Curve A-3. Worksheet for cast Alnico 5DG.

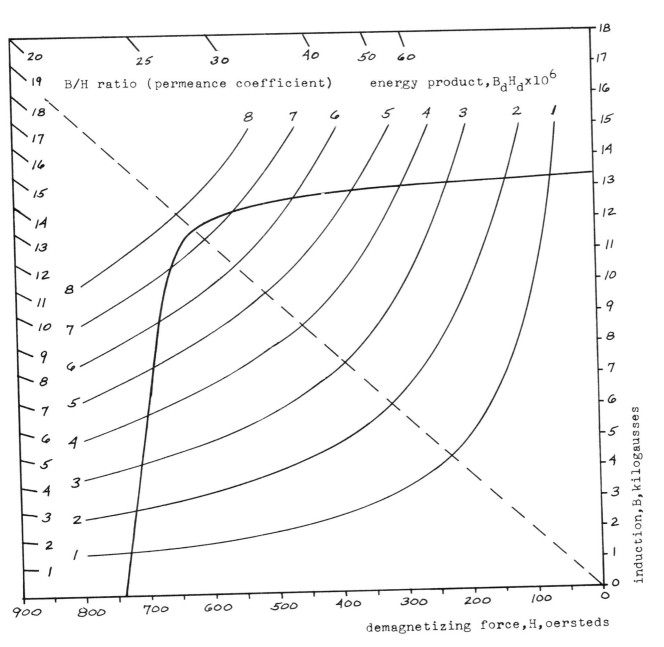

Curve A-4. Worksheet for cast Alnico 5-7.

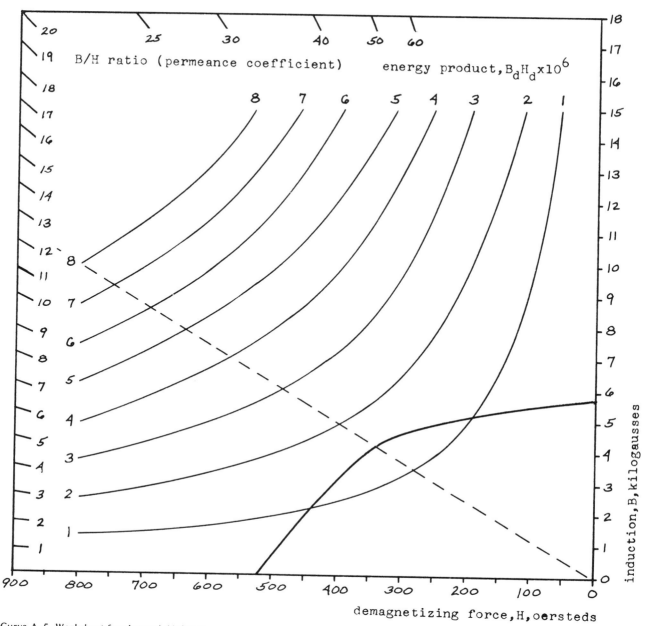

Curve A-5. Worksheet for sintered Alnico 2.

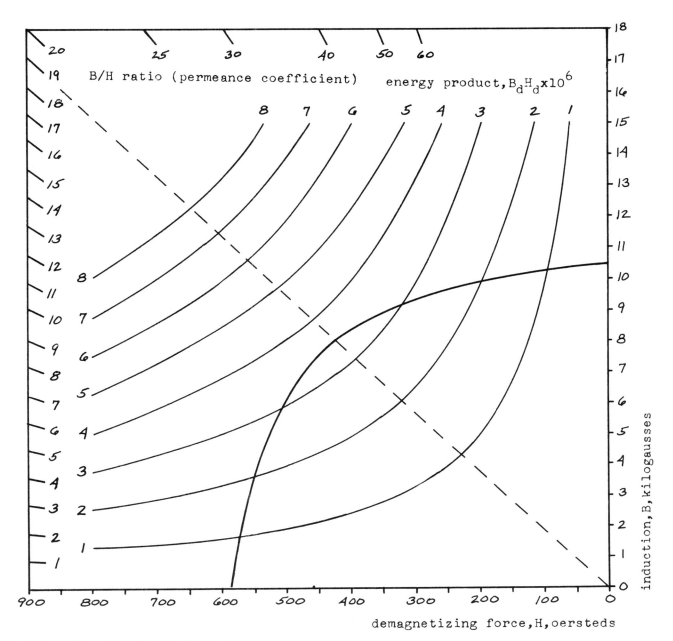

Curve A-6. Worksheet for sintered Alnico 5.

DEMAGNETIZATION CURVES

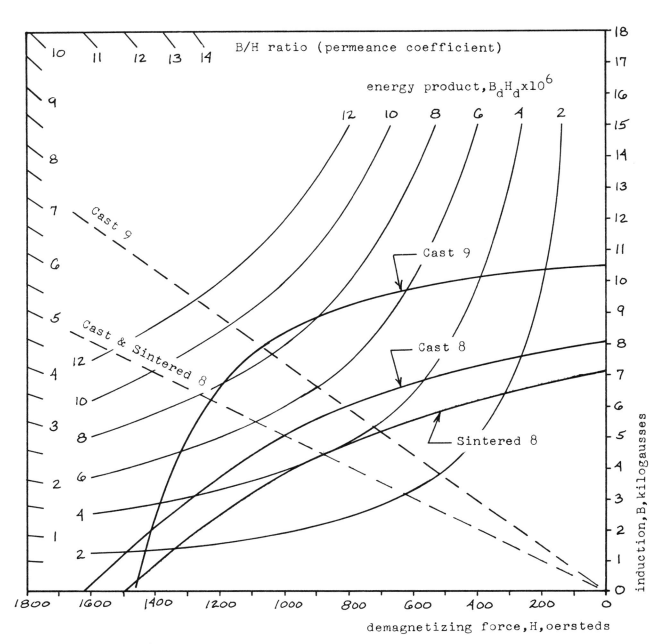

Curve A-7. Demagnetization/energy product curves for cast and sintered Alnico 8 and cast Alnico 9.

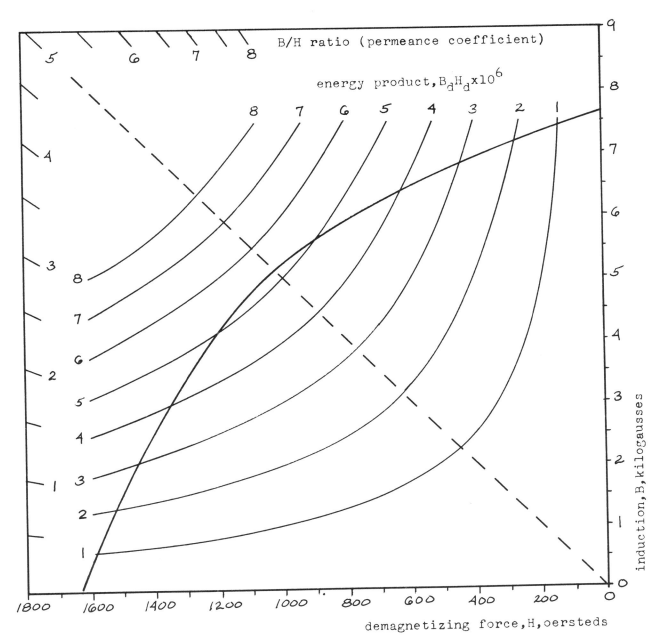

Curve A-8. Worksheet for cast Alnico 8.

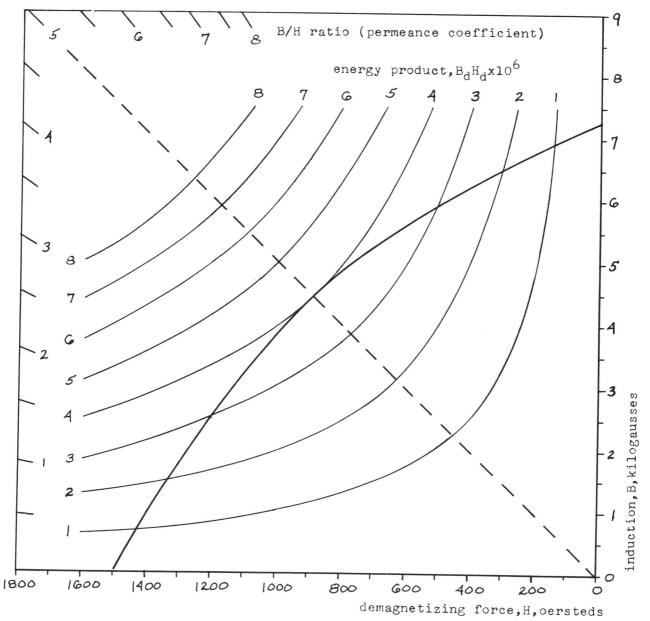

Curve A-9. Worksheet for sintered Alnico 8.

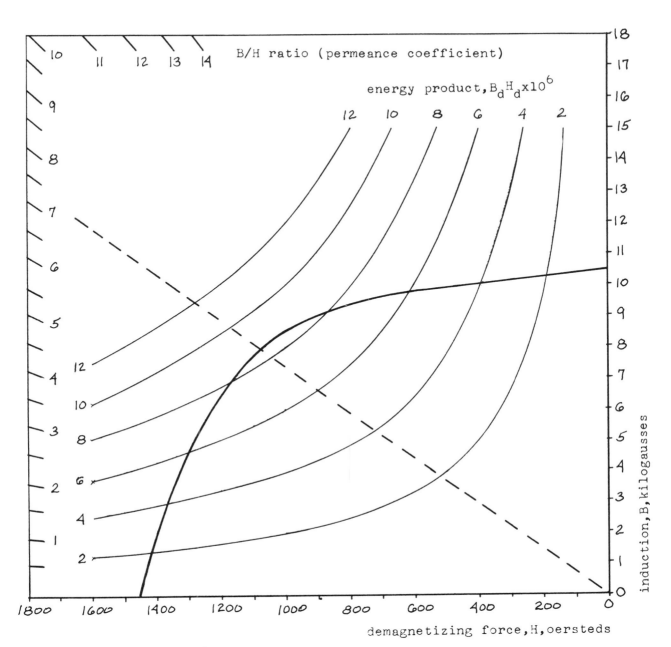

Curve A-10. Worksheet for cast Alnico 9.

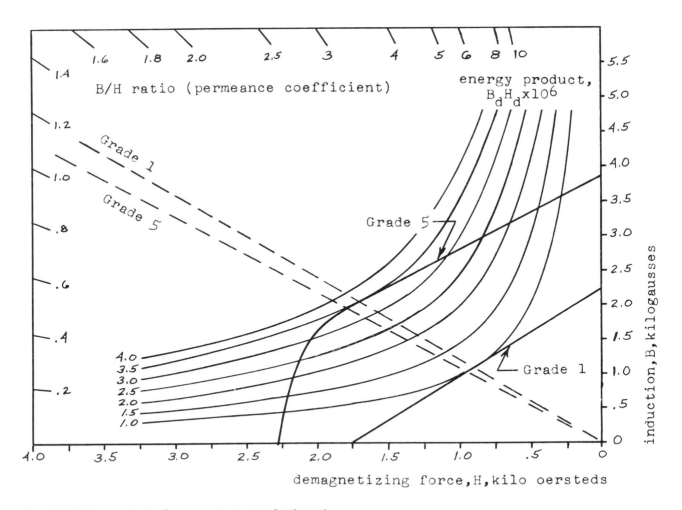

Curve A-11. Demagnetization/energy product curves for sintered ferrites 1 and 5.

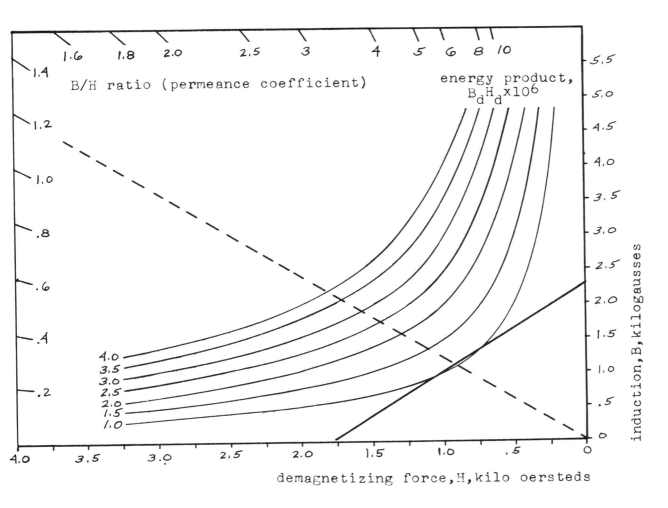

Curve A-12. Worksheet for sintered ferrite 1.

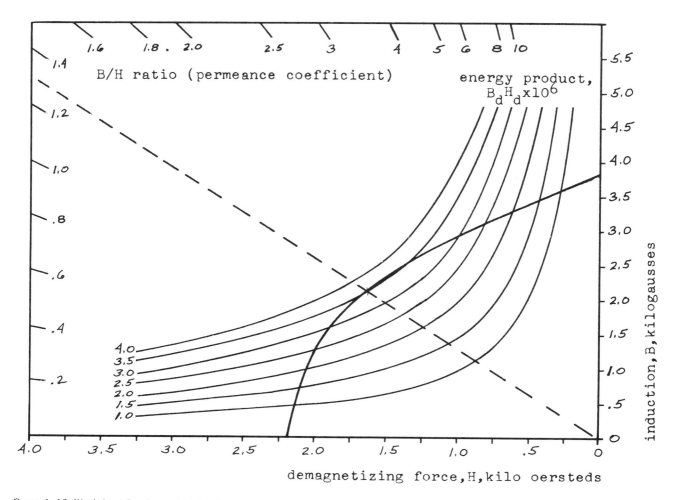

Curve A-13. Worksheet for sintered ferrite 5.

Appendix 2

Data Sheets

Magnetic & Physical Properties

Note: Demagnetization Curves are not included for these materials. See Appendix 1 for typical Demagnetizayion Curves or contact producer.

SOURCE(S):	**A-20 (segments)**
	(Alloy or Grade)
Stackpole Corp.	Sintered Ferrite
	(Type of Material)
USA	1.8/3.2
	(MMPA Brief Designation)
	-
	(IEC Code Reference)

MAGNETIC PROPERTIES		PHYSICAL PROPERTIES	
Residual Induction B_r (G)	2850	Density (lbs./in.^{3})	Est. .1734
Coercive Force (Oe) Normal (H_c)	2400	General Mechanical Properties	VHB
Coercive Force (Oe) Intrinsic (H_{ci})	3100	Tensile Strength (psi)	-
Maximum Energy Product BH_{max} (MGO)	1.8	Transverse Modulus of Rupture (psi)	-
B/H Ratio (Load Line) at BH_{max}	-	Coeff. of Therm. Expansion inches x 10^{-6}/°C	Orientation: Par. - Rt.Angl. -
Flux Density in Magnet B_m at BH_{max} (G)	-	Resistivity (micro-ohms/cm/cm^2)	1.0 @ 25 °C
Coercive Force in Magnet H_m at BH_{max} (Oe)	-	Hardness 6.5	Scale Moh's
Energy Loss W_c (watt-sec/cycle/lb)	-	Basic Forming Method	PF
Importance of Operating Magnet at BH_{max}	E	Basic Finishing Method	WGD
Average Recoil Permeability	Est. 1.1	Relative Corrosion Resistance	Ex
Field Strength Required to Fully Saturate Magnet (Oe)	10,000	Primary Chemical Composition	Ba or Sr & Fe
Ability to Withstand Demagnetizing Fields	Ex	Special Notes:	
Maximum Service Temperature (°F)	482		
Approximate Temperature Permanently Affecting Magnet (°F)	842		
Inherent Orientation of Material	none		

Note: All information based on available producers literature. In some cases of foreign materials interpretation or calculation required.

SOURCE(S):	**A-20 (sleeves)**
	(Alloy or Grade)
Stackpole Corp.	Sintered Ferrite
	(Type of Material)
USA	1.4/3.1
	(MMPA Brief Designation)
	-
	(IEC Code Reference)

MAGNETIC PROPERTIES		PHYSICAL PROPERTIES	
Residual Induction B_r (G)	2575	Density (lbs./in.^{3})	Est. .1734
Coercive Force (Oe) Normal (H_c)	2130	General Mechanical Properties	VHB
Coercive Force (Oe) Intrinsic (H_{ci})	3100	Tensile Strength (psi)	-
Maximum Energy Product BH_{max} (MGO)	1.35	Transverse Modulus of Rupture (psi)	-
B/H Ratio (Load Line) at BH_{max}	-	Coeff. of Therm. Expansion inches x 10^{-6}/°C	Orientation: Par. - Rt.Angl. -
Flux Density in Magnet B_m at BH_{max} (G)	-	Resistivity (micro-ohms/cm/cm)	21.0 @ 25 °C
Coercive Force in Magnet H_m at BH_{max} (Oe)	-	Hardness Scale	6.5 Moh's
Energy Loss W_c (watt-sec/cycle/lb)	-	Basic Forming Method	PF
Importance of Operating Magnet at BH_{max}	E	Basic Finishing Method	WGD
Average Recoil Permeability	Est. 1.1	Relative Corrosion Resistance	Ex
Field Strength Required to Fully Saturate Magnet (Oe)	10,000	Primary Chemical Composition	Ba or Sr & Fe
Ability to Withstand Demagnetizing Fields	Ex	Special Notes:	
Maximum Service Temperature (°F)	482		
Approximate Temperature Permanently Affecting Magnet (°F)	842		
Inherent Orientation of Material	none		

Note: All information based on available producers literature. In some cases of foreign materials interpretation or calculation required.

SOURCE(S):	**Alcomax 3A**
	(Alloy or Grade)
SG Magnets, Ltd.	Sintered Alnico
	(Type of Material)
USA	4.9/.67
	(MMPA Brief Designation)
	-
	(IEC Code Reference)

MAGNETIC PROPERTIES		PHYSICAL PROPERTIES	
Residual Induction B_r (G)	11,700	Density (lbs./in.^{3})	.2570
Coercive Force (Oe) — Normal (H_c)	650	General Mechanical Properties	HT
Coercive Force (Oe) — Intrinsic (H_{ci})	670	Tensile Strength (psi)	27,560
Maximum Energy Product BH_{max} (MGO)	4.9	Transverse Modulus of Rupture (psi)	43,560
B/H Ratio (Load Line) at BH_{max}	18.0	Coeff. of Therm. Expansion inches x 10^{-6}/°C	Orientation: Par. 10.04 Rt.Angl. -
Flux Density in Magnet B_m at BH_{max} (G)	9350	Resistivity (micro-ohms/cm/cm^2)	52 @ 25 °C
Coercive Force in Magnet H_m at BH_{max} (Oe)	520	Hardness Scale	47 Rockwell C
Energy Loss W_c (watt-sec/cycle/lb)	-	Basic Forming Method	DPM
Importance of Operating Magnet at BH_{max}	A	Basic Finishing Method	WGR
Average Recoil Permeability	3.8	Relative Corrosion Resistance	G
Field Strength Required to Fully Saturate Magnet (Oe)	3000	Primary Chemical Composition	Al-Ni-Co-Fe
Ability to Withstand Demagnetizing Fields	F	Special Notes:	
Maximum Service Temperature (°F)	842		
Approximate Temperature Permanently Affecting Magnet (°F)	1436		
Inherent Orientation of Material	MP		

Note: All information based on available producers literature. In some cases of foreign materials interpretation or calculation required.

SOURCE(S):	
SG Magnets, Ltd.	
USA	

Alcomax 4
(Alloy or Grade)

Sintered Alnico
(Type of Material)

4.1/.73
(MMPA Brief Designation)

-
(IEC Code Reference)

MAGNETIC PROPERTIES			PHYSICAL PROPERTIES	
Residual Induction B_r (G)		10,600	Density (lbs./in.^{3})	.2570
Coercive Force (Oe)	Normal (H_c)	720	General Mechanical Properties	HT
	Intrinsic (H_{ci})	730	Tensile Strength (psi)	27,560
Maximum Energy Product BH_{max} (MGO)		4.1	Transverse Modulus of Rupture (psi)	43,500
B/H Ratio (Load Line) at BH_{max}		14.6	Coeff. of Therm. Expansion inches x 10^{-6}/°C	Orientation: Par. 11.0 Rt.Angl. -
Flux Density in Magnet B_m at BH_{max} (G)		7800	Resistivity (micro-ohms/cm/cm^2)	52 @ 25 °C
Coercive Force in Magnet H_m at BH_{max} (Oe)		535	Hardness Scale	48 Rockwell C
Energy Loss W_c (watt-sec/cycle/lb)		-	Basic Forming Method	DPM
Importance of Operating Magnet at BH_{max}		A	Basic Finishing Method	WGR
Average Recoil Permeability		4.3	Relative Corrosion Resistance	G
Field Strength Required to Fully Saturate Magnet (Oe)		3000	Primary Chemical Composition	Al-Ni-Co-Fe
Ability to Withstand Demagnetizing Fields		F	Special Notes:	
Maximum Service Temperature (°F)		842		
Approximate Temperature Permanently Affecting Magnet (°F)		1436		
Inherent Orientation of Material		MP		

Note: All information based on available producers literature. In some cases of foreign materials interpretation or calculation required.

SOURCE(S):	Alni 120
Magnetfabrik-Bonn	(Alloy or Grade)
	Sintered Alnico
	(Type of Material)
Germany	1.1/.58
	(MMPA Brief Designation)
	-
	(IEC Code Reference)

MAGNETIC PROPERTIES		PHYSICAL PROPERTIES	
Residual Induction B_r (G)	5500	Density (lbs./in.^{3})	.2457
Coercive Force (Oe) — Normal (H_c)	550	General Mechanical Properties	HT
Coercive Force (Oe) — Intrinsic (H_{ci})	575	Tensile Strength (psi)	Approx. 28,000
Maximum Energy Product BH_{max} (MGO)	1.1	Transverse Modulus of Rupture (psi)	Approx. 44,000
B/H Ratio (Load Line) at BH_{max}	10.3	Coeff. of Therm. Expansion inches x 10^{-6}/°C — Orientation: Par. - Rt.Angl. -	
Flux Density in Magnet B_m at BH_{max} (G)	3300	Resistivity (micro-ohms/cm/cm^2)	Approx. 50 @ 25 °C
Coercive Force in Magnet H_m at BH_{max} (Oe)	320	Hardness Scale	Approx. 50 Rockwell C
Energy Loss W_c (watt-sec/cycle/lb)	-	Basic Forming Method	DP
Importance of Operating Magnet at BH_{max}	A	Basic Finishing Method	WGR
Average Recoil Permeability	5.0	Relative Corrosion Resistance	G
Field Strength Required to Fully Saturate Magnet (Oe)	3000	Primary Chemical Composition	Al-Ni-Co-Cu-Fe
Ability to Withstand Demagnetizing Fields	F	Special Notes:	
Maximum Service Temperature (°F)	842		
Approximate Temperature Permanently Affecting Magnet (°F)	1436		
Inherent Orientation of Material	none		

Note: All information based on available producers literature. In some cases of foreign materials interpretation or calculation required.

SOURCE(S):

SG Magnets, Ltd.

USA

Alnico F
(Alloy or Grade)

Sintered Alnico
(Type of Material)

1.5/.63
(MMPA Brief Designation)

-
(IEC Code Reference)

MAGNETIC PROPERTIES		PHYSICAL PROPERTIES	
Residual Induction B_r (G)	6200	Density (lbs./in.^{3})	.2460
Coercive Force (Oe) Normal (H_c)	620	General Mechanical Properties	HT
Coercive Force (Oe) Intrinsic (H_{ci})	630	Tensile Strength (psi)	27,560
Maximum Energy Product BH_{max} (MGO)	1.5	Transverse Modulus of Rupture (psi)	43.500
B/H Ratio (Load Line) at BH_{max}	9.5	Coeff. of Therm. Expansion inches x 10^{-6}/°C	Orientation: Par. 12.0 Rt.Angl. _
Flux Density in Magnet B_m at BH_{max} (G)	3700	Resistivity (micro-ohms/cm/cm^2)	65 @ 25 °C
Coercive Force in Magnet H_m at BH_{max} (Oe)	390	Hardness 40	Scale Rockwell C
Energy Loss W_c (watt-sec/cycle/lb)	-	Basic Forming Method	DP
Importance of Operating Magnet at BH_{max}	A	Basic Finishing Method	WGR
Average Recoil Permeability	5.0	Relative Corrosion Resistance	G
Field Strength Required to Fully Saturate Magnet (Oe)	3000	Primary Chemical Composition	Al-Ni-Co-Fe
Ability to Withstand Demagnetizing Fields	F	Special Notes:	
Maximum Service Temperature (°F)	842		
Approximate Temperature Permanently Affecting Magnet (°F)	1436		
Inherent Orientation of Material	none		

Note: All information based on available producers literature. In some cases of foreign materials interpretation or calculation required.

SOURCE(S):

SG Magnets, Ltd.

USA

Alnico HRF
(Alloy or Grade)

Sintered Alnico
(Type of Material)

2.6/.48
(MMPA Brief Designation)

-
(IEC Code Reference)

MAGNETIC PROPERTIES	
Residual Induction B_r (G)	7600
Coercive Force (Oe) — Normal (H_c)	470
Coercive Force (Oe) — Intrinsic (H_{ci})	480
Maximum Energy Product BH_{max} (MGO)	1.6
B/H Ratio (Load Line) at BH_{max}	15.9
Flux Density in Magnet B_m at BH_{max} (G)	5100
Coercive Force in Magnet H_m at BH_{max} (Oe)	320
Energy Loss W_c (watt-sec/cycle/lb)	-
Importance of Operating Magnet at BH_{max}	A
Average Recoil Permeability	6.7
Field Strength Required to Fully Saturate Magnet (Oe)	3000
Ability to Withstand Demagnetizing Fields	F
Maximum Service Temperature (°F)	842
Approximate Temperature Permanently Affecting Magnet (°F)	1436
Inherent Orientation of Material	none

PHYSICAL PROPERTIES	
Density (lbs./in.^{3})	.2530
General Mechanical Properties	HT
Tensile Strength (psi)	27,560
Transverse Modulus of Rupture (psi)	43,500
Coeff. of Therm. Expansion inches x 10^{-6}/°C	Orientation: Par. 12.0 Rt.Angl. -
Resistivity (micro-ohms/cm/cm^2)	65 @ 25 °C
Hardness 43	Scale Rockwell C
Basic Forming Method	DP
Basic Finishing Method	WGR
Relative Corrosion Resistance	G
Primary Chemical Composition	Al-Ni-Co-Fe
Special Notes:	

Note: All information based on available producers literature. In some cases of foreign materials interpretation or calculation required.

SOURCE(S): Hitachi Magnetics Corp. USA

Alnico ISO.8
(Alloy or Grade)

Cast Alnico
(Type of Material)

2.1/5.0
(MMPA Brief Designation)

-
(IEC Code Reference)

MAGNETIC PROPERTIES		PHYSICAL PROPERTIES	
Residual Induction B_r (G)	5800	Density (lbs./in.^{3})	.252
Coercive Force (Oe) Normal (H_c)	1200	General Mechanical Properties	HB
Coercive Force (Oe) Intrinsic (H_{ci})	5000	Tensile Strength (psi)	39,500
Maximum Energy Product BH_{max} (MGO)	2.1	Transverse Modulus of Rupture (psi)	30,000
B/H Ratio (Load Line) at BH_{max}	4.3	Coeff. of Therm. Expansion inches x 10^{-6}/°C	Orientation: Par. 11.6 Rt.Angl. -
Flux Density in Magnet B_m at BH_{max} (G)	3000	Resistivity (micro-ohms/cm/cm^2)	50 @ 25 °C
Coercive Force in Magnet H_m at BH_{max} (Oe)	700	Hardness 43	Scale Rockwell C
Energy Loss W_c (watt-sec/cycle/lb)	15.3	Basic Forming Method	C
Importance of Operating Magnet at BH_{max}	C	Basic Finishing Method	WGR
Average Recoil Permeability	1.9	Relative Corrosion Resistance	G
Field Strength Required to Fully Saturate Magnet (Oe)	3000	Primary Chemical Composition	Al-Ni-Co-Cu-Fe
Ability to Withstand Demagnetizing Fields	Ex	Special Notes: (1) Lowest temperature coefficient of all materials. Little irreversible loss even at high temperatures. (2) Should be magnetized in final circuit.	
Maximum Service Temperature (°F)	900		
Approximate Temperature Permanently Affecting Magnet (°F)	1580		
Inherent Orientation of Material	none		

Note: All information based on available producers literature. In some cases of foreign materials interpretation or calculation required.

SOURCE(S):	**Alnico 5cc**
	(Alloy or Grade)
Arnold Engineering Co.	Cast Alnico
	(Type of Material)
USA	6.5/.67
	(MMPA Brief Designation)
	R1-1-2
	(IEC Code Reference)

MAGNETIC PROPERTIES		PHYSICAL PROPERTIES	
Residual Induction B_r (G)	13,200	Density (lbs./in.^{3})	.2640
Coercive Force (Oe) — Normal (H_c)	675	General Mechanical Properties	HB
Coercive Force (Oe) — Intrinsic (H_{ci})	Est. 670	Tensile Strength (psi)	5200
Maximum Energy Product BH_{max} (MGO)	6.5	Transverse Modulus of Rupture (psi)	9000
B/H Ratio (Load Line) at BH_{max}	18.5	Coeff. of Therm. Expansion inches x 10^{-6}/°C	Orientation: Par. 11.5 Rt.Angl. —
Flux Density in Magnet B_m at BH_{max} (G)	11,000	Resistivity (micro-ohms/cm/cm^2)	47 @ 25 °C
Coercive Force in Magnet H_m at BH_{max} (Oe)	595	Hardness 50	Scale Rockwell C
Energy Loss W_c (watt-sec/cycle/lb)	15.3	Basic Forming Method	CM
Importance of Operating Magnet at BH_{max}	A	Basic Finishing Method	WGR
Average Recoil Permeability	2.4	Relative Corrosion Resistance	G
Field Strength Required to Fully Saturate Magnet (Oe)	3000	Primary Chemical Composition	Al-Ni-Co-Cu-Fe
Ability to Withstand Demagnetizing Fields	G	Special Notes:	
Maximum Service Temperature (°F)	900		
Approximate Temperature Permanently Affecting Magnet (°F)	1580		
Inherent Orientation of Material	CHT		

Note: All information based on available producers literature. In some cases of foreign materials interpretation or calculation required.

SOURCE(S):	**Alnico 8B**
	(Alloy or Grade)
Arnold Engineering Co.	Cast Alnico
	(Type of Material)
USA	-
	(MMPA Brief Designation)
	-
	(IEC Code Reference)

MAGNETIC PROPERTIES		PHYSICAL PROPERTIES	
Residual Induction B_r (G)	8300	Density (lbs./in.^{3})	.2620
Coercive Force (Oe) — Normal (H_c)	1650	General Mechanical Properties	HT
Coercive Force (Oe) — Intrinsic (H_{ci})	-	Tensile Strength (psi)	9000
Maximum Energy Product BH_{max} (MGO)	5.5	Transverse Modulus of Rupture (psi)	-
B/H Ratio (Load Line) at BH_{max}	4.5	Coeff. of Therm. Expansion inches x 10^{-6}/°C	Orientation: Par. 11.0 Rt.Angl. -
Flux Density in Magnet B_m at BH_{max} (G)	5000	Resistivity (micro-ohms/cm/cm^2)	50 @ 25 °C
Coercive Force in Magnet H_m at BH_{max} (Oe)	1100	Hardness Scale	56 Rockwell C
Energy Loss W_c (watt-sec/cycle/lb)	-	Basic Forming Method	C
Importance of Operating Magnet at BH_{max}	B	Basic Finishing Method	WGR
Average Recoil Permeability	2.0	Relative Corrosion Resistance	G
Field Strength Required to Fully Saturate Magnet (Oe)	6000	Primary Chemical Composition	Al-Ni-Co-Cu-Fe
Ability to Withstand Demagnetizing Fields	G	Special Notes:	
Maximum Service Temperature (°F)	-		
Approximate Temperature Permanently Affecting Magnet (°F)	1598		
Inherent Orientation of Material	HT		

Note: All information based on available producers literature. In some cases of foreign materials interpretation or calculation required.

SOURCE(S):

Arnold Engineering Co.

USA

Alnico 8H
(Alloy or Grade)

Cast Alnico
(Type of Material)

-
(MMPA Brief Designation)

-
(IEC Code Reference)

MAGNETIC PROPERTIES		PHYSICAL PROPERTIES	
Residual Induction B_r (G)	7400	Density (lbs./in.^{3})	.262
Coercive Force (Oe) — Normal (H_c)	1900	General Mechanical Properties	HT
Coercive Force (Oe) — Intrinsic (H_{ci})	-	Tensile Strength (psi)	8500
Maximum Energy Product BH_{max} (MGO)	5.5	Transverse Modulus of Rupture (psi)	-
B/H Ratio (Load Line) at BH_{max}	3.5	Coeff. of Therm. Expansion inches x 10^{-6}/°C	Orientation: Par. 11.0 Rt.Angl. -
Flux Density in Magnet B_m at BH_{max} (G)	4400	Resistivity (micro-ohms/cm/cm^2)	50 @ 25 °C
Coercive Force in Magnet H_m at BH_{max} (Oe)	1257	Hardness 56	Scale Rockwell C
Energy Loss W_c (watt-sec/cycle/lb)	-	Basic Forming Method	C
Importance of Operating Magnet at BH_{max}	B	Basic Finishing Method	WGR
Average Recoil Permeability	2.0	Relative Corrosion Resistance	G
Field Strength Required to Fully Saturate Magnet (Oe)	6000	Primary Chemical Composition	Al-Ni-Co-Cu-Fe
Ability to Withstand Demagnetizing Fields	G	Special Notes:	
Maximum Service Temperature (°F)	-		
Approximate Temperature Permanently Affecting Magnet (°F)	1598		
Inherent Orientation of Material	HT		

Note: All information based on available producers literature. In some cases of foreign materials interpretation or calculation required.

SOURCE(S):	**Alnico 8HE**
	(Alloy or Grade)
Arnold Engineering Co.	Cast Alnico
	(Type of Material)
USA	5.5/1.5
	(MMPA Brief Designation)
	R1-1-5
	(IEC Code Reference)

MAGNETIC PROPERTIES		PHYSICAL PROPERTIES	
Residual Induction B_r (G)	9300	Density (lbs./in.^{3})	.2620
Coercive Force (Oe) Normal (H_c)	1550	General Mechanical Properties	HB
Coercive Force (Oe) Intrinsic (H_{ci})	Est. 1500	Tensile Strength (psi)	10,000
Maximum Energy Product BH_{max} (MGO)	6.0	Transverse Modulus of Rupture (psi)	–
B/H Ratio (Load Line) at BH_{max}	5.5	Coeff. of Therm. Expansion inches x 10^{-6}/°C	Orientation: Par. 11.0 Rt.Angl. –
Flux Density in Magnet B_m at BH_{max} (G)	5750	Resistivity (micro-ohms/cm/cm^2)	47 @ 25 °C
Coercive Force in Magnet H_m at BH_{max} (Oe)	1070	Hardness 56 Scale Rockwell C	
Energy Loss W_c (watt-sec/cycle/lb)	15.3	Basic Forming Method	CM
Importance of Operating Magnet at BH_{max}	C	Basic Finishing Method	WGR
Average Recoil Permeability	2.0	Relative Corrosion Resistance	G
Field Strength Required to Fully Saturate Magnet (Oe)	3000	Primary Chemical Composition	Al-Ni-Co-Cu-Fe
Ability to Withstand Demagnetizing Fields	E	Special Notes:	
Maximum Service Temperature (°F)	900		
Approximate Temperature Permanently Affecting Magnet (°F)	1580		
Inherent Orientation of Material	CHT		

Note: All information based on available producers literature. In some cases of foreign materials interpretation or calculation required.

SOURCE(S):

Arnold Engineering Co.

USA

Alnico 9
(Alloy or Grade)

Cast Alnico
(Type of Material)

-
(MMPA Brief Designation)

-
(IEC Code Reference)

MAGNETIC PROPERTIES		PHYSICAL PROPERTIES	
Residual Induction B_r (G)	11,000	Density (lbs./in.^{3})	.2620
Coercive Force (Oe) — Normal (H_c)	1500	General Mechanical Properties	HT
Coercive Force (Oe) — Intrinsic (H_{ci})	-	Tensile Strength (psi)	7000
Maximum Energy Product BH_{max} (MGO)	10.5	Transverse Modulus of Rupture (psi)	8000
B/H Ratio (Load Line) at BH_{max}	6.5	Coeff. of Therm. Expansion inches x 10^{-6}/°C	Orientation: Par. 11.0 Rt.Angl. -
Flux Density in Magnet B_m at BH_{max} (G)	8250	Resistivity (micro-ohms/cm/cm^2)	50 @ 25 °C
Coercive Force in Magnet H_m at BH_{max} (Oe)	1270	Hardness Scale 56	Rockwell C
Energy Loss W_c (watt-sec/cycle/lb)	-	Basic Forming Method	C
Importance of Operating Magnet at BH_{max}	B	Basic Finishing Method	WGR
Average Recoil Permeability	1.3	Relative Corrosion Resistance	G
Field Strength Required to Fully Saturate Magnet (Oe)	6000	Primary Chemical Composition	Al-Ni-Co-Cu-Fe
Ability to Withstand Demagnetizing Fields	G	Special Notes:	
Maximum Service Temperature (°F)	-		
Approximate Temperature Permanently Affecting Magnet (°F)	1598		
Inherent Orientation of Material	HT		

Note: All information based on available producers literature. In some cases of foreign materials interpretation or calculation required.

SOURCE(S):	Alnico 35/5
	(Alloy or Grade)
Magnetfabrik-Bonn	Cast Alnico
	(Type of Material)
Germany	4.4/.60
	(MMPA Brief Designation)
	R1-1-1
	(IEC Code Reference)

MAGNETIC PROPERTIES		PHYSICAL PROPERTIES	
Residual Induction B_r (G)	11,200	Density (lbs./in.^{3})	.2601
Coercive Force (Oe) — Normal (H_c)	590	General Mechanical Properties	HB
Coercive Force (Oe) — Intrinsic (H_{ci})	630	Tensile Strength (psi)	–
Maximum Energy Product BH_{max} (MGO)	4.4	Transverse Modulus of Rupture (psi)	–
B/H Ratio (Load Line) at BH_{max}	18.0	Coeff. of Therm. Expansion inches x 10^{-6}/°C	Orientation: Par. – Rt.Angl. –
Flux Density in Magnet B_m at BH_{max} (G)	9000	Resistivity (micro-ohms/cm/cm^2)	50 @ 25 °C
Coercive Force in Magnet H_m at BH_{max} (Oe)	500	Hardness Scale	Approx. 50 Rockwell C
Energy Loss W_c (watt-sec/cycle/lb)	15.3	Basic Forming Method	CM
Importance of Operating Magnet at BH_{max}	A	Basic Finishing Method	WGR
Average Recoil Permeability	4.0	Relative Corrosion Resistance	G
Field Strength Required to Fully Saturate Magnet (Oe)	3000	Primary Chemical Composition	Al-Ni-Co-Cu-Fe
Ability to Withstand Demagnetizing Fields	F	Special Notes:	
Maximum Service Temperature (°F)	900		
Approximate Temperature Permanently Affecting Magnet (°F)	1580		
Inherent Orientation of Material	CHT		

Note: All information based on available producers literature. In some cases of foreign materials interpretation or calculation required.

SOURCE(S):

Magnetfabrik-Bonn

Germany

Alnico 40/12
(Alloy or Grade)

Sintered Alnico
(Type of Material)

5/1.6
(MMPA Brief Designation)

-
(IEC Code Reference)

MAGNETIC PROPERTIES			PHYSICAL PROPERTIES	
Residual Induction B_r (G)		8500	Density (lbs./in.^{3})	.2637
Coercive Force (Oe)	Normal (H_c)	1500	General Mechanical Properties	HT
	Intrinsic (H_{ci})	1600	Tensile Strength (psi)	Approx. 28,000
Maximum Energy Product BH_{max} (MGO)		5.0	Transverse Modulus of Rupture (psi)	Approx. 44,000
B/H Ratio (Load Line) at BH_{max}		5.1	Coeff. of Therm. Expansion inches x 10^{-6}/°C	Orientation: Par. - Rt.Angl. -
Flux Density in Magnet B_m at BH_{max} (G)		5000	Resistivity (micro-ohms/cm/cm^2)	Approx. 50 @ 25 °C
Coercive Force in Magnet H_m at BH_{max} (Oe)		990	Hardness Approx. 50	Scale Rockwell C
Energy Loss W_c (watt-sec/cycle/lb)		-	Basic Forming Method	DPM
Importance of Operating Magnet at BH_{max}		B	Basic Finishing Method	WGR
Average Recoil Permeability		2.0	Relative Corrosion Resistance	G
Field Strength Required to Fully Saturate Magnet (Oe)		3000	Primary Chemical Composition	Al-Ni-Co-Cu-Ti-Fe
Ability to Withstand Demagnetizing Fields		G	Special Notes:	
Maximum Service Temperature (°F)		842		
Approximate Temperature Permanently Affecting Magnet (°F)		1436		
Inherent Orientation of Material		MP		

Note: All information based on available producers literature. In some cases of foreign materials interpretation or calculation required.

SOURCE(S):	
Magnetfabrik-Bonn	**Alnico 40/15**
	(Alloy or Grade)
	Cast Alnico
	(Type of Material)
Germany	5.0/2.0
	(MMPA Brief Designation)
	R1-1-5
	(IEC Code Reference)

MAGNETIC PROPERTIES		PHYSICAL PROPERTIES	
Residual Induction B_r (G)	7500	Density (lbs./in.^{3})	.2637
Coercive Force (Oe) — Normal (H_c)	1800	General Mechanical Properties	HB
Coercive Force (Oe) — Intrinsic (H_{ci})	1980	Tensile Strength (psi)	-
Maximum Energy Product BH_{max} (MGO)	5.0	Transverse Modulus of Rupture (psi)	-
B/H Ratio (Load Line) at BH_{max}	4.0	Coeff. of Therm. Expansion inches x 10^{-6}/°C	Orientation: Par. - Rt.Angl. -
Flux Density in Magnet B_m at BH_{max} (G)	4000	Resistivity (micro-ohms/cm/cm^2)	50 @ 25 °C
Coercive Force in Magnet H_m at BH_{max} (Oe)	1000	Hardness Approx. 50	Scale Rockwell C
Energy Loss W_c (watt-sec/cycle/lb)	15.3	Basic Forming Method	CM
Importance of Operating Magnet at BH_{max}	C	Basic Finishing Method	WGR
Average Recoil Permeability	2.0	Relative Corrosion Resistance	G
Field Strength Required to Fully Saturate Magnet (Oe)	3000	Primary Chemical Composition	Al-Ni-Co-Cu Ti-Fe
Ability to Withstand Demagnetizing Fields	E	Special Notes:	
Maximum Service Temperature (°F)	900		
Approximate Temperature Permanently Affecting Magnet (°F)	1580		
Inherent Orientation of Material	HT		

Note: All information based on available producers literature. In some cases of foreign materials interpretation or calculation required.

SOURCE(S):

Magnetfabrik-Bonn

Germany

Alnico 160
(Alloy or Grade)

Sintered Alnico
(Type of Material)

1.5/.73
(MMPA Brief Designation)

-
(IEC Code Reference)

MAGNETIC PROPERTIES		PHYSICAL PROPERTIES	
Residual Induction B_r (G)	6500	Density (lbs./in.^{3})	.2565
Coercive Force (Oe) Normal (H_c)	680	General Mechanical Properties	HT
Coercive Force (Oe) Intrinsic (H_{ci})	730	Tensile Strength (psi)	Approx. 28,000
Maximum Energy Product BH_{max} (MGO)	1.5	Transverse Modulus of Rupture (psi)	Approx. 44,000
B/H Ratio (Load Line) at BH_{max}	11.1	Coeff. of Therm. Expansion inches x 10^{-6}/°C	Orientation: Par. - Rt.Angl. -
Flux Density in Magnet B_m at BH_{max} (G)	4000	Resistivity (micro-ohms/cm/cm^2)	Approx. 50 @ 25 °C
Coercive Force in Magnet H_m at BH_{max} (Oe)	360	Hardness Scale	Approx. 50 Rockwell C
Energy Loss W_c (watt-sec/cycle/lb)	-	Basic Forming Method	DP
Importance of Operating Magnet at BH_{max}	A	Basic Finishing Method	WGR
Average Recoil Permeability	4.0	Relative Corrosion Resistance	G
Field Strength Required to Fully Saturate Magnet (Oe)	3000	Primary Chemical Composition	Al-Ni-Co-Cu-Fe
Ability to Withstand Demagnetizing Fields	F	Special Notes:	
Maximum Service Temperature (°F)	842		
Approximate Temperature Permanently Affecting Magnet (°F)	1436		
Inherent Orientation of Material	none		

Note: All information based on available producers literature. In some cases of foreign materials interpretation or calculation required.

SOURCE(S):

Magnetfabrik-Bonn

Germany

Alnico 190
(Alloy or Grade)

Sintered Alnico
(Type of Material)

1.8/.7
(MMPA Brief Designation)

-
(IEC Code Reference)

MAGNETIC PROPERTIES	
Residual Induction B_r (G)	7100
Coercive Force (Oe) Normal (H_c)	680
Coercive Force (Oe) Intrinsic (H_{ci})	700
Maximum Energy Product BH_{max} (MGO)	1.8
B/H Ratio (Load Line) at BH_{max}	11.8
Flux Density in Magnet B_m at BH_{max} (G)	4500
Coercive Force in Magnet H_m at BH_{max} (Oe)	380
Energy Loss W_c (watt-sec/cycle/lb)	-
Importance of Operating Magnet at BH_{max}	A
Average Recoil Permeability	4.0
Field Strength Required to Fully Saturate Magnet (Oe)	3000
Ability to Withstand Demagnetizing Fields	F
Maximum Service Temperature (°F)	842
Approximate Temperature Permanently Affecting Magnet (°F)	1436
Inherent Orientation of Material	none

PHYSICAL PROPERTIES	
Density (lbs./in.^{3})	.2565
General Mechanical Properties	HT
Tensile Strength (psi)	Approx. 28,000
Transverse Modulus of Rupture (psi)	Approx. 44,000
Coeff. of Therm. Expansion inches x 10^{-6}/°C	Orientation: Par. - Rt.Angl. -
Resistivity (micro-ohms/cm/cm^2)	Approx. 50 @ 25 °C
Hardness	Scale Approx. 50 Rockwell C
Basic Forming Method	DP
Basic Finishing Method	WGR
Relative Corrosion Resistance	G
Primary Chemical Composition	Al-Ni-Co-Cu-Fe
Special Notes:	

Note: All information based on available producers literature. In some cases of foreign materials interpretation or calculation required.

SOURCE(S):

Magnetfabrik-Bonn

Germany

Alnico 260
(Alloy or Grade)

Sintered Alnico
(Type of Material)

2.2/1.1
(MMPA Brief Designation)

-
(IEC Code Reference)

MAGNETIC PROPERTIES		PHYSICAL PROPERTIES	
Residual Induction B_r (G)	6000	Density (lbs./in.^{3})	.2601
Coercive Force (Oe) — Normal (H_c)	1000	General Mechanical Properties	HT
Coercive Force (Oe) — Intrinsic (H_{ci})	1080	Tensile Strength (psi)	Approx. 28,000
Maximum Energy Product BH_{max} (MGO)	2.2	Transverse Modulus of Rupture (psi)	Approx. 44,000
B/H Ratio (Load Line) at BH_{max}	5.8	Coeff. of Therm. Expansion inches x 10^{-6}/°C	Orientation: Par. - Rt.Angl. -
Flux Density in Magnet B_m at BH_{max} (G)	3500	Resistivity (micro-ohms/cm/cm^2)	Approx. 50 @ 25 °C
Coercive Force in Magnet H_m at BH_{max} (Oe)	600	Hardness — Scale	Approx. 50 Rockwell C
Energy Loss W_c (watt-sec/cycle/lb)	-	Basic Forming Method	DPM
Importance of Operating Magnet at BH_{max}	B	Basic Finishing Method	WGR
Average Recoil Permeability	3.0	Relative Corrosion Resistance	G
Field Strength Required to Fully Saturate Magnet (Oe)	3000	Primary Chemical Composition	Al-Ni-Co-Cu-Ti-Fe
Ability to Withstand Demagnetizing Fields	G	Special Notes:	
Maximum Service Temperature (°F)	842		
Approximate Temperature Permanently Affecting Magnet (°F)	1436		
Inherent Orientation of Material	HT		

Note: All information based on available producers literature. In some cases of foreign materials interpretation or calculation required.

SOURCE(S):		Alnico 350	
		(Alloy or Grade)	
Magnetfabrik-Bonn		Sintered Alnico	
		(Type of Material)	
Germany		3.3/1.1	
		(MMPA Brief Designation)	
		-	
		(IEC Code Reference)	

MAGNETIC PROPERTIES		PHYSICAL PROPERTIES	
Residual Induction B_r (G)	8000	Density (lbs./in.^{3})	.2565
Coercive Force (Oe) — Normal (H_c)	1050	General Mechanical Properties	HT
Coercive Force (Oe) — Intrinsic (H_{ci})	1100	Tensile Strength (psi)	Approx. 28,000
Maximum Energy Product BH_{max} (MGO)	3.3	Transverse Modulus of Rupture (psi)	Approx. 44,000
B/H Ratio (Load Line) at BH_{max}	7.7	Coeff. of Therm. Expansion inches x 10^{-6}/°C	Orientation: Par. – Rt.Angl. –
Flux Density in Magnet B_m at BH_{max} (G)	5000	Resistivity (micro-ohms/cm/cm^2)	Approx. 50 @ 25 °C
Coercive Force in Magnet H_m at BH_{max} (Oe)	650	Hardness Scale	Approx. 50 Rockwell C
Energy Loss W_c (watt-sec/cycle/lb)	-	Basic Forming Method	DPM
Importance of Operating Magnet at BH_{max}	B	Basic Finishing Method	WGR
Average Recoil Permeability	3.0	Relative Corrosion Resistance	G
Field Strength Required to Fully Saturate Magnet (Oe)	3000	Primary Chemical Composition	Al-Ni-Co-Cu-Ti-Fe
Ability to Withstand Demagnetizing Fields	G	Special Notes:	
Maximum Service Temperature (°F)	842		
Approximate Temperature Permanently Affecting Magnet (°F)	1436		
Inherent Orientation of Material	MP		

Note: All information based on available producers literature. In some cases of foreign materials interpretation or calculation required.

SOURCE(S):	Alnico 450
	(Alloy or Grade)
Magnetfabrik-Bonn	Cast Alnico
	(Type of Material)
Germany	4.5/1.40
	(MMPA Brief Designation)
	-
	(IEC Code Reference)

MAGNETIC PROPERTIES		PHYSICAL PROPERTIES	
Residual Induction B_r (G)	8500	Density (lbs./in.^{3})	.2601
Coercive Force (Oe) — Normal (H_c)	1300	General Mechanical Properties	HT
Coercive Force (Oe) — Intrinsic (H_{ci})	1380	Tensile Strength (psi)	-
Maximum Energy Product BH_{max}(MGO)	4.5	Transverse Modulus of Rupture (psi)	-
B/H Ratio (Load Line) at BH_{max}	5.6	Coeff. of Therm. Expansion inches x 10^{-6}/°C	Orientation: Par. - Rt.Angl. -
Flux Density in Magnet B_m at BH_{max} (G)	5000	Resistivity (micro-ohms/cm/cm^2)	50 @ 25 °C
Coercive Force in Magnet H_m at BH_{max} (Oe)	900	Hardness 6.5	Scale Moh's
Energy Loss W_c (watt-sec/cycle/lb)	15.3	Basic Forming Method	CM
Importance of Operating Magnet at BH_{max}	C	Basic Finishing Method	WGR
Average Recoil Permeability	3.0	Relative Corrosion Resistance	G
Field Strength Required to Fully Saturate Magnet (Oe)	3,000	Primary Chemical Composition	Al-Ni-Co-Cu-Ti-Fe
Ability to Withstand Demagnetizing Fields	G	Special Notes:	
Maximum Service Temperature (°F)	900		
Approximate Temperature Permanently Affecting Magnet (°F)	1580		
Inherent Orientation of Material	HT		

Note: All information based on available producers literature. In some cases of foreign materials interpretation or calculation required.

SOURCE(S):

Magnetfabrik-Bonn

Germany

Alnico 500
(Alloy or Grade)

Cast Alnico
(Type of Material)

4.4/6.3
(MMPA Brief Designation)

-
(IEC Code Reference)

MAGNETIC PROPERTIES		PHYSICAL PROPERTIES	
Residual Induction B_r (G)	11,200	Density (lbs./in.^{3})	.2601
Coercive Force (Oe) Normal (H_c)	590	General Mechanical Properties	HT
Coercive Force (Oe) Intrinsic (H_{ci})	630	Tensile Strength (psi)	-
Maximum Energy Product BH_{max} (MGO)	4.4	Transverse Modulus of Rupture (psi)	-
B/H Ratio (Load Line) at BH_{max}	18.0	Coeff. of Therm. Expansion inches x 10^{-6}/°C	Orientation: Par. - Rt.Angl. -
Flux Density in Magnet B_m at BH_{max} (G)	9000	Resistivity (micro-ohms/cm/cm^2)	50 @ 25 °C
Coercive Force in Magnet H_m at BH_{max} (Oe)	500	Hardness 56	Scale Rockwell C
Energy Loss W_c (watt-sec/cycle/lb)	-	Basic Forming Method	C
Importance of Operating Magnet at BH_{max}	A	Basic Finishing Method	WGR
Average Recoil Permeability	4.0	Relative Corrosion Resistance	E
Field Strength Required to Fully Saturate Magnet (Oe)	3000	Primary Chemical Composition	Al-Ni-Co-Cu-Fe
Ability to Withstand Demagnetizing Fields	F	Special Notes:	
Maximum Service Temperature (°F)	-		
Approximate Temperature Permanently Affecting Magnet (°F)	900		
Inherent Orientation of Material	HT		

Note: All information based on available producers literature. In some cases of foreign materials interpretation or calculation required.

SOURCE(S):	**Alnico 500S**
	(Alloy or Grade)
Magnetfabrik-Bonn	Cast Alnico
	(Type of Material)
Germany	4.0/.63
	(MMPA Brief Designation)
	R1-1-1
	(IEC Code Reference)

MAGNETIC PROPERTIES		PHYSICAL PROPERTIES	
Residual Induction B_r (G)	11,200	Density (lbs./in.^{3})	.2601
Coercive Force (Oe) — Normal (H_c)	600	General Mechanical Properties	HB
Coercive Force (Oe) — Intrinsic (H_{ci})	630	Tensile Strength (psi)	-
Maximum Energy Product BH_{max} (MGO)	4.0	Transverse Modulus of Rupture (psi)	-
B/H Ratio (Load Line) at BH_{max}	18.9	Coeff. of Therm. Expansion inches x 10^{-6}/°C	Orientation: Par. _ Rt.Angl. _
Flux Density in Magnet B_m at BH_{max} (G)	8700	Resistivity (micro-ohms/cm/cm^2)	50 @ 25 °C
Coercive Force in Magnet H_m at BH_{max} (Oe)	460	Hardness Approx. 50	Scale Rockwell C
Energy Loss W_c (watt-sec/cycle/lb)	15.3	Basic Forming Method	CM
Importance of Operating Magnet at BH_{max}	A	Basic Finishing Method	WGR
Average Recoil Permeability	4.0	Relative Corrosion Resistance	G
Field Strength Required to Fully Saturate Magnet (Oe)	3000	Primary Chemical Composition	Al-Ni-Co-Cu-Fe
Ability to Withstand Demagnetizing Fields	F	Special Notes:	
Maximum Service Temperature (°F)	900		
Approximate Temperature Permanently Affecting Magnet (°F)	1580		
Inherent Orientation of Material	CHT		

Note: All information based on available producers literature. In some cases of foreign materials interpretation or calculation required.

SOURCE(S):

UGIMag, Inc.

USA

Alnico 550
(Alloy or Grade)

Cast Alnico
(Type of Material)

5.6/.60
(MMPA Brief Designation)

-
(IEC Code Reference)

MAGNETIC PROPERTIES		PHYSICAL PROPERTIES	
Residual Induction B_r (G)	13,500	Density (lbs./in.^{3})	.2619
Coercive Force (Oe) Normal (H_c)	590	General Mechanical Properties	HB
Coercive Force (Oe) Intrinsic (H_{ci})	Est. 600	Tensile Strength (psi)	-
Maximum Energy Product BH_{max} (MGO)	5.6	Transverse Modulus of Rupture (psi)	-
B/H Ratio (Load Line) at BH_{max}	Est. 20.0	Coeff. of Therm. Expansion inches x 10^{-6}/°C	Orientation: Par. 12.0 Rt.Angl. -
Flux Density in Magnet B_m at BH_{max} (G)	Est. 10,000	Resistivity (micro-ohms/cm/cm^2)	47 @ 25 °C
Coercive Force in Magnet H_m at BH_{max} (Oe)	Est. 500	Hardness Scale	50 Rockwell C
Energy Loss W_c (watt-sec/cycle/lb)	15.3	Basic Forming Method	CM
Importance of Operating Magnet at BH_{max}	A	Basic Finishing Method	WGR
Average Recoil Permeability	2.5	Relative Corrosion Resistance	G
Field Strength Required to Fully Saturate Magnet (Oe)	3000	Primary Chemical Composition	Al-Ni-Co-Fe-Cu
Ability to Withstand Demagnetizing Fields	F	Special Notes:	
Maximum Service Temperature (°F)	900		
Approximate Temperature Permanently Affecting Magnet (°F)	1580		
Inherent Orientation of Material	CHT		

Note: All information based on available producers literature. In some cases of foreign materials interpretation or calculation required.

SOURCE(S):	Alnico 550 UGIMAX
	(Alloy or Grade)
UGIMag, Inc.	Cast Alnico
	(Type of Material)
USA	6.0/.62
	(MMPA Brief Designation)
	R1-1-2
	(IEC Code Reference)

MAGNETIC PROPERTIES		PHYSICAL PROPERTIES	
Residual Induction B_r (G)	13,500	Density (lbs./in.^{3})	.2619
Coercive Force (Oe) — Normal (H_c)	620	General Mechanical Properties	HB
Coercive Force (Oe) — Intrinsic (H_{ci})	Est. 620	Tensile Strength (psi)	—
Maximum Energy Product BH_{max} (MGO)	6.0	Transverse Modulus of Rupture (psi)	—
B/H Ratio (Load Line) at BH_{max}	Est. 17.8	Coeff. of Therm. Expansion inches x 10^{-6}/°C	Orientation: Par. 12.0 Rt.Angl. —
Flux Density in Magnet B_m at BH_{max} (G)	Est. 10,000	Resistivity (micro-ohms/cm/cm^2)	47 @ 25 °C
Coercive Force in Magnet H_m at BH_{max} (Oe)	Est. 560	Hardness 50 Scale Rockwell C	
Energy Loss W_c (watt-sec/cycle/lb)	15.3	Basic Forming Method	CM
Importance of Operating Magnet at BH_{max}	A	Basic Finishing Method	WGR
Average Recoil Permeability	2.3	Relative Corrosion Resistance	G
Field Strength Required to Fully Saturate Magnet (Oe)	3000	Primary Chemical Composition	Al-Ni-Co-Fe-Cu
Ability to Withstand Demagnetizing Fields	F	Special Notes:	
Maximum Service Temperature (°F)	900		
Approximate Temperature Permanently Affecting Magnet (°F)	1580		
Inherent Orientation of Material	CHT		

Note: All information based on available producers literature. In some cases of foreign materials interpretation or calculation required.

SOURCE(S):	Alnico 580
	(Alloy or Grade)
Magnetfabrik-Bonn	Cast Alnico
	(Type of Material)
Germany	5.5/.7
	(MMPA Brief Designation)
	R1-1-1
	(IEC Code Reference)

MAGNETIC PROPERTIES		PHYSICAL PROPERTIES	
Residual Induction B_r (G)	12,500	Density (lbs./in.^{3})	.2601
Coercive Force (Oe) — Normal (H_c)	680	General Mechanical Properties	HB
Coercive Force (Oe) — Intrinsic (H_{ci})	700	Tensile Strength (psi)	−
Maximum Energy Product BH_{max} (MGO)	5.5	Transverse Modulus of Rupture (psi)	−
B/H Ratio (Load Line) at BH_{max}	17.2	Coeff. of Therm. Expansion inches x 10^{-6}/°C	Orientation: Par. − Rt.Angl. −
Flux Density in Magnet B_m at BH_{max} (G)	9300	Resistivity (micro-ohms/cm/cm^2)	50 @ 25 °C
Coercive Force in Magnet H_m at BH_{max} (Oe)	540	Hardness Scale — Approx. 50	Rockwell C
Energy Loss W_c (watt-sec/cycle/lb)	15.3	Basic Forming Method	CM
Importance of Operating Magnet at BH_{max}	A	Basic Finishing Method	WGR
Average Recoil Permeability	5.0	Relative Corrosion Resistance	G
Field Strength Required to Fully Saturate Magnet (Oe)	3000	Primary Chemical Composition	Al-Ni-Co-Cu-Fe
Ability to Withstand Demagnetizing Fields	G	Special Notes:	
Maximum Service Temperature (°F)	900		
Approximate Temperature Permanently Affecting Magnet (°F)	1580		
Inherent Orientation of Material	CHT		

Note: All information based on available producers literature. In some cases of foreign materials interpretation or calculation required.

SOURCE(S):

Magnet Applications, Ltd.
(a.k.a. Cookson Group/Arelec)

England/France

Alnico 600
(Alloy or Grade)

Cast Alnico
(Type of Material)

5.3/.70
(MMPA Brief Designation)

R-1-1-1
(IEC Code Reference)

MAGNETIC PROPERTIES		PHYSICAL PROPERTIES	
Residual Induction B_r (G)	12,000	Density (lbs./in.^{3})	.2637
Coercive Force (Oe) — Normal (H_c)	630	General Mechanical Properties	HB
Coercive Force (Oe) — Intrinsic (H_{ci})	Est. 700	Tensile Strength (psi)	-
Maximum Energy Product BH_{max} (MGO)	5.3	Transverse Modulus of Rupture (psi)	-
B/H Ratio (Load Line) at BH_{max}	18.9	Coeff. of Therm. Expansion inches x 10^{-6}/°C — Orientation: Par. _ Rt.Angl. _	
Flux Density in Magnet B_m at BH_{max} (G)	10,000	Resistivity (micro-ohms/cm/cm^2)	50 @ 25 °C
Coercive Force in Magnet H_m at BH_{max} (Oe)	530	Hardness 650 — Scale Vickers	
Energy Loss W_c (watt-sec/cycle/lb)	15.3	Basic Forming Method	CM
Importance of Operating Magnet at BH_{max}	A	Basic Finishing Method	WGR
Average Recoil Permeability	2.8	Relative Corrosion Resistance	G
Field Strength Required to Fully Saturate Magnet (Oe)	3000	Primary Chemical Composition	Al-Ni-Co-Cu-Fe
Ability to Withstand Demagnetizing Fields	F	Special Notes:	
Maximum Service Temperature (°F)	900		
Approximate Temperature Permanently Affecting Magnet (°F)	1580		
Inherent Orientation of Material	CHT		

Note: All information based on available producers literature. In some cases of foreign materials interpretation or calculation required.

SOURCE(S):

UGIMag, Inc.

USA

Alnico 600
(Alloy or Grade)

Cast Alnico
(Type of Material)

5.3/.63
(MMPA Brief Designation)

R1-1-1
(IEC Code Reference)

MAGNETIC PROPERTIES		PHYSICAL PROPERTIES	
Residual Induction B_r (G)	12,600	Density (lbs./in.^{3})	.2637
Coercive Force (Oe) — Normal (H_c)	630	General Mechanical Properties	HB
Coercive Force (Oe) — Intrinsic (H_{ci})	Est. 630	Tensile Strength (psi)	-
Maximum Energy Product BH_{max} (MGO)	5.3	Transverse Modulus of Rupture (psi)	-
B/H Ratio (Load Line) at BH_{max}	Est. 20.0	Coeff. of Therm. Expansion inches x 10^{-6}/°C	Orientation: Par. 12.0 Rt.Angl. _
Flux Density in Magnet B_m at BH_{max} (G)	Est. 10,000	Resistivity (micro-ohms/cm/cm^2)	47 @ 25 °C
Coercive Force in Magnet H_m at BH_{max} (Oe)	Est. 500	Hardness Scale	50 Rockwell C
Energy Loss W_c (watt-sec/cycle/lb)	15.3	Basic Forming Method	CM
Importance of Operating Magnet at BH_{max}	A	Basic Finishing Method	WGR
Average Recoil Permeability	-	Relative Corrosion Resistance	G
Field Strength Required to Fully Saturate Magnet (Oe)	3000	Primary Chemical Composition	Al-Ni-Co-Fe-Cu
Ability to Withstand Demagnetizing Fields	E	Special Notes:	
Maximum Service Temperature (°F)	900		
Approximate Temperature Permanently Affecting Magnet (°F)	1580		
Inherent Orientation of Material	CHT		

Note: All information based on available producers literature. In some cases of foreign materials interpretation or calculation required.

SOURCE(S):

UGIMag, Inc.

USA

Alnico 600 Super UGIMAX
(Alloy or Grade)

Cast Alnico
(Type of Material)

7.5/.74
(MMPA Brief Designation)

R1-1-3
(IEC Code Reference)

MAGNETIC PROPERTIES	
Residual Induction B_r (G)	13,500
Coercive Force (Oe) — Normal (H_c)	740
Coercive Force (Oe) — Intrinsic (H_{ci})	Est. 740
Maximum Energy Product BH_{max} (MGO)	7.5
B/H Ratio (Load Line) at BH_{max}	Est. 20.0
Flux Density in Magnet B_m at BH_{max} (G)	Est. 1190
Coercive Force in Magnet H_m at BH_{max} (Oe)	Est. 590
Energy Loss W_c (watt-sec/cycle/lb)	15.3
Importance of Operating Magnet at BH_{max}	A
Average Recoil Permeability	2.0
Field Strength Required to Fully Saturate Magnet (Oe)	3000
Ability to Withstand Demagnetizing Fields	G
Maximum Service Temperature (°F)	900
Approximate Temperature Permanently Affecting Magnet (°F)	1580
Inherent Orientation of Material	CHT

PHYSICAL PROPERTIES	
Density (lbs./in.^{3})	.2637
General Mechanical Properties	HB
Tensile Strength (psi)	—
Transverse Modulus of Rupture (psi)	—
Coeff. of Therm. Expansion inches x 10^{-6}/°C — Orientation: Par.	12.0
Coeff. of Therm. Expansion — Rt. Angl.	—
Resistivity (micro-ohms/cm/cm^2)	50 @ 25 °C
Hardness — Scale	56 Rockwell C
Basic Forming Method	CM
Basic Finishing Method	WGR
Relative Corrosion Resistance	G
Primary Chemical Composition	Al-Ni-Co-Fe-Cu
Special Notes:	

Note: All information based on available producers literature. In some cases of foreign materials interpretation or calculation required.

SOURCE(S):

UGIMag, Inc.

USA

Alnico 600 UGIMAX
(Alloy or Grade)

Cast Alnico
(Type of Material)

6.2/.70
(MMPA Brief Designation)

R1-1-3
(IEC Code Reference)

MAGNETIC PROPERTIES	
Residual Induction B_r (G)	13,200
Coercive Force (Oe) — Normal (H_c)	700
Coercive Force (Oe) — Intrinsic (H_{ci})	Est. 700
Maximum Energy Product BH_{max} (MGO)	6.2
B/H Ratio (Load Line) at BH_{max}	Est. 7.2
Flux Density in Magnet B_m at BH_{max} (G)	Est. 10,000
Coercive Force in Magnet H_m at BH_{max} (Oe)	Est. 580
Energy Loss W_c (watt-sec/cycle/lb)	15.3
Importance of Operating Magnet at BH_{max}	A
Average Recoil Permeability	2.5
Field Strength Required to Fully Saturate Magnet (Oe)	3000
Ability to Withstand Demagnetizing Fields	G
Maximum Service Temperature (°F)	900
Approximate Temperature Permanently Affecting Magnet (°F)	1580
Inherent Orientation of Material	CHT

PHYSICAL PROPERTIES	
Density (lbs./in.^{3})	.2637
General Mechanical Properties	HB
Tensile Strength (psi)	—
Transverse Modulus of Rupture (psi)	—
Coeff. of Therm. Expansion inches x 10^{-6}/°C	Orientation: Par. 12.0 Rt.Angl. —
Resistivity (micro-ohms/cm/cm^2)	47 @ 25 °C
Hardness Scale	50 Rockwell C
Basic Forming Method	CM
Basic Finishing Method	WGR
Relative Corrosion Resistance	G
Primary Chemical Composition	Al-Ni-Co-Fe-Cu
Special Notes:	

Note: All information based on available producers literature. In some cases of foreign materials interpretation or calculation required.

SOURCE(S):	Alnico 700
	(Alloy or Grade)
Magnetfabrik-Bonn	Cast Alnico
	(Type of Material)
Germany	6.0/.7
	(MMPA Brief Designation)
	R1-1-1
	(IEC Code Reference)

MAGNETIC PROPERTIES		PHYSICAL PROPERTIES	
Residual Induction B_r (G)	13,000	Density (lbs./in.^{3})	.2601
Coercive Force (Oe) Normal (H_c)	680	General Mechanical Properties	HB
Coercive Force (Oe) Intrinsic (H_{ci})	700	Tensile Strength (psi)	–
Maximum Energy Product BH_{max} (MGO)	6.5	Transverse Modulus of Rupture (psi)	–
B/H Ratio (Load Line) at BH_{max}	20.0	Coeff. of Therm. Expansion inches x 10^{-6}/°C	Orientation: Par. – Rt.Angl. –
Flux Density in Magnet B_m at BH_{max} (G)	11,000	Resistivity (micro-ohms/cm/cm^2)	50 @ 25 °C
Coercive Force in Magnet H_m at BH_{max} (Oe)	550	Hardness Scale Approx. 50	Rockwell C
Energy Loss W_c (watt-sec/cycle/lb)	15.3	Basic Forming Method	CM
Importance of Operating Magnet at BH_{max}	A	Basic Finishing Method	WGR
Average Recoil Permeability	5.0	Relative Corrosion Resistance	G
Field Strength Required to Fully Saturate Magnet (Oe)	3000	Primary Chemical Composition	Al-Ni-Co-Cu-Fe
Ability to Withstand Demagnetizing Fields	G	Special Notes:	
Maximum Service Temperature (°F)	900		
Approximate Temperature Permanently Affecting Magnet (°F)	1580		
Inherent Orientation of Material	CHT		

Note: All information based on available producers literature. In some cases of foreign materials interpretation or calculation required.

SOURCE(S):

UGIMag, Inc.

USA

Alnico 700
(Alloy or Grade)

Cast Alnico
(Type of Material)

5.2/.70
(MMPA Brief Designation)

R1-1-1
(IEC Code Reference)

MAGNETIC PROPERTIES	
Residual Induction B_r (G)	12,000
Coercive Force (Oe) — Normal (H_c)	700
Coercive Force (Oe) — Intrinsic (H_{ci})	Est. 700
Maximum Energy Product BH_{max} (MGO)	5.2
B/H Ratio (Load Line) at BH_{max}	Est. 18.5
Flux Density in Magnet B_m at BH_{max} (G)	Est. 10,000
Coercive Force in Magnet H_m at BH_{max} (Oe)	Est. 540
Energy Loss W_c (watt-sec/cycle/lb)	15.3
Importance of Operating Magnet at BH_{max}	A
Average Recoil Permeability	-
Field Strength Required to Fully Saturate Magnet (Oe)	3000
Ability to Withstand Demagnetizing Fields	E
Maximum Service Temperature (°F)	900
Approximate Temperature Permanently Affecting Magnet (°F)	1580
Inherent Orientation of Material	CHT

PHYSICAL PROPERTIES	
Density (lbs./in.^{3})	.2637
General Mechanical Properties	HB
Tensile Strength (psi)	-
Transverse Modulus of Rupture (psi)	-
Coeff. of Therm. Expansion inches x 10^{-6}/°C — Orientation: Par.	12.0
Coeff. of Therm. Expansion — Rt. Angl.	-
Resistivity (micro-ohms/cm/cm^2)	47 @ 25 °C
Hardness — Scale Rockwell C	50
Basic Forming Method	CM
Basic Finishing Method	WGR
Relative Corrosion Resistance	G
Primary Chemical Composition	Al-Ni-Co-Fe-Cu
Special Notes:	

Note: All information based on available producers literature. In some cases of foreign materials interpretation or calculation required.

SOURCE(S):	**Alnico 800**
	(Alloy or Grade)
UGIMag, Inc.	Cast Alnico
	(Type of Material)
USA	4.3/.80
	(MMPA Brief Designation)
	-
	(IEC Code Reference)

MAGNETIC PROPERTIES		PHYSICAL PROPERTIES	
Residual Induction B_r (G)	11,000	Density (lbs./in.^{3})	.2637
Coercive Force (Oe) — Normal (H_c)	800	General Mechanical Properties	HB
Coercive Force (Oe) — Intrinsic (H_{ci})	Est. 800	Tensile Strength (psi)	-
Maximum Energy Product BH_{max} (MGO)	4.3	Transverse Modulus of Rupture (psi)	-
B/H Ratio (Load Line) at BH_{max}	Est. 13.0	Coeff. of Therm. Expansion inches x 10^{-6}/°C	Orientation: Par. 12.0 Rt.Angl. -
Flux Density in Magnet B_m at BH_{max} (G)	Est. 7500	Resistivity (micro-ohms/cm/cm^2)	50 @ 25 °C
Coercive Force in Magnet H_m at BH_{max} (Oe)	Est. 575	Hardness Scale	56 Rockwell C
Energy Loss W_c (watt-sec/cycle/lb)	15.3	Basic Forming Method	CM
Importance of Operating Magnet at BH_{max}	B	Basic Finishing Method	WGR
Average Recoil Permeability	-	Relative Corrosion Resistance	G
Field Strength Required to Fully Saturate Magnet (Oe)	3000	Primary Chemical Composition	Al-Ni-Co-Fe-Cu
Ability to Withstand Demagnetizing Fields	E	Special Notes:	
Maximum Service Temperature (°F)	900		
Approximate Temperature Permanently Affecting Magnet (°F)	1580		
Inherent Orientation of Material	CHT		

Note: All information based on available producers literature. In some cases of foreign materials interpretation or calculation required.

SOURCE(S):

Alnico 800 UGIMAX
(Alloy or Grade)

UGIMag, Inc.

USA

Cast Alnico
(Type of Material)

6.0/.85
(MMPA Brief Designation)

-
(IEC Code Reference)

MAGNETIC PROPERTIES	
Residual Induction B_r (G)	12,000
Coercive Force (Oe) — Normal (H_c)	850
Coercive Force (Oe) — Intrinsic (H_{ci})	Est. 850
Maximum Energy Product BH_{max} (MGO)	6.0
B/H Ratio (Load Line) at BH_{max}	Est. 16.4
Flux Density in Magnet B_m at BH_{max} (G)	Est. 10,000
Coercive Force in Magnet H_m at BH_{max} (Oe)	Est. 610
Energy Loss W_c (watt-sec/cycle/lb)	15.3
Importance of Operating Magnet at BH_{max}	B
Average Recoil Permeability	3.0
Field Strength Required to Fully Saturate Magnet (Oe)	3000
Ability to Withstand Demagnetizing Fields	G
Maximum Service Temperature (°F)	900
Approximate Temperature Permanently Affecting Magnet (°F)	1580
Inherent Orientation of Material	CHT

PHYSICAL PROPERTIES	
Density (lbs./in.^{3})	.2637
General Mechanical Properties	HB
Tensile Strength (psi)	-
Transverse Modulus of Rupture (psi)	-
Coeff. of Therm. Expansion inches x 10^{-6}/°C	Orientation: Par. 12.0 Rt.Angl. -
Resistivity (micro-ohms/cm/cm^2)	50 @ 25 °C
Hardness — Scale	56 Rockwell C
Basic Forming Method	CM
Basic Finishing Method	WGR
Relative Corrosion Resistance	G
Primary Chemical Composition	Al-Ni-Co-Fe-Cu
Special Notes:	

Note: All information based on available producers literature. In some cases of foreign materials interpretation or calculation required.

SOURCE(S):	Alnico 1300
	(Alloy or Grade)
UGIMag, Inc.	Cast Alnico
	(Type of Material)
USA	5.0/1.4
	(MMPA Brief Designation)
	-
	(IEC Code Reference)

MAGNETIC PROPERTIES		PHYSICAL PROPERTIES	
Residual Induction B_r (G)	8800	Density (lbs./in.^{3})	.2637
Coercive Force (Oe) — Normal (H_c)	1400	General Mechanical Properties	HB
Coercive Force (Oe) — Intrinsic (H_{ci})	Est. 1600	Tensile Strength (psi)	-
Maximum Energy Product BH_{max} (MGO)	5.0	Transverse Modulus of Rupture (psi)	-
B/H Ratio (Load Line) at BH_{max}	Est. 5.1	Coeff. of Therm. Expansion inches x 10^{-6}/°C	Orientation: Par. 12.0 Rt.Angl. -
Flux Density in Magnet B_m at BH_{max} (G)	Est. 5100	Resistivity (micro-ohms/cm/cm^2)	50 @ 25 °C
Coercive Force in Magnet H_m at BH_{max} (Oe)	Est. 1000	Hardness Scale 58	Rockwell C
Energy Loss W_c (watt-sec/cycle/lb)	15.3	Basic Forming Method	CM
Importance of Operating Magnet at BH_{max}	C	Basic Finishing Method	WGR
Average Recoil Permeability	-	Relative Corrosion Resistance	G
Field Strength Required to Fully Saturate Magnet (Oe)	3000	Primary Chemical Composition	Al-Ni-Co-Fe-Cu
Ability to Withstand Demagnetizing Fields	E	Special Notes:	
Maximum Service Temperature (°F)	900		
Approximate Temperature Permanently Affecting Magnet (°F)	1580		
Inherent Orientation of Material	HT		

Note: All information based on available producers literature. In some cases of foreign materials interpretation or calculation required.

SOURCE(S):

UGIMag, Inc.

USA

Alnico 1500
(Alloy or Grade)

Cast Alnico
(Type of Material)

5.7/1.8
(MMPA Brief Designation)

R1-1-5
(IEC Code Reference)

MAGNETIC PROPERTIES	
Residual Induction B_r (G)	9000
Coercive Force (Oe) — Normal (H_c)	1570
Coercive Force (Oe) — Intrinsic (H_{ci})	Est. 1800
Maximum Energy Product BH_{max} (MGO)	5.7
B/H Ratio (Load Line) at BH_{max}	Est. 6.6
Flux Density in Magnet B_m at BH_{max} (G)	Est. 6300
Coercive Force in Magnet H_m at BH_{max} (Oe)	Est. 950
Energy Loss W_c (watt-sec/cycle/lb)	15.3
Importance of Operating Magnet at BH_{max}	C
Average Recoil Permeability	2.5
Field Strength Required to Fully Saturate Magnet (Oe)	3000
Ability to Withstand Demagnetizing Fields	E
Maximum Service Temperature (°F)	900
Approximate Temperature Permanently Affecting Magnet (°F)	1580
Inherent Orientation of Material	HT

PHYSICAL PROPERTIES	
Density (lbs./in.^{3})	2637
General Mechanical Properties	HB
Tensile Strength (psi)	-
Transverse Modulus of Rupture (psi)	-
Coeff. of Therm. Expansion inches x 10^{-6}/°C — Orientation: Par. / Rt.Angl.	12.0 / -
Resistivity (micro-ohms/cm/cm^2)	50 @ 25 °C
Hardness — Scale Rockwell C	58
Basic Forming Method	CM
Basic Finishing Method	WGR
Relative Corrosion Resistance	G
Primary Chemical Composition	Al-Ni-Co-Fe-Cu
Special Notes:	

Note: All information based on available producers literature. In some cases of foreign materials interpretation or calculation required.

SOURCE(S):	Alnico 1500 Super UGIMAX
UGIMag, Inc.	(Alloy or Grade)
	Cast Alnico
	(Type of Material)
USA	10.3/1.5
	(MMPA Brief Designation)
	R1-1-6
	(IEC Code Reference)

MAGNETIC PROPERTIES		PHYSICAL PROPERTIES	
Residual Induction B_r (G)	10,800	Density (lbs./in.^{3})	2637
Coercive Force (Oe) Normal (H_c)	1550	General Mechanical Properties	HB
Coercive Force (Oe) Intrinsic (H_{ci})	Est. 1500	Tensile Strength (psi)	−
Maximum Energy Product BH_{max} (MGO)	10.3	Transverse Modulus of Rupture (psi)	−
B/H Ratio (Load Line) at BH_{max}	Est. 6.7	Coeff. of Therm. Expansion inches x 10^{-6}/°C	Orientation: Par. − Rt.Angl. −
Flux Density in Magnet B_m at BH_{max} (G)	Est. 8000	Resistivity (micro-ohms/cm/cm^2)	50 @ 25 °C
Coercive Force in Magnet H_m at BH_{max} (Oe)	Est. 1200	Hardness Scale	58 Rockwell C
Energy Loss W_c (watt-sec/cycle/lb)	15.3	Basic Forming Method	CM
Importance of Operating Magnet at BH_{max}	C	Basic Finishing Method	WGR
Average Recoil Permeability	1.5	Relative Corrosion Resistance	G
Field Strength Required to Fully Saturate Magnet (Oe)	3000	Primary Chemical Composition	Al-Ni-Co-Fe-Cu
Ability to Withstand Demagnetizing Fields	E	Special Notes:	
Maximum Service Temperature (°F)	900		
Approximate Temperature Permanently Affecting Magnet (°F)	1580		
Inherent Orientation of Material	HT		

Note: All information based on available producers literature. In some cases of foreign materials interpretation or calculation required.

SOURCE(S):

Alnico 2000
(Alloy or Grade)

UGIMag, Inc.

Cast Alnico
(Type of Material)

USA

6.5/2.0
(MMPA Brief Designation)

-
(IEC Code Reference)

MAGNETIC PROPERTIES		PHYSICAL PROPERTIES	
Residual Induction B_r (G)	8000	Density (lbs./in.^{3})	.2637
Coercive Force (Oe) Normal (H_c)	2000	General Mechanical Properties	HB
Coercive Force (Oe) Intrinsic (H_{ci})	Est. 2000	Tensile Strength (psi)	-
Maximum Energy Product BH_{max} (MGO)	6.5	Transverse Modulus of Rupture (psi)	-
B/H Ratio (Load Line) at BH_{max}	Est. 3.8	Coeff. of Therm. Expansion inches x $10^{-6}/°C$	Orientation: Par. 10.2 Rt.Angl. -
Flux Density in Magnet B_m at BH_{max} (G)	Est. 4800	Resistivity (micro-ohms/cm/cm^2	50 @ 25 °C
Coercive Force in Magnet H_m at BH_{max} (Oe)	Est. 1250	Hardness Scale	60 Rockwell C
Energy Loss W_c (watt-sec/cycle/lb)	15.3	Basic Forming Method	CM
Importance of Operating Magnet at BH_{max}	C	Basic Finishing Method	WGR
Average Recoil Permeability	1.3	Relative Corrosion Resistance	G
Field Strength Required to Fully Saturate Magnet (Oe)	3000	Primary Chemical Composition	Al-Ni-Co-Fe-Cu
Ability to Withstand Demagnetizing Fields	E	Special Notes:	
Maximum Service Temperature (°F)	900		
Approximate Temperature Permanently Affecting Magnet (°F)	1580		
Inherent Orientation of Material	HT		

Note: All information based on available producers literature. In some cases of foreign materials interpretation or calculation required.

SOURCE(S):

UGIMag, Inc.

USA

Alnico 2200
(Alloy or Grade)

Cast Alnico
(Type of Material)

5.5/2.4
(MMPA Brief Designation)

R1-1-7
(IEC Code Reference)

MAGNETIC PROPERTIES		PHYSICAL PROPERTIES	
Residual Induction B_r (G)	7400	Density (lbs./in.^{3})	.2673
Coercive Force (Oe) — Normal (H_c)	2200	General Mechanical Properties	HB
Coercive Force (Oe) — Intrinsic (H_{ci})	Est. 2400	Tensile Strength (psi)	—
Maximum Energy Product BH_{max} (MGO)	5.5	Transverse Modulus of Rupture (psi)	—
B/H Ratio (Load Line) at BH_{max}	5.2	Coeff. of Therm. Expansion inches x 10^{-6}/°C — Orientation: Par. / Rt.Angl.	10.2 / —
Flux Density in Magnet B_m at BH_{max} (G)	Est. 5800	Resistivity (micro-ohms/cm/cm^2)	58 @ 25 °C
Coercive Force in Magnet H_m at BH_{max} (Oe)	Est. 1100	Hardness / Scale	60 Rockwell C
Energy Loss W_c (watt-sec/cycle/lb)	15.3	Basic Forming Method	C
Importance of Operating Magnet at BH_{max}	C	Basic Finishing Method	WGR
Average Recoil Permeability	1.3	Relative Corrosion Resistance	G
Field Strength Required to Fully Saturate Magnet (Oe)	3000	Primary Chemical Composition	Al-Ni-Co-Fe-Cu
Ability to Withstand Demagnetizing Fields	E	Special Notes:	
Maximum Service Temperature (°F)	900		
Approximate Temperature Permanently Affecting Magnet (°F)	1580		
Inherent Orientation of Material	none		

Note: All information based on available producers literature. In some cases of foreign materials interpretation or calculation required.

SOURCE(S): **Arnold Engineering Co.**

USA

ArKomax 800
(Alloy or Grade)

Cast Alnico
(Type of Material)

8.1/.74
(MMPA Brief Designation)

R1-1-3
(IEC Code Reference)

MAGNETIC PROPERTIES	
Residual Induction B_r (G)	13,700
Coercive Force (Oe) Normal (H_c)	740
Coercive Force (Oe) Intrinsic (H_{ci})	Est. 740
Maximum Energy Product BH_{max} (MGO)	8.1
B/H Ratio (Load Line) at BH_{max}	18.0
Flux Density in Magnet B_m at BH_{max} (G)	12,000
Coercive Force in Magnet H_m at BH_{max} (Oe)	670
Energy Loss W_c (watt-sec/cycle/lb)	15.3
Importance of Operating Magnet at BH_{max}	A
Average Recoil Permeability	2.0
Field Strength Required to Fully Saturate Magnet (Oe)	3000
Ability to Withstand Demagnetizing Fields	G
Maximum Service Temperature (°F)	900
Approximate Temperature Permanently Affecting Magnet (°F)	1580
Inherent Orientation of Material	CHT

PHYSICAL PROPERTIES	
Density (lbs./in.^{3})	.264
General Mechanical Properties	HB
Tensile Strength (psi)	5000
Transverse Modulus of Rupture (psi)	8000
Coeff. of Therm. Expansion inches x 10^{-6}/°C	Orientation: Par. 11.5 Rt.Angl. _
Resistivity (micro-ohms/cm/cm^2)	47 @ 25 °C
Hardness 50	Scale Rockwell C
Basic Forming Method	CM
Basic Finishing Method	WGR
Relative Corrosion Resistance	G
Primary Chemical Composition	Al-Ni-Co-Cu-Fe
Special Notes:	

Note: All information based on available producers literature. In some cases of foreign materials interpretation or calculation required.

SOURCE(S):

Arnold Engineering Co.

USA

ArKomax 800 Hi-Hc
(Alloy or Grade)

Cast Alnico
(Type of Material)

8.1/.80
(MMPA Brief Designation)

R1-1-4
(IEC Code Reference)

MAGNETIC PROPERTIES		PHYSICAL PROPERTIES	
Residual Induction B_r (G)	13,200	Density (lbs./in.^{3})	.264
Coercive Force (Oe) Normal (H_c)	810	General Mechanical Properties	HB
Coercive Force (Oe) Intrinsic (H_{ci})	Est. 800	Tensile Strength (psi)	5000
Maximum Energy Product BH_{max} (MGO)	8.1	Transverse Modulus of Rupture (psi)	8000
B/H Ratio (Load Line) at BH_{max}	15.5	Coeff. of Therm. Expansion inches x 10^{-6}/°C	Orientation: Par. 11.5 Rt.Angl. –
Flux Density in Magnet B_m at BH_{max} (G)	11,200	Resistivity (micro-ohms/cm/cm^2	47 @ 25 °C
Coercive Force in Magnet H_m at BH_{max} (Oe)	720	Hardness 50	Scale Rockwell C
Energy Loss W_c (watt-sec/cycle/lb)	15.3	Basic Forming Method	CM
Importance of Operating Magnet at BH_{max}	B	Basic Finishing Method	WGR
Average Recoil Permeability	2.0	Relative Corrosion Resistance	G
Field Strength Required to Fully Saturate Magnet (Oe)	3000	Primary Chemical Composition	Al-Ni-Co-Cu-Fe
Ability to Withstand Demagnetizing Fields	G	Special Notes:	
Maximum Service Temperature (°F)	900		
Approximate Temperature Permanently Affecting Magnet (°F)	1580		
Inherent Orientation of Material	CHT		

Note: All information based on available producers literature. In some cases of foreign materials interpretation or calculation required.

SOURCE(S):	Armax 18
Arnold Engineering Co.	(Alloy or Grade)
	Sintered Rare Earth
	(Type of Material)
USA	18/20
	(MMPA Brief Designation)
	R5-1
	(IEC Code Reference)

MAGNETIC PROPERTIES		PHYSICAL PROPERTIES	
Residual Induction B_r (G)	8800	Density (lbs./in.^{3})	.305
Coercive Force (Oe) — Normal (H_c)	8500	General Mechanical Properties	HB
Coercive Force (Oe) — Intrinsic (H_{ci})	16,000	Tensile Strength (psi)	6000
Maximum Energy Product BH_{max} (MGO)	18.0	Transverse Modulus of Rupture (psi)	–
B/H Ratio (Load Line) at BH_{max}	Est. 1.0	Coeff. of Therm. Expansion inches x 10^{-6}/°C	Orientation: Par. 7.0 Rt.Angl. 13.0
Flux Density in Magnet B_m at BH_{max} (G)	Est. 4200	Resistivity (micro-ohms/cm/cm^2)	53 @ 25 °C
Coercive Force in Magnet H_m at BH_{max} (Oe)	Est. 4200	Hardness Scale	– –
Energy Loss W_c (watt-sec/cycle/lb)	–	Basic Forming Method	DPM
Importance of Operating Magnet at BH_{max}	E	Basic Finishing Method	WGR
Average Recoil Permeability	1.02	Relative Corrosion Resistance	G
Field Strength Required to Fully Saturate Magnet (Oe)	32000	Primary Chemical Composition	Sm-Co
Ability to Withstand Demagnetizing Fields	0	Special Notes:	
Maximum Service Temperature (°F)	523		
Approximate Temperature Permanently Affecting Magnet (°F)	993		
Inherent Orientation of Material	MP		

Note: All information based on available producers literature. In some cases of foreign materials interpretation or calculation required.

SOURCE(S):

Arnold Engineering Co.

USA

Armax 20
(Alloy or Grade)

Sintered Rare Earth
(Type of Material)

20/15
(MMPA Brief Designation)

R5-1
(IEC Code Reference)

MAGNETIC PROPERTIES	
Residual Induction B_r (G)	9000
Coercive Force (Oe) — Normal (H_c)	8900
Coercive Force (Oe) — Intrinsic (H_{ci})	15,000
Maximum Energy Product BH_{max} (MGO)	20.0
B/H Ratio (Load Line) at BH_{max}	Est. 1.0
Flux Density in Magnet B_m at BH_{max} (G)	Est. 4250
Coercive Force in Magnet H_m at BH_{max} (Oe)	Est. 4250
Energy Loss W_c (watt-sec/cycle/lb)	–
Importance of Operating Magnet at BH_{max}	E
Average Recoil Permeability	1.02
Field Strength Required to Fully Saturate Magnet (Oe)	32,000
Ability to Withstand Demagnetizing Fields	0
Maximum Service Temperature (°F)	523
Approximate Temperature Permanently Affecting Magnet (°F)	993
Inherent Orientation of Material	MP

PHYSICAL PROPERTIES	
Density (lbs./in.^{3})	.3053
General Mechanical Properties	HB
Tensile Strength (psi)	6000
Transverse Modulus of Rupture (psi)	–
Coeff. of Therm. Expansion inches x 10^{-6}/°C Orientation: Par. 7.0 Rt.Angl. 13.0	
Resistivity (micro-ohms/cm/cm^2)	53 @ 25 °C
Hardness Scale	– –
Basic Forming Method	DPM
Basic Finishing Method	WGR
Relative Corrosion Resistance	G
Primary Chemical Composition	Sm-Co
Special Notes:	

Note: All information based on available producers literature. In some cases of foreign materials interpretation or calculation required.

SOURCE(S):	
Arnold Engineering Co. USA	**Armax 22H** (Alloy or Grade)
	Sintered Rare Earth (Type of Material)
	22/15 (MMPA Brief Designation)
	R5-1 (IEC Code Reference)

MAGNETIC PROPERTIES		PHYSICAL PROPERTIES	
Residual Induction B_r (G)	9800	Density (lbs./in.^{3})	.3017
Coercive Force (Oe) — Normal (H_c)	8700	General Mechanical Properties	HB
Coercive Force (Oe) — Intrinsic (H_{ci})	16,000	Tensile Strength (psi)	5000
Maximum Energy Product BH_{max} (MGO)	22.0	Transverse Modulus of Rupture (psi)	–
B/H Ratio (Load Line) at BH_{max}	Est. 1.1	Coeff. of Therm. Expansion inches x 10^{-6}/°C	Orientation: Par. 10.0 Rt.Angl. 12.0
Flux Density in Magnet B_m at BH_{max} (G)	Est. 5100	Resistivity (micro-ohms/cm/cm^2)	86 @ 25 °C
Coercive Force in Magnet H_m at BH_{max} (Oe)	Est. 4700	Hardness Scale	
Energy Loss W_c (watt-sec/cycle/lb)	–	Basic Forming Method	DPM
Importance of Operating Magnet at BH_{max}	E	Basic Finishing Method	WGR
Average Recoil Permeability	1.08	Relative Corrosion Resistance	G
Field Strength Required to Fully Saturate Magnet (Oe)	40,000	Primary Chemical Composition	Sm-Co
Ability to Withstand Demagnetizing Fields	0	Special Notes:	
Maximum Service Temperature (°F)	572		
Approximate Temperature Permanently Affecting Magnet (°F)	1490		
Inherent Orientation of Material	MP		

Note: All information based on available producers literature. In some cases of foreign materials interpretation or calculation required.

SOURCE(S):

Arnold Engineering Co.

USA

Armax 26H
(Alloy or Grade)

Sintered Rare Earth
(Type of Material)

26/20
(MMPA Brief Designation)

R7-3
(IEC Code Reference)

MAGNETIC PROPERTIES		PHYSICAL PROPERTIES	
Residual Induction B_r (G)	10,600	Density (lbs./in.^{3})	.3017
Coercive Force (Oe) Normal (H_c)	9700	General Mechanical Properties	HB
Coercive Force (Oe) Intrinsic (H_{ci})	16,000	Tensile Strength (psi)	5000
Maximum Energy Product BH_{max} (MGO)	26.0	Transverse Modulus of Rupture (psi)	-
B/H Ratio (Load Line) at BH_{max}	Est. 1.14	Coeff. of Therm. Expansion inches x 10^{-6}/°C	Orientation: Par. 10.0 Rt.Angl. 12.0
Flux Density in Magnet B_m at BH_{max} (G)	Est. 5600	Resistivity (micro-ohms/cm/cm^2)	86 @ 25 °C
Coercive Force in Magnet H_m at BH_{max} (Oe)	Est. 4900	Hardness Scale	- -
Energy Loss W_c (watt-sec/cycle/lb)	-	Basic Forming Method	DPM
Importance of Operating Magnet at BH_{max}	E	Basic Finishing Method	WGR
Average Recoil Permeability	1.08	Relative Corrosion Resistance	G
Field Strength Required to Fully Saturate Magnet (Oe)	40,000	Primary Chemical Composition	Sm-Co
Ability to Withstand Demagnetizing Fields	0	Special Notes:	
Maximum Service Temperature (°F)	572		
Approximate Temperature Permanently Affecting Magnet (°F)	1490		
Inherent Orientation of Material	MP		

Note: All information based on available producers literature. In some cases of foreign materials interpretation or calculation required.

SOURCE(S):

Arnold Engineering Co.

USA

Arnife 1
(Alloy or Grade)

Wrought
(Type of Material)

-
(MMPA Brief Designation)

-
(IEC Code Reference)

MAGNETIC PROPERTIES		PHYSICAL PROPERTIES	
Residual Induction B_r (G)	10,000	Density (lbs./in.^{3})	.2580
Coercive Force (Oe) — Normal (H_c)	170	General Mechanical Properties	HT
Coercive Force (Oe) — Intrinsic (H_{ci})	-	Tensile Strength (psi)	23,000
Maximum Energy Product BH_{max} (MGO)	.9	Transverse Modulus of Rupture (psi)	-
B/H Ratio (Load Line) at BH_{max}	58	Coeff. of Therm. Expansion inches x 10^{-6}/°C	Orientation: Par. 12.8 Rt.Angl. -
Flux Density in Magnet B_m at BH_{max} (G)	7300	Resistivity (micro-ohms/cm/cm^2)	56 @ 25 °C
Coercive Force in Magnet H_m at BH_{max} (Oe)	126	Hardness 47 Scale Rockwell C	
Energy Loss W_c (watt-sec/cycle/lb)	-	Basic Forming Method	W
Importance of Operating Magnet at BH_{max}	A	Basic Finishing Method	WGR
Average Recoil Permeability	17.0	Relative Corrosion Resistance	E
Field Strength Required to Fully Saturate Magnet (Oe)	300	Primary Chemical Composition	-
Ability to Withstand Demagnetizing Fields	P	Special Notes: Discontinued material	
Maximum Service Temperature (°F)	900		
Approximate Temperature Permanently Affecting Magnet (°F)	1200		
Inherent Orientation of Material	-		

Note: All information based on available producers literature. In some cases of foreign materials interpretation or calculation required.

SOURCE(S):

Arnold Engineering Co.

USA

Arnife 2
(Alloy or Grade)

Wrought
(Type of Material)

-
(MMPA Brief Designation)

-
(IEC Code Reference)

MAGNETIC PROPERTIES	
Residual Induction B_r (G)	9300
Coercive Force (Oe) Normal (H_c)	230
Coercive Force (Oe) Intrinsic (H_{ci})	-
Maximum Energy Product BH_{max} (MGO)	1.1
B/H Ratio (Load Line) at BH_{max}	35
Flux Density in Magnet B_m at BH_{max} (G)	6200
Coercive Force in Magnet H_m at BH_{max} (Oe)	175
Energy Loss W_c (watt-sec/cycle/lb)	-
Importance of Operating Magnet at BH_{max}	A
Average Recoil Permeability	12.0
Field Strength Required to Fully Saturate Magnet (Oe)	300
Ability to Withstand Demagnetizing Fields	P
Maximum Service Temperature (°F)	900
Approximate Temperature Permanently Affecting Magnet (°F)	1200
Inherent Orientation of Material	-

PHYSICAL PROPERTIES	
Density (lbs./in.^{3})	.2580
General Mechanical Properties	HT
Tensile Strength (psi)	23,000
Transverse Modulus of Rupture (psi)	-
Coeff. of Therm. Expansion inches x 10^{-6}/°C	Orientation: Par. 12.8 Rt.Angl. -
Resistivity (micro-ohms/cm/cm^2)	56 @ 25 °C
Hardness 47	Scale Rockwell C
Basic Forming Method	W
Basic Finishing Method	WGR
Relative Corrosion Resistance	E
Primary Chemical Composition	-
Special Notes: Discontinued material	

Note: All information based on available producers literature. In some cases of foreign materials interpretation or calculation required.

SOURCE(S): **Arnife 3**
(Alloy or Grade)

Arnold Engineering Co.

Wrought
(Type of Material)

USA

-
(MMPA Brief Designation)

-
(IEC Code Reference)

MAGNETIC PROPERTIES		PHYSICAL PROPERTIES	
Residual Induction B_r (G)	9500	Density (lbs./in.^{3})	.2560
Coercive Force (Oe) — Normal (H_c)	75	General Mechanical Properties	HT
Coercive Force (Oe) — Intrinsic (H_{ci})	-	Tensile Strength (psi)	39,000
Maximum Energy Product BH_{max} (MGO)	.4	Transverse Modulus of Rupture (psi)	-
B/H Ratio (Load Line) at BH_{max}	125	Coeff. of Therm. Expansion inches x 10^{-6}/°C	Orientation: Par. 12.8 Rt.Angl. _
Flux Density in Magnet B_m at BH_{max} (G)	7000	Resistivity (micro-ohms/cm/cm^2)	56 @ 25 °C
Coercive Force in Magnet H_m at BH_{max} (Oe)	56	Hardness 43	Scale Rockwell C
Energy Loss W_c (watt-sec/cycle/lb)	-	Basic Forming Method	W
Importance of Operating Magnet at BH_{max}	A	Basic Finishing Method	WGR
Average Recoil Permeability	19.0	Relative Corrosion Resistance	E
Field Strength Required to Fully Saturate Magnet (Oe)	300	Primary Chemical Composition	-
Ability to Withstand Demagnetizing Fields	P	Special Notes: Discontinued material	
Maximum Service Temperature (°F)	900		
Approximate Temperature Permanently Affecting Magnet (°F)	1200		
Inherent Orientation of Material	-		

Note: All information based on available producers literature. In some cases of foreign materials interpretation or calculation required.

SOURCE(S):	
Arnold Engineering Co.	**Arnife 4** (Alloy or Grade)
	Wrought (Type of Material)
USA	- (MMPA Brief Designation)
	- (IEC Code Reference)

MAGNETIC PROPERTIES		PHYSICAL PROPERTIES	
Residual Induction B_r (G)	9700	Density (lbs./in.^{3})	.2560
Coercive Force (Oe) — Normal (H_c)	125	General Mechanical Properties	HT
Coercive Force (Oe) — Intrinsic (H_{ci})	-	Tensile Strength (psi)	26,000
Maximum Energy Product BH_{max} (MGO)	.7	Transverse Modulus of Rupture (psi)	-
B/H Ratio (Load Line) at BH_{max}	72	Coeff. of Therm. Expansion inches x 10^{-6}/°C	Orientation: Par. 12.8 Rt.Angl. -
Flux Density in Magnet B_m at BH_{max} (G)	7000	Resistivity (micro-ohms/cm/cm^2)	56 @ 25 °C
Coercive Force in Magnet H_m at BH_{max} (Oe)	97	Hardness Scale	46 Rockwell C
Energy Loss W_c (watt-sec/cycle/lb)	-	Basic Forming Method	W
Importance of Operating Magnet at BH_{max}	A	Basic Finishing Method	WGR
Average Recoil Permeability	15	Relative Corrosion Resistance	E
Field Strength Required to Fully Saturate Magnet (Oe)	300	Primary Chemical Composition	-
Ability to Withstand Demagnetizing Fields	P	Special Notes: Discontinued material	
Maximum Service Temperature (°F)	900		
Approximate Temperature Permanently Affecting Magnet (°F)	1200		
Inherent Orientation of Material	-		

Note: All information based on available producers literature. In some cases of foreign materials interpretation or calculation required.

SOURCE(S):	Arnokrome III
	(Alloy or Grade)
Arnold Engineering Co.	Wrought
	(Type of Material)
USA	-
	(MMPA Brief Designation)
	-
	(IEC Code Reference)

MAGNETIC PROPERTIES		PHYSICAL PROPERTIES	
Residual Induction B_r (G)	9000 to 12,000	Density (lbs./in.^{3})	.2770
Coercive Force (Oe) — Normal (H_c)	50 to 300	General Mechanical Properties	HT
Coercive Force (Oe) — Intrinsic (H_{ci})	-	Tensile Strength (psi)	75,000-120,000
Maximum Energy Product BH_{max} (MGO)	.4 to 1.2	Transverse Modulus of Rupture (psi)	-
B/H Ratio (Load Line) at BH_{max}	-	Coeff. of Therm. Expansion inches x 10^{-6}/°C	Orientation: Par. 8.67 Rt.Angl. -
Flux Density in Magnet B_m at BH_{max} (G)	-	Resistivity (micro-ohms/cm/cm^2)	69 @ 25 °C
Coercive Force in Magnet H_m at BH_{max} (Oe)	-	Hardness Scale	75-25 Rockwell C
Energy Loss W_c (watt-sec/cycle/lb)	-	Basic Forming Method	W
Importance of Operating Magnet at BH_{max}	A or B	Basic Finishing Method	WGR
Average Recoil Permeability	-	Relative Corrosion Resistance	E
Field Strength Required to Fully Saturate Magnet (Oe)	3000	Primary Chemical Composition	Fe-Co-Cr
Ability to Withstand Demagnetizing Fields	P to F	Special Notes: (1) All properties vary with heat treatment method	
Maximum Service Temperature (°F)	900		
Approximate Temperature Permanently Affecting Magnet (°F)	1200		
Inherent Orientation of Material	none		

Note: All information based on available producers literature. In some cases of foreign materials interpretation or calculation required.

SOURCE(S):	Arnox 7
Arnold Engineering Co.	(Alloy or Grade)
	Sintered Ferrite
	(Type of Material)
USA	2.7/4.2
	(MMPA Brief Designation)
	-
	(IEC Code Reference)

MAGNETIC PROPERTIES		PHYSICAL PROPERTIES	
Residual Induction B_r (G)	3450	Density (lbs./in.^{3})	.1720
Coercive Force (Oe) — Normal (H_c)	3200	General Mechanical Properties	VHB
Coercive Force (Oe) — Intrinsic (H_{ci})	4200	Tensile Strength (psi)	-
Maximum Energy Product BH_{max} (MGO)	2.7	Transverse Modulus of Rupture (psi)	-
B/H Ratio (Load Line) at BH_{max}	.83	Coeff. of Therm. Expansion inches x 10^{-6}/°C	Orientation: Par. - Rt.Angl. -
Flux Density in Magnet B_m at BH_{max} (G)	2500	Resistivity (micro-ohms/cm/cm^2)	1.0 @ 25 °C
Coercive Force in Magnet H_m at BH_{max} (Oe)	1800	Hardness Scale	6.5 Moh's
Energy Loss W_c (watt-sec/cycle/lb)	-	Basic Forming Method	PMF
Importance of Operating Magnet at BH_{max}	E	Basic Finishing Method	WGD
Average Recoil Permeability	1.05-1.20	Relative Corrosion Resistance	Ex
Field Strength Required to Fully Saturate Magnet (Oe)	15,000	Primary Chemical Composition	Sr-Fe
Ability to Withstand Demagnetizing Fields	Ex	Special Notes:	
Maximum Service Temperature (°F)	482		
Approximate Temperature Permanently Affecting Magnet (°F)	842		
Inherent Orientation of Material	MP		

Note: All information based on available producers literature. In some cases of foreign materials interpretation or calculation required.

SOURCE(S):

Arnold Engineering Co.

USA

Arnox 8
(Alloy or Grade)

Sintered Ferrite
(Type of Material)

3.5/3.2
(MMPA Brief Designation)

-
(IEC Code Reference)

MAGNETIC PROPERTIES		PHYSICAL PROPERTIES	
Residual Induction B_r (G)	3850	Density (lbs./in.^{3})	.1770
Coercive Force (Oe) — Normal (H_c)	2900	General Mechanical Properties	VHB
Coercive Force (Oe) — Intrinsic (H_{ci})	3150	Tensile Strength (psi)	-
Maximum Energy Product BH_{max} (MGO)	3.5	Transverse Modulus of Rupture (psi)	-
B/H Ratio (Load Line) at BH_{max}	1.14	Coeff. of Therm. Expansion inches x 10^{-6}/°C	Orientation: Par. - Rt.Angl. -
Flux Density in Magnet B_m at BH_{max} (G)	2000	Resistivity (micro-ohms/cm/cm^2)	1.0 @ 25 °C
Coercive Force in Magnet H_m at BH_{max} (Oe)	1750	Hardness 6.5	Scale Moh's
Energy Loss W_c (watt-sec/cycle/lb)	-	Basic Forming Method	PMF
Importance of Operating Magnet at BH_{max}	E	Basic Finishing Method	WGD
Average Recoil Permeability	1.02-1.20	Relative Corrosion Resistance	Ex
Field Strength Required to Fully Saturate Magnet (Oe)	15,000	Primary Chemical Composition	Sr-Fe
Ability to Withstand Demagnetizing Fields	Ex	Special Notes:	
Maximum Service Temperature (°F)	482		
Approximate Temperature Permanently Affecting Magnet (°F)	842		
Inherent Orientation of Material	MP		

Note: All information based on available producers literature. In some cases of foreign materials interpretation or calculation required.

SOURCE(S):

Arnold Engineering Co.

USA

Arnox 8B
(Alloy or Grade)

Sintered Ferrite
(Type of Material)

4.0/3.0
(MMPA Brief Designation)

-
(IEC Code Reference)

MAGNETIC PROPERTIES			PHYSICAL PROPERTIES	
Residual Induction B_r (G)		4100	Density (lbs./in.^{3})	.1750
Coercive Force (Oe)	Normal (H_c)	2900	General Mechanical Properties	VHB
	Intrinsic (H_{ci})	3000	Tensile Strength (psi)	-
Maximum Energy Product BH_{max} (MGO)		4.0	Transverse Modulus of Rupture (psi)	-
B/H Ratio (Load Line) at BH_{max}		.95	Coeff. of Therm. Expansion inches x 10^{-6}/°C	Orientation: Par. - Rt.Angl. -
Flux Density in Magnet B_m at BH_{max} (G)		1950	Resistivity (micro-ohms/cm/cm^2)	1.0 @ 25 °C
Coercive Force in Magnet H_m at BH_{max} (Oe)		2050	Hardness Scale	6.5 Moh's
Energy Loss W_c (watt-sec/cycle/lb)		-	Basic Forming Method	PMF
Importance of Operating Magnet at BH_{max}		E	Basic Finishing Method	WGD
Average Recoil Permeability		1.02-1.2	Relative Corrosion Resistance	Ex
Field Strength Required to Fully Saturate Magnet (Oe)		15,000	Primary Chemical Composition	Sr-Fe
Ability to Withstand Demagnetizing Fields		Ex	Special Notes:	
Maximum Service Temperature (°F)		482		
Approximate Temperature Permanently Affecting Magnet (°F)		842		
Inherent Orientation of Material		MP		

Note: All information based on available producers literature. In some cases of foreign materials interpretation or calculation required.

SOURCE(S):	Arnox 8H
Arnold Engineering Co.	(Alloy or Grade)
	Sintered Ferrite
	(Type of Material)
USA	3.4/4.0
	(MMPA Brief Designation)
	-
	(IEC Code Reference)

MAGNETIC PROPERTIES		PHYSICAL PROPERTIES	
Residual Induction B_r (G)	3800	Density (lbs./in.^{3})	.1750
Coercive Force (Oe) Normal (H_c)	3600	General Mechanical Properties	VHB
Coercive Force (Oe) Intrinsic (H_{ci})	4000	Tensile Strength (psi)	-
Maximum Energy Product BH_{max} (MGO)	3.4	Transverse Modulus of Rupture (psi)	-
B/H Ratio (Load Line) at BH_{max}	1.06	Coeff. of Therm. Expansion inches x 10^{-6}/°C	Orientation: Par. - Rt.Angl. -
Flux Density in Magnet B_m at BH_{max} (G)	1900	Resistivity (micro-ohms/cm/cm^2	1.0 @ 25 °C
Coercive Force in Magnet H_m at BH_{max} (Oe)	1800	Hardness 6.5 Scale Moh's	
Energy Loss W_c (watt-sec/cycle/lb)	-	Basic Forming Method	PMF
Importance of Operating Magnet at BH_{max}	E	Basic Finishing Method	WGD
Average Recoil Permeability	1.02-1.2	Relative Corrosion Resistance	Ex
Field Strength Required to Fully Saturate Magnet (Oe)	15,000	Primary Chemical Composition	Sr-Fe
Ability to Withstand Demagnetizing Fields	Ex	Special Notes:	
Maximum Service Temperature (°F)	482		
Approximate Temperature Permanently Affecting Magnet (°F)	842		
Inherent Orientation of Material	MP		

Note: All information based on available producers literature. In some cases of foreign materials interpretation or calculation required.

SOURCE(S):

Arnold Engineering Co.

USA

Arnox 9
(Alloy or Grade)

Sintered Ferrite
(Type of Material)

3.3/4.3
(MMPA Brief Designation)

-
(IEC Code Reference)

MAGNETIC PROPERTIES		PHYSICAL PROPERTIES	
Residual Induction B_r (G)	3700	Density (lbs./in.^{3})	.1730
Coercive Force (Oe) Normal (H_c)	3500	General Mechanical Properties	VHB
Coercive Force (Oe) Intrinsic (H_{ci})	4300	Tensile Strength (psi)	-
Maximum Energy Product BH_{max} (MGO)	3.25	Transverse Modulus of Rupture (psi)	-
B/H Ratio (Load Line) at BH_{max}	1.12	Coeff. of Therm. Expansion inches x 10^{-6}/°C Orientation: Par. - Rt.Angl. -	
Flux Density in Magnet B_m at BH_{max} (G)	1900	Resistivity (micro-ohms/cm/cm^2)	1.0 @ 25 °C
Coercive Force in Magnet H_m at BH_{max} (Oe)	1700	Hardness Scale 6.5 Moh's	
Energy Loss W_c (watt-sec/cycle/lb)	-	Basic Forming Method	PMF
Importance of Operating Magnet at BH_{max}	E	Basic Finishing Method	WGD
Average Recoil Permeability	Est. 1.1	Relative Corrosion Resistance	Ex
Field Strength Required to Fully Saturate Magnet (Oe)	15,000	Primary Chemical Composition	Sr-Fe
Ability to Withstand Demagnetizing Fields	Ex	Special Notes:	
Maximum Service Temperature (°F)	482		
Approximate Temperature Permanently Affecting Magnet (°F)	842		
Inherent Orientation of Material	MP		

Note: All information based on available producers literature. In some cases of foreign materials interpretation or calculation required.

SOURCE(S):	Arnox 9H
Arnold Engineering Co.	(Alloy or Grade)
	Sintered Ferrite
	(Type of Material)
USA	3.3/4.8
	(MMPA Brief Designation)
	-
	(IEC Code Reference)

MAGNETIC PROPERTIES		PHYSICAL PROPERTIES	
Residual Induction B_r (G)	3700	Density (lbs./in.^{3})	.1730
Coercive Force (Oe) Normal (H_c)	3500	General Mechanical Properties	VHB
Coercive Force (Oe) Intrinsic (H_{ci})	4800	Tensile Strength (psi)	-
Maximum Energy Product BH_{max}(MGO)	3.25	Transverse Modulus of Rupture (psi)	-
B/H Ratio (Load Line) at BH_{max}	1.12	Coeff. of Therm. Expansion inches x 10^{-6}/°C	Orientation: Par. - Rt.Angl. -
Flux Density in Magnet B_m at BH_{max} (G)	1900	Resistivity (micro-ohms/cm/cm^2)	1.0 @ 25 °C
Coercive Force in Magnet H_m at BH_{max} (Oe)	1700	Hardness 6.5	Scale Moh's
Energy Loss W_c (watt-sec/cycle/lb)	-	Basic Forming Method	PMF
Importance of Operating Magnet at BH_{max}	E	Basic Finishing Method	WGD
Average Recoil Permeability	1.02-1.20	Relative Corrosion Resistance	Ex
Field Strength Required to Fully Saturate Magnet (Oe)	15,000	Primary Chemical Composition	Sr-Fe
Ability to Withstand Demagnetizing Fields	Ex	Special Notes:	
Maximum Service Temperature (°F)	482		
Approximate Temperature Permanently Affecting Magnet (°F)	842		
Inherent Orientation of Material	MP		

Note: All information based on available producers literature. In some cases of foreign materials interpretation or calculation required.

SOURCE(S):	**Arnox 4000**
	(Alloy or Grade)
Arnold Engineering Co.	Sintered Ferrite
	(Type of Material)
USA	3.9/4.0
	(MMPA Brief Designation)
	-
	(IEC Code Reference)

MAGNETIC PROPERTIES		PHYSICAL PROPERTIES	
Residual Induction B_r (G)	4000	Density (lbs./in.^{3})	.1750
Coercive Force (Oe) — Normal (H_c)	3700	General Mechanical Properties	VHB
Coercive Force (Oe) — Intrinsic (H_{ci})	4000	Tensile Strength (psi)	-
Maximum Energy Product BH_{max} (MGO)	3.9	Transverse Modulus of Rupture (psi)	-
B/H Ratio (Load Line) at BH_{max}	1.07	Coeff. of Therm. Expansion inches x 10^{-6}/°C Orientation: Par. - Rt.Angl. -	
Flux Density in Magnet B_m at BH_{max} (G)	2050	Resistivity (micro-ohms/cm/cm^2)	1.0 @ 25 °C
Coercive Force in Magnet H_m at BH_{max} (Oe)	1920	Hardness Scale	6.5 Moh's
Energy Loss W_c (watt-sec/cycle/lb)	-	Basic Forming Method	PMF
Importance of Operating Magnet at BH_{max}	E	Basic Finishing Method	WGD
Average Recoil Permeability	1.02-1.20	Relative Corrosion Resistance	Ex
Field Strength Required to Fully Saturate Magnet (Oe)	15,000	Primary Chemical Composition	Sr-Fe
Ability to Withstand Demagnetizing Fields	Ex	Special Notes:	
Maximum Service Temperature (°F)	482		
Approximate Temperature Permanently Affecting Magnet (°F)	842		
Inherent Orientation of Material	MP		

Note: All information based on available producers literature. In some cases of foreign materials interpretation or calculation required.

SOURCE(S):

Arnold Engineering Co.

USA

B-1013
(Alloy or Grade)

Bonded Ferrite
(Type of Material)

1.4/3.3
(MMPA Brief Designation)

-
(IEC Code Reference)

MAGNETIC PROPERTIES		PHYSICAL PROPERTIES	
Residual Induction B_r (G)	2450	Density (lbs./in.^{3})	.1340
Coercive Force (Oe) Normal (H_c)	2200	General Mechanical Properties	FL
Coercive Force (Oe) Intrinsic (H_{ci})	3250	Tensile Strength (psi)	640
Maximum Energy Product BH_{max} (MGO)	1.4	Transverse Modulus of Rupture (psi)	-
B/H Ratio (Load Line) at BH_{max}	-	Coeff. of Therm. Expansion inches x 10^{-6}/°C	Orientation: Par. 9.8 Rt.Angl. -
Flux Density in Magnet B_m at BH_{max} (G)	-	Resistivity (micro-ohms/cm/cm^2)	10 @ 25 °C
Coercive Force in Magnet H_m at BH_{max} (Oe)	-	Hardness 55 Scale Shore D	
Energy Loss W_c (watt-sec/cycle/lb)	-	Basic Forming Method	E
Importance of Operating Magnet at BH_{max}	E	Basic Finishing Method	SCD/SRD
Average Recoil Permeability	-	Relative Corrosion Resistance	E
Field Strength Required to Fully Saturate Magnet (Oe)	10,000	Primary Chemical Composition	Ba-Fe + binder
Ability to Withstand Demagnetizing Fields	Ex	Special Notes: (1) Flexible material (2) Best for general fabricated parts (3) Binder is vulcanized nitrile rubber	
Maximum Service Temperature (°F)	250		
Approximate Temperature Permanently Affecting Magnet (°F)	-		
Inherent Orientation of Material	none		

Note: All information based on available producers literature. In some cases of foreign materials interpretation or calculation required.

SOURCE(S):	B-1030
Arnold Engineering Co. USA	(Alloy or Grade) Bonded Ferrite (Type of Material) 1.4/3.3 (MMPA Brief Designation) - (IEC Code Reference)

MAGNETIC PROPERTIES		PHYSICAL PROPERTIES	
Residual Induction B_r (G)	2450	Density (lbs./in.^{3})	.1340
Coercive Force (Oe) — Normal (H_c)	2200	General Mechanical Properties	FL
Coercive Force (Oe) — Intrinsic (H_{ci})	3250	Tensile Strength (psi)	640
Maximum Energy Product BH_{max} (MGO)	1.4	Transverse Modulus of Rupture (psi)	-
B/H Ratio (Load Line) at BH_{max}	-	Coeff. of Therm. Expansion inches x $10^{-6}/°C$ — Orientation: Par. / Rt.Angl.	9.8 / -
Flux Density in Magnet B_m at BH_{max} (G)	-	Resistivity (micro-ohms/cm/cm^2) 10 @ 25 °C	
Coercive Force in Magnet H_m at BH_{max} (Oe)	-	Hardness Scale — 55 Shore D	
Energy Loss W_c (watt-sec/cycle/lb)	-	Basic Forming Method	E
Importance of Operating Magnet at BH_{max}	E	Basic Finishing Method	SCD/SRD
Average Recoil Permeability	-	Relative Corrosion Resistance	E
Field Strength Required to Fully Saturate Magnet (Oe)	10,000	Primary Chemical Composition	Ba-Fe + binder
Ability to Withstand Demagnetizing Fields	Ex	Special Notes: (1) Flexible material (2) Designed for DC motors where greater flexuval capabilities required. (3) Binder is vulcanized nitrile rubber	
Maximum Service Temperature (°F)	250		
Approximate Temperature Permanently Affecting Magnet (°F)	-		
Inherent Orientation of Material	none		

Note: All information based on available producers literature. In some cases of foreign materials interpretation or calculation required.

SOURCE(S):

Arnold Engineering Co.

USA

B-1033
(Alloy or Grade)

Bonded Ferrite
(Type of Material)

1.6/3.6
(MMPA Brief Designation)

-
(IEC Code Reference)

MAGNETIC PROPERTIES		PHYSICAL PROPERTIES	
Residual Induction B_r (G)	2500	Density (lbs./in.^{3})	.137
Coercive Force (Oe) Normal (H_c)	2300	General Mechanical Properties	FL
Coercive Force (Oe) Intrinsic (H_{ci})	3600	Tensile Strength (psi)	640
Maximum Energy Product BH_{max} (MGO)	1.6	Transverse Modulus of Rupture (psi)	-
B/H Ratio (Load Line) at BH_{max}	-	Coeff. of Therm. Expansion inches x 10^{-6}/°C	Orientation: Par. 9.8 Rt.Angl. -
Flux Density in Magnet B_m at BH_{max} (G)	-	Resistivity (micro-ohms/cm/cm^2)	10 @ 25 °C
Coercive Force in Magnet H_m at BH_{max} (Oe)	-	Hardness 55	Scale Shore D
Energy Loss W_c (watt-sec/cycle/lb)	-	Basic Forming Method	E
Importance of Operating Magnet at BH_{max}	E	Basic Finishing Method	SCD/SRD
Average Recoil Permeability	-	Relative Corrosion Resistance	E
Field Strength Required to Fully Saturate Magnet (Oe)	10,000	Primary Chemical Composition	Ba-Fe + binder
Ability to Withstand Demagnetizing Fields	Ex	Special Notes:	
Maximum Service Temperature (°F)	250		
Approximate Temperature Permanently Affecting Magnet (°F)	-		
Inherent Orientation of Material	none		

Note: All information based on available producers literature. In some cases of foreign materials interpretation or calculation required.

SOURCE(S):

Arnold Engineering Co.

USA

B-1035/B-1044/B-1046
(Alloy or Grade)

Bonded Ferrite
(Type of Material)

-
(MMPA Brief Designation)

-
(IEC Code Reference)

MAGNETIC PROPERTIES		PHYSICAL PROPERTIES	
Residual Induction B_r (G)	See	Density (lbs./in.^{3})	-
Coercive Force (Oe) — Normal (H_c)	Note	General Mechanical Properties	FL
Coercive Force (Oe) — Intrinsic (H_{ci})	(1)	Tensile Strength (psi)	-
Maximum Energy Product BH_{max} (MGO)	1.4	Transverse Modulus of Rupture (psi)	-
B/H Ratio (Load Line) at BH_{max}	-	Coeff. of Therm. Expansion inches x 10^{-6}/°C	Orientation: Par. - Rt.Angl. -
Flux Density in Magnet B_m at BH_{max} (G)	-	Resistivity (micro-ohms/cm/cm^2)	- @ - °C
Coercive Force in Magnet H_m at BH_{max} (Oe)	-	Hardness Scale	- -
Energy Loss W_c (watt-sec/cycle/lb)	-	Basic Forming Method	E
Importance of Operating Magnet at BH_{max}	E	Basic Finishing Method	SCD/SRD
Average Recoil Permeability	-	Relative Corrosion Resistance	E
Field Strength Required to Fully Saturate Magnet (Oe)	10,000	Primary Chemical Composition	Ba-Fe + binder
Ability to Withstand Demagnetizing Fields	Ex	Special Notes: (1) Contact manufacturer for properties (2) Flexible material	
Maximum Service Temperature (°F)	250		
Approximate Temperature Permanently Affecting Magnet (°F)	-		
Inherent Orientation of Material	-		

Note: All information based on available producers literature. In some cases of foreign materials interpretation or calculation required.

SOURCE(S):

Arnold Engineering Co.

USA

B-1037
(Alloy or Grade)

Bonded Ferrite
(Type of Material)

1.0/2.2
(MMPA Brief Designation)

-
(IEC Code Reference)

MAGNETIC PROPERTIES	
Residual Induction B_r (G)	2150
Coercive Force (Oe) — Normal (H_c)	1060
Coercive Force (Oe) — Intrinsic (H_{ci})	2150
Maximum Energy Product BH_{max} (MGO)	1.0
B/H Ratio (Load Line) at BH_{max}	-
Flux Density in Magnet B_m at BH_{max} (G)	-
Coercive Force in Magnet H_m at BH_{max} (Oe)	-
Energy Loss W_c (watt-sec/cycle/lb)	-
Importance of Operating Magnet at BH_{max}	E
Average Recoil Permeability	1.08
Field Strength Required to Fully Saturate Magnet (Oe)	10,000
Ability to Withstand Demagnetizing Fields	Ex
Maximum Service Temperature (°F)	200
Approximate Temperature Permanently Affecting Magnet (°F)	-
Inherent Orientation of Material	none

PHYSICAL PROPERTIES	
Density (lbs./in.^{3})	.134
General Mechanical Properties	FL
Tensile Strength (psi)	640
Transverse Modulus of Rupture (psi)	-
Coeff. of Therm. Expansion inches x 10^{-6}/°C	Orientation: Par. 9.8 Rt.Angl. -
Resistivity (micro-ohms/cm/cm^2)	10 @ 15 °C
Hardness	55 Scale Shore D
Basic Forming Method	E
Basic Finishing Method	SCD/SRD
Relative Corrosion Resistance	E
Primary Chemical Composition	Ba-Fe + binder
Special Notes: (1) Flexible material (2) Rubber bonded	

Note: All information based on available producers literature. In some cases of foreign materials interpretation or calculation required.

SOURCE(S):

Arnold Engineering Co.

USA

B-1060

(Alloy or Grade)

Bonded Ferrite

(Type of Material)

1.7/4.1

(MMPA Brief Designation)

-

(IEC Code Reference)

MAGNETIC PROPERTIES	
Residual Induction B_r (G)	2650
Coercive Force (Oe) Normal (H_c)	2425
Coercive Force (Oe) Intrinsic (H_{ci})	4075
Maximum Energy Product BH_{max} (MGO)	1.7
B/H Ratio (Load Line) at BH_{max}	-
Flux Density in Magnet B_m at BH_{max} (G)	-
Coercive Force in Magnet H_m at BH_{max} (Oe)	-
Energy Loss W_c (watt-sec/cycle/lb)	-
Importance of Operating Magnet at BH_{max}	E
Average Recoil Permeability	-
Field Strength Required to Fully Saturate Magnet (Oe)	10,000
Ability to Withstand Demagnetizing Fields	Ex
Maximum Service Temperature (°F)	300
Approximate Temperature Permanently Affecting Magnet (°F)	-
Inherent Orientation of Material	none

PHYSICAL PROPERTIES	
Density (lbs./in.^{3})	.1300
General Mechanical Properties	HT
Tensile Strength (psi)	2200
Transverse Modulus of Rupture (psi)	-
Coeff. of Therm. Expansion inches x 10^{-6}/°C	Orientation: Par. .50 Rt.Angl. -
Resistivity (micro-ohms/cm/cm^2)	- @ - °C
Hardness 85	Scale Shore D
Basic Forming Method	E
Basic Finishing Method	WGD
Relative Corrosion Resistance	E
Primary Chemical Composition	Ba-Fe + binder
Special Notes: (1) Thermoplastic binder (2) Rigid bonded material	

Note: All information based on available producers literature. In some cases of foreign materials interpretation or calculation required.

SOURCE(S):

Arnold Engineering Co.

USA

B-1061
(Alloy or Grade)

Bonded Ferrite
(Type of Material)

1.9/3.0
(MMPA Brief Designation)

-
(IEC Code Reference)

MAGNETIC PROPERTIES		PHYSICAL PROPERTIES	
Residual Induction B_r (G)	2800	Density (lbs./in.^{3})	.1300
Coercive Force (Oe) Normal (H_c)	2250	General Mechanical Properties	HT
Coercive Force (Oe) Intrinsic (H_{ci})	3000	Tensile Strength (psi)	2200
Maximum Energy Product BH_{max} (MGO)	1.9	Transverse Modulus of Rupture (psi)	-
B/H Ratio (Load Line) at BH_{max}	-	Coeff. of Therm. Expansion inches $\times 10^{-6}$/°C	Orientation: Par. .50 Rt.Angl. -
Flux Density in Magnet B_m at BH_{max} (G)	-	Resistivity (micro-ohms/cm/cm^2)	- @ - °C
Coercive Force in Magnet H_m at BH_{max} (Oe)	-	Hardness Scale 85	Shore D
Energy Loss W_c (watt-sec/cycle/lb)	-	Basic Forming Method	E
Importance of Operating Magnet at BH_{max}	E	Basic Finishing Method	WGD
Average Recoil Permeability	-	Relative Corrosion Resistance	E
Field Strength Required to Fully Saturate Magnet (Oe)	10,000	Primary Chemical Composition	Ba-Fe + binder
Ability to Withstand Demagnetizing Fields	Ex	Special Notes: (1) Thermoplastic binder (2) Rigid bonded material	
Maximum Service Temperature (°F)	300		
Approximate Temperature Permanently Affecting Magnet (°F)	-		
Inherent Orientation of Material	none		

Note: All information based on available producers literature. In some cases of foreign materials interpretation or calculation required.

SOURCE(S):	B-1062
	(Alloy or Grade)
Arnold Engineering Co.	Bonded Ferrite
	(Type of Material)
USA	1.9/5.0
	(MMPA Brief Designation)
	-
	(IEC Code Reference)

MAGNETIC PROPERTIES		PHYSICAL PROPERTIES	
Residual Induction B_r (G)	2760	Density (lbs./in.^{3})	.1300
Coercive Force (Oe) — Normal (H_c)	2650	General Mechanical Properties	HT
Coercive Force (Oe) — Intrinsic (H_{ci})	5000	Tensile Strength (psi)	2200
Maximum Energy Product BH_{max} (MGO)	1.9	Transverse Modulus of Rupture (psi)	-
B/H Ratio (Load Line) at BH_{max}	-	Coeff. of Therm. Expansion inches x 10^{-6}/°C Orientation: Par. - Rt.Angl. -	
Flux Density in Magnet B_m at BH_{max} (G)	-	Resistivity (micro-ohms/cm/cm^2) - @ - °C	
Coercive Force in Magnet H_m at BH_{max} (Oe)	-	Hardness Scale 85 Shore D	
Energy Loss W_c (watt-sec/cycle/lb)	-	Basic Forming Method	E
Importance of Operating Magnet at BH_{max}	E	Basic Finishing Method	WGD
Average Recoil Permeability	-	Relative Corrosion Resistance	E
Field Strength Required to Fully Saturate Magnet (Oe)	10,000	Primary Chemical Composition	Ba-Fe + binder
Ability to Withstand Demagnetizing Fields	E	Special Notes: (1) Thermoplastic binder (2) Rigid bonded material	
Maximum Service Temperature (°F)	300		
Approximate Temperature Permanently Affecting Magnet (°F)	-		
Inherent Orientation of Material	MT		

Note: All information based on available producers literature. In some cases of foreign materials interpretation or calculation required.

SOURCE(S):

Arnold Engineering Co.

USA

B-1316/B-1317
(Alloy or Grade)

Bonded Ferrite
(Type of Material)

-
(MMPA Brief Designation)

-
(IEC Code Reference)

MAGNETIC PROPERTIES	
Residual Induction B_r (G)	See
Coercive Force (Oe) — Normal (H_c)	Note
Coercive Force (Oe) — Intrinsic (H_{ci})	(2)
Maximum Energy Product BH_{max} (MGO)	-
B/H Ratio (Load Line) at BH_{max}	-
Flux Density in Magnet B_m at BH_{max} (G)	-
Coercive Force in Magnet H_m at BH_{max} (Oe)	-
Energy Loss W_c (watt-sec/cycle/lb)	-
Importance of Operating Magnet at BH_{max}	E
Average Recoil Permeability	-
Field Strength Required to Fully Saturate Magnet (Oe)	10,000
Ability to Withstand Demagnetizing Fields	Ex
Maximum Service Temperature (°F)	250
Approximate Temperature Permanently Affecting Magnet (°F)	-
Inherent Orientation of Material	-

PHYSICAL PROPERTIES	
Density (lbs./in.^{3})	.1340
General Mechanical Properties	FL
Tensile Strength (psi)	640
Transverse Modulus of Rupture (psi)	-
Coeff. of Therm. Expansion inches x 10^{-6}/°C	Orientation: Par. 9.8 Rt.Angl. -
Resistivity (micro-ohms/cm/cm^2)	10 @ 25 °C
Hardness	55 Scale Shore D
Basic Forming Method	E
Basic Finishing Method	SCD/SRD
Relative Corrosion Resistance	E
Primary Chemical Composition	Ba or Sr-Fe + binder

Special Notes:

(1) Adhesive backed material
(2) Contact manufacturer for magnetic & other physical properties
(3) Polymer binder

Note: All information based on available producers literature. In some cases of foreign materials interpretation or calculation required.

SOURCE(S):

TDK Corp. of America

USA

BPA16
(Alloy or Grade)

Bonded Ferrite
(Type of Material)

1.6/2.8
(MMPA Brief Designation)

-
(IEC Code Reference)

MAGNETIC PROPERTIES		PHYSICAL PROPERTIES	
Residual Induction B_r (G)	2550	Density (lbs./in.^{3})	.1301
Coercive Force (Oe) Normal (H_c)	2380	General Mechanical Properties	HT
Coercive Force (Oe) Intrinsic (H_{ci})	2800	Tensile Strength (psi)	-
Maximum Energy Product BH_{max} (MGO)	1.6	Transverse Modulus of Rupture (psi)	-
B/H Ratio (Load Line) at BH_{max}	-	Coeff. of Therm. Expansion inches x 10^{-6}/°C	Orientation: Par. 65.0 Rt.Angl. -
Flux Density in Magnet B_m at BH_{max} (G)	-	Resistivity (micro-ohms/cm/cm^2)	- @ - °C
Coercive Force in Magnet H_m at BH_{max} (Oe)	-	Hardness Scale	- -
Energy Loss W_c (watt-sec/cycle/lb)	-	Basic Forming Method	E
Importance of Operating Magnet at BH_{max}	E	Basic Finishing Method	WGD
Average Recoil Permeability	-	Relative Corrosion Resistance	E
Field Strength Required to Fully Saturate Magnet (Oe)	10,000	Primary Chemical Composition	Sr-Fe + binder
Ability to Withstand Demagnetizing Fields	Ex	Special Notes: (1) Nylon 12 (polymide resin) binder (2) Injection molded	
Maximum Service Temperature (°F)	266		
Approximate Temperature Permanently Affecting Magnet (°F)	-		
Inherent Orientation of Material	none		

Note: All information based on available producers literature. In some cases of foreign materials interpretation or calculation required.

SOURCE(S): **BPB04**
(Alloy or Grade)

TDK Corp. of America

Bonded Ferrite
(Type of Material)

USA

.4/3.1
(MMPA Brief Designation)

-
(IEC Code Reference)

MAGNETIC PROPERTIES		PHYSICAL PROPERTIES	
Residual Induction B_r (G)	1300	Density (lbs./in.^{3})	.1192
Coercive Force (Oe) Normal (H_c)	1150	General Mechanical Properties	HT
Coercive Force (Oe) Intrinsic (H_{ci})	3100	Tensile Strength (psi)	-
Maximum Energy Product BH_{max} (MGO)	.40	Transverse Modulus of Rupture (psi)	-
B/H Ratio (Load Line) at BH_{max}	-	Coeff. of Therm. Expansion inches x 10^{-6}/°C	Orientation: Par. 45.0 Rt.Angl. -
Flux Density in Magnet B_m at BH_{max} (G)	-	Resistivity (micro-ohms/cm/cm^2	- @ - °C
Coercive Force in Magnet H_m at BH_{max} (Oe)	-	Hardness Scale	- -
Energy Loss W_c (watt-sec/cycle/lb)	-	Basic Forming Method	E
Importance of Operating Magnet at BH_{max}	E	Basic Finishing Method	WGD
Average Recoil Permeability	-	Relative Corrosion Resistance	E
Field Strength Required to Fully Saturate Magnet (Oe)	10,000	Primary Chemical Composition	Ba or Sr-Fe + binder
Ability to Withstand Demagnetizing Fields	Ex	Special Notes: (1) Nylon 12 (polymide resin) binder (2) Injection molded	
Maximum Service Temperature (°F)	266		
Approximate Temperature Permanently Affecting Magnet (°F)	-		
Inherent Orientation of Material	none		

Note: All information based on available producers literature. In some cases of foreign materials interpretation or calculation required.

SOURCE(S):	BPB12
	(Alloy or Grade)
TDK Corp. of America	Bonded Ferrite
	(Type of Material)
USA	1.3/3.4
	(MMPA Brief Designation)
	-
	(IEC Code Reference)

MAGNETIC PROPERTIES			PHYSICAL PROPERTIES	
Residual Induction B_r (G)		2250	Density (lbs./in.^{3})	.1192
Coercive Force (Oe)	Normal (H_c)	2100	General Mechanical Properties	HT
	Intrinsic (H_{ci})	3400	Tensile Strength (psi)	-
Maximum Energy Product BH_{max} (MGO)		1.3	Transverse Modulus of Rupture (psi)	-
B/H Ratio (Load Line) at BH_{max}		.80	Coeff. of Therm. Expansion inches x 10^{-6}/°C	Orientation: Par. 45.0 Rt.Angl. -
Flux Density in Magnet B_m at BH_{max} (G)		1000	Resistivity (micro-ohms/cm/cm^2)	- @ - °C
Coercive Force in Magnet H_m at BH_{max} (Oe)		1250	Hardness Scale	- -
Energy Loss W_c (watt-sec/cycle/lb)		-	Basic Forming Method	E
Importance of Operating Magnet at BH_{max}		E	Basic Finishing Method	WGD
Average Recoil Permeability		-	Relative Corrosion Resistance	E
Field Strength Required to Fully Saturate Magnet (Oe)		10,000	Primary Chemical Composition	Ba or Sr-Fe + binder
Ability to Withstand Demagnetizing Fields		Ex	Special Notes: (1) Nylon 12 (polymide resin) binder (2) Injection molded	
Maximum Service Temperature (°F)		266		
Approximate Temperature Permanently Affecting Magnet (°F)		-		
Inherent Orientation of Material		none		

Note: All information based on available producers literature. In some cases of foreign materials interpretation or calculation required.

SOURCE(S):

TDK Corp. of America

USA

BPC05
(Alloy or Grade)

Bonded Ferrite
(Type of Material)

.5/1.1
(MMPA Brief Designation)

-
(IEC Code Reference)

MAGNETIC PROPERTIES		PHYSICAL PROPERTIES	
Residual Induction B_r (G)	1600	Density (lbs./in.^{3})	-
Coercive Force (Oe) Normal (H_c)	800	General Mechanical Properties	HT
Coercive Force (Oe) Intrinsic (H_{ci})	1100	Tensile Strength (psi)	-
Maximum Energy Product BH_{max} (MGO)	.50	Transverse Modulus of Rupture (psi)	-
B/H Ratio (Load Line) at BH_{max}	-	Coeff. of Therm. Expansion inches x 10^{-6}/°C	Orientation: Par. - Rt.Angl. -
Flux Density in Magnet B_m at BH_{max} (G)	-	Resistivity (micro-ohms/cm/cm^2)	- @ - °C
Coercive Force in Magnet H_m at BH_{max} (Oe)	-	Hardness Scale	- -
Energy Loss W_c (watt-sec/cycle/lb)	-	Basic Forming Method	E
Importance of Operating Magnet at BH_{max}	E	Basic Finishing Method	WGD
Average Recoil Permeability	-	Relative Corrosion Resistance	E
Field Strength Required to Fully Saturate Magnet (Oe)	10,000	Primary Chemical Composition	Ba or Sr-Fe + binder
Ability to Withstand Demagnetizing Fields	Ex	Special Notes: (1) Nylon 12 (polymide resin) binder (2) Injection molded	
Maximum Service Temperature (°F)	266		
Approximate Temperature Permanently Affecting Magnet (°F)	-		
Inherent Orientation of Material	none		

Note: All information based on available producers literature. In some cases of foreign materials interpretation or calculation required.

SOURCE(S):	BPC16
	(Alloy or Grade)
TDK Corp. of America	Bonded Ferrite
	(Type of Material)
USA	1.6/3.1
	(MMPA Brief Designation)
	-
	(IEC Code Reference)

MAGNETIC PROPERTIES		PHYSICAL PROPERTIES	
Residual Induction B_r (G)	2600	Density (lbs./in.^{3})	.1301
Coercive Force (Oe) — Normal (H_c)	2325	General Mechanical Properties	HT
Coercive Force (Oe) — Intrinsic (H_{ci})	3100	Tensile Strength (psi)	-
Maximum Energy Product BH_{max} (MGO)	1.6	Transverse Modulus of Rupture (psi)	-
B/H Ratio (Load Line) at BH_{max}	-	Coeff. of Therm. Expansion inches x 10^{-6}/°C	Orientation: Par. 65.0 Rt.Angl. -
Flux Density in Magnet B_m at BH_{max} (G)	-	Resistivity (micro-ohms/cm/cm^2	- @ - °C
Coercive Force in Magnet H_m at BH_{max} (Oe)	-	Hardness - Scale -	
Energy Loss W_c (watt-sec/cycle/lb)	-	Basic Forming Method	E
Importance of Operating Magnet at BH_{max}	E	Basic Finishing Method	WGD
Average Recoil Permeability	-	Relative Corrosion Resistance	E
Field Strength Required to Fully Saturate Magnet (Oe)	10,000	Primary Chemical Composition	Ba or Sr-Fe + binder
Ability to Withstand Demagnetizing Fields	Ex	Special Notes: (1) Nylon 6 (polymide resin) binder (2) Injection molded	
Maximum Service Temperature (°F)	266		
Approximate Temperature Permanently Affecting Magnet (°F)	-		
Inherent Orientation of Material	none		

Note: All information based on available producers literature. In some cases of foreign materials interpretation or calculation required.

SOURCE(S):

TDK Corp. of America

USA

BQA14
(Alloy or Grade)

Bonded Ferrite
(Type of Material)

1.4/3.1
(MMPA Brief Designation)

-
(IEC Code Reference)

MAGNETIC PROPERTIES	
Residual Induction B_r (G)	2400
Coercive Force (Oe) — Normal (H_c)	2225
Coercive Force (Oe) — Intrinsic (H_{ci})	3100
Maximum Energy Product BH_{max} (MGO)	1.4
B/H Ratio (Load Line) at BH_{max}	.88
Flux Density in Magnet B_m at BH_{max} (G)	1100
Coercive Force in Magnet H_m at BH_{max} (Oe)	1250
Energy Loss W_c (watt-sec/cycle/lb)	-
Importance of Operating Magnet at BH_{max}	E
Average Recoil Permeability	-
Field Strength Required to Fully Saturate Magnet (Oe)	10,000
Ability to Withstand Demagnetizing Fields	Ex
Maximum Service Temperature (°F)	212
Approximate Temperature Permanently Affecting Magnet (°F)	See Note (2)
Inherent Orientation of Material	none

PHYSICAL PROPERTIES	
Density (lbs./in.^{3})	.1301–.1337
General Mechanical Properties	FL
Tensile Strength (psi)	-
Transverse Modulus of Rupture (psi)	-
Coeff. of Therm. Expansion inches x 10^{-6}/°C	Orientation: Par. 150 Rt.Angl. -
Resistivity (micro-ohms/cm/cm^2)	- @ - °C
Hardness Scale	30–40 Shore D
Basic Forming Method	E
Basic Finishing Method	SRD
Relative Corrosion Resistance	E
Primary Chemical Composition	Ba or Sr-Fe + binder
Special Notes: (1) NBR rubber binder (2) Injection molded	

Note: All information based on available producers literature. In some cases of foreign materials interpretation or calculation required.

SOURCE(S):	BQB14
	(Alloy or Grade)
TDK Corp. of America	Bonded Ferrite
	(Type of Material)
USA	1.4/3.4
	(MMPA Brief Designation)
	-
	(IEC Code Reference)

MAGNETIC PROPERTIES		PHYSICAL PROPERTIES	
Residual Induction B_r (G)	2400	Density (lbs./in.^{3})	.1301–.1337
Coercive Force (Oe) — Normal (H_c)	2225	General Mechanical Properties	FL
Coercive Force (Oe) — Intrinsic (H_{ci})	3350	Tensile Strength (psi)	-
Maximum Energy Product BH_{max} (MGO)	1.4	Transverse Modulus of Rupture (psi)	-
B/H Ratio (Load Line) at BH_{max}	1.0	Coeff. of Therm. Expansion inches x 10^{-6}/°C	Orientation: Par. 150 Rt.Angl. -
Flux Density in Magnet B_m at BH_{max} (G)	1200	Resistivity (micro-ohms/cm/cm^2)	- @ - °C
Coercive Force in Magnet H_m at BH_{max} (Oe)	1200	Hardness Scale 30-50	Shore D
Energy Loss W_c (watt-sec/cycle/lb)	-	Basic Forming Method	E
Importance of Operating Magnet at BH_{max}	E	Basic Finishing Method	SRD
Average Recoil Permeability	-	Relative Corrosion Resistance	E
Field Strength Required to Fully Saturate Magnet (Oe)	10,000	Primary Chemical Composition	Ba or Sr-Fe + binder
Ability to Withstand Demagnetizing Fields	Ex	Special Notes: (1) NBR rubber bonded (2) Injection molded	
Maximum Service Temperature (°F)	176		
Approximate Temperature Permanently Affecting Magnet (°F)	See Note (2)		
Inherent Orientation of Material	none		

Note: All information based on available producers literature. In some cases of foreign materials interpretation or calculation required.

SOURCE(S):

TDK Corp. of America

USA

BQC14
(Alloy or Grade)

Bonded Ferrite
(Type of Material)

1.4/3.8
(MMPA Brief Designation)

-
(IEC Code Reference)

MAGNETIC PROPERTIES		PHYSICAL PROPERTIES	
Residual Induction B_r (G)	2400	Density (lbs./in.^{3})	.1301-.1337
Coercive Force (Oe) Normal (H_c)	2225	General Mechanical Properties	FL
Coercive Force (Oe) Intrinsic (H_{ci})	3850	Tensile Strength (psi)	-
Maximum Energy Product BH_{max} (MGO)	1.4	Transverse Modulus of Rupture (psi)	-
B/H Ratio (Load Line) at BH_{max}	1.0	Coeff. of Therm. Expansion inches x 10^{-6}/°C	Orientation: Par. - Rt.Angl. -
Flux Density in Magnet B_m at BH_{max} (G)	1200	Resistivity (micro-ohms/cm/cm^2	- @ - °C
Coercive Force in Magnet H_m at BH_{max} (Oe)	1200	Hardness Scale 28-35	Shore D
Energy Loss W_c (watt-sec/cycle/lb)	-	Basic Forming Method	E
Importance of Operating Magnet at BH_{max}	E	Basic Finishing Method	SRD
Average Recoil Permeability	-	Relative Corrosion Resistance	E
Field Strength Required to Fully Saturate Magnet (Oe)	10,000	Primary Chemical Composition	Ba or Sr-Fe + binder
Ability to Withstand Demagnetizing Fields	Ex	Special Notes: (1) NBR rubber binder (2) Injection molded	
Maximum Service Temperature (°F)	See Note (2)		
Approximate Temperature Permanently Affecting Magnet (°F)	-		
Inherent Orientation of Material	none		

Note: All information based on available producers literature. In some cases of foreign materials interpretation or calculation required.

SOURCE(S):	BQK12
	(Alloy or Grade)
TDK Corp. of America	Bonded Ferrite
	(Type of Material)
USA	1.2/2.5
	(MMPA Brief Designation)
	-
	(IEC Code Reference)

MAGNETIC PROPERTIES		PHYSICAL PROPERTIES	
Residual Induction B_r (G)	2300	Density (lbs./in.^{3})	.1337
Coercive Force (Oe) — Normal (H_c)	2090	General Mechanical Properties	FL
Coercive Force (Oe) — Intrinsic (H_{ci})	2500	Tensile Strength (psi)	-
Maximum Energy Product BH_{max} (MGO)	1.23	Transverse Modulus of Rupture (psi)	-
B/H Ratio (Load Line) at BH_{max}	-	Coeff. of Therm. Expansion inches x 10^{-6}/°C	Orientation: Par. 44.0 Rt.Angl. -
Flux Density in Magnet B_m at BH_{max} (G)	-	Resistivity (micro-ohms/cm/cm^2)	- @ - °C
Coercive Force in Magnet H_m at BH_{max} (Oe)	-	Hardness Scale	- -
Energy Loss W_c (watt-sec/cycle/lb)	-	Basic Forming Method	E
Importance of Operating Magnet at BH_{max}	E	Basic Finishing Method	SRD
Average Recoil Permeability	-	Relative Corrosion Resistance	E
Field Strength Required to Fully Saturate Magnet (Oe)	10,000	Primary Chemical Composition	Ba or Sr-Fe + binder
Ability to Withstand Demagnetizing Fields	Ex	Special Notes: (1) Chlorinate polyethelene rubber binder (2) Injection molded	
Maximum Service Temperature (°F)	See Note (2)		
Approximate Temperature Permanently Affecting Magnet (°F)	-		
Inherent Orientation of Material	none		

Note: All information based on available producers literature. In some cases of foreign materials interpretation or calculation required.

SOURCE(S):	**BRA70**
	(Alloy or Grade)
TDK Corp. of America	Bonded Rare Earth
	(Type of Material)
USA	6.9/6.4
	(MMPA Brief Designation)
	-
	(IEC Code Reference)

MAGNETIC PROPERTIES		PHYSICAL PROPERTIES	
Residual Induction B_r (G)	5600	Density (lbs./in.^{3})	.2023–.2132
Coercive Force (Oe) — Normal (H_c)	4300	General Mechanical Properties	HT
Coercive Force (Oe) — Intrinsic (H_{ci})	6390	Tensile Strength (psi)	-
Maximum Energy Product BH_{max} (MGO)	6.9	Transverse Modulus of Rupture (psi)	-
B/H Ratio (Load Line) at BH_{max}	-	Coeff. of Therm. Expansion inches x 10^{-6}/°C	Orientation: Par. - Rt.Angl. -
Flux Density in Magnet B_m at BH_{max} (G)	-	Resistivity (micro-ohms/cm/cm^2)	- @ - °C
Coercive Force in Magnet H_m at BH_{max} (Oe)	-	Hardness - Scale -	
Energy Loss W_c (watt-sec/cycle/lb)	-	Basic Forming Method	E
Importance of Operating Magnet at BH_{max}	E	Basic Finishing Method	SCD/SRD
Average Recoil Permeability	-	Relative Corrosion Resistance	G
Field Strength Required to Fully Saturate Magnet (Oe)	30,000	Primary Chemical Composition	Sm-Co + binder (See Note 2)
Ability to Withstand Demagnetizing Fields	0	Special Notes: (1) Nylon 12 (polymide resin) bonded (2) Injection molded	
Maximum Service Temperature (°F)	248		
Approximate Temperature Permanently Affecting Magnet (°F)	594		
Inherent Orientation of Material	MP		

Note: All information based on available producers literature. In some cases of foreign materials interpretation or calculation required.

SOURCE(S):		
TDK Corp. of America		**BRA90**
		(Alloy or Grade)
		Bonded Rare Earth
		(Type of Material)
USA		8.4/6.4
		(MMPA Brief Designation)
		-
		(IEC Code Reference)

MAGNETIC PROPERTIES		PHYSICAL PROPERTIES	
Residual Induction B_r (G)	6100	Density (lbs./in.^{3})	.2023-.2132
Coercive Force (Oe) — Normal (H_c)	4640	General Mechanical Properties	HT
Coercive Force (Oe) — Intrinsic (H_{ci})	6390	Tensile Strength (psi)	-
Maximum Energy Product BH_{max} (MGO)	8.4	Transverse Modulus of Rupture (psi)	-
B/H Ratio (Load Line) at BH_{max}	-	Coeff. of Therm. Expansion inches x 10^{-6}/°C — Orientation: Par. - Rt.Angl. -	
Flux Density in Magnet B_m at BH_{max} (G)	-	Resistivity (micro-ohms/cm/cm^2)	- @ - °C
Coercive Force in Magnet H_m at BH_{max} (Oe)	-	Hardness Scale	- -
Energy Loss W_c (watt-sec/cycle/lb)	-	Basic Forming Method	E
Importance of Operating Magnet at BH_{max}	E	Basic Finishing Method	SCD/SRD
Average Recoil Permeability	-	Relative Corrosion Resistance	G
Field Strength Required to Fully Saturate Magnet (Oe)	30,000	Primary Chemical Composition	Sm-Co + binder (See Note 2)
Ability to Withstand Demagnetizing Fields	0	Special Notes: (1) Nylon 12 (polymide resin) bonded (2) Injection Molded	
Maximum Service Temperature (°F)	248		
Approximate Temperature Permanently Affecting Magnet (°F)	594		
Inherent Orientation of Material	none		

Note: All information based on available producers literature. In some cases of foreign materials interpretation or calculation required.

SOURCE(S):

Magnet Applications, Ltd.
(a.k.a. Cookson Group/Arelec)

England/France

Bremag 4NF
(Alloy or Grade)

Bonded Rare Earth
(Type of Material)

-
(MMPA Brief Designation)

-
(IEC Code Reference)

MAGNETIC PROPERTIES		PHYSICAL PROPERTIES	
Residual Induction B_r (G)	See	Density (lbs./in.^{3})	-
Coercive Force (Oe) — Normal (H_c)	Note	General Mechanical Properties	-
Coercive Force (Oe) — Intrinsic (H_{ci})	(1)	Tensile Strength (psi)	-
Maximum Energy Product BH_{max} (MGO)	-	Transverse Modulus of Rupture (psi)	-
B/H Ratio (Load Line) at BH_{max}	-	Coeff. of Therm. Expansion inches x 10^{-6}/°C Orientation: Par. - Rt.Angl. -	
Flux Density in Magnet B_m at BH_{max} (G)	-	Resistivity (micro-ohms/cm/cm^2) - @ - °C	
Coercive Force in Magnet H_m at BH_{max} (Oe)	-	Hardness Scale - -	
Energy Loss W_c (watt-sec/cycle/lb)	-	Basic Forming Method	E
Importance of Operating Magnet at BH_{max}	E	Basic Finishing Method	SCD/SRD
Average Recoil Permeability	-	Relative Corrosion Resistance	G
Field Strength Required to Fully Saturate Magnet (Oe)	-	Primary Chemical Composition	Nd-Fe-B + binder See Note (2)
Ability to Withstand Demagnetizing Fields	0	Special Notes:	
Maximum Service Temperature (°F)	Approx. 250	(1) No additional date available (1993). Material not released commercially as of date of publication.	
Approximate Temperature Permanently Affecting Magnet (°F)	594	(2) Elasticmeric polymer bonded.	
Inherent Orientation of Material	-		

Note: All information based on available producers literature. In some cases of foreign materials interpretation or calculation required.

SOURCE(S):	
	Bremag 5N
	(Alloy or Grade)
Magnet Applications, Ltd.	Bonded Rare Earth
(a.k.a. Cookson Group/Arelec)	(Type of Material)
	-
England/France	(MMPA Brief Designation)
	-
	(IEC Code Reference)

MAGNETIC PROPERTIES		PHYSICAL PROPERTIES	
Residual Induction B_r (G)	4300	Density (lbs./in.^{3})	.1626
Coercive Force (Oe) — Normal (H_c)	4000	General Mechanical Properties	HT
Coercive Force (Oe) — Intrinsic (H_{ci})	6300	Tensile Strength (psi)	-
Maximum Energy Product BH_{max} (MGO)	3.8	Transverse Modulus of Rupture (psi)	-
B/H Ratio (Load Line) at BH_{max}	-	Coeff. of Therm. Expansion inches x 10^{-6}/°C Orientation: Par. - Rt.Angl. -	
Flux Density in Magnet B_m at BH_{max} (G)	-	Resistivity (micro-ohms/cm/cm^2)	750 @ 25 °C
Coercive Force in Magnet H_m at BH_{max} (Oe)	-	Hardness Scale 34 Vickers	
Energy Loss W_c (watt-sec/cycle/lb)	-	Basic Forming Method	E
Importance of Operating Magnet at BH_{max}	E	Basic Finishing Method	WGR
Average Recoil Permeability	1.25	Relative Corrosion Resistance	G
Field Strength Required to Fully Saturate Magnet (Oe)	22,600	Primary Chemical Composition	Nd-Fe-B + binder (See Note)
Ability to Withstand Demagnetizing Fields	0	Special Notes: (1) Nylon or other plastic bonded	
Maximum Service Temperature (°F)	212		
Approximate Temperature Permanently Affecting (°F)	594		
In... of M...	none		

Note: ...sed on available producers literature. ...oreign materials interpretation or calcula... required.

SOURCE(S):

Magnet Applications, Ltd.
(a.k.a. Cookson Group/Arelec)

England/France

Bremag 10N
(Alloy or Grade)

Bonded Rare Earth
(Type of Material)

-
(MMPA Brief Designation)

-
(IEC Code Reference)

MAGNETIC PROPERTIES		PHYSICAL PROPERTIES	
Residual Induction B_r (G)	6800	Density (lbs./in.^{3})	.2168
Coercive Force (Oe) — Normal (H_c)	5780	General Mechanical Properties	HT
Coercive Force (Oe) — Intrinsic (H_{ci})	10,300	Tensile Strength (psi)	-
Maximum Energy Product BH_{max} (MGO)	10.0	Transverse Modulus of Rupture (psi)	-
B/H Ratio (Load Line) at BH_{max}	-	Coeff. of Therm. Expansion inches x 10^{-6}/°C	Orientation: Par. - Rt.Angl. -
Flux Density in Magnet B_m at BH_{max} (G)	-	Resistivity (micro-ohms/cm/cm^2)	- @ - °C
Coercive Force in Magnet H_m at BH_{max} (Oe)	-	Hardness Scale	45 Vickers
Energy Loss W_c (watt-sec/cycle/lb)	-	Basic Forming Method	E
Importance of Operating Magnet at BH_{max}	E	Basic Finishing Method	WGR
Average Recoil Permeability	-	Relative Corrosion Resistance	G
Field Strength Required to Fully Saturate Magnet (Oe)	22,600	Primary Chemical Composition	Nb-Fe-B + binder
Ability to Withstand Demagnetizing Fields	0	Special Notes: (1) Bonded with proprietary resin	
Maximum Service Temperature (°F)	300		
Approximate Temperature Permanently Affecting Magnet (°F)	594		
Inherent Orientation of Material	none		

Note: All information based on available producers literature. In some cases of foreign materials interpretation or calculation required.

SOURCE(S):	Bremag 18
	(Alloy or Grade)
Magnet Applications, Ltd. (a.k.a. Cookson Group/Arelec)	Sintered Rare Earth
	(Type of Material)
	20/18
	(MMPA Brief Designation)
England/France	R5-1
	(IEC Code Reference)

MAGNETIC PROPERTIES		PHYSICAL PROPERTIES	
Residual Induction B_r (G)	9000	Density (lbs./in.^{3})	.2962
Coercive Force (Oe) — Normal (H_c)	8500	General Mechanical Properties	HB
Coercive Force (Oe) — Intrinsic (H_{ci})	18,000	Tensile Strength (psi)	Est. 5000
Maximum Energy Product BH_{max} (MGO)	20.0	Transverse Modulus of Rupture (psi)	–
B/H Ratio (Load Line) at BH_{max}	Est. 1.14	Coeff. of Therm. Expansion inches x 10^{-6}/°C — Orientation: Par. / Rt.Angl.	– / –
Flux Density in Magnet B_m at BH_{max} (G)	Est. 4600	Resistivity (micro-ohms/cm/cm^2)	50 @ 25 °C
Coercive Force in Magnet H_m at BH_{max} (Oe)	Est. 4000	Hardness / Scale	600 Vickers
Energy Loss W_c (watt-sec/cycle/lb)	–	Basic Forming Method	DPM
Importance of Operating Magnet at BH_{max}	E	Basic Finishing Method	WGR
Average Recoil Permeability	1.05	Relative Corrosion Resistance	G
Field Strength Required to Fully Saturate Magnet (Oe)	35,000	Primary Chemical Composition	Sm-Co
Ability to Withstand Demagnetizing Fields	0	Special Notes:	
Maximum Service Temperature (°F)	392		
Approximate Temperature Permanently Affecting Magnet (°F)	Est. 1328		
Inherent Orientation of Material	MP		

Note: All information based on available producers literature. In some cases of foreign materials interpretation or calculation required.

SOURCE(S):

Magnet Applications, Ltd.
(a.k.a. Cookson Group/Arelec)

England/France

Bremag 26
(Alloy or Grade)

Sintered Rare Earth
(Type of Material)

20/18
(MMPA Brief Designation)

R5-1
(IEC Code Reference)

MAGNETIC PROPERTIES	
Residual Induction B_r (G)	10,300
Coercive Force (Oe) Normal (H_c)	8000
Coercive Force (Oe) Intrinsic (H_{ci})	18,000
Maximum Energy Product BH_{max} (MGO)	20.0
B/H Ratio (Load Line) at BH_{max}	Est. .8
Flux Density in Magnet B_m at BH_{max} (G)	Est. 4000
Coercive Force in Magnet H_m at BH_{max} (Oe)	Est. 5000
Energy Loss W_c (watt-sec/cycle/lb)	-
Importance of Operating Magnet at BH_{max}	E
Average Recoil Permeability	1.05
Field Strength Required to Fully Saturate Magnet (Oe)	35,000
Ability to Withstand Demagnetizing Fields	0
Maximum Service Temperature (°F)	572
Approximate Temperature Permanently Affecting Magnet (°F)	Est. 1472
Inherent Orientation of Material	MP

PHYSICAL PROPERTIES	
Density (lbs./in.^{3})	.2962
General Mechanical Properties	HB
Tensile Strength (psi)	Est. 5000
Transverse Modulus of Rupture (psi)	-
Coeff. of Therm. Expansion inches x 10^{-6}/°C	Orientation: Par. - Rt.Angl. -
Resistivity (micro-ohms/cm/cm^2)	80 @ 25 °C
Hardness 600	Scale Vickers
Basic Forming Method	DPM
Basic Finishing Method	WGR
Relative Corrosion Resistance	G
Primary Chemical Composition	Sm-Co
Special Notes:	

Note: All information based on available producers literature. In some cases of foreign materials interpretation or calculation required.

SOURCE(S):

**Magnet Applications, Ltd.
(a.k.a. Cookson Group/Arelec)**

England/France

Bremag 27N
(Alloy or Grade)

Sintered Rare Earth
(Type of Material)

27/13
(MMPA Brief Designation)

R7-3
(IEC Code Reference)

MAGNETIC PROPERTIES		PHYSICAL PROPERTIES	
Residual Induction B_r (G)	11,000	Density (lbs./in.^{3})	.2673
Coercive Force (Oe) — Normal (H_c)	8,750	General Mechanical Properties	HB
Coercive Force (Oe) — Intrinsic (H_{ci})	12,500	Tensile Strength (psi)	-
Maximum Energy Product BH_{max} (MGO)	27.0	Transverse Modulus of Rupture (psi)	-
B/H Ratio (Load Line) at BH_{max}	Est. 1.1	Coeff. of Therm. Expansion inches x 10^{-6}/°C — Orientation: Par. _ Rt.Angl. _	
Flux Density in Magnet B_m at BH_{max} (G)	Est. 5500	Resistivity (micro-ohms/cm/cm^2)	1.5 @ 25 °C
Coercive Force in Magnet H_m at BH_{max} (Oe)	Est. 4800	Hardness — Scale	500 Vickers
Energy Loss W_c (watt-sec/cycle/lb)	-	Basic Forming Method	DPM
Importance of Operating Magnet at BH_{max}	E	Basic Finishing Method	WGR
Average Recoil Permeability	1.05	Relative Corrosion Resistance	P
Field Strength Required to Fully Saturate Magnet (Oe)	35,000	Primary Chemical Composition	Nd-Fe-B
Ability to Withstand Demagnetizing Fields	0	Special Notes: (1) Anti-corrosion coating recommended	
Maximum Service Temperature (°F)	248		
Approximate Temperature Permanently Affecting Magnet (°F)	Est. 590		
Inherent Orientation of Material	MP		

Note: All information based on available producers literature. In some cases of foreign materials interpretation or calculation required.

SOURCE(S):	Bremag 35N
	(Alloy or Grade)
Magnet Applications, Ltd.	Sintered Rare Earth
(a.k.a. Cookson Group/Arelec)	(Type of Material)
	35/17
	(MMPA Brief Designation)
England/France	R7-3
	(IEC Code Reference)

MAGNETIC PROPERTIES		PHYSICAL PROPERTIES	
Residual Induction B_r (G)	12,100	Density (lbs./in.^{3})	.2673
Coercive Force (Oe) — Normal (H_c)	11,600	General Mechanical Properties	HB
Coercive Force (Oe) — Intrinsic (H_{ci})	17,000	Tensile Strength (psi)	-
Maximum Energy Product BH_{max} (MGO)	35.0	Transverse Modulus of Rupture (psi)	-
B/H Ratio (Load Line) at BH_{max}	Est. 1.1	Coeff. of Therm. Expansion inches x 10^{-6}/°C	Orientation: Par. - Rt.Angl. -
Flux Density in Magnet B_m at BH_{max} (G)	Est. 6200	Resistivity (micro-ohms/cm/cm^2)	1.5 @ 25 °C
Coercive Force in Magnet H_m at BH_{max} (Oe)	Est. 5600	Hardness Scale	600 Vickers
Energy Loss W_c (watt-sec/cycle/lb)	-	Basic Forming Method	DPM
Importance of Operating Magnet at BH_{max}	E	Basic Finishing Method	WGR
Average Recoil Permeability	1.05	Relative Corrosion Resistance	P
Field Strength Required to Fully Saturate Magnet (Oe)	40,000	Primary Chemical Composition	Nd-Fe-B
Ability to Withstand Demagnetizing Fields	0	Special Notes: (1) Anti-corrosion coating recommended	
Maximum Service Temperature (°F)	284		
Approximate Temperature Permanently Affecting Magnet (°F)	Est. 590		
Inherent Orientation of Material	MP		

Note: All information based on available producers literature. In some cases of foreign materials interpretation or calculation required.

SOURCE(S):	CM8B
TDK Corp. of America	(Alloy or Grade)
	Bonded Rare Earth
	(Type of Material)
USA	-
	(MMPA Brief Designation)
	-
	(IEC Code Reference)

MAGNETIC PROPERTIES		PHYSICAL PROPERTIES	
Residual Induction B_r (G)	6400	Density (lbs./in.^{3})	.2132
Coercive Force (Oe) — Normal (H_c)	8700	General Mechanical Properties	HT
Coercive Force (Oe) — Intrinsic (H_{ci})	-	Tensile Strength (psi)	-
Maximum Energy Product BH_{max} (MGO)	8.4	Transverse Modulus of Rupture (psi)	-
B/H Ratio (Load Line) at BH_{max}	-	Coeff. of Therm. Expansion inches x 10^{-6}/°C	Orientation: Par. 10.5 Rt.Angl. -
Flux Density in Magnet B_m at BH_{max} (G)	-	Resistivity (micro-ohms/cm/cm^2)	- @ - °C
Coercive Force in Magnet H_m at BH_{max} (Oe)	-	Hardness - Scale -	
Energy Loss W_c (watt-sec/cycle/lb)	-	Basic Forming Method	E
Importance of Operating Magnet at BH_{max}	E	Basic Finishing Method	SGC/SRD
Average Recoil Permeability	1.08	Relative Corrosion Resistance	G
Field Strength Required to Fully Saturate Magnet (Oe)	30,000	Primary Chemical Composition	Nd-Fe-B + binder (See Note 2)
Ability to Withstand Demagnetizing Fields	0	Special Notes: (1) Compression molded (2) Binder not indicated	
Maximum Service Temperature (°F)	-		
Approximate Temperature Permanently Affecting Magnet (°F)	594		
Inherent Orientation of Material	none		

Note: All information based on available producers literature. In some cases of foreign materials interpretation or calculation required.

SOURCE(S):

Stackpole Corp.

USA

Ceramag A
(Alloy or Grade)

Sintered Ferrite
(Type of Material)

1.1/3.2
(MMPA Brief Designation)

S1-0-1
(IEC Code Reference)

MAGNETIC PROPERTIES		PHYSICAL PROPERTIES	
Residual Induction B_r (G)	2100	Density (lbs./in.^{3})	Est. .1731
Coercive Force (Oe) Normal (H_c)	1700	General Mechanical Properties	VHB
Coercive Force (Oe) Intrinsic (H_{ci})	3160	Tensile Strength (psi)	-
Maximum Energy Product BH_{max} (MGO)	.9	Transverse Modulus of Rupture (psi)	-
B/H Ratio (Load Line) at BH_{max}	Est. 1.29	Coeff. of Therm. Expansion inches x 10^{-6}/°C	Orientation: Par. - Rt.Angl. -
Flux Density in Magnet B_m at BH_{max} (G)	Est. 1100	Resistivity (micro-ohms/cm/cm^2)	1.0 @ 25 °C
Coercive Force in Magnet H_m at BH_{max} (Oe)	Est. 850	Hardness Scale	6.5 Moh's
Energy Loss W_c (watt-sec/cycle/lb)	-	Basic Forming Method	PF
Importance of Operating Magnet at BH_{max}	E	Basic Finishing Method	WGD
Average Recoil Permeability	Est. 1.1	Relative Corrosion Resistance	Ex
Field Strength Required to Fully Saturate Magnet (Oe)	10,000	Primary Chemical Composition	Ba or Sr & Fe
Ability to Withstand Demagnetizing Fields	Ex	Special Notes:	
Maximum Service Temperature (°F)	482		
Approximate Temperature Permanently Affecting Magnet (°F)	842		
Inherent Orientation of Material	none		

Note: All information based on available producers literature. In some cases of foreign materials interpretation or calculation required.

SOURCE(S): **Vacuumschmelze GMBH Germany**

Crovac 10/130
(Alloy or Grade)

Wrought
(Type of Material)

-
(MMPA Brief Designation)

-
(IEC Code Reference)

MAGNETIC PROPERTIES		PHYSICAL PROPERTIES	
Residual Induction B_r (G)	9500	Density (lbs./in.^{3})	.2746
Coercive Force (Oe) — Normal (H_c)	445	General Mechanical Properties	See Notes
Coercive Force (Oe) — Intrinsic (H_{ci})	-	Tensile Strength (psi)	-
Maximum Energy Product BH_{max} (MGO)	1.6	Transverse Modulus of Rupture (psi)	-
B/H Ratio (Load Line) at BH_{max}	32.7	Coeff. of Therm. Expansion inches x 10^{-6}/°C	Orientation: Par. 10.0 Rt.Angl. -
Flux Density in Magnet B_m at BH_{max} (G)	7200	Resistivity (micro-ohms/cm/cm^2)	- @ - °C
Coercive Force in Magnet H_m at BH_{max} (Oe)	220	Hardness Scale	See Notes
Energy Loss W_c (watt-sec/cycle/lb)	-	Basic Forming Method	C+W
Importance of Operating Magnet at BH_{max}	B	Basic Finishing Method	See Notes
Average Recoil Permeability	-	Relative Corrosion Resistance	G
Field Strength Required to Fully Saturate Magnet (Oe)	3000	Primary Chemical Composition	Al-Ni-Co-Fe
Ability to Withstand Demagnetizing Fields	F	Special Notes: (1) Vickers Hardness: (a) cold worked = 310-340 (b) sol. annealed = 210-240 (c) after H.T. = 440-520 (2) Must be grind finished after H.T.	
Maximum Service Temperature (°F)	896		
Approximate Temperature Permanently Affecting Magnet (°F)	1184		
Inherent Orientation of Material	none		

Note: All information based on available producers literature. In some cases of foreign materials interpretation or calculation required.

SOURCE(S):

Vacuumschmelze GMBH

Germany

Crovac 10/380
(Alloy or Grade)

Wrought
(Type of Material)

-
(MMPA Brief Designation)

-
(IEC Code Reference)

MAGNETIC PROPERTIES	
Residual Induction B_r (G)	12,500
Coercive Force (Oe) — Normal (H_c)	580
Coercive Force (Oe) — Intrinsic (H_{ci})	-
Maximum Energy Product BH_{max} (MGO)	4.1
B/H Ratio (Load Line) at BH_{max}	16.4
Flux Density in Magnet B_m at BH_{max} (G)	8200
Coercive Force in Magnet H_m at BH_{max} (Oe)	500
Energy Loss W_c (watt-sec/cycle/lb)	-
Importance of Operating Magnet at BH_{max}	B
Average Recoil Permeability	-
Field Strength Required to Fully Saturate Magnet (Oe)	3000
Ability to Withstand Demagnetizing Fields	F
Maximum Service Temperature (°F)	896
Approximate Temperature Permanently Affecting Magnet (°F)	1184
Inherent Orientation of Material	CR

PHYSICAL PROPERTIES	
Density (lbs./in.^{3})	.2746
General Mechanical Properties	See Notes
Tensile Strength (psi)	-
Transverse Modulus of Rupture (psi)	-
Coeff. of Therm. Expansion inches x 10^{-6}/°C	Orientation: Par. 10.0 Rt.Angl. -
Resistivity (micro-ohms/cm/cm^2	- @ - °C
Hardness Scale	See Notes
Basic Forming Method	C+W
Basic Finishing Method	See Notes
Relative Corrosion Resistance	G
Primary Chemical Composition	Al-Ni-Co-Fe

Special Notes:

(1) Vickers Hardness:
   (a) cold worked = 310-340
   (b) sol. annealed = 210-240
   (b) after H.T. = 440-520
(2) Must be grind finished after H.T.

Note: All information based on available producers literature. In some cases of foreign materials interpretation or calculation required.

SOURCE(S):	Crovac 15/150
	(Alloy or Grade)
Vacuumschmelze GMBH	Wrought
	(Type of Material)
Germany	-
	(MMPA Brief Designation)
	-
	(IEC Code Reference)

MAGNETIC PROPERTIES		PHYSICAL PROPERTIES	
Residual Induction B_r (G)	9000	Density (lbs./in.^{3})	.2746
Coercive Force (Oe) — Normal (H_c)	450	General Mechanical Properties	See Notes
Coercive Force (Oe) — Intrinsic (H_{ci})	-	Tensile Strength (psi)	-
Maximum Energy Product BH_{max} (MGO)	1.7	Transverse Modulus of Rupture (psi)	-
B/H Ratio (Load Line) at BH_{max}	12.3	Coeff. of Therm. Expansion inches x 10^{-6}/°C	Orientation: Par. 10.0 Rt.Angl. -
Flux Density in Magnet B_m at BH_{max} (G)	4300	Resistivity (micro-ohms/cm/cm^2)	- @ - °C
Coercive Force in Magnet H_m at BH_{max} (Oe)	350	Hardness Scale	See Notes
Energy Loss W_c (watt-sec/cycle/lb)	-	Basic Forming Method	C+W
Importance of Operating Magnet at BH_{max}	B	Basic Finishing Method	See Notes
Average Recoil Permeability	-	Relative Corrosion Resistance	G
Field Strength Required to Fully Saturate Magnet (Oe)	3000	Primary Chemical Composition	Al-Ni-Co-Fe
Ability to Withstand Demagnetizing Fields	F	Special Notes: (1) Vickers Hardness: (a) cold worked = 310-340 (b) sol. anneled = 210-240 (c) after H.T. = 440-520 (2) Must be grind finished after H.T.	
Maximum Service Temperature (°F)	896		
Approximate Temperature Permanently Affecting Magnet (°F)	1184		
Inherent Orientation of Material	none		

Note: All information based on available producers literature. In some cases of foreign materials interpretation or calculation required.

SOURCE(S):

Vacuumschmelze GMBH

Germany

Crovac 15/400
(Alloy or Grade)

Wrought
(Type of Material)

-
(MMPA Brief Designation)

-
(IEC Code Reference)

MAGNETIC PROPERTIES	
Residual Induction B_r (G)	12,000
Coercive Force (Oe) — Normal (H_c)	615
Coercive Force (Oe) — Intrinsic (H_{ci})	-
Maximum Energy Product BH_{max} (MGO)	4.3
B/H Ratio (Load Line) at BH_{max}	14.1
Flux Density in Magnet B_m at BH_{max} (G)	7600
Coercive Force in Magnet H_m at BH_{max} (Oe)	540
Energy Loss W_c (watt-sec/cycle/lb)	-
Importance of Operating Magnet at BH_{max}	B
Average Recoil Permeability	-
Field Strength Required to Fully Saturate Magnet (Oe)	3000
Ability to Withstand Demagnetizing Fields	F
Maximum Service Temperature (°F)	896
Approximate Temperature Permanently Affecting Magnet (°F)	1184
Inherent Orientation of Material	none

PHYSICAL PROPERTIES	
Density (lbs./in.^{3})	.2746
General Mechanical Properties	See Notes
Tensile Strength (psi)	-
Transverse Modulus of Rupture (psi)	-
Coeff. of Therm. Expansion inches x 10^{-6}/°C	Orientation: Par. 10.0 Rt.Angl. -
Resistivity (micro-ohms/cm/cm^2)	- @ - °C
Hardness Scale	See Notes
Basic Forming Method	C+W
Basic Finishing Method	See Notes
Relative Corrosion Resistance	G
Primary Chemical Composition	Al-Ni-Co-Fe
Special Notes: (1) Vickers Hardness: (a) cold worked = 310-340 (b) sol. anneled = 210-240 (c) after H.T. = 440-520 (2) Must be grind finished after H.T.	

Note: All information based on available producers literature. In some cases of foreign materials interpretation or calculation required.

SOURCE(S):	Crovac 23/250
	(Alloy or Grade)
Vacuumschmelze GMBH	Wrought
	(Type of Material)
Germany	-
	(MMPA Brief Designation)
	-
	(IEC Code Reference)

MAGNETIC PROPERTIES			PHYSICAL PROPERTIES	
Residual Induction B_r (G)		8500	Density (lbs./in.^{3})	.2746
Coercive Force (Oe)	Normal (H_c)	640	General Mechanical Properties	See Notes
	Intrinsic (H_{ci})	-	Tensile Strength (psi)	-
Maximum Energy Product BH_{max} (MGO)		2.3	Transverse Modulus of Rupture (psi)	-
B/H Ratio (Load Line) at BH_{max}		9.1	Coeff. of Therm. Expansion inches x 10^{-6}/°C	Orientation: Par. 10.0 Rt.Angl. -
Flux Density in Magnet B_m at BH_{max} (G)		4500	Resistivity (micro-ohms/cm/cm^2)	- @ - °C
Coercive Force in Magnet H_m at BH_{max} (Oe)		495	Hardness Scale	See Notes
Energy Loss W_c (watt-sec/cycle/lb)		-	Basic Forming Method	C+W
Importance of Operating Magnet at BH_{max}		B	Basic Finishing Method	See Notes
Average Recoil Permeability		-	Relative Corrosion Resistance	G
Field Strength Required to Fully Saturate Magnet (Oe)		3000	Primary Chemical Composition	Al-Ni-Co-Fe
Ability to Withstand Demagnetizing Fields		F	Special Notes: (1) Obsolete material (2) Vickers Hardness: (a) cold worked = 310-340 (b) sol. annealed = 210-240 (c) after H.T. = 440-520 (3) Must be grind finished after H.T.	
Maximum Service Temperature (°F)		896		
Approximate Temperature Permanently Affecting Magnet (°F)		1184		
Inherent Orientation of Material		none		

Note: All information based on available producers literature. In some cases of foreign materials interpretation or calculation required.

SOURCE(S):

Vacuumschmelze GMBH

Germany

Crovac 23/500
(Alloy or Grade)

Wrought
(Type of Material)

-
(MMPA Brief Designation)

-
(IEC Code Reference)

MAGNETIC PROPERTIES	
Residual Induction B_r (G)	11,500
Coercive Force (Oe) — Normal (H_c)	700
Coercive Force (Oe) — Intrinsic (H_{ci})	-
Maximum Energy Product BH_{max} (MGO)	4.7
B/H Ratio (Load Line) at BH_{max}	10.9
Flux Density in Magnet B_m at BH_{max} (G)	7000
Coercive Force in Magnet H_m at BH_{max} (Oe)	640
Energy Loss W_c (watt-sec/cycle/lb)	-
Importance of Operating Magnet at BH_{max}	B
Average Recoil Permeability	-
Field Strength Required to Fully Saturate Magnet (Oe)	3000
Ability to Withstand Demagnetizing Fields	F
Maximum Service Temperature (°F)	896
Approximate Temperature Permanently Affecting Magnet (°F)	1184
Inherent Orientation of Material	CR

PHYSICAL PROPERTIES	
Density (lbs./in.^{3})	.2746
General Mechanical Properties	See Notes
Tensile Strength (psi)	-
Transverse Modulus of Rupture (psi)	-
Coeff. of Therm. Expansion inches x 10^{-6}/°C — Orientation: Par. 10.0, Rt.Angl. -	
Resistivity (micro-ohms/cm/cm^2 - @ - °C)	
Hardness Scale	See Notes
Basic Forming Method	C+W
Basic Finishing Method	See Notes
Relative Corrosion Resistance	G
Primary Chemical Composition	Al-Ni-Co-Fe

Special Notes:

(1) Obsolete material
(2) Vickers Hardness:
   (a) cold worked = 310-340
   (b) Sol. anneled = 210-240
   (c) after H.T. = 440-520
(3) Must be grind finished after H.T.

Note: All information based on available producers literature. In some cases of foreign materials interpretation or calculation required.

SOURCE(S):	EEC 1:5-18
	(Alloy or Grade)
Electron Energy Corp.	Sintered Rare Earth
	(Type of Material)
USA	18/30
	(MMPA Brief Designation)
	R5-1
	(IEC Code Reference)

MAGNETIC PROPERTIES		PHYSICAL PROPERTIES	
Residual Induction B_r (G)	8600	Density (lbs./in.^{3})	.2962
Coercive Force (Oe) — Normal (H_c)	8400	General Mechanical Properties	HB
Coercive Force (Oe) — Intrinsic (H_{ci})	30,000	Tensile Strength (psi)	Est. 5000
Maximum Energy Product BH_{max} (MGO)	18.0	Transverse Modulus of Rupture (psi)	-
B/H Ratio (Load Line) at BH_{max}	1.13	Coeff. of Therm. Expansion inches x 10^{-6}/°C	Orientation: Par. 7.0 Rt.Angl. 13.0
Flux Density in Magnet B_m at BH_{max} (G)	4500	Resistivity (micro-ohms/cm/cm^2)	5.2 @ 25 °C
Coercive Force in Magnet H_m at BH_{max} (Oe)	4000	Hardness Scale	56-59 Rockwell C
Energy Loss W_c (watt-sec/cycle/lb)	-	Basic Forming Method	DPM
Importance of Operating Magnet at BH_{max}	E	Basic Finishing Method	WGR
Average Recoil Permeability	1.05	Relative Corrosion Resistance	G
Field Strength Required to Fully Saturate Magnet (Oe)	50,000	Primary Chemical Composition	Sm-Co
Ability to Withstand Demagnetizing Fields	0	Special Notes:	
Maximum Service Temperature (°F)	572		
Approximate Temperature Permanently Affecting Magnet (°F)	Est. 1328		
Inherent Orientation of Material	MP		

Note: All information based on available producers literature. In some cases of foreign materials interpretation or calculation required.

SOURCE(S):

Electron Energy Corp.

USA

EEC 1:5TC-9
(Alloy or Grade)

Sintered Rare Earth
(Type of Material)

9/30
(MMPA Brief Designation)

R5-1
(IEC Code Reference)

MAGNETIC PROPERTIES		PHYSICAL PROPERTIES	
Residual Induction B_r (G)	6100	Density (lbs./in.^{3})	.2962
Coercive Force (Oe) Normal (H_c)	6000	General Mechanical Properties	HB
Coercive Force (Oe) Intrinsic (H_{ci})	30,000	Tensile Strength (psi)	Est. 5000
Maximum Energy Product BH_{max} (MGO)	9.0	Transverse Modulus of Rupture (psi)	-
B/H Ratio (Load Line) at BH_{max}	1.14	Coeff. of Therm. Expansion inches x 10^{-6}/°C	Orientation: Par. 7.0 Rt.Angl. 13.0
Flux Density in Magnet B_m at BH_{max} (G)	3200	Resistivity (micro-ohms/cm/cm^2	5.2 @ 25 °C
Coercive Force in Magnet H_m at BH_{max} (Oe)	2800	Hardness Scale 56-59	Rockwell C
Energy Loss W_c (watt-sec/cycle/lb)	-	Basic Forming Method	DPM
Importance of Operating Magnet at BH_{max}	E	Basic Finishing Method	WGR
Average Recoil Permeability	1.05	Relative Corrosion Resistance	G
Field Strength Required to Fully Saturate Magnet (Oe)	50,000	Primary Chemical Composition	Sm-Co
Ability to Withstand Demagnetizing Fields	O	Special Notes:	
Maximum Service Temperature (°F)	572		
Approximate Temperature Permanently Affecting Magnet (°F)	Est. 1328		
Inherent Orientation of Material	MP		

Note: All information based on available producers literature. In some cases of foreign materials interpretation or calculation required.

SOURCE(S):	EEC 1:5TC-13
	(Alloy or Grade)
Electron Energy Corp.	Sintered Rare Earth
	(Type of Material)
USA	13/30
	(MMPA Brief Designation)
	R5-1
	(IEC Code Reference)

MAGNETIC PROPERTIES		PHYSICAL PROPERTIES	
Residual Induction B_r (G)	7300	Density (lbs./in.^{3})	.2962
Coercive Force (Oe) Normal (H_c)	7200	General Mechanical Properties	HB
Coercive Force (Oe) Intrinsic (H_{ci})	30,000	Tensile Strength (psi)	Est. 5000
Maximum Energy Product BH_{max} (MGO)	13.0	Transverse Modulus of Rupture (psi)	-
B/H Ratio (Load Line) at BH_{max}	.85	Coeff. of Therm. Expansion inches x 10^{-6}/°C	Orientation: Par. 7.0 Rt.Angl. 13.0
Flux Density in Magnet B_m at BH_{max} (G)	3300	Resistivity (micro-ohms/cm/cm^2)	5.2 @ 25 °C
Coercive Force in Magnet H_m at BH_{max} (Oe)	3900	Hardness Scale	56-59 Rockwell C
Energy Loss W_c (watt-sec/cycle/lb)	-	Basic Forming Method	DPM
Importance of Operating Magnet at BH_{max}	E	Basic Finishing Method	WGR
Average Recoil Permeability	1.05	Relative Corrosion Resistance	G
Field Strength Required to Fully Saturate Magnet (Oe)	50,000	Primary Chemical Composition	Sm-Co
Ability to Withstand Demagnetizing Fields	0	Special Notes:	
Maximum Service Temperature (°F)	572		
Approximate Temperature Permanently Affecting Magnet (°F)	Est. 1328		
Inherent Orientation of Material	MP		

Note: All information based on available producers literature. In some cases of foreign materials interpretation or calculation required.

SOURCE(S):

Electron Energy Corp.

USA

EEC 1:5TC-15
(Alloy or Grade)

Sintered Rare Earth
(Type of Material)

15/30
(MMPA Brief Designation)

R5-1
(IEC Code Reference)

MAGNETIC PROPERTIES		PHYSICAL PROPERTIES	
Residual Induction B_r (G)	7800	Density (lbs./in.^{3})	.2962
Coercive Force (Oe) — Normal (H_c)	7700	General Mechanical Properties	HB
Coercive Force (Oe) — Intrinsic (H_{ci})	30,000	Tensile Strength (psi)	Est. 5000
Maximum Energy Product BH_{max} (MGO)	15.0	Transverse Modulus of Rupture (psi)	–
B/H Ratio (Load Line) at BH_{max}	1.04	Coeff. of Therm. Expansion inches x 10^{-6}/°C	Orientation: Par. 7.0 Rt.Angl. 13.0
Flux Density in Magnet B_m at BH_{max} (G)	3950	Resistivity (micro-ohms/cm/cm^2)	5.2 @ 25 °C
Coercive Force in Magnet H_m at BH_{max} (Oe)	3800	Hardness Scale	56-59 Rockwell C
Energy Loss W_c (watt-sec/cycle/lb)	–	Basic Forming Method	DPM
Importance of Operating Magnet at BH_{max}	E	Basic Finishing Method	WGR
Average Recoil Permeability	1.05	Relative Corrosion Resistance	G
Field Strength Required to Fully Saturate Magnet (Oe)	50,000	Primary Chemical Composition	Sm-Co
Ability to Withstand Demagnetizing Fields	0	Special Notes:	
Maximum Service Temperature (°F)	572		
Approximate Temperature Permanently Affecting Magnet (°F)	Est. 1328		
Inherent Orientation of Material	MP		

Note: All information based on available producers literature. In some cases of foreign materials interpretation or calculation required.

SOURCE(S): **Electron Energy Corp.** USA

EEC 2:17-15	(Alloy or Grade)
Sintered Rare Earth	(Type of Material)
14/20	(MMPA Brief Designation)
R5-1	(IEC Code Reference)

MAGNETIC PROPERTIES		PHYSICAL PROPERTIES	
Residual Induction B_r (G)	8000	Density (lbs./in.^{3})	.2962
Coercive Force (Oe) — Normal (H_c)	7200	General Mechanical Properties	HB
Coercive Force (Oe) — Intrinsic (H_{ci})	20,000	Tensile Strength (psi)	Est. 5000
Maximum Energy Product BH_{max} (MGO)	14.5	Transverse Modulus of Rupture (psi)	–
B/H Ratio (Load Line) at BH_{max}	.90	Coeff. of Therm. Expansion inches x 10^{-6}/°C	Orientation: Par. 8.0 / Rt.Angl. 11.0
Flux Density in Magnet B_m at BH_{max} (G)	3600	Resistivity (micro-ohms/cm/cm^2)	86 @ 25 °C
Coercive Force in Magnet H_m at BH_{max} (Oe)	4000	Hardness / Scale	49-55 Rockwell C
Energy Loss W_c (watt-sec/cycle/lb)	–	Basic Forming Method	DPM
Importance of Operating Magnet at BH_{max}	E	Basic Finishing Method	WGR
Average Recoil Permeability	1.05	Relative Corrosion Resistance	G
Field Strength Required to Fully Saturate Magnet (Oe)	40,000	Primary Chemical Composition	Sm-Co
Ability to Withstand Demagnetizing Fields	0	Special Notes:	
Maximum Service Temperature (°F)	572		
Approximate Temperature Permanently Affecting Magnet (°F)	Est. 1328		
Inherent Orientation of Material	MP		

Note: All information based on available producers literature. In some cases of foreign materials interpretation or calculation required.

SOURCE(S):

Electron Energy Corp.

USA

EEC 2:17-24
(Alloy or Grade)

Sintered Rare Earth
(Type of Material)

24/15
(MMPA Brief Designation)

R5-2
(IEC Code Reference)

MAGNETIC PROPERTIES		PHYSICAL PROPERTIES	
Residual Induction B_r (G)	10,100	Density (lbs./in.^{3})	.2962
Coercive Force (Oe) — Normal (H_c)	9300	General Mechanical Properties	HB
Coercive Force (Oe) — Intrinsic (H_{ci})	25,000	Tensile Strength (psi)	Est. 5000
Maximum Energy Product BH_{max} (MGO)	24.0	Transverse Modulus of Rupture (psi)	-
B/H Ratio (Load Line) at BH_{max}	1.09	Coeff. of Therm. Expansion inches x 10^{-6}/°C	Orientation: Par. 8.0 Rt.Angl. 11.0
Flux Density in Magnet B_m at BH_{max} (G)	5100	Resistivity (micro-ohms/cm/cm^2)	86 @ 25 °C
Coercive Force in Magnet H_m at BH_{max} (Oe)	4700	Hardness 49-55	Scale Rockwell C
Energy Loss W_c (watt-sec/cycle/lb)	-	Basic Forming Method	DPM
Importance of Operating Magnet at BH_{max}	E	Basic Finishing Method	WGR
Average Recoil Permeability	1.05	Relative Corrosion Resistance	G
Field Strength Required to Fully Saturate Magnet (Oe)	40,000	Primary Chemical Composition	Sm-Co
Ability to Withstand Demagnetizing Fields	0	Special Notes:	
Maximum Service Temperature (°F)	572		
Approximate Temperature Permanently Affecting Magnet (°F)	Est. 1328		
Inherent Orientation of Material	MP		

Note: All information based on available producers literature. In some cases of foreign materials interpretation or calculation required.

SOURCE(S): **Electron Energy Corp.** USA

(Alloy or Grade): EEC 2:17-27
(Type of Material): Sintered Rare Earth
(MMPA Brief Designation): 27/25
(IEC Code Reference): R5-2

MAGNETIC PROPERTIES	
Residual Induction B_r (G)	10,800
Coercive Force (Oe) Normal (H_c)	10,100
Coercive Force (Oe) Intrinsic (H_{ci})	25,000
Maximum Energy Product BH_{max} (MGO)	27.5
B/H Ratio (Load Line) at BH_{max}	1.06
Flux Density in Magnet B_m at BH_{max} (G)	5400
Coercive Force in Magnet H_m at BH_{max} (Oe)	5100
Energy Loss W_c (watt-sec/cycle/lb)	-
Importance of Operating Magnet at BH_{max}	E
Average Recoil Permeability	1.05
Field Strength Required to Fully Saturate Magnet (Oe)	40,000
Ability to Withstand Demagnetizing Fields	0
Maximum Service Temperature (°F)	572
Approximate Temperature Permanently Affecting Magnet (°F)	Est. 1328
Inherent Orientation of Material	MP

PHYSICAL PROPERTIES	
Density (lbs./in.^{3})	.2962
General Mechanical Properties	HB
Tensile Strength (psi)	Est. 5000
Transverse Modulus of Rupture (psi)	-
Coeff. of Therm. Expansion inches x 10^{-6}/°C	Orientation: Par. 8.0 Rt.Angl. 11.0
Resistivity (micro-ohms/cm/cm^2)	86 @ 25 °C
Hardness 49-55	Scale Rockwell C
Basic Forming Method	DPM
Basic Finishing Method	WGR
Relative Corrosion Resistance	G
Primary Chemical Composition	Sm-Co
Special Notes:	

Note: All information based on available producers literature. In some cases of foreign materials interpretation or calculation required.

SOURCE(S):

Electron Energy Corp.

USA

EEC 2:17TC-15
(Alloy or Grade)

Sintered Rare Earth
(Type of Material)

14/20
(MMPA Brief Designation)

R5-1
(IEC Code Reference)

MAGNETIC PROPERTIES		PHYSICAL PROPERTIES	
Residual Induction B_r (G)	8000	Density (lbs./in.^{3})	.2962
Coercive Force (Oe) Normal (H_c)	7200	General Mechanical Properties	HB
Coercive Force (Oe) Intrinsic (H_{ci})	20,000	Tensile Strength (psi)	Est. 5000
Maximum Energy Product BH_{max} (MGO)	14.5	Transverse Modulus of Rupture (psi)	-
B/H Ratio (Load Line) at BH_{max}	.90	Coeff. of Therm. Expansion inches x 10^{-6}/°C	Orientation: Par. 8.0 Rt.Angl. 11.0
Flux Density in Magnet B_m at BH_{max} (G)	3600	Resistivity (micro-ohms/cm/cm^2)	86 @ 25 °C
Coercive Force in Magnet H_m at BH_{max} (Oe)	4000	Hardness Scale	49-55 Rockwell C
Energy Loss W_c (watt-sec/cycle/lb)	-	Basic Forming Method	DPM
Importance of Operating Magnet at BH_{max}	E	Basic Finishing Method	WGR
Average Recoil Permeability	1.05	Relative Corrosion Resistance	G
Field Strength Required to Fully Saturate Magnet (Oe)	40,000	Primary Chemical Composition	Sm-Co
Ability to Withstand Demagnetizing Fields	O	Special Notes:	
Maximum Service Temperature (°F)	572		
Approximate Temperature Permanently Affecting Magnet (°F)	Est. 1328		
Inherent Orientation of Material	MP		

Note: All information based on available producers literature. In some cases of foreign materials interpretation or calculation required.

SOURCE(S):	EEC 2:17TC-18
	(Alloy or Grade)
Electron Energy Corp.	Sintered Rare Earth
	(Type of Material)
USA	18/25
	(MMPA Brief Designation)
	R5-1
	(IEC Code Reference)

MAGNETIC PROPERTIES		PHYSICAL PROPERTIES	
Residual Induction B_r (G)	9000	Density (lbs./in.^{3})	.2962
Coercive Force (Oe) — Normal (H_c)	8200	General Mechanical Properties	HB
Coercive Force (Oe) — Intrinsic (H_{ci})	25,000	Tensile Strength (psi)	Est. 5000
Maximum Energy Product BH_{max} (MGO)	18.5	Transverse Modulus of Rupture (psi)	-
B/H Ratio (Load Line) at BH_{max}	.95	Coeff. of Therm. Expansion inches x 10^{-6}/°C	Orientation: Par. 8.0 Rt.Angl. 11.0
Flux Density in Magnet B_m at BH_{max} (G)	4200	Resistivity (micro-ohms/cm/cm^2)	86 @ 25 °C
Coercive Force in Magnet H_m at BH_{max} (Oe)	4400	Hardness 49-55	Scale Rockwell C
Energy Loss W_c (watt-sec/cycle/lb)	-	Basic Forming Method	DPM
Importance of Operating Magnet at BH_{max}	E	Basic Finishing Method	WGR
Average Recoil Permeability	1.05	Relative Corrosion Resistance	G
Field Strength Required to Fully Saturate Magnet (Oe)	40,000	Primary Chemical Composition	Sm-Co
Ability to Withstand Demagnetizing Fields	0	Special Notes:	
Maximum Service Temperature (°F)	Est. 482		
Approximate Temperature Permanently Affecting Magnet (°F)	Est. 1328		
Inherent Orientation of Material	MP		

Note: All information based on available producers literature. In some cases of foreign materials interpretation or calculation required.

SOURCE(S):	EEC NEO 27
Electron Energy Corp.	(Alloy or Grade)
	Sintered Rare Earth
	(Type of Material)
USA	27/15
	(MMPA Brief Designation)
	R7-3
	(IEC Code Reference)

MAGNETIC PROPERTIES		PHYSICAL PROPERTIES	
Residual Induction B_r (G)	10,800	Density (lbs./in.^{3})	.2673
Coercive Force (Oe) — Normal (H_c)	10,200	General Mechanical Properties	HB
Coercive Force (Oe) — Intrinsic (H_{ci})	15,000	Tensile Strength (psi)	-
Maximum Energy Product BH_{max} (MGO)	27.5	Transverse Modulus of Rupture (psi)	-
B/H Ratio (Load Line) at BH_{max}	1.14	Coeff. of Therm. Expansion inches x 10^{-6}/°C	Orientation: Par. 3.4 Rt.Angl. 4.8
Flux Density in Magnet B_m at BH_{max} (G)	5600	Resistivity (micro-ohms/cm/cm^2)	1.4 @ 25 °C
Coercive Force in Magnet H_m at BH_{max} (Oe)	4925	Hardness 55	Scale Rockwell C
Energy Loss W_c (watt-sec/cycle/lb)	-	Basic Forming Method	DPM
Importance of Operating Magnet at BH_{max}	E	Basic Finishing Method	WGR
Average Recoil Permeability	1.05	Relative Corrosion Resistance	P
Field Strength Required to Fully Saturate Magnet (Oe)	32,000	Primary Chemical Composition	Nd-Fe-B
Ability to Withstand Demagnetizing Fields	0	Special Notes:	
Maximum Service Temperature (°F)	Est. 245		
Approximate Temperature Permanently Affecting Magnet (°F)	Est. 590		
Inherent Orientation of Material	MP		

Note: All information based on available producers literature. In some cases of foreign materials interpretation or calculation required.

SOURCE(S):

Electron Energy Corp.

USA

EEC NEO 33
(Alloy or Grade)

Sintered Rare Earth
(Type of Material)

33/15
(MMPA Brief Designation)

R7-3
(IEC Code Reference)

MAGNETIC PROPERTIES	
Residual Induction B_r (G)	12,000
Coercive Force (Oe) — Normal (H_c)	10,600
Coercive Force (Oe) — Intrinsic (H_{ci})	15,000
Maximum Energy Product BH_{max} (MGO)	33.5
B/H Ratio (Load Line) at BH_{max}	1.01
Flux Density in Magnet B_m at BH_{max} (G)	5825
Coercive Force in Magnet H_m at BH_{max} (Oe)	5750
Energy Loss W_c (watt-sec/cycle/lb)	-
Importance of Operating Magnet at BH_{max}	E
Average Recoil Permeability	1.05
Field Strength Required to Fully Saturate Magnet (Oe)	32,000
Ability to Withstand Demagnetizing Fields	0
Maximum Service Temperature (°F)	Est. 245
Approximate Temperature Permanently Affecting Magnet (°F)	Est. 590
Inherent Orientation of Material	MP

PHYSICAL PROPERTIES	
Density (lbs./in.^{3})	.2673
General Mechanical Properties	HB
Tensile Strength (psi)	-
Transverse Modulus of Rupture (psi)	-
Coeff. of Therm. Expansion inches x 10^{-6}/°C	Orientation: Par. 3.4 Rt.Angl. 4.8
Resistivity (micro-ohms/cm/cm^2)	1.4 @ 25 °C
Hardness	55 Scale Rockwell C
Basic Forming Method	DPM
Basic Finishing Method	WGR
Relative Corrosion Resistance	P
Primary Chemical Composition	Nd-Fe-B
Special Notes:	

Note: All information based on available producers literature. In some cases of foreign materials interpretation or calculation required.

SOURCE(S):

TDK Corp. of America

USA

FB1
(Alloy or Grade)

Sintered Ferrite
(Type of Material)

1.1/3.5
(MMPA Brief Designation)

-
(IEC Code Reference)

MAGNETIC PROPERTIES		PHYSICAL PROPERTIES	
Residual Induction B_r (G)	2200	Density (lbs./in.^{3})	.1662–.1770
Coercive Force (Oe) — Normal (H_c)	1900	General Mechanical Properties	VHB
Coercive Force (Oe) — Intrinsic (H_{ci})	3500	Tensile Strength (psi)	-
Maximum Energy Product BH_{max} (MGO)	1.1	Transverse Modulus of Rupture (psi)	-
B/H Ratio (Load Line) at BH_{max}	1.0	Coeff. of Therm. Expansion inches x 10^{-6}/°C Orientation: Par. - Rt.Angl. -	
Flux Density in Magnet B_m at BH_{max} (G)	1000	Resistivity (micro-ohms/cm/cm^2)	1.0 @ 25 °C
Coercive Force in Magnet H_m at BH_{max} (Oe)	1000	Hardness 6.5 Scale Moh's	
Energy Loss W_c (watt-sec/cycle/lb)	-	Basic Forming Method	PF
Importance of Operating Magnet at BH_{max}	E	Basic Finishing Method	WGD
Average Recoil Permeability	1.1 to 1.2	Relative Corrosion Resistance	Ex
Field Strength Required to Fully Saturate Magnet (Oe)	10,000	Primary Chemical Composition	Ba-Fe
Ability to Withstand Demagnetizing Fields	Ex	Special Notes:	
Maximum Service Temperature (°F)	482		
Approximate Temperature Permanently Affecting Magnet (°F)	842		
Inherent Orientation of Material	none		

Note: All information based on available producers literature. In some cases of foreign materials interpretation or calculation required.

SOURCE(S):	
	FB2
	(Alloy or Grade)
TDK Corp. of America	Sintered Ferrite
	(Type of Material)
USA	3.5/2.0
	(MMPA Brief Designation)
	-
	(IEC Code Reference)

MAGNETIC PROPERTIES		PHYSICAL PROPERTIES	
Residual Induction B_r (G)	3850	Density (lbs./in.^{3})	.1734-.1806
Coercive Force (Oe) Normal (H_c)	2000	General Mechanical Properties	VHB
Coercive Force (Oe) Intrinsic (H_{ci})	2020	Tensile Strength (psi)	-
Maximum Energy Product BH_{max} (MGO)	3.5	Transverse Modulus of Rupture (psi)	-
B/H Ratio (Load Line) at BH_{max}	1.08	Coeff. of Therm. Expansion inches x 10^{-6}/°C	Orientation: Par. - Rt.Angl. -
Flux Density in Magnet B_m at BH_{max} (G)	1950	Resistivity (micro-ohms/cm/cm^2	1.0 @ 25 °C
Coercive Force in Magnet H_m at BH_{max} (Oe)	1800	Hardness Scale	6.5 Moh's
Energy Loss W_c (watt-sec/cycle/lb)	-	Basic Forming Method	PMF
Importance of Operating Magnet at BH_{max}	E	Basic Finishing Method	WGD
Average Recoil Permeability	1.1 to 1.2	Relative Corrosion Resistance	Ex
Field Strength Required to Fully Saturate Magnet (Oe)	10,000	Primary Chemical Composition	Ba-Fe
Ability to Withstand Demagnetizing Fields	Ex	Special Notes:	
Maximum Service Temperature (°F)	482		
Approximate Temperature Permanently Affecting Magnet (°F)	842		
Inherent Orientation of Material	MP		

Note: All information based on available producers literature. In some cases of foreign materials interpretation or calculation required.

SOURCE(S):	FB3B
	(Alloy or Grade)
TDK Corp. of America	Sintered Ferrite
	(Type of Material)
USA	3.0/3.4
	(MMPA Brief Designation)
	-
	(IEC Code Reference)

MAGNETIC PROPERTIES		PHYSICAL PROPERTIES	
Residual Induction B_r (G)	3550	Density (lbs./in.^{3})	.1662-.1770
Coercive Force (Oe) Normal (H_c)	3000	General Mechanical Properties	VHB
Coercive Force (Oe) Intrinsic (H_{ci})	3400	Tensile Strength (psi)	-
Maximum Energy Product BH_{max} (MGO)	3.0	Transverse Modulus of Rupture (psi)	-
B/H Ratio (Load Line) at BH_{max}	1.06	Coeff. of Therm. Expansion inches x 10^{-6}/°C	Orientation: Par. - Rt.Angl. -
Flux Density in Magnet B_m at BH_{max} (G)	1750	Resistivity (micro-ohms/cm/cm^2)	1.0 @ 25 °C
Coercive Force in Magnet H_m at BH_{max} (Oe)	1650	Hardness Scale	6.5 Moh's
Energy Loss W_c (watt-sec/cycle/lb)	-	Basic Forming Method	PMF
Importance of Operating Magnet at BH_{max}	E	Basic Finishing Method	WGD
Average Recoil Permeability	1.1 to 1.2	Relative Corrosion Resistance	Ex
Field Strength Required to Fully Saturate Magnet (Oe)	10,000	Primary Chemical Composition	Sr-Fe
Ability to Withstand Demagnetizing Fields	Ex	Special Notes:	
Maximum Service Temperature (°F)	482		
Approximate Temperature Permanently Affecting Magnet (°F)	842		
Inherent Orientation of Material	MP		

Note: All information based on available producers literature. In some cases of foreign materials interpretation or calculation required.

SOURCE(S):	FB3K
	(Alloy or Grade)
TDK Corp. of America	Sintered Ferrite
	(Type of Material)
USA	2.1/3.4
	(MMPA Brief Designation)
	-
	(IEC Code Reference)

MAGNETIC PROPERTIES			PHYSICAL PROPERTIES	
Residual Induction B_r (G)		3000	Density (lbs./in.^{3})	.1662-.1770
Coercive Force (Oe)	Normal (H_c)	2750	General Mechanical Properties	VHB
	Intrinsic (H_{ci})	3400	Tensile Strength (psi)	-
Maximum Energy Product BH_{max} (MGO)		2.1	Transverse Modulus of Rupture (psi)	-
B/H Ratio (Load Line) at BH_{max}		.90	Coeff. of Therm. Expansion inches x 10^{-6}/°C	Orientation: Par. - Rt.Angl. -
Flux Density in Magnet B_m at BH_{max} (G)		1400	Resistivity (micro-ohms/cm/cm^2)	1.0 @ 25 °C
Coercive Force in Magnet H_m at BH_{max} (Oe)		1500	Hardness Scale	6.5 Moh's
Energy Loss W_c (watt-sec/cycle/lb)		-	Basic Forming Method	PMF
Importance of Operating Magnet at BH_{max}		E	Basic Finishing Method	WGD
Average Recoil Permeability		1.1-1.2	Relative Corrosion Resistance	Ex
Field Strength Required to Fully Saturate Magnet (Oe)		10,000	Primary Chemical Composition	Sr-Fe
Ability to Withstand Demagnetizing Fields		Ex	Special Notes:	
Maximum Service Temperature (°F)		482		
Approximate Temperature Permanently Affecting Magnet (°F)		842		
Inherent Orientation of Material		MP		

Note: All information based on available producers literature. In some cases of foreign materials interpretation or calculation required.

SOURCE(S):	FB3X
	(Alloy or Grade)
TDK Corp. of America	Sintered Ferrite
	(Type of Material)
USA	3.2/3.0
	(MMPA Brief Designation)
	-
	(IEC Code Reference)

MAGNETIC PROPERTIES		PHYSICAL PROPERTIES	
Residual Induction B_r (G)	3750	Density (lbs./in.^{3})	.1662-.1770
Coercive Force (Oe) — Normal (H_c)	2950	General Mechanical Properties	VHB
Coercive Force (Oe) — Intrinsic (H_{ci})	3000	Tensile Strength (psi)	-
Maximum Energy Product BH_{max} (MGO)	3.2	Transverse Modulus of Rupture (psi)	-
B/H Ratio (Load Line) at BH_{max}	1.07	Coeff. of Therm. Expansion inches x 10^{-6}/°C	Orientation: Par. -, Rt.Angl. -
Flux Density in Magnet B_m at BH_{max} (G)	1850	Resistivity (micro-ohms/cm/cm^2)	1.0 @ 25 °C
Coercive Force in Magnet H_m at BH_{max} (Oe)	1725	Hardness — Scale	6.5 Moh's
Energy Loss W_c (watt-sec/cycle/lb)	-	Basic Forming Method	PMF
Importance of Operating Magnet at BH_{max}	E	Basic Finishing Method	WGD
Average Recoil Permeability	1.1 to 1.2	Relative Corrosion Resistance	Ex
Field Strength Required to Fully Saturate Magnet (Oe)	10,000	Primary Chemical Composition	Sr-Fe
Ability to Withstand Demagnetizing Fields	Ex	Special Notes:	
Maximum Service Temperature (°F)	482		
Approximate Temperature Permanently Affecting Magnet (°F)	842		
Inherent Orientation of Material	MP		

Note: All information based on available producers literature. In some cases of foreign materials interpretation or calculation required.

SOURCE(S):	FB4A
	(Alloy or Grade)
TDK Corp. of America	Sintered Ferrite
	(Type of Material)
USA	4.0/2.2
	(MMPA Brief Designation)
	-
	(IEC Code Reference)

MAGNETIC PROPERTIES		PHYSICAL PROPERTIES	
Residual Induction B_r (G)	4100	Density (lbs./in.^{3})	.1806-.1842
Coercive Force (Oe) — Normal (H_c)	2200	General Mechanical Properties	VHB
Coercive Force (Oe) — Intrinsic (H_{ci})	2220	Tensile Strength (psi)	-
Maximum Energy Product BH_{max} (MGO)	4.0	Transverse Modulus of Rupture (psi)	-
B/H Ratio (Load Line) at BH_{max}	1.05	Coeff. of Therm. Expansion inches x 10^{-6}/°C	Orientation: Par. - Rt.Angl. -
Flux Density in Magnet B_m at BH_{max} (G)	1050	Resistivity (micro-ohms/cm/cm^2)	1.0 @ 25 °C
Coercive Force in Magnet H_m at BH_{max} (Oe)	1950	Hardness 6.5	Scale Moh's
Energy Loss W_c (watt-sec/cycle/lb)	-	Basic Forming Method	PMF
Importance of Operating Magnet at BH_{max}	E	Basic Finishing Method	WGD
Average Recoil Permeability	1.1 to 1.2	Relative Corrosion Resistance	Ex
Field Strength Required to Fully Saturate Magnet (Oe)	10,000	Primary Chemical Composition	Ba-Fe
Ability to Withstand Demagnetizing Fields	Ex	Special Notes:	
Maximum Service Temperature (°F)	482		
Approximate Temperature Permanently Affecting Magnet (°F)	842		
Inherent Orientation of Material	MP		

Note: All information based on available producers literature. In some cases of foreign materials interpretation or calculation required.

SOURCE(S):	
TDK Corp. of America	**FB4B** (Alloy or Grade)
USA	Sintered Ferrite (Type of Material)
	3.8/3.3 (MMPA Brief Designation)
	- (IEC Code Reference)

MAGNETIC PROPERTIES		PHYSICAL PROPERTIES	
Residual Induction B_r (G)	4000	Density (lbs./in.^{3})	.1770-.1806
Coercive Force (Oe) Normal (H_c)	3200	General Mechanical Properties	VHB
Coercive Force (Oe) Intrinsic (H_{ci})	3300	Tensile Strength (psi)	-
Maximum Energy Product BH_{max} (MGO)	3.8	Transverse Modulus of Rupture (psi)	-
B/H Ratio (Load Line) at BH_{max}	1.02	Coeff. of Therm. Expansion inches x 10^{-6}/°C	Orientation: Par. - Rt.Angl. -
Flux Density in Magnet B_m at BH_{max} (G)	1975	Resistivity (micro-ohms/cm/cm^2)	1.0 @ 25 °C
Coercive Force in Magnet H_m at BH_{max} (Oe)	1925	Hardness Scale	6.5 Moh's
Energy Loss W_c (watt-sec/cycle/lb)	-	Basic Forming Method	PMF
Importance of Operating Magnet at BH_{max}	E	Basic Finishing Method	WGD
Average Recoil Permeability	1.05 to 1.1	Relative Corrosion Resistance	Ex
Field Strength Required to Fully Saturate Magnet (Oe)	10,000	Primary Chemical Composition	Sr-Fe
Ability to Withstand Demagnetizing Fields	Ex	Special Notes:	
Maximum Service Temperature (°F)	482		
Approximate Temperature Permanently Affecting Magnet (°F)	842		
Inherent Orientation of Material	MP		

Note: All information based on available producers literature. In some cases of foreign materials interpretation or calculation required.

SOURCE(S):

TDK Corp. of America

USA

FB4H
(Alloy or Grade)

Sintered Ferrite
(Type of Material)

3.2/4.2
(MMPA Brief Designation)

-
(IEC Code Reference)

MAGNETIC PROPERTIES		PHYSICAL PROPERTIES	
Residual Induction B_r (G)	3700	Density (lbs./in.^{3})	.1770-.1806
Coercive Force (Oe) — Normal (H_c)	3450	General Mechanical Properties	VHB
Coercive Force (Oe) — Intrinsic (H_{ci})	4200	Tensile Strength (psi)	-
Maximum Energy Product BH_{max} (MGO)	3.25	Transverse Modulus of Rupture (psi)	-
B/H Ratio (Load Line) at BH_{max}	1.17	Coeff. of Therm. Expansion inches x 10^{-6}/°C Orientation: Par. - Rt.Angl. -	
Flux Density in Magnet B_m at BH_{max} (G)	1925	Resistivity (micro-ohms/cm/cm^2)	1.0 @ 25 °C
Coercive Force in Magnet H_m at BH_{max} (Oe)	1650	Hardness Scale	6.5 Moh's
Energy Loss W_c (watt-sec/cycle/lb)	-	Basic Forming Method	PMF
Importance of Operating Magnet at BH_{max}	E	Basic Finishing Method	WGD
Average Recoil Permeability	1.05 to 1.1	Relative Corrosion Resistance	Ex
Field Strength Required to Fully Saturate Magnet (Oe)	10,000	Primary Chemical Composition	Sr-Fe
Ability to Withstand Demagnetizing Fields	Ex	Special Notes:	
Maximum Service Temperature (°F)	482		
Approximate Temperature Permanently Affecting Magnet (°F)	842		
Inherent Orientation of Material	MP		

Note: All information based on available producers literature. In some cases of foreign materials interpretation or calculation required.

SOURCE(S):	FB4N
	(Alloy or Grade)
TDK Corp. of America	Sintered Ferrite
	(Type of Material)
USA	4.4/2.3
	(MMPA Brief Designation)
	-
	(IEC Code Reference)

MAGNETIC PROPERTIES		PHYSICAL PROPERTIES	
Residual Induction B_r (G)	4300	Density (lbs./in.^{3})	.1770–.1806
Coercive Force (Oe) — Normal (H_c)	2300	General Mechanical Properties	VHB
Coercive Force (Oe) — Intrinsic (H_{ci})	2320	Tensile Strength (psi)	-
Maximum Energy Product BH_{max} (MGO)	4.4	Transverse Modulus of Rupture (psi)	-
B/H Ratio (Load Line) at BH_{max}	1.0	Coeff. of Therm. Expansion inches x 10^{-6}/°C	Orientation: Par. - Rt.Angl. -
Flux Density in Magnet B_m at BH_{max} (G)	2100	Resistivity (micro-ohms/cm/cm^2)	1.0 @ 25 °C
Coercive Force in Magnet H_m at BH_{max} (Oe)	2100	Hardness Scale	6.5 Moh's
Energy Loss W_c (watt-sec/cycle/lb)	-	Basic Forming Method	PMF
Importance of Operating Magnet at BH_{max}	E	Basic Finishing Method	WGD
Average Recoil Permeability	1.05 to 1.1	Relative Corrosion Resistance	Ex
Field Strength Required to Fully Saturate Magnet (Oe)	10,000	Primary Chemical Composition	Sr-Fe
Ability to Withstand Demagnetizing Fields	Ex	Special Notes:	
Maximum Service Temperature (°F)	482		
Approximate Temperature Permanently Affecting Magnet (°F)	842		
Inherent Orientation of Material	MP		

Note: All information based on available producers literature. In some cases of foreign materials interpretation or calculation required.

SOURCE(S):	FB4X
	(Alloy or Grade)
TDK Corp. of America	Sintered Ferrite
	(Type of Material)
USA	4.2/3.0
	(MMPA Brief Designation)
	-
	(IEC Code Reference)

MAGNETIC PROPERTIES		PHYSICAL PROPERTIES	
Residual Induction B_r (G)	4200	Density (lbs./in.^{3})	.1770–.1806
Coercive Force (Oe) — Normal (H_c)	2950	General Mechanical Properties	VHB
Coercive Force (Oe) — Intrinsic (H_{ci})	3000	Tensile Strength (psi)	-
Maximum Energy Product BH_{max} (MGO)	4.2	Transverse Modulus of Rupture (psi)	-
B/H Ratio (Load Line) at BH_{max}	1.05	Coeff. of Therm. Expansion inches x 10^{-6}/°C	Orientation: Par. _ Rt.Angl. _
Flux Density in Magnet B_m at BH_{max} (G)	2100	Resistivity (micro-ohms/cm/cm^2)	1.0 @ 25 °C
Coercive Force in Magnet H_m at BH_{max} (Oe)	2000	Hardness 6.5	Scale Moh's
Energy Loss W_c (watt-sec/cycle/lb)	-	Basic Forming Method	PMF
Importance of Operating Magnet at BH_{max}	E	Basic Finishing Method	WGD
Average Recoil Permeability	1.05 to 1.1	Relative Corrosion Resistance	Ex
Field Strength Required to Fully Saturate Magnet (Oe)	10,000	Primary Chemical Composition	Sr-Fe
Ability to Withstand Demagnetizing Fields	Ex	Special Notes:	
Maximum Service Temperature (°F)	482		
Approximate Temperature Permanently Affecting Magnet (°F)	842		
Inherent Orientation of Material	MP		

Note: All information based on available producers literature. In some cases of foreign materials interpretation or calculation required.

SOURCE(S):

TDK Corp. of America

USA

FB5B
(Alloy or Grade)

Sintered Ferrite
(Type of Material)

4.3/3.4
(MMPA Brief Designation)

-
(IEC Code Reference)

MAGNETIC PROPERTIES	
Residual Induction B_r (G)	4200
Coercive Force (Oe) — Normal (H_c)	3300
Coercive Force (Oe) — Intrinsic (H_{ci})	3350
Maximum Energy Product BH_{max} (MGO)	4.3
B/H Ratio (Load Line) at BH_{max}	1.75
Flux Density in Magnet B_m at BH_{max} (G)	2150
Coercive Force in Magnet H_m at BH_{max} (Oe)	2000
Energy Loss W_c (watt-sec/cycle/lb)	-
Importance of Operating Magnet at BH_{max}	E
Average Recoil Permeability	1.05 to 1.1
Field Strength Required to Fully Saturate Magnet (Oe)	10,000
Ability to Withstand Demagnetizing Fields	Ex
Maximum Service Temperature (°F)	482
Approximate Temperature Permanently Affecting Magnet (°F)	842
Inherent Orientation of Material	MP

PHYSICAL PROPERTIES	
Density (lbs./in.^{3})	.1770-.1806
General Mechanical Properties	VHB
Tensile Strength (psi)	-
Transverse Modulus of Rupture (psi)	-
Coeff. of Therm. Expansion inches x 10^{-6}/°C	Orientation: Par. - Rt.Angl. -
Resistivity (micro-ohms/cm/cm^2)	1.0 @ 25 °C
Hardness	6.5 Moh's Scale
Basic Forming Method	PMF
Basic Finishing Method	WGD
Relative Corrosion Resistance	Ex
Primary Chemical Composition	Sr-Fe
Special Notes:	

Note: All information based on available producers literature. In some cases of foreign materials interpretation or calculation required.

SOURCE(S):

TDK Corp. of America

USA

FB5E
(Alloy or Grade)

Sintered Ferrite
(Type of Material)

3.2/5.0
(MMPA Brief Designation)

-
(IEC Code Reference)

MAGNETIC PROPERTIES		PHYSICAL PROPERTIES	
Residual Induction B_r (G)	3700	Density (lbs./in.^{3})	.1734-.1770
Coercive Force (Oe) — Normal (H_c)	3550	General Mechanical Properties	VHB
Coercive Force (Oe) — Intrinsic (H_{ci})	4950	Tensile Strength (psi)	-
Maximum Energy Product BH_{max} (MGO)	3.2	Transverse Modulus of Rupture (psi)	-
B/H Ratio (Load Line) at BH_{max}	1.12	Coeff. of Therm. Expansion inches x 10^{-6}/°C — Orientation: Par. _ Rt.Angl. _	
Flux Density in Magnet B_m at BH_{max} (G)	1900	Resistivity (micro-ohms/cm/cm^2)	1.0 @ 25 °C
Coercive Force in Magnet H_m at BH_{max} (Oe)	1700	Hardness Scale	6.5 Moh's
Energy Loss W_c (watt-sec/cycle/lb)	-	Basic Forming Method	PMF
Importance of Operating Magnet at BH_{max}	E	Basic Finishing Method	WGD
Average Recoil Permeability	1.05-1.1	Relative Corrosion Resistance	Ex
Field Strength Required to Fully Saturate Magnet (Oe)	10,000	Primary Chemical Composition	Sr-Fe
Ability to Withstand Demagnetizing Fields	Ex	Special Notes:	
Maximum Service Temperature (°F)	482		
Approximate Temperature Permanently Affecting Magnet (°F)	842		
Inherent Orientation of Material	MP		

Note: All information based on available producers literature. In some cases of foreign materials interpretation or calculation required.

SOURCE(S):

TDK Corp. of America

USA

FB5H
(Alloy or Grade)

Sintered Ferrite
(Type of Material)

3.8/4.0
(MMPA Brief Designation)

-
(IEC Code Reference)

MAGNETIC PROPERTIES			PHYSICAL PROPERTIES	
Residual Induction B_r (G)		4050	Density (lbs./in.^{3})	.1752-.1788
Coercive Force (Oe)	Normal (H_c)	3650	General Mechanical Properties	VHB
	Intrinsic (H_{ci})	4050	Tensile Strength (psi)	-
Maximum Energy Product BH_{max} (MGO)		3.8	Transverse Modulus of Rupture (psi)	-
B/H Ratio (Load Line) at BH_{max}		1.11	Coeff. of Therm. Expansion inches x 10^{-6}/°C	Orientation: Par. - Rt.Angl. -
Flux Density in Magnet B_m at BH_{max} (G)		2050	Resistivity (micro-ohms/cm/cm^2	1.0 @ 25 °C
Coercive Force in Magnet H_m at BH_{max} (Oe)		1850	Hardness Scale 6.5 Moh's	
Energy Loss W_c (watt-sec/cycle/lb)		-	Basic Forming Method	PMF
Importance of Operating Magnet at BH_{max}		E	Basic Finishing Method	WGD
Average Recoil Permeability		1.05-1.1	Relative Corrosion Resistance	Ex
Field Strength Required to Fully Saturate Magnet (Oe)		10,000	Primary Chemical Composition	Sr-Fe
Ability to Withstand Demagnetizing Fields		Ex	Special Notes:	
Maximum Service Temperature (°F)		482		
Approximate Temperature Permanently Affecting Magnet (°F)		842		
Inherent Orientation of Material		MP		

Note: All information based on available producers literature. In some cases of foreign materials interpretation or calculation required.

SOURCE(S):		FB5N
		(Alloy or Grade)
TDK Corp. of America		Sintered Ferrite
		(Type of Material)
USA		4.5/2.9
		(MMPA Brief Designation)
		-
		(IEC Code Reference)

MAGNETIC PROPERTIES		PHYSICAL PROPERTIES	
Residual Induction B_r (G)	4400	Density (lbs./in.^{3})	.1770-.1806
Coercive Force (Oe) — Normal (H_c)	2850	General Mechanical Properties	VHB
Coercive Force (Oe) — Intrinsic (H_{ci})	2880	Tensile Strength (psi)	-
Maximum Energy Product BH_{max} (MGO)	4.5	Transverse Modulus of Rupture (psi)	-
B/H Ratio (Load Line) at BH_{max}	1.07	Coeff. of Therm. Expansion inches x 10^{-6}/°C	Orientation: Par. - Rt.Angl. -
Flux Density in Magnet B_m at BH_{max} (G)	2200	Resistivity (micro-ohms/cm/cm^2)	1.0 @ 25 °C
Coercive Force in Magnet H_m at BH_{max} (Oe)	2050	Hardness Scale	6.5 Moh's
Energy Loss W_c (watt-sec/cycle/lb)	-	Basic Forming Method	PMF
Importance of Operating Magnet at BH_{max}	E	Basic Finishing Method	WGD
Average Recoil Permeability	1.05-1.1	Relative Corrosion Resistance	Ex
Field Strength Required to Fully Saturate Magnet (Oe)	10,000	Primary Chemical Composition	Sr-Fe
Ability to Withstand Demagnetizing Fields	Ex	Special Notes:	
Maximum Service Temperature (°F)	482		
Approximate Temperature Permanently Affecting Magnet (°F)	842		
Inherent Orientation of Material	MP		

Note: All information based on available producers literature. In some cases of foreign materials interpretation or calculation required.

SOURCE(S):

Crucible Magnetics

USA

FM5
(Alloy or Grade)

Sintered Ferrite
(Type of Material)

3.6/2.5
(MMPA Brief Designation)

-
(IEC Code Reference)

MAGNETIC PROPERTIES	
Residual Induction B_r (G)	3950
Coercive Force (Oe) — Normal (H_c)	2400
Coercive Force (Oe) — Intrinsic (H_{ci})	2450
Maximum Energy Product BH_{max} (MGO)	3.6
B/H Ratio (Load Line) at BH_{max}	Est. 1.08
Flux Density in Magnet B_m at BH_{max} (G)	Est. 1900
Coercive Force in Magnet H_m at BH_{max} (Oe)	1800
Energy Loss W_c (watt-sec/cycle/lb)	-
Importance of Operating Magnet at BH_{max}	E
Average Recoil Permeability	Est. 1.1
Field Strength Required to Fully Saturate Magnet (Oe)	10,000
Ability to Withstand Demagnetizing Fields	Ex
Maximum Service Temperature (°F)	480
Approximate Temperature Permanently Affecting Magnet (°F)	842
Inherent Orientation of Material	MP

PHYSICAL PROPERTIES	
Density (lbs./in.^{3})	.1780
General Mechanical Properties	VHB
Tensile Strength (psi)	-
Transverse Modulus of Rupture (psi)	-
Coeff. of Therm. Expansion inches x 10^{-6}/°C	Orientation: Par. - Rt.Angl. -
Resistivity (micro-ohms/cm/cm^2)	1.0 @ 25 °C
Hardness	6.5 Scale Moh's
Basic Forming Method	PMF
Basic Finishing Method	WGD
Relative Corrosion Resistance	Ex
Primary Chemical Composition	Sr-Fe
Special Notes:	

Note: All information based on available producers literature. In some cases of foreign materials interpretation or calculation required.

SOURCE(S): **FM7A**
_____
(Alloy or Grade)

Crucible Magnetics

Sintered Ferrite
_____
(Type of Material)

USA

2.8/4.2
_____
(MMPA Brief Designation)

-
_____
(IEC Code Reference)

MAGNETIC PROPERTIES		PHYSICAL PROPERTIES	
Residual Induction B_r (G)	3500	Density (lbs./in.^{3})	.1740
Coercive Force (Oe) — Normal (H_c)	3250	General Mechanical Properties	VHB
Coercive Force (Oe) — Intrinsic (H_{ci})	4200	Tensile Strength (psi)	-
Maximum Energy Product BH_{max} (MGO)	2.8	Transverse Modulus of Rupture (psi)	-
B/H Ratio (Load Line) at BH_{max}	Est. .80	Coeff. of Therm. Expansion inches x 10^{-6}/°C	Orientation: Par. - Rt.Angl. -
Flux Density in Magnet B_m at BH_{max} (G)	Est. 2500	Resistivity (micro-ohms/cm/cm^2)	1.0 @ 25 °C
Coercive Force in Magnet H_m at BH_{max} (Oe)	Est. 1800	Hardness Scale	6.5 Moh's
Energy Loss W_c (watt-sec/cycle/lb)	-	Basic Forming Method	PMF
Importance of Operating Magnet at BH_{max}	E	Basic Finishing Method	WGD
Average Recoil Permeability	Est. 1.1	Relative Corrosion Resistance	Ex
Field Strength Required to Fully Saturate Magnet (Oe)	10,000	Primary Chemical Composition	Sr-Fe
Ability to Withstand Demagnetizing Fields	Ex	Special Notes:	
Maximum Service Temperature (°F)	480		
Approximate Temperature Permanently Affecting Magnet (°F)	842		
Inherent Orientation of Material	MP		

Note: All information based on available producers literature. In some cases of foreign materials interpretation or calculation required.

SOURCE(S):

Crucible Magnetics

USA

FM7B
(Alloy or Grade)

Sintered Ferrite
(Type of Material)

3.3/4.0
(MMPA Brief Designation)

-
(IEC Code Reference)

MAGNETIC PROPERTIES		PHYSICAL PROPERTIES	
Residual Induction B_r (G)	3800	Density (lbs./in.^{3})	.1740
Coercive Force (Oe) Normal (H_c)	3500	General Mechanical Properties	VHB
Coercive Force (Oe) Intrinsic (H_{ci})	4000	Tensile Strength (psi)	-
Maximum Energy Product BH_{max} (MGO)	3.3	Transverse Modulus of Rupture (psi)	-
B/H Ratio (Load Line) at BH_{max}	Est. 1.1	Coeff. of Therm. Expansion inches x 10^{-6}/°C	Orientation: Par. - Rt.Angl. -
Flux Density in Magnet B_m at BH_{max} (G)	Est. 1900	Resistivity (micro-ohms/cm/cm^2)	1.0 @ 25 °C
Coercive Force in Magnet H_m at BH_{max} (Oe)	Est. 1700	Hardness Scale	6.5 Moh's
Energy Loss W_c (watt-sec/cycle/lb)	-	Basic Forming Method	PMF
Importance of Operating Magnet at BH_{max}	E	Basic Finishing Method	WGD
Average Recoil Permeability	Est. 1.1	Relative Corrosion Resistance	Ex
Field Strength Required to Fully Saturate Magnet (Oe)	10,000	Primary Chemical Composition	Sr-Fe
Ability to Withstand Demagnetizing Fields	Ex	Special Notes:	
Maximum Service Temperature (°F)	480		
Approximate Temperature Permanently Affecting Magnet (°F)	842		
Inherent Orientation of Material	MP		

Note: All information based on available producers literature. In some cases of foreign materials interpretation or calculation required.

SOURCE(S):	FM8A
	(Alloy or Grade)
Crucible Magnetics	Sintered Ferrite
	(Type of Material)
USA	3.5/3.3
	(MMPA Brief Designation)
	-
	(IEC Code Reference)

MAGNETIC PROPERTIES		PHYSICAL PROPERTIES	
Residual Induction B_r (G)	3900	Density (lbs./in.^{3})	.1760
Coercive Force (Oe) — Normal (H_c)	3200	General Mechanical Properties	VHB
Coercive Force (Oe) — Intrinsic (H_{ci})	3250	Tensile Strength (psi)	-
Maximum Energy Product BH_{max} (MGO)	3.5	Transverse Modulus of Rupture (psi)	-
B/H Ratio (Load Line) at BH_{max}	Est. 1.14	Coeff. of Therm. Expansion inches x 10^{-6}/°C	Orientation: Par. - Rt.Angl. -
Flux Density in Magnet B_m at BH_{max} (G)	Est. 2000	Resistivity (micro-ohms/cm/cm^2)	1.0 @ 25 °C
Coercive Force in Magnet H_m at BH_{max} (Oe)	Est. 1750	Hardness Scale	6.5 Moh's
Energy Loss W_c (watt-sec/cycle/lb)	-	Basic Forming Method	PMF
Importance of Operating Magnet at BH_{max}	E	Basic Finishing Method	WGD
Average Recoil Permeability	Est. 1.1	Relative Corrosion Resistance	Ex
Field Strength Required to Fully Saturate Magnet (Oe)	10,000	Primary Chemical Composition	Sr-Fe
Ability to Withstand Demagnetizing Fields	Ex	Special Notes:	
Maximum Service Temperature (°F)	480		
Approximate Temperature Permanently Affecting Magnet (°F)	842		
Inherent Orientation of Material	MP		

Note: All information based on available producers literature. In some cases of foreign materials interpretation or calculation required.

SOURCE(S):	**FM8B**
	(Alloy or Grade)
Crucible Magnetics	Sintered Ferrite
	(Type of Material)
USA	4.1/3.0
	(MMPA Brief Designation)
	-
	(IEC Code Reference)

MAGNETIC PROPERTIES		PHYSICAL PROPERTIES	
Residual Induction B_r (G)	4200	Density (lbs./in.^{3})	.1790
Coercive Force (Oe) — Normal (H_c)	2900	General Mechanical Properties	VHB
Coercive Force (Oe) — Intrinsic (H_{ci})	2950	Tensile Strength (psi)	-
Maximum Energy Product BH_{max} (MGO)	4.1	Transverse Modulus of Rupture (psi)	-
B/H Ratio (Load Line) at BH_{max}	Est. 1.05	Coeff. of Therm. Expansion inches x 10^{-6}/°C	Orientation: Par. - Rt.Angl. -
Flux Density in Magnet B_m at BH_{max} (G)	Est. 2100	Resistivity (micro-ohms/cm/cm^2)	1.0 @ 25 °C
Coercive Force in Magnet H_m at BH_{max} (Oe)	Est. 2000	Hardness 6.5 Scale Moh's	
Energy Loss W_c (watt-sec/cycle/lb)	-	Basic Forming Method	PMF
Importance of Operating Magnet at BH_{max}	E	Basic Finishing Method	WGD
Average Recoil Permeability	Est. 1.1	Relative Corrosion Resistance	Ex
Field Strength Required to Fully Saturate Magnet (Oe)	10,000	Primary Chemical Composition	Sr-Fe
Ability to Withstand Demagnetizing Fields	Ex	Special Notes:	
Maximum Service Temperature (°F)	480		
Approximate Temperature Permanently Affecting Magnet (°F)	842		
Inherent Orientation of Material	MP		

Note: All information based on available producers literature. In some cases of foreign materials interpretation or calculation required.

SOURCE(S): **Crucible Magnetics**

USA

FM8C
(Alloy or Grade)

Sintered Ferrite
(Type of Material)

4.3/2.5
(MMPA Brief Designation)

-
(IEC Code Reference)

MAGNETIC PROPERTIES	
Residual Induction B_r (G)	4300
Coercive Force (Oe) — Normal (H_c)	2400
Coercive Force (Oe) — Intrinsic (H_{ci})	2450
Maximum Energy Product BH_{max} (MGO)	4.3
B/H Ratio (Load Line) at BH_{max}	Est. 1.0
Flux Density in Magnet B_m at BH_{max} (G)	Est. 2100
Coercive Force in Magnet H_m at BH_{max} (Oe)	Est. 2100
Energy Loss W_c (watt-sec/cycle/lb)	-
Importance of Operating Magnet at BH_{max}	E
Average Recoil Permeability	Est. 1.1
Field Strength Required to Fully Saturate Magnet (Oe)	10,000
Ability to Withstand Demagnetizing Fields	Ex
Maximum Service Temperature (°F)	480
Approximate Temperature Permanently Affecting Magnet (°F)	842
Inherent Orientation of Material	MP

PHYSICAL PROPERTIES	
Density (lbs./in.^{3})	.1790
General Mechanical Properties	VHB
Tensile Strength (psi)	-
Transverse Modulus of Rupture (psi)	-
Coeff. of Therm. Expansion inches x 10^{-6}/°C — Orientation: Par.	-
Coeff. of Therm. Expansion inches x 10^{-6}/°C — Rt. Angl.	-
Resistivity (micro-ohms/cm/cm^2)	1.0 @ 25 °C
Hardness Scale	6.5 Moh's
Basic Forming Method	PMF
Basic Finishing Method	WGD
Relative Corrosion Resistance	Ex
Primary Chemical Composition	Sr-Fe
Special Notes:	

Note: All information based on available producers literature. In some cases of foreign materials interpretation or calculation required.

SOURCE(S):

Crucible Magnetics

USA

FM8D
(Alloy or Grade)

Sintered Ferrite
(Type of Material)

3.8/3.5
(MMPA Brief Designation)

-
(IEC Code Reference)

MAGNETIC PROPERTIES		PHYSICAL PROPERTIES	
Residual Induction B_r (G)	4000	Density (lbs./in.^{3})	.1770
Coercive Force (Oe) — Normal (H_c)	3500	General Mechanical Properties	VHB
Coercive Force (Oe) — Intrinsic (H_{ci})	3500	Tensile Strength (psi)	-
Maximum Energy Product BH_{max} (MGO)	3.8	Transverse Modulus of Rupture (psi)	-
B/H Ratio (Load Line) at BH_{max}	Est. .56	Coeff. of Therm. Expansion inches x 10^{-6}/°C	Orientation: Par. - Rt.Angl. -
Flux Density in Magnet B_m at BH_{max} (G)	Est. 1500	Resistivity (micro-ohms/cm/cm^2)	1.0 @ 25 °C
Coercive Force in Magnet H_m at BH_{max} (Oe)	Est. 2700	Hardness 6.5	Scale Moh's
Energy Loss W_c (watt-sec/cycle/lb)	-	Basic Forming Method	PMF
Importance of Operating Magnet at BH_{max}	E	Basic Finishing Method	WGD
Average Recoil Permeability	Est. 1.1	Relative Corrosion Resistance	Ex
Field Strength Required to Fully Saturate Magnet (Oe)	10,000	Primary Chemical Composition	Sr-Fe
Ability to Withstand Demagnetizing Fields	Ex	Special Notes:	
Maximum Service Temperature (°F)	480		
Approximate Temperature Permanently Affecting Magnet (°F)	842		
Inherent Orientation of Material	MP		

Note: All information based on available producers literature. In some cases of foreign materials interpretation or calculation required.

SOURCE(S):

Philips Components

The Neatherlands

FXD300
(Alloy or Grade)

Sintered Ferrite
(Type of Material)

3.7/2.0
(MMPA Brief Designation)

-
(IEC Code Reference)

MAGNETIC PROPERTIES	
Residual Induction B_r (G)	4000
Coercive Force (Oe) — Normal (H_c)	2000
Coercive Force (Oe) — Intrinsic (H_{ci})	2050
Maximum Energy Product BH_{max} (MGO)	3.7
B/H Ratio (Load Line) at BH_{max}	1.3
Flux Density in Magnet B_m at BH_{max} (G)	2200
Coercive Force in Magnet H_m at BH_{max} (Oe)	1700
Energy Loss W_c (watt-sec/cycle/lb)	-
Importance of Operating Magnet at BH_{max}	E
Average Recoil Permeability	1.1
Field Strength Required to Fully Saturate Magnet (Oe)	7000
Ability to Withstand Demagnetizing Fields	Ex
Maximum Service Temperature (°F)	482
Approximate Temperature Permanently Affecting Magnet (°F)	842
Inherent Orientation of Material	MP

PHYSICAL PROPERTIES	
Density (lbs./in.^{3})	.177
General Mechanical Properties	VHB
Tensile Strength (psi)	-
Transverse Modulus of Rupture (psi)	-
Coeff. of Therm. Expansion inches x 10^{-6}/°C — Orientation: Par.	-
Coeff. of Therm. Expansion inches x 10^{-6}/°C — Rt.Angl.	-
Resistivity (micro-ohms/cm/cm^2)	1.0 @ 25 °C
Hardness	6.5 Moh's Scale
Basic Forming Method	PMF
Basic Finishing Method	WGD
Relative Corrosion Resistance	Ex
Primary Chemical Composition	Ba-Fe
Special Notes:	

Note: All information based on available producers literature. In some cases of foreign materials interpretation or calculation required.

SOURCE(S):	FXD330
	(Alloy or Grade)
Philips Components	Sintered Ferrite
	(Type of Material)
The Neatherlands	3.2/3.2
	(MMPA Brief Designation)
	-
	(IEC Code Reference)

MAGNETIC PROPERTIES		PHYSICAL PROPERTIES	
Residual Induction B_r (G)	3700	Density (lbs./in.^{3})	.1680
Coercive Force (Oe) — Normal (H_c)	3100	General Mechanical Properties	VHB
Coercive Force (Oe) — Intrinsic (H_{ci})	3200	Tensile Strength (psi)	-
Maximum Energy Product BH_{max} (MGO)	3.2	Transverse Modulus of Rupture (psi)	-
B/H Ratio (Load Line) at BH_{max}	1.03	Coeff. of Therm. Expansion inches x 10^{-6}/°C	Orientation: Par. - Rt.Angl. -
Flux Density in Magnet B_m at BH_{max} (G)	1800	Resistivity (micro-ohms/cm/cm^2)	1.0 @ 25 °C
Coercive Force in Magnet H_m at BH_{max} (Oe)	1750	Hardness 6.5	Scale Moh's
Energy Loss W_c (watt-sec/cycle/lb)	-	Basic Forming Method	PMF
Importance of Operating Magnet at BH_{max}	E	Basic Finishing Method	WGD
Average Recoil Permeability	1.1	Relative Corrosion Resistance	Ex
Field Strength Required to Fully Saturate Magnet (Oe)	11,000	Primary Chemical Composition	Sr-Fe
Ability to Withstand Demagnetizing Fields	Ex	Special Notes:	
Maximum Service Temperature (°F)	482		
Approximate Temperature Permanently Affecting Magnet (°F)	842		
Inherent Orientation of Material	MP		

Note: All information based on available producers literature. In some cases of foreign materials interpretation or calculation required.

SOURCE(S):	FXD380
	(Alloy or Grade)
Philips Components	Sintered Ferrite
	(Type of Material)
The Neatherlands	3.6/3.5
	(MMPA Brief Designation)
	-
	(IEC Code Reference)

MAGNETIC PROPERTIES		PHYSICAL PROPERTIES	
Residual Induction B_r (G)	3900	Density (lbs./in.^{3})	.1716
Coercive Force (Oe) — Normal (H_c)	3300	General Mechanical Properties	VHB
Coercive Force (Oe) — Intrinsic (H_{ci})	3500	Tensile Strength (psi)	-
Maximum Energy Product BH_{max} (MGO)	3.6	Transverse Modulus of Rupture (psi)	-
B/H Ratio (Load Line) at BH_{max}	1.27	Coeff. of Therm. Expansion inches x 10^{-6}/°C — Orientation: Par. - Rt.Angl. -	
Flux Density in Magnet B_m at BH_{max} (G)	1900	Resistivity (micro-ohms/cm/cm^2)	1.0 @ 25 °C
Coercive Force in Magnet H_m at BH_{max} (Oe)	1850	Hardness — Scale	6.5 Moh's
Energy Loss W_c (watt-sec/cycle/lb)	-	Basic Forming Method	PMF
Importance of Operating Magnet at BH_{max}	E	Basic Finishing Method	WGD
Average Recoil Permeability	1.1	Relative Corrosion Resistance	Ex
Field Strength Required to Fully Saturate Magnet (Oe)	10,000	Primary Chemical Composition	Sr-Fe
Ability to Withstand Demagnetizing Fields	Ex	Special Notes:	
Maximum Service Temperature (°F)	482		
Approximate Temperature Permanently Affecting Magnet (°F)	842		
Inherent Orientation of Material	MP		

Note: All information based on available producers literature. In some cases of foreign materials interpretation or calculation required.

SOURCE(S):	FXD400
	(Alloy or Grade)
Philips Components	Sintered Ferrite
	(Type of Material)
The Neatherlands	3.9/3.5
	(MMPA Brief Designation)
	-
	(IEC Code Reference)

MAGNETIC PROPERTIES		PHYSICAL PROPERTIES	
Residual Induction B_r (G)	4100	Density (lbs./in.^{3})	.1734
Coercive Force (Oe) — Normal (H_c)	3300	General Mechanical Properties	VHB
Coercive Force (Oe) — Intrinsic (H_{ci})	3500	Tensile Strength (psi)	-
Maximum Energy Product BH_{max} (MGO)	3.9	Transverse Modulus of Rupture (psi)	-
B/H Ratio (Load Line) at BH_{max}	1.02	Coeff. of Therm. Expansion inches x 10^{-6}/°C — Orientation: Par. - Rt.Angl. -	
Flux Density in Magnet B_m at BH_{max} (G)	2000	Resistivity (micro-ohms/cm/cm^2)	1.0 @ 25 °C
Coercive Force in Magnet H_m at BH_{max} (Oe)	1950	Hardness Scale	6.5 Moh's
Energy Loss W_c (watt-sec/cycle/lb)	-	Basic Forming Method	PMF
Importance of Operating Magnet at BH_{max}	E	Basic Finishing Method	WGD
Average Recoil Permeability	1.1	Relative Corrosion Resistance	Ex
Field Strength Required to Fully Saturate Magnet (Oe)	10,000	Primary Chemical Composition	Sr-Fe
Ability to Withstand Demagnetizing Fields	Ex	Special Notes:	
Maximum Service Temperature (°F)	482		
Approximate Temperature Permanently Affecting Magnet (°F)	842		
Inherent Orientation of Material	MP		

Note: All information based on available producers literature. In some cases of foreign materials interpretation or calculation required.

SOURCE(S): **Philips Components**

The Neatherlands

FXD480
(Alloy or Grade)

Sintered Ferrite
(Type of Material)

3.4/4.0
(MMPA Brief Designation)

-
(IEC Code Reference)

MAGNETIC PROPERTIES		PHYSICAL PROPERTIES	
Residual Induction B_r (G)	3800	Density (lbs./in.^{3})	.1698
Coercive Force (Oe) — Normal (H_c)	3500	General Mechanical Properties	VHB
Coercive Force (Oe) — Intrinsic (H_{ci})	4000	Tensile Strength (psi)	-
Maximum Energy Product BH_{max} (MGO)	3.4	Transverse Modulus of Rupture (psi)	-
B/H Ratio (Load Line) at BH_{max}	1.09	Coeff. of Therm. Expansion inches x 10^{-6}/°C Orientation: Par. - Rt.Angl. -	
Flux Density in Magnet B_m at BH_{max} (G)	1900	Resistivity (micro-ohms/cm/cm^2)	1.0 @ 25 °C
Coercive Force in Magnet H_m at BH_{max} (Oe)	1750	Hardness Scale	6.5 Moh's
Energy Loss W_c (watt-sec/cycle/lb)	-	Basic Forming Method	PMF
Importance of Operating Magnet at BH_{max}	E	Basic Finishing Method	WGD
Average Recoil Permeability	1.1	Relative Corrosion Resistance	Ex
Field Strength Required to Fully Saturate Magnet (Oe)	10,000	Primary Chemical Composition	Sr-Fe
Ability to Withstand Demagnetizing Fields	Ex	Special Notes:	
Maximum Service Temperature (°F)	482		
Approximate Temperature Permanently Affecting Magnet (°F)	842		
Inherent Orientation of Material	MP		

Note: All information based on available producers literature. In some cases of foreign materials interpretation or calculation required.

SOURCE(S):

Philips Components

The Neatherlands

FXD500
(Alloy or Grade)

Sintered Ferrite
(Type of Material)

3.8/4.2
(MMPA Brief Designation)

-
(IEC Code Reference)

MAGNETIC PROPERTIES		PHYSICAL PROPERTIES	
Residual Induction B_r (G)	4000	Density (lbs./in.^{3})	.1752
Coercive Force (Oe) — Normal (H_c)	3700	General Mechanical Properties	VHB
Coercive Force (Oe) — Intrinsic (H_{ci})	4150	Tensile Strength (psi)	-
Maximum Energy Product BH_{max} (MGO)	3.8	Transverse Modulus of Rupture (psi)	-
B/H Ratio (Load Line) at BH_{max}	1.05	Coeff. of Therm. Expansion inches x 10^{-6}/°C — Orientation: Par. − Rt.Angl. −	
Flux Density in Magnet B_m at BH_{max} (G)	2000	Resistivity (micro-ohms/cm/cm^2)	1.0 @ 25 °C
Coercive Force in Magnet H_m at BH_{max} (Oe)	1900	Hardness Scale	6.5 Moh's
Energy Loss W_c (watt-sec/cycle/lb)	-	Basic Forming Method	PMF
Importance of Operating Magnet at BH_{max}	E	Basic Finishing Method	PMF
Average Recoil Permeability	1.1	Relative Corrosion Resistance	Ex
Field Strength Required to Fully Saturate Magnet (Oe)	10,000	Primary Chemical Composition	Sr-Fe
Ability to Withstand Demagnetizing Fields	Ex	Special Notes:	
Maximum Service Temperature (°F)	482		
Approximate Temperature Permanently Affecting Magnet (°F)	842		
Inherent Orientation of Material	MP		

Note: All information based on available producers literature. In some cases of foreign materials interpretation or calculation required.

SOURCE(S):

Philips Components

The Neatherlands

FXD520
(Alloy or Grade)

Sintered Ferrite
(Type of Material)

4.2/3.3
(MMPA Brief Designation)

-
(IEC Code Reference)

MAGNETIC PROPERTIES	
Residual Induction B_r (G)	4250
Coercive Force (Oe) — Normal (H_c)	3100
Coercive Force (Oe) — Intrinsic (H_{ci})	3300
Maximum Energy Product BH_{max} (MGO)	4.2
B/H Ratio (Load Line) at BH_{max}	1.05
Flux Density in Magnet B_m at BH_{max} (G)	2100
Coercive Force in Magnet H_m at BH_{max} (Oe)	2000
Energy Loss W_c (watt-sec/cycle/lb)	-
Importance of Operating Magnet at BH_{max}	E
Average Recoil Permeability	1.1
Field Strength Required to Fully Saturate Magnet (Oe)	10,000
Ability to Withstand Demagnetizing Fields	Ex
Maximum Service Temperature (°F)	482
Approximate Temperature Permanently Affecting Magnet (°F)	842
Inherent Orientation of Material	MP

PHYSICAL PROPERTIES	
Density (lbs./in.^{3})	.1770
General Mechanical Properties	VHB
Tensile Strength (psi)	-
Transverse Modulus of Rupture (psi)	-
Coeff. of Therm. Expansion inches x 10^{-6}/°C	Orientation: Par. - Rt.Angl. -
Resistivity (micro-ohms/cm/cm^2)	1.0 @ 25 °C
Hardness	6.5 Scale Moh's
Basic Forming Method	PMF
Basic Finishing Method	WGD
Relative Corrosion Resistance	Ex
Primary Chemical Composition	Sr-Fe
Special Notes:	

Note: All information based on available producers literature. In some cases of foreign materials interpretation or calculation required.

SOURCE(S):

Philips Components

The Neatherlands

FXD580
(Alloy or Grade)

Sintered Ferrite
(Type of Material)

3.5/4.5
(MMPA Brief Designation)

-
(IEC Code Reference)

MAGNETIC PROPERTIES		PHYSICAL PROPERTIES	
Residual Induction B_r (G)	3850	Density (lbs./in.^{3})	.1752
Coercive Force (Oe) Normal (H_c)	3770	General Mechanical Properties	VHB
Coercive Force (Oe) Intrinsic (H_{ci})	4520	Tensile Strength (psi)	-
Maximum Energy Product BH_{max} (MGO)	3.5	Transverse Modulus of Rupture (psi)	-
B/H Ratio (Load Line) at BH_{max}	1.05	Coeff. of Therm. Expansion inches x 10^{-6}/°C Orientation: Par. - Rt.Angl. -	
Flux Density in Magnet B_m at BH_{max} (G)	1900	Resistivity (micro-ohms/cm/cm^2)	1.0 @ 25 °C
Coercive Force in Magnet H_m at BH_{max} (Oe)	1800	Hardness Scale	6.5 Moh's
Energy Loss W_c (watt-sec/cycle/lb)	-	Basic Forming Method	PMF
Importance of Operating Magnet at BH_{max}	E	Basic Finishing Method	WGD
Average Recoil Permeability	1.1	Relative Corrosion Resistance	Ex
Field Strength Required to Fully Saturate Magnet (Oe)	10,000	Primary Chemical Composition	Sr-Fe
Ability to Withstand Demagnetizing Fields	Ex	Special Notes:	
Maximum Service Temperature (°F)	482		
Approximate Temperature Permanently Affecting Magnet (°F)	842		
Inherent Orientation of Material	MP		

Note: All information based on available producers literature. In some cases of foreign materials interpretation or calculation required.

SOURCE(S):

Magnet Applications, Ltd.
(a.k.a. Cookson Group/Arelec)

England/France

Ferram I
(Alloy or Grade)

Sintered Ferrite
(Type of Material)

1.1/3.4
(MMPA Brief Designation)

-
(IEC Code Reference)

MAGNETIC PROPERTIES		PHYSICAL PROPERTIES	
Residual Induction B_r (G)	2200	Density (lbs./in.^{3})	.2637
Coercive Force (Oe) — Normal (H_c)	2000	General Mechanical Properties	VHB
Coercive Force (Oe) — Intrinsic (H_{ci})	3400	Tensile Strength (psi)	-
Maximum Energy Product BH_{max} (MGO)	1.1	Transverse Modulus of Rupture (psi)	-
B/H Ratio (Load Line) at BH_{max}	Est. 1.0	Coeff. of Therm. Expansion inches $\times 10^{-6}$/°C — Orientation: Par. / Rt.Angl.	- / -
Flux Density in Magnet B_m at BH_{max} (G)	Est. 1000	Resistivity (micro-ohms/cm/cm^2)	1.0 @ 25 °C
Coercive Force in Magnet H_m at BH_{max} (Oe)	Est. 1000	Hardness Scale	530 Vickers
Energy Loss W_c (watt-sec/cycle/lb)	-	Basic Forming Method	PF
Importance of Operating Magnet at BH_{max}	E	Basic Finishing Method	WGD
Average Recoil Permeability	1.2	Relative Corrosion Resistance	Ex
Field Strength Required to Fully Saturate Magnet (Oe)	10,000	Primary Chemical Composition	Ba or Sr-Fe
Ability to Withstand Demagnetizing Fields	Ex	Special Notes:	
Maximum Service Temperature (°F)	840		
Approximate Temperature Permanently Affecting Magnet (°F)	482		
Inherent Orientation of Material	none		

Note: All information based on available producers literature. In some cases of foreign materials interpretation or calculation required.

SOURCE(S):	
Magnet Applications, Ltd. (a.k.a. Cookson Group/Arelec) England/France	**Ferram A-Ba** (Alloy or Grade)
	Sintered Ferrite (Type of Material)
	3.5/1.9 (MMPA Brief Designation)
	- (IEC Code Reference)

MAGNETIC PROPERTIES		PHYSICAL PROPERTIES	
Residual Induction B_r (G)	3900	Density (lbs./in.^{3})	.1770
Coercive Force (Oe) — Normal (H_c)	1800	General Mechanical Properties	VHB
Coercive Force (Oe) — Intrinsic (H_{ci})	1850	Tensile Strength (psi)	-
Maximum Energy Product BH_{max} (MGO)	3.5	Transverse Modulus of Rupture (psi)	-
B/H Ratio (Load Line) at BH_{max}	Est. 1.08	Coeff. of Therm. Expansion inches x 10^{-6}/°C	Orientation: Par. - Rt.Angl. -
Flux Density in Magnet B_m at BH_{max} (G)	Est. 1950	Resistivity (micro-ohms/cm/cm^2)	1.0 @ 25 °C
Coercive Force in Magnet H_m at BH_{max} (Oe)	Est. 1800	Hardness Scale	530 Vickers
Energy Loss W_c (watt-sec/cycle/lb)	-	Basic Forming Method	PMF
Importance of Operating Magnet at BH_{max}	E	Basic Finishing Method	WGD
Average Recoil Permeability	1.1	Relative Corrosion Resistance	Ex
Field Strength Required to Fully Saturate Magnet (Oe)	10,000	Primary Chemical Composition	Ba-Fe
Ability to Withstand Demagnetizing Fields	Ex	Special Notes:	
Maximum Service Temperature (°F)	480		
Approximate Temperature Permanently Affecting Magnet (°F)	842		
Inherent Orientation of Material	MP		

Note: All information based on available producers literature. In some cases of foreign materials interpretation or calculation required.

SOURCE(S):	Ferram A-Str
	(Alloy or Grade)
Magnet Applications, Ltd.	Sintered Ferrite
(a.k.a. Cookson Group/Arelec)	(Type of Material)
	3.2/3.0
England/France	(MMPA Brief Designation)
	-
	(IEC Code Reference)

MAGNETIC PROPERTIES		PHYSICAL PROPERTIES	
Residual Induction B_r (G)	3600	Density (lbs./in.^{3})	.1770
Coercive Force (Oe) — Normal (H_c)	2900	General Mechanical Properties	VHB
Coercive Force (Oe) — Intrinsic (H_{ci})	3000	Tensile Strength (psi)	-
Maximum Energy Product BH_{max} (MGO)	3.2	Transverse Modulus of Rupture (psi)	-
B/H Ratio (Load Line) at BH_{max}	Est. 1.05	Coeff. of Therm. Expansion inches x 10^{-6}/°C	Orientation: Par. - Rt.Angl. -
Flux Density in Magnet B_m at BH_{max} (G)	Est. 1820	Resistivity (micro-ohms/cm/cm^2)	1.0 @ 25 °C
Coercive Force in Magnet H_m at BH_{max} (Oe)	Est. 1734	Hardness 530	Scale Vickers
Energy Loss W_c (watt-sec/cycle/lb)	-	Basic Forming Method	PMF
Importance of Operating Magnet at BH_{max}	E	Basic Finishing Method	WGD
Average Recoil Permeability	1.1	Relative Corrosion Resistance	Ex
Field Strength Required to Fully Saturate Magnet (Oe)	10,000	Primary Chemical Composition	Sr-Fe
Ability to Withstand Demagnetizing Fields	Ex	Special Notes:	
Maximum Service Temperature (°F)	480		
Approximate Temperature Permanently Affecting Magnet (°F)	842		
Inherent Orientation of Material	MP		

Note: All information based on available producers literature. In some cases of foreign materials interpretation or calculation required.

SOURCE(S):

Magnet Applications, Ltd.
(a.k.a. Cookson Group/Arelec)

England/France

Flexam
(Alloy or Grade)

Bonded Ferrite
(Type of Material)

—
(MMPA Brief Designation)

—
(IEC Code Reference)

MAGNETIC PROPERTIES		PHYSICAL PROPERTIES	
Residual Induction B_r (G)	2000	Density (lbs./in.^{3})	.1337
Coercive Force (Oe) Normal (H_c)	1600	General Mechanical Properties	FL
Coercive Force (Oe) Intrinsic (H_{ci})	—	Tensile Strength (psi)	—
Maximum Energy Product BH_{max} (MGO)	.8	Transverse Modulus of Rupture (psi)	—
B/H Ratio (Load Line) at BH_{max}	—	Coeff. of Therm. Expansion inches x 10^{-6}/°C	Orientation: Par. — Rt.Angl. —
Flux Density in Magnet B_m at BH_{max} (G)	—	Resistivity (micro-ohms/cm/cm^2	— @ — °C
Coercive Force in Magnet H_m at BH_{max} (Oe)	—	Hardness Scale 50 Shore D	
Energy Loss W_c (watt-sec/cycle/lb)	—	Basic Forming Method	E
Importance of Operating Magnet at BH_{max}	E	Basic Finishing Method	SCD/SRD
Average Recoil Permeability	1.1	Relative Corrosion Resistance	E
Field Strength Required to Fully Saturate Magnet (Oe)	10,000	Primary Chemical Composition	Ba or Sr-Fe + binder See Note (1)
Ability to Withstand Demagnetizing Fields	Ex	Special Notes: (1) Rubber binder. Special types with other binders to 248° F.	
Maximum Service Temperature (°F)	176 See Note (1)		
Approximate Temperature Permanently Affecting Magnet (°F)	—		
Inherent Orientation of Material	none		

Note: All information based on available producers literature. In some cases of foreign materials interpretation or calculation required.

SOURCE(S):

**Magnet Applications, Ltd.
(a.k.a. Cookson Group/Arelec)**

England/France

Flexam P-5
(Alloy or Grade)

Bonded Ferrite
(Type of Material)

.5/2.3
(MMPA Brief Designation)

-
(IEC Code Reference)

MAGNETIC PROPERTIES	
Residual Induction B_r (G)	1400
Coercive Force (Oe) Normal (H_c)	1200
Coercive Force (Oe) Intrinsic (H_{ci})	2300
Maximum Energy Product BH_{max} (MGO)	.5
B/H Ratio (Load Line) at BH_{max}	-
Flux Density in Magnet B_m at BH_{max} (G)	-
Coercive Force in Magnet H_m at BH_{max} (Oe)	-
Energy Loss W_c (watt-sec/cycle/lb)	-
Importance of Operating Magnet at BH_{max}	E
Average Recoil Permeability	1.1
Field Strength Required to Fully Saturate Magnet (Oe)	10,000
Ability to Withstand Demagnetizing Fields	Ex
Maximum Service Temperature (°F)	212
Approximate Temperature Permanently Affecting Magnet (°F)	-
Inherent Orientation of Material	none See Note (2)

PHYSICAL PROPERTIES	
Density (lbs./in.^{3})	.1373
General Mechanical Properties	HT
Tensile Strength (psi)	-
Transverse Modulus of Rupture (psi)	-
Coeff. of Therm. Expansion inches x 10^{-6}/°C	Orientation: Par. - Rt.Angl. -
Resistivity (micro-ohms/cm/cm^2	- @ - °C
Hardness 86	Scale Shore D
Basic Forming Method	E
Basic Finishing Method	WGD
Relative Corrosion Resistance	E
Primary Chemical Composition	Ba or Sr-Fe + binder See Note (1)
Special Notes: (1) Nylon or other plastic binder (2) Can be made anisotropic on special order with improved magnetic properties	

Note: All information based on available producers literature. In some cases of foreign materials interpretation or calculation required.

SOURCE(S):

Magnet Applications, Ltd.
(a.k.a. Cookson Group/Arelec)

England/France

Flexam P8
(Alloy or Grade)

Bonded Ferrite
(Type of Material)

.8/2.7
(MMPA Brief Designation)

-
(IEC Code Reference)

MAGNETIC PROPERTIES	
Residual Induction B_r (G)	2000
Coercive Force (Oe) Normal (H_c)	1600
Coercive Force (Oe) Intrinsic (H_{ci})	2700
Maximum Energy Product BH_{max} (MGO)	.8
B/H Ratio (Load Line) at BH_{max}	-
Flux Density in Magnet B_m at BH_{max} (G)	-
Coercive Force in Magnet H_m at BH_{max} (Oe)	-
Energy Loss W_c (watt-sec/cycle/lb)	-
Importance of Operating Magnet at BH_{max}	E
Average Recoil Permeability	1.1
Field Strength Required to Fully Saturate Magnet (Oe)	10,000
Ability to Withstand Demagnetizing Fields	Ex
Maximum Service Temperature (°F)	212
Approximate Temperature Permanently Affecting Magnet (°F)	-
Inherent Orientation of Material	none See Note (2)

PHYSICAL PROPERTIES	
Density (lbs./in.^{3})	.1264
General Mechanical Properties	HT
Tensile Strength (psi)	-
Transverse Modulus of Rupture (psi)	-
Coeff. of Therm. Expansion inches x 10^{-6}/°C	Orientation: Par. - Rt.Angl. -
Resistivity (micro-ohms/cm/cm^2)	- @ - °C
Hardness Scale	86 Shore D
Basic Forming Method	E
Basic Finishing Method	WGD
Relative Corrosion Resistance	E
Primary Chemical Composition	Ba or Sr-Fe See Note (1)

Special Notes:

(1) Nylon or other plastic binder
(2) Can be made anisotropic on special order with improved magnetic properties

Note: All information based on available producers literature. In some cases of foreign materials interpretation or calculation required.

SOURCE(S):	Flexor 15
	(Alloy or Grade)
Magnet Applications, Ltd.	Bonded Ferrite
(a.k.a. Cookson Group/Arelec)	(Type of Material)
	1.15/2.6
England/France	(MMPA Brief Designation)
	-
	(IEC Code Reference)

MAGNETIC PROPERTIES		PHYSICAL PROPERTIES	
Residual Induction B_r (G)	2200	Density (lbs./in.^{3})	.1409
Coercive Force (Oe) Normal (H_c)	1950	General Mechanical Properties	FL
Coercive Force (Oe) Intrinsic (H_{ci})	2600	Tensile Strength (psi)	-
Maximum Energy Product BH_{max} (MGO)	1.15	Transverse Modulus of Rupture (psi)	-
B/H Ratio (Load Line) at BH_{max}	-	Coeff. of Therm. Expansion inches x 10^{-6}/°C	Orientation: Par. - Rt.Angl. -
Flux Density in Magnet B_m at BH_{max} (G)	-	Resistivity (micro-ohms/cm/cm^2	- @ - °C
Coercive Force in Magnet H_m at BH_{max} (Oe)	-	Hardness Scale 50	Shore D
Energy Loss W_c (watt-sec/cycle/lb)	-	Basic Forming Method	E
Importance of Operating Magnet at BH_{max}	E	Basic Finishing Method	SCD/SRD
Average Recoil Permeability	1.1	Relative Corrosion Resistance	E
Field Strength Required to Fully Saturate Magnet (Oe)	10,000	Primary Chemical Composition	Ba or Sr-Fe + binder
Ability to Withstand Demagnetizing Fields	Ex	Special Notes:	
Maximum Service Temperature (°F)	176	(1) Rubber binder. Specifics not known.	
Approximate Temperature Permanently Affecting Magnet (°F)	-		
Inherent Orientation of Material	none		

Note: All information based on available producers literature. In some cases of foreign materials interpretation or calculation required.

SOURCE(S):

Magnet Applications, Ltd.
(a.k.a. Cookson Group/Arelec)

England/France

Flexor 45
(Alloy or Grade)

Bonded Ferrite
(Type of Material)

1.45/2.7
(MMPA Brief Designation)

-
(IEC Code Reference)

MAGNETIC PROPERTIES	
Residual Induction B_r (G)	2500
Coercive Force (Oe) Normal (H_c)	2000
Coercive Force (Oe) Intrinsic (H_{ci})	2700
Maximum Energy Product BH_{max} (MGO)	1.45
B/H Ratio (Load Line) at BH_{max}	-
Flux Density in Magnet B_m at BH_{max} (G)	-
Coercive Force in Magnet H_m at BH_{max} (Oe)	-
Energy Loss W_c (watt-sec/cycle/lb)	-
Importance of Operating Magnet at BH_{max}	E
Average Recoil Permeability	1.1
Field Strength Required to Fully Saturate Magnet (Oe)	10,000
Ability to Withstand Demagnetizing Fields	Ex
Maximum Service Temperature (°F)	176
Approximate Temperature Permanently Affecting Magnet (°F)	-
Inherent Orientation of Material	none

PHYSICAL PROPERTIES	
Density (lbs./in.^{3})	.1409
General Mechanical Properties	FL
Tensile Strength (psi)	-
Transverse Modulus of Rupture (psi)	-
Coeff. of Therm. Expansion inches x 10^{-6}/°C	Orientation: Par. - Rt.Angl. -
Resistivity (micro-ohms/cm/cm^2	- @ - °C
Hardness 50	Scale Shore D
Basic Forming Method	E
Basic Finishing Method	SCD/SRD
Relative Corrosion Resistance	E
Primary Chemical Composition	Ba or Sr-Fe + binder
Special Notes: (1) Rubber binder. Specifics not known.	

Note: All information based on available producers literature. In some cases of foreign materials interpretation or calculation required.

SOURCE(S):	G1-1a
	(Alloy or Grade)
Magnetfabrik Schramberg GMBH & Co.	Sintered Rare Earth
	(Type of Material)
	19/25
Germany	(MMPA Brief Designation)
	R5-1
	(IEC Code Reference)

MAGNETIC PROPERTIES		PHYSICAL PROPERTIES	
Residual Induction B_r (G)	8800	Density (lbs./in.^{3})	.2999
Coercive Force (Oe) — Normal (H_c)	8670	General Mechanical Properties	HB
Coercive Force (Oe) — Intrinsic (H_{ci})	25,120	Tensile Strength (psi)	Est. 5000
Maximum Energy Product BH_{max} (MGO)	19.4	Transverse Modulus of Rupture (psi)	–
B/H Ratio (Load Line) at BH_{max}	1.10	Coeff. of Therm. Expansion inches x 10^{-6}/°C Orientation: Par. _ Rt.Angl. _	
Flux Density in Magnet B_m at BH_{max} (G)	4500	Resistivity (micro-ohms/cm/cm^2)	.5 @ 25 °C
Coercive Force in Magnet H_m at BH_{max} (Oe)	4100	Hardness Scale	560 Vickers
Energy Loss W_c (watt-sec/cycle/lb)	–	Basic Forming Method	DPM
Importance of Operating Magnet at BH_{max}	E	Basic Finishing Method	WGR
Average Recoil Permeability	1.04	Relative Corrosion Resistance	G
Field Strength Required to Fully Saturate Magnet (Oe)	32,000	Primary Chemical Composition	Sm-Co
Ability to Withstand Demagnetizing Fields	0	Special Notes:	
Maximum Service Temperature (°F)	482		
Approximate Temperature Permanently Affecting Magnet (°F)	1328		
Inherent Orientation of Material	MP		

Note: All information based on available producers literature. In some cases of foreign materials interpretation or calculation required.

SOURCE(S):	
Magnetfabrik Schramberg GMBH & Co. Germany	**G1-1i** (Alloy or Grade)
	Sintered Rare Earth (Type of Material)
	21/25 (MMPA Brief Designation)
	R5-1 (IEC Code Reference)

MAGNETIC PROPERTIES		PHYSICAL PROPERTIES	
Residual Induction B_r (G)	9400	Density (lbs./in.^{3})	.2999
Coercive Force (Oe) — Normal (H_c)	8920	General Mechanical Properties	HB
Coercive Force (Oe) — Intrinsic (H_{ci})	25,120	Tensile Strength (psi)	Est. 5000
Maximum Energy Product BH_{max} (MGO)	21.2	Transverse Modulus of Rupture (psi)	—
B/H Ratio (Load Line) at BH_{max}	1.16	Coeff. of Therm. Expansion inches x 10^{-6}/°C — Orientation: Par. _ Rt.Angl. _	
Flux Density in Magnet B_m at BH_{max} (G)	5000	Resistivity (micro-ohms/cm/cm^2 .5 @ 25 °C	
Coercive Force in Magnet H_m at BH_{max} (Oe)	4300	Hardness Scale 560 Vickers	
Energy Loss W_c (watt-sec/cycle/lb)	—	Basic Forming Method	DPM
Importance of Operating Magnet at BH_{max}	E	Basic Finishing Method	WGR
Average Recoil Permeability	1.04	Relative Corrosion Resistance	G
Field Strength Required to Fully Saturate Magnet (Oe)	32,000	Primary Chemical Composition	Sm-Co
Ability to Withstand Demagnetizing Fields	0	Special Notes:	
Maximum Service Temperature (°F)	482		
Approximate Temperature Permanently Affecting Magnet (°F)	1328		
Inherent Orientation of Material	MP		

Note: All information based on available producers literature. In some cases of foreign materials interpretation or calculation required.

SOURCE(S):	G2-1a
	(Alloy or Grade)
Magnetfabrik Schramberg	Sintered Rare Earth
GMBH & Co.	(Type of Material)
	24/12
Germany	(MMPA Brief Designation)
	R5-2
	(IEC Code Reference)

MAGNETIC PROPERTIES		PHYSICAL PROPERTIES	
Residual Induction B_r (G)	10,000	Density (lbs./in.^{3})	.3035
Coercive Force (Oe) Normal (H_c)	9000	General Mechanical Properties	HB
Coercive Force (Oe) Intrinsic (H_{ci})	12,000	Tensile Strength (psi)	Est. 5000
Maximum Energy Product BH_{max} (MGO)	24.0	Transverse Modulus of Rupture (psi)	-
B/H Ratio (Load Line) at BH_{max}	1.14	Coeff. of Therm. Expansion inches x 10^{-6}/°C	Orientation: Par. - Rt.Angl. -
Flux Density in Magnet B_m at BH_{max} (G)	5200	Resistivity (micro-ohms/cm/cm^2)	- @ - °C
Coercive Force in Magnet H_m at BH_{max} (Oe)	4550	Hardness 600	Scale Vickers
Energy Loss W_c (watt-sec/cycle/lb)	-	Basic Forming Method	DPM
Importance of Operating Magnet at BH_{max}	E	Basic Finishing Method	WGR
Average Recoil Permeability	1.07	Relative Corrosion Resistance	G
Field Strength Required to Fully Saturate Magnet (Oe)	30,000	Primary Chemical Composition	Sm-Co
Ability to Withstand Demagnetizing Fields	0	Special Notes:	
Maximum Service Temperature (°F)	662		
Approximate Temperature Permanently Affecting Magnet (°F)	1577		
Inherent Orientation of Material	MP		

Note: All information based on available producers literature. In some cases of foreign materials interpretation or calculation required.

SOURCE(S):

Magnetfabrik Schramberg GMBH & Co.

Germany

G2-1i
(Alloy or Grade)

Sintered Rare Earth
(Type of Material)

27/15
(MMPA Brief Designation)

-
(IEC Code Reference)

MAGNETIC PROPERTIES	
Residual Induction B_r (G)	10,500
Coercive Force (Oe) — Normal (H_c)	9500
Coercive Force (Oe) — Intrinsic (H_{ci})	15,000
Maximum Energy Product BH_{max} (MGO)	27.0
B/H Ratio (Load Line) at BH_{max}	1.13
Flux Density in Magnet B_m at BH_{max} (G)	5700
Coercive Force in Magnet H_m at BH_{max} (Oe)	4650
Energy Loss W_c (watt-sec/cycle/lb)	-
Importance of Operating Magnet at BH_{max}	E
Average Recoil Permeability	1.05
Field Strength Required to Fully Saturate Magnet (Oe)	30,000
Ability to Withstand Demagnetizing Fields	O
Maximum Service Temperature (°F)	662
Approximate Temperature Permanently Affecting Magnet (°F)	1517
Inherent Orientation of Material	MP

PHYSICAL PROPERTIES	
Density (lbs./in.^{3})	.3035
General Mechanical Properties	HB
Tensile Strength (psi)	Est. 5000
Transverse Modulus of Rupture (psi)	-
Coeff. of Therm. Expansion inches x 10^{-6}/°C	Orientation: Par. - Rt.Angl. -
Resistivity (micro-ohms/cm/cm^2)	- @ - °C
Hardness — Scale	600 Vickers
Basic Forming Method	DPM
Basic Finishing Method	WGR
Relative Corrosion Resistance	G
Primary Chemical Composition	Sm-Co
Special Notes:	

Note: All information based on available producers literature. In some cases of foreign materials interpretation or calculation required.

SOURCE(S):	G3-1a
	(Alloy or Grade)
Magnetfabrik Schramberg GMBH & Co.	Sintered Rare Earth
	(Type of Material)
	31/13
Germany	(MMPA Brief Designation)
	R7-3
	(IEC Code Reference)

MAGNETIC PROPERTIES			PHYSICAL PROPERTIES	
Residual Induction B_r (G)		11,500	Density (lbs./in.^{3})	.2673
Coercive Force (Oe)	Normal (H_c)	10,500	General Mechanical Properties	HB
	Intrinsic (H_{ci})	13,000	Tensile Strength (psi)	—
Maximum Energy Product BH_{max} (MGO)		31.0	Transverse Modulus of Rupture (psi)	—
B/H Ratio (Load Line) at BH_{max}		1.15	Coeff. of Therm. Expansion inches x 10^{-6}/°C	Orientation: Par. — Rt.Angl. —
Flux Density in Magnet B_m at BH_{max} (G)		6000	Resistivity (micro-ohms/cm/cm^2)	1.6 @ 25 °C
Coercive Force in Magnet H_m at BH_{max} (Oe)		5200	Hardness Scale	— —
Energy Loss W_c (watt-sec/cycle/lb)		—	Basic Forming Method	DPM
Importance of Operating Magnet at BH_{max}		E	Basic Finishing Method	WGR
Average Recoil Permeability		1.08	Relative Corrosion Resistance	P
Field Strength Required to Fully Saturate Magnet (Oe)		30,000	Primary Chemical Composition	Nd-Fe-B
Ability to Withstand Demagnetizing Fields		0	Special Notes:	
Maximum Service Temperature (°F)		248		
Approximate Temperature Permanently Affecting Magnet (°F)		599		
Inherent Orientation of Material		MP		

Note: All information based on available producers literature. In some cases of foreign materials interpretation or calculation required.

SOURCE(S):

Magnetfabrik Schramberg GMBH & Co.

Germany

G3-1i
(Alloy or Grade)

Sintered Rare Earth
(Type of Material)

34/13
(MMPA Brief Designation)

R7-3
(IEC Code Reference)

MAGNETIC PROPERTIES	
Residual Induction B_r (G)	12,200
Coercive Force (Oe) Normal (H_c)	11,000
Coercive Force (Oe) Intrinsic (H_{ci})	13,000
Maximum Energy Product BH_{max} (MGO)	34.0
B/H Ratio (Load Line) at BH_{max}	1.17
Flux Density in Magnet B_m at BH_{max} (G)	63
Coercive Force in Magnet H_m at BH_{max} (Oe)	54
Energy Loss W_c (watt-sec/cycle/lb)	-
Importance of Operating Magnet at BH_{max}	E
Average Recoil Permeability	1.08
Field Strength Required to Fully Saturate Magnet (Oe)	30,000
Ability to Withstand Demagnetizing Fields	0
Maximum Service Temperature (°F)	248
Approximate Temperature Permanently Affecting Magnet (°F)	599
Inherent Orientation of Material	MP

PHYSICAL PROPERTIES	
Density (lbs./in.^{3})	.2673
General Mechanical Properties	HB
Tensile Strength (psi)	-
Transverse Modulus of Rupture (psi)	-
Coeff. of Therm. Expansion inches x 10^{-6}/°C	Orientation: Par. - Rt.Angl. -
Resistivity (micro-ohms/cm/cm^2)	1.6 @ 25 °C
Hardness	Scale — —
Basic Forming Method	DPM
Basic Finishing Method	WGR
Relative Corrosion Resistance	P
Primary Chemical Composition	Nd-Fe-B
Special Notes:	

Note: All information based on available producers literature. In some cases of foreign materials interpretation or calculation required.

SOURCE(S):		
Magnetfabrik Schramberg GMBH & Co. Germany	**G3-2a** (Alloy or Grade)	
	Sintered Rare Earth (Type of Material)	
	26/17 (MMPA Brief Designation)	
	R7-3 (IEC Code Reference)	

MAGNETIC PROPERTIES		PHYSICAL PROPERTIES	
Residual Induction B_r (G)	10,800	Density (lbs./in.^{3})	.2710
Coercive Force (Oe) — Normal (H_c)	10,100	General Mechanical Properties	HB
Coercive Force (Oe) — Intrinsic (H_{ci})	17,000	Tensile Strength (psi)	-
Maximum Energy Product BH_{max} (MGO)	26.0	Transverse Modulus of Rupture (psi)	-
B/H Ratio (Load Line) at BH_{max}	1.14	Coeff. of Therm. Expansion inches x 10^{-6}/°C Orientation: Par. - Rt.Angl. -	
Flux Density in Magnet B_m at BH_{max} (G)	5500	Resistivity (micro-ohms/cm/cm^2)	- @ - °C
Coercive Force in Magnet H_m at BH_{max} (Oe)	4800	Hardness Scale	- -
Energy Loss W_c (watt-sec/cycle/lb)	-	Basic Forming Method	DPM
Importance of Operating Magnet at BH_{max}	E	Basic Finishing Method	WGR
Average Recoil Permeability	1.08	Relative Corrosion Resistance	P
Field Strength Required to Fully Saturate Magnet (Oe)	30,000	Primary Chemical Composition	Nd-Fe-B
Ability to Withstand Demagnetizing Fields	0	Special Notes:	
Maximum Service Temperature (°F)	302		
Approximate Temperature Permanently Affecting Magnet (°F)	608		
Inherent Orientation of Material	MP		

Note: All information based on available producers literature. In some cases of foreign materials interpretation or calculation required.

SOURCE(S):	G3-2i
	(Alloy or Grade)
Magnetfabrik Schramberg GMBH & Co.	Sintered Rare Earth
	(Type of Material)
	32/17
Germany	(MMPA Brief Designation)
	R7-3
	(IEC Code Reference)

MAGNETIC PROPERTIES		PHYSICAL PROPERTIES	
Residual Induction B_r (G)	11,600	Density (lbs./in.^{3})	.2710
Coercive Force (Oe) — Normal (H_c)	10,800	General Mechanical Properties	HB
Coercive Force (Oe) — Intrinsic (H_{ci})	17,000	Tensile Strength (psi)	-
Maximum Energy Product BH_{max} (MGO)	32.0	Transverse Modulus of Rupture (psi)	-
B/H Ratio (Load Line) at BH_{max}	1.13	Coeff. of Therm. Expansion inches x 10^{-6}/°C	Orientation: Par. - Rt.Angl. -
Flux Density in Magnet B_m at BH_{max} (G)	6000	Resistivity (micro-ohms/cm/cm^2)	- @ - °C
Coercive Force in Magnet H_m at BH_{max} (Oe)	5300	Hardness — Scale	- — -
Energy Loss W_c (watt-sec/cycle/lb)	-	Basic Forming Method	DPM
Importance of Operating Magnet at BH_{max}	E	Basic Finishing Method	WGR
Average Recoil Permeability	1.08	Relative Corrosion Resistance	P
Field Strength Required to Fully Saturate Magnet (Oe)	30,000	Primary Chemical Composition	Nd-Fe-B
Ability to Withstand Demagnetizing Fields	0	Special Notes:	
Maximum Service Temperature (°F)	302		
Approximate Temperature Permanently Affecting Magnet (°F)	608		
Inherent Orientation of Material	MP		

Note: All information based on available producers literature. In some cases of foreign materials interpretation or calculation required.

SOURCE(S):

Magnetfabrik Schramberg GMBH & Co.

Germany

G3-3a	(Alloy or Grade)
Sintered Rare Earth	(Type of Material)
25/21	(MMPA Brief Designation)
R7-3	(IEC Code Reference)

MAGNETIC PROPERTIES		PHYSICAL PROPERTIES	
Residual Induction B_r (G)	10,500	Density (lbs./in.^{3})	.2728
Coercive Force (Oe) — Normal (H_c)	10,000	General Mechanical Properties	HB
Coercive Force (Oe) — Intrinsic (H_{ci})	21,000	Tensile Strength (psi)	-
Maximum Energy Product BH_{max} (MGO)	25.0	Transverse Modulus of Rupture (psi)	-
B/H Ratio (Load Line) at BH_{max}	1.14	Coeff. of Therm. Expansion inches x 10^{-6}/°C — Orientation: Par. — Rt.Angl. —	
Flux Density in Magnet B_m at BH_{max} (G)	5300	Resistivity (micro-ohms/cm/cm^2)	1.6 @ 25 °C
Coercive Force in Magnet H_m at BH_{max} (Oe)	4650	Hardness — Scale	-
Energy Loss W_c (watt-sec/cycle/lb)	-	Basic Forming Method	DPM
Importance of Operating Magnet at BH_{max}	E	Basic Finishing Method	WGR
Average Recoil Permeability	1.08	Relative Corrosion Resistance	P
Field Strength Required to Fully Saturate Magnet (Oe)	40,000	Primary Chemical Composition	Nd-Fe-B
Ability to Withstand Demagnetizing Fields	0	Special Notes:	
Maximum Service Temperature (°F)	302		
Approximate Temperature Permanently Affecting Magnet (°F)	608		
Inherent Orientation of Material	MP		

Note: All information based on available producers literature. In some cases of foreign materials interpretation or calculation required.

SOURCE(S):

Magnetfabrik Schramberg GMBH & Co.

Germany

G3-3i
(Alloy or Grade)

Sintered Rare Earth
(Type of Material)

28/21
(MMPA Brief Designation)

R7-3
(IEC Code Reference)

MAGNETIC PROPERTIES		PHYSICAL PROPERTIES	
Residual Induction B_r (G)	11,200	Density (lbs./in.^{3})	.2728
Coercive Force (Oe) Normal (H_c)	10,500	General Mechanical Properties	HB
Coercive Force (Oe) Intrinsic (H_{ci})	21,000	Tensile Strength (psi)	-
Maximum Energy Product BH_{max} (MGO)	28.0	Transverse Modulus of Rupture (psi)	-
B/H Ratio (Load Line) at BH_{max}	1.18	Coeff. of Therm. Expansion inches x 10^{-6}/°C	Orientation: Par. - Rt.Angl. -
Flux Density in Magnet B_m at BH_{max} (G)	5900	Resistivity (micro-ohms/cm/cm^2)	1.6 @ 25 °C
Coercive Force in Magnet H_m at BH_{max} (Oe)	4900	Hardness -	Scale -
Energy Loss W_c (watt-sec/cycle/lb)	-	Basic Forming Method	DPM
Importance of Operating Magnet at BH_{max}	E	Basic Finishing Method	WGR
Average Recoil Permeability	1.08	Relative Corrosion Resistance	P
Field Strength Required to Fully Saturate Magnet (Oe)	40,000	Primary Chemical Composition	Nd-Fe-B
Ability to Withstand Demagnetizing Fields	0	Special Notes:	
Maximum Service Temperature (°F)	302		
Approximate Temperature Permanently Affecting Magnet (°F)	608		
Inherent Orientation of Material	MP		

Note: All information based on available producers literature. In some cases of foreign materials interpretation or calculation required.

SOURCE(S):	G4-1
	(Alloy or Grade)
Magnetfabrik Schramberg GMBH & Co.	Bonded Rare Earth
	(Type of Material)
	9/14
Germany	(MMPA Brief Designation)
	-
	(IEC Code Reference)

MAGNETIC PROPERTIES		PHYSICAL PROPERTIES	
Residual Induction B_r (G)	6000	Density (lbs./in.^{3})	.2216
Coercive Force (Oe) — Normal (H_c)	5000	General Mechanical Properties	HT
Coercive Force (Oe) — Intrinsic (H_{ci})	13,800	Tensile Strength (psi)	-
Maximum Energy Product BH_{max} (MGO)	9.4	Transverse Modulus of Rupture (psi)	-
B/H Ratio (Load Line) at BH_{max}	-	Coeff. of Therm. Expansion inches x 10^{-6}/°C Orientation: Par. - Rt.Angl. -	
Flux Density in Magnet B_m at BH_{max} (G)	-	Resistivity (micro-ohms/cm/cm^2)	3.8 @ 25 °C
Coercive Force in Magnet H_m at BH_{max} (Oe)	-	Hardness Scale	- -
Energy Loss W_c (watt-sec/cycle/lb)	-	Basic Forming Method	E
Importance of Operating Magnet at BH_{max}	E	Basic Finishing Method	WGR
Average Recoil Permeability	1.15	Relative Corrosion Resistance	G
Field Strength Required to Fully Saturate Magnet (Oe)	40,000	Primary Chemical Composition	Nd-Fe-B + binder (See Note)
Ability to Withstand Demagnetizing Fields	0	Special Notes: (1) Epoxy resin binder	
Maximum Service Temperature (°F)	257		
Approximate Temperature Permanently Affecting Magnet (°F)	594		
Inherent Orientation of Material	none		

Note: All information based on available producers literature. In some cases of foreign materials interpretation or calculation required.

SOURCE(S):	**G4-1a**
	(Alloy or Grade)
Magnetfabrik Schramberg GMBH & Co.	Bonded Rare Earth
	(Type of Material)
	9/14
Germany	(MMPA Brief Designation)
	-
	(IEC Code Reference)

MAGNETIC PROPERTIES		PHYSICAL PROPERTIES	
Residual Induction B_r (G)	6000	Density (lbs./in.^{3})	.2216
Coercive Force (Oe) — Normal (H_c)	5000	General Mechanical Properties	HT
Coercive Force (Oe) — Intrinsic (H_{ci})	14,000	Tensile Strength (psi)	-
Maximum Energy Product BH_{max} (MGO)	9.4	Transverse Modulus of Rupture (psi)	-
B/H Ratio (Load Line) at BH_{max}	1.14	Coeff. of Therm. Expansion inches x 10^{-6}/°C	Orientation: Par. - Rt.Angl. -
Flux Density in Magnet B_m at BH_{max} (G)	3275	Resistivity (micro-ohms/cm/cm^2)	3.8 @ 25 °C
Coercive Force in Magnet H_m at BH_{max} (Oe)	2875	Hardness — Scale	- — -
Energy Loss W_c (watt-sec/cycle/lb)	-	Basic Forming Method	E
Importance of Operating Magnet at BH_{max}	E	Basic Finishing Method	WGR
Average Recoil Permeability	1.15	Relative Corrosion Resistance	G
Field Strength Required to Fully Saturate Magnet (Oe)	40,000	Primary Chemical Composition	Nd-Fe-B + binder (See Note)
Ability to Withstand Demagnetizing Fields	0	Special Notes: (1) Epoxy resin binder	
Maximum Service Temperature (°F)	257		
Approximate Temperature Permanently Affecting Magnet (°F)	594		
Inherent Orientation of Material	none		

Note: All information based on available producers literature. In some cases of foreign materials interpretation or calculation required.

SOURCE(S):	G4-2
	(Alloy or Grade)
Magnetfabrik Schramberg GMBH & Co.	Bonded Rare Earth
	(Type of Material)
	10/10
Germany	(MMPA Brief Designation)
	-
	(IEC Code Reference)

MAGNETIC PROPERTIES		PHYSICAL PROPERTIES	
Residual Induction B_r (G)	6400	Density (lbs./in.^{3})	.2216
Coercive Force (Oe) — Normal (H_c)	5300	General Mechanical Properties	HT
Coercive Force (Oe) — Intrinsic (H_{ci})	10,000	Tensile Strength (psi)	-
Maximum Energy Product BH_{max} (MGO)	10.0	Transverse Modulus of Rupture (psi)	-
B/H Ratio (Load Line) at BH_{max}	-	Coeff. of Therm. Expansion inches x 10^{-6}/°C — Orientation: Par. / Rt.Angl.	- / -
Flux Density in Magnet B_m at BH_{max} (G)	-	Resistivity (micro-ohms/cm/cm^2)	3.8 @ 25 °C
Coercive Force in Magnet H_m at BH_{max} (Oe)	-	Hardness / Scale	- / -
Energy Loss W_c (watt-sec/cycle/lb)	-	Basic Forming Method	E
Importance of Operating Magnet at BH_{max}	E	Basic Finishing Method	WGR
Average Recoil Permeability	1.15	Relative Corrosion Resistance	G
Field Strength Required to Fully Saturate Magnet (Oe)	40,000	Primary Chemical Composition (See Note)	Nd-Fe-B + binder
Ability to Withstand Demagnetizing Fields	0	Special Notes: (1) Epoxy resin binder	
Maximum Service Temperature (°F)	257		
Approximate Temperature Permanently Affecting Magnet (°F)	594		
Inherent Orientation of Material	none		

Note: All information based on available producers literature. In some cases of foreign materials interpretation or calculation required.

SOURCE(S):

Magnetfabrik Schramberg GMBH & Co.

Germany

G4-2a
(Alloy or Grade)

Bonded Rare Earth
(Type of Material)

9/9
(MMPA Brief Designation)

-
(IEC Code Reference)

MAGNETIC PROPERTIES		PHYSICAL PROPERTIES	
Residual Induction B_r (G)	6300	Density (lbs./in.^{3})	.2216
Coercive Force (Oe) Normal (H_c)	4800	General Mechanical Properties	HT
Coercive Force (Oe) Intrinsic (H_{ci})	9200	Tensile Strength (psi)	-
Maximum Energy Product BH_{max} (MGO)	9.4	Transverse Modulus of Rupture (psi)	-
B/H Ratio (Load Line) at BH_{max}	1.14	Coeff. of Therm. Expansion inches x 10^{-6}/°C Orientation: Par. Rt.Angl.	- -
Flux Density in Magnet B_m at BH_{max} (G)	3275	Resistivity (micro-ohms/cm/cm^2)	3.8 @ 25 °C
Coercive Force in Magnet H_m at BH_{max} (Oe)	2875	Hardness Scale	- -
Energy Loss W_c (watt-sec/cycle/lb)	-	Basic Forming Method	E
Importance of Operating Magnet at BH_{max}	E	Basic Finishing Method	WGR
Average Recoil Permeability	1.15	Relative Corrosion Resistance	G
Field Strength Required to Fully Saturate Magnet (Oe)	26,600	Primary Chemical Composition (See Note)	Nd-Fe-B + binder
Ability to Withstand Demagnetizing Fields	0	Special Notes: (1) Epoxy resin binder	
Maximum Service Temperature (°F)	212		
Approximate Temperature Permanently Affecting Magnet (°F)	594		
Inherent Orientation of Material	none		

Note: All information based on available producers literature. In some cases of foreign materials interpretation or calculation required.

SOURCE(S):

Magnetfabrik Schramberg GMBH & Co.

Germany

G4-3
(Alloy or Grade)

Bonded Rare Earth
(Type of Material)

9/14
(MMPA Brief Designation)

-
(IEC Code Reference)

MAGNETIC PROPERTIES		PHYSICAL PROPERTIES	
Residual Induction B_r (G)	6000	Density (lbs./in.^{3})	.2216
Coercive Force (Oe) Normal (H_c)	5000	General Mechanical Properties	HT
Coercive Force (Oe) Intrinsic (H_{ci})	13,800	Tensile Strength (psi)	-
Maximum Energy Product BH_{max} (MGO)	9.4	Transverse Modulus of Rupture (psi)	-
B/H Ratio (Load Line) at BH_{max}	-	Coeff. of Therm. Expansion inches x 10^{-6}/°C Orientation: Par. - Rt.Angl. -	
Flux Density in Magnet B_m at BH_{max} (G)	-	Resistivity (micro-ohms/cm/cm^2)	3.8 @ 25 °C
Coercive Force in Magnet H_m at BH_{max} (Oe)	-	Hardness -	Scale -
Energy Loss W_c (watt-sec/cycle/lb)	-	Basic Forming Method	E
Importance of Operating Magnet at BH_{max}	E	Basic Finishing Method	WGR
Average Recoil Permeability	1.15	Relative Corrosion Resistance	G
Field Strength Required to Fully Saturate Magnet (Oe)	40,000	Primary Chemical Composition (See Note)	Nd-Fe-B + binder
Ability to Withstand Demagnetizing Fields	0	Special Notes: (1) Epoxy resin binder	
Maximum Service Temperature (°F)	257		
Approximate Temperature Permanently Affecting Magnet (°F)	594		
Inherent Orientation of Material	none		

Note: All information based on available producers literature. In some cases of foreign materials interpretation or calculation required.

SOURCE(S):

Magnetfabrik Schramberg GMBH & Co.

Germany

G4-4
(Alloy or Grade)

Bonded Rare Earth
(Type of Material)

10/14
(MMPA Brief Designation)

-
(IEC Code Reference)

MAGNETIC PROPERTIES		PHYSICAL PROPERTIES	
Residual Induction B_r (G)	6400	Density (lbs./in.^{3})	.2216
Coercive Force (Oe) Normal (H_c)	5000	General Mechanical Properties	HT
Coercive Force (Oe) Intrinsic (H_{ci})	13,800	Tensile Strength (psi)	-
Maximum Energy Product BH_{max} (MGO)	10.0	Transverse Modulus of Rupture (psi)	-
B/H Ratio (Load Line) at BH_{max}	-	Coeff. of Therm. Expansion inches x 10^{-6}/°C	Orientation: Par. - Rt.Angl. -
Flux Density in Magnet B_m at BH_{max} (G)	-	Resistivity (micro-ohms/cm/cm^2)	3.8 @ 25 °C
Coercive Force in Magnet H_m at BH_{max} (Oe)	-	Hardness Scale	- -
Energy Loss W_c (watt-sec/cycle/lb)	-	Basic Forming Method	E
Importance of Operating Magnet at BH_{max}	E	Basic Finishing Method	WGR
Average Recoil Permeability	1.15	Relative Corrosion Resistance	G
Field Strength Required to Fully Saturate Magnet (Oe)	40,000	Primary Chemical Composition	Nd-Fe-B + binder (See Note)
Ability to Withstand Demagnetizing Fields	0	Special Notes: (1) Epoxy resin binder	
Maximum Service Temperature (°F)	257		
Approximate Temperature Permanently Affecting Magnet (°F)	594		
Inherent Orientation of Material	none		

Note: All information based on available producers literature. In some cases of foreign materials interpretation or calculation required.

SOURCE(S):

General Magnetic Co.

USA

Genox 5
(Alloy or Grade)

Sintered Ferrite
(Type of Material)

3.4/2.4
(MMPA Brief Designation)

S1-1-6
(IEC Code Reference)

MAGNETIC PROPERTIES	
Residual Induction B_r (G)	3900
Coercive Force (Oe) Normal (H_c)	2380
Coercive Force (Oe) Intrinsic (H_{ci})	2440
Maximum Energy Product BH_{max} (MGO)	3.4
B/H Ratio (Load Line) at BH_{max}	1.05
Flux Density in Magnet B_m at BH_{max} (G)	1900
Coercive Force in Magnet H_m at BH_{max} (Oe)	1800
Energy Loss W_c (watt-sec/cycle/lb)	-
Importance of Operating Magnet at BH_{max}	E
Average Recoil Permeability	Est. 1.1
Field Strength Required to Fully Saturate Magnet (Oe)	10,000
Ability to Withstand Demagnetizing Fields	Ex
Maximum Service Temperature (°F)	482
Approximate Temperature Permanently Affecting Magnet (°F)	842
Inherent Orientation of Material	MP

PHYSICAL PROPERTIES	
Density (lbs./in.^{3})	.1626
General Mechanical Properties	HB
Tensile Strength (psi)	-
Transverse Modulus of Rupture (psi)	-
Coeff. of Therm. Expansion inches x 10^{-6}/°C	Orientation: Par. - Rt.Angl. -
Resistivity (micro-ohms/cm/cm^2)	1.0 @ 25 °C
Hardness 6.5	Scale Moh's
Basic Forming Method	PMF
Basic Finishing Method	WGD
Relative Corrosion Resistance	Ex
Primary Chemical Composition	Ba or Sr-Fe
Special Notes:	

Note: All information based on available producers literature. In some cases of foreign materials interpretation or calculation required.

SOURCE(S):	**Genox 8**
	(Alloy or Grade)
General Magnetic Co.	Sintered Ferrite
	(Type of Material)
USA	3.4/3.1
	(MMPA Brief Designation)
	S1-1-5
	(IEC Code Reference)

MAGNETIC PROPERTIES		PHYSICAL PROPERTIES	
Residual Induction B_r (G)	3850	Density (lbs./in.^{3})	.1626
Coercive Force (Oe) Normal (H_c)	2940	General Mechanical Properties	HB
Coercive Force (Oe) Intrinsic (H_{ci})	3080	Tensile Strength (psi)	-
Maximum Energy Product BH_{max} (MGO)	3.4	Transverse Modulus of Rupture (psi)	-
B/H Ratio (Load Line) at BH_{max}	.85	Coeff. of Therm. Expansion inches x 10^{-6}/°C	Orientation: Par. - Rt.Angl. -
Flux Density in Magnet B_m at BH_{max} (G)	1700	Resistivity (micro-ohms/cm/cm^2)	1.0 @ 25 °C
Coercive Force in Magnet H_m at BH_{max} (Oe)	2000	Hardness Scale	6.5 Moh's
Energy Loss W_c (watt-sec/cycle/lb)	-	Basic Forming Method	PMF
Importance of Operating Magnet at BH_{max}	E	Basic Finishing Method	WGD
Average Recoil Permeability	Est. 1.1	Relative Corrosion Resistance	Ex
Field Strength Required to Fully Saturate Magnet (Oe)	10,000	Primary Chemical Composition	Ba or Sr-Fe
Ability to Withstand Demagnetizing Fields	Ex	Special Notes:	
Maximum Service Temperature (°F)	482		
Approximate Temperature Permanently Affecting Magnet (°F)	842		
Inherent Orientation of Material	MP		

Note: All information based on available producers literature. In some cases of foreign materials interpretation or calculation required.

SOURCE(S):

General Magnetic Co.

USA

Genox 8H
(Alloy or Grade)

Sintered Ferrite
(Type of Material)

3.9/2.9
(MMPA Brief Designation)

-
(IEC Code Reference)

MAGNETIC PROPERTIES		PHYSICAL PROPERTIES	
Residual Induction B_r (G)	4100	Density (lbs./in.^{3})	.1626
Coercive Force (Oe) Normal (H_c)	2850	General Mechanical Properties	VHB
Coercive Force (Oe) Intrinsic (H_{ci})	2900	Tensile Strength (psi)	-
Maximum Energy Product BH_{max} (MGO)	3.9	Transverse Modulus of Rupture (psi)	-
B/H Ratio (Load Line) at BH_{max}	1.03	Coeff. of Therm. Expansion inches x 10^{-6}/°C Orientation: Par. Rt.Angl.	- -
Flux Density in Magnet B_m at BH_{max} (G)	2000	Resistivity (micro-ohms/cm/cm^2)	1.0 @ 25 °C
Coercive Force in Magnet H_m at BH_{max} (Oe)	1950	Hardness Scale	6.5 Moh's
Energy Loss W_c (watt-sec/cycle/lb)	-	Basic Forming Method	PMF
Importance of Operating Magnet at BH_{max}	E	Basic Finishing Method	WGD
Average Recoil Permeability	Est. 1.1	Relative Corrosion Resistance	Ex
Field Strength Required to Fully Saturate Magnet (Oe)	10,000	Primary Chemical Composition	Ba or Sr-Fe
Ability to Withstand Demagnetizing Fields	Ex	Special Notes:	
Maximum Service Temperature (°F)	482		
Approximate Temperature Permanently Affecting Magnet (°F)	842		
Inherent Orientation of Material	MP		

Note: All information based on available producers literature. In some cases of foreign materials interpretation or calculation required.

SOURCE(S):

Hitachi Magnetics Corp.

USA

H-13S
(Alloy or Grade)

Sintered Rare Earth
(Type of Material)

12/20
(MMPA Brief Designation)

R5-1
(IEC Code Reference)

MAGNETIC PROPERTIES	
Residual Induction B_r (G)	7000
Coercive Force (Oe) — Normal (H_c)	6000
Coercive Force (Oe) — Intrinsic (H_{ci})	20,000
Maximum Energy Product BH_{max} (MGO)	12.0
B/H Ratio (Load Line) at BH_{max}	.95
Flux Density in Magnet B_m at BH_{max} (G)	4000
Coercive Force in Magnet H_m at BH_{max} (Oe)	4200
Energy Loss W_c (watt-sec/cycle/lb)	-
Importance of Operating Magnet at BH_{max}	E
Average Recoil Permeability	1.05
Field Strength Required to Fully Saturate Magnet (Oe)	32,000
Ability to Withstand Demagnetizing Fields	0
Maximum Service Temperature (°F)	482
Approximate Temperature Permanently Affecting Magnet (°F)	1328
Inherent Orientation of Material	MP

PHYSICAL PROPERTIES	
Density (lbs./in.^{3})	.2999
General Mechanical Properties	HB
Tensile Strength (psi)	5000
Transverse Modulus of Rupture (psi)	-
Coeff. of Therm. Expansion inches x 10^{-6}/°C	Orientation: Par. 7.3, Rt.Angl. 11.7
Resistivity (micro-ohms/cm/cm^2)	50 @ 25 °C
Hardness	600 Scale Vickers
Basic Forming Method	DPM
Basic Finishing Method	WGR
Relative Corrosion Resistance	G
Primary Chemical Composition	Sm-Co
Special Notes:	

Note: All information based on available producers literature. In some cases of foreign materials interpretation or calculation required.

SOURCE(S):

Hitachi Magnetics Corp.

USA

H-18B
(Alloy or Grade)

Sintered Rare Earth
(Type of Material)

18/15
(MMPA Brief Designation)

R5-1
(IEC Code Reference)

MAGNETIC PROPERTIES		PHYSICAL PROPERTIES	
Residual Induction B_r (G)	8500	Density (lbs./in.^{3})	.2999
Coercive Force (Oe) — Normal (H_c)	8000	General Mechanical Properties	HB
Coercive Force (Oe) — Intrinsic (H_{ci})	15,000	Tensile Strength (psi)	5000
Maximum Energy Product BH_{max} (MGO)	17.5	Transverse Modulus of Rupture (psi)	-
B/H Ratio (Load Line) at BH_{max}	.95	Coeff. of Therm. Expansion inches x 10^{-6}/°C	Orientation: Par. 6.6 Rt.Angl. 12.6
Flux Density in Magnet B_m at BH_{max} (G)	4000	Resistivity (micro-ohms/cm/cm^2)	50 @ 25 °C
Coercive Force in Magnet H_m at BH_{max} (Oe)	4200	Hardness Scale	600 Vickers
Energy Loss W_c (watt-sec/cycle/lb)	-	Basic Forming Method	DPM
Importance of Operating Magnet at BH_{max}	E	Basic Finishing Method	WGR
Average Recoil Permeability	1.05	Relative Corrosion Resistance	G
Field Strength Required to Fully Saturate Magnet (Oe)	32,000	Primary Chemical Composition	Sm-Co
Ability to Withstand Demagnetizing Fields	0	Special Notes:	
Maximum Service Temperature (°F)	572		
Approximate Temperature Permanently Affecting Magnet (°F)	1472		
Inherent Orientation of Material	MP		

Note: All information based on available producers literature. In some cases of foreign materials interpretation or calculation required.

SOURCE(S):	H-20SV
	(Alloy or Grade)
Hitachi Magnetics Corp.	Sintered Rare Earth
	(Type of Material)
USA	20/14
	(MMPA Brief Designation)
	R5-1
	(IEC Code Reference)

MAGNETIC PROPERTIES		PHYSICAL PROPERTIES	
Residual Induction B_r (G)	9300	Density (lbs./in.^{3})	.3071
Coercive Force (Oe) — Normal (H_c)	8750	General Mechanical Properties	HB
Coercive Force (Oe) — Intrinsic (H_{ci})	14,000	Tensile Strength (psi)	5000
Maximum Energy Product BH_{max} (MGO)	20.0	Transverse Modulus of Rupture (psi)	–
B/H Ratio (Load Line) at BH_{max}	1.1	Coeff. of Therm. Expansion inches x 10^{-6}/°C — Orientation: Par. – Rt.Angl. –	
Flux Density in Magnet B_m at BH_{max} (G)	4600	Resistivity (micro-ohms/cm/cm^2)	50 @ 25 °C
Coercive Force in Magnet H_m at BH_{max} (Oe)	4300	Hardness Scale	670 Vickers
Energy Loss W_c (watt-sec/cycle/lb)	–	Basic Forming Method	DPM
Importance of Operating Magnet at BH_{max}	E	Basic Finishing Method	WGR
Average Recoil Permeability	1.05	Relative Corrosion Resistance	G
Field Strength Required to Fully Saturate Magnet (Oe)	32,000	Primary Chemical Composition	Sm-Co
Ability to Withstand Demagnetizing Fields	O	Special Notes:	
Maximum Service Temperature (°F)	472		
Approximate Temperature Permanently Affecting Magnet (°F)	1472		
Inherent Orientation of Material	MP		

Note: All information based on available producers literature. In some cases of foreign materials interpretation or calculation required.

SOURCE(S): **Hitachi Magnetics Corp.**

USA

H-21EV
(Alloy or Grade)

Sintered Rare Earth
(Type of Material)

20/7
(MMPA Brief Designation)

R5-1
(IEC Code Reference)

MAGNETIC PROPERTIES		PHYSICAL PROPERTIES	
Residual Induction B_r (G)	9200	Density (lbs./in.^{3})	.3071
Coercive Force (Oe) — Normal (H_c)	6000	General Mechanical Properties	HB
Coercive Force (Oe) — Intrinsic (H_{ci})	7000	Tensile Strength (psi)	5000
Maximum Energy Product BH_{max} (MGO)	20.0	Transverse Modulus of Rupture (psi)	-
B/H Ratio (Load Line) at BH_{max}	1.4	Coeff. of Therm. Expansion inches x 10^{-6}/°C	Orientation: Par. 8.0 Rt.Angl. -
Flux Density in Magnet B_m at BH_{max} (G)	500	Resistivity (micro-ohms/cm/cm^2)	80 @ 25 °C
Coercive Force in Magnet H_m at BH_{max} (Oe)	3500	Hardness 670	Scale Vickers
Energy Loss W_c (watt-sec/cycle/lb)	-	Basic Forming Method	DPM
Importance of Operating Magnet at BH_{max}	E	Basic Finishing Method	WGR
Average Recoil Permeability	1.05	Relative Corrosion Resistance	G
Field Strength Required to Fully Saturate Magnet (Oe)	32,000	Primary Chemical Composition	Sm-Co
Ability to Withstand Demagnetizing Fields	0	Special Notes:	
Maximum Service Temperature (°F)	482		
Approximate Temperature Permanently Affecting Magnet (°F)	1472		
Inherent Orientation of Material	MP		

Note: All information based on available producers literature. In some cases of foreign materials interpretation or calculation required.

SOURCE(S):

Hitachi Magnetics Corp.

USA

H-22A
(Alloy or Grade)

Sintered Rare Earth
(Type of Material)

20/15
(MMPA Brief Designation)

R5-1
(IEC Code Reference)

MAGNETIC PROPERTIES		PHYSICAL PROPERTIES	
Residual Induction B_r (G)	9000	Density (lbs./in.^{3})	.2999
Coercive Force (Oe) Normal (H_c)	8750	General Mechanical Properties	HB
Coercive Force (Oe) Intrinsic (H_{ci})	15,000	Tensile Strength (psi)	5000
Maximum Energy Product BH_{max} (MGO)	20.0	Transverse Modulus of Rupture (psi)	-
B/H Ratio (Load Line) at BH_{max}	1.0	Coeff. of Therm. Expansion inches x 10^{-6}/°C	Orientation: Par. 6.6 Rt.Angl. 12.6
Flux Density in Magnet B_m at BH_{max} (G)	4250	Resistivity (micro-ohms/cm/cm^2)	50 @ 25 °C
Coercive Force in Magnet H_m at BH_{max} (Oe)	4250	Hardness Scale	600 Vickers
Energy Loss W_c (watt-sec/cycle/lb)	-	Basic Forming Method	DPM
Importance of Operating Magnet at BH_{max}	E	Basic Finishing Method	WGR
Average Recoil Permeability	1.05	Relative Corrosion Resistance	G
Field Strength Required to Fully Saturate Magnet (Oe)	32,000	Primary Chemical Composition	Sm-Co
Ability to Withstand Demagnetizing Fields	0	Special Notes:	
Maximum Service Temperature (°F)	482		
Approximate Temperature Permanently Affecting Magnet (°F)	1472		
Inherent Orientation of Material	MP		

Note: All information based on available producers literature. In some cases of foreign materials interpretation or calculation required.

SOURCE(S):

Hitachi Magnetics Corp.

USA

H-23B
(Alloy or Grade)

Sintered Rare Earth
(Type of Material)

21/5
(MMPA Brief Designation)

R5-2
(IEC Code Reference)

MAGNETIC PROPERTIES		PHYSICAL PROPERTIES	
Residual Induction B_r (G)	9900	Density (lbs./in.^{3})	.3071
Coercive Force (Oe) — Normal (H_c)	6750	General Mechanical Properties	HB
Coercive Force (Oe) — Intrinsic (H_{ci})	5300	Tensile Strength (psi)	5000
Maximum Energy Product BH_{max} (MGO)	21.5	Transverse Modulus of Rupture (psi)	-
B/H Ratio (Load Line) at BH_{max}	1.3	Coeff. of Therm. Expansion inches x 10^{-6}/°C	Orientation: Par. 6.6 Rt.Angl. _
Flux Density in Magnet B_m at BH_{max} (G)	5200	Resistivity (micro-ohms/cm/cm^2)	50 @ 25 °C
Coercive Force in Magnet H_m at BH_{max} (Oe)	4125	Hardness Scale	600 Vickers
Energy Loss W_c (watt-sec/cycle/lb)	-	Basic Forming Method	DPM
Importance of Operating Magnet at BH_{max}	E	Basic Finishing Method	WGR
Average Recoil Permeability	1.05	Relative Corrosion Resistance	G
Field Strength Required to Fully Saturate Magnet (Oe)	32,000	Primary Chemical Composition	Sm-Co
Ability to Withstand Demagnetizing Fields	0	Special Notes:	
Maximum Service Temperature (°F)	482		
Approximate Temperature Permanently Affecting Magnet (°F)	1472		
Inherent Orientation of Material	MP		

Note: All information based on available producers literature. In some cases of foreign materials interpretation or calculation required.

SOURCE(S):	H-23CV
Hitachi Magnetics Corp.	(Alloy or Grade)
	Sintered Rare Earth
	(Type of Material)
USA	24/18
	(MMPA Brief Designation)
	R5-2
	(IEC Code Reference)

MAGNETIC PROPERTIES		PHYSICAL PROPERTIES	
Residual Induction B_r (G)	10,000	Density (lbs./in.^{3})	.3071
Coercive Force (Oe) — Normal (H_c)	8750	General Mechanical Properties	HB
Coercive Force (Oe) — Intrinsic (H_{ci})	–	Tensile Strength (psi)	5000
Maximum Energy Product BH_{max} (MGO)	23.0	Transverse Modulus of Rupture (psi)	–
B/H Ratio (Load Line) at BH_{max}	1.1	Coeff. of Therm. Expansion inches x 10^{-6}/°C — Orientation: Par. – Rt.Angl. –	
Flux Density in Magnet B_m at BH_{max} (G)	5100	Resistivity (micro-ohms/cm/cm^2)	50 @ 25 °C
Coercive Force in Magnet H_m at BH_{max} (Oe)	4700	Hardness — Scale	600 Vickers
Energy Loss W_c (watt-sec/cycle/lb)	–	Basic Forming Method	DPM
Importance of Operating Magnet at BH_{max}	E	Basic Finishing Method	WGR
Average Recoil Permeability	1.05	Relative Corrosion Resistance	G
Field Strength Required to Fully Saturate Magnet (Oe)	32,000	Primary Chemical Composition	Sm-Co
Ability to Withstand Demagnetizing Fields	0	Special Notes:	
Maximum Service Temperature (°F)	572		
Approximate Temperature Permanently Affecting Magnet (°F)	1472		
Inherent Orientation of Material	MP		

Note: All information based on available producers literature. In some cases of foreign materials interpretation or calculation required.

SOURCE(S):	H-23EH
Hitachi Magnetics Corp.	(Alloy or Grade)
	Sintered Rare Earth
	(Type of Material)
USA	22/8
	(MMPA Brief Designation)
	R5-2
	(IEC Code Reference)

MAGNETIC PROPERTIES		PHYSICAL PROPERTIES	
Residual Induction B_r (G)	9800	Density (lbs./in.^{3})	.3071
Coercive Force (Oe) Normal (H_c)	7000	General Mechanical Properties	HB
Coercive Force (Oe) Intrinsic (H_{ci})	8000	Tensile Strength (psi)	5000
Maximum Energy Product BH_{max} (MGO)	22.0	Transverse Modulus of Rupture (psi)	-
B/H Ratio (Load Line) at BH_{max}	1.1	Coeff. of Therm. Expansion inches x 10^{-6}/°C	Orientation: Par. 6.6 Rt.Angl. _
Flux Density in Magnet B_m at BH_{max} (G)	5800	Resistivity (micro-ohms/cm/cm^2)	50 @ 25 °C
Coercive Force in Magnet H_m at BH_{max} (Oe)	5200	Hardness 670	Scale Vickers
Energy Loss W_c (watt-sec/cycle/lb)	-	Basic Forming Method	DPM
Importance of Operating Magnet at BH_{max}	E	Basic Finishing Method	WGR
Average Recoil Permeability	1.05	Relative Corrosion Resistance	G
Field Strength Required to Fully Saturate Magnet (Oe)	32,000	Primary Chemical Composition	Sm-Co
Ability to Withstand Demagnetizing Fields	0	Special Notes:	
Maximum Service Temperature (°F)	472		
Approximate Temperature Permanently Affecting Magnet (°F)	1472		
Inherent Orientation of Material	MP		

Note: All information based on available producers literature. In some cases of foreign materials interpretation or calculation required.

SOURCE(S):	H-23EV
	(Alloy or Grade)
Hitachi Magnetics Corp.	Sintered Rare Earth
	(Type of Material)
USA	22/8
	(MMPA Brief Designation)
	R5-2
	(IEC Code Reference)

MAGNETIC PROPERTIES		PHYSICAL PROPERTIES	
Residual Induction B_r (G)	9800	Density (lbs./in.^{3})	.2999
Coercive Force (Oe) — Normal (H_c)	6500	General Mechanical Properties	HB
Coercive Force (Oe) — Intrinsic (H_{ci})	8000	Tensile Strength (psi)	5000
Maximum Energy Product BH_{max} (MGO)	22.0	Transverse Modulus of Rupture (psi)	—
B/H Ratio (Load Line) at BH_{max}	1.4	Coeff. of Therm. Expansion inches x 10^{-6}/°C	Orientation: Par. — Rt.Angl. —
Flux Density in Magnet B_m at BH_{max} (G)	5000	Resistivity (micro-ohms/cm/cm^2	— @ — °C
Coercive Force in Magnet H_m at BH_{max} (Oe)	3500	Hardness Scale 670 Vickers	
Energy Loss W_c (watt-sec/cycle/lb)	—	Basic Forming Method	DPM
Importance of Operating Magnet at BH_{max}	E	Basic Finishing Method	WGR
Average Recoil Permeability	1.05	Relative Corrosion Resistance	G
Field Strength Required to Fully Saturate Magnet (Oe)	32,000	Primary Chemical Composition	Sm-Co
Ability to Withstand Demagnetizing Fields	0	Special Notes:	
Maximum Service Temperature (°F)	572		
Approximate Temperature Permanently Affecting Magnet (°F)	1472		
Inherent Orientation of Material	MP		

Note: All information based on available producers literature. In some cases of foreign materials interpretation or calculation required.

SOURCE(S):	H-25B
	(Alloy or Grade)
Hitachi Magnetics Corp.	Sintered Rare Earth
	(Type of Material)
USA	24/7
	(MMPA Brief Designation)
	R5-2
	(IEC Code Reference)

MAGNETIC PROPERTIES		PHYSICAL PROPERTIES	
Residual Induction B_r (G)	10,600	Density (lbs./in.^{3})	.3071
Coercive Force (Oe) — Normal (H_c)	6750	General Mechanical Properties	HB
Coercive Force (Oe) — Intrinsic (H_{ci})	5300	Tensile Strength (psi)	5000
Maximum Energy Product BH_{max} (MGO)	24.5	Transverse Modulus of Rupture (psi)	-
B/H Ratio (Load Line) at BH_{max}	1.4	Coeff. of Therm. Expansion inches x 10^{-6}/°C	Orientation: Par. - Rt.Angl. -
Flux Density in Magnet B_m at BH_{max} (G)	5600	Resistivity (micro-ohms/cm/cm^2)	50 @ 25 °C
Coercive Force in Magnet H_m at BH_{max} (Oe)	4000	Hardness 600	Scale Vickers
Energy Loss W_c (watt-sec/cycle/lb)	-	Basic Forming Method	DPM
Importance of Operating Magnet at BH_{max}	E	Basic Finishing Method	WGR
Average Recoil Permeability	1.05	Relative Corrosion Resistance	G
Field Strength Required to Fully Saturate Magnet (Oe)	32,000	Primary Chemical Composition	Sm-Co
Ability to Withstand Demagnetizing Fields	0	Special Notes:	
Maximum Service Temperature (°F)	572		
Approximate Temperature Permanently Affecting Magnet (°F)	1472		
Inherent Orientation of Material	MP		

Note: All information based on available producers literature. In some cases of foreign materials interpretation or calculation required.

SOURCE(S):	H-25EH
	(Alloy or Grade)
Hitachi Magnetics Corp.	Sintered Rare Earth
	(Type of Material)
USA	24/9
	(MMPA Brief Designation)
	R5-2
	(IEC Code Reference)

MAGNETIC PROPERTIES		PHYSICAL PROPERTIES	
Residual Induction B_r (G)	10,200	Density (lbs./in.^{3})	.2999
Coercive Force (Oe) — Normal (H_c)	7000	General Mechanical Properties	HB
Coercive Force (Oe) — Intrinsic (H_{ci})	8000	Tensile Strength (psi)	5000
Maximum Energy Product BH_{max} (MGO)	24.5	Transverse Modulus of Rupture (psi)	-
B/H Ratio (Load Line) at BH_{max}	1.05	Coeff. of Therm. Expansion inches x 10^{-6}/°C	Orientation: Par. - Rt.Angl. -
Flux Density in Magnet B_m at BH_{max} (G)	5000	Resistivity (micro-ohms/cm/cm^2)	- @ - °C
Coercive Force in Magnet H_m at BH_{max} (Oe)	4750	Hardness 670 Scale Vickers	
Energy Loss W_c (watt-sec/cycle/lb)	-	Basic Forming Method	DPM
Importance of Operating Magnet at BH_{max}	E	Basic Finishing Method	WGR
Average Recoil Permeability	1.05	Relative Corrosion Resistance	G
Field Strength Required to Fully Saturate Magnet (Oe)	32,000	Primary Chemical Composition	Sm-Co
Ability to Withstand Demagnetizing Fields	0	Special Notes:	
Maximum Service Temperature (°F)	572		
Approximate Temperature Permanently Affecting Magnet (°F)	1472		
Inherent Orientation of Material	MP		

Note: All information based on available producers literature. In some cases of foreign materials interpretation or calculation required.

SOURCE(S):	H-30CH
	(Alloy or Grade)
Hitachi Magnetics Corp.	Sintered Rare Earth
	(Type of Material)
USA	28/9
	(MMPA Brief Designation)
	R5-2
	(IEC Code Reference)

MAGNETIC PROPERTIES		PHYSICAL PROPERTIES	
Residual Induction B_r (G)	10,900	Density (lbs./in.^{3})	.3071
Coercive Force (Oe) — Normal (H_c)	9000	General Mechanical Properties	HB
Coercive Force (Oe) — Intrinsic (H_{ci})	8500	Tensile Strength (psi)	5000
Maximum Energy Product BH_{max} (MGO)	28.0	Transverse Modulus of Rupture (psi)	-
B/H Ratio (Load Line) at BH_{max}	1.1	Coeff. of Therm. Expansion inches x 10^{-6}/°C	Orientation: Par. - Rt.Angl. -
Flux Density in Magnet B_m at BH_{max} (G)	5500	Resistivity (micro-ohms/cm/cm^2)	50 @ 25 °C
Coercive Force in Magnet H_m at BH_{max} (Oe)	5000	Hardness Scale	600 Vickers
Energy Loss W_c (watt-sec/cycle/lb)	-	Basic Forming Method	DPM
Importance of Operating Magnet at BH_{max}	E	Basic Finishing Method	WGR
Average Recoil Permeability	1.05	Relative Corrosion Resistance	G
Field Strength Required to Fully Saturate Magnet (Oe)	32,000	Primary Chemical Composition	Sm-Co
Ability to Withstand Demagnetizing Fields	O	Special Notes:	
Maximum Service Temperature (°F)	572		
Approximate Temperature Permanently Affecting Magnet (°F)	1472		
Inherent Orientation of Material	MP		

Note: All information based on available producers literature. In some cases of foreign materials interpretation or calculation required.

SOURCE(S):

Hitachi Magnetics Corp.

USA

HB-061
(Alloy or Grade)

Bonded Rare Earth
(Type of Material)

-
(MMPA Brief Designation)

-
(IEC Code Reference)

MAGNETIC PROPERTIES		PHYSICAL PROPERTIES	
Residual Induction B_r (G)	5700	Density (lbs./in.^{3})	Approx. .2130
Coercive Force (Oe) Normal (H_c)	4600	General Mechanical Properties	HT
Coercive Force (Oe) Intrinsic (H_{ci})	9000	Tensile Strength (psi)	-
Maximum Energy Product BH_{max} (MGO)	6.4	Transverse Modulus of Rupture (psi)	-
B/H Ratio (Load Line) at BH_{max}	1.2	Coeff. of Therm. Expansion inches x 10^{-6}/°C	Orientation: Par. - Rt.Angl. -
Flux Density in Magnet B_m at BH_{max} (G)	3100	Resistivity (micro-ohms/cm/cm^2)	- @ - °C
Coercive Force in Magnet H_m at BH_{max} (Oe)	2600	Hardness Scale	- -
Energy Loss W_c (watt-sec/cycle/lb)	-	Basic Forming Method	E See Note (2)
Importance of Operating Magnet at BH_{max}	E	Basic Finishing Method	SCD
Average Recoil Permeability	-	Relative Corrosion Resistance	G
Field Strength Required to Fully Saturate Magnet (Oe)	22,600	Primary Chemical Composition	Nd-Fe-B + binder (See Note)
Ability to Withstand Demagnetizing Fields	0	Special Notes: (1) Used for small motors, speakers, pickups, sensors, relays and magnet rolls. (2) Binder not indicated. Injection molded. (3) Surface treatment available on request.	
Maximum Service Temperature (°F)	-		
Approximate Temperature Permanently Affecting Magnet (°F)	594		
Inherent Orientation of Material	none		

Note: All information based on available producers literature. In some cases of foreign materials interpretation or calculation required.

SOURCE(S):

Hitachi Magnetics Corp.

USA

HB-081
(Alloy or Grade)

Bonded Rare Earth
(Type of Material)

-
(MMPA Brief Designation)

-
(IEC Code Reference)

MAGNETIC PROPERTIES		PHYSICAL PROPERTIES	
Residual Induction B_r (G)	6400	Density (lbs./in.^{3})	Approx. .2130
Coercive Force (Oe) Normal (H_c)	5000	General Mechanical Properties	HT
Coercive Force (Oe) Intrinsic (H_{ci})	9000	Tensile Strength (psi)	-
Maximum Energy Product BH_{max} (MGO)	8.3	Transverse Modulus of Rupture (psi)	-
B/H Ratio (Load Line) at BH_{max}	1.5	Coeff. of Therm. Expansion inches x 10^{-6}/°C	Orientation: Par. - Rt.Angl. -
Flux Density in Magnet B_m at BH_{max} (G)	3000	Resistivity (micro-ohms/cm/cm^2	- @ - °C
Coercive Force in Magnet H_m at BH_{max} (Oe)	2000	Hardness Scale	- -
Energy Loss W_c (watt-sec/cycle/lb)	-	Basic Forming Method	E See Note (2)
Importance of Operating Magnet at BH_{max}	E	Basic Finishing Method	SCD/SRD
Average Recoil Permeability	-	Relative Corrosion Resistance	G
Field Strength Required to Fully Saturate Magnet (Oe)	22,600	Primary Chemical Composition	Nd-Fe-B + binder (See Note)
Ability to Withstand Demagnetizing Fields	0	Special Notes:	
Maximum Service Temperature (°F)	-	(1) Used for small motors, speakers, pickups, sensors, relays and magnet rolls.	
Approximate Temperature Permanently Affecting Magnet (°F)	594	(2) Binder not indicated. Compression molded.	
Inherent Orientation of Material	none	(3) Surface treatment available on request.	

Note: All information based on available producers literature. In some cases of foreign materials interpretation or calculation required.

SOURCE(S):

Hitachi Magnetics Corp.

USA

HIMAG
(Alloy or Grade)

Cast Alnico
(Type of Material)

7.5/.75
(MMPA Brief Designation)

R1-1-3
(IEC Code Reference)

MAGNETIC PROPERTIES	
Residual Induction B_r (G)	13,500
Coercive Force (Oe) Normal (H_c)	750
Coercive Force (Oe) Intrinsic (H_{ci})	Approx. 750
Maximum Energy Product BH_{max} (MGO)	7.5
B/H Ratio (Load Line) at BH_{max}	20.0
Flux Density in Magnet B_m at BH_{max} (G)	1190
Coercive Force in Magnet H_m at BH_{max} (Oe)	590
Energy Loss W_c (watt-sec/cycle/lb)	15.3
Importance of Operating Magnet at BH_{max}	A
Average Recoil Permeability	3.8
Field Strength Required to Fully Saturate Magnet (Oe)	3000
Ability to Withstand Demagnetizing Fields	G
Maximum Service Temperature (°F)	900
Approximate Temperature Permanently Affecting Magnet (°F)	1580
Inherent Orientation of Material	CHT

PHYSICAL PROPERTIES	
Density (lbs./in.^{3})	.2640
General Mechanical Properties	HB
Tensile Strength (psi)	-
Transverse Modulus of Rupture (psi)	-
Coeff. of Therm. Expansion inches x 10^{-6}/°C	Orientation: Par. 11.4 Rt.Angl. -
Resistivity (micro-ohms/cm/cm^2)	47 @ 25 °C
Hardness	53 Scale Rockwell
Basic Forming Method	CM
Basic Finishing Method	WGR
Relative Corrosion Resistance	G
Primary Chemical Composition	Al-Ni-Co

Special Notes:

(1) Lowest temperature co-efficient of all materials. Little irreversible loss even at high temperatures.
(2) Should be magnetized in final ciccuit.

Note: All information based on available producers literature. In some cases of foreign materials interpretation or calculation required.

SOURCE(S):

Hitachi Magnetics Corp.

USA

HS-20BR
(Alloy or Grade)

Sintered Rare Earth
(Type of Material)

21/15
(MMPA Brief Designation)

R7-3
(IEC Code Reference)

MAGNETIC PROPERTIES		PHYSICAL PROPERTIES	
Residual Induction B_r (G)	9600	Density (lbs./in.^{3})	.2710
Coercive Force (Oe) Normal (H_c)	9000	General Mechanical Properties	HB
Coercive Force (Oe) Intrinsic (H_{ci})	15,000	Tensile Strength (psi)	-
Maximum Energy Product BH_{max} (MGO)	21.0	Transverse Modulus of Rupture (psi)	-
B/H Ratio (Load Line) at BH_{max}	1.1	Coeff. of Therm. Expansion inches x 10^{-6}/°C	Orientation: Par. -5.0 Rt.Angl. -1.6
Flux Density in Magnet B_m at BH_{max} (G)	5000	Resistivity (micro-ohms/cm/cm^2)	150 @ 25 °C
Coercive Force in Magnet H_m at BH_{max} (Oe)	4500	Hardness Scale	600 Vickers
Energy Loss W_c (watt-sec/cycle/lb)	-	Basic Forming Method	DPM
Importance of Operating Magnet at BH_{max}	E	Basic Finishing Method	WGR
Average Recoil Permeability	1.05	Relative Corrosion Resistance	P
Field Strength Required to Fully Saturate Magnet (Oe)	32,000	Primary Chemical Composition	Nd-Fe-B
Ability to Withstand Demagnetizing Fields	0	Special Notes:	
Maximum Service Temperature (°F)	245		
Approximate Temperature Permanently Affecting Magnet (°F)	500		
Inherent Orientation of Material	Radial oriented ring only		

Note: All information based on available producers literature. In some cases of foreign materials interpretation or calculation required.

SOURCE(S):

Hitachi Magnetics Corp.

USA

HS-25BR
(Alloy or Grade)

Sintered Rare Earth
(Type of Material)

26/15
(MMPA Brief Designation)

R7-3
(IEC Code Reference)

MAGNETIC PROPERTIES		PHYSICAL PROPERTIES	
Residual Induction B_r (G)	10,600	Density (lbs./in.^{3})	.2710
Coercive Force (Oe) Normal (H_c)	10,000	General Mechanical Properties	HB
Coercive Force (Oe) Intrinsic (H_{ci})	15,000	Tensile Strength (psi)	-
Maximum Energy Product BH_{max} (MGO)	26.0	Transverse Modulus of Rupture (psi)	-
B/H Ratio (Load Line) at BH_{max}	1.14	Coeff. of Therm. Expansion inches x 10^{-6}/°C	Orientation: Par. -5.0 Rt.Angl. -1.6
Flux Density in Magnet B_m at BH_{max} (G)	5600	Resistivity (micro-ohms/cm/cm^2)	150 @ 25 °C
Coercive Force in Magnet H_m at BH_{max} (Oe)	4900	Hardness 600	Scale Vickers
Energy Loss W_c (watt-sec/cycle/lb)	-	Basic Forming Method	DPM
Importance of Operating Magnet at BH_{max}	E	Basic Finishing Method	WGR
Average Recoil Permeability	1.05	Relative Corrosion Resistance	P
Field Strength Required to Fully Saturate Magnet (Oe)	32,000	Primary Chemical Composition	Nd-Fe-B
Ability to Withstand Demagnetizing Fields	0	Special Notes:	
Maximum Service Temperature (°F)	245		
Approximate Temperature Permanently Affecting Magnet (°F)	500		
Inherent Orientation of Material	Radial oriented ring only		

Note: All information based on available producers literature. In some cases of foreign materials interpretation or calculation required.

SOURCE(S): **Hitachi Magnetics Corp.**

USA

HS-25CR
(Alloy or Grade)

Sintered Rare Earth
(Type of Material)

24/15
(MMPA Brief Designation)

R7-3
(IEC Code Reference)

MAGNETIC PROPERTIES		PHYSICAL PROPERTIES	
Residual Induction B_r (G)	10,400	Density (lbs./in.^{3})	.2710
Coercive Force (Oe) Normal (H_c)	9800	General Mechanical Properties	HB
Coercive Force (Oe) Intrinsic (H_{ci})	15,000	Tensile Strength (psi)	-
Maximum Energy Product BH_{max} (MGO)	24.0	Transverse Modulus of Rupture (psi)	-
B/H Ratio (Load Line) at BH_{max}	1.3	Coeff. of Therm. Expansion inches x 10^{-6}/°C	Orientation: Par. -5.0 Rt.Angl. -1.6
Flux Density in Magnet B_m at BH_{max} (G)	5800	Resistivity (micro-ohms/cm/cm^2)	150 @ 25 °C
Coercive Force in Magnet H_m at BH_{max} (Oe)	4600	Hardness Scale	600 Vickers
Energy Loss W_c (watt-sec/cycle/lb)	-	Basic Forming Method	DPM
Importance of Operating Magnet at BH_{max}	E	Basic Finishing Method	WGR
Average Recoil Permeability	1.05	Relative Corrosion Resistance	P
Field Strength Required to Fully Saturate Magnet (Oe)	32,000	Primary Chemical Composition	Nd-Fe-B
Ability to Withstand Demagnetizing Fields	0	Special Notes:	
Maximum Service Temperature (°F)	245		
Approximate Temperature Permanently Affecting Magnet (°F)	Est. 590		
Inherent Orientation of Material	Radial oriented ring only		

Note: All information based on available producers literature. In some cases of foreign materials interpretation or calculation required.

SOURCE(S):

Hitachi Magnetics Corp.

USA

HS-27CV
(Alloy or Grade)

Sintered Rare Earth
(Type of Material)

27/20
(MMPA Brief Designation)

R7-3
(IEC Code Reference)

MAGNETIC PROPERTIES		PHYSICAL PROPERTIES	
Residual Induction B_r (G)	10,700	Density (lbs./in.^{3})	.2710
Coercive Force (Oe) Normal (H_c)	9500	General Mechanical Properties	HB
Coercive Force (Oe) Intrinsic (H_{ci})	20,000	Tensile Strength (psi)	-
Maximum Energy Product BH_{max} (MGO)	27.0	Transverse Modulus of Rupture (psi)	-
B/H Ratio (Load Line) at BH_{max}	1.1	Coeff. of Therm. Expansion inches x 10^{-6}/°C	Orientation: Par. -5.0 Rt.Angl. -1.6
Flux Density in Magnet B_m at BH_{max} (G)	5500	Resistivity (micro-ohms/cm/cm^2)	150 @ 25 °C
Coercive Force in Magnet H_m at BH_{max} (Oe)	5000	Hardness Scale	600 Vickers
Energy Loss W_c (watt-sec/cycle/lb)	-	Basic Forming Method	DPM
Importance of Operating Magnet at BH_{max}	E	Basic Finishing Method	WGR
Average Recoil Permeability	1.05	Relative Corrosion Resistance	P
Field Strength Required to Fully Saturate Magnet (Oe)	40,000	Primary Chemical Composition	Nd-Fe-B
Ability to Withstand Demagnetizing Fields	0	Special Notes:	
Maximum Service Temperature (°F)	392		
Approximate Temperature Permanently Affecting Magnet (°F)	500		
Inherent Orientation of Material	MP		

Note: All information based on available producers literature. In some cases of foreign materials interpretation or calculation required.

SOURCE(S): **Hitachi Magnetics Corp.**

USA

HS-30BV
(Alloy or Grade)

Sintered Rare Earth
(Type of Material)

28/15
(MMPA Brief Designation)

R7-3
(IEC Code Reference)

MAGNETIC PROPERTIES	
Residual Induction B_r (G)	10,500
Coercive Force (Oe) — Normal (H_c)	10,500
Coercive Force (Oe) — Intrinsic (H_{ci})	15,000
Maximum Energy Product BH_{max} (MGO)	28.0
B/H Ratio (Load Line) at BH_{max}	1.35
Flux Density in Magnet B_m at BH_{max} (G)	6400
Coercive Force in Magnet H_m at BH_{max} (Oe)	4750
Energy Loss W_c (watt-sec/cycle/lb)	–
Importance of Operating Magnet at BH_{max}	E
Average Recoil Permeability	1.05
Field Strength Required to Fully Saturate Magnet (Oe)	32,000
Ability to Withstand Demagnetizing Fields	0
Maximum Service Temperature (°F)	248
Approximate Temperature Permanently Affecting Magnet (°F)	500
Inherent Orientation of Material	MP

PHYSICAL PROPERTIES	
Density (lbs./in.^{3})	.2710
General Mechanical Properties	HB
Tensile Strength (psi)	–
Transverse Modulus of Rupture (psi)	–
Coeff. of Therm. Expansion inches x 10^{-6}/°C	Orientation: Par. –5.0 Rt.Angl. –1.6
Resistivity (micro-ohms/cm/cm^2)	150 @ 25 °C
Hardness	Scale 600 Vickers
Basic Forming Method	DPM
Basic Finishing Method	WGR
Relative Corrosion Resistance	P
Primary Chemical Composition	Nd-Fe-B
Special Notes:	

Note: All information based on available producers literature. In some cases of foreign materials interpretation or calculation required.

SOURCE(S):

Hitachi Magnetics Corp.

USA

HS-30CH
(Alloy or Grade)

Sintered Rare Earth
(Type of Material)

30/20
(MMPA Brief Designation)

R7-3
(IEC Code Reference)

MAGNETIC PROPERTIES		PHYSICAL PROPERTIES	
Residual Induction B_r (G)	11,300	Density (lbs./in.^{3})	.2710
Coercive Force (Oe) Normal (H_c)	10,800	General Mechanical Properties	HB
Coercive Force (Oe) Intrinsic (H_{ci})	20,000	Tensile Strength (psi)	-
Maximum Energy Product BH_{max} (MGO)	30.0	Transverse Modulus of Rupture (psi)	-
B/H Ratio (Load Line) at BH_{max}	1.35	Coeff. of Therm. Expansion inches x 10^{-6}/°C	Orientation: Par. -5.0 Rt.Angl. -1.6
Flux Density in Magnet B_m at BH_{max} (G)	6400	Resistivity (micro-ohms/cm/cm^2	150 @ 25 °C
Coercive Force in Magnet H_m at BH_{max} (Oe)	4750	Hardness Scale	- -
Energy Loss W_c (watt-sec/cycle/lb)	-	Basic Forming Method	DPM
Importance of Operating Magnet at BH_{max}	E	Basic Finishing Method	WRG
Average Recoil Permeability	1.05	Relative Corrosion Resistance	P
Field Strength Required to Fully Saturate Magnet (Oe)	40,000	Primary Chemical Composition	Nd-Fe-B
Ability to Withstand Demagnetizing Fields	0	Special Notes:	
Maximum Service Temperature (°F)	392		
Approximate Temperature Permanently Affecting Magnet (°F)	500		
Inherent Orientation of Material	MP		

Note: All information based on available producers literature. In some cases of foreign materials interpretation or calculation required.

SOURCE(S):	HS-30CR
	(Alloy or Grade)
Hitachi Magnetics Corp.	Sintered Rare Earth
	(Type of Material)
USA	30/19
	(MMPA Brief Designation)
	R7-3
	(IEC Code Reference)

MAGNETIC PROPERTIES		PHYSICAL PROPERTIES	
Residual Induction B_r (G)	11,200	Density (lbs./in.^{3})	.2710
Coercive Force (Oe) — Normal (H_c)	10,700	General Mechanical Properties	HB
Coercive Force (Oe) — Intrinsic (H_{ci})	19,000	Tensile Strength (psi)	-
Maximum Energy Product BH_{max} (MGO)	30.0	Transverse Modulus of Rupture (psi)	-
B/H Ratio (Load Line) at BH_{max}	1.2	Coeff. of Therm. Expansion inches x 10^{-6}/°C	Orientation: Par. -5.0 Rt.Angl. -1.6
Flux Density in Magnet B_m at BH_{max} (G)	600	Resistivity (micro-ohms/cm/cm^2)	150 @ 25 °C
Coercive Force in Magnet H_m at BH_{max} (Oe)	5000	Hardness 600 Scale Vickers	
Energy Loss W_c (watt-sec/cycle/lb)	-	Basic Forming Method	DPM
Importance of Operating Magnet at BH_{max}	E	Basic Finishing Method	WGR
Average Recoil Permeability	1.05	Relative Corrosion Resistance	P
Field Strength Required to Fully Saturate Magnet (Oe)	40,000	Primary Chemical Composition	Nd-Fe-B
Ability to Withstand Demagnetizing Fields	0	Special Notes:	
Maximum Service Temperature (°F)	302		
Approximate Temperature Permanently Affecting Magnet (°F)	500		
Inherent Orientation of Material	Radial oriented ring only		

Note: All information based on available producers literature. In some cases of foreign materials interpretation or calculation required.

SOURCE(S):

Hitachi Magnetics Corp.

USA

HS-30CV
(Alloy or Grade)

Sintered Rare Earth
(Type of Material)

30/18
(MMPA Brief Designation)

R7-3
(IEC Code Reference)

MAGNETIC PROPERTIES	
Residual Induction B_r (G)	11,200
Coercive Force (Oe) Normal (H_c)	10,700
Coercive Force (Oe) Intrinsic (H_{ci})	19,000
Maximum Energy Product BH_{max} (MGO)	30.0
B/H Ratio (Load Line) at BH_{max}	1.2
Flux Density in Magnet B_m at BH_{max} (G)	6000
Coercive Force in Magnet H_m at BH_{max} (Oe)	5000
Energy Loss W_c (watt-sec/cycle/lb)	-
Importance of Operating Magnet at BH_{max}	E
Average Recoil Permeability	1.05
Field Strength Required to Fully Saturate Magnet (Oe)	40,000
Ability to Withstand Demagnetizing Fields	0
Maximum Service Temperature (°F)	302
Approximate Temperature Permanently Affecting Magnet (°F)	500
Inherent Orientation of Material	MP

PHYSICAL PROPERTIES	
Density (lbs./in.^{3})	.2710
General Mechanical Properties	HB
Tensile Strength (psi)	-
Transverse Modulus of Rupture (psi)	-
Coeff. of Therm. Expansion inches x 10^{-6}/°C	Orientation: Par. -5.0 Rt.Angl. -1.6
Resistivity (micro-ohms/cm/cm^2)	150 @ 25 °C
Hardness 600	Scale Vickers
Basic Forming Method	DPM
Basic Finishing Method	WGR
Relative Corrosion Resistance	P
Primary Chemical Composition	Nd-Fe-B
Special Notes:	

Note: All information based on available producers literature. In some cases of foreign materials interpretation or calculation required.

SOURCE(S):

Hitachi Magnetics Corp.

USA

HS-32BV
(Alloy or Grade)

Sintered Rare Earth
(Type of Material)

32/16
(MMPA Brief Designation)

R7-3
(IEC Code Reference)

MAGNETIC PROPERTIES	
Residual Induction B_r (G)	11,600
Coercive Force (Oe) Normal (H_c)	11,000
Coercive Force (Oe) Intrinsic (H_{ci})	16,000
Maximum Energy Product BH_{max} (MGO)	32.0
B/H Ratio (Load Line) at BH_{max}	1.0
Flux Density in Magnet B_m at BH_{max} (G)	5700
Coercive Force in Magnet H_m at BH_{max} (Oe)	5700
Energy Loss W_c (watt-sec/cycle/lb)	-
Importance of Operating Magnet at BH_{max}	E
Average Recoil Permeability	1.05
Field Strength Required to Fully Saturate Magnet (Oe)	32,000
Ability to Withstand Demagnetizing Fields	0
Maximum Service Temperature (°F)	248
Approximate Temperature Permanently Affecting Magnet (°F)	500
Inherent Orientation of Material	MP

PHYSICAL PROPERTIES	
Density (lbs./in.^{3})	.2710
General Mechanical Properties	HB
Tensile Strength (psi)	-
Transverse Modulus of Rupture (psi)	-
Coeff. of Therm. Expansion inches x 10^{-6}/°C	Orientation: Par. -5.0 Rt.Angl. -1.6
Resistivity (micro-ohms/cm/cm^2)	150 @ 25 °C
Hardness	600 Vickers Scale
Basic Forming Method	DPM
Basic Finishing Method	WGR
Relative Corrosion Resistance	P
Primary Chemical Composition	Nd-Fe-B
Special Notes:	

Note: All information based on available producers literature. In some cases of foreign materials interpretation or calculation required.

SOURCE(S):

Hitachi Magnetics Corp.

USA

HS-35BH
(Alloy or Grade)

Sintered Rare Earth
(Type of Material)

35/15
(MMPA Brief Designation)

R7-3
(IEC Code Reference)

MAGNETIC PROPERTIES		PHYSICAL PROPERTIES	
Residual Induction B_r (G)	12,300	Density (lbs./in.^{3})	.2710
Coercive Force (Oe) — Normal (H_c)	11,500	General Mechanical Properties	HB
Coercive Force (Oe) — Intrinsic (H_{ci})	15,000	Tensile Strength (psi)	-
Maximum Energy Product BH_{max} (MGO)	35.0	Transverse Modulus of Rupture (psi)	-
B/H Ratio (Load Line) at BH_{max}	1.1	Coeff. of Therm. Expansion inches x 10^{-6}/°C	Orientation: Par. _ Rt.Angl. _
Flux Density in Magnet B_m at BH_{max} (G)	6100	Resistivity (micro-ohms/cm/cm^2)	150 @ 25 °C
Coercive Force in Magnet H_m at BH_{max} (Oe)	5700	Hardness Scale	600 Vickers
Energy Loss W_c (watt-sec/cycle/lb)	-	Basic Forming Method	DPM
Importance of Operating Magnet at BH_{max}	E	Basic Finishing Method	WGR
Average Recoil Permeability	1.05	Relative Corrosion Resistance	P
Field Strength Required to Fully Saturate Magnet (Oe)	32,000	Primary Chemical Composition	Nd-Fe-B
Ability to Withstand Demagnetizing Fields	0	Special Notes:	
Maximum Service Temperature (°F)	245		
Approximate Temperature Permanently Affecting Magnet (°F)	500		
Inherent Orientation of Material	MP		

Note: All information based on available producers literature. In some cases of foreign materials interpretation or calculation required.

SOURCE(S):

Hitachi Magnetics Corp.

USA

HS-35CH
(Alloy or Grade)

Sintered Rare Earth
(Type of Material)

35/19
(MMPA Brief Designation)

R7-3
(IEC Code Reference)

MAGNETIC PROPERTIES		PHYSICAL PROPERTIES	
Residual Induction B_r (G)	12,100	Density (lbs./in.^{3})	.2710
Coercive Force (Oe) — Normal (H_c)	11,500	General Mechanical Properties	HB
Coercive Force (Oe) — Intrinsic (H_{ci})	19,000	Tensile Strength (psi)	-
Maximum Energy Product BH_{max} (MGO)	35.0	Transverse Modulus of Rupture (psi)	-
B/H Ratio (Load Line) at BH_{max}	1.0	Coeff. of Therm. Expansion inches x 10^{-6}/°C	Orientation: Par. -5.0 Rt.Angl. -1.6
Flux Density in Magnet B_m at BH_{max} (G)	5900	Resistivity (micro-ohms/cm/cm^2)	150 @ 25 °C
Coercive Force in Magnet H_m at BH_{max} (Oe)	5900	Hardness 60	Scale Vickers
Energy Loss W_c (watt-sec/cycle/lb)	-	Basic Forming Method	DPM
Importance of Operating Magnet at BH_{max}	E	Basic Finishing Method	WGR
Average Recoil Permeability	1.05	Relative Corrosion Resistance	P
Field Strength Required to Fully Saturate Magnet (Oe)	40,000	Primary Chemical Composition	Nd-Fe-B
Ability to Withstand Demagnetizing Fields	0	Special Notes:	
Maximum Service Temperature (°F)	302		
Approximate Temperature Permanently Affecting Magnet (°F)	500		
Inherent Orientation of Material	MP		

Note: All information based on available producers literature. In some cases of foreign materials interpretation or calculation required.

SOURCE(S):

Hitachi Magnetics Corp.

USA

HS-37BH
(Alloy or Grade)

Sintered Rare Earth
(Type of Material)

37/16
(MMPA Brief Designation)

R7-3
(IEC Code Reference)

MAGNETIC PROPERTIES		PHYSICAL PROPERTIES	
Residual Induction B_r (G)	12,500	Density (lbs./in.^{3})	.2710
Coercive Force (Oe) Normal (H_c)	11,800	General Mechanical Properties	HB
Coercive Force (Oe) Intrinsic (H_{ci})	16,000	Tensile Strength (psi)	-
Maximum Energy Product BH_{max} (MGO)	37.0	Transverse Modulus of Rupture (psi)	-
B/H Ratio (Load Line) at BH_{max}	1.0	Coeff. of Therm. Expansion inches x 10^{-6}/°C	Orientation: Par. -5.0 Rt.Angl. -1.6
Flux Density in Magnet B_m at BH_{max} (G)	6000	Resistivity (micro-ohms/cm/cm^2)	150 @ 25 °C
Coercive Force in Magnet H_m at BH_{max} (Oe)	6000	Hardness Scale	600 Vickers
Energy Loss W_c (watt-sec/cycle/lb)	-	Basic Forming Method	DPM
Importance of Operating Magnet at BH_{max}	E	Basic Finishing Method	WGR
Average Recoil Permeability	1.05	Relative Corrosion Resistance	P
Field Strength Required to Fully Saturate Magnet (Oe)	32,000	Primary Chemical Composition	Nd-Fe-B
Ability to Withstand Demagnetizing Fields	0	Special Notes:	
Maximum Service Temperature (°F)	248		
Approximate Temperature Permanently Affecting Magnet (°F)	500		
Inherent Orientation of Material	MP		

Note: All information based on available producers literature. In some cases of foreign materials interpretation or calculation required.

SOURCE(S):	Hycomax 1
	(Alloy or Grade)
SG Magnets, Ltd.	Sintered Alnico
	(Type of Material)
USA	3.0/.95
	(MMPA Brief Designation)
	-
	(IEC Code Reference)

MAGNETIC PROPERTIES		PHYSICAL PROPERTIES	
Residual Induction B_r (G)	7900	Density (lbs./in.^{3})	.2530
Coercive Force (Oe) — Normal (H_c)	910	General Mechanical Properties	HT
Coercive Force (Oe) — Intrinsic (H_{ci})	950	Tensile Strength (psi)	27,560
Maximum Energy Product BH_{max} (MGO)	3.0	Transverse Modulus of Rupture (psi)	43,500
B/H Ratio (Load Line) at BH_{max}	8.6	Coeff. of Therm. Expansion inches x 10^{-6}/°C — Orientation: Par. 11.5 Rt.Angl. -	
Flux Density in Magnet B_m at BH_{max} (G)	5100	Resistivity (micro-ohms/cm/cm^2)	55 @ 25 °C
Coercive Force in Magnet H_m at BH_{max} (Oe)	590	Hardness 47 — Scale Rockwell C	
Energy Loss W_c (watt-sec/cycle/lb)	-	Basic Forming Method	DPM
Importance of Operating Magnet at BH_{max}	B	Basic Finishing Method	WGR
Average Recoil Permeability	4.5	Relative Corrosion Resistance	G
Field Strength Required to Fully Saturate Magnet (Oe)	3000	Primary Chemical Composition	Al-Ni-Co-Fe
Ability to Withstand Demagnetizing Fields	G	Special Notes:	
Maximum Service Temperature (°F)	842		
Approximate Temperature Permanently Affecting Magnet (°F)	1436		
Inherent Orientation of Material	HT		

Note: All information based on available producers literature. In some cases of foreign materials interpretation or calculation required.

SOURCE(S):

SG Magnets, Ltd.

USA

Hycomax 3
(Alloy or Grade)

Sintered Alnico
(Type of Material)

4.8/1.7
(MMPA Brief Designation)

-
(IEC Code Reference)

MAGNETIC PROPERTIES			PHYSICAL PROPERTIES	
Residual Induction B_r (G)		7800	Density (lbs./in.^{3})	.2500
Coercive Force (Oe)	Normal (H_c)	1600	General Mechanical Properties	HT
	Intrinsic (H_{ci})	1700	Tensile Strength (psi)	24,650
Maximum Energy Product BH_{max}(MGO)		4.8	Transverse Modulus of Rupture (psi)	43,500
B/H Ratio (Load Line) at BH_{max}		4.9	Coeff. of Therm. Expansion inches x 10^{-6}/°C	Orientation: Par. 11.5 Rt.Angl. -
Flux Density in Magnet B_m at BH_{max} (G)		5000	Resistivity (micro-ohms/cm/cm^2)	55 @ 25 °C
Coercive Force in Magnet H_m at BH_{max} (Oe)		1020	Hardness 52	Scale Rockwell C
Energy Loss W_c (watt-sec/cycle/lb)		-	Basic Forming Method	DPM
Importance of Operating Magnet at BH_{max}		C	Basic Finishing Method	WGR
Average Recoil Permeability		2.2	Relative Corrosion Resistance	G
Field Strength Required to Fully Saturate Magnet (Oe)		3000	Primary Chemical Composition	Al-Ni-Co-Fe
Ability to Withstand Demagnetizing Fields		G	Special Notes:	
Maximum Service Temperature (°F)		842		
Approximate Temperature Permanently Affecting Magnet (°F)		1436		
Inherent Orientation of Material		MP		

Note: All information based on available producers literature. In some cases of foreign materials interpretation or calculation required.

SOURCE(S):

SG Magnets, Ltd.

USA

Hycomax 4
(Alloy or Grade)

Sintered Alnico
(Type of Material)

5.0/2.0
(MMPA Brief Designation)

-
(IEC Code Reference)

MAGNETIC PROPERTIES		PHYSICAL PROPERTIES	
Residual Induction B_r (G)	6800	Density (lbs./in.^{3})	.2500
Coercive Force (Oe) — Normal (H_c)	1940	General Mechanical Properties	HT
Coercive Force (Oe) — Intrinsic (H_{ci})	2000	Tensile Strength (psi)	24,650
Maximum Energy Product BH_{max} (MGO)	5.0	Transverse Modulus of Rupture (psi)	43,500
B/H Ratio (Load Line) at BH_{max}	3.5	Coeff. of Therm. Expansion inches x 10^{-6}/°C	Orientation: Par. 11.5 Rt.Angl. _
Flux Density in Magnet B_m at BH_{max} (G)	4200	Resistivity (micro-ohms/cm/cm^2)	55 @ 25 °C
Coercive Force in Magnet H_m at BH_{max} (Oe)	1190	Hardness 55 Scale Rockwell C	
Energy Loss W_c (watt-sec/cycle/lb)	-	Basic Forming Method	DPM
Importance of Operating Magnet at BH_{max}	C	Basic Finishing Method	WGR
Average Recoil Permeability	1.9	Relative Corrosion Resistance	G
Field Strength Required to Fully Saturate Magnet (Oe)	3000	Primary Chemical Composition	Al-Ni-Co-Fe
Ability to Withstand Demagnetizing Fields	G	Special Notes:	
Maximum Service Temperature (°F)	842		
Approximate Temperature Permanently Affecting Magnet (°F)	1436		
Inherent Orientation of Material	MP		

Note: All information based on available producers literature. In some cases of foreign materials interpretation or calculation required.

SOURCE(S):	**Hynico**
	(Alloy or Grade)
SG Magnets, Ltd.	Sintered Alnico
	(Type of Material)
USA	2.5/1.2
	(MMPA Brief Designation)
	-
	(IEC Code Reference)

MAGNETIC PROPERTIES		PHYSICAL PROPERTIES	
Residual Induction B_r (G)	6200	Density (lbs./in.^{3})	.2530
Coercive Force (Oe) — Normal (H_c)	1100	General Mechanical Properties	HT
Coercive Force (Oe) — Intrinsic (H_{ci})	1200	Tensile Strength (psi)	27,560
Maximum Energy Product BH_{max} (MGO)	2.5	Transverse Modulus of Rupture (psi)	43,500
B/H Ratio (Load Line) at BH_{max}	5.6	Coeff. of Therm. Expansion inches x 10^{-6}/°C	Orientation: Par. 11.5 Rt.Angl. _
Flux Density in Magnet B_m at BH_{max} (G)	3500	Resistivity (micro-ohms/cm/cm^2)	58 @ 25 °C
Coercive Force in Magnet H_m at BH_{max} (Oe)	630	Hardness Scale	50 Rockwell C
Energy Loss W_c (watt-sec/cycle/lb)	-	Basic Forming Method	DPM
Importance of Operating Magnet at BH_{max}	C	Basic Finishing Method	WGR
Average Recoil Permeability	3.1	Relative Corrosion Resistance	G
Field Strength Required to Fully Saturate Magnet (Oe)	3000	Primary Chemical Composition	Al-Ni-Co-Fe
Ability to Withstand Demagnetizing Fields	G	Special Notes:	
Maximum Service Temperature (°F)	842		
Approximate Temperature Permanently Affecting Magnet (°F)	1436		
Inherent Orientation of Material	HT		

Note: All information based on available producers literature. In some cases of foreign materials interpretation or calculation required.

SOURCE(S):

UGIMag, Inc.

USA

INCOR 18
(Alloy or Grade)

Sintered Rare Earth
(Type of Material)

18/19
(MMPA Brief Designation)

R5-1
(IEC Code Reference)

MAGNETIC PROPERTIES		PHYSICAL PROPERTIES	
Residual Induction B_r (G)	8750	Density (lbs./in.^{3})	.3071
Coercive Force (Oe) — Normal (H_c)	8600	General Mechanical Properties	HB
Coercive Force (Oe) — Intrinsic (H_{ci})	19,000	Tensile Strength (psi)	Est. 5000
Maximum Energy Product BH_{max} (MGO)	19.0	Transverse Modulus of Rupture (psi)	—
B/H Ratio (Load Line) at BH_{max}	.53	Coeff. of Therm. Expansion inches x 10^{-6}/°C	Orientation: Par. — Rt.Angl. —
Flux Density in Magnet B_m at BH_{max} (G)	3200	Resistivity (micro-ohms/cm/cm^2)	55 @ 22 °C
Coercive Force in Magnet H_m at BH_{max} (Oe)	6000	Hardness Scale	600 Vickers
Energy Loss W_c (watt-sec/cycle/lb)	—	Basic Forming Method	DPM
Importance of Operating Magnet at BH_{max}	E	Basic Finishing Method	WGR
Average Recoil Permeability	1.02	Relative Corrosion Resistance	G
Field Strength Required to Fully Saturate Magnet (Oe)	40,000	Primary Chemical Composition	Sm-Co
Ability to Withstand Demagnetizing Fields	0	Special Notes: (1) Axially oriented	
Maximum Service Temperature (°F)	Est. 482		
Approximate Temperature Permanently Affecting Magnet (°F)	Est. 1328		
Inherent Orientation of Material	MP		

Note: All information based on available producers literature. In some cases of foreign materials interpretation or calculation required.

SOURCE(S):

UGIMag, Inc.

USA

INCOR 21
(Alloy or Grade)

Sintered Rare Earth
(Type of Material)

21/19
(MMPA Brief Designation)

R5-1
(IEC Code Reference)

MAGNETIC PROPERTIES		PHYSICAL PROPERTIES	
Residual Induction B_r (G)	9300	Density (lbs./in.^{3})	.3071
Coercive Force (Oe) — Normal (H_c)	9100	General Mechanical Properties	HB
Coercive Force (Oe) — Intrinsic (H_{ci})	19,000	Tensile Strength (psi)	Est. 5000
Maximum Energy Product BH_{max} (MGO)	21.0	Transverse Modulus of Rupture (psi)	–
B/H Ratio (Load Line) at BH_{max}	.49	Coeff. of Therm. Expansion inches x 10^{-6}/°C Orientation: Par. – Rt.Angl. –	
Flux Density in Magnet B_m at BH_{max} (G)	3200	Resistivity (micro-ohms/cm/cm^2)	55 @ 22 °C
Coercive Force in Magnet H_m at BH_{max} (Oe)	6500	Hardness Scale 600 Vickers	
Energy Loss W_c (watt-sec/cycle/lb)	–	Basic Forming Method	DPM
Importance of Operating Magnet at BH_{max}	E	Basic Finishing Method	WGR
Average Recoil Permeability	1.02	Relative Corrosion Resistance	G
Field Strength Required to Fully Saturate Magnet (Oe)	40,000	Primary Chemical Composition	Sm-Co
Ability to Withstand Demagnetizing Fields	0	Special Notes: (1) Transverse oriented	
Maximum Service Temperature (°F)	Est. 482		
Approximate Temperature Permanently Affecting Magnet (°F)	Est. 1328		
Inherent Orientation of Material	MP		

Note: All information based on available producers literature. In some cases of foreign materials interpretation or calculation required.

SOURCE(S):		INCOR 24	
		(Alloy or Grade)	
UGIMag, Inc.		Sintered Rare Earth	
		(Type of Material)	
USA		24/18	
		(MMPA Brief Designation)	
		R5-2	
		(IEC Code Reference)	

MAGNETIC PROPERTIES		PHYSICAL PROPERTIES	
Residual Induction B_r (G)	10,100	Density (lbs./in.^{3})	.2999
Coercive Force (Oe) — Normal (H_c)	9500	General Mechanical Properties	HB
Coercive Force (Oe) — Intrinsic (H_{ci})	18,000	Tensile Strength (psi)	Est. 5000
Maximum Energy Product BH_{max} (MGO)	24.0	Transverse Modulus of Rupture (psi)	–
B/H Ratio (Load Line) at BH_{max}	.63	Coeff. of Therm. Expansion inches x 10^{-6}/°C	Orientation: Par. – Rt.Angl. –
Flux Density in Magnet B_m at BH_{max} (G)	3900	Resistivity (micro-ohms/cm/cm^2)	90 @ 22 °C
Coercive Force in Magnet H_m at BH_{max} (Oe)	6200	Hardness Scale	600 Vickers
Energy Loss W_c (watt-sec/cycle/lb)	–	Basic Forming Method	DPM
Importance of Operating Magnet at BH_{max}	E	Basic Finishing Method	WGR
Average Recoil Permeability	1.05	Relative Corrosion Resistance	G
Field Strength Required to Fully Saturate Magnet (Oe)	40,000	Primary Chemical Composition	Sm-Co
Ability to Withstand Demagnetizing Fields	O	Special Notes: (1) Axially oriented	
Maximum Service Temperature (°F)	Est. 482		
Approximate Temperature Permanently Affecting Magnet (°F)	Est. 1328		
Inherent Orientation of Material	MP		

Note: All information based on available producers literature. In some cases of foreign materials interpretation or calculation required.

SOURCE(S):	**INCOR 26**
	(Alloy or Grade)
UGIMag, Inc.	Sintered Rare Earth
	(Type of Material)
USA	26/18
	(MMPA Brief Designation)
	R5-2
	(IEC Code Reference)

MAGNETIC PROPERTIES		PHYSICAL PROPERTIES	
Residual Induction B_r (G)	10,500	Density (lbs./in.^{3})	.2999
Coercive Force (Oe) — Normal (H_c)	10,100	General Mechanical Properties	HB
Coercive Force (Oe) — Intrinsic (H_{ci})	18,000	Tensile Strength (psi)	Est. 5000
Maximum Energy Product BH_{max} (MGO)	26.0	Transverse Modulus of Rupture (psi)	–
B/H Ratio (Load Line) at BH_{max}	.59	Coeff. of Therm. Expansion inches x 10^{-6}/°C	Orientation: Par. – Rt.Angl. –
Flux Density in Magnet B_m at BH_{max} (G)	3900	Resistivity (micro-ohms/cm/cm^2)	90 @ 22 °C
Coercive Force in Magnet H_m at BH_{max} (Oe)	6600	Hardness Scale	600 Vickers
Energy Loss W_c (watt-sec/cycle/lb)	–	Basic Forming Method	DPM
Importance of Operating Magnet at BH_{max}	E	Basic Finishing Method	WGR
Average Recoil Permeability	1.05	Relative Corrosion Resistance	G
Field Strength Required to Fully Saturate Magnet (Oe)	40,000	Primary Chemical Composition	Sm-Co
Ability to Withstand Demagnetizing Fields	O	Special Notes: (1) Transverse oriented	
Maximum Service Temperature (°F)	Est. 482		
Approximate Temperature Permanently Affecting Magnet (°F)	Est. 1328		
Inherent Orientation of Material	MP		

Note: All information based on available producers literature. In some cases of foreign materials interpretation or calculation required.

SOURCE(S):	**KHJ-1**
	(Alloy or Grade)
Hitachi Magnetics Corp.	Wrought
	(Type of Material)
USA	-
	(MMPA Brief Designation)
	-
	(IEC Code Reference)

MAGNETIC PROPERTIES		PHYSICAL PROPERTIES	
Residual Induction B_r (G)	13,000	Density (lbs./in.^{3})	.2746
Coercive Force (Oe) — Normal (H_c)	600	General Mechanical Properties	HT
Coercive Force (Oe) — Intrinsic (H_{ci})	-	Tensile Strength (psi)	-
Maximum Energy Product BH_{max} (MGO)	5.7	Transverse Modulus of Rupture (psi)	-
B/H Ratio (Load Line) at BH_{max}	21.0	Coeff. of Therm. Expansion inches x 10^{-6}/°C	Orientation: Par. - Rt.Angl. -
Flux Density in Magnet B_m at BH_{max} (G)	10,400	Resistivity (micro-ohms/cm/cm^2)	75 @ 25 °C
Coercive Force in Magnet H_m at BH_{max} (Oe)	500	Hardness Scale	350 Vickers
Energy Loss W_c (watt-sec/cycle/lb)	-	Basic Forming Method	W (See Notes)
Importance of Operating Magnet at BH_{max}	A	Basic Finishing Method	WGR (See Notes)
Average Recoil Permeability	-	Relative Corrosion Resistance	E
Field Strength Required to Fully Saturate Magnet (Oe)	3000	Primary Chemical Composition	Fe-Cr-Co
Ability to Withstand Demagnetizing Fields	F	Special Notes: (1) Can be formed into shapes difficult for Alnico such as wire and thin films. (2) Can also be punched or drawn. (3) Very high mechanical strength.	
Maximum Service Temperature (°F)	900		
Approximate Temperature Permanently Affecting Magnet (°F)	1200 min.		
Inherent Orientation of Material	HT		

Note: All information based on available producers literature. In some cases of foreign materials interpretation or calculation required.

SOURCE(S):

Hitachi Magnetics Corp.

USA

KHJ-1D
(Alloy or Grade)

Wrought
(Type of Material)

–
(MMPA Brief Designation)

–
(IEC Code Reference)

MAGNETIC PROPERTIES	
Residual Induction B_r (G)	12,000
Coercive Force (Oe) — Normal (H_c)	550
Coercive Force (Oe) — Intrinsic (H_{ci})	–
Maximum Energy Product BH_{max} (MGO)	4.5
B/H Ratio (Load Line) at BH_{max}	22.2
Flux Density in Magnet B_m at BH_{max} (G)	10,000
Coercive Force in Magnet H_m at BH_{max} (Oe)	450
Energy Loss W_c (watt-sec/cycle/lb)	–
Importance of Operating Magnet at BH_{max}	A
Average Recoil Permeability	–
Field Strength Required to Fully Saturate Magnet (Oe)	3000
Ability to Withstand Demagnetizing Fields	F
Maximum Service Temperature (°F)	900
Approximate Temperature Permanently Affecting Magnet (°F)	1200 min.
Inherent Orientation of Material	CR

PHYSICAL PROPERTIES	
Density (lbs./in.^{3})	.2746
General Mechanical Properties	HT
Tensile Strength (psi)	–
Transverse Modulus of Rupture (psi)	–
Coeff. of Therm. Expansion inches x 10^{-6}/°C — Orientation: Par. / Rt.Angl.	– / –
Resistivity (micro-ohms/cm/cm^2)	75 @ 25 °C
Hardness — Scale Vickers	315
Basic Forming Method	W (See Notes)
Basic Finishing Method	WGR (See Notes)
Relative Corrosion Resistance	E
Primary Chemical Composition	Fe-Cr-Co

Special Notes:

(1) Can be formed into shapes difficult for Alnico such as wire and thin films.
(2) Can also be punched or drawn.
(3) Very high mechanical strength.

Note: All information based on available producers literature. In some cases of foreign materials interpretation or calculation required.

SOURCE(S):

Hitachi Magnetics Corp.

USA

KHJ-3A
(Alloy or Grade)

Wrought
(Type of Material)

-
(MMPA Brief Designation)

-
(IEC Code Reference)

MAGNETIC PROPERTIES		PHYSICAL PROPERTIES	
Residual Induction B_r (G)	13,500	Density (lbs./in.^{3})	.2746
Coercive Force (Oe) — Normal (H_c)	650	General Mechanical Properties	HT
Coercive Force (Oe) — Intrinsic (H_{ci})	-	Tensile Strength (psi)	-
Maximum Energy Product BH_{max} (MGO)	6.5	Transverse Modulus of Rupture (psi)	-
B/H Ratio (Load Line) at BH_{max}	20.0	Coeff. of Therm. Expansion inches x 10^{-6}/°C	Orientation: Par. - Rt.Angl. -
Flux Density in Magnet B_m at BH_{max} (G)	11,400	Resistivity (micro-ohms/cm/cm^2)	75 @ 25 °C
Coercive Force in Magnet H_m at BH_{max} (Oe)	570	Hardness Scale	350 Vickers
Energy Loss W_c (watt-sec/cycle/lb)	-	Basic Forming Method	W (See Notes)
Importance of Operating Magnet at BH_{max}	B	Basic Finishing Method	WGR (See Notes)
Average Recoil Permeability	-	Relative Corrosion Resistance	E
Field Strength Required to Fully Saturate Magnet (Oe)	3000	Primary Chemical Composition	Fe-Cr-Co
Ability to Withstand Demagnetizing Fields	F	Special Notes:	
Maximum Service Temperature (°F)	900	(1) Can be formed into shapes difficult for Alnico such as wire and thin films.	
Approximate Temperature Permanently Affecting Magnet (°F)	1200 min.	(2) Can also be punched or drawn. (3) Very high mechanical strength.	
Inherent Orientation of Material	HT		

Note: All information based on available producers literature. In some cases of foreign materials interpretation or calculation required.

SOURCE(S):

Hitachi Magnetics Corp.

USA

KHJ-4DA
(Alloy or Grade)

Wrought
(Type of Material)

-
(MMPA Brief Designation)

-
(IEC Code Reference)

MAGNETIC PROPERTIES		PHYSICAL PROPERTIES	
Residual Induction B_r (G)	9000	Density (lbs./in.^{3})	.2746
Coercive Force (Oe) — Normal (H_c)	250	General Mechanical Properties	HT
Coercive Force (Oe) — Intrinsic (H_{ci})	-	Tensile Strength (psi)	-
Maximum Energy Product BH_{max} (MGO)	.30	Transverse Modulus of Rupture (psi)	-
B/H Ratio (Load Line) at BH_{max}	-	Coeff. of Therm. Expansion inches x 10^{-6}/°C	Orientation: Par. - Rt.Angl. -
Flux Density in Magnet B_m at BH_{max} (G)	-	Resistivity (micro-ohms/cm/cm^2	- @ - °C
Coercive Force in Magnet H_m at BH_{max} (Oe)	-	Hardness -	Scale -
Energy Loss W_c (watt-sec/cycle/lb)	-	Basic Forming Method	W (See Notes)
Importance of Operating Magnet at BH_{max}	A	Basic Finishing Method	WGR (See Notes)
Average Recoil Permeability	-	Relative Corrosion Resistance	E
Field Strength Required to Fully Saturate Magnet (Oe)	1000	Primary Chemical Composition	Fe-Cr-Co
Ability to Withstand Demagnetizing Fields	P	Special Notes: (1) Can be formed into shapes difficult for Alnico such as wire and thin films. (2) Can also be punched or drawn. (3) Very high mechanical strength.	
Maximum Service Temperature (°F)	900		
Approximate Temperature Permanently Affecting Magnet (°F)	1200 min.		
Inherent Orientation of Material	none		

Note: All information based on available producers literature. In some cases of foreign materials interpretation or calculation required.

SOURCE(S): **Hitachi Magnetics Corp.**

USA

KHJ-4DB
(Alloy or Grade)

Wrought
(Type of Material)

-
(MMPA Brief Designation)

-
(IEC Code Reference)

MAGNETIC PROPERTIES	
Residual Induction B_r (G)	9000
Coercive Force (Oe) — Normal (H_c)	250
Coercive Force (Oe) — Intrinsic (H_{ci})	-
Maximum Energy Product BH_{max} (MGO)	.90
B/H Ratio (Load Line) at BH_{max}	22.5
Flux Density in Magnet B_m at BH_{max} (G)	4500
Coercive Force in Magnet H_m at BH_{max} (Oe)	200
Energy Loss W_c (watt-sec/cycle/lb)	-
Importance of Operating Magnet at BH_{max}	A
Average Recoil Permeability	-
Field Strength Required to Fully Saturate Magnet (Oe)	1000
Ability to Withstand Demagnetizing Fields	P
Maximum Service Temperature (°F)	900
Approximate Temperature Permanently Affecting Magnet (°F)	1200 min.
Inherent Orientation of Material	CR

PHYSICAL PROPERTIES	
Density (lbs./in.^{3})	.2746
General Mechanical Properties	HT
Tensile Strength (psi)	-
Transverse Modulus of Rupture (psi)	-
Coeff. of Therm. Expansion inches x 10^{-6}/°C — Orientation: Par.	-
Coeff. of Therm. Expansion inches x 10^{-6}/°C — Rt.Angl.	-
Resistivity (micro-ohms/cm/cm^2)	75 @ 25 °C
Hardness — Scale Vickers	315
Basic Forming Method	W (See Notes)
Basic Finishing Method	WGR (See Notes)
Relative Corrosion Resistance	E
Primary Chemical Composition	Fe-Cr-Co

Special Notes:

(1) Can be formed into shapes difficult for Alnico such as wire and thin films.
(2) Can also be punched or drawn.
(3) Very high mechanical strength.

Note: All information based on available producers literature. In some cases of foreign materials interpretation or calculation required.

SOURCE(S):

Hitachi Magnetics Corp.

USA

KHJ-4DC
(Alloy or Grade)

Wrought
(Type of Material)

-
(MMPA Brief Designation)

-
(IEC Code Reference)

MAGNETIC PROPERTIES	
Residual Induction B_r (G)	9000
Coercive Force (Oe) Normal (H_c)	400
Coercive Force (Oe) Intrinsic (H_{ci})	-
Maximum Energy Product BH_{max} (MGO)	1.5
B/H Ratio (Load Line) at BH_{max}	20
Flux Density in Magnet B_m at BH_{max} (G)	5500
Coercive Force in Magnet H_m at BH_{max} (Oe)	275
Energy Loss W_c (watt-sec/cycle/lb)	-
Importance of Operating Magnet at BH_{max}	B
Average Recoil Permeability	-
Field Strength Required to Fully Saturate Magnet (Oe)	3000
Ability to Withstand Demagnetizing Fields	F
Maximum Service Temperature (°F)	900
Approximate Temperature Permanently Affecting Magnet (°F)	1200 min.
Inherent Orientation of Material	CR

PHYSICAL PROPERTIES	
Density (lbs./in.^{3})	.2746
General Mechanical Properties	HT
Tensile Strength (psi)	-
Transverse Modulus of Rupture (psi)	-
Coeff. of Therm. Expansion inches x 10^{-6}/°C	Orientation: Par. - Rt.Angl. -
Resistivity (micro-ohms/cm/cm^2)	75 @ 25 °C
Hardness 315	Scale Vickers
Basic Forming Method	W (See Notes)
Basic Finishing Method	WGR (See Notes)
Relative Corrosion Resistance	E
Primary Chemical Composition	Fe-Cr-Co

Special Notes:

(1) Can be formed into shapes difficult for Alnico such as wire and thin films.
(2) Can also be punched or drawn.
(3) Very high mechanical strength.

Note: All information based on available producers literature. In some cases of foreign materials interpretation or calculation required.

SOURCE(S):	KHJ-4DD
	(Alloy or Grade)
Hitachi Magnetics Corp.	Wrought
	(Type of Material)
USA	-
	(MMPA Brief Designation)
	-
	(IEC Code Reference)

MAGNETIC PROPERTIES		PHYSICAL PROPERTIES	
Residual Induction B_r (G)	8000	Density (lbs./in.^{3})	.2746
Coercive Force (Oe) — Normal (H_c)	550	General Mechanical Properties	HT
Coercive Force (Oe) — Intrinsic (H_{ci})	-	Tensile Strength (psi)	-
Maximum Energy Product BH_{max} (MGO)	2.0	Transverse Modulus of Rupture (psi)	-
B/H Ratio (Load Line) at BH_{max}	17.1	Coeff. of Therm. Expansion inches x 10^{-6}/°C	Orientation: Par. _ Rt.Angl. _
Flux Density in Magnet B_m at BH_{max} (G)	6000	Resistivity (micro-ohms/cm/cm^2)	75 @ 25 °C
Coercive Force in Magnet H_m at BH_{max} (Oe)	350	Hardness Scale	315 Vickers
Energy Loss W_c (watt-sec/cycle/lb)	-	Basic Forming Method	W (See Notes)
Importance of Operating Magnet at BH_{max}	B	Basic Finishing Method	WGR (See Notes)
Average Recoil Permeability	-	Relative Corrosion Resistance	E
Field Strength Required to Fully Saturate Magnet (Oe)	3000	Primary Chemical Composition	Fe-Cr-Co
Ability to Withstand Demagnetizing Fields	F	Special Notes:	
Maximum Service Temperature (°F)	900	(1) Can be formed into shapes difficult for Alnico such as wire and thin films.	
Approximate Temperature Permanently Affecting Magnet (°F)	1200 min.	(2) Can also be punched or drawn. (3) Very high mechanical strength.	
Inherent Orientation of Material	CR		

Note: All information based on available producers literature. In some cases of foreign materials interpretation or calculation required.

SOURCE(S):

Hitachi Magnetics Corp.

USA

KHJ-7DA
(Alloy or Grade)

Wrought
(Type of Material)

-
(MMPA Brief Designation)

-
(IEC Code Reference)

MAGNETIC PROPERTIES	
Residual Induction B_r (G)	13,000
Coercive Force (Oe) — Normal (H_c)	120
Coercive Force (Oe) — Intrinsic (H_{ci})	-
Maximum Energy Product BH_{max} (MGO)	1.0
B/H Ratio (Load Line) at BH_{max}	100
Flux Density in Magnet B_m at BH_{max} (G)	10,000
Coercive Force in Magnet H_m at BH_{max} (Oe)	100
Energy Loss W_c (watt-sec/cycle/lb)	-
Importance of Operating Magnet at BH_{max}	A
Average Recoil Permeability	-
Field Strength Required to Fully Saturate Magnet (Oe)	1000
Ability to Withstand Demagnetizing Fields	P
Maximum Service Temperature (°F)	900
Approximate Temperature Permanently Affecting Magnet (°F)	1200 min.
Inherent Orientation of Material	none

PHYSICAL PROPERTIES	
Density (lbs./in.^{3})	.2746
General Mechanical Properties	HT
Tensile Strength (psi)	-
Transverse Modulus of Rupture (psi)	-
Coeff. of Therm. Expansion inches x 10^{-6}/°C	Orientation: Par. - Rt.Angl. -
Resistivity (micro-ohms/cm/cm^2)	75 @ 25 °C
Hardness	Scale 315 Vickers
Basic Forming Method	W (See Notes)
Basic Finishing Method	WGR (See Notes)
Relative Corrosion Resistance	E
Primary Chemical Composition	Fe-Cr-Co

Special Notes:

(1) Can be formed into shapes difficult for Alnico such as wire and thin films.
(2) Can also be punched or drawn.
(3) Very high mechanical strength.

Note: All information based on available producers literature. In some cases of foreign materials interpretation or calculation required.

SOURCE(S):

Hitachi Magnetics Corp.

USA

KPM-2A
(Alloy or Grade)

Bonded Ferrite
(Type of Material)

1.5/2.9
(MMPA Brief Designation)

-
(IEC Code Reference)

MAGNETIC PROPERTIES		PHYSICAL PROPERTIES	
Residual Induction B_r (G)	2400	Density (lbs./in.^{3})	-
Coercive Force (Oe) — Normal (H_c)	2100	General Mechanical Properties	HT
Coercive Force (Oe) — Intrinsic (H_{ci})	2900	Tensile Strength (psi)	-
Maximum Energy Product BH_{max} (MGO)	1.45	Transverse Modulus of Rupture (psi)	-
B/H Ratio (Load Line) at BH_{max}	1.1	Coeff. of Therm. Expansion inches x 10^{-6}/°C	Orientation: Par. - Rt.Angl. -
Flux Density in Magnet B_m at BH_{max} (G)	1270	Resistivity (micro-ohms/cm/cm^2	- @ - °C
Coercive Force in Magnet H_m at BH_{max} (Oe)	1150	Hardness Scale	- -
Energy Loss W_c (watt-sec/cycle/lb)	-	Basic Forming Method	VM
Importance of Operating Magnet at BH_{max}	E	Basic Finishing Method	SRD
Average Recoil Permeability	-	Relative Corrosion Resistance	E
Field Strength Required to Fully Saturate Magnet (Oe)	10,000	Primary Chemical Composition	Ba-Fe + binder
Ability to Withstand Demagnetizing Fields	Ex	Special Notes: (1) Nylon binder, injection molded. (2) Maximum service temperature depends on grade of nylon binder used.	
Maximum Service Temperature (°F)	See Note		
Approximate Temperature Permanently Affecting Magnet (°F)	200		
Inherent Orientation of Material	none		

Note: All information based on available producers literature. In some cases of foreign materials interpretation or calculation required.

SOURCE(S):	KPM-3
	(Alloy or Grade)
Hitachi Magnetics Corp.	Bonded Ferrite
	(Type of Material)
USA	.4/2.0
	(MMPA Brief Designation)
	-
	(IEC Code Reference)

MAGNETIC PROPERTIES		PHYSICAL PROPERTIES	
Residual Induction B_r (G)	1400	Density (lbs./in.^{3})	-
Coercive Force (Oe) — Normal (H_c)	1200	General Mechanical Properties	HT
Coercive Force (Oe) — Intrinsic (H_{ci})	1950	Tensile Strength (psi)	-
Maximum Energy Product BH_{max} (MGO)	.40	Transverse Modulus of Rupture (psi)	-
B/H Ratio (Load Line) at BH_{max}	1.5	Coeff. of Therm. Expansion inches x 10^{-6}/°C	Orientation: Par. - Rt.Angl. -
Flux Density in Magnet B_m at BH_{max} (G)	750	Resistivity (micro-ohms/cm/cm^2)	- @ - °C
Coercive Force in Magnet H_m at BH_{max} (Oe)	500	Hardness Scale	- -
Energy Loss W_c (watt-sec/cycle/lb)	-	Basic Forming Method	E
Importance of Operating Magnet at BH_{max}	E	Basic Finishing Method	SRD
Average Recoil Permeability	-	Relative Corrosion Resistance	E
Field Strength Required to Fully Saturate Magnet (Oe)	10,000	Primary Chemical Composition	Ba-Fe + binder
Ability to Withstand Demagnetizing Fields	Ex	Special Notes: (1) Nylon binder, injection molded. (2) Maximum service temperature depends on grade of nylon binder used.	
Maximum Service Temperature (°F)	See Note		
Approximate Temperature Permanently Affecting Magnet (°F)	200		
Inherent Orientation of Material	none		

Note: All information based on available producers literature. In some cases of foreign materials interpretation or calculation required.

SOURCE(S):

Krupp Widia GMBH

Germany

Koerdym 32P
(Alloy or Grade)

Bonded Rare Earth
(Type of Material)

28/90P
(MMPA Brief Designation)

-
(IEC Code Reference)

MAGNETIC PROPERTIES		PHYSICAL PROPERTIES	
Residual Induction B_r (G)	4500	Density (lbs./in.^{3})	.1590
Coercive Force (Oe) — Normal (H_c)	3770	General Mechanical Properties	HT
Coercive Force (Oe) — Intrinsic (H_{ci})	12,750	Tensile Strength (psi)	-
Maximum Energy Product BH_{max} (MGO)	4.0	Transverse Modulus of Rupture (psi)	-
B/H Ratio (Load Line) at BH_{max}	-	Coeff. of Therm. Expansion inches x 10^{-6}/°C — Orientation: Par. - Rt.Angl. -	
Flux Density in Magnet B_m at BH_{max} (G)	-	Resistivity (micro-ohms/cm/cm^2)	- @ - °C
Coercive Force in Magnet H_m at BH_{max} (Oe)	-	Hardness — Scale	- -
Energy Loss W_c (watt-sec/cycle/lb)	-	Basic Forming Method	E
Importance of Operating Magnet at BH_{max}	E	Basic Finishing Method	MCT
Average Recoil Permeability	1.15	Relative Corrosion Resistance	G
Field Strength Required to Fully Saturate Magnet (Oe)	30,000	Primary Chemical Composition	Nd-Fe-B + binder (See Note)
Ability to Withstand Demagnetizing Fields	0	Special Notes:	
Maximum Service Temperature (°F)	-	(1) Injection Molded. Binder not indicated.	
Approximate Temperature Permanently Affecting Magnet (°F)	594		
Inherent Orientation of Material	none		

Note: All information based on available producers literature. In some cases of foreign materials interpretation or calculation required.

SOURCE(S):

Krupp Widia GMBH

Germany

Koerdym 35P
(Alloy or Grade)

Bonded Rare Earth
(Type of Material)

30/54P
(MMPA Brief Designation)

-
(IEC Code Reference)

MAGNETIC PROPERTIES	
Residual Induction B_r (G)	4800
Coercive Force (Oe) Normal (H_c)	3770
Coercive Force (Oe) Intrinsic (H_{ci})	7540
Maximum Energy Product BH_{max} (MGO)	4.4
B/H Ratio (Load Line) at BH_{max}	-
Flux Density in Magnet B_m at BH_{max} (G)	-
Coercive Force in Magnet H_m at BH_{max} (Oe)	-
Energy Loss W_c (watt-sec/cycle/lb)	-
Importance of Operating Magnet at BH_{max}	E
Average Recoil Permeability	1.25
Field Strength Required to Fully Saturate Magnet (Oe)	30,000
Ability to Withstand Demagnetizing Fields	0
Maximum Service Temperature (°F)	-
Approximate Temperature Permanently Affecting Magnet (°F)	594
Inherent Orientation of Material	none

PHYSICAL PROPERTIES	
Density (lbs./in.^{3})	.1590
General Mechanical Properties	HT
Tensile Strength (psi)	-
Transverse Modulus of Rupture (psi)	-
Coeff. of Therm. Expansion inches x 10^{-6}/°C	Orientation: Par. - Rt.Angl. -
Resistivity (micro-ohms/cm/cm^2)	- @ - °C
Hardness -	Scale -
Basic Forming Method	E
Basic Finishing Method	MCT
Relative Corrosion Resistance	G
Primary Chemical Composition	Nd-Fe-B + binder (See Note)
Special Notes: (1) Injection Molded. Binder not indicated.	

Note: All information based on available producers literature. In some cases of foreign materials interpretation or calculation required.

SOURCE(S):

Krupp Widia GMBH

Germany

Koerdym 38P
(Alloy or Grade)

Bonded Rare Earth
(Type of Material)

34/90P
(MMPA Brief Designation)

-
(IEC Code Reference)

MAGNETIC PROPERTIES	
Residual Induction B_r (G)	4900
Coercive Force (Oe) — Normal (H_c)	4020
Coercive Force (Oe) — Intrinsic (H_{ci})	12,700
Maximum Energy Product BH_{max} (MGO)	4.8
B/H Ratio (Load Line) at BH_{max}	1.2
Flux Density in Magnet B_m at BH_{max} (G)	2,400
Coercive Force in Magnet H_m at BH_{max} (Oe)	2,000
Energy Loss W_c (watt-sec/cycle/lb)	-
Importance of Operating Magnet at BH_{max}	E
Average Recoil Permeability	1.15
Field Strength Required to Fully Saturate Magnet (Oe)	30,000
Ability to Withstand Demagnetizing Fields	0
Maximum Service Temperature (°F)	-
Approximate Temperature Permanently Affecting Magnet (°F)	594
Inherent Orientation of Material	none

PHYSICAL PROPERTIES	
Density (lbs./in.^{3})	.1734
General Mechanical Properties	HT
Tensile Strength (psi)	-
Transverse Modulus of Rupture (psi)	-
Coeff. of Therm. Expansion inches x 10^{-6}/°C	Orientation: Par. - Rt.Angl. -
Resistivity (micro-ohms/cm/cm^2)	- @ - °C
Hardness / Scale	- / -
Basic Forming Method	E
Basic Finishing Method	MCT
Relative Corrosion Resistance	G
Primary Chemical Composition	Nd-Fe-B + binder (See Note)
Special Notes: (1) Injection Molded. Binder not indicated.	

Note: All information based on available producers literature. In some cases of foreign materials interpretation or calculation required.

SOURCE(S):	Koerdym 42P
	(Alloy or Grade)
Krupp Widia GMBH	Bonded Rare Earth
	(Type of Material)
Germany	35/54P
	(MMPA Brief Designation)
	-
	(IEC Code Reference)

MAGNETIC PROPERTIES		PHYSICAL PROPERTIES	
Residual Induction B_r (G)	5200	Density (lbs./in.^{3})	.1734
Coercive Force (Oe) — Normal (H_c)	4020	General Mechanical Properties	HT
Coercive Force (Oe) — Intrinsic (H_{ci})	7540	Tensile Strength (psi)	-
Maximum Energy Product BH_{max} (MGO)	5.3	Transverse Modulus of Rupture (psi)	-
B/H Ratio (Load Line) at BH_{max}	1.19	Coeff. of Therm. Expansion inches x 10^{-6}/°C	Orientation: Par. - Rt.Angl. -
Flux Density in Magnet B_m at BH_{max} (G)	2500	Resistivity (micro-ohms/cm/cm^2)	- @ - °C
Coercive Force in Magnet H_m at BH_{max} (Oe)	2100	Hardness - Scale -	
Energy Loss W_c (watt-sec/cycle/lb)	-	Basic Forming Method	E
Importance of Operating Magnet at BH_{max}	E	Basic Finishing Method	MCT
Average Recoil Permeability	1.25	Relative Corrosion Resistance	G
Field Strength Required to Fully Saturate Magnet (Oe)	30,000	Primary Chemical Composition	Nd-Fe-B + binder (See Note)
Ability to Withstand Demagnetizing Fields	0	Special Notes: (1) Injection Molded. Binder not indicated.	
Maximum Service Temperature (°F)	-		
Approximate Temperature Permanently Affecting Magnet (°F)	594		
Inherent Orientation of Material	none		

Note: All information based on available producers literature. In some cases of foreign materials interpretation or calculation required.

SOURCE(S):	Koerdym 55/100P
Krupp Widia GMBH	(Alloy or Grade)
	Bonded Rare Earth
	(Type of Material)
Germany	55/100P
	(MMPA Brief Designation)
	-
	(IEC Code Reference)

MAGNETIC PROPERTIES		PHYSICAL PROPERTIES	
Residual Induction B_r (G)	6000	Density (lbs./in.^{3})	.2168
Coercive Force (Oe) — Normal (H_c)	5100	General Mechanical Properties	HT
Coercive Force (Oe) — Intrinsic (H_{ci})	14,400	Tensile Strength (psi)	-
Maximum Energy Product BH_{max} (MGO)	7.5	Transverse Modulus of Rupture (psi)	-
B/H Ratio (Load Line) at BH_{max}	1.27	Coeff. of Therm. Expansion inches x 10^{-6}/°C Orientation: Par. — Rt.Angl. —	
Flux Density in Magnet B_m at BH_{max} (G)	3500	Resistivity (micro-ohms/cm/cm^2)	- @ - °C
Coercive Force in Magnet H_m at BH_{max} (Oe)	2760	Hardness Scale	- -
Energy Loss W_c (watt-sec/cycle/lb)	-	Basic Forming Method	E
Importance of Operating Magnet at BH_{max}	E	Basic Finishing Method	MCT
Average Recoil Permeability	1.15	Relative Corrosion Resistance	G
Field Strength Required to Fully Saturate Magnet (Oe)	35,000	Primary Chemical Composition	Nd + Fe-B + binder (See Note)
Ability to Withstand Demagnetizing Fields	0	Special Notes:	
Maximum Service Temperature (°F)	-	(1) Obsolete material - see other Koerdym grades of plastic bonded rare earth.	
Approximate Temperature Permanently Affecting Magnet (°F)	594	(2) Injection or compression molded. Binder not indicated.	
Inherent Orientation of Material	none		

Note: All information based on available producers literature. In some cases of foreign materials interpretation or calculation required.

SOURCE(S):

Krupp Widia GMBH

Germany

Koerdym 60P
(Alloy or Grade)

Bonded Rare Earth
(Type of Material)

55/95P
(MMPA Brief Designation)

-
(IEC Code Reference)

MAGNETIC PROPERTIES	
Residual Induction B_r (G)	5700
Coercive Force (Oe) Normal (H_c)	4780
Coercive Force (Oe) Intrinsic (H_{ci})	11,940
Maximum Energy Product BH_{max} (MGO)	6.9
B/H Ratio (Load Line) at BH_{max}	1.18
Flux Density in Magnet B_m at BH_{max} (G)	2850
Coercive Force in Magnet H_m at BH_{max} (Oe)	425
Energy Loss W_c (watt-sec/cycle/lb)	-
Importance of Operating Magnet at BH_{max}	E
Average Recoil Permeability	1.15
Field Strength Required to Fully Saturate Magnet (Oe)	30,000
Ability to Withstand Demagnetizing Fields	0
Maximum Service Temperature (°F)	-
Approximate Temperature Permanently Affecting Magnet (°F)	594
Inherent Orientation of Material	none

PHYSICAL PROPERTIES	
Density (lbs./in.^{3})	.2132
General Mechanical Properties	HT
Tensile Strength (psi)	-
Transverse Modulus of Rupture (psi)	-
Coeff. of Therm. Expansion inches x 10^{-6}/°C	Orientation: Par. - Rt.Angl. -
Resistivity (micro-ohms/cm/cm^2)	- @ - °C
Hardness -	Scale -
Basic Forming Method	E
Basic Finishing Method	MCT
Relative Corrosion Resistance	G
Primary Chemical Composition	Nd-Fe-B + binder (See Note)

Special Notes:

(1) Compression Molded. Binder not indicated.

Note: All information based on available producers literature. In some cases of foreign materials interpretation or calculation required.

SOURCE(S):

Koerdym 75P
(Alloy or Grade)

Krupp Widia GMBH

Bonded Rare Earth
(Type of Material)

Germany

68/62P
(MMPA Brief Designation)

-
(IEC Code Reference)

MAGNETIC PROPERTIES		PHYSICAL PROPERTIES	
Residual Induction B_r (G)	6900	Density (lbs./in.^{3})	.2132
Coercive Force (Oe) — Normal (H_c)	5030	General Mechanical Properties	HT
Coercive Force (Oe) — Intrinsic (H_{ci})	8550	Tensile Strength (psi)	-
Maximum Energy Product BH_{max} (MGO)	9.4	Transverse Modulus of Rupture (psi)	-
B/H Ratio (Load Line) at BH_{max}	1.63	Coeff. of Therm. Expansion inches x 10^{-6}/°C Orientation: Par. - Rt.Angl. -	
Flux Density in Magnet B_m at BH_{max} (G)	3900	Resistivity (micro-ohms/cm/cm^2 - @ - °C	
Coercive Force in Magnet H_m at BH_{max} (Oe)	2400	Hardness Scale	- -
Energy Loss W_c (watt-sec/cycle/lb)	-	Basic Forming Method	E
Importance of Operating Magnet at BH_{max}	E	Basic Finishing Method	MCT
Average Recoil Permeability	1.25	Relative Corrosion Resistance	G
Field Strength Required to Fully Saturate Magnet (Oe)	30,000	Primary Chemical Composition	Nd-Fe-B + binder (See Note)
Ability to Withstand Demagnetizing Fields	0	Special Notes: (1) Compression Molded. Binder not indicated.	
Maximum Service Temperature (°F)	-		
Approximate Temperature Permanently Affecting Magnet (°F)	594		
Inherent Orientation of Material	none		

Note: All information based on available producers literature. In some cases of foreign materials interpretation or calculation required.

SOURCE(S):

Koerdym 190
(Alloy or Grade)

Krupp Widia GMBH

Sintered Rare Earth
(Type of Material)

Germany

24/21
(MMPA Brief Designation)

R7-3
(IEC Code Reference)

MAGNETIC PROPERTIES		PHYSICAL PROPERTIES	
Residual Induction B_r (G)	10,000	Density (lbs./in.^{3})	.2673
Coercive Force (Oe) — Normal (H_c)	9,420	General Mechanical Properties	HB
Coercive Force (Oe) — Intrinsic (H_{ci})	21,350	Tensile Strength (psi)	-
Maximum Energy Product BH_{max} (MGO)	23.9	Transverse Modulus of Rupture (psi)	-
B/H Ratio (Load Line) at BH_{max}	1.25	Coeff. of Therm. Expansion inches x 10^{-6}/°C	Orientation: Par. 3.4 Rt.Angl. -4.8
Flux Density in Magnet B_m at BH_{max} (G)	5500	Resistivity (micro-ohms/cm/cm^2)	140 @ 25 °C
Coercive Force in Magnet H_m at BH_{max} (Oe)	4400	Hardness Scale	500-600 Vickers
Energy Loss W_c (watt-sec/cycle/lb)	-	Basic Forming Method	DPM
Importance of Operating Magnet at BH_{max}	E	Basic Finishing Method	WGR
Average Recoil Permeability	1.1	Relative Corrosion Resistance	P
Field Strength Required to Fully Saturate Magnet (Oe)	40,000	Primary Chemical Composition	Nd-Fe-B
Ability to Withstand Demagnetizing Fields	0	Special Notes:	
Maximum Service Temperature (°F)	Est. 392		
Approximate Temperature Permanently Affecting Magnet (°F)	Est. 698		
Inherent Orientation of Material	MP		

Note: All information based on available producers literature. In some cases of foreign materials interpretation or calculation required.

SOURCE(S):

Krupp Widia GMBH

Germany

Koerdym 210
(Alloy or Grade)

Sintered Rare Earth
(Type of Material)

26/18
(MMPA Brief Designation)

R7-3
(IEC Code Reference)

MAGNETIC PROPERTIES	
Residual Induction B_r (G)	10,500
Coercive Force (Oe) — Normal (H_c)	9800
Coercive Force (Oe) — Intrinsic (H_{ci})	17,580
Maximum Energy Product BH_{max} (MGO)	26.4
B/H Ratio (Load Line) at BH_{max}	1.28
Flux Density in Magnet B_m at BH_{max} (G)	5800
Coercive Force in Magnet H_m at BH_{max} (Oe)	4520
Energy Loss W_c (watt-sec/cycle/lb)	-
Importance of Operating Magnet at BH_{max}	E
Average Recoil Permeability	1.1
Field Strength Required to Fully Saturate Magnet (Oe)	32,000
Ability to Withstand Demagnetizing Fields	0
Maximum Service Temperature (°F)	Est. 356
Approximate Temperature Permanently Affecting Magnet (°F)	Est. 698
Inherent Orientation of Material	MP

PHYSICAL PROPERTIES	
Density (lbs./in.^{3})	.2673
General Mechanical Properties	HB
Tensile Strength (psi)	-
Transverse Modulus of Rupture (psi)	-
Coeff. of Therm. Expansion inches x 10^{-6}/°C	Orientation: Par. 3.4 Rt.Angl. -4.8
Resistivity (micro-ohms/cm/cm^2)	140 @ 25 °C
Hardness	Scale 500-600 Vickers
Basic Forming Method	DPM
Basic Finishing Method	WGR
Relative Corrosion Resistance	P
Primary Chemical Composition	Nd-Fe-B
Special Notes:	

Note: All information based on available producers literature. In some cases of foreign materials interpretation or calculation required.

SOURCE(S):	Koerdym 230
	(Alloy or Grade)
Krupp Widia GMBH	Sintered Rare Earth
	(Type of Material)
Germany	29/13
	(MMPA Brief Designation)
	R7-3
	(IEC Code Reference)

MAGNETIC PROPERTIES		PHYSICAL PROPERTIES	
Residual Induction B_r (G)	10,800	Density (lbs./in.^{3})	.2673
Coercive Force (Oe) — Normal (H_c)	10,420	General Mechanical Properties	HB
Coercive Force (Oe) — Intrinsic (H_{ci})	12,560	Tensile Strength (psi)	-
Maximum Energy Product BH_{max} (MGO)	28.9	Transverse Modulus of Rupture (psi)	-
B/H Ratio (Load Line) at BH_{max}	1.29	Coeff. of Therm. Expansion inches x 10^{-6}/°C	Orientation: Par. 3.4 Rt.Angl. -4.8
Flux Density in Magnet B_m at BH_{max} (G)	6000	Resistivity (micro-ohms/cm/cm^2)	140 @ 25 °C
Coercive Force in Magnet H_m at BH_{max} (Oe)	4650	Hardness Scale	500-600 Vickers
Energy Loss W_c (watt-sec/cycle/lb)	-	Basic Forming Method	DPM
Importance of Operating Magnet at BH_{max}	E	Basic Finishing Method	WGR
Average Recoil Permeability	1.1	Relative Corrosion Resistance	P
Field Strength Required to Fully Saturate Magnet (Oe)	32,000	Primary Chemical Composition	Nd-Fe-B
Ability to Withstand Demagnetizing Fields	0	Special Notes:	
Maximum Service Temperature (°F)	Est. 240		
Approximate Temperature Permanently Affecting Magnet (°F)	Est. 590		
Inherent Orientation of Material	MP		

Note: All information based on available producers literature. In some cases of foreign materials interpretation or calculation required.

SOURCE(S): **Krupp Widia GMBH** **Germany**

Koerdym 240
(Alloy or Grade)

Sintered Rare Earth
(Type of Material)

30/21
(MMPA Brief Designation)

R7-3
(IEC Code Reference)

MAGNETIC PROPERTIES	
Residual Induction B_r (G)	11,000
Coercive Force (Oe) — Normal (H_c)	10,550
Coercive Force (Oe) — Intrinsic (H_{ci})	21,350
Maximum Energy Product BH_{max} (MGO)	30.1
B/H Ratio (Load Line) at BH_{max}	1.29
Flux Density in Magnet B_m at BH_{max} (G)	6000
Coercive Force in Magnet H_m at BH_{max} (Oe)	4650
Energy Loss W_c (watt-sec/cycle/lb)	-
Importance of Operating Magnet at BH_{max}	E
Average Recoil Permeability	1.1
Field Strength Required to Fully Saturate Magnet (Oe)	40,000
Ability to Withstand Demagnetizing Fields	0
Maximum Service Temperature (°F)	Est. 392
Approximate Temperature Permanently Affecting Magnet (°F)	Est. 698
Inherent Orientation of Material	MP

PHYSICAL PROPERTIES	
Density (lbs./in.^{3})	.2673
General Mechanical Properties	HB
Tensile Strength (psi)	-
Transverse Modulus of Rupture (psi)	-
Coeff. of Therm. Expansion inches x 10^{-6}/°C	Orientation: Par. 3.4 Rt.Angl. -4.8
Resistivity (micro-ohms/cm/cm^2)	140 @ 25 °C
Hardness Scale	500-600 Vickers
Basic Forming Method	DPM
Basic Finishing Method	WGR
Relative Corrosion Resistance	P
Primary Chemical Composition	Nd-Fe-B
Special Notes:	

Note: All information based on available producers literature. In some cases of foreign materials interpretation or calculation required.

SOURCE(S):

Krupp Widia GMBH

Germany

Koerdym 260
(Alloy or Grade)

Sintered Rare Earth
(Type of Material)

33/18
(MMPA Brief Designation)

R7-3
(IEC Code Reference)

MAGNETIC PROPERTIES	
Residual Induction B_r (G)	11,500
Coercive Force (Oe) Normal (H_c)	11,050
Coercive Force (Oe) Intrinsic (H_{ci})	17,580
Maximum Energy Product BH_{max} (MGO)	32.6
B/H Ratio (Load Line) at BH_{max}	1.27
Flux Density in Magnet B_m at BH_{max} (G)	6200
Coercive Force in Magnet H_m at BH_{max} (Oe)	4900
Energy Loss W_c (watt-sec/cycle/lb)	-
Importance of Operating Magnet at BH_{max}	E
Average Recoil Permeability	1.1
Field Strength Required to Fully Saturate Magnet (Oe)	32,000
Ability to Withstand Demagnetizing Fields	0
Maximum Service Temperature (°F)	Est. 248
Approximate Temperature Permanently Affecting Magnet (°F)	Est. 590
Inherent Orientation of Material	MP

PHYSICAL PROPERTIES	
Density (lbs./in.^{3})	.2673
General Mechanical Properties	HB
Tensile Strength (psi)	-
Transverse Modulus of Rupture (psi)	-
Coeff. of Therm. Expansion inches x 10^{-6}/°C	Orientation: Par. 3.4 Rt.Angl. -4.8
Resistivity (micro-ohms/cm/cm^2)	140 @ 25 °C
Hardness	Scale 500-600 Vickers
Basic Forming Method	DPM
Basic Finishing Method	WGR
Relative Corrosion Resistance	P
Primary Chemical Composition	Nd-Fe-B
Special Notes:	

Note: All information based on available producers literature. In some cases of foreign materials interpretation or calculation required.

SOURCE(S):	**Koerdym 280**
	(Alloy or Grade)
Krupp Widia GMBH	Sintered Rare Earth
	(Type of Material)
Germany	35/13
	(MMPA Brief Designation)
	R7-3
	(IEC Code Reference)

MAGNETIC PROPERTIES		PHYSICAL PROPERTIES	
Residual Induction B_r (G)	12,000	Density (lbs./in.^{3})	.2673
Coercive Force (Oe) — Normal (H_c)	11,560	General Mechanical Properties	HB
Coercive Force (Oe) — Intrinsic (H_{ci})	12,560	Tensile Strength (psi)	-
Maximum Energy Product BH_{max} (MGO)	35.1	Transverse Modulus of Rupture (psi)	-
B/H Ratio (Load Line) at BH_{max}	1.27	Coeff. of Therm. Expansion inches x 10^{-6}/°C	Orientation: Par. 3.4 Rt.Angl. -4.8
Flux Density in Magnet B_m at BH_{max} (G)	6700	Resistivity (micro-ohms/cm/cm^2)	140 @ 25 °C
Coercive Force in Magnet H_m at BH_{max} (Oe)	5280	Hardness Scale	500-600 Vickers
Energy Loss W_c (watt-sec/cycle/lb)	-	Basic Forming Method	DPM
Importance of Operating Magnet at BH_{max}	E	Basic Finishing Method	WGR
Average Recoil Permeability	1.1	Relative Corrosion Resistance	P
Field Strength Required to Fully Saturate Magnet (Oe)	32,000	Primary Chemical Composition	Nd-Fe-B
Ability to Withstand Demagnetizing Fields	O	Special Notes:	
Maximum Service Temperature (°F)	Est. 212		
Approximate Temperature Permanently Affecting Magnet (°F)	Est. 590		
Inherent Orientation of Material	MP		

Note: All information based on available producers literature. In some cases of foreign materials interpretation or calculation required.

SOURCE(S):

Krupp Widia GMBH

Germany

Koerflex 160 [8/2]
(Alloy or Grade)

Wrought
(Type of Material)

1.3/.31
(MMPA Brief Designation)

(IEC Code Reference)

MAGNETIC PROPERTIES	
Residual Induction B_r (G)	9400
Coercive Force (Oe) — Normal (H_c)	300
Coercive Force (Oe) — Intrinsic (H_{ci})	310
Maximum Energy Product BH_{max} (MGO)	1.3
B/H Ratio (Load Line) at BH_{max}	36.8
Flux Density in Magnet B_m at BH_{max} (G)	7000
Coercive Force in Magnet H_m at BH_{max} (Oe)	190
Energy Loss W_c (watt-sec/cycle/lb)	-
Importance of Operating Magnet at BH_{max}	A
Average Recoil Permeability	5.0
Field Strength Required to Fully Saturate Magnet (Oe)	1600
Ability to Withstand Demagnetizing Fields	F
Maximum Service Temperature (°F)	900
Approximate Temperature Permanently Affecting Magnet (°F)	1175
Inherent Orientation of Material	CR

PHYSICAL PROPERTIES	
Density (lbs./in.^{3})	.2728
General Mechanical Properties	HT
Tensile Strength (psi)	-
Transverse Modulus of Rupture (psi)	-
Coeff. of Therm. Expansion inches x 10^{-6}/°C	Orientation: Par. 12.0 Rt.Angl. _
Resistivity (micro-ohms/cm/cm^2)	70 @ 25 °C
Hardness Scale	220-320 Vickers
Basic Forming Method	W
Basic Finishing Method	WGR
Relative Corrosion Resistance	E
Primary Chemical Composition	Fe-Cr-Co
Special Notes:	

Note: All information based on available producers literature. In some cases of foreign materials interpretation or calculation required.

SOURCE(S): **Krupp Widia GMBH** Germany

Koerflex 160 [10/4]
(Alloy or Grade)

Wrought
(Type of Material)

1.5/.51
(MMPA Brief Designation)

(IEC Code Reference)

MAGNETIC PROPERTIES	
Residual Induction B_r (G)	8400
Coercive Force (Oe) Normal (H_c)	500
Coercive Force (Oe) Intrinsic (H_{ci})	510
Maximum Energy Product BH_{max} (MGO)	1.5
B/H Ratio (Load Line) at BH_{max}	24.0
Flux Density in Magnet B_m at BH_{max} (G)	6000
Coercive Force in Magnet H_m at BH_{max} (Oe)	250
Energy Loss W_c (watt-sec/cycle/lb)	-
Importance of Operating Magnet at BH_{max}	B
Average Recoil Permeability	3.0
Field Strength Required to Fully Saturate Magnet (Oe)	1600
Ability to Withstand Demagnetizing Fields	F
Maximum Service Temperature (°F)	900
Approximate Temperature Permanently Affecting Magnet (°F)	1175
Inherent Orientation of Material	CR

PHYSICAL PROPERTIES	
Density (lbs./in.^{3})	.2728
General Mechanical Properties	HT
Tensile Strength (psi)	-
Transverse Modulus of Rupture (psi)	-
Coeff. of Therm. Expansion inches x 10^{-6}/°C	Orientation: Par. - Rt.Angl. -
Resistivity (micro-ohms/cm/cm^2)	70 @ 25 °C
Hardness 220-320	Scale Vickers
Basic Forming Method	W
Basic Finishing Method	WGR
Relative Corrosion Resistance	E
Primary Chemical Composition	Fe-Cr-Co
Special Notes:	

Note: All information based on available producers literature. In some cases of foreign materials interpretation or calculation required.

SOURCE(S):

Krupp Widia GMBH

Germany

Koerflex 300 (strip)
(Alloy or Grade)

Wrought
(Type of Material)

1.8/.34
(MMPA Brief Designation)

_____
(IEC Code Reference)

MAGNETIC PROPERTIES			PHYSICAL PROPERTIES	
Residual Induction B_r (G)		9500	Density (lbs./in.^{3})	.2962
Coercive Force (Oe)	Normal (H_c)	330	General Mechanical Properties	HT
	Intrinsic (H_{ci})	340	Tensile Strength (psi)	-
Maximum Energy Product BH_{max} (MGO)		1.8	Transverse Modulus of Rupture (psi)	-
B/H Ratio (Load Line) at BH_{max}		30.0	Coeff. of Therm. Expansion inches x 10^{-6}/°C	Orientation: Par. 12.0 Rt.Angl. _
Flux Density in Magnet B_m at BH_{max} (G)		7500	Resistivity (micro-ohms/cm/cm^2 -	@ - °C
Coercive Force in Magnet H_m at BH_{max} (Oe)		250	Hardness Scale 500-950	Vickers
Energy Loss W_c (watt-sec/cycle/lb)		-	Basic Forming Method	W
Importance of Operating Magnet at BH_{max}		A	Basic Finishing Method	WGR
Average Recoil Permeability		5.0	Relative Corrosion Resistance	E
Field Strength Required to Fully Saturate Magnet (Oe)		1600	Primary Chemical Composition	Fe-Cr-Co-V
Ability to Withstand Demagnetizing Fields		F	Special Notes:	
Maximum Service Temperature (°F)		900		
Approximate Temperature Permanently Affecting Magnet (°F)		1337		
Inherent Orientation of Material		CR		

Note: All information based on available producers literature. In some cases of foreign materials interpretation or calculation required.

SOURCE(S):

Krupp Widia GMBH

Germany

Koerflex 300 (wire)
(Alloy or Grade)

Wrought
(Type of Material)

2.5/.39
(MMPA Brief Designation)

(IEC Code Reference)

MAGNETIC PROPERTIES	
Residual Induction B_r (G)	11,000
Coercive Force (Oe) — Normal (H_c)	380
Coercive Force (Oe) — Intrinsic (H_{ci})	390
Maximum Energy Product BH_{max} (MGO)	2.5
B/H Ratio (Load Line) at BH_{max}	32.1
Flux Density in Magnet B_m at BH_{max} (G)	9000
Coercive Force in Magnet H_m at BH_{max} (Oe)	280
Energy Loss W_c (watt-sec/cycle/lb)	-
Importance of Operating Magnet at BH_{max}	B
Average Recoil Permeability	3.0
Field Strength Required to Fully Saturate Magnet (Oe)	1600
Ability to Withstand Demagnetizing Fields	F
Maximum Service Temperature (°F)	900
Approximate Temperature Permanently Affecting Magnet (°F)	1337
Inherent Orientation of Material	CR

PHYSICAL PROPERTIES	
Density (lbs./in.^{3})	.2962
General Mechanical Properties	HT
Tensile Strength (psi)	-
Transverse Modulus of Rupture (psi)	-
Coeff. of Therm. Expansion inches x 10^{-6}/°C — Orientation: Par.	12.0
Coeff. of Therm. Expansion inches x 10^{-6}/°C — Rt.Angl.	-
Resistivity (micro-ohms/cm/cm^2)	55 @ 25 °C
Hardness — Scale Vickers	500-950
Basic Forming Method	W
Basic Finishing Method	WGR
Relative Corrosion Resistance	E
Primary Chemical Composition	Fe-Cr-Co-V
Special Notes:	

Note: All information based on available producers literature. In some cases of foreign materials interpretation or calculation required.

SOURCE(S):

Krupp Widia GMBH

Germany

Koermax 150
(Alloy or Grade)

Sintered Rare Earth
(Type of Material)

18/6
(MMPA Brief Designation)

R5-1
(IEC Code Reference)

MAGNETIC PROPERTIES	
Residual Induction B_r (G)	8900
Coercive Force (Oe) Normal (H_c)	5400
Coercive Force (Oe) Intrinsic (H_{ci})	6280
Maximum Energy Product BH_{max} (MGO)	17.6
B/H Ratio (Load Line) at BH_{max}	1.59
Flux Density in Magnet B_m at BH_{max} (G)	5200
Coercive Force in Magnet H_m at BH_{max} (Oe)	3270
Energy Loss W_c (watt-sec/cycle/lb)	-
Importance of Operating Magnet at BH_{max}	E
Average Recoil Permeability	1.1
Field Strength Required to Fully Saturate Magnet (Oe)	15,000
Ability to Withstand Demagnetizing Fields	0
Maximum Service Temperature (°F)	Est. 482
Approximate Temperature Permanently Affecting Magnet (°F)	Est. 1328
Inherent Orientation of Material	MP

PHYSICAL PROPERTIES	
Density (lbs./in.^{3})	.2999
General Mechanical Properties	HB
Tensile Strength (psi)	Est. 5000
Transverse Modulus of Rupture (psi)	-
Coeff. of Therm. Expansion inches x 10^{-6}/°C	Orientation: Par. _ Rt.Angl. _
Resistivity (micro-ohms/cm/cm^2)	85 @ 25 °C
Hardness	Scale 500-600 Vickers
Basic Forming Method	DPM
Basic Finishing Method	WGR
Relative Corrosion Resistance	G
Primary Chemical Composition	Sm-Co
Special Notes:	

Note: All information based on available producers literature. In some cases of foreign materials interpretation or calculation required.

SOURCE(S): **Krupp Widia GMBH** Germany

Koermax 160
(Alloy or Grade)

Sintered Rare Earth
(Type of Material)

20/19
(MMPA Brief Designation)

R5-1
(IEC Code Reference)

MAGNETIC PROPERTIES	
Residual Induction B_r (G)	9400
Coercive Force (Oe) — Normal (H_c)	8540
Coercive Force (Oe) — Intrinsic (H_{ci})	18,840
Maximum Energy Product BH_{max} (MGO)	20.1
B/H Ratio (Load Line) at BH_{max}	1.14
Flux Density in Magnet B_m at BH_{max} (G)	4600
Coercive Force in Magnet H_m at BH_{max} (Oe)	4020
Energy Loss W_c (watt-sec/cycle/lb)	-
Importance of Operating Magnet at BH_{max}	E
Average Recoil Permeability	1.1
Field Strength Required to Fully Saturate Magnet (Oe)	40,000
Ability to Withstand Demagnetizing Fields	0
Maximum Service Temperature (°F)	Est. 482
Approximate Temperature Permanently Affecting Magnet (°F)	Est. 1328
Inherent Orientation of Material	MP

PHYSICAL PROPERTIES	
Density (lbs./in.^{3})	.2999
General Mechanical Properties	HB
Tensile Strength (psi)	Est. 5000
Transverse Modulus of Rupture (psi)	-
Coeff. of Therm. Expansion inches x 10^{-6}/°C	Orientation: Par. - Rt.Angl. -
Resistivity (micro-ohms/cm/cm^2)	85 @ 25 °C
Hardness Scale	500-600 Vickers
Basic Forming Method	DPM
Basic Finishing Method	WGR
Relative Corrosion Resistance	G
Primary Chemical Composition	Sm-Co
Special Notes:	

Note: All information based on available producers literature. In some cases of foreign materials interpretation or calculation required.

SOURCE(S):	
Krupp Widia GMBH Germany	**Koermax 170** (Alloy or Grade)
	Sintered Rare Earth (Type of Material)
	21/6 (MMPA Brief Designation)
	R5-1 (IEC Code Reference)

MAGNETIC PROPERTIES		PHYSICAL PROPERTIES	
Residual Induction B_r (G)	9600	Density (lbs./in.^{3})	.2999
Coercive Force (Oe) — Normal (H_c)	5650	General Mechanical Properties	HB
Coercive Force (Oe) — Intrinsic (H_{ci})	6280	Tensile Strength (psi)	Est. 5000
Maximum Energy Product BH_{max} (MGO)	21.3	Transverse Modulus of Rupture (psi)	-
B/H Ratio (Load Line) at BH_{max}	1.59	Coeff. of Therm. Expansion inches x 10^{-6}/°C	Orientation: Par. - Rt.Angl. -
Flux Density in Magnet B_m at BH_{max} (G)	5800	Resistivity (micro-ohms/cm/cm^2)	85 @ 25 °C
Coercive Force in Magnet H_m at BH_{max} (Oe)	3640	Hardness Scale	500-600 Vickers
Energy Loss W_c (watt-sec/cycle/lb)	-	Basic Forming Method	DPM
Importance of Operating Magnet at BH_{max}	E	Basic Finishing Method	WGR
Average Recoil Permeability	1.1	Relative Corrosion Resistance	G
Field Strength Required to Fully Saturate Magnet (Oe)	15,000	Primary Chemical Composition	Sm-Co
Ability to Withstand Demagnetizing Fields	0	Special Notes:	
Maximum Service Temperature (°F)	Est. 482		
Approximate Temperature Permanently Affecting Magnet (°F)	Est. 1328		
Inherent Orientation of Material	MP		

Note: All information based on available producers literature. In some cases of foreign materials interpretation or calculation required.

SOURCE(S):	Koermax 200
	(Alloy or Grade)
Krupp Widia GMBH	Sintered Rare Earth
	(Type of Material)
Germany	25/19
	(MMPA Brief Designation)
	R5-2
	(IEC Code Reference)

MAGNETIC PROPERTIES		PHYSICAL PROPERTIES	
Residual Induction B_r (G)	10,200	Density (lbs./in.^{3})	.2999
Coercive Force (Oe) — Normal (H_c)	9,290	General Mechanical Properties	HB
Coercive Force (Oe) — Intrinsic (H_{ci})	18,840	Tensile Strength (psi)	Est. 5000
Maximum Energy Product BH_{max} (MGO)	25.1	Transverse Modulus of Rupture (psi)	-
B/H Ratio (Load Line) at BH_{max}	1.02	Coeff. of Therm. Expansion inches x 10^{-6}/°C	Orientation: Par. - Rt.Angl. -
Flux Density in Magnet B_m at BH_{max} (G)	5,000	Resistivity (micro-ohms/cm/cm^2)	85 @ 25 °C
Coercive Force in Magnet H_m at BH_{max} (Oe)	4,900	Hardness	Scale 500-600 Vickers
Energy Loss W_c (watt-sec/cycle/lb)	-	Basic Forming Method	DPM
Importance of Operating Magnet at BH_{max}	E	Basic Finishing Method	WGR
Average Recoil Permeability	1.1	Relative Corrosion Resistance	G
Field Strength Required to Fully Saturate Magnet (Oe)	40,000	Primary Chemical Composition	Sm-Co
Ability to Withstand Demagnetizing Fields	0	Special Notes:	
Maximum Service Temperature (°F)	Est. 482		
Approximate Temperature Permanently Affecting Magnet (°F)	Est. 1328		
Inherent Orientation of Material	MP		

Note: All information based on available producers literature. In some cases of foreign materials interpretation or calculation required.

SOURCE(S):

Krupp Widia GMBH

Germany

Koerox 1/18P
(Alloy or Grade)

Bonded Ferrite
(Type of Material)

.16/2.3
(MMPA Brief Designation)

-
(IEC Code Reference)

MAGNETIC PROPERTIES		PHYSICAL PROPERTIES	
Residual Induction B_r (G)	830	Density (lbs./in.^{3})	.0867
Coercive Force (Oe) — Normal (H_c)	820	General Mechanical Properties	HT
Coercive Force (Oe) — Intrinsic (H_{ci})	2260	Tensile Strength (psi)	-
Maximum Energy Product BH_{max} (MGO)	.16	Transverse Modulus of Rupture (psi)	-
B/H Ratio (Load Line) at BH_{max}	.82	Coeff. of Therm. Expansion inches x 10^{-6}/°C	Orientation: Par. - Rt.Angl. -
Flux Density in Magnet B_m at BH_{max} (G)	360	Resistivity (micro-ohms/cm/cm^2)	- @ - °C
Coercive Force in Magnet H_m at BH_{max} (Oe)	440	Hardness / Scale	- / -
Energy Loss W_c (watt-sec/cycle/lb)	-	Basic Forming Method	E
Importance of Operating Magnet at BH_{max}	E	Basic Finishing Method	WGD
Average Recoil Permeability	1.15	Relative Corrosion Resistance	E
Field Strength Required to Fully Saturate Magnet (Oe)	8000	Primary Chemical Composition	Ba or Sr-Fe + binder
Ability to Withstand Demagnetizing Fields	Ex	Special Notes: (1) Injection molded with thermoplastic binder (2) Maximum service temperature depends on binder used. Binder not identified by manufacturer.	
Maximum Service Temperature (°F)	See Note		
Approximate Temperature Permanently Affecting Magnet (°F)	-		
Inherent Orientation of Material	none		

Note: All information based on available producers literature. In some cases of foreign materials interpretation or calculation required.

SOURCE(S):	Koerox 2/20P
	(Alloy or Grade)
Krupp Widia GMBH	Bonded Ferrite
	(Type of Material)
Germany	.3/2.5
	(MMPA Brief Designation)
	-
	(IEC Code Reference)

MAGNETIC PROPERTIES		PHYSICAL PROPERTIES	
Residual Induction B_r (G)	1180	Density (lbs./in.^{3})	.1192
Coercive Force (Oe) — Normal (H_c)	1010	General Mechanical Properties	HT
Coercive Force (Oe) — Intrinsic (H_{ci})	2510	Tensile Strength (psi)	-
Maximum Energy Product BH_{max} (MGO)	.30	Transverse Modulus of Rupture (psi)	-
B/H Ratio (Load Line) at BH_{max}	1.2	Coeff. of Therm. Expansion inches x 10^{-6}/°C Orientation: Par. - Rt.Angl. -	
Flux Density in Magnet B_m at BH_{max} (G)	600	Resistivity (micro-ohms/cm/cm^2)	- @ - °C
Coercive Force in Magnet H_m at BH_{max} (Oe)	500	Hardness Scale	- -
Energy Loss W_c (watt-sec/cycle/lb)	-	Basic Forming Method	E
Importance of Operating Magnet at BH_{max}	E	Basic Finishing Method	WGD
Average Recoil Permeability	1.15	Relative Corrosion Resistance	E
Field Strength Required to Fully Saturate Magnet (Oe)	8000	Primary Chemical Composition	Ba or Sr-Fe + binder
Ability to Withstand Demagnetizing Fields	Ex	Special Notes: (1) Injection molded with thermoplastic binder. (2) Maximum service temperature depends on binder used. Binder not identified by manufacturer.	
Maximum Service Temperature (°F)	See Note		
Approximate Temperature Permanently Affecting Magnet (°F)	-		
Inherent Orientation of Material	none		

Note: All information based on available producers literature. In some cases of foreign materials interpretation or calculation required.

SOURCE(S):

Krupp Widia GMBH

Germany

Koerox 4/22P
(Alloy or Grade)

Bonded Ferrite
(Type of Material)

.46/2.8
(MMPA Brief Designation)

-
(IEC Code Reference)

MAGNETIC PROPERTIES	
Residual Induction B_r (G)	1500
Coercive Force (Oe) — Normal (H_c)	1190
Coercive Force (Oe) — Intrinsic (H_{ci})	2830
Maximum Energy Product BH_{max} (MGO)	.46
B/H Ratio (Load Line) at BH_{max}	1.19
Flux Density in Magnet B_m at BH_{max} (G)	750
Coercive Force in Magnet H_m at BH_{max} (Oe)	628
Energy Loss W_c (watt-sec/cycle/lb)	-
Importance of Operating Magnet at BH_{max}	E
Average Recoil Permeability	1.15
Field Strength Required to Fully Saturate Magnet (Oe)	10,000
Ability to Withstand Demagnetizing Fields	E
Maximum Service Temperature (°F)	See Note
Approximate Temperature Permanently Affecting Magnet (°F)	-
Inherent Orientation of Material	none

PHYSICAL PROPERTIES	
Density (lbs./in.^{3})	.1373
General Mechanical Properties	HT
Tensile Strength (psi)	-
Transverse Modulus of Rupture (psi)	-
Coeff. of Therm. Expansion inches x 10^{-6}/°C — Orientation: Par.	-
Coeff. of Therm. Expansion — Rt.Angl.	-
Resistivity (micro-ohms/cm/cm^2) @ °C	-
Hardness / Scale	-
Basic Forming Method	E
Basic Finishing Method	WGD
Relative Corrosion Resistance	E
Primary Chemical Composition	Ba or Sr-Fe + binder

Special Notes:

(1) Injection molded with thermoplastic binder.
(2) Maximum service temperature depends on binder used. Binder not identified by manufacturer.

Note: All information based on available producers literature. In some cases of foreign materials interpretation or calculation required.

SOURCE(S):

Krupp Widia GMBH

Germany

Koerox 8/19P
(Alloy or Grade)

Bonded Ferrite
(Type of Material)

1.0/2.4
(MMPA Brief Designation)

-
(IEC Code Reference)

MAGNETIC PROPERTIES		PHYSICAL PROPERTIES	
Residual Induction B_r (G)	2100	Density (lbs./in.^{3})	.1264
Coercive Force (Oe) — Normal (H_c)	1570	General Mechanical Properties	HT
Coercive Force (Oe) — Intrinsic (H_{ci})	2390	Tensile Strength (psi)	-
Maximum Energy Product BH_{max} (MGO)	1.0	Transverse Modulus of Rupture (psi)	-
B/H Ratio (Load Line) at BH_{max}	1.0	Coeff. of Therm. Expansion inches x 10^{-6}/°C	Orientation: Par. - Rt.Angl. -
Flux Density in Magnet B_m at BH_{max} (G)	1000	Resistivity (micro-ohms/cm/cm^2 - @ - °C)	
Coercive Force in Magnet H_m at BH_{max} (Oe)	1000	Hardness - Scale -	
Energy Loss W_c (watt-sec/cycle/lb)	-	Basic Forming Method	E
Importance of Operating Magnet at BH_{max}	E	Basic Finishing Method	WGD
Average Recoil Permeability	1.1	Relative Corrosion Resistance	E
Field Strength Required to Fully Saturate Magnet (Oe)	10,000	Primary Chemical Composition	Ba or Sr-Fe + binder
Ability to Withstand Demagnetizing Fields	E	Special Notes:	
Maximum Service Temperature (°F)	See Note	(1) Injection molded with thermoplastic binder. (2) Maximum service temperature depends on binder used. Binder not identified by manufacturer.	
Approximate Temperature Permanently Affecting Magnet (°F)	-		
Inherent Orientation of Material	none		

Note: All information based on available producers literature. In some cases of foreign materials interpretation or calculation required.

SOURCE(S):	**Koerox 10/22P**
	(Alloy or Grade)
Krupp Widia GMBH	Bonded Ferrite
	(Type of Material)
Germany	1.3/2.8
	(MMPA Brief Designation)
	-
	(IEC Code Reference)

MAGNETIC PROPERTIES		PHYSICAL PROPERTIES	
Residual Induction B_r (G)	2300	Density (lbs./in.^{3})	.1156
Coercive Force (Oe) — Normal (H_c)	1950	General Mechanical Properties	HT
Coercive Force (Oe) — Intrinsic (H_{ci})	2830	Tensile Strength (psi)	-
Maximum Energy Product BH_{max} (MGO)	1.31	Transverse Modulus of Rupture (psi)	-
B/H Ratio (Load Line) at BH_{max}	1.09	Coeff. of Therm. Expansion inches x 10^{-6}/°C	Orientation: Par. - Rt.Angl. -
Flux Density in Magnet B_m at BH_{max} (G)	1200	Resistivity (micro-ohms/cm/cm^2)	- @ - °C
Coercive Force in Magnet H_m at BH_{max} (Oe)	1100	Hardness -	Scale -
Energy Loss W_c (watt-sec/cycle/lb)	-	Basic Forming Method	E
Importance of Operating Magnet at BH_{max}	E	Basic Finishing Method	WGD
Average Recoil Permeability	1.1	Relative Corrosion Resistance	E
Field Strength Required to Fully Saturate Magnet (Oe)	10,000	Primary Chemical Composition	Ba or Sr-Fe + binder
Ability to Withstand Demagnetizing Fields	E	Special Notes: (1) Injection molded with thermoplastic binder. (2) Maximum service temperature depends on binder used. Binder not identified by manufacturer.	
Maximum Service Temperature (°F)	See Note		
Approximate Temperature Permanently Affecting Magnet (°F)	-		
Inherent Orientation of Material	none		

Note: All information based on available producers literature. In some cases of foreign materials interpretation or calculation required.

SOURCE(S):	
Krupp Widia GMBH	Koerox 12/22P
	(Alloy or Grade)
	Bonded Ferrite
	(Type of Material)
Germany	1.6/2.8
	(MMPA Brief Designation)
	-
	(IEC Code Reference)

MAGNETIC PROPERTIES		PHYSICAL PROPERTIES	
Residual Induction B_r (G)	2600	Density (lbs./in.^{3})	.1228
Coercive Force (Oe) — Normal (H_c)	2260	General Mechanical Properties	HT
Coercive Force (Oe) — Intrinsic (H_{ci})	2830	Tensile Strength (psi)	-
Maximum Energy Product BH_{max} (MGO)	1.63	Transverse Modulus of Rupture (psi)	-
B/H Ratio (Load Line) at BH_{max}	.96	Coeff. of Therm. Expansion inches x 10^{-6}/°C	Orientation: Par. - Rt.Angl. -
Flux Density in Magnet B_m at BH_{max} (G)	1250	Resistivity (micro-ohms/cm/cm^2)	- @ - °C
Coercive Force in Magnet H_m at BH_{max} (Oe)	1300	Hardness Scale	- -
Energy Loss W_c (watt-sec/cycle/lb)	-	Basic Forming Method	E
Importance of Operating Magnet at BH_{max}	E	Basic Finishing Method	WGD
Average Recoil Permeability	1.1	Relative Corrosion Resistance	E
Field Strength Required to Fully Saturate Magnet (Oe)	10,000	Primary Chemical Composition	Ba or Sr-Fe + binder
Ability to Withstand Demagnetizing Fields	E	Special Notes: (1) Injection molded with thermpolastic binder. (2) Maximum service temperature depends on binder used. Binder not identified by manufacturer.	
Maximum Service Temperature (°F)	See Note		
Approximate Temperature Permanently Affecting Magnet (°F)	-		
Inherent Orientation of Material	none		

Note: All information based on available producers literature. In some cases of foreign materials interpretation or calculation required.

SOURCE(S):		**Koerox 100**	
		(Alloy or Grade)	
Krupp Widia GMBH		Sintered Ferrite	
		(Type of Material)	
Germany		1.1/3.6	
		(MMPA Brief Designation)	
		-	
		(IEC Code Reference)	

MAGNETIC PROPERTIES		PHYSICAL PROPERTIES	
Residual Induction B_r (G)	2300	Density (lbs./in.^{3})	.1734
Coercive Force (Oe) — Normal (H_c)	1950	General Mechanical Properties	VHB
Coercive Force (Oe) — Intrinsic (H_{ci})	3640	Tensile Strength (psi)	-
Maximum Energy Product BH_{max} (MGO)	1.13	Transverse Modulus of Rupture (psi)	-
B/H Ratio (Load Line) at BH_{max}	1.1	Coeff. of Therm. Expansion inches x 10^{-6}/°C	Orientation: Par. - Rt.Angl. -
Flux Density in Magnet B_m at BH_{max} (G)	1125	Resistivity (micro-ohms/cm/cm^2)	1.0 @ 25 °C
Coercive Force in Magnet H_m at BH_{max} (Oe)	1000	Hardness Scale 8	Vickers HV5
Energy Loss W_c (watt-sec/cycle/lb)	-	Basic Forming Method	PF
Importance of Operating Magnet at BH_{max}	E	Basic Finishing Method	WGD
Average Recoil Permeability	1.2	Relative Corrosion Resistance	Ex
Field Strength Required to Fully Saturate Magnet (Oe)	10,000	Primary Chemical Composition	Ba or Sr-Fe
Ability to Withstand Demagnetizing Fields	Ex	Special Notes:	
Maximum Service Temperature (°F)	482		
Approximate Temperature Permanently Affecting Magnet (°F)	842		
Inherent Orientation of Material	none		

Note: All information based on available producers literature. In some cases of foreign materials interpretation or calculation required.

SOURCE(S):	Koerox 150
	(Alloy or Grade)
Krupp Widia GMBH	Sintered Ferrite
	(Type of Material)
Germany	1.5/3.6
	(MMPA Brief Designation)
	-
	(IEC Code Reference)

MAGNETIC PROPERTIES		PHYSICAL PROPERTIES	
Residual Induction B_r (G)	2700	Density (lbs./in.^{3})	.1770
Coercive Force (Oe) — Normal (H_c)	2200	General Mechanical Properties	VHB
Coercive Force (Oe) — Intrinsic (H_{ci})	3640	Tensile Strength (psi)	-
Maximum Energy Product BH_{max} (MGO)	1.5	Transverse Modulus of Rupture (psi)	-
B/H Ratio (Load Line) at BH_{max}	1.27	Coeff. of Therm. Expansion inches x 10^{-6}/°C Orientation: Par. _ Rt.Angl. _	
Flux Density in Magnet B_m at BH_{max} (G)	1400	Resistivity (micro-ohms/cm/cm^2)	1.0 @ 25 °C
Coercive Force in Magnet H_m at BH_{max} (Oe)	1100	Hardness Scale 8 Vickers HV5	
Energy Loss W_c (watt-sec/cycle/lb)	-	Basic Forming Method	PF
Importance of Operating Magnet at BH_{max}	E	Basic Finishing Method	WGD
Average Recoil Permeability	1.2	Relative Corrosion Resistance	Ex
Field Strength Required to Fully Saturate Magnet (Oe)	10,000	Primary Chemical Composition	Ba or Sr-Fe
Ability to Withstand Demagnetizing Fields	Ex	Special Notes:	
Maximum Service Temperature (°F)	482		
Approximate Temperature Permanently Affecting Magnet (°F)	842		
Inherent Orientation of Material	partial MP		

Note: All information based on available producers literature. In some cases of foreign materials interpretation or calculation required.

SOURCE(S): **Krupp Widia GMBH** Germany

Koerox 300
(Alloy or Grade)

Sintered Ferrite
(Type of Material)

3.0/3.0
(MMPA Brief Designation)

-
(IEC Code Reference)

MAGNETIC PROPERTIES		PHYSICAL PROPERTIES	
Residual Induction B_r (G)	3650	Density (lbs./in.^{3})	.1734
Coercive Force (Oe) Normal (H_c)	2640	General Mechanical Properties	VHB
Coercive Force (Oe) Intrinsic (H_{ci})	3020	Tensile Strength (psi)	-
Maximum Energy Product BH_{max} (MGO)	3.0	Transverse Modulus of Rupture (psi)	-
B/H Ratio (Load Line) at BH_{max}	1.12	Coeff. of Therm. Expansion inches x 10^{-6}/°C Orientation: Par. - Rt.Angl. -	
Flux Density in Magnet B_m at BH_{max} (G)	1850	Resistivity (micro-ohms/cm/cm^2)	1.0 @ 25 °C
Coercive Force in Magnet H_m at BH_{max} (Oe)	2650	Hardness Scale 8 Vickers HV5	
Energy Loss W_c (watt-sec/cycle/lb)	-	Basic Forming Method	PMF
Importance of Operating Magnet at BH_{max}	E	Basic Finishing Method	WGD
Average Recoil Permeability	1.1	Relative Corrosion Resistance	Ex
Field Strength Required to Fully Saturate Magnet (Oe)	10,000	Primary Chemical Composition	Ba or Sr-Fe
Ability to Withstand Demagnetizing Fields	Ex	Special Notes:	
Maximum Service Temperature (°F)	482		
Approximate Temperature Permanently Affecting Magnet (°F)	842		
Inherent Orientation of Material	MP		

Note: All information based on available producers literature. In some cases of foreign materials interpretation or calculation required.

SOURCE(S):

Krupp Widia GMBH

Germany

Koerox 330
(Alloy or Grade)

Sintered Ferrite
(Type of Material)

3.3/3.3
(MMPA Brief Designation)

-
(IEC Code Reference)

MAGNETIC PROPERTIES		PHYSICAL PROPERTIES	
Residual Induction B_r (G)	3700	Density (lbs./in.^{3})	.1734
Coercive Force (Oe) — Normal (H_c)	3020	General Mechanical Properties	VHB
Coercive Force (Oe) — Intrinsic (H_{ci})	3270	Tensile Strength (psi)	-
Maximum Energy Product BH_{max} (MGO)	3.3	Transverse Modulus of Rupture (psi)	-
B/H Ratio (Load Line) at BH_{max}	.95	Coeff. of Therm. Expansion inches x 10^{-6}/°C	Orientation: Par. - Rt.Angl. -
Flux Density in Magnet B_m at BH_{max} (G)	1775	Resistivity (micro-ohms/cm/cm^2)	1.0 @ 25 °C
Coercive Force in Magnet H_m at BH_{max} (Oe)	1875	Hardness 8	Scale Vickers HV5
Energy Loss W_c (watt-sec/cycle/lb)	-	Basic Forming Method	PMF
Importance of Operating Magnet at BH_{max}	E	Basic Finishing Method	WGD
Average Recoil Permeability	1.1	Relative Corrosion Resistance	Ex
Field Strength Required to Fully Saturate Magnet (Oe)	10,000	Primary Chemical Composition	Ba or Sr-Fe
Ability to Withstand Demagnetizing Fields	Ex	Special Notes:	
Maximum Service Temperature (°F)	482		
Approximate Temperature Permanently Affecting Magnet (°F)	842		
Inherent Orientation of Material	MP		

Note: All information based on available producers literature. In some cases of foreign materials interpretation or calculation required.

SOURCE(S):	Koerox 350
	(Alloy or Grade)
Krupp Widia GMBH	Sintered Ferrite
	(Type of Material)
Germany	3.3/3.6
	(MMPA Brief Designation)
	-
	(IEC Code Reference)

MAGNETIC PROPERTIES		PHYSICAL PROPERTIES	
Residual Induction B_r (G)	3700	Density (lbs./in.^{3})	.1734
Coercive Force (Oe) — Normal (H_c)	3140	General Mechanical Properties	VHB
Coercive Force (Oe) — Intrinsic (H_{ci})	3640	Tensile Strength (psi)	-
Maximum Energy Product BH_{max} (MGO)	3.3	Transverse Modulus of Rupture (psi)	-
B/H Ratio (Load Line) at BH_{max}	1.09	Coeff. of Therm. Expansion inches x 10^{-6}/°C	Orientation: Par. - Rt.Angl. -
Flux Density in Magnet B_m at BH_{max} (G)	1900	Resistivity (micro-ohms/cm/cm^2)	1.0 @ 25 °C
Coercive Force in Magnet H_m at BH_{max} (Oe)	1750	Hardness 8	Scale Vickers HV5
Energy Loss W_c (watt-sec/cycle/lb)	-	Basic Forming Method	PMF
Importance of Operating Magnet at BH_{max}	E	Basic Finishing Method	WGD
Average Recoil Permeability	1.1	Relative Corrosion Resistance	Ex
Field Strength Required to Fully Saturate Magnet (Oe)	10,000	Primary Chemical Composition	Ba or Sr-Fe
Ability to Withstand Demagnetizing Fields	Ex	Special Notes:	
Maximum Service Temperature (°F)	482		
Approximate Temperature Permanently Affecting Magnet (°F)	842		
Inherent Orientation of Material	MP		

Note: All information based on available producers literature. In some cases of foreign materials interpretation or calculation required.

SOURCE(S):

Krupp Widia GMBH

Germany

Koerox 360
(Alloy or Grade)

Sintered Ferrite
(Type of Material)

3.8/2.2
(MMPA Brief Designation)

-
(IEC Code Reference)

MAGNETIC PROPERTIES	
Residual Induction B_r (G)	4000
Coercive Force (Oe) Normal (H_c)	2140
Coercive Force (Oe) Intrinsic (H_{ci})	2200
Maximum Energy Product BH_{max} (MGO)	3.8
B/H Ratio (Load Line) at BH_{max}	1.17
Flux Density in Magnet B_m at BH_{max} (G)	2100
Coercive Force in Magnet H_m at BH_{max} (Oe)	1800
Energy Loss W_c (watt-sec/cycle/lb)	-
Importance of Operating Magnet at BH_{max}	E
Average Recoil Permeability	1.1
Field Strength Required to Fully Saturate Magnet (Oe)	10,000
Ability to Withstand Demagnetizing Fields	Ex
Maximum Service Temperature (°F)	482
Approximate Temperature Permanently Affecting Magnet (°F)	842
Inherent Orientation of Material	MP

PHYSICAL PROPERTIES	
Density (lbs./in.^{3})	.1770
General Mechanical Properties	VHB
Tensile Strength (psi)	-
Transverse Modulus of Rupture (psi)	-
Coeff. of Therm. Expansion inches x 10^{-6}/°C	Orientation: Par. - Rt.Angl. -
Resistivity (micro-ohms/cm/cm^2)	1.0 @ 25 °C
Hardness 8	Scale Vickers HV5
Basic Forming Method	PMF
Basic Finishing Method	WGD
Relative Corrosion Resistance	Ex
Primary Chemical Composition	Ba or Sr-Fe
Special Notes:	

Note: All information based on available producers literature. In some cases of foreign materials interpretation or calculation required.

SOURCE(S):

Krupp Widia GMBH

Germany

Koerox 400
(Alloy or Grade)

Sintered Ferrite
(Type of Material)

3.9/3.3
(MMPA Brief Designation)

-
(IEC Code Reference)

MAGNETIC PROPERTIES	
Residual Induction B_r (G)	4000
Coercive Force (Oe) Normal (H_c)	3200
Coercive Force (Oe) Intrinsic (H_{ci})	3270
Maximum Energy Product BH_{max} (MGO)	3.9
B/H Ratio (Load Line) at BH_{max}	1.03
Flux Density in Magnet B_m at BH_{max} (G)	2000
Coercive Force in Magnet H_m at BH_{max} (Oe)	1950
Energy Loss W_c (watt-sec/cycle/lb)	-
Importance of Operating Magnet at BH_{max}	E
Average Recoil Permeability	1.1
Field Strength Required to Fully Saturate Magnet (Oe)	10,000
Ability to Withstand Demagnetizing Fields	Ex
Maximum Service Temperature (°F)	482
Approximate Temperature Permanently Affecting Magnet (°F)	842
Inherent Orientation of Material	MP

PHYSICAL PROPERTIES	
Density (lbs./in.^{3})	.1770
General Mechanical Properties	VHB
Tensile Strength (psi)	-
Transverse Modulus of Rupture (psi)	-
Coeff. of Therm. Expansion inches x 10^{-6}/°C	Orientation: Par. - Rt.Angl. -
Resistivity (micro-ohms/cm/cm^2)	1.0 @ 25 °C
Hardness	8 Scale Vickers HV5
Basic Forming Method	PMF
Basic Finishing Method	WGD
Relative Corrosion Resistance	Ex
Primary Chemical Composition	Ba or Sr-Fe
Special Notes:	

Note: All information based on available producers literature. In some cases of foreign materials interpretation or calculation required.

SOURCE(S):	Koerox 420
	(Alloy or Grade)
Krupp Widia GMBH	Sintered Ferrite
	(Type of Material)
Germany	4.3/3.4
	(MMPA Brief Designation)
	-
	(IEC Code Reference)

MAGNETIC PROPERTIES		PHYSICAL PROPERTIES	
Residual Induction B_r (G)	4200	Density (lbs./in.^{3})	.1770
Coercive Force (Oe) — Normal (H_c)	3330	General Mechanical Properties	VHB
Coercive Force (Oe) — Intrinsic (H_{ci})	3390	Tensile Strength (psi)	-
Maximum Energy Product BH_{max} (MGO)	4.3	Transverse Modulus of Rupture (psi)	-
B/H Ratio (Load Line) at BH_{max}	1.11	Coeff. of Therm. Expansion inches x 10^{-6}/°C	Orientation: Par. _ Rt.Angl. _
Flux Density in Magnet B_m at BH_{max} (G)	2200	Resistivity (micro-ohms/cm/cm^2)	1.0 @ 25 °C
Coercive Force in Magnet H_m at BH_{max} (Oe)	1975	Hardness 8	Scale Vickers HV
Energy Loss W_c (watt-sec/cycle/lb)	-	Basic Forming Method	PMF
Importance of Operating Magnet at BH_{max}	E	Basic Finishing Method	WGD
Average Recoil Permeability	1.1	Relative Corrosion Resistance	Ex
Field Strength Required to Fully Saturate Magnet (Oe)	10,000	Primary Chemical Composition	Ba or Sr-Fe
Ability to Withstand Demagnetizing Fields	Ex	Special Notes:	
Maximum Service Temperature (°F)	482		
Approximate Temperature Permanently Affecting Magnet (°F)	842		
Inherent Orientation of Material	MP		

Note: All information based on available producers literature. In some cases of foreign materials interpretation or calculation required.

SOURCE(S):	**Koerzit 3/.6**
	(Alloy or Grade)
Krupp Widia GMBH	Sintered Alnico
	(Type of Material)
Germany	.4/.80
	(MMPA Brief Designation)
	(IEC Code Reference)

MAGNETIC PROPERTIES		PHYSICAL PROPERTIES	
Residual Induction B_r (G)	9900	Density (lbs./in.^{3})	.2565
Coercive Force (Oe) — Normal (H_c)	80	General Mechanical Properties	HT
Coercive Force (Oe) — Intrinsic (H_{ci})	80	Tensile Strength (psi)	Approx. 28,000
Maximum Energy Product BH_{max} (MGO)	.4	Transverse Modulus of Rupture (psi)	Approx. 44,000
B/H Ratio (Load Line) at BH_{max}	14.2	Coeff. of Therm. Expansion inches x 10^{-6}/°C	Orientation: Par. — Rt.Angl. —
Flux Density in Magnet B_m at BH_{max} (G)	710	Resistivity (micro-ohms/cm/cm^2)	55 @ 25 °C
Coercive Force in Magnet H_m at BH_{max} (Oe)	50	Hardness Scale	300-500 Vickers
Energy Loss W_c (watt-sec/cycle/lb)	–	Basic Forming Method	DP
Importance of Operating Magnet at BH_{max}	A	Basic Finishing Method	WGR
Average Recoil Permeability	–	Relative Corrosion Resistance	G
Field Strength Required to Fully Saturate Magnet (Oe)	1000	Primary Chemical Composition	Al-Ni-Co-Fe
Ability to Withstand Demagnetizing Fields	P	Special Notes: (1) Primarily used in hysteresis applications such as motor and couplings. (2) Magnetic properties strongly shape dependant.	
Maximum Service Temperature (°F)	842		
Approximate Temperature Permanently Affecting Magnet (°F)	1436		
Inherent Orientation of Material	none		

Note: All information based on available producers literature. In some cases of foreign materials interpretation or calculation required.

SOURCE(S):

Krupp Widia GMBH

Germany

Koerzit 4/.9
(Alloy or Grade)

Sintered Alnico
(Type of Material)

.5/.11
(MMPA Brief Designation)

(IEC Code Reference)

MAGNETIC PROPERTIES	
Residual Induction B_r (G)	9000
Coercive Force (Oe) — Normal (H_c)	110
Coercive Force (Oe) — Intrinsic (H_{ci})	110
Maximum Energy Product BH_{max} (MGO)	.5
B/H Ratio (Load Line) at BH_{max}	81.0
Flux Density in Magnet B_m at BH_{max} (G)	6100
Coercive Force in Magnet H_m at BH_{max} (Oe)	75
Energy Loss W_c (watt-sec/cycle/lb)	-
Importance of Operating Magnet at BH_{max}	A
Average Recoil Permeability	-
Field Strength Required to Fully Saturate Magnet (Oe)	1000
Ability to Withstand Demagnetizing Fields	P
Maximum Service Temperature (°F)	842
Approximate Temperature Permanently Affecting Magnet (°F)	1436
Inherent Orientation of Material	none

PHYSICAL PROPERTIES	
Density (lbs./in.^{3})	.2565
General Mechanical Properties	HT
Tensile Strength (psi)	Approx. 28,000
Transverse Modulus of Rupture (psi)	Approx. 44,000
Coeff. of Therm. Expansion inches x 10^{-6}/°C	Orientation: Par. - Rt.Angl. -
Resistivity (micro-ohms/cm/cm^2)	55 @ 25 °C
Hardness	Scale 300-500 Vickers
Basic Forming Method	DP
Basic Finishing Method	WGR
Relative Corrosion Resistance	G
Primary Chemical Composition	Al-Ni-Co-Fe

Special Notes:

(1) Primarily used in hysteresis applications such as motor and couplings.
(2) Magnetic properties strongly shape dependant.

Note: All information based on available producers literature. In some cases of foreign materials interpretation or calculation required.

SOURCE(S):	Koerzit 4/1.0
	(Alloy or Grade)
Krupp Widia GMBH	Sintered Alnico
	(Type of Material)
Germany	4/1
	(MMPA Brief Designation)
	(IEC Code Reference)

MAGNETIC PROPERTIES		PHYSICAL PROPERTIES	
Residual Induction B_r (G)	8800	Density (lbs./in.^{3})	.2565
Coercive Force (Oe) — Normal (H_c)	125	General Mechanical Properties	HT
Coercive Force (Oe) — Intrinsic (H_{ci})	125	Tensile Strength (psi)	–
Maximum Energy Product BH_{max} (MGO)	.55	Transverse Modulus of Rupture (psi)	–
B/H Ratio (Load Line) at BH_{max}	67.8	Coeff. of Therm. Expansion inches x 10^{-6}/°C	Orientation: Par. – Rt.Angl. –
Flux Density in Magnet B_m at BH_{max} (G)	6100	Resistivity (micro-ohms/cm/cm^2)	55 @ 25 °C
Coercive Force in Magnet H_m at BH_{max} (Oe)	90	Hardness Scale	300-500 Vickers
Energy Loss W_c (watt-sec/cycle/lb)	–	Basic Forming Method	W
Importance of Operating Magnet at BH_{max}	A	Basic Finishing Method	WGR
Average Recoil Permeability	–	Relative Corrosion Resistance	G
Field Strength Required to Fully Saturate Magnet (Oe)	1000	Primary Chemical Composition	Al-Ni-Co-Fe
Ability to Withstand Demagnetizing Fields	P	Special Notes: (1) Primarily used in hysteresis applications such as motor and couplings. (2) Magnetic properties strongly shape dependant.	
Maximum Service Temperature (°F)	–		
Approximate Temperature Permanently Affecting Magnet (°F)	1382-1652		
Inherent Orientation of Material	none		

Note: All information based on available producers literature. In some cases of foreign materials interpretation or calculation required.

SOURCE(S):	**Koerzit 5/1.4**	
	(Alloy or Grade)	
Krupp Widia GMBH	Sintered Alnico	
	(Type of Material)	
Germany	.65/.18	
	(MMPA Brief Designation)	
	(IEC Code Reference)	

MAGNETIC PROPERTIES		PHYSICAL PROPERTIES	
Residual Induction B_r (G)	8400	Density (lbs./in.^{3})	.2529
Coercive Force (Oe) — Normal (H_c)	180	General Mechanical Properties	HT
Coercive Force (Oe) — Intrinsic (H_{ci})	180	Tensile Strength (psi)	Approx. 28,000
Maximum Energy Product BH_{max} (MGO)	.65	Transverse Modulus of Rupture (psi)	Approx. 44,000
B/H Ratio (Load Line) at BH_{max}	49.1	Coeff. of Therm. Expansion inches x 10^{-6}/°C Orientation: Par. — Rt.Angl. —	
Flux Density in Magnet B_m at BH_{max} (G)	5650	Resistivity (micro-ohms/cm/cm^2)	55 @ 25 °C
Coercive Force in Magnet H_m at BH_{max} (Oe)	115	Hardness Scale 300-500	Vickers
Energy Loss W_c (watt-sec/cycle/lb)	—	Basic Forming Method	DP
Importance of Operating Magnet at BH_{max}	A	Basic Finishing Method	WGR
Average Recoil Permeability	—	Relative Corrosion Resistance	G
Field Strength Required to Fully Saturate Magnet (Oe)	1000	Primary Chemical Composition	Al-Ni-Co-Fe
Ability to Withstand Demagnetizing Fields	P	Special Notes: (1) Primarily used in hysteresis applications such as motor and couplings. (2) Magnetic properties strongly shape dependant.	
Maximum Service Temperature (°F)	842		
Approximate Temperature Permanently Affecting Magnet (°F)	1436		
Inherent Orientation of Material	none		

Note: All information based on available producers literature. In some cases of foreign materials interpretation or calculation required.

SOURCE(S):

Koerzit 6/1.6
(Alloy or Grade)

Krupp Widia GMBH

Sintered Alnico
(Type of Material)

Germany

.80/.20
(MMPA Brief Designation)

(IEC Code Reference)

MAGNETIC PROPERTIES	
Residual Induction B_r (G)	8200
Coercive Force (Oe) Normal (H_c)	200
Coercive Force (Oe) Intrinsic (H_{ci})	200
Maximum Energy Product BH_{max} (MGO)	.8
B/H Ratio (Load Line) at BH_{max}	38.9
Flux Density in Magnet B_m at BH_{max} (G)	5600
Coercive Force in Magnet H_m at BH_{max} (Oe)	145
Energy Loss W_c (watt-sec/cycle/lb)	-
Importance of Operating Magnet at BH_{max}	A
Average Recoil Permeability	-
Field Strength Required to Fully Saturate Magnet (Oe)	1000
Ability to Withstand Demagnetizing Fields	P
Maximum Service Temperature (°F)	842
Approximate Temperature Permanently Affecting Magnet (°F)	1436
Inherent Orientation of Material	none

PHYSICAL PROPERTIES	
Density (lbs./in.^{3})	.2529
General Mechanical Properties	HT
Tensile Strength (psi)	Approx. 28,000
Transverse Modulus of Rupture (psi)	Approx. 44,000
Coeff. of Therm. Expansion inches x 10^{-6}/°C	Orientation: Par. _ Rt.Angl. _
Resistivity (micro-ohms/cm/cm^2)	55 @ 25 °C
Hardness	Scale 300-500 Vickers
Basic Forming Method	DP
Basic Finishing Method	WGR
Relative Corrosion Resistance	G
Primary Chemical Composition	Al-Ni-Co-Fe

Special Notes:

(1) Primarily used in hysteresis applications such as motor and couplings.
(2) Magnetic properties strongly shape dependant.

Note: All information based on available producers literature. In some cases of foreign materials interpretation or calculation required.

SOURCE(S):	**Koerzit 120**
Krupp Widia GMBH	(Alloy or Grade)
	Sintered Alnico
	(Type of Material)
Germany	1.2/.65
	(MMPA Brief Designation)
	(IEC Code Reference)

MAGNETIC PROPERTIES		PHYSICAL PROPERTIES	
Residual Induction B_r (G)	5300	Density (lbs./in.^{3})	.2457
Coercive Force (Oe) — Normal (H_c)	650	General Mechanical Properties	HT
Coercive Force (Oe) — Intrinsic (H_{ci})	680	Tensile Strength (psi)	Approx. 28,000
Maximum Energy Product BH_{max} (MGO)	1.2	Transverse Modulus of Rupture (psi)	Approx. 44,000
B/H Ratio (Load Line) at BH_{max}	8.6	Coeff. of Therm. Expansion inches x 10^{-6}/°C	Orientation: Par. — Rt.Angl. —
Flux Density in Magnet B_m at BH_{max} (G)	3200	Resistivity Approx. (micro-ohms/cm/cm^2)	50 @ 25 °C
Coercive Force in Magnet H_m at BH_{max} (Oe)	370	Hardness Scale	Approx. 50 Rockwell C
Energy Loss W_c (watt-sec/cycle/lb)	–	Basic Forming Method	DP
Importance of Operating Magnet at BH_{max}	A	Basic Finishing Method	WGR
Average Recoil Permeability	4.5	Relative Corrosion Resistance	G
Field Strength Required to Fully Saturate Magnet (Oe)	3000	Primary Chemical Composition	Al-Ni-Co-Cu-Nb-Ti-Fe
Ability to Withstand Demagnetizing Fields	F	Special Notes:	
Maximum Service Temperature (°F)	842		
Approximate Temperature Permanently Affecting Magnet (°F)	1436		
Inherent Orientation of Material	none		

Note: All information based on available producers literature. In some cases of foreign materials interpretation or calculation required.

SOURCE(S):

Krupp Widia GMBH

Germany

Koerzit 130K
(Alloy or Grade)

Sintered Alnico
(Type of Material)

1.3/.72
(MMPA Brief Designation)

(IEC Code Reference)

MAGNETIC PROPERTIES	
Residual Induction B_r (G)	5500
Coercive Force (Oe) — Normal (H_c)	680
Coercive Force (Oe) — Intrinsic (H_{ci})	720
Maximum Energy Product BH_{max} (MGO)	1.3
B/H Ratio (Load Line) at BH_{max}	8.0
Flux Density in Magnet B_m at BH_{max} (G)	3200
Coercive Force in Magnet H_m at BH_{max} (Oe)	400
Energy Loss W_c (watt-sec/cycle/lb)	-
Importance of Operating Magnet at BH_{max}	A
Average Recoil Permeability	4.0
Field Strength Required to Fully Saturate Magnet (Oe)	3000
Ability to Withstand Demagnetizing Fields	F
Maximum Service Temperature (°F)	842
Approximate Temperature Permanently Affecting Magnet (°F)	1436
Inherent Orientation of Material	none

PHYSICAL PROPERTIES	
Density (lbs./in.^{3})	.2493
General Mechanical Properties	HT
Tensile Strength (psi)	Approx. 28,000
Transverse Modulus of Rupture (psi)	Approx. 44,000
Coeff. of Therm. Expansion inches x 10^{-6}/°C	Orientation: Par. — Rt.Angl. —
Resistivity (micro-ohms/cm/cm^2)	Approx. 50 @ 25 °C
Hardness — Approx. 50	Scale Rockwell C
Basic Forming Method	DP
Basic Finishing Method	WGR
Relative Corrosion Resistance	G
Primary Chemical Composition	Al-Ni-Co-Cu-Nb-Ti-Fe
Special Notes:	

Note: All information based on available producers literature. In some cases of foreign materials interpretation or calculation required.

SOURCE(S):

Krupp Widia GMBH

Germany

Koerzit 160
(Alloy or Grade)

Sintered Alnico
(Type of Material)

1.7/.76
(MMPA Brief Designation)

(IEC Code Reference)

MAGNETIC PROPERTIES		PHYSICAL PROPERTIES	
Residual Induction B_r (G)	6500	Density (lbs./in.^{3})	.2565
Coercive Force (Oe) — Normal (H_c)	720	General Mechanical Properties	HT
Coercive Force (Oe) — Intrinsic (H_{ci})	760	Tensile Strength (psi)	Approx. 28,000
Maximum Energy Product BH_{max} (MGO)	1.7	Transverse Modulus of Rupture (psi)	Approx. 44,000
B/H Ratio (Load Line) at BH_{max}	8.6	Coeff. of Therm. Expansion inches x 10^{-6}/°C	Orientation: Par. _ Rt.Angl. _
Flux Density in Magnet B_m at BH_{max} (G)	3800	Resistivity (micro-ohms/cm/cm^2)	Approx. 50 @ 25 °C
Coercive Force in Magnet H_m at BH_{max} (Oe)	440	Hardness Scale	Approx. 50 Rockwell C
Energy Loss W_c (watt-sec/cycle/lb)	-	Basic Forming Method	DP
Importance of Operating Magnet at BH_{max}	A	Basic Finishing Method	WGR
Average Recoil Permeability	4.7	Relative Corrosion Resistance	G
Field Strength Required to Fully Saturate Magnet (Oe)	3000	Primary Chemical Composition	Al-Ni-Co-Cu-Nb-T-Fe
Ability to Withstand Demagnetizing Fields	F	Special Notes:	
Maximum Service Temperature (°F)	842		
Approximate Temperature Permanently Affecting Magnet (°F)	1436		
Inherent Orientation of Material	none		

Note: All information based on available producers literature. In some cases of foreign materials interpretation or calculation required.

SOURCE(S): **Krupp Widia GMBH**

Germany

Koerzit 190
(Alloy or Grade)

Sintered Alnico
(Type of Material)

2.1/.78
(MMPA Brief Designation)

(IEC Code Reference)

MAGNETIC PROPERTIES	
Residual Induction B_r (G)	7500
Coercive Force (Oe) — Normal (H_c)	750
Coercive Force (Oe) — Intrinsic (H_{ci})	780
Maximum Energy Product BH_{max} (MGO)	2.1
B/H Ratio (Load Line) at BH_{max}	9.6
Flux Density in Magnet B_m at BH_{max} (G)	4500
Coercive Force in Magnet H_m at BH_{max} (Oe)	470
Energy Loss W_c (watt-sec/cycle/lb)	-
Importance of Operating Magnet at BH_{max}	A
Average Recoil Permeability	4.5
Field Strength Required to Fully Saturate Magnet (Oe)	3000
Ability to Withstand Demagnetizing Fields	F
Maximum Service Temperature (°F)	842
Approximate Temperature Permanently Affecting Magnet (°F)	1436
Inherent Orientation of Material	HT

PHYSICAL PROPERTIES	
Density (lbs./in.^{3})	.2565
General Mechanical Properties	HT
Tensile Strength (psi)	Approx. 28,000
Transverse Modulus of Rupture (psi)	Approx. 44,000
Coeff. of Therm. Expansion inches x 10^{-6}/°C — Orientation: Par. / Rt.Angl.	- / -
Resistivity (micro-ohms/cm/cm^2)	Approx. 50 @ 25 °C
Hardness — Scale	Approx. 50 / Rockwell C
Basic Forming Method	DPM
Basic Finishing Method	WGR
Relative Corrosion Resistance	G
Primary Chemical Composition	Al-Ni-Co-Cu-Nd-T-Fe
Special Notes:	

Note: All information based on available producers literature. In some cases of foreign materials interpretation or calculation required.

SOURCE(S):	Koerzit 260
Krupp Widia GMBH	(Alloy or Grade)
	Sintered Alnico
	(Type of Material)
Germany	2.8/1.4
	(MMPA Brief Designation)
	(IEC Code Reference)

MAGNETIC PROPERTIES		PHYSICAL PROPERTIES	
Residual Induction B_r (G)	6400	Density (lbs./in.^{3})	.2637
Coercive Force (Oe) — Normal (H_c)	1320	General Mechanical Properties	HT
Coercive Force (Oe) — Intrinsic (H_{ci})	1430	Tensile Strength (psi)	Approx. 28,000
Maximum Energy Product BH_{max} (MGO)	2.8	Transverse Modulus of Rupture (psi)	Approx. 44,000
B/H Ratio (Load Line) at BH_{max}	4.6	Coeff. of Therm. Expansion inches x 10^{-6}/°C	Orientation: Par. _ Rt.Angl. _
Flux Density in Magnet B_m at BH_{max} (G)	3600	Resistivity (micro-ohms/cm/cm^2)	Approx. 50 @ 25 °C
Coercive Force in Magnet H_m at BH_{max} (Oe)	780	Hardness Scale	Approx. 50 Rockwell C
Energy Loss W_c (watt-sec/cycle/lb)	-	Basic Forming Method	DPM
Importance of Operating Magnet at BH_{max}	C	Basic Finishing Method	WGR
Average Recoil Permeability	3.8	Relative Corrosion Resistance	G
Field Strength Required to Fully Saturate Magnet (Oe)	3000	Primary Chemical Composition	Al-Ni-Co-Cu-Nb-T-Fe
Ability to Withstand Demagnetizing Fields	G	Special Notes:	
Maximum Service Temperature (°F)	842		
Approximate Temperature Permanently Affecting Magnet (°F)	1436		
Inherent Orientation of Material	HT		

Note: All information based on available producers literature. In some cases of foreign materials interpretation or calculation required.

SOURCE(S):	**Koerzit 400K**
	(Alloy or Grade)
Krupp Widia GMBH	Sintered Alnico
	(Type of Material)
Germany	4.0/.82
	(MMPA Brief Designation)
	(IEC Code Reference)

MAGNETIC PROPERTIES		PHYSICAL PROPERTIES	
Residual Induction B_r (G)	11,000	Density (lbs./in.^{3})	.2637
Coercive Force (Oe) — Normal (H_c)	800	General Mechanical Properties	HT
Coercive Force (Oe) — Intrinsic (H_{ci})	815	Tensile Strength (psi)	Approx. 28,000
Maximum Energy Product BH_{max} (MGO)	4.0	Transverse Modulus of Rupture (psi)	Approx. 44,000
B/H Ratio (Load Line) at BH_{max}	13.1	Coeff. of Therm. Expansion inches x 10^{-6}/°C	Orientation: Par. _ Rt.Angl. _
Flux Density in Magnet B_m at BH_{max} (G)	7200	Resistivity (micro-ohms/cm/cm^2)	Approx. 50 @ 25 °C
Coercive Force in Magnet H_m at BH_{max} (Oe)	550	Hardness Scale	Approx. 50 Rockwell C
Energy Loss W_c (watt-sec/cycle/lb)	-	Basic Forming Method	DPM
Importance of Operating Magnet at BH_{max}	A	Basic Finishing Method	WGR
Average Recoil Permeability	4.5	Relative Corrosion Resistance	G
Field Strength Required to Fully Saturate Magnet (Oe)	3000	Primary Chemical Composition	Al-Ni-Co-Cu-Nd-T-Fe
Ability to Withstand Demagnetizing Fields	F	Special Notes:	
Maximum Service Temperature (°F)	842		
Approximate Temperature Permanently Affecting Magnet (°F)	1436		
Inherent Orientation of Material	MP		

Note: All information based on available producers literature. In some cases of foreign materials interpretation or calculation required.

SOURCE(S):

Krupp Widia GMBH

Germany

Koerzit 420
(Alloy or Grade)

Sintered Alnico
(Type of Material)

.55/.13
(MMPA Brief Designation)

(IEC Code Reference)

MAGNETIC PROPERTIES		PHYSICAL PROPERTIES	
Residual Induction B_r (G)	8800	Density (lbs./in.^{3})	.2565
Coercive Force (Oe) — Normal (H_c)	125	General Mechanical Properties	HT
Coercive Force (Oe) — Intrinsic (H_{ci})	125	Tensile Strength (psi)	Approx. 28,000
Maximum Energy Product BH_{max} (MGO)	.55	Transverse Modulus of Rupture (psi)	Approx. 44,000
B/H Ratio (Load Line) at BH_{max}	67.8	Coeff. of Therm. Expansion inches x 10^{-6}/°C	Orientation: Par. — Rt.Angl. —
Flux Density in Magnet B_m at BH_{max} (G)	6100	Resistivity (micro-ohms/cm/cm^2)	55 @ 25 °C
Coercive Force in Magnet H_m at BH_{max} (Oe)	90	Hardness Scale	300-500 Vickers
Energy Loss W_c (watt-sec/cycle/lb)	-	Basic Forming Method	DP
Importance of Operating Magnet at BH_{max}	A	Basic Finishing Method	WGR
Average Recoil Permeability	-	Relative Corrosion Resistance	G
Field Strength Required to Fully Saturate Magnet (Oe)	1000	Primary Chemical Composition	Al-Ni-Co-Fe
Ability to Withstand Demagnetizing Fields	P	Special Notes:	
Maximum Service Temperature (°F)	842	(1) Primarily used in hysteresis applications such as motor and couplings.	
Approximate Temperature Permanently Affecting Magnet (°F)	1436	(2) Magnetic properties strongly shape dependant.	
Inherent Orientation of Material	none		

Note: All information based on available producers literature. In some cases of foreign materials interpretation or calculation required.

SOURCE(S):

Koerzit 450
(Alloy or Grade)

Krupp Widia GMBH

Sintered Alnico
(Type of Material)

Germany

5.5/1.5
(MMPA Brief Designation)

(IEC Code Reference)

MAGNETIC PROPERTIES	
Residual Induction B_r (G)	8800
Coercive Force (Oe) — Normal (H_c)	1450
Coercive Force (Oe) — Intrinsic (H_{ci})	1500
Maximum Energy Product BH_{max} (MGO)	5.5
B/H Ratio (Load Line) at BH_{max}	5.5
Flux Density in Magnet B_m at BH_{max} (G)	5500
Coercive Force in Magnet H_m at BH_{max} (Oe)	1000
Energy Loss W_c (watt-sec/cycle/lb)	-
Importance of Operating Magnet at BH_{max}	B
Average Recoil Permeability	3.0
Field Strength Required to Fully Saturate Magnet (Oe)	3000
Ability to Withstand Demagnetizing Fields	G
Maximum Service Temperature (°F)	842
Approximate Temperature Permanently Affecting Magnet (°F)	1436
Inherent Orientation of Material	MP

PHYSICAL PROPERTIES	
Density (lbs./in.^{3})	.2637
General Mechanical Properties	HT
Tensile Strength (psi)	Approx. 28,000
Transverse Modulus of Rupture (psi)	Approx. 44,000
Coeff. of Therm. Expansion inches x 10^{-6}/°C	Orientation: Par. _ Rt.Angl. _
Resistivity (micro-ohms/cm/cm^2)	Approx. 50 @ 25 °C
Hardness — Approx. 50	Scale Rockwell C
Basic Forming Method	DPM
Basic Finishing Method	WGR
Relative Corrosion Resistance	G
Primary Chemical Composition	Al-Ni-Co-Cu-Nb-T-Fe
Special Notes:	

Note: All information based on available producers literature. In some cases of foreign materials interpretation or calculation required.

SOURCE(S):

Krupp Widia GMBH

Germany

Koerzit 500
(Alloy or Grade)

Sintered Alnico
(Type of Material)

5.2/.65
(MMPA Brief Designation)

(IEC Code Reference)

MAGNETIC PROPERTIES	
Residual Induction B_r (G)	12,400
Coercive Force (Oe) — Normal (H_c)	640
Coercive Force (Oe) — Intrinsic (H_{ci})	645
Maximum Energy Product BH_{max} (MGO)	5.2
B/H Ratio (Load Line) at BH_{max}	19.2
Flux Density in Magnet B_m at BH_{max} (G)	10,000
Coercive Force in Magnet H_m at BH_{max} (Oe)	520
Energy Loss W_c (watt-sec/cycle/lb)	-
Importance of Operating Magnet at BH_{max}	A
Average Recoil Permeability	3.75
Field Strength Required to Fully Saturate Magnet (Oe)	3000
Ability to Withstand Demagnetizing Fields	F
Maximum Service Temperature (°F)	842
Approximate Temperature Permanently Affecting Magnet (°F)	1436
Inherent Orientation of Material	MP

PHYSICAL PROPERTIES	
Density (lbs./in.^{3})	.2637
General Mechanical Properties	HT
Tensile Strength (psi)	Approx. 28,000
Transverse Modulus of Rupture (psi)	Approx. 44,000
Coeff. of Therm. Expansion inches x 10^{-6}/°C	Orientation: Par. _ Rt.Angl. _
Resistivity (micro-ohms/cm/cm^2)	Approx. 50 @ 25 °C
Hardness	Approx. 50 Scale Rockwell C
Basic Forming Method	DPM
Basic Finishing Method	WGR
Relative Corrosion Resistance	G
Primary Chemical Composition	Al-Ni-Co-Cu-Nb-T-Fe
Special Notes:	

Note: All information based on available producers literature. In some cases of foreign materials interpretation or calculation required.

SOURCE(S):	Koerzit 700
	(Alloy or Grade)
Krupp Widia GMBH	Cast Alnico
	(Type of Material)
Germany	7.8/.74
	(MMPA Brief Designation)
	R1-1-3
	(IEC Code Reference)

MAGNETIC PROPERTIES		PHYSICAL PROPERTIES	
Residual Induction B_r (G)	13,500	Density (lbs./in.^{3})	.2637
Coercive Force (Oe) — Normal (H_c)	730	General Mechanical Properties HB	
Coercive Force (Oe) — Intrinsic (H_{ci})	740	Tensile Strength (psi)	-
Maximum Energy Product BH_{max} (MGO)	7.8	Transverse Modulus of Rupture (psi)	-
B/H Ratio (Load Line) at BH_{max}	18.5	Coeff. of Therm. Expansion inches x 10^{-6}/°C	Orientation: Par. - Rt.Angl. -
Flux Density in Magnet B_m at BH_{max} (G)	12,000	Resistivity (micro-ohms/cm/cm^2)	50 @ 25 °C
Coercive Force in Magnet H_m at BH_{max} (Oe)	650	Hardness Scale Approx. 50	Rockwell C
Energy Loss W_c (watt-sec/cycle/lb)	15.3	Basic Forming Method	CM
Importance of Operating Magnet at BH_{max}	A	Basic Finishing Method	WGR
Average Recoil Permeability	1.5-3.0	Relative Corrosion Resistance	G
Field Strength Required to Fully Saturate Magnet (Oe)	3000	Primary Chemical Composition	Al-Ni-Co-Cu-Nb-T-Fe
Ability to Withstand Demagnetizing Fields	G	Special Notes:	
Maximum Service Temperature (°F)	900		
Approximate Temperature Permanently Affecting Magnet (°F)	1580		
Inherent Orientation of Material	CHT		

Note: All information based on available producers literature. In some cases of foreign materials interpretation or calculation required.

SOURCE(S):	Koerzit 1800
	(Alloy or Grade)
Krupp Widia GMBH	Sintered Alnico
	(Type of Material)
Germany	5.5/2.0
	(MMPA Brief Designation)
	(IEC Code Reference)

MAGNETIC PROPERTIES		PHYSICAL PROPERTIES	
Residual Induction B_r (G)	7400	Density (lbs./in.^{3})	.2601
Coercive Force (Oe) — Normal (H_c)	1880	General Mechanical Properties	HT
Coercive Force (Oe) — Intrinsic (H_{ci})	2010	Tensile Strength (psi)	Approx. 28,000
Maximum Energy Product BH_{max} (MGO)	5.5	Transverse Modulus of Rupture (psi)	Approx. 44,000
B/H Ratio (Load Line) at BH_{max}	3.5	Coeff. of Therm. Expansion inches x 10^{-6}/°C	Orientation: Par. — Rt.Angl. —
Flux Density in Magnet B_m at BH_{max} (G)	4400	Resistivity (micro-ohms/cm/cm^2)	Approx. 50 @ 25 °C
Coercive Force in Magnet H_m at BH_{max} (Oe)	1250	Hardness — Approx. 50	Scale Rockwell C
Energy Loss W_c (watt-sec/cycle/lb)	—	Basic Forming Method	DPM
Importance of Operating Magnet at BH_{max}	B	Basic Finishing Method	WGR
Average Recoil Permeability	2.0	Relative Corrosion Resistance	G
Field Strength Required to Fully Saturate Magnet (Oe)	3000	Primary Chemical Composition	Al-Ni-Co-Cu-Nb-T-Fe
Ability to Withstand Demagnetizing Fields	G	Special Notes:	
Maximum Service Temperature (°F)	842		
Approximate Temperature Permanently Affecting Magnet (°F)	1436		
Inherent Orientation of Material	MP		

Note: All information based on available producers literature. In some cases of foreign materials interpretation or calculation required.

SOURCE(S):

Arnold Engineering Co.

USA

MGO-1016
(Alloy or Grade)

Bonded Ferrite
(Type of Material)

-
(MMPA Brief Designation)

-
(IEC Code Reference)

MAGNETIC PROPERTIES		PHYSICAL PROPERTIES	
Residual Induction B_r (G)	See	Density (lbs./in.^{3})	-
Coercive Force (Oe) Normal (H_c)	Note	General Mechanical Properties	FL
Coercive Force (Oe) Intrinsic (H_{ci})	(2)	Tensile Strength (psi)	-
Maximum Energy Product BH_{max} (MGO)	-	Transverse Modulus of Rupture (psi)	-
B/H Ratio (Load Line) at BH_{max}	-	Coeff. of Therm. Expansion inches x 10^{-6}/°C	Orientation: Par. - Rt.Angl. -
Flux Density in Magnet B_m at BH_{max} (G)	-	Resistivity (micro-ohms/cm/cm^2)	- @ - °C
Coercive Force in Magnet H_m at BH_{max} (Oe)	-	Hardness Scale	- -
Energy Loss W_c (watt-sec/cycle/lb)	-	Basic Forming Method	E
Importance of Operating Magnet at BH_{max}	E	Basic Finishing Method	SCD/SRD
Average Recoil Permeability	-	Relative Corrosion Resistance	E
Field Strength Required to Fully Saturate Magnet (Oe)	10,000	Primary Chemical Composition	Ba-Fe + binder
Ability to Withstand Demagnetizing Fields	Ex	Special Notes: (1) Available as strips or sheeting (2) Contact manufacturer for properties not shown	
Maximum Service Temperature (°F)	-		
Approximate Temperature Permanently Affecting Magnet (°F)	-		
Inherent Orientation of Material	-		

Note: All information based on available producers literature. In some cases of foreign materials interpretation or calculation required.

SOURCE(S):

Vacuumschmelze GMBH

Germany

Magnetoflex 35 (strip)
(Alloy or Grade)

Wrought
(Type of Material)

-
(MMPA Brief Designation)

-
(IEC Code Reference)

MAGNETIC PROPERTIES		PHYSICAL PROPERTIES	
Residual Induction B_r (G)	8750	Density (lbs./in.^{3})	.2926
Coercive Force (Oe) — Normal (H_c)	340	General Mechanical Properties	See Notes
Coercive Force (Oe) — Intrinsic (H_{ci})	-	Tensile Strength (psi)	-
Maximum Energy Product BH_{max} (MGO)	1.7	Transverse Modulus of Rupture (psi)	-
B/H Ratio (Load Line) at BH_{max}	24.0	Coeff. of Therm. Expansion inches x 10^{-6}/°C	Orientation: Par. 11.0 Rt.Angl. -
Flux Density in Magnet B_m at BH_{max} (G)	6000	Resistivity (micro-ohms/cm/cm^2)	- @ - °C
Coercive Force in Magnet H_m at BH_{max} (Oe)	250	Hardness Scale	See Notes
Energy Loss W_c (watt-sec/cycle/lb)	-	Basic Forming Method	W
Importance of Operating Magnet at BH_{max}	B	Basic Finishing Method	See Notes
Average Recoil Permeability	-	Relative Corrosion Resistance	G
Field Strength Required to Fully Saturate Magnet (Oe)	2500	Primary Chemical Composition	Fe-Co-V
Ability to Withstand Demagnetizing Fields	F	Special Notes:	
Maximum Service Temperature (°F)	932	(1) Vickers Hardness: (a) cold worked = 400 (b) after H.T. = 900	
Approximate Temperature Permanently Affecting Magnet (°F)	1292	(2) May be formed before H.T. Must be ground after H.T.	
Inherent Orientation of Material	CR		

Note: All information based on available producers literature. In some cases of foreign materials interpretation or calculation required.

SOURCE(S):

Vacuumschmelze GMBH

Germany

Magnetoflex 35 (wire)
(Alloy or Grade)

Wrought
(Type of Material)

-
(MMPA Brief Designation)

-
(IEC Code Reference)

MAGNETIC PROPERTIES		PHYSICAL PROPERTIES	
Residual Induction B_r (G)	9750	Density (lbs./in.^{3})	.2926
Coercive Force (Oe) — Normal (H_c)	340	General Mechanical Properties	See Notes
Coercive Force (Oe) — Intrinsic (H_{ci})		Tensile Strength (psi)	-
Maximum Energy Product BH_{max} (MGO)	2.5	Transverse Modulus of Rupture (psi)	-
B/H Ratio (Load Line) at BH_{max}	30.3	Coeff. of Therm. Expansion inches x 10^{-6}/°C	Orientation: Par. 11.0 Rt.Angl. -
Flux Density in Magnet B_m at BH_{max} (G)	8800	Resistivity (micro-ohms/cm/cm^2 - @ - °C)	
Coercive Force in Magnet H_m at BH_{max} (Oe)	290	Hardness Scale	See Notes
Energy Loss W_c (watt-sec/cycle/lb)	-	Basic Forming Method	W
Importance of Operating Magnet at BH_{max}	A	Basic Finishing Method	See Notes
Average Recoil Permeability	-	Relative Corrosion Resistance	G
Field Strength Required to Fully Saturate Magnet (Oe)	2500	Primary Chemical Composition	Fe-Co-V
Ability to Withstand Demagnetizing Fields	P	Special Notes: (1) Vickers Hardness: (a) cold worked = 400 (b) after H.T. = 900 (2) May be formed before H.T. Must be ground after H.T.	
Maximum Service Temperature (°F)	932		
Approximate Temperature Permanently Affecting Magnet (°F)	1292		
Inherent Orientation of Material	CR		

Note: All information based on available producers literature. In some cases of foreign materials interpretation or calculation required.

SOURCE(S):

Vacuumschmelze GMBH

Germany

Magnetoflex 40 (anisotropic)
(Alloy or Grade)

Wrought
(Type of Material)

-
(MMPA Brief Designation)

-
(IEC Code Reference)

MAGNETIC PROPERTIES	
Residual Induction B_r (G)	10,000-12,000
Coercive Force (Oe) — Normal (H_c)	110-140
Coercive Force (Oe) — Intrinsic (H_{ci})	-
Maximum Energy Product BH_{max} (MGO)	.5-1.0
B/H Ratio (Load Line) at BH_{max}	220.0
Flux Density in Magnet B_m at BH_{max} (G)	13,200
Coercive Force in Magnet H_m at BH_{max} (Oe)	60
Energy Loss W_c (watt-sec/cycle/lb)	-
Importance of Operating Magnet at BH_{max}	A
Average Recoil Permeability	-
Field Strength Required to Fully Saturate Magnet (Oe)	2500
Ability to Withstand Demagnetizing Fields	P
Maximum Service Temperature (°F)	932
Approximate Temperature Permanently Affecting Magnet (°F)	1292
Inherent Orientation of Material	CR

PHYSICAL PROPERTIES	
Density (lbs./in.^{3})	.2926
General Mechanical Properties	See Notes
Tensile Strength (psi)	-
Transverse Modulus of Rupture (psi)	-
Coeff. of Therm. Expansion inches x 10^{-6}/°C	Orientation: Par. 11.0 Rt.Angl. -
Resistivity (micro-ohms/cm/cm^2)	- @ - °C
Hardness Scale	See Notes
Basic Forming Method	W
Basic Finishing Method	See Notes
Relative Corrosion Resistance	G
Primary Chemical Composition	Fe-Co-V

Special Notes:

(1) Vickers Hardness
   (a) cold worked = 400
   (b) after H.T. = 900
(2) May be formed before H.T. Must be ground after H.T.

Note: All information based on available producers literature. In some cases of foreign materials interpretation or calculation required.

SOURCE(S):

Vacuumschmelze GMBH

Germany

Magnetoflex 40 (isotropic)
(Alloy or Grade)

Wrought
(Type of Material)

-
(MMPA Brief Designation)

-
(IEC Code Reference)

MAGNETIC PROPERTIES	
Residual Induction B_r (G)	8000
Coercive Force (Oe) Normal (H_c)	150
Coercive Force (Oe) Intrinsic (H_{ci})	-
Maximum Energy Product BH_{max} (MGO)	.80
B/H Ratio (Load Line) at BH_{max}	36.0
Flux Density in Magnet B_m at BH_{max} (G)	5400
Coercive Force in Magnet H_m at BH_{max} (Oe)	150
Energy Loss W_c (watt-sec/cycle/lb)	-
Importance of Operating Magnet at BH_{max}	A
Average Recoil Permeability	-
Field Strength Required to Fully Saturate Magnet (Oe)	2500
Ability to Withstand Demagnetizing Fields	P
Maximum Service Temperature (°F)	932
Approximate Temperature Permanently Affecting Magnet (°F)	1292
Inherent Orientation of Material	none

PHYSICAL PROPERTIES	
Density (lbs./in.^{3})	.2926
General Mechanical Properties	See Notes
Tensile Strength (psi)	-
Transverse Modulus of Rupture (psi)	-
Coeff. of Therm. Expansion inches x 10^{-6}/°C	Orientation: Par. 11.0 Rt.Angl. -
Resistivity (micro-ohms/cm/cm^2)	- @ - °C
Hardness Scale	See Notes
Basic Forming Method	W
Basic Finishing Method	See Notes
Relative Corrosion Resistance	G
Primary Chemical Composition	Fe-Co-V

Special Notes:

(1) Vickers Hardness:
   (a) cold worked = 400
   (b) after H.T. = 900
(2) May be formed before H.T. Must be ground after H.T.

Note: All information based on available producers literature. In some cases of foreign materials interpretation or calculation required.

SOURCE(S):

Hitachi Magnetics Corp.

USA

Nd94-EA
(Alloy or Grade)

Sintered Rare Earth
(Type of Material)

28/15
(MMPA Brief Designation)

R7-3
(IEC Code Reference)

MAGNETIC PROPERTIES	
Residual Induction B_r (G)	10,800
Coercive Force (Oe) — Normal (H_c)	10,000
Coercive Force (Oe) — Intrinsic (H_{ci})	15,000
Maximum Energy Product BH_{max} (MGO)	28.0
B/H Ratio (Load Line) at BH_{max}	.77
Flux Density in Magnet B_m at BH_{max} (G)	4600
Coercive Force in Magnet H_m at BH_{max} (Oe)	6000
Energy Loss W_c (watt-sec/cycle/lb)	-
Importance of Operating Magnet at BH_{max}	E
Average Recoil Permeability	1.05
Field Strength Required to Fully Saturate Magnet (Oe)	32,000
Ability to Withstand Demagnetizing Fields	0
Maximum Service Temperature (°F)	Est. 245
Approximate Temperature Permanently Affecting Magnet (°F)	500
Inherent Orientation of Material	MP

PHYSICAL PROPERTIES	
Density (lbs./in.^{3})	.2710
General Mechanical Properties	HB
Tensile Strength (psi)	-
Transverse Modulus of Rupture (psi)	-
Coeff. of Therm. Expansion inches x 10^{-6}/°C — Orientation: Par. _ Rt.Angl. _	
Resistivity (micro-ohms/cm/cm^2)	- @ - °C
Hardness / Scale	- / -
Basic Forming Method	DPM
Basic Finishing Method	WRG
Relative Corrosion Resistance	P
Primary Chemical Composition	Nd-Fe-B
Special Notes:	

Note: All information based on available producers literature. In some cases of foreign materials interpretation or calculation required.

SOURCE(S):	**Nd94-EB**
	(Alloy or Grade)
Hitachi Magnetics Corp.	Sintered Rare Earth
	(Type of Material)
USA	31/15
	(MMPA Brief Designation)
	R7-3
	(IEC Code Reference)

MAGNETIC PROPERTIES		PHYSICAL PROPERTIES	
Residual Induction B_r (G)	11,500	Density (lbs./in.^{3})	.2710
Coercive Force (Oe) — Normal (H_c)	10,700	General Mechanical Properties	HB
Coercive Force (Oe) — Intrinsic (H_{ci})	15,000	Tensile Strength (psi)	-
Maximum Energy Product BH_{max} (MGO)	31.0	Transverse Modulus of Rupture (psi)	-
B/H Ratio (Load Line) at BH_{max}	.9	Coeff. of Therm. Expansion inches x 10^{-6}/°C	Orientation: Par. - Rt.Angl. -
Flux Density in Magnet B_m at BH_{max} (G)	5200	Resistivity (micro-ohms/cm/cm^2)	- @ - °C
Coercive Force in Magnet H_m at BH_{max} (Oe)	6000	Hardness -	Scale -
Energy Loss W_c (watt-sec/cycle/lb)	-	Basic Forming Method	DPM
Importance of Operating Magnet at BH_{max}	E	Basic Finishing Method	WGR
Average Recoil Permeability	1.05	Relative Corrosion Resistance	P
Field Strength Required to Fully Saturate Magnet (Oe)	32,000	Primary Chemical Composition	Nd-Fe-B
Ability to Withstand Demagnetizing Fields	0	Special Notes:	
Maximum Service Temperature (°F)	Est. 245		
Approximate Temperature Permanently Affecting Magnet (°F)	500		
Inherent Orientation of Material	MP		

Note: All information based on available producers literature. In some cases of foreign materials interpretation or calculation required.

SOURCE(S):	Nd94-FA
	(Alloy or Grade)
Hitachi Magnetics Corp.	Sintered Rare Earth
	(Type of Material)
USA	24/10
	(MMPA Brief Designation)
	R7-3
	(IEC Code Reference)

MAGNETIC PROPERTIES		PHYSICAL PROPERTIES	
Residual Induction B_r (G)	10,300	Density (lbs./in.^{3})	.2710
Coercive Force (Oe) — Normal (H_c)	9500	General Mechanical Properties	HB
Coercive Force (Oe) — Intrinsic (H_{ci})	10,000	Tensile Strength (psi)	-
Maximum Energy Product BH_{max} (MGO)	24.5	Transverse Modulus of Rupture (psi)	-
B/H Ratio (Load Line) at BH_{max}	.7	Coeff. of Therm. Expansion inches x 10^{-6}/°C	Orientation: Par. - Rt.Angl. -
Flux Density in Magnet B_m at BH_{max} (G)	4000	Resistivity (micro-ohms/cm/cm^2	- @ - °C
Coercive Force in Magnet H_m at BH_{max} (Oe)	6000	Hardness -	Scale -
Energy Loss W_c (watt-sec/cycle/lb)	-	Basic Forming Method	DPM
Importance of Operating Magnet at BH_{max}	E	Basic Finishing Method	WGR
Average Recoil Permeability	1.05	Relative Corrosion Resistance	P
Field Strength Required to Fully Saturate Magnet (Oe)	32,000	Primary Chemical Composition	Nd-Fe-B
Ability to Withstand Demagnetizing Fields	0	Special Notes:	
Maximum Service Temperature (°F)	257		
Approximate Temperature Permanently Affecting Magnet (°F)	500		
Inherent Orientation of Material	MP		

Note: All information based on available producers literature. In some cases of foreign materials interpretation or calculation required.

SOURCE(S):

Hitachi Magnetics Corp.

USA

Nd97-EA
(Alloy or Grade)

Sintered Rare Earth
(Type of Material)

30/11
(MMPA Brief Designation)

R7-3
(IEC Code Reference)

MAGNETIC PROPERTIES		PHYSICAL PROPERTIES	
Residual Induction B_r (G)	11,500	Density (lbs./in.^{3})	.2710
Coercive Force (Oe) Normal (H_c)	10,000	General Mechanical Properties	HB
Coercive Force (Oe) Intrinsic (H_{ci})	11,250	Tensile Strength (psi)	-
Maximum Energy Product BH_{max} (MGO)	30.0	Transverse Modulus of Rupture (psi)	-
B/H Ratio (Load Line) at BH_{max}	1.2	Coeff. of Therm. Expansion inches x 10^{-6}/°C	Orientation: Par. - Rt.Angl. -
Flux Density in Magnet B_m at BH_{max} (G)	6000	Resistivity (micro-ohms/cm/cm^2	- @ - °C
Coercive Force in Magnet H_m at BH_{max} (Oe)	5000	Hardness -	Scale -
Energy Loss W_c (watt-sec/cycle/lb)	-	Basic Forming Method	DPM
Importance of Operating Magnet at BH_{max}	E	Basic Finishing Method	WGR
Average Recoil Permeability	1.05	Relative Corrosion Resistance	P
Field Strength Required to Fully Saturate Magnet (Oe)	32,000	Primary Chemical Composition	Nd-Fe-B
Ability to Withstand Demagnetizing Fields	0	Special Notes:	
Maximum Service Temperature (°F)	257		
Approximate Temperature Permanently Affecting Magnet (°F)	500		
Inherent Orientation of Material	MP		

Note: All information based on available producers literature. In some cases of foreign materials interpretation or calculation required.

SOURCE(S):

Hitachi Magnetics Corp.

USA

Nd97-EB
(Alloy or Grade)

Sintered Rare Earth
(Type of Material)

35/11
(MMPA Brief Designation)

R7-3
(IEC Code Reference)

MAGNETIC PROPERTIES	
Residual Induction B_r (G)	13,400
Coercive Force (Oe) — Normal (H_c)	10,100
Coercive Force (Oe) — Intrinsic (H_{ci})	11,250
Maximum Energy Product BH_{max} (MGO)	34.5
B/H Ratio (Load Line) at BH_{max}	1.0
Flux Density in Magnet B_m at BH_{max} (G)	5800
Coercive Force in Magnet H_m at BH_{max} (Oe)	5800
Energy Loss W_c (watt-sec/cycle/lb)	-
Importance of Operating Magnet at BH_{max}	E
Average Recoil Permeability	1.05
Field Strength Required to Fully Saturate Magnet (Oe)	32,000
Ability to Withstand Demagnetizing Fields	0
Maximum Service Temperature (°F)	257
Approximate Temperature Permanently Affecting Magnet (°F)	500
Inherent Orientation of Material	MP

PHYSICAL PROPERTIES	
Density (lbs./in.^{3})	.2710
General Mechanical Properties	HB
Tensile Strength (psi)	-
Transverse Modulus of Rupture (psi)	-
Coeff. of Therm. Expansion inches x 10^{-6}/°C — Orientation: Par. / Rt.Angl.	- / -
Resistivity (micro-ohms/cm/cm^2) @ °C	- @ -
Hardness / Scale	- / -
Basic Forming Method	DPM
Basic Finishing Method	WGR
Relative Corrosion Resistance	P
Primary Chemical Composition	Nd-Fe-B
Special Notes:	

Note: All information based on available producers literature. In some cases of foreign materials interpretation or calculation required.

SOURCE(S):	**NEO 2001A**
	(Alloy or Grade)
Stackpole Corp.	Bonded Rare Earth
	(Type of Material)
USA	4.5/15
	(MMPA Brief Designation)
	R5-1
	(IEC Code Reference)

MAGNETIC PROPERTIES		PHYSICAL PROPERTIES	
Residual Induction B_r (G)	4600	Density (lbs./in.^{3})	.1824
Coercive Force (Oe) — Normal (H_c)	4100	General Mechanical Properties	HT
Coercive Force (Oe) — Intrinsic (H_{ci})	15,000	Tensile Strength (psi)	-
Maximum Energy Product BH_{max} (MGO)	4.5	Transverse Modulus of Rupture (psi)	-
B/H Ratio (Load Line) at BH_{max}	-	Coeff. of Therm. Expansion inches x 10^{-6}/°C — Orientation: Par. / Rt.Angl.	- / -
Flux Density in Magnet B_m at BH_{max} (G)	-	Resistivity (micro-ohms/cm/cm^2) @ °C	- / -
Coercive Force in Magnet H_m at BH_{max} (Oe)	-	Hardness / Scale	- / -
Energy Loss W_c (watt-sec/cycle/lb)	-	Basic Forming Method	E
Importance of Operating Magnet at BH_{max}	E	Basic Finishing Method	WGR
Average Recoil Permeability	1.12	Relative Corrosion Resistance	G
Field Strength Required to Fully Saturate Magnet (Oe)	35,000	Primary Chemical Composition	Nd-Fe-B + binder (See Note)
Ability to Withstand Demagnetizing Fields	0	Special Notes: (1) Binder not indicated	
Maximum Service Temperature (°F)	-		
Approximate Temperature Permanently Affecting Magnet (°F)	594		
Inherent Orientation of Material	none		

Note: All information based on available producers literature. In some cases of foreign materials interpretation or calculation required.

SOURCE(S):	NEO 2001B
	(Alloy or Grade)
Stackpole Corp.	Bonded Rare Earth
	(Type of Material)
USA	5.5/9.5
	(MMPA Brief Designation)
	-
	(IEC Code Reference)

MAGNETIC PROPERTIES		PHYSICAL PROPERTIES	
Residual Induction B_r (G)	5200	Density (lbs./in.^{3})	.1824
Coercive Force (Oe) — Normal (H_c)	4300	General Mechanical Properties	HT
Coercive Force (Oe) — Intrinsic (H_{ci})	9500	Tensile Strength (psi)	-
Maximum Energy Product BH_{max} (MGO)	5.5	Transverse Modulus of Rupture (psi)	-
B/H Ratio (Load Line) at BH_{max}	-	Coeff. of Therm. Expansion inches x 10^{-6}/°C	Orientation: Par. - Rt.Angl. -
Flux Density in Magnet B_m at BH_{max} (G)	-	Resistivity (micro-ohms/cm/cm^2)	- @ - °C
Coercive Force in Magnet H_m at BH_{max} (Oe)	-	Hardness — Scale	- — -
Energy Loss W_c (watt-sec/cycle/lb)	-	Basic Forming Method	E
Importance of Operating Magnet at BH_{max}	E	Basic Finishing Method	WGR
Average Recoil Permeability	1.2	Relative Corrosion Resistance	G
Field Strength Required to Fully Saturate Magnet (Oe)	30,000	Primary Chemical Composition	Nd-Fe-B + binder (See Note)
Ability to Withstand Demagnetizing Fields	0	Special Notes: (1) Binder not indicated	
Maximum Service Temperature (°F)	-		
Approximate Temperature Permanently Affecting Magnet (°F)	594		
Inherent Orientation of Material	none		

Note: All information based on available producers literature. In some cases of foreign materials interpretation or calculation required.

SOURCE(S):

Magnetfabrik-Bonn

Germany

Neofer 31/100p
(Alloy or Grade)

Bonded Rare Earth
(Type of Material)

3.8/12.5   [31/100p]
(MMPA Brief Designation)

-
(IEC Code Reference)

MAGNETIC PROPERTIES			PHYSICAL PROPERTIES	
Residual Induction B_r (G)		4100	Density (lbs./in.^{3})	.1662
Coercive Force (Oe)	Normal (H_c)	3640	General Mechanical Properties	FL
	Intrinsic (H_{ci})	12,500	Tensile Strength (psi)	-
Maximum Energy Product BH_{max} (MGO)		3.8	Transverse Modulus of Rupture (psi)	-
B/H Ratio (Load Line) at BH_{max}		.93	Coeff. of Therm. Expansion inches x 10^{-6}/°C	Orientation: Par. - Rt.Angl. -
Flux Density in Magnet B_m at BH_{max} (G)		2000	Resistivity (micro-ohms/cm/cm^2	- @ - °C
Coercive Force in Magnet H_m at BH_{max} (Oe)		2150	Hardness Scale	- -
Energy Loss W_c (watt-sec/cycle/lb)		-	Basic Forming Method	E
Importance of Operating Magnet at BH_{max}		E	Basic Finishing Method	SCD/SRD
Average Recoil Permeability		1.15	Relative Corrosion Resistance	G
Field Strength Required to Fully Saturate Magnet (Oe)		26,600	Primary Chemical Composition	Nd-Fe-B + binder (See Note)
Ability to Withstand Demagnetizing Fields		0	Special Notes: (1) Binder not indicated	
Maximum Service Temperature (°F)		-		
Approximate Temperature Permanently Affecting Magnet (°F)		594		
Inherent Orientation of Material		none		

Note: All information based on available producers literature. In some cases of foreign materials interpretation or calculation required.

SOURCE(S):	Neofer 35/100p
	(Alloy or Grade)
Magnetfabrik-Bonn	Bonded Rare Earth
	(Type of Material)
Germany	4.3/12 [35/100p]
	(MMPA Brief Designation)
	-
	(IEC Code Reference)

MAGNETIC PROPERTIES		PHYSICAL PROPERTIES	
Residual Induction B_r (G)	4500	Density (lbs./in.^{3})	.1806
Coercive Force (Oe) — Normal (H_c)	3760	General Mechanical Properties	HT
Coercive Force (Oe) — Intrinsic (H_{ci})	12,000	Tensile Strength (psi)	-
Maximum Energy Product BH_{max} (MGO)	4.3	Transverse Modulus of Rupture (psi)	-
B/H Ratio (Load Line) at BH_{max}	1.07	Coeff. of Therm. Expansion inches x 10^{-6}/°C Orientation: Par. - Rt.Angl. -	
Flux Density in Magnet B_m at BH_{max} (G)	2300	Resistivity (micro-ohms/cm/cm^2 - @ - °C	
Coercive Force in Magnet H_m at BH_{max} (Oe)	2150	Hardness Scale - -	
Energy Loss W_c (watt-sec/cycle/lb)	-	Basic Forming Method	E
Importance of Operating Magnet at BH_{max}	E	Basic Finishing Method	WGR
Average Recoil Permeability	1.15	Relative Corrosion Resistance	G
Field Strength Required to Fully Saturate Magnet (Oe)	26,600	Primary Chemical Composition	Nd-Fe-B + binder (See Note)
Ability to Withstand Demagnetizing Fields	0	Special Notes: (1) Binder not indicated	
Maximum Service Temperature (°F)	-		
Approximate Temperature Permanently Affecting Magnet (°F)	594		
Inherent Orientation of Material	none		

Note: All information based on available producers literature. In some cases of foreign materials interpretation or calculation required.

SOURCE(S):

Magnetfabrik-Bonn

Germany

Neofer 37/60p
(Alloy or Grade)

Bonded Rare Earth
(Type of Material)

4.6/7.5 [37/60p]
(MMPA Brief Designation)

-
(IEC Code Reference)

MAGNETIC PROPERTIES		PHYSICAL PROPERTIES	
Residual Induction B_r (G)	4800	Density (lbs./in.^{3})	.1662
Coercive Force (Oe) Normal (H_c)	3760	General Mechanical Properties	FL
Coercive Force (Oe) Intrinsic (H_{ci})	7500	Tensile Strength (psi)	-
Maximum Energy Product BH_{max} (MGO)	4.6	Transverse Modulus of Rupture (psi)	-
B/H Ratio (Load Line) at BH_{max}	1.02	Coeff. of Therm. Expansion inches x 10^{-6}/°C Orientation: Par. / Rt.Angl.	- / -
Flux Density in Magnet B_m at BH_{max} (G)	2200	Resistivity (micro-ohms/cm/cm^2 - @ - °C	
Coercive Force in Magnet H_m at BH_{max} (Oe)	2150	Hardness Scale	- -
Energy Loss W_c (watt-sec/cycle/lb)	-	Basic Forming Method	E
Importance of Operating Magnet at BH_{max}	E	Basic Finishing Method	SCD/SRD
Average Recoil Permeability	1.2	Relative Corrosion Resistance	G
Field Strength Required to Fully Saturate Magnet (Oe)	26,600	Primary Chemical Composition	Nd-Fe-B + binder (See Note)
Ability to Withstand Demagnetizing Fields	0	Special Notes: (1) Binder not indicated	
Maximum Service Temperature (°F)	-		
Approximate Temperature Permanently Affecting Magnet (°F)	594		
Inherent Orientation of Material	none		

Note: All information based on available producers literature. In some cases of foreign materials interpretation or calculation required.

SOURCE(S):

Magnetfabrik-Bonn

Germany

Neofer 44/60p
(Alloy or Grade)

Bonded Rare Earth
(Type of Material)

5.5/7.5   [44/60p]
(MMPA Brief Designation)

-
(IEC Code Reference)

MAGNETIC PROPERTIES		PHYSICAL PROPERTIES	
Residual Induction B_r (G)	5400	Density (lbs./in.^{3})	.1806
Coercive Force (Oe) Normal (H_c)	4000	General Mechanical Properties	FL
Coercive Force (Oe) Intrinsic (H_{ci})	7500	Tensile Strength (psi)	-
Maximum Energy Product BH_{max} (MGO)	5.5	Transverse Modulus of Rupture (psi)	-
B/H Ratio (Load Line) at BH_{max}	1.1	Coeff. of Therm. Expansion inches x 10^{-6}/°C Orientation: Par. - Rt.Angl. -	
Flux Density in Magnet B_m at BH_{max} (G)	2500	Resistivity (micro-ohms/cm/cm^2) - @ - °C	
Coercive Force in Magnet H_m at BH_{max} (Oe)	2250	Hardness Scale - -	
Energy Loss W_c (watt-sec/cycle/lb)	-	Basic Forming Method	E
Importance of Operating Magnet at BH_{max}	E	Basic Finishing Method	SCD/SRD
Average Recoil Permeability	1.2	Relative Corrosion Resistance	G
Field Strength Required to Fully Saturate Magnet (Oe)	22,600	Primary Chemical Composition (See Note)	Nd-Fe-B + binder
Ability to Withstand Demagnetizing Fields	0	Special Notes: (1) Binder not indicated	
Maximum Service Temperature (°F)	26,600		
Approximate Temperature Permanently Affecting Magnet (°F)	594		
Inherent Orientation of Material	none		

Note: All information based on available producers literature. In some cases of foreign materials interpretation or calculation required.

SOURCE(S):

Magnetfabrik-Bonn

Germany

Neofer 55/100p
(Alloy or Grade)

Bonded Rare Earth
(Type of Material)

6.9/12.5   [55/100p]
(MMPA Brief Designation)

-
(IEC Code Reference)

MAGNETIC PROPERTIES		PHYSICAL PROPERTIES	
Residual Induction B_r (G)	5800	Density (lbs./in.^{3})	.2132
Coercive Force (Oe) — Normal (H_c)	5000	General Mechanical Properties	HT
Coercive Force (Oe) — Intrinsic (H_{ci})	12,500	Tensile Strength (psi)	-
Maximum Energy Product BH_{max} (MGO)	6.9	Transverse Modulus of Rupture (psi)	-
B/H Ratio (Load Line) at BH_{max}	1.3	Coeff. of Therm. Expansion inches x 10^{-6}/°C	Orientation: Par. - Rt.Angl. -
Flux Density in Magnet B_m at BH_{max} (G)	3200	Resistivity (micro-ohms/cm/cm^2)	- @ - °C
Coercive Force in Magnet H_m at BH_{max} (Oe)	2500	Hardness Scale	- -
Energy Loss W_c (watt-sec/cycle/lb)	-	Basic Forming Method	E
Importance of Operating Magnet at BH_{max}	E	Basic Finishing Method	WGR
Average Recoil Permeability	1.15	Relative Corrosion Resistance	G
Field Strength Required to Fully Saturate Magnet (Oe)	26,600	Primary Chemical Composition	Nd-Fe-B + binder (See Note)
Ability to Withstand Demagnetizing Fields	0	Special Notes: (1) Binder not indicated	
Maximum Service Temperature (°F)	-		
Approximate Temperature Permanently Affecting Magnet (°F)	594		
Inherent Orientation of Material	none		

Note: All information based on available producers literature. In some cases of foreign materials interpretation or calculation required.

SOURCE(S):	Neofer 62/60p
	(Alloy or Grade)
Magnetfabrik-Bonn	Bonded Rare Earth
	(Type of Material)
Germany	7.7/7.5 [62/60p]
	(MMPA Brief Designation)
	-
	(IEC Code Reference)

MAGNETIC PROPERTIES		PHYSICAL PROPERTIES	
Residual Induction B_r (G)	6500	Density (lbs./in.^{3})	.2132
Coercive Force (Oe) — Normal (H_c)	4770	General Mechanical Properties	HT
Coercive Force (Oe) — Intrinsic (H_{ci})	7500	Tensile Strength (psi)	-
Maximum Energy Product BH_{max} (MGO)	7.7	Transverse Modulus of Rupture (psi)	-
B/H Ratio (Load Line) at BH_{max}	1.5	Coeff. of Therm. Expansion inches x 10^{-6}/°C Orientation: Par. _ Rt.Angl. _	
Flux Density in Magnet B_m at BH_{max} (G)	3300	Resistivity (micro-ohms/cm/cm^2 - @ - °C	
Coercive Force in Magnet H_m at BH_{max} (Oe)	2500	Hardness Scale	- -
Energy Loss W_c (watt-sec/cycle/lb)	-	Basic Forming Method	E
Importance of Operating Magnet at BH_{max}	E	Basic Finishing Method	MCT
Average Recoil Permeability	1.2	Relative Corrosion Resistance	G
Field Strength Required to Fully Saturate Magnet (Oe)	26,600	Primary Chemical Composition	Nd-Fe-B + binder (See Note)
Ability to Withstand Demagnetizing Fields	0	Special Notes: (1) Binder not indicated	
Maximum Service Temperature (°F)	-		
Approximate Temperature Permanently Affecting Magnet (°F)	594		
Inherent Orientation of Material	none		

Note: All information based on available producers literature. In some cases of foreign materials interpretation or calculation required.

SOURCE(S):	**Neofer 230/80**	
	(Alloy or Grade)	
Magnetfabrik-Bonn	Sintered Rare Earth	
	(Type of Material)	
Germany	29/10 [230/80]	
	(MMPA Brief Designation)	
	R7-3	
	(IEC Code Reference)	

MAGNETIC PROPERTIES		PHYSICAL PROPERTIES	
Residual Induction B_r (G)	11,000	Density (lbs./in.^{3})	.2673
Coercive Force (Oe) Normal (H_c)	9500	General Mechanical Properties	HB
Coercive Force (Oe) Intrinsic (H_{ci})	10,000	Tensile Strength (psi)	-
Maximum Energy Product BH_{max} (MGO)	29.0	Transverse Modulus of Rupture (psi)	-
B/H Ratio (Load Line) at BH_{max}	1.2	Coeff. of Therm. Expansion inches x 10^{-6}/°C	Orientation: Par. - Rt.Angl. -
Flux Density in Magnet B_m at BH_{max} (G)	6000	Resistivity (micro-ohms/cm/cm^2 -	@ - °C
Coercive Force in Magnet H_m at BH_{max} (Oe)	5000	Hardness Scale - -	
Energy Loss W_c (watt-sec/cycle/lb)	-	Basic Forming Method	DPM
Importance of Operating Magnet at BH_{max}	E	Basic Finishing Method	WGR
Average Recoil Permeability	1.05	Relative Corrosion Resistance	P
Field Strength Required to Fully Saturate Magnet (Oe)	35,000	Primary Chemical Composition	Nd-Fe-B
Ability to Withstand Demagnetizing Fields	0	Special Notes:	
Maximum Service Temperature (°F)	Est. 212		
Approximate Temperature Permanently Affecting Magnet (°F)	Est. 590		
Inherent Orientation of Material	MP		

Note: All information based on available producers literature. In some cases of foreign materials interpretation or calculation required.

SOURCE(S):

Magnetfabrik-Bonn

Germany

Neofer 230/120
(Alloy or Grade)

Sintered Rare Earth
(Type of Material)

29/15   [230/120]
(MMPA Brief Designation)

R7-3
(IEC Code Reference)

MAGNETIC PROPERTIES		PHYSICAL PROPERTIES	
Residual Induction B_r (G)	11,000	Density (lbs./in.^{3})	.2673
Coercive Force (Oe) — Normal (H_c)	9800	General Mechanical Properties	HB
Coercive Force (Oe) — Intrinsic (H_{ci})	15,000	Tensile Strength (psi)	-
Maximum Energy Product BH_{max} (MGO)	29.0	Transverse Modulus of Rupture (psi)	-
B/H Ratio (Load Line) at BH_{max}	1.2	Coeff. of Therm. Expansion inches x 10^{-6}/°C	Orientation: Par. - Rt.Angl. -
Flux Density in Magnet B_m at BH_{max} (G)	6000	Resistivity (micro-ohms/cm/cm^2)	- @ - °C
Coercive Force in Magnet H_m at BH_{max} (Oe)	5000	Hardness Scale	- -
Energy Loss W_c (watt-sec/cycle/lb)	-	Basic Forming Method	DPM
Importance of Operating Magnet at BH_{max}	E	Basic Finishing Method	WGR
Average Recoil Permeability	1.05	Relative Corrosion Resistance	P
Field Strength Required to Fully Saturate Magnet (Oe)	35,000	Primary Chemical Composition	Nd-Fe-B
Ability to Withstand Demagnetizing Fields	0	Special Notes:	
Maximum Service Temperature (°F)	Est. 212		
Approximate Temperature Permanently Affecting Magnet (°F)	Est. 590		
Inherent Orientation of Material	MP		

Note: All information based on available producers literature. In some cases of foreign materials interpretation or calculation required.

SOURCE(S):

Thyssen Magnettechnik GMBH

Germany

Neolit NQ1C
(Alloy or Grade)

Sintered Rare Earth
(Type of Material)

9/16
(MMPA Brief Designation)

R7-3
(IEC Code Reference)

MAGNETIC PROPERTIES	
Residual Induction B_r (G)	6400
Coercive Force (Oe) — Normal (H_c)	5700
Coercive Force (Oe) — Intrinsic (H_{ci})	16,000
Maximum Energy Product BH_{max} (MGO)	8.8
B/H Ratio (Load Line) at BH_{max}	.97
Flux Density in Magnet B_m at BH_{max} (G)	2900
Coercive Force in Magnet H_m at BH_{max} (Oe)	3000
Energy Loss W_c (watt-sec/cycle/lb)	-
Importance of Operating Magnet at BH_{max}	E
Average Recoil Permeability	1.15
Field Strength Required to Fully Saturate Magnet (Oe)	32,000
Ability to Withstand Demagnetizing Fields	0
Maximum Service Temperature (°F)	257
Approximate Temperature Permanently Affecting Magnet (°F)	878
Inherent Orientation of Material	none

PHYSICAL PROPERTIES	
Density (lbs./in.^{3})	.2204
General Mechanical Properties	HB
Tensile Strength (psi)	-
Transverse Modulus of Rupture (psi)	-
Coeff. of Therm. Expansion inches x 10^{-6}/°C — Orientation: Par. / Rt.Angl.	- / -
Resistivity (micro-ohms/cm/cm^2)	- @ - °C
Hardness 30	Scale Rockwell C
Basic Forming Method	DP
Basic Finishing Method	WGR
Relative Corrosion Resistance	P
Primary Chemical Composition	Nd-Fe-B
Special Notes:	

Note: All information based on available producers literature. In some cases of foreign materials interpretation or calculation required.

SOURCE(S):

Thyssen Magnettechnik GMBH

Germany

Neolit NQ1D
(Alloy or Grade)

Sintered Rare Earth
(Type of Material)

10/11
(MMPA Brief Designation)

R7-3
(IEC Code Reference)

MAGNETIC PROPERTIES		PHYSICAL PROPERTIES	
Residual Induction B_r (G)	7000	Density (lbs./in.^{3})	.2204
Coercive Force (Oe) Normal (H_c)	5700	General Mechanical Properties	HB
Coercive Force (Oe) Intrinsic (H_{ci})	11,000	Tensile Strength (psi)	-
Maximum Energy Product BH_{max} (MGO)	9.5	Transverse Modulus of Rupture (psi)	-
B/H Ratio (Load Line) at BH_{max}	1.0	Coeff. of Therm. Expansion inches x 10^{-6}/°C	Orientation: Par. - Rt.Angl. -
Flux Density in Magnet B_m at BH_{max} (G)	3100	Resistivity (micro-ohms/cm/cm^2	- @ - °C
Coercive Force in Magnet H_m at BH_{max} (Oe)	3100	Hardness Scale	30 Rockwell C
Energy Loss W_c (watt-sec/cycle/lb)	-	Basic Forming Method	DP
Importance of Operating Magnet at BH_{max}	E	Basic Finishing Method	WGR
Average Recoil Permeability	1.15	Relative Corrosion Resistance	P
Field Strength Required to Fully Saturate Magnet (Oe)	32,000	Primary Chemical Composition	Nd-Fe-B
Ability to Withstand Demagnetizing Fields	0	Special Notes:	
Maximum Service Temperature (°F)	230		
Approximate Temperature Permanently Affecting Magnet (°F)	878		
Inherent Orientation of Material	none		

Note: All information based on available producers literature. In some cases of foreign materials interpretation or calculation required.

SOURCE(S):	Neolit NQ2E
Thyssen Magnettechnik GMBH	(Alloy or Grade)
	Sintered Rare Earth
	(Type of Material)
Germany	14/18
	(MMPA Brief Designation)
	R7-3
	(IEC Code Reference)

MAGNETIC PROPERTIES		PHYSICAL PROPERTIES	
Residual Induction B_r (G)	8000	Density (lbs./in.^{3})	.2746
Coercive Force (Oe) — Normal (H_c)	7000	General Mechanical Properties	HB
Coercive Force (Oe) — Intrinsic (H_{ci})	18,000	Tensile Strength (psi)	-
Maximum Energy Product BH_{max} (MGO)	14.0	Transverse Modulus of Rupture (psi)	-
B/H Ratio (Load Line) at BH_{max}	1.08	Coeff. of Therm. Expansion inches x 10^{-6}/°C Orientation: Par. - Rt.Angl. -	
Flux Density in Magnet B_m at BH_{max} (G)	3900	Resistivity (micro-ohms/cm/cm^2)	- @ - °C
Coercive Force in Magnet H_m at BH_{max} (Oe)	3600	Hardness 60 Scale Rockwell C	
Energy Loss W_c (watt-sec/cycle/lb)	-	Basic Forming Method	DP
Importance of Operating Magnet at BH_{max}	E	Basic Finishing Method	WGR
Average Recoil Permeability	1.15	Relative Corrosion Resistance	P
Field Strength Required to Fully Saturate Magnet (Oe)	32,000	Primary Chemical Composition	Nd-Fe-B
Ability to Withstand Demagnetizing Fields	0	Special Notes:	
Maximum Service Temperature (°F)	338		
Approximate Temperature Permanently Affecting Magnet (°F)	635		
Inherent Orientation of Material	none		

Note: All information based on available producers literature. In some cases of foreign materials interpretation or calculation required.

SOURCE(S):	Neolit NQ2F
	(Alloy or Grade)
Thyssen Magnettechnik GMBH	Sintered Rare Earth
	(Type of Material)
Germany	14/20
	(MMPA Brief Designation)
	R7-3
	(IEC Code Reference)

MAGNETIC PROPERTIES		PHYSICAL PROPERTIES	
Residual Induction B_r (G)	8000	Density (lbs./in.^{3})	.2746
Coercive Force (Oe) — Normal (H_c)	7500	General Mechanical Properties	HB
Coercive Force (Oe) — Intrinsic (H_{ci})	20,000	Tensile Strength (psi)	-
Maximum Energy Product BH_{max} (MGO)	14.0	Transverse Modulus of Rupture (psi)	-
B/H Ratio (Load Line) at BH_{max}	1.05	Coeff. of Therm. Expansion inches x 10^{-6}/°C Orientation: Par. _ Rt.Angl. _	
Flux Density in Magnet B_m at BH_{max} (G)	3900	Resistivity (micro-ohms/cm/cm^2)	- @ - °C
Coercive Force in Magnet H_m at BH_{max} (Oe)	3700	Hardness 60 Scale Rockwell C	
Energy Loss W_c (watt-sec/cycle/lb)	-	Basic Forming Method	DP
Importance of Operating Magnet at BH_{max}	E	Basic Finishing Method	WGR
Average Recoil Permeability	1.14	Relative Corrosion Resistance	P
Field Strength Required to Fully Saturate Magnet (Oe)	40,000	Primary Chemical Composition	Nd-Fe-B
Ability to Withstand Demagnetizing Fields	0	Special Notes:	
Maximum Service Temperature (°F)	392		
Approximate Temperature Permanently Affecting Magnet (°F)	698		
Inherent Orientation of Material	none		

Note: All information based on available producers literature. In some cases of foreign materials interpretation or calculation required.

SOURCE(S):

Thyssen Magnettechnik GMBH

Germany

Neolit NQ3E
(Alloy or Grade)

Sintered Rare Earth
(Type of Material)

36/13
(MMPA Brief Designation)

R7-3
(IEC Code Reference)

MAGNETIC PROPERTIES		PHYSICAL PROPERTIES	
Residual Induction B_r (G)	12,500	Density (lbs./in.^{3})	.2746
Coercive Force (Oe) — Normal (H_c)	11,200	General Mechanical Properties	HB
Coercive Force (Oe) — Intrinsic (H_{ci})	13,000	Tensile Strength (psi)	-
Maximum Energy Product BH_{max} (MGO)	36.0	Transverse Modulus of Rupture (psi)	-
B/H Ratio (Load Line) at BH_{max}	.96	Coeff. of Therm. Expansion inches x 10^{-6}/°C Orientation: Par. – Rt.Angl. –	
Flux Density in Magnet B_m at BH_{max} (G)	5900	Resistivity (micro-ohms/cm/cm^2) – @ – °C	
Coercive Force in Magnet H_m at BH_{max} (Oe)	6150	Hardness Scale 60 Rockwell C	
Energy Loss W_c (watt-sec/cycle/lb)	-	Basic Forming Method	DPM
Importance of Operating Magnet at BH_{max}	E	Basic Finishing Method	WGR
Average Recoil Permeability	1.09	Relative Corrosion Resistance	P
Field Strength Required to Fully Saturate Magnet (Oe)	32,000	Primary Chemical Composition	Nd-Fe-B
Ability to Withstand Demagnetizing Fields	0	Special Notes:	
Maximum Service Temperature (°F)	302		
Approximate Temperature Permanently Affecting Magnet (°F)	635		
Inherent Orientation of Material	MP		

Note: All information based on available producers literature. In some cases of foreign materials interpretation or calculation required.

SOURCE(S):

Thyssen Magnettechnik GMBH

Germany

Neolit NQ3F
(Alloy or Grade)

Sintered Rare Earth
(Type of Material)

35/17
(MMPA Brief Designation)

R7-3
(IEC Code Reference)

MAGNETIC PROPERTIES		PHYSICAL PROPERTIES	
Residual Induction B_r (G)	12,300	Density (lbs./in.^{3})	.2746
Coercive Force (Oe) — Normal (H_c)	11,500	General Mechanical Properties	HB
Coercive Force (Oe) — Intrinsic (H_{ci})	17,000	Tensile Strength (psi)	-
Maximum Energy Product BH_{max} (MGO)	35.0	Transverse Modulus of Rupture (psi)	-
B/H Ratio (Load Line) at BH_{max}	1.11	Coeff. of Therm. Expansion inches x 10^{-6}/°C	Orientation: Par. — Rt.Angl. —
Flux Density in Magnet B_m at BH_{max} (G)	6200	Resistivity (micro-ohms/cm/cm^2)	- @ - °C
Coercive Force in Magnet H_m at BH_{max} (Oe)	5600	Hardness Scale 60	Rockwell C
Energy Loss W_c (watt-sec/cycle/lb)	-	Basic Forming Method	DPM
Importance of Operating Magnet at BH_{max}	E	Basic Finishing Method	WGR
Average Recoil Permeability	1.07	Relative Corrosion Resistance	P
Field Strength Required to Fully Saturate Magnet (Oe)	32,000	Primary Chemical Composition	Nd-Fe-B
Ability to Withstand Demagnetizing Fields	0	Special Notes:	
Maximum Service Temperature (°F)	356		
Approximate Temperature Permanently Affecting Magnet (°F)	698		
Inherent Orientation of Material	MP		

Note: All information based on available producers literature. In some cases of foreign materials interpretation or calculation required.

SOURCE(S):

Thyssen Magnettechnik GMBH

Germany

Neolit NQ3G
(Alloy or Grade)

Sintered Rare Earth
(Type of Material)

31/23
(MMPA Brief Designation)

R7-3
(IEC Code Reference)

MAGNETIC PROPERTIES		PHYSICAL PROPERTIES	
Residual Induction B_r (G)	11,400	Density (lbs./in.^{3})	.2746
Coercive Force (Oe) — Normal (H_c)	10,700	General Mechanical Properties	HB
Coercive Force (Oe) — Intrinsic (H_{ci})	23,000	Tensile Strength (psi)	-
Maximum Energy Product BH_{max} (MGO)	31.0	Transverse Modulus of Rupture (psi)	-
B/H Ratio (Load Line) at BH_{max}	1.02	Coeff. of Therm. Expansion inches x 10^{-6}/°C	Orientation: Par. - Rt.Angl. -
Flux Density in Magnet B_m at BH_{max} (G)	5600	Resistivity (micro-ohms/cm/cm^2)	- @ - °C
Coercive Force in Magnet H_m at BH_{max} (Oe)	5500	Hardness 60	Scale Rockwell C
Energy Loss W_c (watt-sec/cycle/lb)	-	Basic Forming Method	DPM
Importance of Operating Magnet at BH_{max}	E	Basic Finishing Method	WGR
Average Recoil Permeability	1.05	Relative Corrosion Resistance	P
Field Strength Required to Fully Saturate Magnet (Oe)	32,000	Primary Chemical Composition	Nd-Fe-B
Ability to Withstand Demagnetizing Fields	0	Special Notes:	
Maximum Service Temperature (°F)	392		
Approximate Temperature Permanently Affecting Magnet (°F)	698		
Inherent Orientation of Material	MP		

Note: All information based on available producers literature. In some cases of foreign materials interpretation or calculation required.

SOURCE(S):   **NEOREC-27SH**
                                                (Alloy or Grade)

TDK Corp. of America   Sintered Rare Earth
                                                (Type of Material)

          USA   27/21
                                              (MMPA Brief Designation)

                                                 R7-3
                                               (IEC Code Reference)

MAGNETIC PROPERTIES		PHYSICAL PROPERTIES	
Residual Induction B_r (G)	10,600	Density (lbs./in.^{3})	.2673
Coercive Force (Oe) — Normal (H_c)	10,000	General Mechanical Properties	HB
Coercive Force (Oe) — Intrinsic (H_{ci})	21,000	Tensile Strength (psi)	-
Maximum Energy Product BH_{max} (MGO)	27.0	Transverse Modulus of Rupture (psi)	-
B/H Ratio (Load Line) at BH_{max}	1.0	Coeff. of Therm. Expansion inches x 10^{-6}/°C	Orientation: Par. - Rt.Angl. -
Flux Density in Magnet B_m at BH_{max} (G)	5200	Resistivity (micro-ohms/cm/cm^2)	130 @ 20 °C
Coercive Force in Magnet H_m at BH_{max} (Oe)	5200	Hardness Scale	550-600 Vickers
Energy Loss W_c (watt-sec/cycle/lb)	-	Basic Forming Method	DPM
Importance of Operating Magnet at BH_{max}	E	Basic Finishing Method	WGR
Average Recoil Permeability	1.05	Relative Corrosion Resistance	P
Field Strength Required to Fully Saturate Magnet (Oe)	32,000	Primary Chemical Composition	Nd-Fe-B
Ability to Withstand Demagnetizing Fields	0	Special Notes:	
Maximum Service Temperature (°F)	Est. 302		
Approximate Temperature Permanently Affecting Magnet (°F)	Est. 590		
Inherent Orientation of Material	MP		

Note: All information based on available producers literature. In some cases of foreign materials interpretation or calculation required.

SOURCE(S):

TDK Corp. of America

USA

NEOREC-30SH
(Alloy or Grade)

Sintered Rare Earth
(Type of Material)

30/21
(MMPA Brief Designation)

R7-3
(IEC Code Reference)

MAGNETIC PROPERTIES		PHYSICAL PROPERTIES	
Residual Induction B_r (G)	11,200	Density (lbs./in.^{3})	.2673
Coercive Force (Oe) — Normal (H_c)	10,700	General Mechanical Properties	HB
Coercive Force (Oe) — Intrinsic (H_{ci})	21,000	Tensile Strength (psi)	-
Maximum Energy Product BH_{max} (MGO)	30.0	Transverse Modulus of Rupture (psi)	-
B/H Ratio (Load Line) at BH_{max}	1.17	Coeff. of Therm. Expansion inches x 10^{-6}/°C — Orientation: Par. / Rt.Angl.	- / -
Flux Density in Magnet B_m at BH_{max} (G)	5950	Resistivity (micro-ohms/cm/cm^2)	130 @ 20 °C
Coercive Force in Magnet H_m at BH_{max} (Oe)	5100	Hardness Scale	550-600 Vickers
Energy Loss W_c (watt-sec/cycle/lb)	-	Basic Forming Method	DPM
Importance of Operating Magnet at BH_{max}	E	Basic Finishing Method	WGR
Average Recoil Permeability	1.05	Relative Corrosion Resistance	P
Field Strength Required to Fully Saturate Magnet (Oe)	32,000	Primary Chemical Composition	Nd-Fe-B
Ability to Withstand Demagnetizing Fields	0	Special Notes:	
Maximum Service Temperature (°F)	Est. 302		
Approximate Temperature Permanently Affecting Magnet (°F)	608		
Inherent Orientation of Material	MP		

Note: All information based on available producers literature. In some cases of foreign materials interpretation or calculation required.

SOURCE(S):

TDK Corp. of America

USA

NEOREC-32H
(Alloy or Grade)

Sintered Rare Earth
(Type of Material)

31/17
(MMPA Brief Designation)

R7-3
(IEC Code Reference)

MAGNETIC PROPERTIES	
Residual Induction B_r (G)	11,500
Coercive Force (Oe) — Normal (H_c)	10,200
Coercive Force (Oe) — Intrinsic (H_{ci})	17,000+
Maximum Energy Product BH_{max} (MGO)	31.0
B/H Ratio (Load Line) at BH_{max}	1.07
Flux Density in Magnet B_m at BH_{max} (G)	5800
Coercive Force in Magnet H_m at BH_{max} (Oe)	5400
Energy Loss W_c (watt-sec/cycle/lb)	-
Importance of Operating Magnet at BH_{max}	E
Average Recoil Permeability	1.05
Field Strength Required to Fully Saturate Magnet (Oe)	32,000
Ability to Withstand Demagnetizing Fields	0
Maximum Service Temperature (°F)	Est. 250
Approximate Temperature Permanently Affecting Magnet (°F)	608
Inherent Orientation of Material	MP

PHYSICAL PROPERTIES	
Density (lbs./in.^{3})	.2673
General Mechanical Properties	HB
Tensile Strength (psi)	-
Transverse Modulus of Rupture (psi)	-
Coeff. of Therm. Expansion inches x 10^{-6}/°C — Orientation: Par. / Rt.Angl.	- / -
Resistivity (micro-ohms/cm/cm^2)	130 @ 20 °C
Hardness Scale	550-600 Vickers
Basic Forming Method	DPM
Basic Finishing Method	WGR
Relative Corrosion Resistance	P
Primary Chemical Composition	Nd-Fe-B
Special Notes:	

Note: All information based on available producers literature. In some cases of foreign materials interpretation or calculation required.

SOURCE(S):	NEOREC-33
	(Alloy or Grade)
TDK Corp. of America	Sintered Rare Earth
	(Type of Material)
USA	32/11
	(MMPA Brief Designation)
	R7-3
	(IEC Code Reference)

MAGNETIC PROPERTIES		PHYSICAL PROPERTIES	
Residual Induction B_r (G)	11,700	Density (lbs./in.^{3})	.2673
Coercive Force (Oe) — Normal (H_c)	10,500	General Mechanical Properties	HB
Coercive Force (Oe) — Intrinsic (H_{ci})	11,000	Tensile Strength (psi)	-
Maximum Energy Product BH_{max} (MGO)	32.0	Transverse Modulus of Rupture (psi)	-
B/H Ratio (Load Line) at BH_{max}	1.05	Coeff. of Therm. Expansion inches x 10^{-6}/°C	Orientation: Par. - Rt.Angl. -
Flux Density in Magnet B_m at BH_{max} (G)	5800	Resistivity (micro-ohms/cm/cm^2)	130 @ 20 °C
Coercive Force in Magnet H_m at BH_{max} (Oe)	5500	Hardness 550-600 Scale	Vickers
Energy Loss W_c (watt-sec/cycle/lb)	-	Basic Forming Method	DPM
Importance of Operating Magnet at BH_{max}	E	Basic Finishing Method	WGR
Average Recoil Permeability	1.05	Relative Corrosion Resistance	P
Field Strength Required to Fully Saturate Magnet (Oe)	32,000	Primary Chemical Composition	Nd-Fe-B
Ability to Withstand Demagnetizing Fields	0	Special Notes:	
Maximum Service Temperature (°F)	Est. 212		
Approximate Temperature Permanently Affecting Magnet (°F)	608		
Inherent Orientation of Material	MP		

Note: All information based on available producers literature. In some cases of foreign materials interpretation or calculation required.

SOURCE(S): **NEOREC-35H**
(Alloy or Grade)

TDK Corp. of America

Sintered Rare Earth
(Type of Material)

USA

37/17
(MMPA Brief Designation)

R7-3
(IEC Code Reference)

MAGNETIC PROPERTIES		PHYSICAL PROPERTIES	
Residual Induction B_r (G)	12,100	Density (lbs./in.^{3})	.2673
Coercive Force (Oe) — Normal (H_c)	11,500	General Mechanical Properties	HB
Coercive Force (Oe) — Intrinsic (H_{ci})	17,000	Tensile Strength (psi)	-
Maximum Energy Product BH_{max} (MGO)	37.0	Transverse Modulus of Rupture (psi)	-
B/H Ratio (Load Line) at BH_{max}	1.02	Coeff. of Therm. Expansion inches x 10^{-6}/°C	Orientation: Par. _ Rt.Angl. _
Flux Density in Magnet B_m at BH_{max} (G)	6100	Resistivity (micro-ohms/cm/cm^2)	130 @ 20 °C
Coercive Force in Magnet H_m at BH_{max} (Oe)	6000	Hardness Scale	550-600 Vickers
Energy Loss W_c (watt-sec/cycle/lb)	-	Basic Forming Method	DPM
Importance of Operating Magnet at BH_{max}	E	Basic Finishing Method	WGR
Average Recoil Permeability	1.05	Relative Corrosion Resistance	P
Field Strength Required to Fully Saturate Magnet (Oe)	32,000	Primary Chemical Composition	Nd-Fe-B
Ability to Withstand Demagnetizing Fields	0	Special Notes:	
Maximum Service Temperature (°F)	Est. 250		
Approximate Temperature Permanently Affecting Magnet (°F)	608		
Inherent Orientation of Material	MP		

Note: All information based on available producers literature. In some cases of foreign materials interpretation or calculation required.

SOURCE(S):

NEOREC-38
(Alloy or Grade)

TDK Corp. of America

Sintered Rare Earth
(Type of Material)

USA

37/11
(MMPA Brief Designation)

R7-3
(IEC Code Reference)

MAGNETIC PROPERTIES		PHYSICAL PROPERTIES	
Residual Induction B_r (G)	12,600	Density (lbs./in.^{3})	.2673
Coercive Force (Oe) Normal (H_c)	11,500	General Mechanical Properties	HB
Coercive Force (Oe) Intrinsic (H_{ci})	11,000	Tensile Strength (psi)	-
Maximum Energy Product BH_{max} (MGO)	37.0	Transverse Modulus of Rupture (psi)	-
B/H Ratio (Load Line) at BH_{max}	1.03	Coeff. of Therm. Expansion inches x 10^{-6}/°C	Orientation: Par. - Rt.Angl. -
Flux Density in Magnet B_m at BH_{max} (G)	6200	Resistivity (micro-ohms/cm/cm^2)	130 @ 20 °C
Coercive Force in Magnet H_m at BH_{max} (Oe)	600	Hardness Scale	500-600 Vickers
Energy Loss W_c (watt-sec/cycle/lb)	-	Basic Forming Method	DPM
Importance of Operating Magnet at BH_{max}	E	Basic Finishing Method	WGR
Average Recoil Permeability	1.05	Relative Corrosion Resistance	P
Field Strength Required to Fully Saturate Magnet (Oe)	32,000	Primary Chemical Composition	Nd-Fe-B
Ability to Withstand Demagnetizing Fields	0	Special Notes:	
Maximum Service Temperature (°F)	Est. 212		
Approximate Temperature Permanently Affecting Magnet (°F)	608		
Inherent Orientation of Material	MP		

Note: All information based on available producers literature. In some cases of foreign materials interpretation or calculation required.

SOURCE(S):

UGIMag, Inc.

USA

NIALCO 1
(Alloy or Grade)

Bonded Alnico
(Type of Material)

.6/.54
(MMPA Brief Designation)

-
(IEC Code Reference)

MAGNETIC PROPERTIES		PHYSICAL PROPERTIES	
Residual Induction B_r (G)	3200	Density (lbs./in.^{3})	.1861
Coercive Force (Oe) — Normal (H_c)	500	General Mechanical Properties	FL
Coercive Force (Oe) — Intrinsic (H_{ci})	540	Tensile Strength (psi)	-
Maximum Energy Product BH_{max} (MGO)	.60	Transverse Modulus of Rupture (psi)	-
B/H Ratio (Load Line) at BH_{max}	-	Coeff. of Therm. Expansion inches x 10^{-6}/°C	Orientation: Par. - Rt.Angl. -
Flux Density in Magnet B_m at BH_{max} (G)	-	Resistivity (micro-ohms/cm/cm^2)	- @ - °C
Coercive Force in Magnet H_m at BH_{max} (Oe)	-	Hardness Scale	- -
Energy Loss W_c (watt-sec/cycle/lb)	-	Basic Forming Method	E
Importance of Operating Magnet at BH_{max}	B	Basic Finishing Method	SCD/SRD
Average Recoil Permeability	3.0	Relative Corrosion Resistance	G
Field Strength Required to Fully Saturate Magnet (Oe)	3000	Primary Chemical Composition	Al-Ni-Co-Fe-Cu + binder
Ability to Withstand Demagnetizing Fields	G	Special Notes:	
Maximum Service Temperature (°F)	-		
Approximate Temperature Permanently Affecting Magnet (°F)	-		
Inherent Orientation of Material	none		

Note: All information based on available producers literature. In some cases of foreign materials interpretation or calculation required.

SOURCE(S):	**NIALCO 1**
	(Alloy or Grade)
UGIMag, Inc.	Cast Alnico
	(Type of Material)
USA	1.5/.55
	(MMPA Brief Designation)
	-
	(IEC Code Reference)

MAGNETIC PROPERTIES		PHYSICAL PROPERTIES	
Residual Induction B_r (G)	7000	Density (lbs./in.^{3})	.2529
Coercive Force (Oe) Normal (H_c)	550	General Mechanical Properties	HB
Coercive Force (Oe) Intrinsic (H_{ci})	Est. 550	Tensile Strength (psi)	-
Maximum Energy Product BH_{max} (MGO)	1.5	Transverse Modulus of Rupture (psi)	-
B/H Ratio (Load Line) at BH_{max}	Est. 12.5	Coeff. of Therm. Expansion inches x 10^{-6}/°C	Orientation: Par. - Rt.Angl. -
Flux Density in Magnet B_m at BH_{max} (G)	Est. 4700	Resistivity (micro-ohms/cm/cm^2)	65 @ 25 °C
Coercive Force in Magnet H_m at BH_{max} (Oe)	Est. 375	Hardness Scale	45 Rockwell C
Energy Loss W_c (watt-sec/cycle/lb)	15.3	Basic Forming Method	C
Importance of Operating Magnet at BH_{max}	A	Basic Finishing Method	WGR
Average Recoil Permeability	4.0	Relative Corrosion Resistance	G
Field Strength Required to Fully Saturate Magnet (Oe)	3000	Primary Chemical Composition	Al-Ni-Co-Fe-Cu
Ability to Withstand Demagnetizing Fields	F	Special Notes:	
Maximum Service Temperature (°F)	900		
Approximate Temperature Permanently Affecting Magnet (°F)	1580		
Inherent Orientation of Material	none		

Note: All information based on available producers literature. In some cases of foreign materials interpretation or calculation required.

SOURCE(S):	NIALCO 3
	(Alloy or Grade)
UGIMag, Inc.	Cast Alnico
	(Type of Material)
USA	1.8/.70
	(MMPA Brief Designation)
	-
	(IEC Code Reference)

MAGNETIC PROPERTIES		PHYSICAL PROPERTIES	
Residual Induction B_r (G)	6900	Density (lbs./in.^{3})	.2565
Coercive Force (Oe) — Normal (H_c)	700	General Mechanical Properties	HB
Coercive Force (Oe) — Intrinsic (H_{ci})	Est. 700	Tensile Strength (psi)	-
Maximum Energy Product BH_{max} (MGO)	Est. 1.8	Transverse Modulus of Rupture (psi)	-
B/H Ratio (Load Line) at BH_{max}	Est. 8.4	Coeff. of Therm. Expansion inches x 10^{-6}/°C	Orientation: Par. 12.5 Rt.Angl. -
Flux Density in Magnet B_m at BH_{max} (G)	Est. 4000	Resistivity (micro-ohms/cm/cm^2)	65 @ 25 °C
Coercive Force in Magnet H_m at BH_{max} (Oe)	Est. 475	Hardness Scale	45 Rockwell C
Energy Loss W_c (watt-sec/cycle/lb)	15.3	Basic Forming Method	C
Importance of Operating Magnet at BH_{max}	A	Basic Finishing Method	WGR
Average Recoil Permeability	4.5	Relative Corrosion Resistance	G
Field Strength Required to Fully Saturate Magnet (Oe)	3000	Primary Chemical Composition	Al-Ni-Co-Fe-Cu
Ability to Withstand Demagnetizing Fields	G	Special Notes:	
Maximum Service Temperature (°F)	900		
Approximate Temperature Permanently Affecting Magnet (°F)	1580		
Inherent Orientation of Material	none		

Note: All information based on available producers literature. In some cases of foreign materials interpretation or calculation required.

SOURCE(S):

UGIMag, Inc.

USA

NIALCO 4
(Alloy or Grade)

Bonded Alnico
(Type of Material)

1.0/1.1
(MMPA Brief Designation)

-
(IEC Code Reference)

MAGNETIC PROPERTIES		PHYSICAL PROPERTIES	
Residual Induction B_r (G)	3000	Density (lbs./in.^{3})	-
Coercive Force (Oe) Normal (H_c)	1000	General Mechanical Properties	FL
Coercive Force (Oe) Intrinsic (H_{ci})	1100	Tensile Strength (psi)	-
Maximum Energy Product BH_{max} (MGO)	1.0	Transverse Modulus of Rupture (psi)	-
B/H Ratio (Load Line) at BH_{max}	-	Coeff. of Therm. Expansion inches x $10^{-6}/°C$	Orientation: Par. - Rt.Angl. -
Flux Density in Magnet B_m at BH_{max} (G)	-	Resistivity (micro-ohms/cm/cm^2	- @ - °C
Coercive Force in Magnet H_m at BH_{max} (Oe)	-	Hardness Scale	- -
Energy Loss W_c (watt-sec/cycle/lb)	-	Basic Forming Method	E
Importance of Operating Magnet at BH_{max}	B	Basic Finishing Method	SCD/SRD
Average Recoil Permeability	2.5	Relative Corrosion Resistance	G
Field Strength Required to Fully Saturate Magnet (Oe)	3000	Primary Chemical Composition	Al-Ni-Co-Fe-Cu + binder
Ability to Withstand Demagnetizing Fields	G	Special Notes:	
Maximum Service Temperature (°F)	-		
Approximate Temperature Permanently Affecting Magnet (°F)	-		
Inherent Orientation of Material	none		

Note: All information based on available producers literature. In some cases of foreign materials interpretation or calculation required.

SOURCE(S):	NIALCO 5
	(Alloy or Grade)
UGIMag, Inc.	Cast Alnico
	(Type of Material)
USA	2.9/.90
	(MMPA Brief Designation)
	-
	(IEC Code Reference)

MAGNETIC PROPERTIES		PHYSICAL PROPERTIES	
Residual Induction B_r (G)	8500	Density (lbs./in.^{3})	.2601
Coercive Force (Oe) Normal (H_c)	800	General Mechanical Properties	HB
Coercive Force (Oe) Intrinsic (H_{ci})	Est. 900	Tensile Strength (psi)	-
Maximum Energy Product BH_{max} (MGO)	2.9	Transverse Modulus of Rupture (psi)	-
B/H Ratio (Load Line) at BH_{max}	10.0	Coeff. of Therm. Expansion inches x 10^{-6}/°C	Orientation: Par. 12.5 Rt.Angl. -
Flux Density in Magnet B_m at BH_{max} (G)	Est. 4500	Resistivity (micro-ohms/cm/cm^2	65 @ 25 °C
Coercive Force in Magnet H_m at BH_{max} (Oe)	Est. 450	Hardness Scale 45	Rockwell C
Energy Loss W_c (watt-sec/cycle/lb)	15.3	Basic Forming Method	C
Importance of Operating Magnet at BH_{max}	B	Basic Finishing Method	WGR
Average Recoil Permeability	3.7	Relative Corrosion Resistance	G
Field Strength Required to Fully Saturate Magnet (Oe)	3000	Primary Chemical Composition	Al-Ni-Co-Fe-Cu
Ability to Withstand Demagnetizing Fields	G	Special Notes:	
Maximum Service Temperature (°F)	900		
Approximate Temperature Permanently Affecting Magnet (°F)	1580		
Inherent Orientation of Material	none		

Note: All information based on available producers literature. In some cases of foreign materials interpretation or calculation required.

SOURCE(S):

UGIMag, Inc.

USA

NIALCO 6
(Alloy or Grade)

Bonded Alnico
(Type of Material)

1.1/1.2
(MMPA Brief Designation)

-
(IEC Code Reference)

MAGNETIC PROPERTIES		PHYSICAL PROPERTIES	
Residual Induction B_r (G)	3600	Density (lbs./in.^{3})	-
Coercive Force (Oe) — Normal (H_c)	1100	General Mechanical Properties	FL
Coercive Force (Oe) — Intrinsic (H_{ci})	1200	Tensile Strength (psi)	-
Maximum Energy Product BH_{max} (MGO)	1.10	Transverse Modulus of Rupture (psi)	-
B/H Ratio (Load Line) at BH_{max}	-	Coeff. of Therm. Expansion inches x 10^{-6}/°C	Orientation: Par. - Rt.Angl. -
Flux Density in Magnet B_m at BH_{max} (G)	-	Resistivity (micro-ohms/cm/cm^2	- @ - °C
Coercive Force in Magnet H_m at BH_{max} (Oe)	-	Hardness Scale	- -
Energy Loss W_c (watt-sec/cycle/lb)	-	Basic Forming Method	E
Importance of Operating Magnet at BH_{max}	B	Basic Finishing Method	SCD/SRD
Average Recoil Permeability	2.5	Relative Corrosion Resistance	G
Field Strength Required to Fully Saturate Magnet (Oe)	3000	Primary Chemical Composition	Al-Ni-Co-Fe-Cu + binder
Ability to Withstand Demagnetizing Fields	G	Special Notes:	
Maximum Service Temperature (°F)	-		
Approximate Temperature Permanently Affecting Magnet (°F)	-		
Inherent Orientation of Material	none		

Note: All information based on available producers literature. In some cases of foreign materials interpretation or calculation required.

SOURCE(S):	NIALCO 7
	(Alloy or Grade)
UGIMag, Inc.	Cast Alnico
	(Type of Material)
USA	2.3/1.3
	(MMPA Brief Designation)
	-
	(IEC Code Reference)

MAGNETIC PROPERTIES		PHYSICAL PROPERTIES	
Residual Induction B_r (G)	6400	Density (lbs./in.^{3})	Est. .2650
Coercive Force (Oe) — Normal (H_c)	1150	General Mechanical Properties	HB
Coercive Force (Oe) — Intrinsic (H_{ci})	Est. 1300	Tensile Strength (psi)	-
Maximum Energy Product BH_{max} (MGO)	2.3	Transverse Modulus of Rupture (psi)	-
B/H Ratio (Load Line) at BH_{max}	Est. 5.3	Coeff. of Therm. Expansion inches x 10^{-6}/°C	Orientation: Par. 12.0 Rt.Angl. -
Flux Density in Magnet B_m at BH_{max} (G)	Est. 3700	Resistivity (micro-ohms/cm/cm^2)	50 @ 25 °C
Coercive Force in Magnet H_m at BH_{max} (Oe)	Est. 700	Hardness Approx. 50	Scale Rockwell C
Energy Loss W_c (watt-sec/cycle/lb)	15.3	Basic Forming Method	C
Importance of Operating Magnet at BH_{max}	C	Basic Finishing Method	WGR
Average Recoil Permeability	-	Relative Corrosion Resistance	G
Field Strength Required to Fully Saturate Magnet (Oe)	3000	Primary Chemical Composition	Al-Ni-Co-Fe-Cu
Ability to Withstand Demagnetizing Fields	E	Special Notes:	
Maximum Service Temperature (°F)	900		
Approximate Temperature Permanently Affecting Magnet (°F)	1580		
Inherent Orientation of Material	none		

Note: All information based on available producers literature. In some cases of foreign materials interpretation or calculation required.

SOURCE(S):

UGIMag, Inc.

USA

NIALCO 8
(Alloy or Grade)

Cast Alnico
(Type of Material)

2.8/1.9
(MMPA Brief Designation)

-
(IEC Code Reference)

MAGNETIC PROPERTIES		PHYSICAL PROPERTIES	
Residual Induction B_r (G)	5,800	Density (lbs./in.^{3})	Est. .2650
Coercive Force (Oe) Normal (H_c)	1650	General Mechanical Properties	HB
Coercive Force (Oe) Intrinsic (H_{ci})	Est. 1860	Tensile Strength (psi)	-
Maximum Energy Product BH_{max} (MGO)	2.8	Transverse Modulus of Rupture (psi)	-
B/H Ratio (Load Line) at BH_{max}	Est. 4.5	Coeff. of Therm. Expansion inches x 10^{-6}/°C	Orientation: Par. 10.2 Rt.Angl. -
Flux Density in Magnet B_m at BH_{max} (G)	Est. 3500	Resistivity (micro-ohms/cm/cm^2	50 @ 25 °C
Coercive Force in Magnet H_m at BH_{max} (Oe)	Est. 775	Hardness Scale	Approx. 50 Rockwell C
Energy Loss W_c (watt-sec/cycle/lb)	15.3	Basic Forming Method	C
Importance of Operating Magnet at BH_{max}	C	Basic Finishing Method	WGR
Average Recoil Permeability	-	Relative Corrosion Resistance	G
Field Strength Required to Fully Saturate Magnet (Oe)	3000	Primary Chemical Composition	Al-Ni-Co-Fe-Cu
Ability to Withstand Demagnetizing Fields	E	Special Notes:	
Maximum Service Temperature (°F)	900		
Approximate Temperature Permanently Affecting Magnet (°F)	1580		
Inherent Orientation of Material	none		

Note: All information based on available producers literature. In some cases of foreign materials interpretation or calculation required.

SOURCE(S):	Oerstit 160
	(Alloy or Grade)
Thyssen Magnettechnik GMBH	Cast Alnico
	(Type of Material)
Germany	1.8/.80
	(MMPA Brief Designation)
	-
	(IEC Code Reference)

MAGNETIC PROPERTIES		PHYSICAL PROPERTIES	
Residual Induction B_r (G)	7000	Density (lbs./in.^{3})	.1565
Coercive Force (Oe) — Normal (H_c)	800	General Mechanical Properties	HB
Coercive Force (Oe) — Intrinsic (H_{ci})	800	Tensile Strength (psi)	-
Maximum Energy Product BH_{max} (MGO)	1.8	Transverse Modulus of Rupture (psi)	-
B/H Ratio (Load Line) at BH_{max}	8.9	Coeff. of Therm. Expansion inches x 10^{-6}/°C	Orientation: Par. _ Rt.Angl. _
Flux Density in Magnet B_m at BH_{max} (G)	4000	Resistivity (micro-ohms/cm/cm^2)	50 @ 25 °C
Coercive Force in Magnet H_m at BH_{max} (Oe)	450	Hardness Scale	50 Rockwell C
Energy Loss W_c (watt-sec/cycle/lb)	15.3	Basic Forming Method	CM
Importance of Operating Magnet at BH_{max}	B	Basic Finishing Method	WGR
Average Recoil Permeability	4.5	Relative Corrosion Resistance	G
Field Strength Required to Fully Saturate Magnet (Oe)	3000	Primary Chemical Composition	Al-Ni-Co-Fe-Cu
Ability to Withstand Demagnetizing Fields	G	Special Notes:	
Maximum Service Temperature (°F)	900		
Approximate Temperature Permanently Affecting Magnet (°F)	1580		
Inherent Orientation of Material	HT		

Note: All information based on available producers literature. In some cases of foreign materials interpretation or calculation required.

SOURCE(S):	**Oerstit 260**
(Alloy or Grade)
Thyssen Magnettechnik GMBH	Cast Alnico
(Type of Material)
Germany	2.6/1.3
(MMPA Brief Designation)
-
(IEC Code Reference)

MAGNETIC PROPERTIES		PHYSICAL PROPERTIES	
Residual Induction B_r (G)	6100	Density (lbs./in.^{3})	.2565
Coercive Force (Oe) — Normal (H_c)	1200	General Mechanical Properties	HB
Coercive Force (Oe) — Intrinsic (H_{ci})	1300	Tensile Strength (psi)	-
Maximum Energy Product BH_{max} (MGO)	2.6	Transverse Modulus of Rupture (psi)	-
B/H Ratio (Load Line) at BH_{max}	4.97	Coeff. of Therm. Expansion inches x 10^{-6}/°C	Orientation: Par. - Rt.Angl. -
Flux Density in Magnet B_m at BH_{max} (G)	3600	Resistivity (micro-ohms/cm/cm^2)	50 @ 25 °C
Coercive Force in Magnet H_m at BH_{max} (Oe)	725	Hardness Scale	50 Rockwell C
Energy Loss W_c (watt-sec/cycle/lb)	15.3	Basic Forming Method	CM
Importance of Operating Magnet at BH_{max}	C	Basic Finishing Method	WGR
Average Recoil Permeability	2.8	Relative Corrosion Resistance	G
Field Strength Required to Fully Saturate Magnet (Oe)	3000	Primary Chemical Composition	Al-Ni-Co-Fe-Cu
Ability to Withstand Demagnetizing Fields	E	Special Notes:	
Maximum Service Temperature (°F)	900		
Approximate Temperature Permanently Affecting Magnet (°F)	1580		
Inherent Orientation of Material	HT		

Note: All information based on available producers literature. In some cases of foreign materials interpretation or calculation required.

SOURCE(S):

Thyssen Magnettechnik GMBH

Germany

Oerstit 400
(Alloy or Grade)

Cast Alnico
(Type of Material)

3.8/.80
(MMPA Brief Designation)

-
(IEC Code Reference)

MAGNETIC PROPERTIES		
Residual Induction B_r (G)		10,000
Coercive Force (Oe)	Normal (H_c)	700
	Intrinsic (H_{ci})	800
Maximum Energy Product BH_{max} (MGO)		3.8
B/H Ratio (Load Line) at BH_{max}		13.4
Flux Density in Magnet B_m at BH_{max} (G)		7100
Coercive Force in Magnet H_m at BH_{max} (Oe)		530
Energy Loss W_c (watt-sec/cycle/lb)		15.3
Importance of Operating Magnet at BH_{max}		A
Average Recoil Permeability		4.3
Field Strength Required to Fully Saturate Magnet (Oe)		3000
Ability to Withstand Demagnetizing Fields		G
Maximum Service Temperature (°F)		900
Approximate Temperature Permanently Affecting Magnet (°F)		1580
Inherent Orientation of Material		HT

PHYSICAL PROPERTIES	
Density (lbs./in.^{3})	.2565
General Mechanical Properties	HB
Tensile Strength (psi)	-
Transverse Modulus of Rupture (psi)	-
Coeff. of Therm. Expansion inches x 10^{-6}/°C	Orientation: Par. - Rt.Angl. -
Resistivity (micro-ohms/cm/cm^2)	50 @ 25 °C
Hardness 50	Scale Rockwell C
Basic Forming Method	CM
Basic Finishing Method	WGR
Relative Corrosion Resistance	G
Primary Chemical Composition	Al-Ni-Co-Fe-Cu
Special Notes:	

Note: All information based on available producers literature. In some cases of foreign materials interpretation or calculation required.

SOURCE(S):	Oerstit 450
Thyssen Magnettechnik GMBH Germany	(Alloy or Grade)
	Cast Alnico
	(Type of Material)
	5.2/1.6
	(MMPA Brief Designation)
	R1-1-5
	(IEC Code Reference)

MAGNETIC PROPERTIES		PHYSICAL PROPERTIES	
Residual Induction B_r (G)	8300	Density (lbs./in.^{3})	.2565
Coercive Force (Oe) — Normal (H_c)	1500	General Mechanical Properties	HB
Coercive Force (Oe) — Intrinsic (H_{ci})	1600	Tensile Strength (psi)	-
Maximum Energy Product BH_{max} (MGO)	5.2	Transverse Modulus of Rupture (psi)	-
B/H Ratio (Load Line) at BH_{max}	5.2	Coeff. of Therm. Expansion inches x 10^{-6}/°C	Orientation: Par. - Rt.Angl. -
Flux Density in Magnet B_m at BH_{max} (G)	5200	Resistivity (micro-ohms/cm/cm^2)	50 @ 25 °C
Coercive Force in Magnet H_m at BH_{max} (Oe)	1000	Hardness 50	Scale Rockwell C
Energy Loss W_c (watt-sec/cycle/lb)	15.3	Basic Forming Method	CM
Importance of Operating Magnet at BH_{max}	C	Basic Finishing Method	WGR
Average Recoil Permeability	2.2	Relative Corrosion Resistance	G
Field Strength Required to Fully Saturate Magnet (Oe)	3000	Primary Chemical Composition	Al-Ni-Co-Fe-Cu
Ability to Withstand Demagnetizing Fields	E	Special Notes:	
Maximum Service Temperature (°F)	900		
Approximate Temperature Permanently Affecting Magnet (°F)	1580		
Inherent Orientation of Material	HT		

Note: All information based on available producers literature. In some cases of foreign materials interpretation or calculation required.

SOURCE(S):

Thyssen Magnettechnik GMBH

Germany

Oerstit 500G
(Alloy or Grade)

Cast Alnico
(Type of Material)

5.0/.60
(MMPA Brief Designation)

R1-1-1
(IEC Code Reference)

MAGNETIC PROPERTIES		PHYSICAL PROPERTIES	
Residual Induction B_r (G)	12,400	Density (lbs./in.^{3})	.2565
Coercive Force (Oe) — Normal (H_c)	600	General Mechanical Properties	HB
Coercive Force (Oe) — Intrinsic (H_{ci})	600	Tensile Strength (psi)	—
Maximum Energy Product BH_{max} (MGO)	5.0	Transverse Modulus of Rupture (psi)	—
B/H Ratio (Load Line) at BH_{max}	20.5	Coeff. of Therm. Expansion inches x 10^{-6}/°C	Orientation: Par. _ Rt.Angl. _
Flux Density in Magnet B_m at BH_{max} (G)	10,200	Resistivity (micro-ohms/cm/cm^2)	50 @ 25 °C
Coercive Force in Magnet H_m at BH_{max} (Oe)	500	Hardness Scale	50 Rockwell C
Energy Loss W_c (watt-sec/cycle/lb)	15.3	Basic Forming Method	CM
Importance of Operating Magnet at BH_{max}	A	Basic Finishing Method	WRG
Average Recoil Permeability	3.2	Relative Corrosion Resistance	G
Field Strength Required to Fully Saturate Magnet (Oe)	3000	Primary Chemical Composition	Al-Ni-Co-Fe-Cu
Ability to Withstand Demagnetizing Fields	F	Special Notes:	
Maximum Service Temperature (°F)	900		
Approximate Temperature Permanently Affecting Magnet (°F)	1580		
Inherent Orientation of Material	CHT		

Note: All information based on available producers literature. In some cases of foreign materials interpretation or calculation required.

SOURCE(S):	
Thyssen Magnettechnik GMBH	**Oerstit 500S**
	(Alloy or Grade)
	Cast Alnico
	(Type of Material)
Germany	4.5/.60
	(MMPA Brief Designation)
	R1-1-1
	(IEC Code Reference)

MAGNETIC PROPERTIES		PHYSICAL PROPERTIES	
Residual Induction B_r (G)	11,600	Density (lbs./in.^{3})	.2565
Coercive Force (Oe) — Normal (H_c)	600	General Mechanical Properties	HB
Coercive Force (Oe) — Intrinsic (H_{ci})	600	Tensile Strength (psi)	-
Maximum Energy Product BH_{max} (MGO)	4.5	Transverse Modulus of Rupture (psi)	-
B/H Ratio (Load Line) at BH_{max}	17.8	Coeff. of Therm. Expansion inches x 10^{-6}/°C	Orientation: Par. _ Rt.Angl. _
Flux Density in Magnet B_m at BH_{max} (G)	8900	Resistivity (micro-ohms/cm/cm^2)	50 @ 25 °C
Coercive Force in Magnet H_m at BH_{max} (Oe)	500	Hardness 50	Scale Rockwell C
Energy Loss W_c (watt-sec/cycle/lb)	15.3	Basic Forming Method	CM
Importance of Operating Magnet at BH_{max}	A	Basic Finishing Method	WGR
Average Recoil Permeability	3.7	Relative Corrosion Resistance	G
Field Strength Required to Fully Saturate Magnet (Oe)	3000	Primary Chemical Composition	Al-Ni-Co-Fe-Cu
Ability to Withstand Demagnetizing Fields	F	Special Notes:	
Maximum Service Temperature (°F)	900		
Approximate Temperature Permanently Affecting Magnet (°F)	1580		
Inherent Orientation of Material	CHT		

Note: All information based on available producers literature. In some cases of foreign materials interpretation or calculation required.

SOURCE(S):	Ox 100
	(Alloy or Grade)
Magnetfabrik-Bonn	Sintered Ferrite
	(Type of Material)
Germany	1.0/1.8
	(MMPA Brief Designation)
	-
	(IEC Code Reference)

MAGNETIC PROPERTIES		PHYSICAL PROPERTIES	
Residual Induction B_r (G)	2100	Density (lbs./in.^{3})	.1770
Coercive Force (Oe) — Normal (H_c)	1700	General Mechanical Properties	VHB
Coercive Force (Oe) — Intrinsic (H_{ci})	1750	Tensile Strength (psi)	-
Maximum Energy Product BH_{max} (MGO)	1.0	Transverse Modulus of Rupture (psi)	-
B/H Ratio (Load Line) at BH_{max}	1.56	Coeff. of Therm. Expansion inches x 10^{-6}/°C	Orientation: Par. - Rt.Angl. -
Flux Density in Magnet B_m at BH_{max} (G)	1250	Resistivity (micro-ohms/cm/cm^2)	1.0 @ 25 °C
Coercive Force in Magnet H_m at BH_{max} (Oe)	800	Hardness Scale 6.5	Moh's
Energy Loss W_c (watt-sec/cycle/lb)	-	Basic Forming Method	PF
Importance of Operating Magnet at BH_{max}	E	Basic Finishing Method	WGD
Average Recoil Permeability	1.1	Relative Corrosion Resistance	Ex
Field Strength Required to Fully Saturate Magnet (Oe)	10,000	Primary Chemical Composition	Ba or Sr or Pb-Fe
Ability to Withstand Demagnetizing Fields	Ex	Special Notes:	
Maximum Service Temperature (°F)	482		
Approximate Temperature Permanently Affecting Magnet (°F)	842		
Inherent Orientation of Material	none		

Note: All information based on available producers literature. In some cases of foreign materials interpretation or calculation required.

SOURCE(S):	
	Ox 300
	(Alloy or Grade)
Magnetfabrik-Bonn	Sintered Ferrite
	(Type of Material)
Germany	2.8/1.9
	(MMPA Brief Designation)
	-
	(IEC Code Reference)

MAGNETIC PROPERTIES		PHYSICAL PROPERTIES	
Residual Induction B_r (G)	3600	Density (lbs./in.^{3})	.1734
Coercive Force (Oe) — Normal (H_c)	1800	General Mechanical Properties	VHB
Coercive Force (Oe) — Intrinsic (H_{ci})	1885	Tensile Strength (psi)	-
Maximum Energy Product BH_{max} (MGO)	2.8	Transverse Modulus of Rupture (psi)	-
B/H Ratio (Load Line) at BH_{max}	1.83	Coeff. of Therm. Expansion inches x 10^{-6}/°C — Orientation: Par. / Rt.Angl.	- / -
Flux Density in Magnet B_m at BH_{max} (G)	2200	Resistivity (micro-ohms/cm/cm^2)	1.0 @ 25 °C
Coercive Force in Magnet H_m at BH_{max} (Oe)	1200	Hardness Scale	6.5 Moh's
Energy Loss W_c (watt-sec/cycle/lb)	-	Basic Forming Method	PMF
Importance of Operating Magnet at BH_{max}	E	Basic Finishing Method	WGD
Average Recoil Permeability	1.1	Relative Corrosion Resistance	Ex
Field Strength Required to Fully Saturate Magnet (Oe)	10,000	Primary Chemical Composition	Ba or Sr or Pb-Fe
Ability to Withstand Demagnetizing Fields	Ex	Special Notes:	
Maximum Service Temperature (°F)	482		
Approximate Temperature Permanently Affecting Magnet (°F)	842		
Inherent Orientation of Material	MP		

Note: All information based on available producers literature. In some cases of foreign materials interpretation or calculation required.

SOURCE(S):	Ox 330
	(Alloy or Grade)
Magnetfabrik-Bonn	Sintered Ferrite
	(Type of Material)
Germany	3.0/2.9
	(MMPA Brief Designation)
	-
	(IEC Code Reference)

MAGNETIC PROPERTIES		PHYSICAL PROPERTIES	
Residual Induction B_r (G)	3500	Density (lbs./in.^{3})	.1662
Coercive Force (Oe) — Normal (H_c)	2700	General Mechanical Properties	VHB
Coercive Force (Oe) — Intrinsic (H_{ci})	2900	Tensile Strength (psi)	-
Maximum Energy Product BH_{max} (MGO)	3.0	Transverse Modulus of Rupture (psi)	-
B/H Ratio (Load Line) at BH_{max}	.75	Coeff. of Therm. Expansion inches x 10^{-6}/°C	Orientation: Par. - Rt.Angl. -
Flux Density in Magnet B_m at BH_{max} (G)	1500	Resistivity (micro-ohms/cm/cm^2)	1.0 @ 25 °C
Coercive Force in Magnet H_m at BH_{max} (Oe)	2000	Hardness 6.5	Scale Moh's
Energy Loss W_c (watt-sec/cycle/lb)	-	Basic Forming Method	PMF
Importance of Operating Magnet at BH_{max}	E	Basic Finishing Method	WGD
Average Recoil Permeability	1.1	Relative Corrosion Resistance	Ex
Field Strength Required to Fully Saturate Magnet (Oe)	10,000	Primary Chemical Composition	Ba or Sr or Pb-Fe
Ability to Withstand Demagnetizing Fields	Ex	Special Notes:	
Maximum Service Temperature (°F)	482		
Approximate Temperature Permanently Affecting Magnet (°F)	842		
Inherent Orientation of Material	MP		

Note: All information based on available producers literature. In some cases of foreign materials interpretation or calculation required.

SOURCE(S):

Magnetfabrik-Bonn

Germany

Ox 380
(Alloy or Grade)

Sintered Ferrite
(Type of Material)

3.4/2.3
(MMPA Brief Designation)

-
(IEC Code Reference)

MAGNETIC PROPERTIES		PHYSICAL PROPERTIES	
Residual Induction B_r (G)	3800	Density (lbs./in.^{3})	.1734
Coercive Force (Oe) Normal (H_c)	2200	General Mechanical Properties	VHB
Coercive Force (Oe) Intrinsic (H_{ci})	2320	Tensile Strength (psi)	-
Maximum Energy Product BH_{max} (MGO)	3.4	Transverse Modulus of Rupture (psi)	-
B/H Ratio (Load Line) at BH_{max}	1.38	Coeff. of Therm. Expansion inches x 10^{-6}/°C	Orientation: Par. - Rt.Angl. -
Flux Density in Magnet B_m at BH_{max} (G)	2200	Resistivity (micro-ohms/cm/cm^2)	1.0 @ 25 °C
Coercive Force in Magnet H_m at BH_{max} (Oe)	1600	Hardness Scale	6.5 Moh's
Energy Loss W_c (watt-sec/cycle/lb)	-	Basic Forming Method	PMF
Importance of Operating Magnet at BH_{max}	E	Basic Finishing Method	WGD
Average Recoil Permeability	1.1	Relative Corrosion Resistance	Ex
Field Strength Required to Fully Saturate Magnet (Oe)	10,000	Primary Chemical Composition	Ba or Sr or Pb-Fe
Ability to Withstand Demagnetizing Fields	Ex	Special Notes:	
Maximum Service Temperature (°F)	482		
Approximate Temperature Permanently Affecting Magnet (°F)	842		
Inherent Orientation of Material	MP		

Note: All information based on available producers literature. In some cases of foreign materials interpretation or calculation required.

SOURCE(S):	
Magnetfabrik-Bonn	**Ox 400** (Alloy or Grade)
	Sintered Ferrite (Type of Material)
Germany	3.8/3.2 (MMPA Brief Designation)
	- (IEC Code Reference)

MAGNETIC PROPERTIES		PHYSICAL PROPERTIES	
Residual Induction B_r (G)	3800	Density (lbs./in.^{3})	.1734
Coercive Force (Oe) — Normal (H_c)	3050	General Mechanical Properties	VHB
Coercive Force (Oe) — Intrinsic (H_{ci})	3200	Tensile Strength (psi)	-
Maximum Energy Product BH_{max} (MGO)	3.8	Transverse Modulus of Rupture (psi)	-
B/H Ratio (Load Line) at BH_{max}	.9	Coeff. of Therm. Expansion inches x 10^{-6}/°C	Orientation: Par. - Rt.Angl. -
Flux Density in Magnet B_m at BH_{max} (G)	1800	Resistivity (micro-ohms/cm/cm^2)	1.0 @ 25 °C
Coercive Force in Magnet H_m at BH_{max} (Oe)	2000	Hardness 6.5	Scale Moh's
Energy Loss W_c (watt-sec/cycle/lb)	-	Basic Forming Method	PMF
Importance of Operating Magnet at BH_{max}	E	Basic Finishing Method	WGD
Average Recoil Permeability	1.1	Relative Corrosion Resistance	Ex
Field Strength Required to Fully Saturate Magnet (Oe)	10,000	Primary Chemical Composition	Ba or Sr or Pb-Fe
Ability to Withstand Demagnetizing Fields	Ex	Special Notes:	
Maximum Service Temperature (°F)	482		
Approximate Temperature Permanently Affecting Magnet (°F)	842		
Inherent Orientation of Material	MP		

Note: All information based on available producers literature. In some cases of foreign materials interpretation or calculation required.

SOURCE(S):

Thyssen Magnettechnik GMBH

Germany

Oxit 100
(Alloy or Grade)

Sintered Ferrite
(Type of Material)

.9/3.1
(MMPA Brief Designation)

S1-0-1
(IEC Code Reference)

MAGNETIC PROPERTIES		PHYSICAL PROPERTIES	
Residual Induction B_r (G)	2100	Density (lbs./in.^{3})	.1734
Coercive Force (Oe) — Normal (H_c)	1700	General Mechanical Properties	VHB
Coercive Force (Oe) — Intrinsic (H_{ci})	3100	Tensile Strength (psi)	-
Maximum Energy Product BH_{max} (MGO)	.9	Transverse Modulus of Rupture (psi)	-
B/H Ratio (Load Line) at BH_{max}	1.29	Coeff. of Therm. Expansion inches x 10^{-6}/°C Orientation: Par. − Rt.Angl. −	
Flux Density in Magnet B_m at BH_{max} (G)	1100	Resistivity (micro-ohms/cm/cm^2)	1.0 @ 25 °C
Coercive Force in Magnet H_m at BH_{max} (Oe)	850	Hardness Scale 6.5 Moh's	
Energy Loss W_c (watt-sec/cycle/lb)	-	Basic Forming Method	PF
Importance of Operating Magnet at BH_{max}	E	Basic Finishing Method	WGD
Average Recoil Permeability	1.1	Relative Corrosion Resistance	Ex
Field Strength Required to Fully Saturate Magnet (Oe)	10,000	Primary Chemical Composition	Sr-Fe
Ability to Withstand Demagnetizing Fields	Ex	Special Notes:	
Maximum Service Temperature (°F)	482		
Approximate Temperature Permanently Affecting Magnet (°F)	842		
Inherent Orientation of Material	none		

Note: All information based on available producers literature. In some cases of foreign materials interpretation or calculation required.

SOURCE(S):

Thyssen Magnettechnik GMBH

Germany

Oxit 360
(Alloy or Grade)

Sintered Ferrite
(Type of Material)

3.2/3.3
(MMPA Brief Designation)

-
(IEC Code Reference)

MAGNETIC PROPERTIES	
Residual Induction B_r (G)	3800
Coercive Force (Oe) — Normal (H_c)	2900
Coercive Force (Oe) — Intrinsic (H_{ci})	3300
Maximum Energy Product BH_{max} (MGO)	3.2
B/H Ratio (Load Line) at BH_{max}	1.0
Flux Density in Magnet B_m at BH_{max} (G)	1800
Coercive Force in Magnet H_m at BH_{max} (Oe)	1800
Energy Loss W_c (watt-sec/cycle/lb)	-
Importance of Operating Magnet at BH_{max}	E
Average Recoil Permeability	1.05
Field Strength Required to Fully Saturate Magnet (Oe)	10,000
Ability to Withstand Demagnetizing Fields	Ex
Maximum Service Temperature (°F)	482
Approximate Temperature Permanently Affecting Magnet (°F)	842
Inherent Orientation of Material	MP

PHYSICAL PROPERTIES	
Density (lbs./in.^{3})	.1734
General Mechanical Properties	VHB
Tensile Strength (psi)	-
Transverse Modulus of Rupture (psi)	-
Coeff. of Therm. Expansion inches x 10^{-6}/°C	Orientation: Par. - Rt.Angl. -
Resistivity (micro-ohms/cm/cm^2)	1.0 @ 25 °C
Hardness	6.5 Scale Moh's
Basic Forming Method	PMF
Basic Finishing Method	WGD
Relative Corrosion Resistance	Ex
Primary Chemical Composition	Sr-Fe
Special Notes:	

Note: All information based on available producers literature. In some cases of foreign materials interpretation or calculation required.

SOURCE(S):

Thyssen Magnettechnik GMBH

Germany

Oxit 380
(Alloy or Grade)

Sintered Ferrite
(Type of Material)

3.6/3.8
(MMPA Brief Designation)

-
(IEC Code Reference)

MAGNETIC PROPERTIES	
Residual Induction B_r (G)	3900
Coercive Force (Oe) — Normal (H_c)	3600
Coercive Force (Oe) — Intrinsic (H_{ci})	3800
Maximum Energy Product BH_{max} (MGO)	3.6
B/H Ratio (Load Line) at BH_{max}	1.05
Flux Density in Magnet B_m at BH_{max} (G)	1950
Coercive Force in Magnet H_m at BH_{max} (Oe)	2850
Energy Loss W_c (watt-sec/cycle/lb)	-
Importance of Operating Magnet at BH_{max}	E
Average Recoil Permeability	1.05
Field Strength Required to Fully Saturate Magnet (Oe)	10,000
Ability to Withstand Demagnetizing Fields	Ex
Maximum Service Temperature (°F)	482
Approximate Temperature Permanently Affecting Magnet (°F)	842
Inherent Orientation of Material	MP

PHYSICAL PROPERTIES	
Density (lbs./in.^{3})	.1734
General Mechanical Properties	VHB
Tensile Strength (psi)	-
Transverse Modulus of Rupture (psi)	-
Coeff. of Therm. Expansion inches x 10^{-6}/°C — Orientation: Par.	-
Coeff. of Therm. Expansion — Rt.Angl.	-
Resistivity (micro-ohms/cm/cm^2)	1.0 @ 25 °C
Hardness Scale	6.5 Moh's
Basic Forming Method	PMF
Basic Finishing Method	WGD
Relative Corrosion Resistance	Ex
Primary Chemical Composition	Sr-Fe
Special Notes:	

Note: All information based on available producers literature. In some cases of foreign materials interpretation or calculation required.

SOURCE(S):	Oxit 380C
	(Alloy or Grade)
Thyssen Magnettechnik GMBH	Sintered Ferrite
	(Type of Material)
Germany	3.2/4.0
	(MMPA Brief Designation)
	-
	(IEC Code Reference)

MAGNETIC PROPERTIES		PHYSICAL PROPERTIES	
Residual Induction B_r (G)	3900	Density (lbs./in.^{3})	.1734
Coercive Force (Oe) — Normal (H_c)	3600	General Mechanical Properties	VHB
Coercive Force (Oe) — Intrinsic (H_{ci})	4000	Tensile Strength (psi)	-
Maximum Energy Product BH_{max} (MGO)	3.2	Transverse Modulus of Rupture (psi)	-
B/H Ratio (Load Line) at BH_{max}	1.12	Coeff. of Therm. Expansion inches x 10^{-6}/°C	Orientation: Par. - Rt.Angl. -
Flux Density in Magnet B_m at BH_{max} (G)	1900	Resistivity (micro-ohms/cm/cm^2)	1.0 @ 25 °C
Coercive Force in Magnet H_m at BH_{max} (Oe)	1700	Hardness Scale	6.5 Moh's
Energy Loss W_c (watt-sec/cycle/lb)	-	Basic Forming Method	PMF
Importance of Operating Magnet at BH_{max}	E	Basic Finishing Method	WGD
Average Recoil Permeability	1.05	Relative Corrosion Resistance	Ex
Field Strength Required to Fully Saturate Magnet (Oe)	10,000	Primary Chemical Composition	Sr-Fe
Ability to Withstand Demagnetizing Fields	Ex	Special Notes:	
Maximum Service Temperature (°F)	482		
Approximate Temperature Permanently Affecting Magnet (°F)	842		
Inherent Orientation of Material	MP		

Note: All information based on available producers literature. In some cases of foreign materials interpretation or calculation required.

SOURCE(S):

Thyssen Magnettechnik GMBH

Germany

Oxit 400
(Alloy or Grade)

Sintered Ferrite
(Type of Material)

4.0/3.5
(MMPA Brief Designation)

-
(IEC Code Reference)

MAGNETIC PROPERTIES		PHYSICAL PROPERTIES	
Residual Induction B_r (G)	4100	Density (lbs./in.^{3})	.1734
Coercive Force (Oe) Normal (H_c)	3300	General Mechanical Properties	VHB
Coercive Force (Oe) Intrinsic (H_{ci})	3500	Tensile Strength (psi)	-
Maximum Energy Product BH_{max} (MGO)	4.0	Transverse Modulus of Rupture (psi)	-
B/H Ratio (Load Line) at BH_{max}	1.0	Coeff. of Therm. Expansion inches x 10^{-6}/°C	Orientation: Par. - Rt.Angl. -
Flux Density in Magnet B_m at BH_{max} (G)	2000	Resistivity (micro-ohms/cm/cm^2)	1.0 @ 25 °C
Coercive Force in Magnet H_m at BH_{max} (Oe)	2000	Hardness Scale	6.5 Moh's
Energy Loss W_c (watt-sec/cycle/lb)	-	Basic Forming Method	PMF
Importance of Operating Magnet at BH_{max}	E	Basic Finishing Method	WGD
Average Recoil Permeability	1.05	Relative Corrosion Resistance	Ex
Field Strength Required to Fully Saturate Magnet (Oe)	10,000	Primary Chemical Composition	Sr-Fe
Ability to Withstand Demagnetizing Fields	Ex	Special Notes:	
Maximum Service Temperature (°F)	482		
Approximate Temperature Permanently Affecting Magnet (°F)	842		
Inherent Orientation of Material	MP		

Note: All information based on available producers literature. In some cases of foreign materials interpretation or calculation required.

SOURCE(S):	Oxit 400C
	(Alloy or Grade)
Thyssen Magnettechnik GMBH	Sintered Ferrite
	(Type of Material)
Germany	4.0/4.0
	(MMPA Brief Designation)
	-
	(IEC Code Reference)

MAGNETIC PROPERTIES		PHYSICAL PROPERTIES	
Residual Induction B_r (G)	4100	Density (lbs./in.^{3})	.1734
Coercive Force (Oe) — Normal (H_c)	3600	General Mechanical Properties	VHB
Coercive Force (Oe) — Intrinsic (H_{ci})	4000	Tensile Strength (psi)	-
Maximum Energy Product BH_{max} (MGO)	4.0	Transverse Modulus of Rupture (psi)	-
B/H Ratio (Load Line) at BH_{max}	.95	Coeff. of Therm. Expansion inches x 10^{-6}/°C — Orientation: Par. / Rt.Angl.	- / -
Flux Density in Magnet B_m at BH_{max} (G)	1950	Resistivity (micro-ohms/cm/cm^2)	1.0 @ 25 °C
Coercive Force in Magnet H_m at BH_{max} (Oe)	2050	Hardness / Scale	6.5 Moh's
Energy Loss W_c (watt-sec/cycle/lb)	-	Basic Forming Method	PMF
Importance of Operating Magnet at BH_{max}	E	Basic Finishing Method	WGD
Average Recoil Permeability	1.05	Relative Corrosion Resistance	Ex
Field Strength Required to Fully Saturate Magnet (Oe)	10,000	Primary Chemical Composition	Sr-Fe
Ability to Withstand Demagnetizing Fields	Ex	Special Notes:	
Maximum Service Temperature (°F)	482		
Approximate Temperature Permanently Affecting Magnet (°F)	842		
Inherent Orientation of Material	MP		

Note: All information based on available producers literature. In some cases of foreign materials interpretation or calculation required.

SOURCE(S):

Thyssen Magnettechnik GMBH

Germany

Oxit 400HC
(Alloy or Grade)

Sintered Ferrite
(Type of Material)

3.8/4.5
(MMPA Brief Designation)

-
(IEC Code Reference)

MAGNETIC PROPERTIES	
Residual Induction B_r (G)	4000
Coercive Force (Oe) Normal (H_c)	3800
Coercive Force (Oe) Intrinsic (H_{ci})	4500
Maximum Energy Product BH_{max} (MGO)	3.8
B/H Ratio (Load Line) at BH_{max}	1.04
Flux Density in Magnet B_m at BH_{max} (G)	1975
Coercive Force in Magnet H_m at BH_{max} (Oe)	1900
Energy Loss W_c (watt-sec/cycle/lb)	-
Importance of Operating Magnet at BH_{max}	E
Average Recoil Permeability	1.05
Field Strength Required to Fully Saturate Magnet (Oe)	10,000
Ability to Withstand Demagnetizing Fields	Ex
Maximum Service Temperature (°F)	482
Approximate Temperature Permanently Affecting Magnet (°F)	842
Inherent Orientation of Material	MP

PHYSICAL PROPERTIES	
Density (lbs./in.^{3})	.1734
General Mechanical Properties	VHB
Tensile Strength (psi)	-
Transverse Modulus of Rupture (psi)	-
Coeff. of Therm. Expansion inches x 10^{-6}/°C Orientation: Par. / Rt.Angl.	- / -
Resistivity (micro-ohms/cm/cm^2)	1.0 @ 25 °C
Hardness Scale	6.5 Moh's
Basic Forming Method	PMF
Basic Finishing Method	WGD
Relative Corrosion Resistance	Ex
Primary Chemical Composition	Sr-Fe
Special Notes:	

Note: All information based on available producers literature. In some cases of foreign materials interpretation or calculation required.

SOURCE(S):	**Oxit 420**
	(Alloy or Grade)
Thyssen Magnettechnik GMBH	Sintered Ferrite
	(Type of Material)
Germany	4.3/3.8
	(MMPA Brief Designation)
	-
	(IEC Code Reference)

MAGNETIC PROPERTIES		PHYSICAL PROPERTIES	
Residual Induction B_r (G)	4300	Density (lbs./in.^{3})	.1734
Coercive Force (Oe) Normal (H_c)	3300	General Mechanical Properties	HB
Coercive Force (Oe) Intrinsic (H_{ci})	3800	Tensile Strength (psi)	-
Maximum Energy Product BH_{max} (MGO)	4.3	Transverse Modulus of Rupture (psi)	-
B/H Ratio (Load Line) at BH_{max}	1.13	Coeff. of Therm. Expansion inches x 10^{-6}/°C	Orientation: Par. - Rt.Angl. -
Flux Density in Magnet B_m at BH_{max} (G)	2200	Resistivity (micro-ohms/cm/cm^2	- @ - °C
Coercive Force in Magnet H_m at BH_{max} (Oe)	1950	Hardness -	Scale -
Energy Loss W_c (watt-sec/cycle/lb)	-	Basic Forming Method	PMF
Importance of Operating Magnet at BH_{max}	E	Basic Finishing Method	WGD
Average Recoil Permeability	1.05	Relative Corrosion Resistance	Ex
Field Strength Required to Fully Saturate Magnet (Oe)	10,000	Primary Chemical Composition	Sr-Fe
Ability to Withstand Demagnetizing Fields	E	Special Notes: (1) Magnetic Properties (2) Temperature Sensitive	
Maximum Service Temperature (°F)	482		
Approximate Temperature Permanently Affecting Magnet (°F)	842		
Inherent Orientation of Material	MP		

Note: All information based on available producers literature. In some cases of foreign materials interpretation or calculation required.

SOURCE(S):

Philips Components

The Neatherlands

P30
(Alloy or Grade)

Bonded Ferrite
(Type of Material)

.35/2.4
(MMPA Brief Designation)

-
(IEC Code Reference)

MAGNETIC PROPERTIES		PHYSICAL PROPERTIES	
Residual Induction B_r (G)	1250	Density (lbs./in.^{3})	.1120
Coercive Force (Oe) — Normal (H_c)	1110	General Mechanical Properties	FL
Coercive Force (Oe) — Intrinsic (H_{ci})	2390	Tensile Strength (psi)	-
Maximum Energy Product BH_{max} (MGO)	.35	Transverse Modulus of Rupture (psi)	-
B/H Ratio (Load Line) at BH_{max}	.96	Coeff. of Therm. Expansion inches x 10^{-6}/°C Orientation: Par. / Rt.Angl.	- / -
Flux Density in Magnet B_m at BH_{max} (G)	600	Resistivity (micro-ohms/cm/cm^2)	10^9 @ 25 °C
Coercive Force in Magnet H_m at BH_{max} (Oe)	628	Hardness Scale 55°C	Shore C
Energy Loss W_c (watt-sec/cycle/lb)	-	Basic Forming Method	E
Importance of Operating Magnet at BH_{max}	E	Basic Finishing Method	SRD
Average Recoil Permeability	-	Relative Corrosion Resistance	E
Field Strength Required to Fully Saturate Magnet (Oe)	10,000	Primary Chemical Composition	Ba or Sr-Fe + binder
Ability to Withstand Demagnetizing Fields	Ex	Special Notes: (1) Maximum service temperature depends on binder used. Binder not identified by manufacturer.	
Maximum Service Temperature (°F)	200		
Approximate Temperature Permanently Affecting Magnet (°F)	-		
Inherent Orientation of Material	none		

Note: All information based on available producers literature. In some cases of foreign materials interpretation or calculation required.

SOURCE(S):

Philips Components

The Neatherlands

P40B
(Alloy or Grade)

Bonded Ferrite
(Type of Material)

.45/2.4
(MMPA Brief Designation)

—
(IEC Code Reference)

MAGNETIC PROPERTIES		PHYSICAL PROPERTIES	
Residual Induction B_r (G)	1450	Density (lbs./in.^{3})	.1337
Coercive Force (Oe) — Normal (H_c)	1210	General Mechanical Properties	FL
Coercive Force (Oe) — Intrinsic (H_{ci})	2390	Tensile Strength (psi)	—
Maximum Energy Product BH_{max} (MGO)	.45	Transverse Modulus of Rupture (psi)	—
B/H Ratio (Load Line) at BH_{max}	.87	Coeff. of Therm. Expansion inches x 10^{-6}/°C	Orientation: Par. — Rt.Angl. —
Flux Density in Magnet B_m at BH_{max} (G)	600	Resistivity (micro-ohms/cm/cm^2)	10^9 @ 25 °C
Coercive Force in Magnet H_m at BH_{max} (Oe)	691	Hardness 55	Scale Shore C
Energy Loss W_c (watt-sec/cycle/lb)	—	Basic Forming Method	E
Importance of Operating Magnet at BH_{max}	E	Basic Finishing Method	SRD
Average Recoil Permeability	—	Relative Corrosion Resistance	E
Field Strength Required to Fully Saturate Magnet (Oe)	10,000	Primary Chemical Composition	Ba or Sr-Fe + binder
Ability to Withstand Demagnetizing Fields	Ex	Special Notes: (1) Maximum service temperature depends on binder used. Binder not identified by manufacturer.	
Maximum Service Temperature (°F)	200		
Approximate Temperature Permanently Affecting Magnet (°F)	—		
Inherent Orientation of Material	none		

Note: All information based on available producers literature. In some cases of foreign materials interpretation or calculation required.

SOURCE(S):	Platinum Cobalt 77/23
	(Alloy or Grade)
Vacuumschmelze GMBH	Platinum Cobalt
	(Type of Material)
Germany	9.5/5.3
	(MMPA Brief Designation)
	-
	(IEC Code Reference)

MAGNETIC PROPERTIES		PHYSICAL PROPERTIES	
Residual Induction B_r (G)	6400	Density (lbs./in.^{3})	.5744
Coercive Force (Oe) — Normal (H_c)	4300	General Mechanical Properties	D
Coercive Force (Oe) — Intrinsic (H_{ci})	5300	Tensile Strength (psi)	200,000
Maximum Energy Product BH_{max} (MGO)	9.5	Transverse Modulus of Rupture (psi)	230,000
B/H Ratio (Load Line) at BH_{max}	1.14	Coeff. of Therm. Expansion inches x 10^{-6}/°C	Orientation: Par. 11.4 Rt.Angl. -
Flux Density in Magnet B_m at BH_{max} (G)	3250	Resistivity (micro-ohms/cm/cm^2)	28.0 @ 25 °C
Coercive Force in Magnet H_m at BH_{max} (Oe)	2850	Hardness Scale	340 Vickers
Energy Loss W_c (watt-sec/cycle/lb)	-	Basic Forming Method	W
Importance of Operating Magnet at BH_{max}	E	Basic Finishing Method	MCT
Average Recoil Permeability	1.2	Relative Corrosion Resistance	Ex
Field Strength Required to Fully Saturate Magnet (Oe)	20,000	Primary Chemical Composition	Pt-Co
Ability to Withstand Demagnetizing Fields	0	Special Notes:	
Maximum Service Temperature (°F)	752	(1) Because of limited availability and high cost, used only for exceptional, small magnet applications.	
Approximate Temperature Permanently Affecting Magnet (°F)	932		
Inherent Orientation of Material	CR		

Note: All information based on available producers literature. In some cases of foreign materials interpretation or calculation required.

SOURCE(S):

Magnetfabrik-Bonn

Germany

Prac 120
(Alloy or Grade)

Alni
(Type of Material)

.39/520
(MMPA Brief Designation)

-
(IEC Code Reference)

MAGNETIC PROPERTIES		PHYSICAL PROPERTIES	
Residual Induction B_r (G)	2800	Density (lbs./in.^{3})	.1879
Coercive Force (Oe) — Normal (H_c)	470	General Mechanical Properties	HT
Coercive Force (Oe) — Intrinsic (H_{ci})	520	Tensile Strength (psi)	-
Maximum Energy Product BH_{max} (MGO)	.39	Transverse Modulus of Rupture (psi)	-
B/H Ratio (Load Line) at BH_{max}	10.0	Coeff. of Therm. Expansion inches x 10^{-6}/°C	Orientation: Par. - Rt.Angl. -
Flux Density in Magnet B_m at BH_{max} (G)	2000	Resistivity (micro-ohms/cm/cm^2)	- @ - °C
Coercive Force in Magnet H_m at BH_{max} (Oe)	200	Hardness Scale	- -
Energy Loss W_c (watt-sec/cycle/lb)	-	Basic Forming Method	E
Importance of Operating Magnet at BH_{max}	A	Basic Finishing Method	WGR
Average Recoil Permeability	3.0	Relative Corrosion Resistance	G
Field Strength Required to Fully Saturate Magnet (Oe)	3000	Primary Chemical Composition	Alni 120 + binder
Ability to Withstand Demagnetizing Fields	F	Special Notes: (1) Temperatures vary with binder used. Consult manufacturer.	
Maximum Service Temperature (°F)	See Note		
Approximate Temperature Permanently Affecting Magnet (°F)	-		
Inherent Orientation of Material	none		

Note: All information based on available producers literature. In some cases of foreign materials interpretation or calculation required.

SOURCE(S):

Magnetfabrik-Bonn

Germany

Prac 160
(Alloy or Grade)

Bonded Alnico
(Type of Material)

.65/6.3
(MMPA Brief Designation)

-
(IEC Code Reference)

MAGNETIC PROPERTIES		PHYSICAL PROPERTIES	
Residual Induction B_r (G)	3200	Density (lbs./in.^{3})	.1915
Coercive Force (Oe) Normal (H_c)	580	General Mechanical Properties	HT
Coercive Force (Oe) Intrinsic (H_{ci})	630	Tensile Strength (psi)	-
Maximum Energy Product BH_{max} (MGO)	.65	Transverse Modulus of Rupture (psi)	-
B/H Ratio (Load Line) at BH_{max}	7.3	Coeff. of Therm. Expansion inches x 10^{-6}/°C	Orientation: Par. - Rt.Angl. -
Flux Density in Magnet B_m at BH_{max} (G)	2200	Resistivity (micro-ohms/cm/cm^2	- @ - °C
Coercive Force in Magnet H_m at BH_{max} (Oe)	300	Hardness Scale	- -
Energy Loss W_c (watt-sec/cycle/lb)	-	Basic Forming Method	E
Importance of Operating Magnet at BH_{max}	A	Basic Finishing Method	WGR
Average Recoil Permeability	3.0	Relative Corrosion Resistance	G
Field Strength Required to Fully Saturate Magnet (Oe)	3000	Primary Chemical Composition	Alnico 160 + binder
Ability to Withstand Demagnetizing Fields	F	Special Notes: (1) Temperatures vary with binder used. Consult manufacturer.	
Maximum Service Temperature (°F)	See Note		
Approximate Temperature Permanently Affecting Magnet (°F)	-		
Inherent Orientation of Material	none		

Note: All information based on available producers literature. In some cases of foreign materials interpretation or calculation required.

SOURCE(S):

Magnetfabrik-Bonn

Germany

Prac 260K
(Alloy or Grade)

Bonded Alnico
(Type of Material)

1.1/1.2
(MMPA Brief Designation)

-
(IEC Code Reference)

MAGNETIC PROPERTIES	
Residual Induction B_r (G)	3800
Coercive Force (Oe) — Normal (H_c)	1000
Coercive Force (Oe) — Intrinsic (H_{ci})	1150
Maximum Energy Product BH_{max} (MGO)	1.1
B/H Ratio (Load Line) at BH_{max}	4.4
Flux Density in Magnet B_m at BH_{max} (G)	2200
Coercive Force in Magnet H_m at BH_{max} (Oe)	500
Energy Loss W_c (watt-sec/cycle/lb)	-
Importance of Operating Magnet at BH_{max}	B
Average Recoil Permeability	3.0
Field Strength Required to Fully Saturate Magnet (Oe)	3000
Ability to Withstand Demagnetizing Fields	G
Maximum Service Temperature (°F)	See Note
Approximate Temperature Permanently Affecting Magnet (°F)	-
Inherent Orientation of Material	none

PHYSICAL PROPERTIES	
Density (lbs./in.^{3})	.1951
General Mechanical Properties	HT
Tensile Strength (psi)	
Transverse Modulus of Rupture (psi)	
Coeff. of Therm. Expansion inches x 10^{-6}/°C	Orientation: Par. - Rt.Angl. -
Resistivity (micro-ohms/cm/cm^2)	- @ - °C
Hardness / Scale	- / -
Basic Forming Method	E
Basic Finishing Method	WGR
Relative Corrosion Resistance	G
Primary Chemical Composition	Alnico 260K + binder
Special Notes: (1) Temperatures vary with binder used. Consult manufacturer.	

Note: All information based on available producers literature. In some cases of foreign materials interpretation or calculation required.

SOURCE(S):	**Prac 260u.260T**
Magnetfabrik-Bonn	(Alloy or Grade)
	Bonded Alnico
	(Type of Material)
Germany	.9/1.1
	(MMPA Brief Designation)
	-
	(IEC Code Reference)

MAGNETIC PROPERTIES		PHYSICAL PROPERTIES	
Residual Induction B_r (G)	3600	Density (lbs./in.^{3})	.1951
Coercive Force (Oe) — Normal (H_c)	900	General Mechanical Properties	HT
Coercive Force (Oe) — Intrinsic (H_{ci})	1050	Tensile Strength (psi)	-
Maximum Energy Product BH_{max} (MGO)	.90	Transverse Modulus of Rupture (psi)	-
B/H Ratio (Load Line) at BH_{max}	3.6	Coeff. of Therm. Expansion inches x 10^{-6}/°C	Orientation: Par. - Rt.Angl. -
Flux Density in Magnet B_m at BH_{max} (G)	1800	Resistivity (micro-ohms/cm/cm^2	- @ - °C
Coercive Force in Magnet H_m at BH_{max} (Oe)	500	Hardness Scale	- -
Energy Loss W_c (watt-sec/cycle/lb)	-	Basic Forming Method	E
Importance of Operating Magnet at BH_{max}	B	Basic Finishing Method	WGR
Average Recoil Permeability	3.0	Relative Corrosion Resistance	G
Field Strength Required to Fully Saturate Magnet (Oe)	3000	Primary Chemical Composition	Alnico 260 + binder
Ability to Withstand Demagnetizing Fields	G	Special Notes: (1) Temperatures vary with binder used. Consult manufacturer.	
Maximum Service Temperature (°F)	See Note		
Approximate Temperature Permanently Affecting Magnet (°F)	-		
Inherent Orientation of Material	none		

Note: All information based on available producers literature. In some cases of foreign materials interpretation or calculation required.

SOURCE(S):

TDK Corp. of America

USA

REC-18
(Alloy or Grade)

Sintered Rare Earth
(Type of Material)

18/16
(MMPA Brief Designation)

R5-1
(IEC Code Reference)

MAGNETIC PROPERTIES	
Residual Induction B_r (G)	8500
Coercive Force (Oe) — Normal (H_c)	8000
Coercive Force (Oe) — Intrinsic (H_{ci})	16,000
Maximum Energy Product BH_{max} (MGO)	18.0
B/H Ratio (Load Line) at BH_{max}	1.0
Flux Density in Magnet B_m at BH_{max} (G)	4200
Coercive Force in Magnet H_m at BH_{max} (Oe)	4200
Energy Loss W_c (watt-sec/cycle/lb)	-
Importance of Operating Magnet at BH_{max}	E
Average Recoil Permeability	1.05
Field Strength Required to Fully Saturate Magnet (Oe)	32,000
Ability to Withstand Demagnetizing Fields	0
Maximum Service Temperature (°F)	Est. 482
Approximate Temperature Permanently Affecting Magnet (°F)	Est. 1328
Inherent Orientation of Material	MP

PHYSICAL PROPERTIES	
Density (lbs./in.^{3})	.2926-.2999
General Mechanical Properties	HB
Tensile Strength (psi)	Est. 5000
Transverse Modulus of Rupture (psi)	-
Coeff. of Therm. Expansion inches x 10^{-6}/°C	Orientation: Par. 6.0 Rt.Angl. 16.0
Resistivity (micro-ohms/cm/cm^2)	53 @ 20 °C
Hardness	Scale 450-500 Vickers
Basic Forming Method	DPM
Basic Finishing Method	WGR
Relative Corrosion Resistance	G
Primary Chemical Composition	Sm-Co
Special Notes:	

Note: All information based on available producers literature. In some cases of foreign materials interpretation or calculation required.

SOURCE(S):

TDK Corp. of America

USA

REC-18B
(Alloy or Grade)

Sintered Rare Earth
(Type of Material)

18/9
(MMPA Brief Designation)

R5-2
(IEC Code Reference)

MAGNETIC PROPERTIES	
Residual Induction B_r (G)	8600
Coercive Force (Oe) — Normal (H_c)	7200
Coercive Force (Oe) — Intrinsic (H_{ci})	9000
Maximum Energy Product BH_{max} (MGO)	18.0
B/H Ratio (Load Line) at BH_{max}	1.13
Flux Density in Magnet B_m at BH_{max} (G)	4500
Coercive Force in Magnet H_m at BH_{max} (Oe)	4000
Energy Loss W_c (watt-sec/cycle/lb)	-
Importance of Operating Magnet at BH_{max}	E
Average Recoil Permeability	1.05
Field Strength Required to Fully Saturate Magnet (Oe)	32,000
Ability to Withstand Demagnetizing Fields	0
Maximum Service Temperature (°F)	Est. 375
Approximate Temperature Permanently Affecting Magnet (°F)	Est. 1328
Inherent Orientation of Material	MP

PHYSICAL PROPERTIES	
Density (lbs./in.^{3})	.2962-.3035
General Mechanical Properties	HB
Tensile Strength (psi)	Est. 5000
Transverse Modulus of Rupture (psi)	-
Coeff. of Therm. Expansion inches x 10^{-6}/°C	Orientation: Par. 8.0 Rt.Angl. 8.0
Resistivity (micro-ohms/cm/cm^2)	53 @ 20 °C
Hardness 500-600	Scale Vickers
Basic Forming Method	DPM
Basic Finishing Method	WGR
Relative Corrosion Resistance	G
Primary Chemical Composition	Sm-Co
Special Notes:	

Note: All information based on available producers literature. In some cases of foreign materials interpretation or calculation required.

SOURCE(S):		REC-20
		(Alloy or Grade)
TDK Corp. of America		Sintered Rare Earth
		(Type of Material)
USA		20/9
		(MMPA Brief Designation)
		R5-2
		(IEC Code Reference)

MAGNETIC PROPERTIES		PHYSICAL PROPERTIES	
Residual Induction B_r (G)	9000	Density (lbs./in.^{3})	.2926-.2999
Coercive Force (Oe) — Normal (H_c)	8750	General Mechanical Properties	HB
Coercive Force (Oe) — Intrinsic (H_{ci})	9000	Tensile Strength (psi)	Est. 5000
Maximum Energy Product BH_{max} (MGO)	20.0	Transverse Modulus of Rupture (psi)	-
B/H Ratio (Load Line) at BH_{max}	1.07	Coeff. of Therm. Expansion inches x 10^{-6}/°C	Orientation: Par. - Rt.Angl. -
Flux Density in Magnet B_m at BH_{max} (G)	4600	Resistivity (micro-ohms/cm/cm^2)	53 @ 20 °C
Coercive Force in Magnet H_m at BH_{max} (Oe)	4300	Hardness Scale	450-500 Vickers
Energy Loss W_c (watt-sec/cycle/lb)	-	Basic Forming Method	DPM
Importance of Operating Magnet at BH_{max}	E	Basic Finishing Method	WGR
Average Recoil Permeability	1.05	Relative Corrosion Resistance	G
Field Strength Required to Fully Saturate Magnet (Oe)	32,000	Primary Chemical Composition	Sm-Co
Ability to Withstand Demagnetizing Fields	0	Special Notes:	
Maximum Service Temperature (°F)	Est. 375		
Approximate Temperature Permanently Affecting Magnet (°F)	Est. 1328		
Inherent Orientation of Material	MP		

Note: All information based on available producers literature. In some cases of foreign materials interpretation or calculation required.

SOURCE(S):	REC-22
	(Alloy or Grade)
TDK Corp. of America	Sintered Rare Earth
	(Type of Material)
USA	22/9
	(MMPA Brief Designation)
	R5-2
	(IEC Code Reference)

MAGNETIC PROPERTIES		PHYSICAL PROPERTIES	
Residual Induction B_r (G)	9550	Density (lbs./in.^{3})	.2999-.3071
Coercive Force (Oe) — Normal (H_c)	8600	General Mechanical Properties	HB
Coercive Force (Oe) — Intrinsic (H_{ci})	9000	Tensile Strength (psi)	Est. 5000
Maximum Energy Product BH_{max} (MGO)	22.0	Transverse Modulus of Rupture (psi)	-
B/H Ratio (Load Line) at BH_{max}	.96	Coeff. of Therm. Expansion inches x 10^{-6}/°C	Orientation: Par. 8.0 Rt.Angl. 8.0
Flux Density in Magnet B_m at BH_{max} (G)	4600	Resistivity (micro-ohms/cm/cm^2)	86 @ 20 °C
Coercive Force in Magnet H_m at BH_{max} (Oe)	4775	Hardness Scale	500-600 Vickers
Energy Loss W_c (watt-sec/cycle/lb)	-	Basic Forming Method	DPM
Importance of Operating Magnet at BH_{max}	E	Basic Finishing Method	WGR
Average Recoil Permeability	1.05	Relative Corrosion Resistance	G
Field Strength Required to Fully Saturate Magnet (Oe)	32,000	Primary Chemical Composition	Sm-Co
Ability to Withstand Demagnetizing Fields	O	Special Notes:	
Maximum Service Temperature (°F)	Est. 375		
Approximate Temperature Permanently Affecting Magnet (°F)	Est. 1328		
Inherent Orientation of Material	MP		

Note: All information based on available producers literature. In some cases of foreign materials interpretation or calculation required.

SOURCE(S):	
TDK Corp. of America	**REC-22B** (Alloy or Grade)
	Sintered Rare Earth (Type of Material)
USA	22/10 (MMPA Brief Designation)
	R5-2 (IEC Code Reference)

MAGNETIC PROPERTIES		PHYSICAL PROPERTIES	
Residual Induction B_r (G)	9600	Density (lbs./in.^{3})	.2962-.3035
Coercive Force (Oe) — Normal (H_c)	7800	General Mechanical Properties	HB
Coercive Force (Oe) — Intrinsic (H_{ci})	10,000	Tensile Strength (psi)	Est. 5000
Maximum Energy Product BH_{max} (MGO)	22.0	Transverse Modulus of Rupture (psi)	-
B/H Ratio (Load Line) at BH_{max}	1.14	Coeff. of Therm. Expansion inches x 10^{-6}/°C	Orientation: Par. 8.0 Rt.Angl. 11.0
Flux Density in Magnet B_m at BH_{max} (G)	5000	Resistivity (micro-ohms/cm/cm^2)	90 @ 20 °C
Coercive Force in Magnet H_m at BH_{max} (Oe)	4400	Hardness Scale	500-600 Vickers
Energy Loss W_c (watt-sec/cycle/lb)	-	Basic Forming Method	DPM
Importance of Operating Magnet at BH_{max}	E	Basic Finishing Method	WGR
Average Recoil Permeability	1.05	Relative Corrosion Resistance	G
Field Strength Required to Fully Saturate Magnet (Oe)	32,000	Primary Chemical Composition	Sm-Co
Ability to Withstand Demagnetizing Fields	0	Special Notes:	
Maximum Service Temperature (°F)	Est. 400		
Approximate Temperature Permanently Affecting Magnet (°F)	Est. 1328		
Inherent Orientation of Material	MP		

Note: All information based on available producers literature. In some cases of foreign materials interpretation or calculation required.

SOURCE(S):	REC-24
	(Alloy or Grade)
TDK Corp. of America	Sintered Rare Earth
	(Type of Material)
USA	23/7
	(MMPA Brief Designation)
	R5-2
	(IEC Code Reference)

MAGNETIC PROPERTIES		PHYSICAL PROPERTIES	
Residual Induction B_r (G)	10,000	Density (lbs./in.^{3})	.2999-.3107
Coercive Force (Oe) — Normal (H_c)	6400	General Mechanical Properties	HB
Coercive Force (Oe) — Intrinsic (H_{ci})	6600	Tensile Strength (psi)	Est. 5000
Maximum Energy Product BH_{max} (MGO)	23.0	Transverse Modulus of Rupture (psi)	-
B/H Ratio (Load Line) at BH_{max}	1.31	Coeff. of Therm. Expansion inches x 10^{-6}/°C	Orientation: Par. 8.0 Rt.Angl. 11.0
Flux Density in Magnet B_m at BH_{max} (G)	5500	Resistivity (micro-ohms/cm/cm^2)	86 @ 20 °C
Coercive Force in Magnet H_m at BH_{max} (Oe)	4200	Hardness Scale	500-600 Vickers
Energy Loss W_c (watt-sec/cycle/lb)	-	Basic Forming Method	DPM
Importance of Operating Magnet at BH_{max}	E	Basic Finishing Method	WGR
Average Recoil Permeability	1.05	Relative Corrosion Resistance	G
Field Strength Required to Fully Saturate Magnet (Oe)	22,600	Primary Chemical Composition	Sm-Co
Ability to Withstand Demagnetizing Fields	0	Special Notes:	
Maximum Service Temperature (°F)	Est. 300		
Approximate Temperature Permanently Affecting Magnet (°F)	Est. 1328		
Inherent Orientation of Material	MP		

Note: All information based on available producers literature. In some cases of foreign materials interpretation or calculation required.

SOURCE(S):

TDK Corp. of America

USA

REC-26
(Alloy or Grade)

Sintered Rare Earth
(Type of Material)

26/10
(MMPA Brief Designation)

R5-2
(IEC Code Reference)

MAGNETIC PROPERTIES	
Residual Induction B_r (G)	10,500
Coercive Force (Oe) — Normal (H_c)	9200
Coercive Force (Oe) — Intrinsic (H_{ci})	10,000
Maximum Energy Product BH_{max} (MGO)	26.0
B/H Ratio (Load Line) at BH_{max}	1.0
Flux Density in Magnet B_m at BH_{max} (G)	5100
Coercive Force in Magnet H_m at BH_{max} (Oe)	5100
Energy Loss W_c (watt-sec/cycle/lb)	-
Importance of Operating Magnet at BH_{max}	E
Average Recoil Permeability	1.05
Field Strength Required to Fully Saturate Magnet (Oe)	32,000
Ability to Withstand Demagnetizing Fields	0
Maximum Service Temperature (°F)	Est. 400
Approximate Temperature Permanently Affecting Magnet (°F)	Est. 1328
Inherent Orientation of Material	MP

PHYSICAL PROPERTIES	
Density (lbs./in.^{3})	.2999-.3107
General Mechanical Properties	HB
Tensile Strength (psi)	Est. 5000
Transverse Modulus of Rupture (psi)	-
Coeff. of Therm. Expansion inches x 10^{-6}/°C	Orientation: Par. 8.0 Rt.Angl. 11.0
Resistivity (micro-ohms/cm/cm^2)	86 @ 20 °C
Hardness	Scale 500-600 Vickers
Basic Forming Method	DPM
Basic Finishing Method	WGR
Relative Corrosion Resistance	G
Primary Chemical Composition	Sm-Co
Special Notes:	

Note: All information based on available producers literature. In some cases of foreign materials interpretation or calculation required.

SOURCE(S):

TDK Corp. of America

USA

REC-26A
(Alloy or Grade)

Sintered Rare Earth
(Type of Material)

25/8
(MMPA Brief Designation)

R5-2
(IEC Code Reference)

MAGNETIC PROPERTIES		PHYSICAL PROPERTIES	
Residual Induction B_r (G)	10,400	Density (lbs./in.^{3})	.2999-.3107
Coercive Force (Oe) — Normal (H_c)	8000	General Mechanical Properties	HB
Coercive Force (Oe) — Intrinsic (H_{ci})	8000	Tensile Strength (psi)	Est. 5000
Maximum Energy Product BH_{max} (MGO)	25.0	Transverse Modulus of Rupture (psi)	-
B/H Ratio (Load Line) at BH_{max}	1.20	Coeff. of Therm. Expansion inches x 10^{-6}/°C	Orientation: Par. 8.0 Rt.Angl. 11.0
Flux Density in Magnet B_m at BH_{max} (G)	5500	Resistivity (micro-ohms/cm/cm^2)	86 @ 20 °C
Coercive Force in Magnet H_m at BH_{max} (Oe)	4600	Hardness Scale	500-600 Vickers
Energy Loss W_c (watt-sec/cycle/lb)	-	Basic Forming Method	DPM
Importance of Operating Magnet at BH_{max}	E	Basic Finishing Method	WGR
Average Recoil Permeability	1.05	Relative Corrosion Resistance	G
Field Strength Required to Fully Saturate Magnet (Oe)	32,000	Primary Chemical Composition	Sm-Co
Ability to Withstand Demagnetizing Fields	0	Special Notes:	
Maximum Service Temperature (°F)	Est. 350		
Approximate Temperature Permanently Affecting Magnet (°F)	Est. 1328		
Inherent Orientation of Material	MP		

Note: All information based on available producers literature. In some cases of foreign materials interpretation or calculation required.

SOURCE(S):	REC-30
TDK Corp. of America	(Alloy or Grade)
	Sintered Rare Earth
	(Type of Material)
USA	30/7
	(MMPA Brief Designation)
	R5-2
	(IEC Code Reference)

MAGNETIC PROPERTIES		PHYSICAL PROPERTIES	
Residual Induction B_r (G)	11,400	Density (lbs./in.^{3})	.2999-.3107
Coercive Force (Oe) — Normal (H_c)	6400	General Mechanical Properties	HB
Coercive Force (Oe) — Intrinsic (H_{ci})	6600	Tensile Strength (psi)	Est. 5000
Maximum Energy Product BH_{max} (MGO)	30.0	Transverse Modulus of Rupture (psi)	-
B/H Ratio (Load Line) at BH_{max}	1.27	Coeff. of Therm. Expansion inches x 10^{-6}/°C	Orientation: Par. 8.0 Rt.Angl. 11.0
Flux Density in Magnet B_m at BH_{max} (G)	6200	Resistivity (micro-ohms/cm/cm^2)	86 @ 20 °C
Coercive Force in Magnet H_m at BH_{max} (Oe)	4900	Hardness Scale	500-600 Vickers
Energy Loss W_c (watt-sec/cycle/lb)	-	Basic Forming Method	DPM
Importance of Operating Magnet at BH_{max}	E	Basic Finishing Method	WGR
Average Recoil Permeability	1.05	Relative Corrosion Resistance	G
Field Strength Required to Fully Saturate Magnet (Oe)	22,600	Primary Chemical Composition	Sm-Co
Ability to Withstand Demagnetizing Fields	0	Special Notes:	
Maximum Service Temperature (°F)	Est. 300		
Approximate Temperature Permanently Affecting Magnet (°F)	Est. 1328		
Inherent Orientation of Material	MP		

Note: All information based on available producers literature. In some cases of foreign materials interpretation or calculation required.

SOURCE(S): **TDK Corp. of America**

USA

REC-32A
(Alloy or Grade)

Sintered Rare Earth
(Type of Material)

30/8
(MMPA Brief Designation)

R5-2
(IEC Code Reference)

MAGNETIC PROPERTIES	
Residual Induction B_r (G)	11,300
Coercive Force (Oe) — Normal (H_c)	8000
Coercive Force (Oe) — Intrinsic (H_{ci})	8000
Maximum Energy Product BH_{max} (MGO)	30.0
B/H Ratio (Load Line) at BH_{max}	1.16
Flux Density in Magnet B_m at BH_{max} (G)	5900
Coercive Force in Magnet H_m at BH_{max} (Oe)	5100
Energy Loss W_c (watt-sec/cycle/lb)	-
Importance of Operating Magnet at BH_{max}	E
Average Recoil Permeability	1.05
Field Strength Required to Fully Saturate Magnet (Oe)	22,600
Ability to Withstand Demagnetizing Fields	0
Maximum Service Temperature (°F)	Est. 350
Approximate Temperature Permanently Affecting Magnet (°F)	Est. 1328
Inherent Orientation of Material	MP

PHYSICAL PROPERTIES	
Density (lbs./in.^{3})	.2999-.3107
General Mechanical Properties	HB
Tensile Strength (psi)	Est. 5000
Transverse Modulus of Rupture (psi)	-
Coeff. of Therm. Expansion inches x 10^{-6}/°C	Orientation: Par. 8.0 Rt.Angl. 11.0
Resistivity (micro-ohms/cm/cm^2)	86 @ 20 °C
Hardness Scale	500-600 Vickers
Basic Forming Method	DPM
Basic Finishing Method	WGR
Relative Corrosion Resistance	G
Primary Chemical Composition	Sm-Co
Special Notes:	

Note: All information based on available producers literature. In some cases of foreign materials interpretation or calculation required.

SOURCE(S):	RECOMA 22
	(Alloy or Grade)
UGIMag, Inc.	Sintered Rare Earth
	(Type of Material)
USA	22/18
	(MMPA Brief Designation)
	R5-1
	(IEC Code Reference)

MAGNETIC PROPERTIES		PHYSICAL PROPERTIES	
Residual Induction B_r (G)	9400	Density (lbs./in.^{3})	.2962
Coercive Force (Oe) — Normal (H_c)	9300	General Mechanical Properties	HB
Coercive Force (Oe) — Intrinsic (H_{ci})	18,000	Tensile Strength (psi)	Est. 5000
Maximum Energy Product BH_{max} (MGO)	22.0	Transverse Modulus of Rupture (psi)	-
B/H Ratio (Load Line) at BH_{max}	.47	Coeff. of Therm. Expansion inches x 10^{-6}/°C	Orientation: Par. - Rt.Angl. -
Flux Density in Magnet B_m at BH_{max} (G)	3200	Resistivity (micro-ohms/cm/cm^2)	- @ - °C
Coercive Force in Magnet H_m at BH_{max} (Oe)	6800	Hardness 600	Scale Vickers
Energy Loss W_c (watt-sec/cycle/lb)	-	Basic Forming Method	DP
Importance of Operating Magnet at BH_{max}	E	Basic Finishing Method	WGR
Average Recoil Permeability	1.02	Relative Corrosion Resistance	G
Field Strength Required to Fully Saturate Magnet (Oe)	40,000	Primary Chemical Composition	Sm-Co
Ability to Withstand Demagnetizing Fields	0	Special Notes:	
Maximum Service Temperature (°F)	Est. 482		
Approximate Temperature Permanently Affecting Magnet (°F)	Est. 1328		
Inherent Orientation of Material	none		

Note: All information based on available producers literature. In some cases of foreign materials interpretation or calculation required.

SOURCE(S): **UGIMag, Inc.** USA

RECOMA 25
(Alloy or Grade)

Sintered Rare Earth
(Type of Material)

25/19
(MMPA Brief Designation)

R5-2
(IEC Code Reference)

MAGNETIC PROPERTIES	
Residual Induction B_r (G)	10,000
Coercive Force (Oe) — Normal (H_c)	9500
Coercive Force (Oe) — Intrinsic (H_{ci})	19,000
Maximum Energy Product BH_{max} (MGO)	25.0
B/H Ratio (Load Line) at BH_{max}	-
Flux Density in Magnet B_m at BH_{max} (G)	-
Coercive Force in Magnet H_m at BH_{max} (Oe)	-
Energy Loss W_c (watt-sec/cycle/lb)	-
Importance of Operating Magnet at BH_{max}	E
Average Recoil Permeability	1.05
Field Strength Required to Fully Saturate Magnet (Oe)	30,000
Ability to Withstand Demagnetizing Fields	0
Maximum Service Temperature (°F)	Est. 482
Approximate Temperature Permanently Affecting Magnet (°F)	Est. 1328
Inherent Orientation of Material	none

PHYSICAL PROPERTIES	
Density (lbs./in.^{3})	Est. .3000
General Mechanical Properties	HB
Tensile Strength (psi)	Est. 5000
Transverse Modulus of Rupture (psi)	-
Coeff. of Therm. Expansion inches x 10^{-6}/°C	Orientation: Par. - Rt.Angl. -
Resistivity (micro-ohms/cm/cm^2)	- @ - °C
Hardness Scale	- -
Basic Forming Method	DP
Basic Finishing Method	WGR
Relative Corrosion Resistance	G
Primary Chemical Composition	Sm-Co
Special Notes:	

Note: All information based on available producers literature. In some cases of foreign materials interpretation or calculation required.

SOURCE(S):

UGIMag, Inc.

USA

RECOMA 28
(Alloy or Grade)

Sintered Rare Earth
(Type of Material)

28/18
(MMPA Brief Designation)

R5-2
(IEC Code Reference)

MAGNETIC PROPERTIES	
Residual Induction B_r (G)	10,700
Coercive Force (Oe) — Normal (H_c)	10,300
Coercive Force (Oe) — Intrinsic (H_{ci})	18,000
Maximum Energy Product BH_{max} (MGO)	28.0
B/H Ratio (Load Line) at BH_{max}	.77
Flux Density in Magnet B_m at BH_{max} (G)	4600
Coercive Force in Magnet H_m at BH_{max} (Oe)	6000
Energy Loss W_c (watt-sec/cycle/lb)	-
Importance of Operating Magnet at BH_{max}	E
Average Recoil Permeability	1.05
Field Strength Required to Fully Saturate Magnet (Oe)	40,000
Ability to Withstand Demagnetizing Fields	0
Maximum Service Temperature (°F)	Est. 482
Approximate Temperature Permanently Affecting Magnet (°F)	Est. 1328
Inherent Orientation of Material	none

PHYSICAL PROPERTIES	
Density (lbs./in.^{3})	.2999
General Mechanical Properties	HB
Tensile Strength (psi)	Est. 5000
Transverse Modulus of Rupture (psi)	-
Coeff. of Therm. Expansion inches x 10^{-6}/°C	Orientation: Par. - Rt.Angl. -
Resistivity (micro-ohms/cm/cm^2)	90 @ 22 °C
Hardness Scale	600 Vickers
Basic Forming Method	DP
Basic Finishing Method	WGR
Relative Corrosion Resistance	G
Primary Chemical Composition	Sm-Co
Special Notes:	

Note: All information based on available producers literature. In some cases of foreign materials interpretation or calculation required.

SOURCE(S):

Philips Components

The Neatherlands

RES190
(Alloy or Grade)

Sintered Rare Earth
(Type of Material)

20/15
(MMPA Brief Designation)

R5-1
(IEC Code Reference)

MAGNETIC PROPERTIES		PHYSICAL PROPERTIES	
Residual Induction B_r (G)	8900	Density (lbs./in.^{3})	.2999
Coercive Force (Oe) — Normal (H_c)	8420	General Mechanical Properties	HB
Coercive Force (Oe) — Intrinsic (H_{ci})	15,000	Tensile Strength (psi)	Est. 5000
Maximum Energy Product BH_{max} (MGO)	19.4	Transverse Modulus of Rupture (psi)	-
B/H Ratio (Load Line) at BH_{max}	1.0	Coeff. of Therm. Expansion inches x 10^{-6}/°C	Orientation: Par. -6.0 Rt.Angl. -12.0
Flux Density in Magnet B_m at BH_{max} (G)	4400	Resistivity (micro-ohms/cm/cm^2)	50 @ 25 °C
Coercive Force in Magnet H_m at BH_{max} (Oe)	4400	Hardness Scale	500 Vickers
Energy Loss W_c (watt-sec/cycle/lb)	-	Basic Forming Method	DPM
Importance of Operating Magnet at BH_{max}	E	Basic Finishing Method	WGR
Average Recoil Permeability	1.05	Relative Corrosion Resistance	G
Field Strength Required to Fully Saturate Magnet (Oe)	32,000	Primary Chemical Composition	Sm-Co
Ability to Withstand Demagnetizing Fields	0	Special Notes:	
Maximum Service Temperature (°F)	Est. 482		
Approximate Temperature Permanently Affecting Magnet (°F)	500		
Inherent Orientation of Material	MP		

Note: All information based on available producers literature. In some cases of foreign materials interpretation or calculation required.

SOURCE(S):	**RES195**
Philips Components	(Alloy or Grade)
	Sintered Rare Earth
	(Type of Material)
The Neatherlands	20/20
	(MMPA Brief Designation)
	R5-1
	(IEC Code Reference)

MAGNETIC PROPERTIES		PHYSICAL PROPERTIES	
Residual Induction B_r (G)	8900	Density (lbs./in.^{3})	.2999
Coercive Force (Oe) — Normal (H_c)	8480	General Mechanical Properties	HB
Coercive Force (Oe) — Intrinsic (H_{ci})	20,015	Tensile Strength (psi)	Est. 5000
Maximum Energy Product BH_{max} (MGO)	19.4	Transverse Modulus of Rupture (psi)	-
B/H Ratio (Load Line) at BH_{max}	Est. 1.0	Coeff. of Therm. Expansion inches x 10^{-6}/°C	Orientation: Par. 6.0 Rt.Angl. 12.0
Flux Density in Magnet B_m at BH_{max} (G)	Est. 4400	Resistivity (micro-ohms/cm/cm^2)	50 @ 25 °C
Coercive Force in Magnet H_m at BH_{max} (Oe)	Est. 4400	Hardness 500	Scale Vickers
Energy Loss W_c (watt-sec/cycle/lb)	-	Basic Forming Method	DPM
Importance of Operating Magnet at BH_{max}	E	Basic Finishing Method	WGR
Average Recoil Permeability	1.05	Relative Corrosion Resistance	G
Field Strength Required to Fully Saturate Magnet (Oe)	32,000	Primary Chemical Composition	Sm-Co
Ability to Withstand Demagnetizing Fields	0	Special Notes:	
Maximum Service Temperature (°F)	Est. 482		
Approximate Temperature Permanently Affecting Magnet (°F)	500		
Inherent Orientation of Material	MP		

Note: All information based on available producers literature. In some cases of foreign materials interpretation or calculation required.

SOURCE(S):

Philips Components

The Neatherlands

RES230
(Alloy or Grade)

Sintered Rare Earth
(Type of Material)

23/9
(MMPA Brief Designation)

R5-2
(IEC Code Reference)

MAGNETIC PROPERTIES	
Residual Induction B_r (G)	10,000
Coercive Force (Oe) — Normal (H_c)	8170
Coercive Force (Oe) — Intrinsic (H_{ci})	9425
Maximum Energy Product BH_{max} (MGO)	23.2
B/H Ratio (Load Line) at BH_{max}	Est. 1.1
Flux Density in Magnet B_m at BH_{max} (G)	Est. 5100
Coercive Force in Magnet H_m at BH_{max} (Oe)	Est. 4800
Energy Loss W_c (watt-sec/cycle/lb)	-
Importance of Operating Magnet at BH_{max}	E
Average Recoil Permeability	1.05
Field Strength Required to Fully Saturate Magnet (Oe)	32,000
Ability to Withstand Demagnetizing Fields	0
Maximum Service Temperature (°F)	Est. 572
Approximate Temperature Permanently Affecting Magnet (°F)	500
Inherent Orientation of Material	MP

PHYSICAL PROPERTIES	
Density (lbs./in.^{3})	.3040
General Mechanical Properties	HB
Tensile Strength (psi)	Est. 5000
Transverse Modulus of Rupture (psi)	-
Coeff. of Therm. Expansion inches x 10^{-6}/°C	Orientation: Par. 8.0 Rt.Angl. 12.0
Resistivity (micro-ohms/cm/cm^2)	80 @ 25 °C
Hardness	550 Scale Vickers
Basic Forming Method	DPM
Basic Finishing Method	WGR
Relative Corrosion Resistance	G
Primary Chemical Composition	Sm-Co
Special Notes:	

Note: All information based on available producers literature. In some cases of foreign materials interpretation or calculation required.

SOURCE(S):

Philips Components

The Neatherlands

RES239
(Alloy or Grade)

Sintered Rare Earth
(Type of Material)

23/38
(MMPA Brief Designation)

R7-3
(IEC Code Reference)

MAGNETIC PROPERTIES	
Residual Induction B_r (G)	9600
Coercive Force (Oe) — Normal (H_c)	9400
Coercive Force (Oe) — Intrinsic (H_{ci})	37,700
Maximum Energy Product BH_{max} (MGO)	23.0
B/H Ratio (Load Line) at BH_{max}	1.04
Flux Density in Magnet B_m at BH_{max} (G)	5200
Coercive Force in Magnet H_m at BH_{max} (Oe)	5000
Energy Loss W_c (watt-sec/cycle/lb)	-
Importance of Operating Magnet at BH_{max}	E
Average Recoil Permeability	1.05
Field Strength Required to Fully Saturate Magnet (Oe)	32,000
Ability to Withstand Demagnetizing Fields	O
Maximum Service Temperature (°F)	Est. 356
Approximate Temperature Permanently Affecting Magnet (°F)	Est. 590
Inherent Orientation of Material	MP

PHYSICAL PROPERTIES	
Density (lbs./in.^{3})	.2673
General Mechanical Properties	HB
Tensile Strength (psi)	-
Transverse Modulus of Rupture (psi)	-
Coeff. of Therm. Expansion inches x 10^{-6}/°C	Orientation: Par. 5.7 Rt.Angl. -.5
Resistivity (micro-ohms/cm/cm^2)	1.4 @ 25 °C
Hardness	Scale 500 Vickers
Basic Forming Method	DPM
Basic Finishing Method	WGR
Relative Corrosion Resistance	P
Primary Chemical Composition	Nd-Fe-B
Special Notes:	

Note: All information based on available producers literature. In some cases of foreign materials interpretation or calculation required.

SOURCE(S):

Philips Components

The Neatherlands

RES255
(Alloy or Grade)

Sintered Rare Earth
(Type of Material)

25/19
(MMPA Brief Designation)

R7-3
(IEC Code Reference)

MAGNETIC PROPERTIES	
Residual Induction B_r (G)	10,500
Coercive Force (Oe) — Normal (H_c)	9425
Coercive Force (Oe) — Intrinsic (H_{ci})	18,850
Maximum Energy Product BH_{max} (MGO)	25.1
B/H Ratio (Load Line) at BH_{max}	1.0
Flux Density in Magnet B_m at BH_{max} (G)	5000
Coercive Force in Magnet H_m at BH_{max} (Oe)	5000
Energy Loss W_c (watt-sec/cycle/lb)	-
Importance of Operating Magnet at BH_{max}	E
Average Recoil Permeability	1.05
Field Strength Required to Fully Saturate Magnet (Oe)	32,000
Ability to Withstand Demagnetizing Fields	0
Maximum Service Temperature (°F)	Est. 302
Approximate Temperature Permanently Affecting Magnet (°F)	300
Inherent Orientation of Material	MP

PHYSICAL PROPERTIES	
Density (lbs./in.^{3})	.2673
General Mechanical Properties	HB
Tensile Strength (psi)	-
Transverse Modulus of Rupture (psi)	-
Coeff. of Therm. Expansion inches x 10^{-6}/°C	Orientation: Par. 5.7 Rt.Angl. -.5
Resistivity (micro-ohms/cm/cm^2)	1.4 @ 25 °C
Hardness	Scale 500 Vickers
Basic Forming Method	DPM
Basic Finishing Method	WGR
Relative Corrosion Resistance	P
Primary Chemical Composition	Nd-Fe-B
Special Notes:	

Note: All information based on available producers literature. In some cases of foreign materials interpretation or calculation required.

SOURCE(S):	RES257
	(Alloy or Grade)
Philips Components	Sintered Rare Earth
	(Type of Material)
The Neatherlands	25/25
	(MMPA Brief Designation)
	R7-3
	(IEC Code Reference)

MAGNETIC PROPERTIES		PHYSICAL PROPERTIES	
Residual Induction B_r (G)	10,200	Density (lbs./in.^{3})	.2673
Coercive Force (Oe) — Normal (H_c)	9400	General Mechanical Properties	HB
Coercive Force (Oe) — Intrinsic (H_{ci})	25,150	Tensile Strength (psi)	–
Maximum Energy Product BH_{max} (MGO)	25.0	Transverse Modulus of Rupture (psi)	–
B/H Ratio (Load Line) at BH_{max}	1.01	Coeff. of Therm. Expansion inches x 10^{-6}/°C	Orientation: Par. 5.7 Rt.Angl. .5
Flux Density in Magnet B_m at BH_{max} (G)	4900	Resistivity (micro-ohms/cm/cm^2)	1.4 @ 25 °C
Coercive Force in Magnet H_m at BH_{max} (Oe)	4850	Hardness 500	Scale Vickers
Energy Loss W_c (watt-sec/cycle/lb)	–	Basic Forming Method	DPM
Importance of Operating Magnet at BH_{max}	E	Basic Finishing Method	WGR
Average Recoil Permeability	1.05	Relative Corrosion Resistance	P
Field Strength Required to Fully Saturate Magnet (Oe)	32,000	Primary Chemical Composition	Nd-Fe-B
Ability to Withstand Demagnetizing Fields	0	Special Notes:	
Maximum Service Temperature (°F)	Est. 356		
Approximate Temperature Permanently Affecting Magnet (°F)	300		
Inherent Orientation of Material	MP		

Note: All information based on available producers literature. In some cases of foreign materials interpretation or calculation required.

SOURCE(S):

Philips Components

The Neatherlands

RES270
(Alloy or Grade)

Sintered Rare Earth
(Type of Material)

30/12
(MMPA Brief Designation)

R7-3
(IEC Code Reference)

MAGNETIC PROPERTIES		PHYSICAL PROPERTIES	
Residual Induction B_r (G)	11,200	Density (lbs./in.^{3})	.2673
Coercive Force (Oe) Normal (H_c)	10,700	General Mechanical Properties	HB
Coercive Force (Oe) Intrinsic (H_{ci})	12,550	Tensile Strength (psi)	-
Maximum Energy Product BH_{max} (MGO)	30.0	Transverse Modulus of Rupture (psi)	-
B/H Ratio (Load Line) at BH_{max}	1.0	Coeff. of Therm. Expansion inches x 10^{-6}/°C	Orientation: Par. 5.7 Rt.Angl. -.5
Flux Density in Magnet B_m at BH_{max} (G)	5200	Resistivity (micro-ohms/cm/cm^2)	1.4 @ 25 °C
Coercive Force in Magnet H_m at BH_{max} (Oe)	5200	Hardness 500	Scale Vickers
Energy Loss W_c (watt-sec/cycle/lb)	-	Basic Forming Method	DPM
Importance of Operating Magnet at BH_{max}	E	Basic Finishing Method	WGR
Average Recoil Permeability	1.05	Relative Corrosion Resistance	P
Field Strength Required to Fully Saturate Magnet (Oe)	32,000	Primary Chemical Composition	Nd-Fe-B
Ability to Withstand Demagnetizing Fields	0	Special Notes:	
Maximum Service Temperature (°F)	Est. 212		
Approximate Temperature Permanently Affecting Magnet (°F)	500		
Inherent Orientation of Material	MP		

Note: All information based on available producers literature. In some cases of foreign materials interpretation or calculation required.

SOURCE(S):

Philips Components

The Neatherlands

RES275
(Alloy or Grade)

Sintered Rare Earth
(Type of Material)

31/19
(MMPA Brief Designation)

R7-3
(IEC Code Reference)

MAGNETIC PROPERTIES		PHYSICAL PROPERTIES	
Residual Induction B_r (G)	11,500	Density (lbs./in.^{3})	.2673
Coercive Force (Oe) — Normal (H_c)	10,700	General Mechanical Properties	HB
Coercive Force (Oe) — Intrinsic (H_{ci})	18,850	Tensile Strength (psi)	-
Maximum Energy Product BH_{max} (MGO)	31.0	Transverse Modulus of Rupture (psi)	-
B/H Ratio (Load Line) at BH_{max}	1.0	Coeff. of Therm. Expansion inches x 10^{-6}/°C	Orientation: Par. 5.7 Rt.Angl. 1.5
Flux Density in Magnet B_m at BH_{max} (G)	5200	Resistivity (micro-ohms/cm/cm^2)	1.4 @ 25 °C
Coercive Force in Magnet H_m at BH_{max} (Oe)	5200	Hardness Scale	500 Vickers
Energy Loss W_c (watt-sec/cycle/lb)	-	Basic Forming Method	DPM
Importance of Operating Magnet at BH_{max}	E	Basic Finishing Method	WGR
Average Recoil Permeability	1.05	Relative Corrosion Resistance	P
Field Strength Required to Fully Saturate Magnet (Oe)	32,000	Primary Chemical Composition	Nd-Fe-B
Ability to Withstand Demagnetizing Fields	0	Special Notes:	
Maximum Service Temperature (°F)	Est. 302		
Approximate Temperature Permanently Affecting Magnet (°F)	500		
Inherent Orientation of Material	MP		

Note: All information based on available producers literature. In some cases of foreign materials interpretation or calculation required.

SOURCE(S):

Philips Components

The Neatherlands

RES300
(Alloy or Grade)

Sintered Rare Earth
(Type of Material)

34/12
(MMPA Brief Designation)

R7-3
(IEC Code Reference)

MAGNETIC PROPERTIES		PHYSICAL PROPERTIES	
Residual Induction B_r (G)	12,000	Density (lbs./in.^{3})	.2673
Coercive Force (Oe) Normal (H_c)	11,300	General Mechanical Properties	HB
Coercive Force (Oe) Intrinsic (H_{ci})	12,550	Tensile Strength (psi)	-
Maximum Energy Product BH_{max} (MGO)	34.0	Transverse Modulus of Rupture (psi)	-
B/H Ratio (Load Line) at BH_{max}	1.0	Coeff. of Therm. Expansion inches x 10^{-6}/°C	Orientation: Par. 5.7 Rt.Angl. .5
Flux Density in Magnet B_m at BH_{max} (G)	5500	Resistivity (micro-ohms/cm/cm^2)	1.4 @ 25 °C
Coercive Force in Magnet H_m at BH_{max} (Oe)	5500	Hardness Scale	500 Vickers
Energy Loss W_c (watt-sec/cycle/lb)	-	Basic Forming Method	DPM
Importance of Operating Magnet at BH_{max}	E	Basic Finishing Method	WGR
Average Recoil Permeability	1.05	Relative Corrosion Resistance	P
Field Strength Required to Fully Saturate Magnet (Oe)	32,000	Primary Chemical Composition	Nd-Fe-B
Ability to Withstand Demagnetizing Fields	O	Special Notes:	
Maximum Service Temperature (°F)	Est. 212		
Approximate Temperature Permanently Affecting Magnet (°F)	500		
Inherent Orientation of Material	MP		

Note: All information based on available producers literature. In some cases of foreign materials interpretation or calculation required.

SOURCE(S):

Philips Components

The Neatherlands

RES302
(Alloy or Grade)

Sintered Rare Earth
(Type of Material)

33/15
(MMPA Brief Designation)

R7-3
(IEC Code Reference)

MAGNETIC PROPERTIES		PHYSICAL PROPERTIES	
Residual Induction B_r (G)	11,900	Density (lbs./in.^{3})	.2673
Coercive Force (Oe) — Normal (H_c)	11,300	General Mechanical Properties	HB
Coercive Force (Oe) — Intrinsic (H_{ci})	15,100	Tensile Strength (psi)	-
Maximum Energy Product BH_{max} (MGO)	33.0	Transverse Modulus of Rupture (psi)	-
B/H Ratio (Load Line) at BH_{max}	1.0	Coeff. of Therm. Expansion inches $\times 10^{-6}$/°C	Orientation: Par. 5.7 Rt.Angl. -.5
Flux Density in Magnet B_m at BH_{max} (G)	5200	Resistivity (micro-ohms/cm/cm^2)	1.4 @ 25 °C
Coercive Force in Magnet H_m at BH_{max} (Oe)	5200	Hardness 500	Scale Vickers
Energy Loss W_c (watt-sec/cycle/lb)	-	Basic Forming Method	DPM
Importance of Operating Magnet at BH_{max}	E	Basic Finishing Method	WGR
Average Recoil Permeability	1.05	Relative Corrosion Resistance	P
Field Strength Required to Fully Saturate Magnet (Oe)	32,000	Primary Chemical Composition	Nd-Fe-B
Ability to Withstand Demagnetizing Fields	0	Special Notes:	
Maximum Service Temperature (°F)	Est. 248		
Approximate Temperature Permanently Affecting Magnet (°F)	500		
Inherent Orientation of Material	MP		

Note: All information based on available producers literature. In some cases of foreign materials interpretation or calculation required.

SOURCE(S):

Philips Components

The Neatherlands

RES303
(Alloy or Grade)

Sintered Rare Earth
(Type of Material)

32/16
(MMPA Brief Designation)

R7-3
(IEC Code Reference)

MAGNETIC PROPERTIES	
Residual Induction B_r (G)	11,600
Coercive Force (Oe) Normal (H_c)	11,000
Coercive Force (Oe) Intrinsic (H_{ci})	16,350
Maximum Energy Product BH_{max} (MGO)	32.0
B/H Ratio (Load Line) at BH_{max}	1.0
Flux Density in Magnet B_m at BH_{max} (G)	5200
Coercive Force in Magnet H_m at BH_{max} (Oe)	5200
Energy Loss W_c (watt-sec/cycle/lb)	-
Importance of Operating Magnet at BH_{max}	E
Average Recoil Permeability	1.05
Field Strength Required to Fully Saturate Magnet (Oe)	32,000
Ability to Withstand Demagnetizing Fields	0
Maximum Service Temperature (°F)	Est. 248
Approximate Temperature Permanently Affecting Magnet (°F)	500
Inherent Orientation of Material	MP

PHYSICAL PROPERTIES	
Density (lbs./in.^{3})	.2673
General Mechanical Properties	HB
Tensile Strength (psi)	-
Transverse Modulus of Rupture (psi)	-
Coeff. of Therm. Expansion inches x 10^{-6}/°C	Orientation: Par. 5.7 Rt.Angl. -.5
Resistivity (micro-ohms/cm/cm^2)	1.4 @ 25 °C
Hardness	Scale 500 Vickers
Basic Forming Method	DPM
Basic Finishing Method	WGR
Relative Corrosion Resistance	P
Primary Chemical Composition	Nd-Fe-B
Special Notes:	

Note: All information based on available producers literature. In some cases of foreign materials interpretation or calculation required.

SOURCE(S):

Philips Components

The Neatherlands

RES305
(Alloy or Grade)

Sintered Rare Earth
(Type of Material)

30/19
(MMPA Brief Designation)

R7-3
(IEC Code Reference)

MAGNETIC PROPERTIES		PHYSICAL PROPERTIES	
Residual Induction B_r (G)	11,500	Density (lbs./in.^{3})	.2673
Coercive Force (Oe) — Normal (H_c)	10,680	General Mechanical Properties	HB
Coercive Force (Oe) — Intrinsic (H_{ci})	18,850	Tensile Strength (psi)	-
Maximum Energy Product BH_{max} (MGO)	30.2	Transverse Modulus of Rupture (psi)	-
B/H Ratio (Load Line) at BH_{max}	1.0	Coeff. of Therm. Expansion inches x 10^{-6}/°C	Orientation: Par. 5.7 Rt.Angl. -.5
Flux Density in Magnet B_m at BH_{max} (G)	5000	Resistivity (micro-ohms/cm/cm^2)	1.4 @ 25 °C
Coercive Force in Magnet H_m at BH_{max} (Oe)	5000	Hardness 500	Scale Vickers
Energy Loss W_c (watt-sec/cycle/lb)	-	Basic Forming Method	DPM
Importance of Operating Magnet at BH_{max}	E	Basic Finishing Method	WGR
Average Recoil Permeability	1.05	Relative Corrosion Resistance	P
Field Strength Required to Fully Saturate Magnet (Oe)	32,000	Primary Chemical Composition	Nd-Fe-B
Ability to Withstand Demagnetizing Fields	0	Special Notes:	
Maximum Service Temperature (°F)	Est. 302		
Approximate Temperature Permanently Affecting Magnet (°F)	500		
Inherent Orientation of Material	MP		

Note: All information based on available producers literature. In some cases of foreign materials interpretation or calculation required.

SOURCE(S):

Philips Components

The Neatherlands

RES350
(Alloy or Grade)

Sintered Rare Earth
(Type of Material)

35/13
(MMPA Brief Designation)

R7-3
(IEC Code Reference)

MAGNETIC PROPERTIES		PHYSICAL PROPERTIES	
Residual Induction B_r (G)	12,000	Density (lbs./in.^{3})	.2673
Coercive Force (Oe) — Normal (H_c)	10,680	General Mechanical Properties	HB
Coercive Force (Oe) — Intrinsic (H_{ci})	12,565	Tensile Strength (psi)	-
Maximum Energy Product BH_{max} (MGO)	35.2	Transverse Modulus of Rupture (psi)	-
B/H Ratio (Load Line) at BH_{max}	.92	Coeff. of Therm. Expansion inches x 10^{-6}/°C	Orientation: Par. 5.7 Rt.Angl. -.5
Flux Density in Magnet B_m at BH_{max} (G)	5700	Resistivity (micro-ohms/cm/cm^2)	1.4 @ 25 °C
Coercive Force in Magnet H_m at BH_{max} (Oe)	6200	Hardness Scale	500 Vickers
Energy Loss W_c (watt-sec/cycle/lb)	-	Basic Forming Method	DPM
Importance of Operating Magnet at BH_{max}	E	Basic Finishing Method	WGR
Average Recoil Permeability	1.05	Relative Corrosion Resistance	P
Field Strength Required to Fully Saturate Magnet (Oe)	32,000	Primary Chemical Composition	Nd-Fe-B
Ability to Withstand Demagnetizing Fields	0	Special Notes:	
Maximum Service Temperature (°F)	Est. 212		
Approximate Temperature Permanently Affecting Magnet (°F)	500		
Inherent Orientation of Material	MP		

Note: All information based on available producers literature. In some cases of foreign materials interpretation or calculation required.

SOURCE(S):

Philips Components

The Neatherlands

RES421
(Alloy or Grade)

Sintered Rare Earth
(Type of Material)

40/14
(MMPA Brief Designation)

R7-3
(IEC Code Reference)

MAGNETIC PROPERTIES	
Residual Induction B_r (G)	13,000
Coercive Force (Oe) Normal (H_c)	11,550
Coercive Force (Oe) Intrinsic (H_{ci})	13,800
Maximum Energy Product BH_{max} (MGO)	40
B/H Ratio (Load Line) at BH_{max}	1.0
Flux Density in Magnet B_m at BH_{max} (G)	5400
Coercive Force in Magnet H_m at BH_{max} (Oe)	5400
Energy Loss W_c (watt-sec/cycle/lb)	-
Importance of Operating Magnet at BH_{max}	E
Average Recoil Permeability	1.05
Field Strength Required to Fully Saturate Magnet (Oe)	32,000
Ability to Withstand Demagnetizing Fields	0
Maximum Service Temperature (°F)	Est. 220
Approximate Temperature Permanently Affecting Magnet (°F)	500
Inherent Orientation of Material	MP

PHYSICAL PROPERTIES	
Density (lbs./in.^{3})	.2673
General Mechanical Properties	HB
Tensile Strength (psi)	-
Transverse Modulus of Rupture (psi)	-
Coeff. of Therm. Expansion inches x 10^{-6}/°C	Orientation: Par. 5.7 Rt.Angl. -.5
Resistivity (micro-ohms/cm/cm^2)	1.4 @ 25 °C
Hardness	500 Scale Vickers
Basic Forming Method	DPM
Basic Finishing Method	WGR
Relative Corrosion Resistance	P
Primary Chemical Composition	Nd-Fe-B
Special Notes:	

Note: All information based on available producers literature. In some cases of foreign materials interpretation or calculation required.

SOURCE(S):

Philips Components

The Neatherlands

SP02
(Alloy or Grade)

Bonded Ferrite
(Type of Material)

.25/1.6
(MMPA Brief Designation)

-
(IEC Code Reference)

MAGNETIC PROPERTIES		PHYSICAL PROPERTIES	
Residual Induction B_r (G)	1000	Density (lbs./in.^{3})	.0957
Coercive Force (Oe) — Normal (H_c)	1011	General Mechanical Properties	FL
Coercive Force (Oe) — Intrinsic (H_{ci})	1634	Tensile Strength (psi)	-
Maximum Energy Product BH_{max} (MGO)	.25	Transverse Modulus of Rupture (psi)	-
B/H Ratio (Load Line) at BH_{max}	1.0	Coeff. of Therm. Expansion inches x 10^{-6}/°C	Orientation: Par. 7.5 Rt.Angl. -
Flux Density in Magnet B_m at BH_{max} (G)	500	Resistivity (micro-ohms/cm/cm^2	10^{10} @ 25 °C
Coercive Force in Magnet H_m at BH_{max} (Oe)	500	Hardness Scale	- -
Energy Loss W_c (watt-sec/cycle/lb)	-	Basic Forming Method	E
Importance of Operating Magnet at BH_{max}	E	Basic Finishing Method	SRD
Average Recoil Permeability	-	Relative Corrosion Resistance	E
Field Strength Required to Fully Saturate Magnet (Oe)	10,000	Primary Chemical Composition	Ba or Sr-Fe + binder
Ability to Withstand Demagnetizing Fields	Ex	Special Notes: (1) Maximum service temperature depends on binder used. Binder not identified by manufacturer.	
Maximum Service Temperature (°F)	200		
Approximate Temperature Permanently Affecting Magnet (°F)	-		
Inherent Orientation of Material	none		

Note: All information based on available producers literature. In some cases of foreign materials interpretation or calculation required.

SOURCE(S):

Philips Components

The Neatherlands

SP10
(Alloy or Grade)

Bonded Ferrite
(Type of Material)

.11/2.4
(MMPA Brief Designation)

-
(IEC Code Reference)

MAGNETIC PROPERTIES		PHYSICAL PROPERTIES	
Residual Induction B_r (G)	800	Density (lbs./in.^{3})	.0903
Coercive Force (Oe) — Normal (H_c)	729	General Mechanical Properties	FL
Coercive Force (Oe) — Intrinsic (H_{ci})	2390	Tensile Strength (psi)	-
Maximum Energy Product BH_{max} (MGO)	.11	Transverse Modulus of Rupture (psi)	-
B/H Ratio (Load Line) at BH_{max}	1.2	Coeff. of Therm. Expansion inches x 10^{-6}/°C	Orientation: Par. 5.0 Rt.Angl. -
Flux Density in Magnet B_m at BH_{max} (G)	3700	Resistivity (micro-ohms/cm/cm^2)	10^{10} @ 25 °C
Coercive Force in Magnet H_m at BH_{max} (Oe)	3000	Hardness Scale	- -
Energy Loss W_c (watt-sec/cycle/lb)	-	Basic Forming Method	E
Importance of Operating Magnet at BH_{max}	E	Basic Finishing Method	SRD
Average Recoil Permeability	-	Relative Corrosion Resistance	E
Field Strength Required to Fully Saturate Magnet (Oe)	10,000	Primary Chemical Composition	Ba or Sr-Fe + binder
Ability to Withstand Demagnetizing Fields	Ex	Special Notes: (1) Maximum service temperature depends on binder used. Binder not identified by manufacturer.	
Maximum Service Temperature (°F)	200		
Approximate Temperature Permanently Affecting Magnet (°F)	-		
Inherent Orientation of Material	none		

Note: All information based on available producers literature. In some cases of foreign materials interpretation or calculation required.

SOURCE(S):

Philips Components

The Neatherlands

SP160
(Alloy or Grade)

Bonded Ferrite
(Type of Material)

1.5/3.3
(MMPA Brief Designation)

-
(IEC Code Reference)

MAGNETIC PROPERTIES	
Residual Induction B_r (G)	2450
Coercive Force (Oe) Normal (H_c)	2260
Coercive Force (Oe) Intrinsic (H_{ci})	3270
Maximum Energy Product BH_{max} (MGO)	1.5
B/H Ratio (Load Line) at BH_{max}	-
Flux Density in Magnet B_m at BH_{max} (G)	-
Coercive Force in Magnet H_m at BH_{max} (Oe)	-
Energy Loss W_c (watt-sec/cycle/lb)	-
Importance of Operating Magnet at BH_{max}	E
Average Recoil Permeability	-
Field Strength Required to Fully Saturate Magnet (Oe)	10,000
Ability to Withstand Demagnetizing Fields	Ex
Maximum Service Temperature (°F)	200
Approximate Temperature Permanently Affecting Magnet (°F)	-
Inherent Orientation of Material	none

PHYSICAL PROPERTIES	
Density (lbs./in.^{3})	.1264
General Mechanical Properties	FL
Tensile Strength (psi)	-
Transverse Modulus of Rupture (psi)	-
Coeff. of Therm. Expansion inches x 10^{-6}/°C	Orientation: Par. 15.0 Rt.Angl. -
Resistivity (micro-ohms/cm/cm^2)	10^7 @ 25 °C
Hardness -	Scale -
Basic Forming Method	E
Basic Finishing Method	SRD
Relative Corrosion Resistance	E
Primary Chemical Composition	Sr-Fe + binder

Special Notes:

(1) Maximum service temperature depends on binder used. Binder not identified by manufacturer.

Note: All information based on available producers literature. In some cases of foreign materials interpretation or calculation required.

SOURCE(S):

Philips Components

The Neatherlands

SP170
(Alloy or Grade)

Bonded Ferrite
(Type of Material)

1.8/3.3
(MMPA Brief Designation)

-
(IEC Code Reference)

MAGNETIC PROPERTIES		PHYSICAL PROPERTIES	
Residual Induction B_r (G)	2700	Density (lbs./in.^{3})	.1409
Coercive Force (Oe) — Normal (H_c)	2460	General Mechanical Properties	FL
Coercive Force (Oe) — Intrinsic (H_{ci})	3270	Tensile Strength (psi)	-
Maximum Energy Product BH_{max} (MGO)	1.75	Transverse Modulus of Rupture (psi)	-
B/H Ratio (Load Line) at BH_{max}	-	Coeff. of Therm. Expansion inches x 10^{-6}/°C	Orientation: Par. 15.0 Rt.Angl. _
Flux Density in Magnet B_m at BH_{max} (G)	-	Resistivity (micro-ohms/cm/cm^2)	10^7 @ 25 °C
Coercive Force in Magnet H_m at BH_{max} (Oe)	-	Hardness — Scale	- — -
Energy Loss W_c (watt-sec/cycle/lb)	-	Basic Forming Method	E
Importance of Operating Magnet at BH_{max}	E	Basic Finishing Method	SRD
Average Recoil Permeability	-	Relative Corrosion Resistance	E
Field Strength Required to Fully Saturate Magnet (Oe)	10,000	Primary Chemical Composition	Sr-Fe + binder
Ability to Withstand Demagnetizing Fields	Ex	Special Notes: (1) Maximum service temperature depends on binder used. Binder not identified by manufacturer.	
Maximum Service Temperature (°F)	200		
Approximate Temperature Permanently Affecting Magnet (°F)	-		
Inherent Orientation of Material	none		

Note: All information based on available producers literature. In some cases of foreign materials interpretation or calculation required.

SOURCE(S):

Magnetfabrik-Bonn

Germany

Seco 50/60p
(Alloy or Grade)

Bonded Rare Earth
(Type of Material)

6/8   [50/60p]
(MMPA Brief Designation)

-
(IEC Code Reference)

MAGNETIC PROPERTIES		PHYSICAL PROPERTIES	
Residual Induction B_r (G)	5000	Density (lbs./in.^{3})	.1879
Coercive Force (Oe) Normal (H_c)	4000	General Mechanical Properties	HT
Coercive Force (Oe) Intrinsic (H_{ci})	8000	Tensile Strength (psi)	-
Maximum Energy Product BH_{max} (MGO)	6.0	Transverse Modulus of Rupture (psi)	-
B/H Ratio (Load Line) at BH_{max}	1.2	Coeff. of Therm. Expansion inches x 10^{-6}/°C	Orientation: Par. - Rt.Angl. -
Flux Density in Magnet B_m at BH_{max} (G)	3000	Resistivity (micro-ohms/cm/cm^2	- @ - °C
Coercive Force in Magnet H_m at BH_{max} (Oe)	2500	Hardness Scale	-
Energy Loss W_c (watt-sec/cycle/lb)	-	Basic Forming Method	E
Importance of Operating Magnet at BH_{max}	E	Basic Finishing Method	WGR
Average Recoil Permeability	1.1	Relative Corrosion Resistance	G
Field Strength Required to Fully Saturate Magnet (Oe)	35,000	Primary Chemical Composition	Sm-Co + binder (See Note)
Ability to Withstand Demagnetizing Fields	0	Special Notes: (1) Binder not indicated	
Maximum Service Temperature (°F)	-		
Approximate Temperature Permanently Affecting Magnet (°F)	594		
Inherent Orientation of Material	MP		

Note: All information based on available producers literature. In some cases of foreign materials interpretation or calculation required.

SOURCE(S):

Magnetfabrik-Bonn

Germany

Seco 140/120
(Alloy or Grade)

Sintered Rare Earth
(Type of Material)

18/10   [140-120]
(MMPA Brief Designation)

R5-1
(IEC Code Reference)

MAGNETIC PROPERTIES		PHYSICAL PROPERTIES	
Residual Induction B_r (G)	8500	Density (lbs./in.^{3})	.2962
Coercive Force (Oe) — Normal (H_c)	8000	General Mechanical Properties	HB
Coercive Force (Oe) — Intrinsic (H_{ci})	10,000	Tensile Strength (psi)	Est. 5000
Maximum Energy Product BH_{max} (MGO)	18.0	Transverse Modulus of Rupture (psi)	–
B/H Ratio (Load Line) at BH_{max}	1.1	Coeff. of Therm. Expansion inches x 10^{-6}/°C	Orientation: Par. – Rt.Angl. –
Flux Density in Magnet B_m at BH_{max} (G)	4300	Resistivity (micro-ohms/cm/cm^2)	– @ – °C
Coercive Force in Magnet H_m at BH_{max} (Oe)	4000	Hardness – Scale –	
Energy Loss W_c (watt-sec/cycle/lb)	–	Basic Forming Method	DPM
Importance of Operating Magnet at BH_{max}	E	Basic Finishing Method	WGR
Average Recoil Permeability	1.02	Relative Corrosion Resistance	G
Field Strength Required to Fully Saturate Magnet (Oe)	35,000	Primary Chemical Composition	Sm-Co
Ability to Withstand Demagnetizing Fields	0	Special Notes:	
Maximum Service Temperature (°F)	Est. 650		
Approximate Temperature Permanently Affecting Magnet (°F)	Est. 1577		
Inherent Orientation of Material	MP		

Note: All information based on available producers literature. In some cases of foreign materials interpretation or calculation required.

SOURCE(S): **Seco 170/120**

(Alloy or Grade)

Magnetfabrik-Bonn

Sintered Rare Earth

(Type of Material)

Germany

21/15   [170/120]

(MMPA Brief Designation)

R5-1

(IEC Code Reference)

MAGNETIC PROPERTIES		PHYSICAL PROPERTIES	
Residual Induction B_r (G)	9500	Density (lbs./in.^{3})	.2962
Coercive Force (Oe) — Normal (H_c)	8000	General Mechanical Properties	HB
Coercive Force (Oe) — Intrinsic (H_{ci})	15,000	Tensile Strength (psi)	Est. 5000
Maximum Energy Product BH_{max} (MGO)	21.0	Transverse Modulus of Rupture (psi)	—
B/H Ratio (Load Line) at BH_{max}	1.1	Coeff. of Therm. Expansion inches x 10^{-6}/°C Orientation: Par. — Rt.Angl. —	
Flux Density in Magnet B_m at BH_{max} (G)	5400	Resistivity (micro-ohms/cm/cm^2) — @ — °C	
Coercive Force in Magnet H_m at BH_{max} (Oe)	5000	Hardness — Scale —	
Energy Loss W_c (watt-sec/cycle/lb)	—	Basic Forming Method	DPM
Importance of Operating Magnet at BH_{max}	E	Basic Finishing Method	WGR
Average Recoil Permeability	1.04	Relative Corrosion Resistance	G
Field Strength Required to Fully Saturate Magnet (Oe)	35,000	Primary Chemical Composition	Sm-Co
Ability to Withstand Demagnetizing Fields	O	Special Notes:	
Maximum Service Temperature (°F)	Est. 523		
Approximate Temperature Permanently Affecting Magnet (°F)	Est. 993		
Inherent Orientation of Material	MP		

Note: All information based on available producers literature. In some cases of foreign materials interpretation or calculation required.

SOURCE(S):

Thyssen Magnettechnik GMBH

Germany

Secolit 100TK
(Alloy or Grade)

Sintered Rare Earth
(Type of Material)

13/30
(MMPA Brief Designation)

R5-1
(IEC Code Reference)

MAGNETIC PROPERTIES	
Residual Induction B_r (G)	7300
Coercive Force (Oe) Normal (H_c)	7200
Coercive Force (Oe) Intrinsic (H_{ci})	30,000
Maximum Energy Product BH_{max} (MGO)	13.0
B/H Ratio (Load Line) at BH_{max}	1.4
Flux Density in Magnet B_m at BH_{max} (G)	4200
Coercive Force in Magnet H_m at BH_{max} (Oe)	3000
Energy Loss W_c (watt-sec/cycle/lb)	-
Importance of Operating Magnet at BH_{max}	E
Average Recoil Permeability	1.04
Field Strength Required to Fully Saturate Magnet (Oe)	32,000
Ability to Withstand Demagnetizing Fields	O
Maximum Service Temperature (°F)	482
Approximate Temperature Permanently Affecting Magnet (°F)	1472
Inherent Orientation of Material	MP

PHYSICAL PROPERTIES	
Density (lbs./in.^{3})	.2999
General Mechanical Properties	HB
Tensile Strength (psi)	Est. 5000
Transverse Modulus of Rupture (psi)	-
Coeff. of Therm. Expansion inches x 10^{-6}/°C	Orientation: Par. - Rt.Angl. -
Resistivity (micro-ohms/cm/cm^2	- @ - °C
Hardness Scale	550-700 Vickers
Basic Forming Method	DPM
Basic Finishing Method	WGR
Relative Corrosion Resistance	G
Primary Chemical Composition	Sm-Co
Special Notes:	

Note: All information based on available producers literature. In some cases of foreign materials interpretation or calculation required.

SOURCE(S):

Thyssen Magnettechnik GMBH

Germany

Secolit 170
(Alloy or Grade)

Sintered Rare Earth
(Type of Material)

23/15
(MMPA Brief Designation)

R5-1
(IEC Code Reference)

MAGNETIC PROPERTIES	
Residual Induction B_r (G)	9500
Coercive Force (Oe) Normal (H_c)	9200
Coercive Force (Oe) Intrinsic (H_{ci})	15,000
Maximum Energy Product BH_{max} (MGO)	23.0
B/H Ratio (Load Line) at BH_{max}	1.02
Flux Density in Magnet B_m at BH_{max} (G)	4800
Coercive Force in Magnet H_m at BH_{max} (Oe)	4700
Energy Loss W_c (watt-sec/cycle/lb)	-
Importance of Operating Magnet at BH_{max}	E
Average Recoil Permeability	1.04
Field Strength Required to Fully Saturate Magnet (Oe)	32,000
Ability to Withstand Demagnetizing Fields	0
Maximum Service Temperature (°F)	482
Approximate Temperature Permanently Affecting Magnet (°F)	1472
Inherent Orientation of Material	MP

PHYSICAL PROPERTIES	
Density (lbs./in.^{3})	.3035
General Mechanical Properties	HB
Tensile Strength (psi)	Est. 5000
Transverse Modulus of Rupture (psi)	-
Coeff. of Therm. Expansion inches x 10^{-6}/°C	Orientation: Par. - Rt.Angl. -
Resistivity (micro-ohms/cm/cm^2 - @ - °C	
Hardness Scale 550-700	Vickers
Basic Forming Method	DPM
Basic Finishing Method	WGR
Relative Corrosion Resistance	G
Primary Chemical Composition	Sm-Co
Special Notes:	

Note: All information based on available producers literature. In some cases of foreign materials interpretation or calculation required.

SOURCE(S):		
Thyssen Magnettechnik GMBH		**Secolit 215**
		(Alloy or Grade)
		Sintered Rare Earth
		(Type of Material)
Germany		26/26
		(MMPA Brief Designation)
		R5-2
		(IEC Code Reference)

MAGNETIC PROPERTIES		PHYSICAL PROPERTIES	
Residual Induction B_r (G)	11,000	Density (lbs./in.^{3})	.2999
Coercive Force (Oe) — Normal (H_c)	10,200	General Mechanical Properties	HB
Coercive Force (Oe) — Intrinsic (H_{ci})	26,000	Tensile Strength (psi)	Est. 5000
Maximum Energy Product BH_{max} (MGO)	26.0	Transverse Modulus of Rupture (psi)	-
B/H Ratio (Load Line) at BH_{max}	1.15	Coeff. of Therm. Expansion inches x 10^{-6}/°C	Orientation: Par. - Rt.Angl. -
Flux Density in Magnet B_m at BH_{max} (G)	5500	Resistivity (micro-ohms/cm/cm^2	- @ - °C
Coercive Force in Magnet H_m at BH_{max} (Oe)	4800	Hardness Scale	600-750 Vickers
Energy Loss W_c (watt-sec/cycle/lb)	-	Basic Forming Method	DPM
Importance of Operating Magnet at BH_{max}	E	Basic Finishing Method	WGR
Average Recoil Permeability	1.07	Relative Corrosion Resistance	G
Field Strength Required to Fully Saturate Magnet (Oe)	40,000	Primary Chemical Composition	Sm-Co
Ability to Withstand Demagnetizing Fields	0	Special Notes:	
Maximum Service Temperature (°F)	572		
Approximate Temperature Permanently Affecting Magnet (°F)	1472		
Inherent Orientation of Material	MP		

Note: All information based on available producers literature. In some cases of foreign materials interpretation or calculation required.

SOURCE(S):	Secolit 215N
	(Alloy or Grade)
Thyssen Magnettechnik GMBH	Sintered Rare Earth
	(Type of Material)
Germany	26/10
	(MMPA Brief Designation)
	R5-2
	(IEC Code Reference)

MAGNETIC PROPERTIES		PHYSICAL PROPERTIES	
Residual Induction B_r (G)	11,000	Density (lbs./in.^{3})	.2999
Coercive Force (Oe) — Normal (H_c)	7500	General Mechanical Properties	HB
Coercive Force (Oe) — Intrinsic (H_{ci})	10,000	Tensile Strength (psi)	Est. 5000
Maximum Energy Product BH_{max} (MGO)	26.0	Transverse Modulus of Rupture (psi)	—
B/H Ratio (Load Line) at BH_{max}	1.15	Coeff. of Therm. Expansion inches x 10^{-6}/°C	Orientation: Par. — Rt.Angl. —
Flux Density in Magnet B_m at BH_{max} (G)	5500	Resistivity (micro-ohms/cm/cm^2	— @ — °C
Coercive Force in Magnet H_m at BH_{max} (Oe)	4800	Hardness Scale	600-750 Vickers
Energy Loss W_c (watt-sec/cycle/lb)	—	Basic Forming Method	DPM
Importance of Operating Magnet at BH_{max}	E	Basic Finishing Method	WGR
Average Recoil Permeability	1.07	Relative Corrosion Resistance	G
Field Strength Required to Fully Saturate Magnet (Oe)	25,000	Primary Chemical Composition	Sm-Co
Ability to Withstand Demagnetizing Fields	0	Special Notes:	
Maximum Service Temperature (°F)	572		
Approximate Temperature Permanently Affecting Magnet (°F)	1472		
Inherent Orientation of Material	MP		

Note: All information based on available producers literature. In some cases of foreign materials interpretation or calculation required.

SOURCE(S):		**Sprox HT3/19P**
		(Alloy or Grade)
Magnetfabrik-Bonn		Bonded Ferrite
		(Type of Material)
Germany		.35/2.8
		(MMPA Brief Designation)
		-
		(IEC Code Reference)

MAGNETIC PROPERTIES		PHYSICAL PROPERTIES	
Residual Induction B_r (G)	1350	Density (lbs./in.^{3})	.1301
Coercive Force (Oe) — Normal (H_c)	1000	General Mechanical Properties	FL
Coercive Force (Oe) — Intrinsic (H_{ci})	2300	Tensile Strength (psi)	-
Maximum Energy Product BH_{max} (MGO)	.35	Transverse Modulus of Rupture (psi)	-
B/H Ratio (Load Line) at BH_{max}	1.14	Coeff. of Therm. Expansion inches x 10^{-6}/°C	Orientation: Par. - Rt.Angl. -
Flux Density in Magnet B_m at BH_{max} (G)	640	Resistivity (micro-ohms/cm/cm^2)	- @ - °C
Coercive Force in Magnet H_m at BH_{max} (Oe)	560	Hardness Scale	- -
Energy Loss W_c (watt-sec/cycle/lb)	-	Basic Forming Method	E
Importance of Operating Magnet at BH_{max}	E	Basic Finishing Method	SCD/SRD
Average Recoil Permeability	1.05	Relative Corrosion Resistance	E
Field Strength Required to Fully Saturate Magnet (Oe)	10,000	Primary Chemical Composition	Ba, Sr or Pb-Fe + binder
Ability to Withstand Demagnetizing Fields	Ex	Special Notes: (1) Maximum service temperature depends on binder used. Binder not identified by manufacturer.	
Maximum Service Temperature (°F)	See Note		
Approximate Temperature Permanently Affecting Magnet (°F)	-		
Inherent Orientation of Material	none		

Note: All information based on available producers literature. In some cases of foreign materials interpretation or calculation required.

SOURCE(S):

Magnetfabrik-Bonn

Germany

Sprox HT11/21P
(Alloy or Grade)

Bonded Ferrite
(Type of Material)

1.3/2.6
(MMPA Brief Designation)

-
(IEC Code Reference)

MAGNETIC PROPERTIES	
Residual Induction B_r (G)	2300
Coercive Force (Oe) — Normal (H_c)	2000
Coercive Force (Oe) — Intrinsic (H_{ci})	2600
Maximum Energy Product BH_{max} (MGO)	1.3
B/H Ratio (Load Line) at BH_{max}	.96
Flux Density in Magnet B_m at BH_{max} (G)	1100
Coercive Force in Magnet H_m at BH_{max} (Oe)	1150
Energy Loss W_c (watt-sec/cycle/lb)	-
Importance of Operating Magnet at BH_{max}	E
Average Recoil Permeability	1.05
Field Strength Required to Fully Saturate Magnet (Oe)	10,000
Ability to Withstand Demagnetizing Fields	Ex
Maximum Service Temperature (°F)	See Note
Approximate Temperature Permanently Affecting Magnet (°F)	-
Inherent Orientation of Material	MP

PHYSICAL PROPERTIES	
Density (lbs./in.^{3})	.1156
General Mechanical Properties	FL
Tensile Strength (psi)	-
Transverse Modulus of Rupture (psi)	-
Coeff. of Therm. Expansion inches x 10^{-6}/°C — Orientation: Par.	-
Coeff. of Therm. Expansion inches x 10^{-6}/°C — Rt.Angl.	-
Resistivity (micro-ohms/cm/cm^2) @ °C	-
Hardness / Scale	- / -
Basic Forming Method	E
Basic Finishing Method	SCD/SRD
Relative Corrosion Resistance	E
Primary Chemical Composition	Ba, Sr or Pb-Fe + binder

Special Notes:

(1) Maximum service temperature depends on binder used. Binder not identified by manufacturer.

Note: All information based on available producers literature. In some cases of foreign materials interpretation or calculation required.

SOURCE(S):

Magnetfabrik-Bonn

Germany

Sprox HT14/21P
(Alloy or Grade)

Bonded Ferrite
(Type of Material)

1.8/2.6
(MMPA Brief Designation)

-
(IEC Code Reference)

MAGNETIC PROPERTIES		PHYSICAL PROPERTIES	
Residual Induction B_r (G)	2600	Density (lbs./in.^{3})	.1264
Coercive Force (Oe) Normal (H_c)	2200	General Mechanical Properties	FL
Coercive Force (Oe) Intrinsic (H_{ci})	2600	Tensile Strength (psi)	-
Maximum Energy Product BH_{max} (MGO)	1.75	Transverse Modulus of Rupture (psi)	-
B/H Ratio (Load Line) at BH_{max}	1.25	Coeff. of Therm. Expansion inches x 10^{-6}/°C	Orientation: Par. - Rt.Angl. -
Flux Density in Magnet B_m at BH_{max} (G)	1500	Resistivity (micro-ohms/cm/cm^2	- @ - °C
Coercive Force in Magnet H_m at BH_{max} (Oe)	1200	Hardness Scale	- -
Energy Loss W_c (watt-sec/cycle/lb)	-	Basic Forming Method	E
Importance of Operating Magnet at BH_{max}	E	Basic Finishing Method	SCD/SRD
Average Recoil Permeability	1.05	Relative Corrosion Resistance	E
Field Strength Required to Fully Saturate Magnet (Oe)	10,000	Primary Chemical Composition	Ba, Sr or Pb-Fe + binder
Ability to Withstand Demagnetizing Fields	Ex	Special Notes: (1) Maximum service temperature depends on binder used. Binder not identified by manufacturer.	
Maximum Service Temperature (°F)	See Note		
Approximate Temperature Permanently Affecting Magnet (°F)	-		
Inherent Orientation of Material	none		

Note: All information based on available producers literature. In some cases of foreign materials interpretation or calculation required.

SOURCE(S):

Magnetfabrik-Bonn

Germany

Sprox HT16/19P
(Alloy or Grade)

Bonded Ferrite
(Type of Material)

1.9/2.4
(MMPA Brief Designation)

-
(IEC Code Reference)

MAGNETIC PROPERTIES	
Residual Induction B_r (G)	2800
Coercive Force (Oe) — Normal (H_c)	2200
Coercive Force (Oe) — Intrinsic (H_{ci})	2400
Maximum Energy Product BH_{max} (MGO)	1.9
B/H Ratio (Load Line) at BH_{max}	1.28
Flux Density in Magnet B_m at BH_{max} (G)	1600
Coercive Force in Magnet H_m at BH_{max} (Oe)	1250
Energy Loss W_c (watt-sec/cycle/lb)	-
Importance of Operating Magnet at BH_{max}	E
Average Recoil Permeability	1.05
Field Strength Required to Fully Saturate Magnet (Oe)	10,000
Ability to Withstand Demagnetizing Fields	Ex
Maximum Service Temperature (°F)	See Note
Approximate Temperature Permanently Affecting Magnet (°F)	-
Inherent Orientation of Material	none

PHYSICAL PROPERTIES	
Density (lbs./in.^{3})	.1337
General Mechanical Properties	FL
Tensile Strength (psi)	-
Transverse Modulus of Rupture (psi)	-
Coeff. of Therm. Expansion inches x 10^{-6}/°C — Orientation: Par.	-
Coeff. of Therm. Expansion inches x 10^{-6}/°C — Rt.Angl.	-
Resistivity (micro-ohms/cm/cm^2) @ °C	-
Hardness / Scale	-
Basic Forming Method	E
Basic Finishing Method	SCD/SRD
Relative Corrosion Resistance	E
Primary Chemical Composition	Ba, Sr or Pb-Fe + binder

Special Notes:

(1) Maximum service temperature depends on binder used. Binder not identified by manufacturer.

Note: All information based on available producers literature. In some cases of foreign materials interpretation or calculation required.

SOURCE(S):	
Magnetfabrik-Bonn Germany	**Sprox 2FE** (Alloy or Grade)
	Bonded Ferrite (Type of Material)
	.5/2.4 (MMPA Brief Designation)
	- (IEC Code Reference)

MAGNETIC PROPERTIES		PHYSICAL PROPERTIES	
Residual Induction B_r (G)	1500	Density (lbs./in.^{3})	.1337
Coercive Force (Oe) Normal (H_c)	1300	General Mechanical Properties	FL
Coercive Force (Oe) Intrinsic (H_{ci})	2400	Tensile Strength (psi)	-
Maximum Energy Product BH_{max} (MGO)	.5	Transverse Modulus of Rupture (psi)	-
B/H Ratio (Load Line) at BH_{max}	1.33	Coeff. of Therm. Expansion inches x 10^{-6}/°C	Orientation: Par. - Rt.Angl. -
Flux Density in Magnet B_m at BH_{max} (G)	800	Resistivity (micro-ohms/cm/cm^2)	- @ - °C
Coercive Force in Magnet H_m at BH_{max} (Oe)	600	Hardness Scale	- -
Energy Loss W_c (watt-sec/cycle/lb)	-	Basic Forming Method	E
Importance of Operating Magnet at BH_{max}	E	Basic Finishing Method	SCD/SRD
Average Recoil Permeability	1.05	Relative Corrosion Resistance	E
Field Strength Required to Fully Saturate Magnet (Oe)	10,000	Primary Chemical Composition	Ba, Sr or Pb-Fe + binder
Ability to Withstand Demagnetizing Fields	Ex	Special Notes:	
Maximum Service Temperature (°F)	See Note	(1) Maximum service temperature depends on binder used. Binder not identified by manufacturer.	
Approximate Temperature Permanently Affecting Magnet (°F)	-		
Inherent Orientation of Material	none		

Note: All information based on available producers literature. In some cases of foreign materials interpretation or calculation required.

SOURCE(S): **Magnetfabrik-Bonn** Germany	**Sprox 4Fu.4FE** (Alloy or Grade)
	Bonded Ferrite (Type of Material)
	11.5/2.6 (MMPA Brief Designation)
	- (IEC Code Reference)

MAGNETIC PROPERTIES		PHYSICAL PROPERTIES	
Residual Induction B_r (G)	2200	Density (lbs./in.^{3})	.1337
Coercive Force (Oe) — Normal (H_c)	1900	General Mechanical Properties	FL
Coercive Force (Oe) — Intrinsic (H_{ci})	2600	Tensile Strength (psi)	-
Maximum Energy Product BH_{max} (MGO)	1.15	Transverse Modulus of Rupture (psi)	-
B/H Ratio (Load Line) at BH_{max}	1.15	Coeff. of Therm. Expansion inches x 10^{-6}/°C	Orientation: Par. - Rt.Angl. -
Flux Density in Magnet B_m at BH_{max} (G)	1150	Resistivity (micro-ohms/cm/cm^2)	- @ - °C
Coercive Force in Magnet H_m at BH_{max} (Oe)	1000	Hardness -	Scale -
Energy Loss W_c (watt-sec/cycle/lb)	-	Basic Forming Method	E
Importance of Operating Magnet at BH_{max}	E	Basic Finishing Method	SCD/SRD
Average Recoil Permeability	1.05	Relative Corrosion Resistance	E
Field Strength Required to Fully Saturate Magnet (Oe)	10,000	Primary Chemical Composition	Ba, Sr or Pb-Fe + binder
Ability to Withstand Demagnetizing Fields	Ex	Special Notes: (1) Maximum service temperature depends on binder used. Binder not identified by manufacturer.	
Maximum Service Temperature (°F)	See Note		
Approximate Temperature Permanently Affecting Magnet (°F)	-		
Inherent Orientation of Material	none		

Note: All information based on available producers literature. In some cases of foreign materials interpretation or calculation required.

SOURCE(S):	Sprox 5F
Magnetfabrik-Bonn	(Alloy or Grade)
	Bonded Ferrite
	(Type of Material)
Germany	14/28
	(MMPA Brief Designation)
	-
	(IEC Code Reference)

MAGNETIC PROPERTIES		PHYSICAL PROPERTIES	
Residual Induction B_r (G)	2400	Density (lbs./in.^{3})	.1337
Coercive Force (Oe) Normal (H_c)	2100	General Mechanical Properties	FL
Coercive Force (Oe) Intrinsic (H_{ci})	2800	Tensile Strength (psi)	-
Maximum Energy Product BH_{max} (MGO)	1.4	Transverse Modulus of Rupture (psi)	-
B/H Ratio (Load Line) at BH_{max}	1.04	Coeff. of Therm. Expansion inches x 10^{-6}/°C	Orientation: Par. - Rt.Angl. -
Flux Density in Magnet B_m at BH_{max} (G)	1250	Resistivity (micro-ohms/cm/cm^2)	- @ - °C
Coercive Force in Magnet H_m at BH_{max} (Oe)	1200	Hardness Scale	- -
Energy Loss W_c (watt-sec/cycle/lb)	-	Basic Forming Method	E
Importance of Operating Magnet at BH_{max}	E	Basic Finishing Method	SCD/SRD
Average Recoil Permeability	1.05	Relative Corrosion Resistance	E
Field Strength Required to Fully Saturate Magnet (Oe)	10,000	Primary Chemical Composition	Ba, Sr or Pb-Fe + binder
Ability to Withstand Demagnetizing Fields	Ex	Special Notes: (1) Maximum service temperature depends on binder used. Binder not identified by manufacturer.	
Maximum Service Temperature (°F)	See Note		
Approximate Temperature Permanently Affecting Magnet (°F)	-		
Inherent Orientation of Material	none		

Note: All information based on available producers literature. In some cases of foreign materials interpretation or calculation required.

SOURCE(S):	STAB 0
	(Alloy or Grade)
UGIMag, Inc.	Sintered Rare Earth
	(Type of Material)
USA	10/16
	(MMPA Brief Designation)
	R5-1
	(IEC Code Reference)

MAGNETIC PROPERTIES		PHYSICAL PROPERTIES	
Residual Induction B_r (G)	6400	Density (lbs./in.^{3})	Est. .3000
Coercive Force (Oe) — Normal (H_c)	5900	General Mechanical Properties	HB
Coercive Force (Oe) — Intrinsic (H_{ci})	16,000	Tensile Strength (psi)	Est. 5000
Maximum Energy Product BH_{max} (MGO)	10.0	Transverse Modulus of Rupture (psi)	-
B/H Ratio (Load Line) at BH_{max}	-	Coeff. of Therm. Expansion inches x 10^{-6}/°C	Orientation: Par. - Rt.Angl. -
Flux Density in Magnet B_m at BH_{max} (G)	-	Resistivity (micro-ohms/cm/cm^2)	- @ - °C
Coercive Force in Magnet H_m at BH_{max} (Oe)	-	Hardness -	Scale -
Energy Loss W_c (watt-sec/cycle/lb)	-	Basic Forming Method	DP
Importance of Operating Magnet at BH_{max}	E	Basic Finishing Method	WGR
Average Recoil Permeability	1.05	Relative Corrosion Resistance	G
Field Strength Required to Fully Saturate Magnet (Oe)	40,000	Primary Chemical Composition	Sm-Co
Ability to Withstand Demagnetizing Fields	0	Special Notes:	
Maximum Service Temperature (°F)	Est. 482		
Approximate Temperature Permanently Affecting Magnet (°F)	Est. 1328		
Inherent Orientation of Material	none		

Note: All information based on available producers literature. In some cases of foreign materials interpretation or calculation required.

SOURCE(S):

UGIMag, Inc.

USA

STAB 0.02
(Alloy or Grade)

Sintered Rare Earth
(Type of Material)

15/16
(MMPA Brief Designation)

-
(IEC Code Reference)

MAGNETIC PROPERTIES		PHYSICAL PROPERTIES	
Residual Induction B_r (G)	7700	Density (lbs./in.^{3})	Est. .3000
Coercive Force (Oe) Normal (H_c)	7200	General Mechanical Properties	HB
Coercive Force (Oe) Intrinsic (H_{ci})	16,000	Tensile Strength (psi)	Est. 5000
Maximum Energy Product BH_{max} (MGO)	15.0	Transverse Modulus of Rupture (psi)	-
B/H Ratio (Load Line) at BH_{max}	-	Coeff. of Therm. Expansion inches x 10^{-6}/°C	Orientation: Par. - Rt.Angl. -
Flux Density in Magnet B_m at BH_{max} (G)	-	Resistivity (micro-ohms/cm/cm^2)	- @ - °C
Coercive Force in Magnet H_m at BH_{max} (Oe)	-	Hardness Scale	- -
Energy Loss W_c (watt-sec/cycle/lb)	-	Basic Forming Method	DP
Importance of Operating Magnet at BH_{max}	E	Basic Finishing Method	WGR
Average Recoil Permeability	1.05	Relative Corrosion Resistance	G
Field Strength Required to Fully Saturate Magnet (Oe)	40,000	Primary Chemical Composition	Sm-Co
Ability to Withstand Demagnetizing Fields	0	Special Notes:	
Maximum Service Temperature (°F)	Est. 482		
Approximate Temperature Permanently Affecting Magnet (°F)	Est. 1328		
Inherent Orientation of Material	none		

Note: All information based on available producers literature. In some cases of foreign materials interpretation or calculation required.

SOURCE(S): **Stackpole Corp.** USA

Stabon E15N
(Alloy or Grade)

Bonded Ferrite
(Type of Material)

1.6/3.1
(MMPA Brief Designation)

-
(IEC Code Reference)

MAGNETIC PROPERTIES	
Residual Induction B_r (G)	2600
Coercive Force (Oe) Normal (H_c)	2100
Coercive Force (Oe) Intrinsic (H_{ci})	3100
Maximum Energy Product BH_{max} (MGO)	1.55
B/H Ratio (Load Line) at BH_{max}	-
Flux Density in Magnet B_m at BH_{max} (G)	-
Coercive Force in Magnet H_m at BH_{max} (Oe)	-
Energy Loss W_c (watt-sec/cycle/lb)	-
Importance of Operating Magnet at BH_{max}	E
Average Recoil Permeability	-
Field Strength Required to Fully Saturate Magnet (Oe)	10,000
Ability to Withstand Demagnetizing Fields	Ex
Maximum Service Temperature (°F)	See Note
Approximate Temperature Permanently Affecting Magnet (°F)	-
Inherent Orientation of Material	none

PHYSICAL PROPERTIES	
Density (lbs./in.^{3})	-
General Mechanical Properties	HT
Tensile Strength (psi)	-
Transverse Modulus of Rupture (psi)	-
Coeff. of Therm. Expansion inches x 10^{-6}/°C	Orientation: Par. - Rt.Angl. -
Resistivity (micro-ohms/cm/cm^2)	- @ - °C
Hardness -	Scale -
Basic Forming Method	E
Basic Finishing Method	SRD
Relative Corrosion Resistance	E
Primary Chemical Composition	Ba or Sr + binder

Special Notes:

(1) Maximum service temperature depends on binder used. Binder not identified by manufacturer.

Note: All information based on available producers literature. In some cases of foreign materials interpretation or calculation required.

SOURCE(S):	**Stabon E120**
	(Alloy or Grade)
Stackpole Corp.	Bonded Ferrite
	(Type of Material)
USA	1.2/2.8
	(MMPA Brief Designation)
	-
	(IEC Code Reference)

MAGNETIC PROPERTIES		PHYSICAL PROPERTIES	
Residual Induction B_r (G)	2250	Density (lbs./in.^{3})	.1337
Coercive Force (Oe) — Normal (H_c)	2000	General Mechanical Properties	FL
Coercive Force (Oe) — Intrinsic (H_{ci})	2800	Tensile Strength (psi)	-
Maximum Energy Product BH_{max} (MGO)	1.2	Transverse Modulus of Rupture (psi)	-
B/H Ratio (Load Line) at BH_{max}	-	Coeff. of Therm. Expansion inches x 10^{-6}/°C	Orientation: Par. - Rt.Angl. -
Flux Density in Magnet B_m at BH_{max} (G)	-	Resistivity (micro-ohms/cm/cm^2)	- @ - °C
Coercive Force in Magnet H_m at BH_{max} (Oe)	-	Hardness Scale 65	Shore D
Energy Loss W_c (watt-sec/cycle/lb)	-	Basic Forming Method	E
Importance of Operating Magnet at BH_{max}	E	Basic Finishing Method	SRD
Average Recoil Permeability	-	Relative Corrosion Resistance	E
Field Strength Required to Fully Saturate Magnet (Oe)	10,000	Primary Chemical Composition	Ba or Sr-Fe with binder
Ability to Withstand Demagnetizing Fields	Ex	Special Notes: (1) Extruded material (2) Maximum service temperature depends on binder used. Binder not identified by manufacturer.	
Maximum Service Temperature (°F)	200		
Approximate Temperature Permanently Affecting Magnet (°F)	-		
Inherent Orientation of Material	none		

Note: All information based on available producers literature. In some cases of foreign materials interpretation or calculation required.

SOURCE(S):	**Stabon E140**
	(Alloy or Grade)
Stackpole Corp.	Bonded Ferrite
	(Type of Material)
USA	1.4/3.0
	(MMPA Brief Designation)
	-
	(IEC Code Reference)

MAGNETIC PROPERTIES		PHYSICAL PROPERTIES	
Residual Induction B_r (G)	2450	Density (lbs./in.^{3})	.1337
Coercive Force (Oe) — Normal (H_c)	2200	General Mechanical Properties	FL
Coercive Force (Oe) — Intrinsic (H_{ci})	3000	Tensile Strength (psi)	-
Maximum Energy Product BH_{max} (MGO)	1.4	Transverse Modulus of Rupture (psi)	-
B/H Ratio (Load Line) at BH_{max}	-	Coeff. of Therm. Expansion inches x 10^{-6}/°C — Orientation: Par. — Rt.Angl.	- / -
Flux Density in Magnet B_m at BH_{max} (G)	-	Resistivity (micro-ohms/cm/cm^2 - @ - °C)	
Coercive Force in Magnet H_m at BH_{max} (Oe)	-	Hardness — Scale	- / -
Energy Loss W_c (watt-sec/cycle/lb)	-	Basic Forming Method	E
Importance of Operating Magnet at BH_{max}	E	Basic Finishing Method	SRD
Average Recoil Permeability	-	Relative Corrosion Resistance	E
Field Strength Required to Fully Saturate Magnet (Oe)	10,000	Primary Chemical Composition	Ba or Sr-Fe
Ability to Withstand Demagnetizing Fields	Ex	Special Notes: (1) Maximum service temperature depends on binder used. Binder not identified by manufacturer.	
Maximum Service Temperature (°F)	200		
Approximate Temperature Permanently Affecting Magnet (°F)	-		
Inherent Orientation of Material	none		

Note: All information based on available producers literature. In some cases of foreign materials interpretation or calculation required.

SOURCE(S):

Stackpole Corp.

USA

Stabon IM140
(Alloy or Grade)

Bonded Ferrite
(Type of Material)

1.4/3.0
(MMPA Brief Designation)

-
(IEC Code Reference)

MAGNETIC PROPERTIES		PHYSICAL PROPERTIES	
Residual Induction B_r (G)	2450	Density (lbs./in.^{3})	.1228
Coercive Force (Oe) — Normal (H_c)	2200	General Mechanical Properties	HT
Coercive Force (Oe) — Intrinsic (H_{ci})	3000	Tensile Strength (psi)	-
Maximum Energy Product BH_{max} (MGO)	1.4	Transverse Modulus of Rupture (psi)	-
B/H Ratio (Load Line) at BH_{max}	-	Coeff. of Therm. Expansion inches x 10^{-6}/°C — Orientation: Par. / Rt.Angl.	- / -
Flux Density in Magnet B_m at BH_{max} (G)	-	Resistivity (micro-ohms/cm/cm^2) @ - °C	-
Coercive Force in Magnet H_m at BH_{max} (Oe)	-	Hardness Scale	- / -
Energy Loss W_c (watt-sec/cycle/lb)	-	Basic Forming Method	E
Importance of Operating Magnet at BH_{max}	E	Basic Finishing Method	WGR
Average Recoil Permeability	-	Relative Corrosion Resistance	E
Field Strength Required to Fully Saturate Magnet (Oe)	10,000	Primary Chemical Composition	Ba or Sr-Fe
Ability to Withstand Demagnetizing Fields	Ex	Special Notes:	
Maximum Service Temperature (°F)	See Note	(1) Injection molded with nylon or PPS binder	
Approximate Temperature Permanently Affecting Magnet (°F)	-	(2) Maximum service temperature depends on binder used. Binder not identified by manufacturer.	
Inherent Orientation of Material	none		

Note: All information based on available producers literature. In some cases of foreign materials interpretation or calculation required.

SOURCE(S):

Stackpole Corp.

USA

Stabon IM160
(Alloy or Grade)

Bonded Ferrite
(Type of Material)

1.6/2.8
(MMPA Brief Designation)

-
(IEC Code Reference)

MAGNETIC PROPERTIES		PHYSICAL PROPERTIES	
Residual Induction B_r (G)	2550	Density (lbs./in.^{3})	.1264
Coercive Force (Oe) — Normal (H_c)	2300	General Mechanical Properties	HT
Coercive Force (Oe) — Intrinsic (H_{ci})	2800	Tensile Strength (psi)	-
Maximum Energy Product BH_{max} (MGO)	1.6	Transverse Modulus of Rupture (psi)	-
B/H Ratio (Load Line) at BH_{max}	-	Coeff. of Therm. Expansion inches x 10^{-6}/°C — Orientation: Par. _ Rt.Angl. _	
Flux Density in Magnet B_m at BH_{max} (G)	-	Resistivity (micro-ohms/cm/cm^2) - @ - °C	
Coercive Force in Magnet H_m at BH_{max} (Oe)	-	Hardness - Scale -	
Energy Loss W_c (watt-sec/cycle/lb)	-	Basic Forming Method	E
Importance of Operating Magnet at BH_{max}	E	Basic Finishing Method	WGR
Average Recoil Permeability	-	Relative Corrosion Resistance	E
Field Strength Required to Fully Saturate Magnet (Oe)	10,000	Primary Chemical Composition	Ba or Sr-Fe
Ability to Withstand Demagnetizing Fields	Ex	Special Notes: (1) Injection molded with nylon or PPS binder. (2) Maximum service temperature depends on binder used. Binder not identified by manufacturer.	
Maximum Service Temperature (°F)	See Note		
Approximate Temperature Permanently Affecting Magnet (°F)	-		
Inherent Orientation of Material	none		

Note: All information based on available producers literature. In some cases of foreign materials interpretation or calculation required.

SOURCE(S):	**Stabon IM180**
	(Alloy or Grade)
Stackpole Corp.	Bonded Ferrite
	(Type of Material)
USA	1.8/2.5
	(MMPA Brief Designation)
	-
	(IEC Code Reference)

MAGNETIC PROPERTIES		PHYSICAL PROPERTIES	
Residual Induction B_r (G)	2700	Density (lbs./in.^{3})	.1337
Coercive Force (Oe) — Normal (H_c)	2280	General Mechanical Properties	-
Coercive Force (Oe) — Intrinsic (H_{ci})	2480	Tensile Strength (psi)	-
Maximum Energy Product BH_{max} (MGO)	1.8	Transverse Modulus of Rupture (psi)	-
B/H Ratio (Load Line) at BH_{max}	-	Coeff. of Therm. Expansion inches x 10^{-6}/°C	Orientation: Par. - Rt.Angl. -
Flux Density in Magnet B_m at BH_{max} (G)	-	Resistivity (micro-ohms/cm/cm^2)	- @ - °C
Coercive Force in Magnet H_m at BH_{max} (Oe)	-	Hardness - Scale -	
Energy Loss W_c (watt-sec/cycle/lb)	-	Basic Forming Method	E
Importance of Operating Magnet at BH_{max}	E	Basic Finishing Method	WGR
Average Recoil Permeability	-	Relative Corrosion Resistance	E
Field Strength Required to Fully Saturate Magnet (Oe)	10,000	Primary Chemical Composition	Ba or Sr-Fe
Ability to Withstand Demagnetizing Fields	Ex	Special Notes: (1) Injection molded with nylon bond. (2) Maximum service temperature depends on binder used. Binder not identified by manufacturer.	
Maximum Service Temperature (°F)	See Note		
Approximate Temperature Permanently Affecting Magnet (°F)	-		
Inherent Orientation of Material	none		

Note: All information based on available producers literature. In some cases of foreign materials interpretation or calculation required.

SOURCE(S):

Max Baermann GMBH

Germany

Tromadur 3/16P
(Alloy or Grade)

Bonded Ferrite
(Type of Material)

.38/2.0
(MMPA Brief Designation)

-
(IEC Code Reference)

MAGNETIC PROPERTIES			PHYSICAL PROPERTIES	
Residual Induction B_r (G)		1350	Density (lbs./in.^{3})	-
Coercive Force (Oe)	Normal (H_c)	1131	General Mechanical Properties	HT
	Intrinsic (H_{ci})	2011	Tensile Strength (psi)	-
Maximum Energy Product BH_{max}(MGO)		.38	Transverse Modulus of Rupture (psi)	-
B/H Ratio (Load Line) at BH_{max}		-	Coeff. of Therm. Expansion inches x 10^{-6}/°C	Orientation: Par. - Rt.Angl. -
Flux Density in Magnet B_m at BH_{max} (G)		-	Resistivity (micro-ohms/cm/cm^2)	- @ - °C
Coercive Force in Magnet H_m at BH_{max} (Oe)		-	Hardness Scale	-
Energy Loss W_c (watt-sec/cycle/lb)		-	Basic Forming Method	E
Importance of Operating Magnet at BH_{max}		D	Basic Finishing Method	WGD
Average Recoil Permeability		-	Relative Corrosion Resistance	E
Field Strength Required to Fully Saturate Magnet (Oe)		10,000	Primary Chemical Composition	Ba or Sr-Fe + binder
Ability to Withstand Demagnetizing Fields		E	Special Notes: (1) Thermoplastic Binder (2) Injection molded	
Maximum Service Temperature (°F)		-		
Approximate Temperature Permanently Affecting Magnet (°F)		-		
Inherent Orientation of Material		none		

Note: All information based on available producers literature. In some cases of foreign materials interpretation or calculation required.

SOURCE(S):	Tromadur 9/21P
	(Alloy or Grade)
Max Baermann GMBH	Bonded Ferrite
	(Type of Material)
Germany	1.1/2.7
	(MMPA Brief Designation)
	-
	(IEC Code Reference)

MAGNETIC PROPERTIES		PHYSICAL PROPERTIES	
Residual Induction B_r (G)	2200	Density (lbs./in.^{3})	-
Coercive Force (Oe) Normal (H_c)	1847	General Mechanical Properties	HT
Coercive Force (Oe) Intrinsic (H_{ci})	2700	Tensile Strength (psi)	-
Maximum Energy Product BH_{max} (MGO)	1.1	Transverse Modulus of Rupture (psi)	-
B/H Ratio (Load Line) at BH_{max}	-	Coeff. of Therm. Expansion inches x 10^{-6}/°C Orientation: Par. - Rt.Angl. -	
Flux Density in Magnet B_m at BH_{max} (G)	-	Resistivity (micro-ohms/cm/cm^2 - @ - °C	
Coercive Force in Magnet H_m at BH_{max} (Oe)	-	Hardness Scale - -	
Energy Loss W_c (watt-sec/cycle/lb)	-	Basic Forming Method	E
Importance of Operating Magnet at BH_{max}	E	Basic Finishing Method	WGD
Average Recoil Permeability	-	Relative Corrosion Resistance	E
Field Strength Required to Fully Saturate Magnet (Oe)	10,000	Primary Chemical Composition	Ba or Sr-Fe + binder
Ability to Withstand Demagnetizing Fields	E	Special Notes: (1) Thermoplastic Binder (2) Injection molded	
Maximum Service Temperature (°F)	-		
Approximate Temperature Permanently Affecting Magnet (°F)	-		
Inherent Orientation of Material	MP		

Note: All information based on available producers literature. In some cases of foreign materials interpretation or calculation required.

SOURCE(S):	
Max Baermann GMBH	**Tromadur 9/28P** (Alloy or Grade)
	Bonded Ferrite (Type of Material)
Germany	1.1/3.5 (MMPA Brief Designation)
	- (IEC Code Reference)

MAGNETIC PROPERTIES		PHYSICAL PROPERTIES	
Residual Induction B_r (G)	2200	Density (lbs./in.^{3})	.1264
Coercive Force (Oe) — Normal (H_c)	2136	General Mechanical Properties	FL
Coercive Force (Oe) — Intrinsic (H_{ci})	3519	Tensile Strength (psi)	-
Maximum Energy Product BH_{max} (MGO)	1.1	Transverse Modulus of Rupture (psi)	-
B/H Ratio (Load Line) at BH_{max}	-	Coeff. of Therm. Expansion inches x 10^{-6}/°C	Orientation: Par. - Rt.Angl. -
Flux Density in Magnet B_m at BH_{max} (G)	-	Resistivity (micro-ohms/cm/cm^2	- @ - °C
Coercive Force in Magnet H_m at BH_{max} (Oe)	-	Hardness Scale	- -
Energy Loss W_c (watt-sec/cycle/lb)	-	Basic Forming Method	E
Importance of Operating Magnet at BH_{max}	E	Basic Finishing Method	SCD/SRD
Average Recoil Permeability	-	Relative Corrosion Resistance	E
Field Strength Required to Fully Saturate Magnet (Oe)	10,000	Primary Chemical Composition	Ba or Sr-Fe + binder
Ability to Withstand Demagnetizing Fields	E	Special Notes: (1) Highly flexible material	
Maximum Service Temperature (°F)	-		
Approximate Temperature Permanently Affecting Magnet (°F)	-		
Inherent Orientation of Material	none		

Note: All information based on available producers literature. In some cases of foreign materials interpretation or calculation required.

SOURCE(S):

Max Baermann GMBH

Germany

Tromadur 13/22P
(Alloy or Grade)

Bonded Ferrite
(Type of Material)

1.6/2.8
(MMPA Brief Designation)

-
(IEC Code Reference)

MAGNETIC PROPERTIES		PHYSICAL PROPERTIES	
Residual Induction B_r (G)	2600	Density (lbs./in.^{3})	-
Coercive Force (Oe) Normal (H_c)	2325	General Mechanical Properties	HT
Coercive Force (Oe) Intrinsic (H_{ci})	2765	Tensile Strength (psi)	-
Maximum Energy Product BH_{max} (MGO)	1.6	Transverse Modulus of Rupture (psi)	-
B/H Ratio (Load Line) at BH_{max}	-	Coeff. of Therm. Expansion inches x 10^{-6}/°C Orientation: Par. - Rt.Angl. -	
Flux Density in Magnet B_m at BH_{max} (G)	-	Resistivity (micro-ohms/cm/cm^2) - @ - °C	
Coercive Force in Magnet H_m at BH_{max} (Oe)	-	Hardness - Scale -	
Energy Loss W_c (watt-sec/cycle/lb)	-	Basic Forming Method	E
Importance of Operating Magnet at BH_{max}	E	Basic Finishing Method	WGD
Average Recoil Permeability	-	Relative Corrosion Resistance	E
Field Strength Required to Fully Saturate Magnet (Oe)	10,000	Primary Chemical Composition	Ba or Sr-Fe + binder
Ability to Withstand Demagnetizing Fields	E	Special Notes: (1) Thermoplastic Binder (2) Injection molded	
Maximum Service Temperature (°F)	-		
Approximate Temperature Permanently Affecting Magnet (°F)	-		
Inherent Orientation of Material	MP		

Note: All information based on available producers literature. In some cases of foreign materials interpretation or calculation required.

SOURCE(S):

Max Baermann GMBH

Germany

Tromadur 16/25P
(Alloy or Grade)

Bonded Ferrite
(Type of Material)

2.0/3.1
(MMPA Brief Designation)

-
(IEC Code Reference)

MAGNETIC PROPERTIES		PHYSICAL PROPERTIES	
Residual Induction B_r (G)	2950	Density (lbs./in.^{3})	-
Coercive Force (Oe) Normal (H_c)	2639	General Mechanical Properties	HT
Coercive Force (Oe) Intrinsic (H_{ci})	3142	Tensile Strength (psi)	-
Maximum Energy Product BH_{max} (MGO)	2.0	Transverse Modulus of Rupture (psi)	-
B/H Ratio (Load Line) at BH_{max}	-	Coeff. of Therm. Expansion inches x 10^{-6}/°C	Orientation: Par. - Rt.Angl. -
Flux Density in Magnet B_m at BH_{max} (G)	-	Resistivity (micro-ohms/cm/cm^2 - @ - °C)	
Coercive Force in Magnet H_m at BH_{max} (Oe)	-	Hardness Scale	- -
Energy Loss W_c (watt-sec/cycle/lb)	-	Basic Forming Method	E
Importance of Operating Magnet at BH_{max}	E	Basic Finishing Method	WGD
Average Recoil Permeability	-	Relative Corrosion Resistance	E
Field Strength Required to Fully Saturate Magnet (Oe)	10,000	Primary Chemical Composition	Ba or Sr-Fe + binder
Ability to Withstand Demagnetizing Fields	E	Special Notes: (1) Thermoplastic Binder (2) Injection molded	
Maximum Service Temperature (°F)	-		
Approximate Temperature Permanently Affecting Magnet (°F)	-		
Inherent Orientation of Material	MP		

Note: All information based on available producers literature. In some cases of foreign materials interpretation or calculation required.

SOURCE(S): **Max Baermann GMBH** Germany

Tromadym 35/70P
(Alloy or Grade)

Bonded Rare Earth
(Type of Material)

4.4/8.8
(MMPA Brief Designation)

-
(IEC Code Reference)

MAGNETIC PROPERTIES	
Residual Induction B_r (G)	4700
Coercive Force (Oe) — Normal (H_c)	3770
Coercive Force (Oe) — Intrinsic (H_{ci})	8796
Maximum Energy Product BH_{max} (MGO)	4.4
B/H Ratio (Load Line) at BH_{max}	-
Flux Density in Magnet B_m at BH_{max} (G)	-
Coercive Force in Magnet H_m at BH_{max} (Oe)	-
Energy Loss W_c (watt-sec/cycle/lb)	-
Importance of Operating Magnet at BH_{max}	E
Average Recoil Permeability	-
Field Strength Required to Fully Saturate Magnet (Oe)	35,000
Ability to Withstand Demagnetizing Fields	0
Maximum Service Temperature (°F)	See Note
Approximate Temperature Permanently Affecting Magnet (°F)	Est. 594
Inherent Orientation of Material	none

PHYSICAL PROPERTIES	
Density (lbs./in.^{3})	.1698
General Mechanical Properties	HT
Tensile Strength (psi)	-
Transverse Modulus of Rupture (psi)	-
Coeff. of Therm. Expansion inches x 10^{-6}/°C	Orientation: Par. - Rt.Angl. -
Resistivity (micro-ohms/cm/cm^2)	- @ - °C
Hardness / Scale	- / -
Basic Forming Method	E
Basic Finishing Method	WGR
Relative Corrosion Resistance	G
Primary Chemical Composition	Nd-Fe-B + binder

Special Notes:

(1) Binder is thermoplastic
(2) Injection molded
(3) Maximum service temperature varies with specific binder. Binder not indicated by producer.

Note: All information based on available producers literature. In some cases of foreign materials interpretation or calculation required.

SOURCE(S):

Max Baermann GMBH

Germany

Tromadym 50/70P
(Alloy or Grade)

Bonded Rare Earth
(Type of Material)

6.3/8.8
(MMPA Brief Designation)

-
(IEC Code Reference)

MAGNETIC PROPERTIES		PHYSICAL PROPERTIES	
Residual Induction B_r (G)	5500	Density (lbs./in.^{3})	.1806
Coercive Force (Oe) — Normal (H_c)	4524	General Mechanical Properties	HT
Coercive Force (Oe) — Intrinsic (H_{ci})	8796	Tensile Strength (psi)	-
Maximum Energy Product BH_{max} (MGO)	6.3	Transverse Modulus of Rupture (psi)	-
B/H Ratio (Load Line) at BH_{max}	-	Coeff. of Therm. Expansion inches x 10^{-6}/°C — Orientation: Par. — Rt.Angl.	- / -
Flux Density in Magnet B_m at BH_{max} (G)	-	Resistivity (micro-ohms/cm/cm^2) — @ — °C	-
Coercive Force in Magnet H_m at BH_{max} (Oe)	-	Hardness — Scale	- / -
Energy Loss W_c (watt-sec/cycle/lb)	-	Basic Forming Method	E
Importance of Operating Magnet at BH_{max}	E	Basic Finishing Method	WGR
Average Recoil Permeability	-	Relative Corrosion Resistance	G
Field Strength Required to Fully Saturate Magnet (Oe)	35,000	Primary Chemical Composition	Nd-Fe-B + binder
Ability to Withstand Demagnetizing Fields	0	Special Notes: (1) Thermoplastic bonded (2) Injection molded (3) Maximum service temperature varies with specific binder. Binder not indicated by producer.	
Maximum Service Temperature (°F)	See Note		
Approximate Temperature Permanently Affecting Magnet (°F)	Est. 594		
Inherent Orientation of Material	none		

Note: All information based on available producers literature. In some cases of foreign materials interpretation or calculation required.

SOURCE(S):

Max Baermann GMBH

Germany

Tromadym 60/70P
(Alloy or Grade)

Bonded Rare Earth
(Type of Material)

7.5/8.8
(MMPA Brief Designation)

-
(IEC Code Reference)

MAGNETIC PROPERTIES		PHYSICAL PROPERTIES	
Residual Induction B_r (G)	6000	Density (lbs./in.^{3})	.2095
Coercive Force (Oe) — Normal (H_c)	5027	General Mechanical Properties	HT
Coercive Force (Oe) — Intrinsic (H_{ci})	8796	Tensile Strength (psi)	-
Maximum Energy Product BH_{max} (MGO)	7.5	Transverse Modulus of Rupture (psi)	-
B/H Ratio (Load Line) at BH_{max}	-	Coeff. of Therm. Expansion inches x 10^{-6}/°C Orientation: Par. - Rt.Angl. -	
Flux Density in Magnet B_m at BH_{max} (G)	-	Resistivity (micro-ohms/cm/cm^2)	- @ - °C
Coercive Force in Magnet H_m at BH_{max} (Oe)	-	Hardness Scale	- -
Energy Loss W_c (watt-sec/cycle/lb)	-	Basic Forming Method	E
Importance of Operating Magnet at BH_{max}	E	Basic Finishing Method	WGR
Average Recoil Permeability	-	Relative Corrosion Resistance	G
Field Strength Required to Fully Saturate Magnet (Oe)	35,000	Primary Chemical Composition	Nd-Fe-B + binder
Ability to Withstand Demagnetizing Fields	0	Special Notes: (1) Thermoset bonded (2) Pressure molded (3) Maximum service temperature varies with specific binder. Binder not indicated by producer.	
Maximum Service Temperature (°F)	See Note		
Approximate Temperature Permanently Affecting Magnet (°F)	Est. 594		
Inherent Orientation of Material	none		

Note: All information based on available producers literature. In some cases of foreign materials interpretation or calculation required.

SOURCE(S):

Max Baermann GMBH

Germany

Tromadym 60/110P
(Alloy or Grade)

Bonded Rare Earth
(Type of Material)

7.5/13.8
(MMPA Brief Designation)

-
(IEC Code Reference)

MAGNETIC PROPERTIES		PHYSICAL PROPERTIES	
Residual Induction B_r (G)	6000	Density (lbs./in.^{3})	.2095
Coercive Force (Oe) — Normal (H_c)	5027	General Mechanical Properties	HT
Coercive Force (Oe) — Intrinsic (H_{ci})	13,820	Tensile Strength (psi)	-
Maximum Energy Product BH_{max} (MGO)	7.5	Transverse Modulus of Rupture (psi)	-
B/H Ratio (Load Line) at BH_{max}	-	Coeff. of Therm. Expansion inches x 10^{-6}/°C — Orientation: Par. -, Rt.Angl. -	
Flux Density in Magnet B_m at BH_{max} (G)	-	Resistivity (micro-ohms/cm/cm^2)	- @ - °C
Coercive Force in Magnet H_m at BH_{max} (Oe)	-	Hardness — Scale	- — -
Energy Loss W_c (watt-sec/cycle/lb)	-	Basic Forming Method	E
Importance of Operating Magnet at BH_{max}	E	Basic Finishing Method	WGR
Average Recoil Permeability	-	Relative Corrosion Resistance	G
Field Strength Required to Fully Saturate Magnet (Oe)	35,000	Primary Chemical Composition	Nd-Fe-B + binder
Ability to Withstand Demagnetizing Fields	0	Special Notes: (1) Thermoset bonded (2) Pressure molded (3) Maximum service temperature varies with specific binder. Binder not indicated by producer.	
Maximum Service Temperature (°F)	See Note		
Approximate Temperature Permanently Affecting Magnet (°F)	Est. 594		
Inherent Orientation of Material	none		

Note: All information based on available producers literature. In some cases of foreign materials interpretation or calculation required.

SOURCE(S):	Tromaflex 3/24P
	(Alloy or Grade)
Max Baermann GMBH	Bonded Ferrite
	(Type of Material)
Germany	.38/3.0
	(MMPA Brief Designation)
	-
	(IEC Code Reference)

MAGNETIC PROPERTIES		PHYSICAL PROPERTIES	
Residual Induction B_r (G)	1270	Density (lbs./in.^{3})	.1228
Coercive Force (Oe) — Normal (H_c)	1144	General Mechanical Properties	FL
Coercive Force (Oe) — Intrinsic (H_{ci})	3016	Tensile Strength (psi)	-
Maximum Energy Product BH_{max} (MGO)	.38	Transverse Modulus of Rupture (psi)	-
B/H Ratio (Load Line) at BH_{max}	-	Coeff. of Therm. Expansion inches x 10^{-6}/°C — Orientation: Par. / Rt.Angl.	- / -
Flux Density in Magnet B_m at BH_{max} (G)	-	Resistivity (micro-ohms/cm/cm^2)	- @ - °C
Coercive Force in Magnet H_m at BH_{max} (Oe)	-	Hardness / Scale	- / -
Energy Loss W_c (watt-sec/cycle/lb)	-	Basic Forming Method	E
Importance of Operating Magnet at BH_{max}	E	Basic Finishing Method	SCD/SRD
Average Recoil Permeability	-	Relative Corrosion Resistance	E
Field Strength Required to Fully Saturate Magnet (Oe)	10,000	Primary Chemical Composition	Ba or Sr-Fe + binder
Ability to Withstand Demagnetizing Fields	E	Special Notes: (1) Highly flexible material	
Maximum Service Temperature (°F)	-		
Approximate Temperature Permanently Affecting Magnet (°F)	-		
Inherent Orientation of Material	none		

Note: All information based on available producers literature. In some cases of foreign materials interpretation or calculation required.

SOURCE(S):

Max Baermann GMBH

Germany

Tromaflex 3/25P
(Alloy or Grade)

Bonded Ferrite
(Type of Material)

.43/3.1
(MMPA Brief Designation)

-
(IEC Code Reference)

MAGNETIC PROPERTIES	
Residual Induction B_r (G)	1420
Coercive Force (Oe) Normal (H_c)	1232
Coercive Force (Oe) Intrinsic (H_{ci})	3142
Maximum Energy Product BH_{max} (MGO)	.43
B/H Ratio (Load Line) at BH_{max}	-
Flux Density in Magnet B_m at BH_{max} (G)	-
Coercive Force in Magnet H_m at BH_{max} (Oe)	-
Energy Loss W_c (watt-sec/cycle/lb)	-
Importance of Operating Magnet at BH_{max}	E
Average Recoil Permeability	-
Field Strength Required to Fully Saturate Magnet (Oe)	10,000
Ability to Withstand Demagnetizing Fields	E
Maximum Service Temperature (°F)	-
Approximate Temperature Permanently Affecting Magnet (°F)	-
Inherent Orientation of Material	none

PHYSICAL PROPERTIES	
Density (lbs./in.^{3})	.1337
General Mechanical Properties	FL
Tensile Strength (psi)	-
Transverse Modulus of Rupture (psi)	-
Coeff. of Therm. Expansion inches x 10^{-6}/°C	Orientation: Par. - Rt.Angl. -
Resistivity (micro-ohms/cm/cm^2)	- @ - °C
Hardness / Scale	- / -
Basic Forming Method	E
Basic Finishing Method	SCD/SRD
Relative Corrosion Resistance	E
Primary Chemical Composition	Ba or Sr-Fe + binder

Special Notes:

(1) Highly flexible material

Note: All information based on available producers literature. In some cases of foreign materials interpretation or calculation required.

SOURCE(S): **Max Baermann GMBH** Germany

Tromaflex 4/24P
(Alloy or Grade)

Bonded Ferrite
(Type of Material)

.57/3.0
(MMPA Brief Designation)

-
(IEC Code Reference)

MAGNETIC PROPERTIES		PHYSICAL PROPERTIES	
Residual Induction B_r (G)	1650	Density (lbs./in.^{3})	.1373
Coercive Force (Oe) — Normal (H_c)	1382	General Mechanical Properties	FL
Coercive Force (Oe) — Intrinsic (H_{ci})	3016	Tensile Strength (psi)	-
Maximum Energy Product BH_{max} (MGO)	.57	Transverse Modulus of Rupture (psi)	-
B/H Ratio (Load Line) at BH_{max}	-	Coeff. of Therm. Expansion inches x 10^{-6}/°C — Orientation: Par. _ Rt.Angl. _	
Flux Density in Magnet B_m at BH_{max} (G)	-	Resistivity (micro-ohms/cm/cm^2 - @ - °C	
Coercive Force in Magnet H_m at BH_{max} (Oe)	-	Hardness Scale - -	
Energy Loss W_c (watt-sec/cycle/lb)	-	Basic Forming Method	E
Importance of Operating Magnet at BH_{max}	E	Basic Finishing Method	SCD/SRD
Average Recoil Permeability	-	Relative Corrosion Resistance	E
Field Strength Required to Fully Saturate Magnet (Oe)	10,000	Primary Chemical Composition	Ba or Sr-Fe + binder
Ability to Withstand Demagnetizing Fields	E	Special Notes: (1) Highly flexible material	
Maximum Service Temperature (°F)	-		
Approximate Temperature Permanently Affecting Magnet (°F)	-		
Inherent Orientation of Material	none		

Note: All information based on available producers literature. In some cases of foreign materials interpretation or calculation required.

SOURCE(S):	Tromaflex 9/28P
	(Alloy or Grade)
Max Baermann GMBH	Bonded Ferrite
	(Type of Material)
Germany	-
	(MMPA Brief Designation)
	-
	(IEC Code Reference)

MAGNETIC PROPERTIES			PHYSICAL PROPERTIES	
Residual Induction B_r (G)		2200	Density (lbs./in.^{3})	.1264
Coercive Force (Oe)	Normal (H_c)	2136	General Mechanical Properties	FL
	Intrinsic (H_{ci})	3519	Tensile Strength (psi)	-
Maximum Energy Product BH_{max} (MGO)		1.1	Transverse Modulus of Rupture (psi)	-
B/H Ratio (Load Line) at BH_{max}		-	Coeff. of Therm. Expansion inches x 10^{-6}/°C	Orientation: Par. - Rt.Angl. -
Flux Density in Magnet B_m at BH_{max} (G)		-	Resistivity (micro-ohms/cm/cm^2 - @ - °C	
Coercive Force in Magnet H_m at BH_{max} (Oe)		-	Hardness - Scale -	
Energy Loss W_c (watt-sec/cycle/lb)		-	Basic Forming Method	E
Importance of Operating Magnet at BH_{max}		E	Basic Finishing Method	SCD/SRD
Average Recoil Permeability			Relative Corrosion Resistance	E
Field Strength Required to Fully Saturate Magnet (Oe)		10,000	Primary Chemical Composition	Ba or Sr-Fe + binder
Ability to Withstand Demagnetizing Fields		E	Special Notes: (1) Flexible material	
Maximum Service Temperature (°F)		-		
Approximate Temperature Permanently Affecting Magnet (°F)		-		
Inherent Orientation of Material		none		

Note: All information based on available producers literature. In some cases of foreign materials interpretation or calculation required.

SOURCE(S):	Tromaflex 11/20P
	(Alloy or Grade)
Max Baermann GMBH	Bonded Ferrite
	(Type of Material)
Germany	1.4/2.5
	(MMPA Brief Designation)
	-
	(IEC Code Reference)

MAGNETIC PROPERTIES		PHYSICAL PROPERTIES	
Residual Induction B_r (G)	2450	Density (lbs./in.^{3})	.1337
Coercive Force (Oe) — Normal (H_c)	2086	General Mechanical Properties	FL
Coercive Force (Oe) — Intrinsic (H_{ci})	2450	Tensile Strength (psi)	-
Maximum Energy Product BH_{max} (MGO)	1.4	Transverse Modulus of Rupture (psi)	-
B/H Ratio (Load Line) at BH_{max}	-	Coeff. of Therm. Expansion inches x 10^{-6}/°C	Orientation: Par. - Rt.Angl. -
Flux Density in Magnet B_m at BH_{max} (G)	-	Resistivity (micro-ohms/cm/cm^2)	- @ - °C
Coercive Force in Magnet H_m at BH_{max} (Oe)	-	Hardness — Scale	- — -
Energy Loss W_c (watt-sec/cycle/lb)	-	Basic Forming Method	E
Importance of Operating Magnet at BH_{max}	E	Basic Finishing Method	SCD/SRD
Average Recoil Permeability	-	Relative Corrosion Resistance	E
Field Strength Required to Fully Saturate Magnet (Oe)	10,000	Primary Chemical Composition	Ba or Sr-Fe + binder
Ability to Withstand Demagnetizing Fields	E	Special Notes: (1) Highly flexible material	
Maximum Service Temperature (°F)	-		
Approximate Temperature Permanently Affecting Magnet (°F)	-		
Inherent Orientation of Material	none		

Note: All information based on available producers literature. In some cases of foreign materials interpretation or calculation required.

SOURCE(S):	Tromaflex 13/23P
Max Baermann GMBH	(Alloy or Grade)
	Bonded Ferrite
	(Type of Material)
Germany	1.6/2.9
	(MMPA Brief Designation)
	-
	(IEC Code Reference)

MAGNETIC PROPERTIES		PHYSICAL PROPERTIES	
Residual Induction B_r (G)	2700	Density (lbs./in.^{3})	.1275
Coercive Force (Oe) — Normal (H_c)	2287	General Mechanical Properties	FL
Coercive Force (Oe) — Intrinsic (H_{ci})	2890	Tensile Strength (psi)	-
Maximum Energy Product BH_{max} (MGO)	1.6	Transverse Modulus of Rupture (psi)	-
B/H Ratio (Load Line) at BH_{max}	-	Coeff. of Therm. Expansion inches x 10^{-6}/°C	Orientation: Par. - Rt.Angl. -
Flux Density in Magnet B_m at BH_{max} (G)	-	Resistivity (micro-ohms/cm/cm^2 - @ - °C	
Coercive Force in Magnet H_m at BH_{max} (Oe)	-	Hardness - Scale -	
Energy Loss W_c (watt-sec/cycle/lb)	-	Basic Forming Method	E
Importance of Operating Magnet at BH_{max}	E	Basic Finishing Method	SCD/SRD
Average Recoil Permeability	-	Relative Corrosion Resistance	E
Field Strength Required to Fully Saturate Magnet (Oe)	10,000	Primary Chemical Composition	Ba or Sr-Fe + binder
Ability to Withstand Demagnetizing Fields	E	Special Notes: (1) Highly flexible material	
Maximum Service Temperature (°F)	-		
Approximate Temperature Permanently Affecting Magnet (°F)	-		
Inherent Orientation of Material	none		

Note: All information based on available producers literature. In some cases of foreign materials interpretation or calculation required.

SOURCE(S):

Max Baermann GMBH

Germany

Tromalit 6/6P
(Alloy or Grade)

Bonded Alnico
(Type of Material)

.75/.70
(MMPA Brief Designation)

-
(IEC Code Reference)

MAGNETIC PROPERTIES		PHYSICAL PROPERTIES	
Residual Induction B_r (G)	4350	Density (lbs./in.^{3})	-
Coercive Force (Oe) — Normal (H_c)	628	General Mechanical Properties	HT
Coercive Force (Oe) — Intrinsic (H_{ci})	704	Tensile Strength (psi)	-
Maximum Energy Product BH_{max} (MGO)	.75	Transverse Modulus of Rupture (psi)	-
B/H Ratio (Load Line) at BH_{max}	-	Coeff. of Therm. Expansion inches x 10^{-6}/°C Orientation: Par. - Rt.Angl. -	
Flux Density in Magnet B_m at BH_{max} (G)	-	Resistivity (micro-ohms/cm/cm^2)	- @ - °C
Coercive Force in Magnet H_m at BH_{max} (Oe)	-	Hardness — Scale	- — -
Energy Loss W_c (watt-sec/cycle/lb)	-	Basic Forming Method	E
Importance of Operating Magnet at BH_{max}	A	Basic Finishing Method	WGR
Average Recoil Permeability	-	Relative Corrosion Resistance	G
Field Strength Required to Fully Saturate Magnet (Oe)	3000	Primary Chemical Composition	Al-Ni-Co-Cu-Fe + binder
Ability to Withstand Demagnetizing Fields	G	Special Notes: (1) Pressure molded	
Maximum Service Temperature (°F)	-		
Approximate Temperature Permanently Affecting Magnet (°F)	900		
Inherent Orientation of Material	none		

Note: All information based on available producers literature. In some cases of foreign materials interpretation or calculation required.

SOURCE(S):

Max Baermann GMBH

Germany

Tromalit 11/9P
(Alloy or Grade)

Bonded Alnico
(Type of Material)

1.4/1.1
(MMPA Brief Designation)

-
(IEC Code Reference)

MAGNETIC PROPERTIES	
Residual Induction B_r (G)	4350
Coercive Force (Oe) Normal (H_c)	1018
Coercive Force (Oe) Intrinsic (H_{ci})	1131
Maximum Energy Product BH_{max} (MGO)	1.38
B/H Ratio (Load Line) at BH_{max}	-
Flux Density in Magnet B_m at BH_{max} (G)	-
Coercive Force in Magnet H_m at BH_{max} (Oe)	-
Energy Loss W_c (watt-sec/cycle/lb)	-
Importance of Operating Magnet at BH_{max}	D
Average Recoil Permeability	-
Field Strength Required to Fully Saturate Magnet (Oe)	3000
Ability to Withstand Demagnetizing Fields	D
Maximum Service Temperature (°F)	-
Approximate Temperature Permanently Affecting Magnet (°F)	900
Inherent Orientation of Material	none

PHYSICAL PROPERTIES		
Density (lbs./in.^{3})	-	
General Mechanical Properties	HT	
Tensile Strength (psi)	-	
Transverse Modulus of Rupture (psi)	-	
Coeff. of Therm. Expansion inches x 10^{-6}/°C	Orientation: Par. Rt.Angl.	- -
Resistivity (micro-ohms/cm/cm^2)	- @ - °C	
Hardness	Scale	
-	-	
Basic Forming Method	E	
Basic Finishing Method	WGR	
Relative Corrosion Resistance	G	
Primary Chemical Composition	Al-Ni-Co-Cu-Fe + binder	
Special Notes: (1) Pressure molded		

Note: All information based on available producers literature. In some cases of foreign materials interpretation or calculation required.

SOURCE(S):

UGIMag, Inc.

USA

UGIMAX 30K
(Alloy or Grade)

Sintered Rare Earth
(Type of Material)

30/19
(MMPA Brief Designation)

R7-3
(IEC Code Reference)

MAGNETIC PROPERTIES		PHYSICAL PROPERTIES	
Residual Induction B_r (G)	11,400	Density (lbs./in.^{3})	.2710
Coercive Force (Oe) — Normal (H_c)	10,800	General Mechanical Properties	HB
Coercive Force (Oe) — Intrinsic (H_{ci})	19,000	Tensile Strength (psi)	-
Maximum Energy Product BH_{max} (MGO)	30.0	Transverse Modulus of Rupture (psi)	-
B/H Ratio (Load Line) at BH_{max}	.03	Coeff. of Therm. Expansion inches x 10^{-6}/°C — Orientation: Par. / Rt.Angl.	- / -
Flux Density in Magnet B_m at BH_{max} (G)	5000	Resistivity (micro-ohms/cm/cm^2)	150 @ 22 °C
Coercive Force in Magnet H_m at BH_{max} (Oe)	6000	Hardness / Scale	600 Vickers
Energy Loss W_c (watt-sec/cycle/lb)	-	Basic Forming Method	DPM
Importance of Operating Magnet at BH_{max}	E	Basic Finishing Method	WGR
Average Recoil Permeability	1.05	Relative Corrosion Resistance	P
Field Strength Required to Fully Saturate Magnet (Oe)	30,000	Primary Chemical Composition	Nd-Fe-B
Ability to Withstand Demagnetizing Fields	0	Special Notes: (1) Axially oriented	
Maximum Service Temperature (°F)	Est. 302		
Approximate Temperature Permanently Affecting Magnet (°F)	Est. 590		
Inherent Orientation of Material	MP		

Note: All information based on available producers literature. In some cases of foreign materials interpretation or calculation required.

SOURCE(S):

UGIMag, Inc.

USA

UGIMAX 31H
(Alloy or Grade)

Sintered Rare Earth
(Type of Material)

31/18
(MMPA Brief Designation)

R7-3
(IEC Code Reference)

MAGNETIC PROPERTIES	
Residual Induction B_r (G)	11,500
Coercive Force (Oe) Normal (H_c)	11,000
Coercive Force (Oe) Intrinsic (H_{ci})	18,000
Maximum Energy Product BH_{max} (MGO)	31.0
B/H Ratio (Load Line) at BH_{max}	2.2
Flux Density in Magnet B_m at BH_{max} (G)	8400
Coercive Force in Magnet H_m at BH_{max} (Oe)	3800
Energy Loss W_c (watt-sec/cycle/lb)	-
Importance of Operating Magnet at BH_{max}	E
Average Recoil Permeability	1.05
Field Strength Required to Fully Saturate Magnet (Oe)	30,000
Ability to Withstand Demagnetizing Fields	0
Maximum Service Temperature (°F)	Est. 300
Approximate Temperature Permanently Affecting Magnet (°F)	Est. 590
Inherent Orientation of Material	MP

PHYSICAL PROPERTIES	
Density (lbs./in.^{3})	.2710
General Mechanical Properties	HB
Tensile Strength (psi)	-
Transverse Modulus of Rupture (psi)	-
Coeff. of Therm. Expansion inches x 10^{-6}/°C	Orientation: Par. - Rt.Angl. -
Resistivity (micro-ohms/cm/cm^2)	150 @ 22 °C
Hardness Scale	600 Vickers
Basic Forming Method	DPM
Basic Finishing Method	WGR
Relative Corrosion Resistance	P
Primary Chemical Composition	Nd-Fe-B
Special Notes: (1) Axially oriented	

Note: All information based on available producers literature. In some cases of foreign materials interpretation or calculation required.

SOURCE(S):

UGIMag, Inc.

USA

UGIMAX 34B
(Alloy or Grade)

Sintered Rare Earth
(Type of Material)

34/15
(MMPA Brief Designation)

R7-3
(IEC Code Reference)

MAGNETIC PROPERTIES		PHYSICAL PROPERTIES	
Residual Induction B_r (G)	12,000	Density (lbs./in.^{3})	.2710
Coercive Force (Oe) Normal (H_c)	11,000	General Mechanical Properties	HB
Coercive Force (Oe) Intrinsic (H_{ci})	15,000	Tensile Strength (psi)	-
Maximum Energy Product BH_{max} (MGO)	34.0	Transverse Modulus of Rupture (psi)	-
B/H Ratio (Load Line) at BH_{max}	.8	Coeff. of Therm. Expansion inches x 10^{-6}/°C	Orientation: Par. - Rt.Angl. -
Flux Density in Magnet B_m at BH_{max} (G)	5200	Resistivity (micro-ohms/cm/cm^2)	150 @ 22 °C
Coercive Force in Magnet H_m at BH_{max} (Oe)	65000	Hardness Scale 600 Vickers	
Energy Loss W_c (watt-sec/cycle/lb)	-	Basic Forming Method	DPM
Importance of Operating Magnet at BH_{max}	E	Basic Finishing Method	WGR
Average Recoil Permeability	1.05	Relative Corrosion Resistance	P
Field Strength Required to Fully Saturate Magnet (Oe)	30,000	Primary Chemical Composition	Nd-Fe-B
Ability to Withstand Demagnetizing Fields	0	Special Notes: (1) Axially oriented	
Maximum Service Temperature (°F)	Est. 248		
Approximate Temperature Permanently Affecting Magnet (°F)	Est. 590		
Inherent Orientation of Material	MP		

Note: All information based on available producers literature. In some cases of foreign materials interpretation or calculation required.

SOURCE(S):	UGIMAX 34K
	(Alloy or Grade)
UGIMag, Inc.	Sintered Rare Earth
	(Type of Material)
USA	34/19
	(MMPA Brief Designation)
	R7-3
	(IEC Code Reference)

MAGNETIC PROPERTIES		PHYSICAL PROPERTIES	
Residual Induction B_r (G)	12,500	Density (lbs./in.^{3})	.2710
Coercive Force (Oe) — Normal (H_c)	11,500	General Mechanical Properties	HB
Coercive Force (Oe) — Intrinsic (H_{ci})	19,000	Tensile Strength (psi)	—
Maximum Energy Product BH_{max} (MGO)	34.0	Transverse Modulus of Rupture (psi)	—
B/H Ratio (Load Line) at BH_{max}	.63	Coeff. of Therm. Expansion inches x 10^{-6}/°C	Orientation: Par. — Rt.Angl. —
Flux Density in Magnet B_m at BH_{max} (G)	4600	Resistivity (micro-ohms/cm/cm^2)	150 @ 22 °C
Coercive Force in Magnet H_m at BH_{max} (Oe)	7300	Hardness Scale	600 Vickers
Energy Loss W_c (watt-sec/cycle/lb)	—	Basic Forming Method	DPM
Importance of Operating Magnet at BH_{max}	E	Basic Finishing Method	WGR
Average Recoil Permeability	1.05	Relative Corrosion Resistance	P
Field Strength Required to Fully Saturate Magnet (Oe)	30,000	Primary Chemical Composition	Nd-Fe-B
Ability to Withstand Demagnetizing Fields	0	Special Notes: (1) Transverse oriented	
Maximum Service Temperature (°F)	Est. 302		
Approximate Temperature Permanently Affecting Magnet (°F)	Est. 590		
Inherent Orientation of Material	MP		

Note: All information based on available producers literature. In some cases of foreign materials interpretation or calculation required.

SOURCE(S):

UGIMag, Inc.

USA

UGIMAX 35H
(Alloy or Grade)

Sintered Rare Earth
(Type of Material)

35/18
(MMPA Brief Designation)

R7-3
(IEC Code Reference)

MAGNETIC PROPERTIES		PHYSICAL PROPERTIES	
Residual Induction B_r (G)	12,100	Density (lbs./in.^{3})	.2710
Coercive Force (Oe) — Normal (H_c)	11,700	General Mechanical Properties	HB
Coercive Force (Oe) — Intrinsic (H_{ci})	18,000	Tensile Strength (psi)	–
Maximum Energy Product BH_{max} (MGO)	35.0	Transverse Modulus of Rupture (psi)	–
B/H Ratio (Load Line) at BH_{max}	.71	Coeff. of Therm. Expansion inches x 10^{-6}/°C	Orientation: Par. – Rt.Angl. –
Flux Density in Magnet B_m at BH_{max} (G)	5000	Resistivity (micro-ohms/cm/cm^2)	150 @ 22 °C
Coercive Force in Magnet H_m at BH_{max} (Oe)	7000	Hardness Scale	600 Vickers
Energy Loss W_c (watt-sec/cycle/lb)	–	Basic Forming Method	DPM
Importance of Operating Magnet at BH_{max}	E	Basic Finishing Method	WGR
Average Recoil Permeability	1.05	Relative Corrosion Resistance	P
Field Strength Required to Fully Saturate Magnet (Oe)	30,000	Primary Chemical Composition	Nd-Fe-B
Ability to Withstand Demagnetizing Fields	0	Special Notes: (1) Transverse oriented	
Maximum Service Temperature (°F)	Est. 300		
Approximate Temperature Permanently Affecting Magnet (°F)	Est. 590		
Inherent Orientation of Material	MP		

Note: All information based on available producers literature. In some cases of foreign materials interpretation or calculation required.

SOURCE(S):	UGIMAX 37B
UGIMag, Inc.	(Alloy or Grade)
	Sintered Rare Earth
	(Type of Material)
USA	37/15
	(MMPA Brief Designation)
	R7-3
	(IEC Code Reference)

MAGNETIC PROPERTIES		PHYSICAL PROPERTIES	
Residual Induction B_r (G)	12,500	Density (lbs./in.^{3})	2710
Coercive Force (Oe) — Normal (H_c)	11,700	General Mechanical Properties	HB
Coercive Force (Oe) — Intrinsic (H_{ci})	15,000	Tensile Strength (psi)	-
Maximum Energy Product BH_{max} (MGO)	37.0	Transverse Modulus of Rupture (psi)	-
B/H Ratio (Load Line) at BH_{max}	.79	Coeff. of Therm. Expansion inches x 10^{-6}/°C Orientation: Par. _ Rt.Angl. _	
Flux Density in Magnet B_m at BH_{max} (G)	5400	Resistivity (micro-ohms/cm/cm^2)	150 @ 22 °C
Coercive Force in Magnet H_m at BH_{max} (Oe)	6800	Hardness Scale	600 Vickers
Energy Loss W_c (watt-sec/cycle/lb)	-	Basic Forming Method	DPM
Importance of Operating Magnet at BH_{max}	E	Basic Finishing Method	WGR
Average Recoil Permeability	1.05	Relative Corrosion Resistance	P
Field Strength Required to Fully Saturate Magnet (Oe)	30,000	Primary Chemical Composition	Nd-Fe-B
Ability to Withstand Demagnetizing Fields	0	Special Notes: (1) Transverse oriented	
Maximum Service Temperature (°F)	Est. 245		
Approximate Temperature Permanently Affecting Magnet (°F)	Est. 590		
Inherent Orientation of Material	MP		

Note: All information based on available producers literature. In some cases of foreign materials interpretation or calculation required.

SOURCE(S):	UGISTAB 26x4
UGIMag, Inc.	(Alloy or Grade)
	Sintered Rare Earth
	(Type of Material)
USA	26/24
	(MMPA Brief Designation)
	R7-3
	(IEC Code Reference)

MAGNETIC PROPERTIES		PHYSICAL PROPERTIES	
Residual Induction B_r (G)	10,400	Density (lbs./in.^{3})	.2710
Coercive Force (Oe) — Normal (H_c)	10,000	General Mechanical Properties	HB
Coercive Force (Oe) — Intrinsic (H_{ci})	24,000	Tensile Strength (psi)	-
Maximum Energy Product BH_{max} (MGO)	26.0	Transverse Modulus of Rupture (psi)	-
B/H Ratio (Load Line) at BH_{max}	.42	Coeff. of Therm. Expansion inches x 10^{-6}/°C	Orientation: Par. - Rt.Angl. -
Flux Density in Magnet B_m at BH_{max} (G)	3300	Resistivity (micro-ohms/cm/cm^2)	150 @ 22 °C
Coercive Force in Magnet H_m at BH_{max} (Oe)	7800	Hardness Scale 600	Vickers
Energy Loss W_c (watt-sec/cycle/lb)	-	Basic Forming Method	DPM
Importance of Operating Magnet at BH_{max}	E	Basic Finishing Method	WGR
Average Recoil Permeability	1.05	Relative Corrosion Resistance	P
Field Strength Required to Fully Saturate Magnet (Oe)	30,000	Primary Chemical Composition	Nd-Fe-B
Ability to Withstand Demagnetizing Fields	0	Special Notes: (1) Axial orientation	
Maximum Service Temperature (°F)	Est. 480		
Approximate Temperature Permanently Affecting Magnet (°F)	Est. 1328		
Inherent Orientation of Material	MP		

Note: All information based on available producers literature. In some cases of foreign materials interpretation or calculation required.

SOURCE(S):

UGIMag, Inc.

USA

UGISTAB 30x4
(Alloy or Grade)

Sintered Rare Earth
(Type of Material)

30/24
(MMPA Brief Designation)

R7-3
(IEC Code Reference)

MAGNETIC PROPERTIES		PHYSICAL PROPERTIES	
Residual Induction B_r (G)	11,100	Density (lbs./in.^{3})	.2710
Coercive Force (Oe) — Normal (H_c)	10,800	General Mechanical Properties	HB
Coercive Force (Oe) — Intrinsic (H_{ci})	24,000	Tensile Strength (psi)	-
Maximum Energy Product BH_{max} (MGO)	30.0	Transverse Modulus of Rupture (psi)	-
B/H Ratio (Load Line) at BH_{max}	.41	Coeff. of Therm. Expansion inches x 10^{-6}/°C	Orientation: Par. - Rt.Angl. -
Flux Density in Magnet B_m at BH_{max} (G)	3500	Resistivity (micro-ohms/cm/cm^2)	150 @ 22 °C
Coercive Force in Magnet H_m at BH_{max} (Oe)	8500	Hardness Scale	600 Vickers
Energy Loss W_c (watt-sec/cycle/lb)	-	Basic Forming Method	DPM
Importance of Operating Magnet at BH_{max}	E	Basic Finishing Method	WGR
Average Recoil Permeability	1.05	Relative Corrosion Resistance	P
Field Strength Required to Fully Saturate Magnet (Oe)	30,000	Primary Chemical Composition	Nd-Fe-B
Ability to Withstand Demagnetizing Fields	0	Special Notes: (1) Transverse orientation	
Maximum Service Temperature (°F)	Est. 480		
Approximate Temperature Permanently Affecting Magnet (°F)	Est. 1328		
Inherent Orientation of Material	MP		

Note: All information based on available producers literature. In some cases of foreign materials interpretation or calculation required.

SOURCE(S):

Vacuumschmelze GMBH

Germany

Vacodym 335HR
(Alloy or Grade)

Sintered Rare Earth
(Type of Material)

37/12
(MMPA Brief Designation)

R7-3
(IEC Code Reference)

MAGNETIC PROPERTIES		PHYSICAL PROPERTIES	
Residual Induction B_r (G)	12,500	Density (lbs./in.^{3})	.2673
Coercive Force (Oe) Normal (H_c)	11,300	General Mechanical Properties	HB
Coercive Force (Oe) Intrinsic (H_{ci})	12,000	Tensile Strength (psi)	-
Maximum Energy Product BH_{max} (MGO)	37.0	Transverse Modulus of Rupture (psi)	-
B/H Ratio (Load Line) at BH_{max}	1.0	Coeff. of Therm. Expansion inches x 10^{-6}/°C	Orientation: Par. 5.0 Rt.Angl. -1.0
Flux Density in Magnet B_m at BH_{max} (G)	6100	Resistivity (micro-ohms/cm/cm^2	- @ - °C
Coercive Force in Magnet H_m at BH_{max} (Oe)	6100	Hardness Scale 570	Vickers
Energy Loss W_c (watt-sec/cycle/lb)	-	Basic Forming Method	DPM
Importance of Operating Magnet at BH_{max}	E	Basic Finishing Method	WGR
Average Recoil Permeability	1.05	Relative Corrosion Resistance	P See Note
Field Strength Required to Fully Saturate Magnet (Oe)	32,000	Primary Chemical Composition	Nd-Fe-B
Ability to Withstand Demagnetizing Fields	0	Special Notes: (1) Oxidizes readily. Surface protection by coating or plating advisable.	
Maximum Service Temperature (°F)	212		
Approximate Temperature Permanently Affecting Magnet (°F)	590		
Inherent Orientation of Material	MP		

Note: All information based on available producers literature. In some cases of foreign materials interpretation or calculation required.

SOURCE(S):

Vacuumschmelze GMBH

Germany

Vacodym 335WZ
(Alloy or Grade)

Sintered Rare Earth
(Type of Material)

32/12
(MMPA Brief Designation)

R7-3
(IEC Code Reference)

MAGNETIC PROPERTIES	
Residual Induction B_r (G)	12,000
Coercive Force (Oe) — Normal (H_c)	10,700
Coercive Force (Oe) — Intrinsic (H_{ci})	12,000
Maximum Energy Product BH_{max} (MGO)	32
B/H Ratio (Load Line) at BH_{max}	.90
Flux Density in Magnet B_m at BH_{max} (G)	5400
Coercive Force in Magnet H_m at BH_{max} (Oe)	6000
Energy Loss W_c (watt-sec/cycle/lb)	-
Importance of Operating Magnet at BH_{max}	E
Average Recoil Permeability	1.05
Field Strength Required to Fully Saturate Magnet (Oe)	32,000
Ability to Withstand Demagnetizing Fields	0
Maximum Service Temperature (°F)	212
Approximate Temperature Permanently Affecting Magnet (°F)	590
Inherent Orientation of Material	MP

PHYSICAL PROPERTIES	
Density (lbs./in.^{3})	.2673
General Mechanical Properties	HB
Tensile Strength (psi)	-
Transverse Modulus of Rupture (psi)	-
Coeff. of Therm. Expansion inches x 10^{-6}/°C	Orientation: Par. 5.0 Rt.Angl. -1.0
Resistivity (micro-ohms/cm/cm^2)	- @ - °C
Hardness 570	Scale Vickers
Basic Forming Method	DPM
Basic Finishing Method	WGR
Relative Corrosion Resistance	P See Note
Primary Chemical Composition	Nd-Fe-B

Special Notes:

(1) Oxidizes readily. Surface protection by coating or plating advisable.

Note: All information based on available producers literature. In some cases of foreign materials interpretation or calculation required.

SOURCE(S):

Vacuumschmelze GMBH

Germany

Vacodym 351HR
(Alloy or Grade)

Sintered Rare Earth
(Type of Material)

36/16
(MMPA Brief Designation)

R7-3
(IEC Code Reference)

MAGNETIC PROPERTIES		PHYSICAL PROPERTIES	
Residual Induction B_r (G)	12,400	Density (lbs./in.^{3})	.2673
Coercive Force (Oe) Normal (H_c)	11,800	General Mechanical Properties	HB
Coercive Force (Oe) Intrinsic (H_{ci})	16,000	Tensile Strength (psi)	-
Maximum Energy Product BH_{max} (MGO)	36.5	Transverse Modulus of Rupture (psi)	-
B/H Ratio (Load Line) at BH_{max}	.74	Coeff. of Therm. Expansion inches x 10^{-6}/°C	Orientation: Par. 5.0 Rt.Angl. -1.0
Flux Density in Magnet B_m at BH_{max} (G)	5200	Resistivity (micro-ohms/cm/cm^2	- @ - °C
Coercive Force in Magnet H_m at BH_{max} (Oe)	7000	Hardness Scale 570	Vickers
Energy Loss W_c (watt-sec/cycle/lb)	-	Basic Forming Method	DPM
Importance of Operating Magnet at BH_{max}	E	Basic Finishing Method	WGR
Average Recoil Permeability	1.05	Relative Corrosion Resistance	P See Note
Field Strength Required to Fully Saturate Magnet (Oe)	32,000	Primary Chemical Composition	Nd-Fe-B
Ability to Withstand Demagnetizing Fields	0	Special Notes: (1) Oxidizes readily. Surface protection by coating or plating advisable.	
Maximum Service Temperature (°F)	248		
Approximate Temperature Permanently Affecting Magnet (°F)	590		
Inherent Orientation of Material	MP		

Note: All information based on available producers literature. In some cases of foreign materials interpretation or calculation required.

SOURCE(S):

Vacuumschmelze GMBH

Germany

Vacodym 351WZ
(Alloy or Grade)

Sintered Rare Earth
(Type of Material)

31/16
(MMPA Brief Designation)

R7-3
(IEC Code Reference)

MAGNETIC PROPERTIES		PHYSICAL PROPERTIES	
Residual Induction B_r (G)	11,500	Density (lbs./in.^{3})	.2673
Coercive Force (Oe) Normal (H_c)	11,000	General Mechanical Properties	HB
Coercive Force (Oe) Intrinsic (H_{ci})	16,000	Tensile Strength (psi)	-
Maximum Energy Product BH_{max} (MGO)	31.5	Transverse Modulus of Rupture (psi)	-
B/H Ratio (Load Line) at BH_{max}	.60	Coeff. of Therm. Expansion inches x 10^{-6}/°C	Orientation: Par. 5.0 Rt.Angl. -1.0
Flux Density in Magnet B_m at BH_{max} (G)	4350	Resistivity (micro-ohms/cm/cm^2)	- @ - °C
Coercive Force in Magnet H_m at BH_{max} (Oe)	7275	Hardness Scale 570	Vickers
Energy Loss W_c (watt-sec/cycle/lb)	-	Basic Forming Method	DPM
Importance of Operating Magnet at BH_{max}	E	Basic Finishing Method	WGR
Average Recoil Permeability	1.05	Relative Corrosion Resistance	P See Note
Field Strength Required to Fully Saturate Magnet (Oe)	32,00	Primary Chemical Composition	Nd-Fe-B
Ability to Withstand Demagnetizing Fields	0	Special Notes: (1) Oxidizes readily. Surface protection by coating or plating advisable.	
Maximum Service Temperature (°F)	248		
Approximate Temperature Permanently Affecting Magnet (°F)	590		
Inherent Orientation of Material	MP		

Note: All information based on available producers literature. In some cases of foreign materials interpretation or calculation required.

SOURCE(S):

Vacuumschmelze GMBH

Germany

Vacodym 362HR
(Alloy or Grade)

Sintered Rare Earth
(Type of Material)

(MMPA Brief Designation)

R7-3
(IEC Code Reference)

MAGNETIC PROPERTIES		PHYSICAL PROPERTIES	
Residual Induction B_r (G)	13,300	Density (lbs./in.^{3})	.2673
Coercive Force (Oe) — Normal (H_c)	12,800	General Mechanical Properties	HB
Coercive Force (Oe) — Intrinsic (H_{ci})	16,000	Tensile Strength (psi)	-
Maximum Energy Product BH_{max} (MGO)	42.5	Transverse Modulus of Rupture (psi)	-
B/H Ratio (Load Line) at BH_{max}	.8	Coeff. of Therm. Expansion inches x 10^{-6}/°C	Orientation: Par. - Rt.Angl. -
Flux Density in Magnet B_m at BH_{max} (G)	5800	Resistivity (micro-ohms/cm/cm^2)	- @ - °C
Coercive Force in Magnet H_m at BH_{max} (Oe)	7300	Hardness 570	Scale Vickers
Energy Loss W_c (watt-sec/cycle/lb)	-	Basic Forming Method	DPM
Importance of Operating Magnet at BH_{max}	E	Basic Finishing Method	WGR
Average Recoil Permeability	1.05	Relative Corrosion Resistance	P See Note
Field Strength Required to Fully Saturate Magnet (Oe)	32,000	Primary Chemical Composition	Nd-Fe-B
Ability to Withstand Demagnetizing Fields	0	Special Notes: (1) Oxidizes readily. Surface protection by coating or plating advisable.	
Maximum Service Temperature (°F)	248		
Approximate Temperature Permanently Affecting Magnet (°F)	590		
Inherent Orientation of Material	MP		

Note: All information based on available producers literature. In some cases of foreign materials interpretation or calculation required.

SOURCE(S):	**Vacodym 362WZ**
	(Alloy or Grade)
Vacuumschmelze GMBH	Sintered Rare Earth
	(Type of Material)
Germany	36/16
	(MMPA Brief Designation)
	R7-3
	(IEC Code Reference)

MAGNETIC PROPERTIES		PHYSICAL PROPERTIES	
Residual Induction B_r (G)	12,300	Density (lbs./in.^{3})	.2673
Coercive Force (Oe) — Normal (H_c)	11,700	General Mechanical Properties	HB
Coercive Force (Oe) — Intrinsic (H_{ci})	16,000	Tensile Strength (psi)	-
Maximum Energy Product BH_{max} (MGO)	36.0	Transverse Modulus of Rupture (psi)	-
B/H Ratio (Load Line) at BH_{max}	.73	Coeff. of Therm. Expansion inches x 10^{-6}/°C	Orientation: Par. 5.0 Rt.Angl. -1.0
Flux Density in Magnet B_m at BH_{max} (G)	5100	Resistivity (micro-ohms/cm/cm^2)	- @ - °C
Coercive Force in Magnet H_m at BH_{max} (Oe)	7000	Hardness Scale 570	Vickers
Energy Loss W_c (watt-sec/cycle/lb)	-	Basic Forming Method	DPM
Importance of Operating Magnet at BH_{max}	E	Basic Finishing Method	WGR
Average Recoil Permeability	1.05	Relative Corrosion Resistance	P See Note
Field Strength Required to Fully Saturate Magnet (Oe)	32,000	Primary Chemical Composition	Nd-Fe-B
Ability to Withstand Demagnetizing Fields	0	Special Notes: (1) Oxidizes readily. Surface protection by coating or plating advisable.	
Maximum Service Temperature (°F)	248		
Approximate Temperature Permanently Affecting Magnet (°F)	590		
Inherent Orientation of Material	MP		

Note: All information based on available producers literature. In some cases of foreign materials interpretation or calculation required.

SOURCE(S):

Vacuumschmelze GMBH

Germany

Vacodym 370HR
(Alloy or Grade)

Sintered Rare Earth
(Type of Material)

34/19
(MMPA Brief Designation)

R7-3
(IEC Code Reference)

MAGNETIC PROPERTIES		PHYSICAL PROPERTIES	
Residual Induction B_r (G)	12,000	Density (lbs./in.^{3})	.2673
Coercive Force (Oe) — Normal (H_c)	11,400	General Mechanical Properties	HB
Coercive Force (Oe) — Intrinsic (H_{ci})	19,000	Tensile Strength (psi)	—
Maximum Energy Product BH_{max} (MGO)	34.5	Transverse Modulus of Rupture (psi)	—
B/H Ratio (Load Line) at BH_{max}	.57	Coeff. of Therm. Expansion inches x 10^{-6}/°C	Orientation: Par. 5.0 Rt.Angl. −1.0
Flux Density in Magnet B_m at BH_{max} (G)	4450	Resistivity (micro-ohms/cm/cm^2	— @ — °C
Coercive Force in Magnet H_m at BH_{max} (Oe)	7800	Hardness 570 Scale Vickers	
Energy Loss W_c (watt-sec/cycle/lb)	—	Basic Forming Method	DPM
Importance of Operating Magnet at BH_{max}	E	Basic Finishing Method	WGR
Average Recoil Permeability	1.05	Relative Corrosion Resistance	P See Note
Field Strength Required to Fully Saturate Magnet (Oe)	40,000	Primary Chemical Composition	Nd-Fe-B
Ability to Withstand Demagnetizing Fields	0	Special Notes: (1) Oxidizes readily. Surface protection by coating or plating advisable.	
Maximum Service Temperature (°F)	302		
Approximate Temperature Permanently Affecting Magnet (°F)	590		
Inherent Orientation of Material	MP		

Note: All information based on available producers literature. In some cases of foreign materials interpretation or calculation required.

SOURCE(S):

Vacuumschmelze GMBH

Germany

Vacodym 370WZ
(Alloy or Grade)

Sintered Rare Earth
(Type of Material)

30/19
(MMPA Brief Designation)

R7-3
(IEC Code Reference)

MAGNETIC PROPERTIES		PHYSICAL PROPERTIES	
Residual Induction B_r (G)	11,200	Density (lbs./in.^{3})	.2673
Coercive Force (Oe) — Normal (H_c)	10,700	General Mechanical Properties	HB
Coercive Force (Oe) — Intrinsic (H_{ci})	19,000	Tensile Strength (psi)	–
Maximum Energy Product BH_{max} (MGO)	30.0	Transverse Modulus of Rupture (psi)	–
B/H Ratio (Load Line) at BH_{max}	.49	Coeff. of Therm. Expansion inches x 10^{-6}/°C	Orientation: Par. 5.0 Rt.Angl. -1.0
Flux Density in Magnet B_m at BH_{max} (G)	3800	Resistivity (micro-ohms/cm/cm^2	– @ – °C
Coercive Force in Magnet H_m at BH_{max} (Oe)	7700	Hardness 570	Scale Vickers
Energy Loss W_c (watt-sec/cycle/lb)	–	Basic Forming Method	DPM
Importance of Operating Magnet at BH_{max}	E	Basic Finishing Method	WGR
Average Recoil Permeability	1.05	Relative Corrosion Resistance	P See Note
Field Strength Required to Fully Saturate Magnet (Oe)	40,000	Primary Chemical Composition	Nd-Fe-Co
Ability to Withstand Demagnetizing Fields	0	Special Notes: (1) Oxidizes readily. Surface protection by coating or plating advisable.	
Maximum Service Temperature (°F)	302		
Approximate Temperature Permanently Affecting Magnet (°F)	590		
Inherent Orientation of Material	MP		

Note: All information based on available producers literature. In some cases of foreign materials interpretation or calculation required.

SOURCE(S):

Vacuumschmelze GMBH

Germany

Vacodym 383HR
(Alloy or Grade)

Sintered Rare Earth
(Type of Material)

39/19
(MMPA Brief Designation)

R7-3
(IEC Code Reference)

MAGNETIC PROPERTIES		PHYSICAL PROPERTIES	
Residual Induction B_r (G)	12,700	Density (lbs./in.^{3})	.2673
Coercive Force (Oe) — Normal (H_c)	12,200	General Mechanical Properties	HB
Coercive Force (Oe) — Intrinsic (H_{ci})	19,000	Tensile Strength (psi)	-
Maximum Energy Product BH_{max} (MGO)	39.0	Transverse Modulus of Rupture (psi)	-
B/H Ratio (Load Line) at BH_{max}	.62	Coeff. of Therm. Expansion inches x 10^{-6}/°C	Orientation: Par. 5.0 Rt.Angl. -1.0
Flux Density in Magnet B_m at BH_{max} (G)	4900	Resistivity (micro-ohms/cm/cm^2	- @ - °C
Coercive Force in Magnet H_m at BH_{max} (Oe)	7900	Hardness 570 Scale Vickers	
Energy Loss W_c (watt-sec/cycle/lb)	-	Basic Forming Method	DPM
Importance of Operating Magnet at BH_{max}	E	Basic Finishing Method	WGR
Average Recoil Permeability	1.05	Relative Corrosion Resistance	P See Note
Field Strength Required to Fully Saturate Magnet (Oe)	40,000	Primary Chemical Composition	Nd-Fe-B
Ability to Withstand Demagnetizing Fields	0	Special Notes: (1) Oxidizes readily. Surface protection by coating or plating advisable.	
Maximum Service Temperature (°F)	302		
Approximate Temperature Permanently Affecting Magnet (°F)	590		
Inherent Orientation of Material	MP		

Note: All information based on available producers literature. In some cases of foreign materials interpretation or calculation required.

SOURCE(S):

Vacuumschmelze GMBH

Germany

Vacodym 383WZ
(Alloy or Grade)

Sintered Rare Earth
(Type of Material)

33/19
(MMPA Brief Designation)

R7-3
(IEC Code Reference)

MAGNETIC PROPERTIES	
Residual Induction B_r (G)	11,800
Coercive Force (Oe) — Normal (H_c)	11,200
Coercive Force (Oe) — Intrinsic (H_{ci})	19,000
Maximum Energy Product BH_{max} (MGO)	33.0
B/H Ratio (Load Line) at BH_{max}	.57
Flux Density in Magnet B_m at BH_{max} (G)	4450
Coercive Force in Magnet H_m at BH_{max} (Oe)	7750
Energy Loss W_c (watt-sec/cycle/lb)	-
Importance of Operating Magnet at BH_{max}	E
Average Recoil Permeability	1.05
Field Strength Required to Fully Saturate Magnet (Oe)	32,000
Ability to Withstand Demagnetizing Fields	0
Maximum Service Temperature (°F)	302
Approximate Temperature Permanently Affecting Magnet (°F)	590
Inherent Orientation of Material	MP

PHYSICAL PROPERTIES	
Density (lbs./in.^{3})	.2673
General Mechanical Properties	HB
Tensile Strength (psi)	-
Transverse Modulus of Rupture (psi)	-
Coeff. of Therm. Expansion inches x 10^{-6}/°C	Orientation: Par. 5.0 Rt.Angl. -1.0
Resistivity (micro-ohms/cm/cm^2 - @ - °C)	
Hardness 570	Scale Vickers
Basic Forming Method	DPM
Basic Finishing Method	WGR
Relative Corrosion Resistance	P See Note
Primary Chemical Composition	Nd-Fe-B
Special Notes: (1) Oxidizes readily. Surface protection by coating or plating advisable.	

Note: All information based on available producers literature. In some cases of foreign materials interpretation or calculation required.

SOURCE(S):		**Vacodym 400HR**
		(Alloy or Grade)
Vacuumschmelze GMBH		Sintered Rare Earth
		(Type of Material)
Germany		31/26
		(MMPA Brief Designation)
		R7-3
		(IEC Code Reference)

MAGNETIC PROPERTIES		PHYSICAL PROPERTIES	
Residual Induction B_r (G)	11,500	Density (lbs./in.^{3})	.2673
Coercive Force (Oe) — Normal (H_c)	11,000	General Mechanical Properties	HB
Coercive Force (Oe) — Intrinsic (H_{ci})	26,000	Tensile Strength (psi)	-
Maximum Energy Product BH_{max} (MGO)	31.5	Transverse Modulus of Rupture (psi)	-
B/H Ratio (Load Line) at BH_{max}	.50	Coeff. of Therm. Expansion inches x 10^{-6}/°C	Orientation: Par. 5.0 Rt.Angl. -1.0
Flux Density in Magnet B_m at BH_{max} (G)	3950	Resistivity (micro-ohms/cm/cm^2 -	@ - °C
Coercive Force in Magnet H_m at BH_{max} (Oe)	7950	Hardness Scale 570	Vickers
Energy Loss W_c (watt-sec/cycle/lb)	-	Basic Forming Method	DPM
Importance of Operating Magnet at BH_{max}	E	Basic Finishing Method	WGR
Average Recoil Permeability	1.05	Relative Corrosion Resistance	P See Note
Field Strength Required to Fully Saturate Magnet (Oe)	50,000	Primary Chemical Composition	Nd-Fe-B
Ability to Withstand Demagnetizing Fields	0	Special Notes: (1) Oxidizes readily. Surface protection by coating or plating advisable.	
Maximum Service Temperature (°F)	356		
Approximate Temperature Permanently Affecting Magnet (°F)	590		
Inherent Orientation of Material	MP		

Note: All information based on available producers literature. In some cases of foreign materials interpretation or calculation required.

SOURCE(S):

Vacuumschmelze GMBH

Germany

Vacodym 400WZ
(Alloy or Grade)

Sintered Rare Earth
(Type of Material)

27/26
(MMPA Brief Designation)

R7-3
(IEC Code Reference)

MAGNETIC PROPERTIES	
Residual Induction B_r (G)	10,600
Coercive Force (Oe) Normal (H_c)	10,100
Coercive Force (Oe) Intrinsic (H_{ci})	26,000
Maximum Energy Product BH_{max} (MGO)	27.0
B/H Ratio (Load Line) at BH_{max}	.57
Flux Density in Magnet B_m at BH_{max} (G)	3900
Coercive Force in Magnet H_m at BH_{max} (Oe)	6800
Energy Loss W_c (watt-sec/cycle/lb)	—
Importance of Operating Magnet at BH_{max}	E
Average Recoil Permeability	1.05
Field Strength Required to Fully Saturate Magnet (Oe)	40,000
Ability to Withstand Demagnetizing Fields	0
Maximum Service Temperature (°F)	356
Approximate Temperature Permanently Affecting Magnet (°F)	590
Inherent Orientation of Material	WP

PHYSICAL PROPERTIES	
Density (lbs./in.^{3})	.2673
General Mechanical Properties	HB
Tensile Strength (psi)	—
Transverse Modulus of Rupture (psi)	—
Coeff. of Therm. Expansion inches x 10^{-6}/°C	Orientation: Par. 5.0 Rt.Angl. -1.0
Resistivity (micro-ohms/cm/cm^2	— @ — °C
Hardness 570	Scale Vickers
Basic Forming Method	DPM
Basic Finishing Method	WGR
Relative Corrosion Resistance	P See Note
Primary Chemical Composition	Nd-Fe-B
Special Notes: (1) Oxidizes readily. Surface protection by coating or plating advisable.	

Note: All information based on available producers literature. In some cases of foreign materials interpretation or calculation required.

SOURCE(S):	Vacomax 145 (Standard)
Vacuumschmelze GMBH	(Alloy or Grade)
	Sintered Rare Earth
	(Type of Material)
Germany	20/22
	(MMPA Brief Designation)
	R5-1
	(IEC Code Reference)

MAGNETIC PROPERTIES		PHYSICAL PROPERTIES	
Residual Induction B_r (G)	9000	Density (lbs./in.^{3})	.3035
Coercive Force (Oe) — Normal (H_c)	8300	General Mechanical Properties	HB
Coercive Force (Oe) — Intrinsic (H_{ci})	22,500	Tensile Strength (psi)	5000
Maximum Energy Product BH_{max} (MGO)	20.0	Transverse Modulus of Rupture (psi)	-
B/H Ratio (Load Line) at BH_{max}	.8	Coeff. of Therm. Expansion inches x 10^{-6}/°C	Orientation: Par. 7.0 Rt.Angl. 13.0
Flux Density in Magnet B_m at BH_{max} (G)	4000	Resistivity (micro-ohms/cm/cm^2)	- @ - °C
Coercive Force in Magnet H_m at BH_{max} (Oe)	5000	Hardness Scale	550 Vickers
Energy Loss W_c (watt-sec/cycle/lb)	-	Basic Forming Method	DPM
Importance of Operating Magnet at BH_{max}	E	Basic Finishing Method	WGR
Average Recoil Permeability	1.05	Relative Corrosion Resistance	G
Field Strength Required to Fully Saturate Magnet (Oe)	40,000	Primary Chemical Composition	Sm-Co
Ability to Withstand Demagnetizing Fields	0	Special Notes: (1) Higher temperature stability than most other rare earth materials.	
Maximum Service Temperature (°F) See Note (1)	482		
Approximate Temperature Permanently Affecting Magnet (°F)	1328		
Inherent Orientation of Material	MP		

Note: All information based on available producers literature. In some cases of foreign materials interpretation or calculation required.

SOURCE(S):	**Vacomax 145 (Superquality)**
	(Alloy or Grade)
Vacuumschmelze GMBH	Sintered Rare Earth
	(Type of Material)
Germany	20/30
	(MMPA Brief Designation)
	R5-1
	(IEC Code Reference)

MAGNETIC PROPERTIES		PHYSICAL PROPERTIES	
Residual Induction B_r (G)	9000	Density (lbs./in.^{3})	.3035
Coercive Force (Oe) — Normal (H_c)	8300	General Mechanical Properties	HB
Coercive Force (Oe) — Intrinsic (H_{ci})	30,000	Tensile Strength (psi)	5000
Maximum Energy Product BH_{max} (MGO)	20.0	Transverse Modulus of Rupture (psi)	-
B/H Ratio (Load Line) at BH_{max}	.8	Coeff. of Therm. Expansion inches x 10^{-6}/°C	Orientation: Par. 7.0 Rt.Angl. 13.0
Flux Density in Magnet B_m at BH_{max} (G)	4000	Resistivity (micro-ohms/cm/cm^2 - @ - °C)	
Coercive Force in Magnet H_m at BH_{max} (Oe)	5000	Hardness Scale	550 Vickers
Energy Loss W_c (watt-sec/cycle/lb)	-	Basic Forming Method	DPM
Importance of Operating Magnet at BH_{max}	E	Basic Finishing Method	WGR
Average Recoil Permeability	1.05	Relative Corrosion Resistance	G
Field Strength Required to Fully Saturate Magnet (Oe)	50,000	Primary Chemical Composition	Sm-Co
Ability to Withstand Demagnetizing Fields	0	Special Notes:	
Maximum Service Temperature (°F) (See Note 1)	482	(1) Higher temperature stability than most other rare earth materials.	
Approximate Temperature Permanently Affecting Magnet (°F)	1328		
Inherent Orientation of Material	MP		

Note: All information based on available producers literature. In some cases of foreign materials interpretation or calculation required.

SOURCE(S):	Vacomax 170
	(Alloy or Grade)
Vacuumschmelze GMBH	Sintered Rare Earth
	(Type of Material)
Germany	23/22
	(MMPA Brief Designation)
	R5-1
	(IEC Code Reference)

MAGNETIC PROPERTIES		PHYSICAL PROPERTIES	
Residual Induction B_r (G)	9500	Density (lbs./in.^{3})	.3035
Coercive Force (Oe) — Normal (H_c)	9000	General Mechanical Properties	HB
Coercive Force (Oe) — Intrinsic (H_{ci})	22,500	Tensile Strength (psi)	5000
Maximum Energy Product BH_{max} (MGO)	23.0	Transverse Modulus of Rupture (psi)	-
B/H Ratio (Load Line) at BH_{max}	.46	Coeff. of Therm. Expansion inches x 10^{-6}/°C	Orientation: Par. 7.0 Rt.Angl. 13.0
Flux Density in Magnet B_m at BH_{max} (G)	3250	Resistivity (micro-ohms/cm/cm^2)	- @ - °C
Coercive Force in Magnet H_m at BH_{max} (Oe)	7025	Hardness Scale	550 Vickers
Energy Loss W_c (watt-sec/cycle/lb)	-	Basic Forming Method	DPM
Importance of Operating Magnet at BH_{max}	E	Basic Finishing Method	WGR
Average Recoil Permeability	1.05	Relative Corrosion Resistance	G
Field Strength Required to Fully Saturate Magnet (Oe)	40,000	Primary Chemical Composition	Sm-Co
Ability to Withstand Demagnetizing Fields	0	Special Notes:	
Maximum Service Temperature (°F)	482 See Note (1)	(1) Higher temperature stability than most other rare earth materials.	
Approximate Temperature Permanently Affecting Magnet (°F)	1328		
Inherent Orientation of Material	MP		

Note: All information based on available producers literature. In some cases of foreign materials interpretation or calculation required.

SOURCE(S): **Vacuumschmelze GMBH** Germany

Vacomax 200
(Alloy or Grade)

Sintered Rare Earth
(Type of Material)

25/19
(MMPA Brief Designation)

R5-2
(IEC Code Reference)

MAGNETIC PROPERTIES	
Residual Induction B_r (G)	10,100
Coercive Force (Oe) — Normal (H_c)	9500
Coercive Force (Oe) — Intrinsic (H_{ci})	19,000
Maximum Energy Product BH_{max} (MGO)	25.0
B/H Ratio (Load Line) at BH_{max}	.58
Flux Density in Magnet B_m at BH_{max} (G)	3800
Coercive Force in Magnet H_m at BH_{max} (Oe)	6600
Energy Loss W_c (watt-sec/cycle/lb)	-
Importance of Operating Magnet at BH_{max}	E
Average Recoil Permeability	1.05
Field Strength Required to Fully Saturate Magnet (Oe)	32,000
Ability to Withstand Demagnetizing Fields	O
Maximum Service Temperature (°F)	482 See Note (1)
Approximate Temperature Permanently Affecting Magnet (°F)	1328
Inherent Orientation of Material	MP

PHYSICAL PROPERTIES	
Density (lbs./in.^{3})	.3035
General Mechanical Properties	HB
Tensile Strength (psi)	5000
Transverse Modulus of Rupture (psi)	-
Coeff. of Therm. Expansion inches x 10^{-6}/°C — Orientation: Par. / Rt.Angl.	7.0 / 13.0
Resistivity (micro-ohms/cm/cm^2 - @ - °C)	
Hardness — Scale	550 Vickers
Basic Forming Method	DPM
Basic Finishing Method	WGR
Relative Corrosion Resistance	G
Primary Chemical Composition	Sm-Co

Special Notes:

(1) Higher temperature stability than most other rare earth materials.

Note: All information based on available producers literature. In some cases of foreign materials interpretation or calculation required.

SOURCE(S):

Vacuumschmelze GMBH

Germany 27/18

Vacomax 225HR
(Alloy or Grade)

Sintered Rare Earth
(Type of Material)

___
(MMPA Brief Designation)

R5-2
(IEC Code Reference)

MAGNETIC PROPERTIES		PHYSICAL PROPERTIES	
Residual Induction B_r (G)	11,000	Density (lbs./in.^{3})	.3035
Coercive Force (Oe) — Normal (H_c)	9400	General Mechanical Properties	HB
Coercive Force (Oe) — Intrinsic (H_{ci})	18,000	Tensile Strength (psi)	5000
Maximum Energy Product BH_{max} (MGO)	27.0	Transverse Modulus of Rupture (psi)	—
B/H Ratio (Load Line) at BH_{max}	.53	Coeff. of Therm. Expansion inches x 10^{-6}/°C	Orientation: Par. 10.0 Rt.Angl. 12.0
Flux Density in Magnet B_m at BH_{max} (G)	3800	Resistivity (micro-ohms/cm/cm^2)	— @ — °C
Coercive Force in Magnet H_m at BH_{max} (Oe)	7200	Hardness Scale	640 Vickers
Energy Loss W_c (watt-sec/cycle/lb)	—	Basic Forming Method	DPM
Importance of Operating Magnet at BH_{max}	E	Basic Finishing Method	WGR
Average Recoil Permeability	1.05	Relative Corrosion Resistance	G
Field Strength Required to Fully Saturate Magnet (Oe)	40,000	Primary Chemical Composition	Sm-Co
Ability to Withstand Demagnetizing Fields	0	Special Notes:	
Maximum Service Temperature (°F)	572 See Note (1)	(1) Higher temperature stability than most other rare earth materials	
Approximate Temperature Permanently Affecting Magnet (°F)	1472		
Inherent Orientation of Material	MP		

Note: All information based on available producers literature. In some cases of foreign materials interpretation or calculation required.

SOURCE(S):	Vacomax 225WZ
Vacuumschmelze GMBH	(Alloy or Grade)
	Sintered Rare Earth
	(Type of Material)
Germany	22/20
	(MMPA Brief Designation)
	R5-2
	(IEC Code Reference)

MAGNETIC PROPERTIES		PHYSICAL PROPERTIES	
Residual Induction B_r (G)	10,000	Density (lbs./in.^{3})	.3035
Coercive Force (Oe) — Normal (H_c)	8800	General Mechanical Properties	HB
Coercive Force (Oe) — Intrinsic (H_{ci})	20,000	Tensile Strength (psi)	5000
Maximum Energy Product BH_{max} (MGO)	22.0	Transverse Modulus of Rupture (psi)	-
B/H Ratio (Load Line) at BH_{max}	.49	Coeff. of Therm. Expansion inches x 10^{-6}/°C	Orientation: Par. 10.0 Rt.Angl. 12.0
Flux Density in Magnet B_m at BH_{max} (G)	3300	Resistivity (micro-ohms/cm/cm^2	- @ - °C
Coercive Force in Magnet H_m at BH_{max} (Oe)	6700	Hardness Scale	640 Vickers
Energy Loss W_c (watt-sec/cycle/lb)	-	Basic Forming Method	DPM
Importance of Operating Magnet at BH_{max}	E	Basic Finishing Method	WGR
Average Recoil Permeability	1.05	Relative Corrosion Resistance	G
Field Strength Required to Fully Saturate Magnet (Oe)	40,000	Primary Chemical Composition	Sm-Co
Ability to Withstand Demagnetizing Fields	0	Special Notes:	
Maximum Service Temperature (°F) See Note (1)	572	(1) Higher temperature stability than most other rare earth materials	
Approximate Temperature Permanently Affecting Magnet (°F)	1472		
Inherent Orientation of Material	MP		

Note: All information based on available producers literature. In some cases of foreign materials interpretation or calculation required.

SOURCE(S):

Vacuumschmelze GMBH

Germany

Vacozet 200
(Alloy or Grade)

Wrought
(Type of Material)

-
(MMPA Brief Designation)

-
(IEC Code Reference)

MAGNETIC PROPERTIES	
Residual Induction B_r (G)	14,500
Coercive Force (Oe) — Normal (H_c)	17.0
Coercive Force (Oe) — Intrinsic (H_{ci})	-
Maximum Energy Product BH_{max} (MGO)	.19
B/H Ratio (Load Line) at BH_{max}	933.0
Flux Density in Magnet B_m at BH_{max} (G)	14,000
Coercive Force in Magnet H_m at BH_{max} (Oe)	15.0
Energy Loss W_c (watt-sec/cycle/lb)	-
Importance of Operating Magnet at BH_{max}	A
Average Recoil Permeability	-
Field Strength Required to Fully Saturate Magnet (Oe)	2500
Ability to Withstand Demagnetizing Fields	P
Maximum Service Temperature (°F)	752
Approximate Temperature Permanently Affecting Magnet (°F)	1832
Inherent Orientation of Material	CR

PHYSICAL PROPERTIES	
Density (lbs./in.^{3})	.3107
General Mechanical Properties	See Notes
Tensile Strength (psi)	-
Transverse Modulus of Rupture (psi)	-
Coeff. of Therm. Expansion inches x 10^{-6}/°C	Orientation: Par. 12.2 Rt.Angl. -
Resistivity (micro-ohms/cm/cm^2)	- @ - °C
Hardness Scale	See Notes
Basic Forming Method	W
Basic Finishing Method	See Notes
Relative Corrosion Resistance	G
Primary Chemical Composition	Fe-Co-Ni + Al-T or Nb

Special Notes:
(1) Vickers Hardness:
  (a) cold worked = 330-380
  (b) after final H.T. = 380-430
(2) Final forming must be done before H.T. with carbide tools. Must be ground after H.T.
(3) Can be used for good glass-to-metal seals.

Note: All information based on available producers literature. In some cases of foreign materials interpretation or calculation required.

SOURCE(S): **Vacuumschmelze GMBH, Germany**

Vacozet 258
(Alloy or Grade)

Wrought
(Type of Material)

-
(MMPA Brief Designation)

-
(IEC Code Reference)

MAGNETIC PROPERTIES	
Residual Induction B_r (G)	13,750
Coercive Force (Oe) Normal (H_c)	34.5
Coercive Force (Oe) Intrinsic (H_{ci})	-
Maximum Energy Product BH_{max} (MGO)	.32
B/H Ratio (Load Line) at BH_{max}	359.0
Flux Density in Magnet B_m at BH_{max} (G)	10,500
Coercive Force in Magnet H_m at BH_{max} (Oe)	32.0
Energy Loss W_c (watt-sec/cycle/lb)	-
Importance of Operating Magnet at BH_{max}	A
Average Recoil Permeability	-
Field Strength Required to Fully Saturate Magnet (Oe)	2500
Ability to Withstand Demagnetizing Fields	P
Maximum Service Temperature (°F)	752
Approximate Temperature Permanently Affecting Magnet (°F)	1472
Inherent Orientation of Material	CR

PHYSICAL PROPERTIES	
Density (lbs./in.^{3})	.2926
General Mechanical Properties	See Notes
Tensile Strength (psi)	-
Transverse Modulus of Rupture (psi)	-
Coeff. of Therm. Expansion inches x 10^{-6}/°C	Orientation: Par. 10.7 Rt.Angl. -
Resistivity (micro-ohms/cm/cm^2	- @ - °C
Hardness Scale	See Notes
Basic Forming Method	W
Basic Finishing Method	See Notes
Relative Corrosion Resistance	G
Primary Chemical Composition	Fe-Co-Ni + Al-T or Nb

Special Notes:
(1) Vickers Hardness:
 (a) cold worked = 380-430
 (b) after final H.T. = 480-630
(2) Final forming must be done before H.T. with carbide tools. Must be ground after H.T.
(3) Can be used for good glass-to-metal seals.

Note: All information based on available producers literature. In some cases of foreign materials interpretation or calculation required.

SOURCE(S):

Vacuumschmelze GMBH

Germany

Vacozet 655
(Alloy or Grade)

Wrought
(Type of Material)

-
(MMPA Brief Designation)

-
(IEC Code Reference)

MAGNETIC PROPERTIES	
Residual Induction B_r (G)	13,500
Coercive Force (Oe) Normal (H_c)	51.0
Coercive Force (Oe) Intrinsic (H_{ci})	-
Maximum Energy Product BH_{max} (MGO)	.51
B/H Ratio (Load Line) at BH_{max}	203
Flux Density in Magnet B_m at BH_{max} (G)	11,000
Coercive Force in Magnet H_m at BH_{max} (Oe)	54.0
Energy Loss W_c (watt-sec/cycle/lb)	-
Importance of Operating Magnet at BH_{max}	A
Average Recoil Permeability	-
Field Strength Required to Fully Saturate Magnet (Oe)	2500
Ability to Withstand Demagnetizing Fields	P
Maximum Service Temperature (°F)	752
Approximate Temperature Permanently Affecting Magnet (°F)	1472
Inherent Orientation of Material	CR

PHYSICAL PROPERTIES	
Density (lbs./in.^{3})	.2926
General Mechanical Properties	See Notes
Tensile Strength (psi)	-
Transverse Modulus of Rupture (psi)	-
Coeff. of Therm. Expansion inches x 10^{-6}/°C	Orientation: Par. 11.0 Rt.Angl. -
Resistivity (micro-ohms/cm/cm^2)	- @ - °C
Hardness Scale	See Notes
Basic Forming Method	W
Basic Finishing Method	See Notes
Relative Corrosion Resistance	G
Primary Chemical Composition	Fe-Co-Ni + Al-T or Nb
Special Notes: (1) Vickers Hardness: (a) cold worked = 400-450 (b) after final H.T. = 600-700 (2) Final forming must be done before H.T. with carbide tools. Must be ground after H.T. (3) Can be used for good glass-to-metal seals.	

Note: All information based on available producers literature. In some cases of foreign materials interpretation or calculation required.

SOURCE(S):	YBM-1A
Hitachi Magnetics Corp.	(Alloy or Grade)
	Sintered Ferrite
	(Type of Material)
USA	3.5/3.0
	(MMPA Brief Designation)
	-
	(IEC Code Reference)

MAGNETIC PROPERTIES		PHYSICAL PROPERTIES	
Residual Induction B_r (G)	3900	Density (lbs./in.^{3})	.1770
Coercive Force (Oe) — Normal (H_c)	1900	General Mechanical Properties	VHB
Coercive Force (Oe) — Intrinsic (H_{ci})	3000	Tensile Strength (psi)	-
Maximum Energy Product BH_{max} (MGO)	3.5	Transverse Modulus of Rupture (psi)	-
B/H Ratio (Load Line) at BH_{max}	1.1	Coeff. of Therm. Expansion inches x 10^{-6}/°C	Orientation: Par. - Rt.Angl. -
Flux Density in Magnet B_m at BH_{max} (G)	2000	Resistivity (micro-ohms/cm/cm^2)	1.0 @ 25 °C
Coercive Force in Magnet H_m at BH_{max} (Oe)	1750	Hardness Scale	6.5 Moh's
Energy Loss W_c (watt-sec/cycle/lb)	-	Basic Forming Method	PMF
Importance of Operating Magnet at BH_{max}	E	Basic Finishing Method	WGD
Average Recoil Permeability	1.1	Relative Corrosion Resistance	Ex
Field Strength Required to Fully Saturate Magnet (Oe)	10,000	Primary Chemical Composition	Ba or Sr-Fe
Ability to Withstand Demagnetizing Fields	Ex	Special Notes: (1) Magnetic properties are temperature sensitive.	
Maximum Service Temperature (°F)	482		
Approximate Temperature Permanently Affecting Magnet (°F)	842		
Inherent Orientation of Material	MP		

Note: All information based on available producers literature. In some cases of foreign materials interpretation or calculation required.

SOURCE(S):	YBM-1B
	(Alloy or Grade)
Hitachi Magnetics Corp.	Sintered Ferrite
	(Type of Material)
USA	3.0/3.2
	(MMPA Brief Designation)
	-
	(IEC Code Reference)

MAGNETIC PROPERTIES		PHYSICAL PROPERTIES	
Residual Induction B_r (G)	4050	Density (lbs./in.^{3})	.1770
Coercive Force (Oe) — Normal (H_c)	2000	General Mechanical Properties	VHB
Coercive Force (Oe) — Intrinsic (H_{ci})	3200	Tensile Strength (psi)	-
Maximum Energy Product BH_{max} (MGO)	3.0	Transverse Modulus of Rupture (psi)	-
B/H Ratio (Load Line) at BH_{max}	1.1	Coeff. of Therm. Expansion inches x 10^{-6}/°C	Orientation: Par. - Rt.Angl. -
Flux Density in Magnet B_m at BH_{max} (G)	2000	Resistivity (micro-ohms/cm/cm^2)	1.0 @ 25 °C
Coercive Force in Magnet H_m at BH_{max} (Oe)	1750	Hardness Scale	6.5 Moh's
Energy Loss W_c (watt-sec/cycle/lb)	-	Basic Forming Method	PMF
Importance of Operating Magnet at BH_{max}	E	Basic Finishing Method	WGD
Average Recoil Permeability	1.1	Relative Corrosion Resistance	Ex
Field Strength Required to Fully Saturate Magnet (Oe)	10,000	Primary Chemical Composition	Ba or Sr-Fe
Ability to Withstand Demagnetizing Fields	Ex	Special Notes: (1) Magnetic properties are temperature sensitive.	
Maximum Service Temperature (°F)	482		
Approximate Temperature Permanently Affecting Magnet (°F)	842		
Inherent Orientation of Material	MP		

Note: All information based on available producers literature. In some cases of foreign materials interpretation or calculation required.

SOURCE(S):	**YBM-1BB**
	(Alloy or Grade)
Hitachi Magnetics Corp.	Sintered Ferrite
	(Type of Material)
USA	3.7/3.4
	(MMPA Brief Designation)
	-
	(IEC Code Reference)

MAGNETIC PROPERTIES		PHYSICAL PROPERTIES	
Residual Induction B_r (G)	3950	Density (lbs./in.^{3})	.1770
Coercive Force (Oe) Normal (H_c)	2400	General Mechanical Properties	VHB
Coercive Force (Oe) Intrinsic (H_{ci})	3400	Tensile Strength (psi)	-
Maximum Energy Product BH_{max} (MGO)	3.7	Transverse Modulus of Rupture (psi)	-
B/H Ratio (Load Line) at BH_{max}	.90	Coeff. of Therm. Expansion inches x 10^{-6}/°C	Orientation: Par. - Rt.Angl. -
Flux Density in Magnet B_m at BH_{max} (G)	1850	Resistivity (micro-ohms/cm/cm^2)	1.0 @ 25 °C
Coercive Force in Magnet H_m at BH_{max} (Oe)	2000	Hardness Scale	6.5 Moh's
Energy Loss W_c (watt-sec/cycle/lb)	-	Basic Forming Method	PMF
Importance of Operating Magnet at BH_{max}	E	Basic Finishing Method	WGD
Average Recoil Permeability	1.1	Relative Corrosion Resistance	Ex
Field Strength Required to Fully Saturate Magnet (Oe)	10,000	Primary Chemical Composition	Ba or Sr-Fe
Ability to Withstand Demagnetizing Fields	Ex	Special Notes: (1) Magnetic properties are temperature sensitive.	
Maximum Service Temperature (°F)	482		
Approximate Temperature Permanently Affecting Magnet (°F)	842		
Inherent Orientation of Material	MP		

Note: All information based on available producers literature. In some cases of foreign materials interpretation or calculation required.

SOURCE(S):	
	YBM-2BE (Alloy or Grade)
Hitachi Magnetics Corp.	Sintered Ferrite (Type of Material)
USA	3.4/4.1 (MMPA Brief Designation)
	- (IEC Code Reference)

MAGNETIC PROPERTIES		PHYSICAL PROPERTIES	
Residual Induction B_r (G)	3800	Density (lbs./in.^{3})	.1770
Coercive Force (Oe) Normal (H_c)	3550	General Mechanical Properties	VHB
Coercive Force (Oe) Intrinsic (H_{ci})	4100	Tensile Strength (psi)	-
Maximum Energy Product BH_{max} (MGO)	3.4	Transverse Modulus of Rupture (psi)	-
B/H Ratio (Load Line) at BH_{max}	1.0	Coeff. of Therm. Expansion inches x 10^{-6}/°C	Orientation: Par. - Rt.Angl. -
Flux Density in Magnet B_m at BH_{max} (G)	1850	Resistivity (micro-ohms/cm/cm^2)	1.0 @ 25 °C
Coercive Force in Magnet H_m at BH_{max} (Oe)	1850	Hardness Scale	6.5 Moh's
Energy Loss W_c (watt-sec/cycle/lb)	-	Basic Forming Method	PMF
Importance of Operating Magnet at BH_{max}	E	Basic Finishing Method	WGD
Average Recoil Permeability	1.1	Relative Corrosion Resistance	Ex
Field Strength Required to Fully Saturate Magnet (Oe)	10,000	Primary Chemical Composition	Ba or Sr-Fe
Ability to Withstand Demagnetizing Fields	Ex	Special Notes: (1) Magnetic properties are temperature sensitive.	
Maximum Service Temperature (°F)	482		
Approximate Temperature Permanently Affecting Magnet (°F)	842		
Inherent Orientation of Material	MP		

Note: All information based on available producers literature. In some cases of foreign materials interpretation or calculation required.

SOURCE(S):	YBM-2BF
Hitachi Magnetics Corp.	(Alloy or Grade)
	Sintered Ferrite
	(Type of Material)
USA	3.3/5.0
	(MMPA Brief Designation)
	-
	(IEC Code Reference)

MAGNETIC PROPERTIES		PHYSICAL PROPERTIES	
Residual Induction B_r (G)	3750	Density (lbs./in.^{3})	.1770
Coercive Force (Oe) — Normal (H_c)	3550	General Mechanical Properties	VHB
Coercive Force (Oe) — Intrinsic (H_{ci})	5000	Tensile Strength (psi)	-
Maximum Energy Product BH_{max} (MGO)	3.3	Transverse Modulus of Rupture (psi)	-
B/H Ratio (Load Line) at BH_{max}	.83	Coeff. of Therm. Expansion inches x 10^{-6}/°C Orientation: Par. - Rt.Angl. -	
Flux Density in Magnet B_m at BH_{max} (G)	1650	Resistivity (micro-ohms/cm/cm^2)	1.0 @ 25 °C
Coercive Force in Magnet H_m at BH_{max} (Oe)	2000	Hardness Scale	6.5 Moh's
Energy Loss W_c (watt-sec/cycle/lb)	-	Basic Forming Method	PMF
Importance of Operating Magnet at BH_{max}	E	Basic Finishing Method	WGD
Average Recoil Permeability	1.1	Relative Corrosion Resistance	Ex
Field Strength Required to Fully Saturate Magnet (Oe)	10,000	Primary Chemical Composition	Ba or Sr-Fe
Ability to Withstand Demagnetizing Fields	Ex	Special Notes: (1) Magnetic properties are temperature sensitive.	
Maximum Service Temperature (°F)	482		
Approximate Temperature Permanently Affecting Magnet (°F)	842		
Inherent Orientation of Material	MP		

Note: All information based on available producers literature. In some cases of foreign materials interpretation or calculation required.

SOURCE(S):	YBM-2C
	(Alloy or Grade)
Hitachi Magnetics Corp.	Sintered Ferrite
	(Type of Material)
USA	3.2/3.3
	(MMPA Brief Designation)
	-
	(IEC Code Reference)

MAGNETIC PROPERTIES		PHYSICAL PROPERTIES	
Residual Induction B_r (G)	3700	Density (lbs./in.^{3})	.1770
Coercive Force (Oe) — Normal (H_c)	3000	General Mechanical Properties	VHB
Coercive Force (Oe) — Intrinsic (H_{ci})	3250	Tensile Strength (psi)	-
Maximum Energy Product BH_{max} (MGO)	3.2	Transverse Modulus of Rupture (psi)	-
B/H Ratio (Load Line) at BH_{max}	1.8	Coeff. of Therm. Expansion inches x 10^{-6}/°C	Orientation: Par. - Rt.Angl. -
Flux Density in Magnet B_m at BH_{max} (G)	2400	Resistivity (micro-ohms/cm/cm^2)	1.0 @ 25 °C
Coercive Force in Magnet H_m at BH_{max} (Oe)	1350	Hardness 6.5	Scale Moh's
Energy Loss W_c (watt-sec/cycle/lb)	-	Basic Forming Method	PMF
Importance of Operating Magnet at BH_{max}	E	Basic Finishing Method	WGD
Average Recoil Permeability	1.1	Relative Corrosion Resistance	Ex
Field Strength Required to Fully Saturate Magnet (Oe)	10,000	Primary Chemical Composition	Ba or Sr-Fe
Ability to Withstand Demagnetizing Fields	Ex	Special Notes: (1) Magnetic properties are temperature sensitive.	
Maximum Service Temperature (°F)	482		
Approximate Temperature Permanently Affecting Magnet (°F)	842		
Inherent Orientation of Material	MP		

Note: All information based on available producers literature. In some cases of foreign materials interpretation or calculation required.

SOURCE(S):	**YBM-2D**
	(Alloy or Grade)
Hitachi Magnetics Corp.	Sintered Ferrite
	(Type of Material)
USA	-
	(MMPA Brief Designation)
	-
	(IEC Code Reference)

MAGNETIC PROPERTIES		PHYSICAL PROPERTIES	
Residual Induction B_r (G)	3500	Density (lbs./in.^{3})	.1770
Coercive Force (Oe) Normal (H_c)	2950	General Mechanical Properties	VHB
Coercive Force (Oe) Intrinsic (H_{ci})	3250	Tensile Strength (psi)	Low
Maximum Energy Product BH_{max} (MGO)	2.9	Transverse Modulus of Rupture (psi)	-
B/H Ratio (Load Line) at BH_{max}	1.03	Coeff. of Therm. Expansion inches x 10^{-6}/°C	Orientation: Par. - Rt.Angl. -
Flux Density in Magnet B_m at BH_{max} (G)	1750	Resistivity (micro-ohms/cm/cm^2 10^6	@ 25 °C
Coercive Force in Magnet H_m at BH_{max} (Oe)	1700	Hardness Scale 6.5 Moh's	
Energy Loss W_c (watt-sec/cycle/lb)	-	Basic Forming Method	PMF
Importance of Operating Magnet at BH_{max}	D	Basic Finishing Method	WGD
Average Recoil Permeability	1.1	Relative Corrosion Resistance	Ex
Field Strength Required to Fully Saturate Magnet (Oe)	7000	Primary Chemical Composition	Ba or Sr-Fe
Ability to Withstand Demagnetizing Fields	E	Special Notes: (1) Magnetic properties are temperature sensitive.	
Maximum Service Temperature (°F)	482		
Approximate Temperature Permanently Affecting Magnet (°F)	842		
Inherent Orientation of Material	MP		

Note: All information based on available producers literature. In some cases of foreign materials interpretation or calculation required.

SOURCE(S):	**YBM-4A**
	(Alloy or Grade)
Hitachi Magnetics Corp.	Sintered Ferrite
	(Type of Material)
USA	-
	(MMPA Brief Designation)
	-
	(IEC Code Reference)

MAGNETIC PROPERTIES		PHYSICAL PROPERTIES	
Residual Induction B_r (G)	2500	Density (lbs./in.^{3})	.1770
Coercive Force (Oe) — Normal (H_c)	1900	General Mechanical Properties	VHB
Coercive Force (Oe) — Intrinsic (H_{ci})	-	Tensile Strength (psi)	Low
Maximum Energy Product BH_{max} (MGO)	1.3	Transverse Modulus of Rupture (psi)	-
B/H Ratio (Load Line) at BH_{max}	1.0	Coeff. of Therm. Expansion inches x 10^{-6}/°C Orientation: Par. - Rt.Angl. -	
Flux Density in Magnet B_m at BH_{max} (G)	2000	Resistivity (micro-ohms/cm/cm^2) 10^6 @ 25 °C	
Coercive Force in Magnet H_m at BH_{max} (Oe)	2000	Hardness Scale 6.5 Moh's	
Energy Loss W_c (watt-sec/cycle/lb)	-	Basic Forming Method	PMF
Importance of Operating Magnet at BH_{max}	D	Basic Finishing Method	WGD
Average Recoil Permeability	1.1	Relative Corrosion Resistance	E
Field Strength Required to Fully Saturate Magnet (Oe)	7000	Primary Chemical Composition	Ba or Sr-Fe
Ability to Withstand Demagnetizing Fields	E	Special Notes: (1) Magnetic properties are temperature sensitive.	
Maximum Service Temperature (°F)	482		
Approximate Temperature Permanently Affecting Magnet (°F)	842		
Inherent Orientation of Material	MP		

Note: All information based on available producers literature. In some cases of foreign materials interpretation or calculation required.

SOURCE(S):	YBM-4B
Hitachi Magnetics Corp. USA	(Alloy or Grade)
	Sintered Ferrite
	(Type of Material)
	-
	(MMPA Brief Designation)
	-
	(IEC Code Reference)

MAGNETIC PROPERTIES		PHYSICAL PROPERTIES	
Residual Induction B_r (G)	2700	Density (lbs./in.^{3})	.1770
Coercive Force (Oe) — Normal (H_c)	2200	General Mechanical Properties	VHB
Coercive Force (Oe) — Intrinsic (H_{ci})	-	Tensile Strength (psi)	Low
Maximum Energy Product BH_{max} (MGO)	1.5	Transverse Modulus of Rupture (psi)	-
B/H Ratio (Load Line) at BH_{max}	1.0	Coeff. of Therm. Expansion inches x 10^{-6}/°C	Orientation: Par. - Rt.Angl. -
Flux Density in Magnet B_m at BH_{max} (G)	2000	Resistivity (micro-ohms/cm/cm^2)	10^6 @ 25 °C
Coercive Force in Magnet H_m at BH_{max} (Oe)	2000	Hardness 6.5	Scale Moh's
Energy Loss W_c (watt-sec/cycle/lb)	-	Basic Forming Method	PMF
Importance of Operating Magnet at BH_{max}	D	Basic Finishing Method	WGD
Average Recoil Permeability	1.1	Relative Corrosion Resistance	E
Field Strength Required to Fully Saturate Magnet (Oe)	7000	Primary Chemical Composition	Ba or Sr-Fe
Ability to Withstand Demagnetizing Fields	E	Special Notes: (1) Magnetic properties are temperature sensitive.	
Maximum Service Temperature (°F)	482		
Approximate Temperature Permanently Affecting Magnet (°F)	842		
Inherent Orientation of Material	MP		

Note: All information based on available producers literature. In some cases of foreign materials interpretation or calculation required.

SOURCE(S): Hitachi Magnetics Corp. USA

YBM-4D
(Alloy or Grade)

Sintered Ferrite
(Type of Material)

-
(MMPA Brief Designation)

-
(IEC Code Reference)

MAGNETIC PROPERTIES		PHYSICAL PROPERTIES	
Residual Induction B_r (G)	2500	Density (lbs./in.^{3})	.1770
Coercive Force (Oe) Normal (H_c)	1990	General Mechanical Properties	VHB
Coercive Force (Oe) Intrinsic (H_{ci})	Est. 3100	Tensile Strength (psi)	-
Maximum Energy Product BH_{max} (MGO)	1.3	Transverse Modulus of Rupture (psi)	-
B/H Ratio (Load Line) at BH_{max}	-	Coeff. of Therm. Expansion inches x 10^{-6}/°C	Orientation: Par. - Rt.Angl. -
Flux Density in Magnet B_m at BH_{max} (G)	-	Resistivity (micro-ohms/cm/cm^2)	1.0 @ 25 °C
Coercive Force in Magnet H_m at BH_{max} (Oe)	-	Hardness 6.5	Scale Moh's
Energy Loss W_c (watt-sec/cycle/lb)	-	Basic Forming Method	PMF
Importance of Operating Magnet at BH_{max}	E	Basic Finishing Method	WGD
Average Recoil Permeability	1.1	Relative Corrosion Resistance	Ex
Field Strength Required to Fully Saturate Magnet (Oe)	10,000	Primary Chemical Composition	Ba or Sr-Fe
Ability to Withstand Demagnetizing Fields	Ex	Special Notes: (1) Magnetic properties are temperature sensitive.	
Maximum Service Temperature (°F)	482		
Approximate Temperature Permanently Affecting Magnet (°F)	842		
Inherent Orientation of Material	MP		

Note: All information based on available producers literature. In some cases of foreign materials interpretation or calculation required.

SOURCE(S):

Hitachi Magnetics Corp.

USA

YBM-4E (rings)
(Alloy or Grade)

Sintered Ferrite
(Type of Material)

−
(MMPA Brief Designation)

−
(IEC Code Reference)

MAGNETIC PROPERTIES		PHYSICAL PROPERTIES	
Residual Induction B_r (G)	See Special Note	Density (lbs./in.^{3})	.1770
Coercive Force (Oe) Normal (H_c)	"	General Mechanical Properties	VHB
Coercive Force (Oe) Intrinsic (H_{ci})	"	Tensile Strength (psi)	−
Maximum Energy Product BH_{max} (MGO)	"	Transverse Modulus of Rupture (psi)	−
B/H Ratio (Load Line) at BH_{max}	"	Coeff. of Therm. Expansion inches x 10^{-6}/°C	Orientation: Par. _ Rt.Angl. _
Flux Density in Magnet B_m at BH_{max} (G)	"	Resistivity (micro-ohms/cm/cm^2)	1.0 @ 25 °C
Coercive Force in Magnet H_m at BH_{max} (Oe)	"	Hardness Scale	6.5 Moh's
Energy Loss W_c (watt-sec/cycle/lb)	−	Basic Forming Method	PMF
Importance of Operating Magnet at BH_{max}	E	Basic Finishing Method	WGD
Average Recoil Permeability	1.1	Relative Corrosion Resistance	Ex
Field Strength Required to Fully Saturate Magnet (Oe)	10,000	Primary Chemical Composition	Ba or Sr-Fe
Ability to Withstand Demagnetizing Fields	Ex	Special Notes: (1) Magnetic properties are temperature sensitive. (2) YBM-4E if only available as a ring-shaped magnet, so it is difficult to measure B_r, H_c & BH_{max}.	
Maximum Service Temperature (°F)	482		
Approximate Temperature Permanently Affecting Magnet (°F)	842		
Inherent Orientation of Material	MP		

Note: All information based on available producers literature. In some cases of foreign materials interpretation or calculation required.

SOURCE(S):

Hitachi Magnetics Corp.

USA

YBM-4F (rings)
(Alloy or Grade)

Sintered Ferrite
(Type of Material)

-
(MMPA Brief Designation)

-
(IEC Code Reference)

MAGNETIC PROPERTIES			PHYSICAL PROPERTIES	
Residual Induction B_r (G)		See Special Note	Density (lbs./in.^{3})	.1770
Coercive Force (Oe)	Normal (H_c)	"	General Mechanical Properties	VHB
	Intrinsic (H_{ci})	"	Tensile Strength (psi)	-
Maximum Energy Product BH_{max} (MGO)		"	Transverse Modulus of Rupture (psi)	-
B/H Ratio (Load Line) at BH_{max}		"	Coeff. of Therm. Expansion inches x 10^{-6}/°C	Orientation: Par. - Rt.Angl. -
Flux Density in Magnet B_m at BH_{max} (G)		"	Resistivity (micro-ohms/cm/cm^2)	1.0 @ 25 °C
Coercive Force in Magnet H_m at BH_{max} (Oe)		"	Hardness 6.5	Scale Moh's
Energy Loss W_c (watt-sec/cycle/lb)		-	Basic Forming Method	PMF
Importance of Operating Magnet at BH_{max}		E	Basic Finishing Method	WGD
Average Recoil Permeability		1.1	Relative Corrosion Resistance	Ex
Field Strength Required to Fully Saturate Magnet (Oe)		10,000	Primary Chemical Composition	Ba or Sr-Fe
Ability to Withstand Demagnetizing Fields		Ex	Special Notes: (1) Magnetic properties are temperature sensitive. (2) YBM-4F is only available as a ring-shaped magnet, so it is difficult to measure B_r, H_c & BH_{max}.	
Maximum Service Temperature (°F)		482		
Approximate Temperature Permanently Affecting Magnet (°F)		842		
Inherent Orientation of Material		MP		

Note: All information based on available producers literature. In some cases of foreign materials interpretation or calculation required.

SOURCE(S):	**YBM-5BB**
	(Alloy or Grade)
Hitachi Magnetics Corp.	Sintered Ferrite
	(Type of Material)
USA	4.4/2.9
	(MMPA Brief Designation)
	-
	(IEC Code Reference)

MAGNETIC PROPERTIES		PHYSICAL PROPERTIES	
Residual Induction B_r (G)	4300	Density (lbs./in.^{3})	.1770
Coercive Force (Oe) — Normal (H_c)	2850	General Mechanical Properties	VHB
Coercive Force (Oe) — Intrinsic (H_{ci})	2900	Tensile Strength (psi)	-
Maximum Energy Product BH_{max} (MGO)	4.4	Transverse Modulus of Rupture (psi)	-
B/H Ratio (Load Line) at BH_{max}	.58	Coeff. of Therm. Expansion inches x 10^{-6}/°C — Orientation: Par. / Rt.Angl.	- / -
Flux Density in Magnet B_m at BH_{max} (G)	1600	Resistivity (micro-ohms/cm/cm^2)	1.0 @ 25 °C
Coercive Force in Magnet H_m at BH_{max} (Oe)	2750	Hardness Scale	6.5 Moh's
Energy Loss W_c (watt-sec/cycle/lb)	-	Basic Forming Method	PMF
Importance of Operating Magnet at BH_{max}	E	Basic Finishing Method	WGD
Average Recoil Permeability	1.1	Relative Corrosion Resistance	Ex
Field Strength Required to Fully Saturate Magnet (Oe)	10,000	Primary Chemical Composition	Ba or Sr-Fe
Ability to Withstand Demagnetizing Fields	Ex	Special Notes: (1) Magnetic properties are temperature sensitive.	
Maximum Service Temperature (°F)	482		
Approximate Temperature Permanently Affecting Magnet (°F)	842		
Inherent Orientation of Material	MP		

Note: All information based on available producers literature. In some cases of foreign materials interpretation or calculation required.

SOURCE(S):

Hitachi Magnetics Corp.

USA

YBM-5BD
(Alloy or Grade)

Sintered Ferrite
(Type of Material)

4.0/3.6
(MMPA Brief Designation)

-
(IEC Code Reference)

MAGNETIC PROPERTIES	
Residual Induction B_r (G)	4100
Coercive Force (Oe) — Normal (H_c)	3450
Coercive Force (Oe) — Intrinsic (H_{ci})	3600
Maximum Energy Product BH_{max} (MGO)	4.0
B/H Ratio (Load Line) at BH_{max}	.56
Flux Density in Magnet B_m at BH_{max} (G)	1500
Coercive Force in Magnet H_m at BH_{max} (Oe)	2700
Energy Loss W_c (watt-sec/cycle/lb)	-
Importance of Operating Magnet at BH_{max}	E
Average Recoil Permeability	1.1
Field Strength Required to Fully Saturate Magnet (Oe)	10,000
Ability to Withstand Demagnetizing Fields	Ex
Maximum Service Temperature (°F)	482
Approximate Temperature Permanently Affecting Magnet (°F)	842
Inherent Orientation of Material	MP

PHYSICAL PROPERTIES	
Density (lbs./in.^{3})	.1770
General Mechanical Properties	VHB
Tensile Strength (psi)	-
Transverse Modulus of Rupture (psi)	-
Coeff. of Therm. Expansion inches x 10^{-6}/°C — Orientation: Par. / Rt.Angl.	- / -
Resistivity (micro-ohms/cm/cm^2)	1.0 @ 25 °C
Hardness Scale	6.5 Moh's
Basic Forming Method	PMF
Basic Finishing Method	WGD
Relative Corrosion Resistance	Ex
Primary Chemical Composition	Ba or Sr-Fe

Special Notes:

(1) Magnetic properties are temperature sensitive.

Note: All information based on available producers literature. In some cases of foreign materials interpretation or calculation required.

SOURCE(S):	YBM-5BE
	(Alloy or Grade)
Hitachi Magnetics Corp.	Sintered Ferrite
	(Type of Material)
USA	3.8/4.1
	(MMPA Brief Designation)
	-
	(IEC Code Reference)

MAGNETIC PROPERTIES		PHYSICAL PROPERTIES	
Residual Induction B_r (G)	4000	Density (lbs./in.^{3})	.1770
Coercive Force (Oe) — Normal (H_c)	3750	General Mechanical Properties	VHB
Coercive Force (Oe) — Intrinsic (H_{ci})	4050	Tensile Strength (psi)	-
Maximum Energy Product BH_{max} (MGO)	3.8	Transverse Modulus of Rupture (psi)	-
B/H Ratio (Load Line) at BH_{max}	.63	Coeff. of Therm. Expansion inches x 10^{-6}/°C	Orientation: Par. - Rt.Angl. -
Flux Density in Magnet B_m at BH_{max} (G)	1600	Resistivity (micro-ohms/cm/cm^2)	1.0 @ 25 °C
Coercive Force in Magnet H_m at BH_{max} (Oe)	2400	Hardness Scale	6.5 Moh's
Energy Loss W_c (watt-sec/cycle/lb)	-	Basic Forming Method	PMF
Importance of Operating Magnet at BH_{max}	E	Basic Finishing Method	WGD
Average Recoil Permeability	1.1	Relative Corrosion Resistance	Ex
Field Strength Required to Fully Saturate Magnet (Oe)	10,000	Primary Chemical Composition	Ba or Sr-Fe
Ability to Withstand Demagnetizing Fields	Ex	Special Notes: (1) Magnetic properties are temperature sensitive.	
Maximum Service Temperature (°F)	482		
Approximate Temperature Permanently Affecting Magnet (°F)	842		
Inherent Orientation of Material	MP		

Note: All information based on available producers literature. In some cases of foreign materials interpretation or calculation required.

SOURCE(S):	**YBM-5BF**
	(Alloy or Grade)
Hitachi Magnetics Corp.	Sintered Ferrite
	(Type of Material)
USA	3.6/4.6
	(MMPA Brief Designation)
	-
	(IEC Code Reference)

MAGNETIC PROPERTIES		PHYSICAL PROPERTIES	
Residual Induction B_r (G)	3900	Density (lbs./in.^{3})	.1770
Coercive Force (Oe) — Normal (H_c)	3650	General Mechanical Properties	VHB
Coercive Force (Oe) — Intrinsic (H_{ci})	4600	Tensile Strength (psi)	-
Maximum Energy Product BH_{max} (MGO)	3.6	Transverse Modulus of Rupture (psi)	-
B/H Ratio (Load Line) at BH_{max}	.625	Coeff. of Therm. Expansion inches x 10^{-6}/°C — Orientation: Par. / Rt.Angl.	- / -
Flux Density in Magnet B_m at BH_{max} (G)	1500	Resistivity (micro-ohms/cm/cm^2)	1.0 @ 25 °C
Coercive Force in Magnet H_m at BH_{max} (Oe)	2600	Hardness Scale	6.5 Moh's
Energy Loss W_c (watt-sec/cycle/lb)	-	Basic Forming Method	PMF
Importance of Operating Magnet at BH_{max}	E	Basic Finishing Method	WGD
Average Recoil Permeability	1.1	Relative Corrosion Resistance	Ex
Field Strength Required to Fully Saturate Magnet (Oe)	10,000	Primary Chemical Composition	Ba or Sr-Fe
Ability to Withstand Demagnetizing Fields	Ex	Special Notes:	
Maximum Service Temperature (°F)	482	(1) Magnetic properties are temperature sensitive.	
Approximate Temperature Permanently Affecting Magnet (°F)	842		
Inherent Orientation of Material	MP		

Note: All information based on available producers literature. In some cases of foreign materials interpretation or calculation required.

SOURCE(S):

Hitachi Magnetics Corp.

USA

YCM-1B
(Alloy or Grade)

Cast Alnico
(Type of Material)

5.2/.63
(MMPA Brief Designation)

R1-1-1
(IEC Code Reference)

MAGNETIC PROPERTIES	
Residual Induction B_r (G)	13,000
Coercive Force (Oe) Normal (H_c)	630
Coercive Force (Oe) Intrinsic (H_{ci})	630
Maximum Energy Product BH_{max} (MGO)	5.2
B/H Ratio (Load Line) at BH_{max}	20.0
Flux Density in Magnet B_m at BH_{max} (G)	10,200
Coercive Force in Magnet H_m at BH_{max} (Oe)	510
Energy Loss W_c (watt-sec/cycle/lb)	15.3
Importance of Operating Magnet at BH_{max}	A
Average Recoil Permeability	4.3
Field Strength Required to Fully Saturate Magnet (Oe)	3000
Ability to Withstand Demagnetizing Fields	G
Maximum Service Temperature (°F)	900
Approximate Temperature Permanently Affecting Magnet (°F)	1580
Inherent Orientation of Material	CHT

PHYSICAL PROPERTIES	
Density (lbs./in.^{3})	.2640
General Mechanical Properties	HB
Tensile Strength (psi)	-
Transverse Modulus of Rupture (psi)	-
Coeff. of Therm. Expansion inches x 10^{-6}/°C	Orientation: Par. 11.6 Rt.Angl. _
Resistivity (micro-ohms/cm/cm^2)	47 @ 25 °C
Hardness 50	Scale Rockwell C
Basic Forming Method	CM
Basic Finishing Method	WGR
Relative Corrosion Resistance	G
Primary Chemical Composition	Al-Ni-Co-Cu-Fe

Special Notes:

(1) Lowest temperature co-efficient of all materials. Little irreversible loss even at high temperatures.
(2) Should be magnetized in final circuit.

Note: All information based on available producers literature. In some cases of foreign materials interpretation or calculation required.

SOURCE(S):	YCM-1D
Hitachi Magnetics Corp.	(Alloy or Grade)
	Cast Alnico
	(Type of Material)
USA	5.8/.69
	(MMPA Brief Designation)
	R1-1-2
	(IEC Code Reference)

MAGNETIC PROPERTIES		PHYSICAL PROPERTIES	
Residual Induction B_r (G)	13,000	Density (lbs./in.^{3})	.2640
Coercive Force (Oe) — Normal (H_c)	690	General Mechanical Properties	HB
Coercive Force (Oe) — Intrinsic (H_{ci})	690	Tensile Strength (psi)	-
Maximum Energy Product BH_{max} (MGO)	5.8	Transverse Modulus of Rupture (psi)	-
B/H Ratio (Load Line) at BH_{max}	17.2	Coeff. of Therm. Expansion inches x 10^{-6}/°C	Orientation: Par. 11.6 Rt.Angl. -
Flux Density in Magnet B_m at BH_{max} (G)	10,000	Resistivity (micro-ohms/cm/cm^2	47 @ 25 °C
Coercive Force in Magnet H_m at BH_{max} (Oe)	580	Hardness 50	Scale Rockwell C
Energy Loss W_c (watt-sec/cycle/lb)	15.3	Basic Forming Method	CM
Importance of Operating Magnet at BH_{max}	A	Basic Finishing Method	WGR
Average Recoil Permeability	4.3	Relative Corrosion Resistance	G
Field Strength Required to Fully Saturate Magnet (Oe)	3000	Primary Chemical Composition	Al-Ni-Co-Cu-Fe
Ability to Withstand Demagnetizing Fields	G	Special Notes:	
Maximum Service Temperature (°F)	900	(1) Lowest temperature co-efficient of all materials. Little irreversible loss even at high temperatures.	
Approximate Temperature Permanently Affecting Magnet (°F)	1580	(2) Should be magnetized in final circuit.	
Inherent Orientation of Material	CHT		

Note: All information based on available producers literature. In some cases of foreign materials interpretation or calculation required.

SOURCE(S):	
Hitachi Magnetics Corp. USA	**YCM-2B** (Alloy or Grade)
	Cast Alnico (Type of Material)
	4.0/.75 (MMPA Brief Designation)
	- (IEC Code Reference)

MAGNETIC PROPERTIES		PHYSICAL PROPERTIES	
Residual Induction B_r (G)	10,750	Density (lbs./in.^{3})	.2640
Coercive Force (Oe) — Normal (H_c)	725	General Mechanical Properties	HB
Coercive Force (Oe) — Intrinsic (H_{ci})	750	Tensile Strength (psi)	-
Maximum Energy Product BH_{max} (MGO)	4.0	Transverse Modulus of Rupture (psi)	-
B/H Ratio (Load Line) at BH_{max}	13.3	Coeff. of Therm. Expansion inches x 10^{-6}/°C — Orientation: Par. / Rt.Angl.	11.6 / _
Flux Density in Magnet B_m at BH_{max} (G)	7300	Resistivity (micro-ohms/cm/cm^2)	47 @ 25 °C
Coercive Force in Magnet H_m at BH_{max} (Oe)	550	Hardness / Scale	50 / Rockwell C
Energy Loss W_c (watt-sec/cycle/lb)	15.3	Basic Forming Method	CM
Importance of Operating Magnet at BH_{max}	A	Basic Finishing Method	WGR
Average Recoil Permeability	4.3	Relative Corrosion Resistance	G
Field Strength Required to Fully Saturate Magnet (Oe)	3000	Primary Chemical Composition	Al-Ni-Co-Cu-Fe
Ability to Withstand Demagnetizing Fields	G	Special Notes: (1) Lowest temperature coefficient of all materials. Little irreversible loss even at high temperatures. (2) Should be magnetized in final circuit.	
Maximum Service Temperature (°F)	900		
Approximate Temperature Permanently Affecting Magnet (°F)	1580		
Inherent Orientation of Material	HT		

Note: All information based on available producers literature. In some cases of foreign materials interpretation or calculation required.

SOURCE(S):	YCM-2C
Hitachi Magnetics Corp.	(Alloy or Grade)
	Cast Alnico
	(Type of Material)
USA	3.5/.80
	(MMPA Brief Designation)
	R1-1-4
	(IEC Code Reference)

MAGNETIC PROPERTIES		PHYSICAL PROPERTIES	
Residual Induction B_r (G)	10,000	Density (lbs./in.^{3})	.2680
Coercive Force (Oe) — Normal (H_c)	800	General Mechanical Properties	HB
Coercive Force (Oe) — Intrinsic (H_{ci})	800	Tensile Strength (psi)	–
Maximum Energy Product BH_{max} (MGO)	3.5	Transverse Modulus of Rupture (psi)	–
B/H Ratio (Load Line) at BH_{max}	11.6	Coeff. of Therm. Expansion inches x 10^{-6}/°C — Orientation: Par. 11.4 Rt.Angl. –	
Flux Density in Magnet B_m at BH_{max} (G)	6400	Resistivity (micro-ohms/cm/cm^2)	50 @ 25 °C
Coercive Force in Magnet H_m at BH_{max} (Oe)	550	Hardness 50 Scale Rockwell C	
Energy Loss W_c (watt-sec/cycle/lb)	15.3	Basic Forming Method	CM
Importance of Operating Magnet at BH_{max}	B	Basic Finishing Method	WGR
Average Recoil Permeability	4.3	Relative Corrosion Resistance	G
Field Strength Required to Fully Saturate Magnet (Oe)	3000	Primary Chemical Composition	Al-Ni-Co-Cu-Fe
Ability to Withstand Demagnetizing Fields	G	Special Notes:	
Maximum Service Temperature (°F)	900	(1) Lowest temperature coefficient of all materials. Little irreversible loss even at high temperatures.	
Approximate Temperature Permanently Affecting Magnet (°F)	1580	(2) Should be magnetized in final circuit.	
Inherent Orientation of Material	HT		

Note: All information based on available producers literature. In some cases of foreign materials interpretation or calculation required.

SOURCE(S):	YCM-4A
	(Alloy or Grade)
Hitachi Magnetics Corp.	Cast Alnico
	(Type of Material)
USA	1.5/.80
	(MMPA Brief Designation)
	-
	(IEC Code Reference)

MAGNETIC PROPERTIES		PHYSICAL PROPERTIES	
Residual Induction B_r (G)	6000	Density (lbs./in.^{3})	.2680
Coercive Force (Oe) — Normal (H_c)	750	General Mechanical Properties	HB
Coercive Force (Oe) — Intrinsic (H_{ci})	Approx. 800	Tensile Strength (psi)	-
Maximum Energy Product BH_{max} (MGO)	1.5	Transverse Modulus of Rupture (psi)	-
B/H Ratio (Load Line) at BH_{max}	7.6	Coeff. of Therm. Expansion inches x 10^{-6}/°C	Orientation: Par. 11.4 Rt.Angl. -
Flux Density in Magnet B_m at BH_{max} (G)	3400	Resistivity (micro-ohms/cm/cm^2)	50 @ 25 °C
Coercive Force in Magnet H_m at BH_{max} (Oe)	450	Hardness 50	Scale Rockwell C
Energy Loss W_c (watt-sec/cycle/lb)	15.3	Basic Forming Method	C
Importance of Operating Magnet at BH_{max}	A	Basic Finishing Method	WGR
Average Recoil Permeability	4.3	Relative Corrosion Resistance	G
Field Strength Required to Fully Saturate Magnet (Oe)	3000	Primary Chemical Composition	Al-Ni-Co-Cu-Fe
Ability to Withstand Demagnetizing Fields	G	Special Notes:	
Maximum Service Temperature (°F)	900	(1) Lowest temperature coefficient of all materials. Little irreversible loss even at high temperatures.	
Approximate Temperature Permanently Affecting Magnet (°F)	1580	(2) Should be magnetized in final circuit.	
Inherent Orientation of Material	none		

Note: All information based on available producers literature. In some cases of foreign materials interpretation or calculation required.

SOURCE(S):

Hitachi Magnetics Corp.

USA

YCM-4B
(Alloy or Grade)

Cast Alnico
(Type of Material)

1.7/.60
(MMPA Brief Designation)

-
(IEC Code Reference)

MAGNETIC PROPERTIES	
Residual Induction B_r (G)	6500
Coercive Force (Oe) — Normal (H_c)	600
Coercive Force (Oe) — Intrinsic (H_{ci})	Approx. 600
Maximum Energy Product BH_{max} (MGO)	1.7
B/H Ratio (Load Line) at BH_{max}	10.8
Flux Density in Magnet B_m at BH_{max} (G)	4300
Coercive Force in Magnet H_m at BH_{max} (Oe)	400
Energy Loss W_c (watt-sec/cycle/lb)	15.3
Importance of Operating Magnet at BH_{max}	A
Average Recoil Permeability	4.3
Field Strength Required to Fully Saturate Magnet (Oe)	3000
Ability to Withstand Demagnetizing Fields	G
Maximum Service Temperature (°F)	900
Approximate Temperature Permanently Affecting Magnet (°F)	1580
Inherent Orientation of Material	none

PHYSICAL PROPERTIES	
Density (lbs./in.^{3})	.2680
General Mechanical Properties	HB
Tensile Strength (psi)	-
Transverse Modulus of Rupture (psi)	-
Coeff. of Therm. Expansion inches x 10^{-6}/°C	Orientation: Par. 11.4 Rt.Angl. -
Resistivity (micro-ohms/cm/cm^2)	50 @ 25 °C
Hardness	50 Scale Rockwell C
Basic Forming Method	C
Basic Finishing Method	WGR
Relative Corrosion Resistance	G
Primary Chemical Composition	Al-Ni-Co-Cu-Fe

Special Notes:

(1) Lowest temperature coefficient of all materials. Little irreversible loss even at high temperatures.
(2) Should be magnetized in final circuit.

Note: All information based on available producers literature. In some cases of foreign materials interpretation or calculation required.

SOURCE(S):	**YCM-4C**
Hitachi Magnetics Corp.	(Alloy or Grade)
	Cast Alnico
	(Type of Material)
USA	1.6/.50
	(MMPA Brief Designation)
	-
	(IEC Code Reference)

MAGNETIC PROPERTIES			PHYSICAL PROPERTIES	
Residual Induction B_r (G)		7500	Density (lbs./in.^{3})	.2680
Coercive Force (Oe)	Normal (H_c)	500	General Mechanical Properties	HB
	Intrinsic (H_{ci})	Approx. 500	Tensile Strength (psi)	-
Maximum Energy Product BH_{max} (MGO)		1.6	Transverse Modulus of Rupture (psi)	-
B/H Ratio (Load Line) at BH_{max}		13.7	Coeff. of Therm. Expansion inches x 10^{-6}/°C	Orientation: Par. 11.4 Rt.Angl. -
Flux Density in Magnet B_m at BH_{max} (G)		4800	Resistivity (micro-ohms/cm/cm^2)	50 @ 25 °C
Coercive Force in Magnet H_m at BH_{max} (Oe)		350	Hardness 50	Scale Rockwell C
Energy Loss W_c (watt-sec/cycle/lb)		15.3	Basic Forming Method	C
Importance of Operating Magnet at BH_{max}		A	Basic Finishing Method	WGR
Average Recoil Permeability		4.3	Relative Corrosion Resistance	G
Field Strength Required to Fully Saturate Magnet (Oe)		3000	Primary Chemical Composition	Al-Ni-Co-Cu-Fe
Ability to Withstand Demagnetizing Fields		F	Special Notes: (1) Lowest temperature co-efficient of all materials. Little irreversible loss even at high temperatures. (2) Should be magnetized in final circuit.	
Maximum Service Temperature (°F)		900		
Approximate Temperature Permanently Affecting Magnet (°F)		1580		
Inherent Orientation of Material		none		

Note: All information based on available producers literature. In some cases of foreign materials interpretation or calculation required.

SOURCE(S):

Hitachi Magnetics Corp.

USA

YCM-4D
(Alloy or Grade)

Cast Alnico
(Type of Material)

2.4/1.1
(MMPA Brief Designation)

-
(IEC Code Reference)

MAGNETIC PROPERTIES	
Residual Induction B_r (G)	6500
Coercive Force (Oe) — Normal (H_c)	1075
Coercive Force (Oe) — Intrinsic (H_{ci})	Approx. 1100
Maximum Energy Product BH_{max} (MGO)	2.4
B/H Ratio (Load Line) at BH_{max}	4.3
Flux Density in Magnet B_m at BH_{max} (G)	3200
Coercive Force in Magnet H_m at BH_{max} (Oe)	750
Energy Loss W_c (watt-sec/cycle/lb)	15.3
Importance of Operating Magnet at BH_{max}	B
Average Recoil Permeability	4.3
Field Strength Required to Fully Saturate Magnet (Oe)	3000
Ability to Withstand Demagnetizing Fields	E
Maximum Service Temperature (°F)	900
Approximate Temperature Permanently Affecting Magnet (°F)	1580
Inherent Orientation of Material	HT

PHYSICAL PROPERTIES	
Density (lbs./in.^{3})	.2680
General Mechanical Properties	HB
Tensile Strength (psi)	-
Transverse Modulus of Rupture (psi)	-
Coeff. of Therm. Expansion inches x 10^{-6}/°C	Orientation: Par. 11.4 Rt.Angl. _
Resistivity (micro-ohms/cm/cm^2)	50 @ 25 °C
Hardness	50 Scale Rockwell C
Basic Forming Method	CM
Basic Finishing Method	WGR
Relative Corrosion Resistance	G
Primary Chemical Composition	Al-Ni-Co-Cu-Fe

Special Notes:

(1) Lowest temperature coefficient of all materials. Little irreversible loss even at high temperatures.
(2) Should be magnetized in final circuit.

Note: All information based on available producers literature. In some cases of foreign materials interpretation or calculation required.

SOURCE(S):

Hitachi Magnetics Corp.

USA

YCM-5AB
(Alloy or Grade)

Cast Alnico
(Type of Material)

-
(MMPA Brief Designation)

-
(IEC Code Reference)

MAGNETIC PROPERTIES			PHYSICAL PROPERTIES	
Residual Induction B_r (G)		3000	Density (lbs./in.^{3})	.2680
Coercive Force (Oe)	Normal (H_c)	150	General Mechanical Properties	HB
	Intrinsic (H_{ci})	-	Tensile Strength (psi)	-
Maximum Energy Product BH_{max} (MGO)		-	Transverse Modulus of Rupture (psi)	-
B/H Ratio (Load Line) at BH_{max}		-	Coeff. of Therm. Expansion inches x 10^{-6}/°C	Orientation: Par. 11.4 Rt.Angl. _
Flux Density in Magnet B_m at BH_{max} (G)		-	Resistivity (micro-ohms/cm/cm^2)	50 @ 25 °C
Coercive Force in Magnet H_m at BH_{max} (Oe)		-	Hardness Scale	Approx. 50 Rockwell C
Energy Loss W_c (watt-sec/cycle/lb)		-	Basic Forming Method	C
Importance of Operating Magnet at BH_{max}		A	Basic Finishing Method	WGR
Average Recoil Permeability		-	Relative Corrosion Resistance	G
Field Strength Required to Fully Saturate Magnet (Oe)		2000	Primary Chemical Composition	Al-Ni-Co-Cu-Fe
Ability to Withstand Demagnetizing Fields		P	Special Notes: (1) Lowest temperature coefficient of all materials. Little irreversible loss even at high temperatures. (2) Should be magnetized in final circuit.	
Maximum Service Temperature (°F)		900		
Approximate Temperature Permanently Affecting Magnet (°F)		1580		
Inherent Orientation of Material		none		

Note: All information based on available producers literature. In some cases of foreign materials interpretation or calculation required.

SOURCE(S):

Hitachi Magnetics Corp.

USA

YCM-5CD
(Alloy or Grade)

Cast Alnico
(Type of Material)

-
(MMPA Brief Designation)

-
(IEC Code Reference)

MAGNETIC PROPERTIES			PHYSICAL PROPERTIES	
Residual Induction B_r (G)		8500	Density (lbs./in.^{3})	.2680
Coercive Force (Oe)	Normal (H_c)	195	General Mechanical Properties	HB
	Intrinsic (H_{ci})	-	Tensile Strength (psi)	-
Maximum Energy Product BH_{max} (MGO)		-	Transverse Modulus of Rupture (psi)	-
B/H Ratio (Load Line) at BH_{max}		-	Coeff. of Therm. Expansion inches x 10^{-6}/°C	Orientation: Par. 11.4 Rt.Angl. -
Flux Density in Magnet B_m at BH_{max} (G)		-	Resistivity (micro-ohms/cm/cm^2	50 @ 25 °C
Coercive Force in Magnet H_m at BH_{max} (Oe)		-	Hardness Scale Approx. 50	Rockwell C
Energy Loss W_c (watt-sec/cycle/lb)		-	Basic Forming Method	C
Importance of Operating Magnet at BH_{max}		A	Basic Finishing Method	WGR
Average Recoil Permeability		-	Relative Corrosion Resistance	G
Field Strength Required to Fully Saturate Magnet (Oe)		2000	Primary Chemical Composition	Al-Ni-Co-Cu-Fe
Ability to Withstand Demagnetizing Fields		P	Special Notes:	
Maximum Service Temperature (°F)		900	(1) Lowest temperature co-efficient of all materials. Little irreversible loss even at high temperatures. (2) Should be magnetized in final circuit.	
Approximate Temperature Permanently Affecting Magnet (°F)		1580		
Inherent Orientation of Material		none		

Note: All information based on available producers literature. In some cases of foreign materials interpretation or calculation required.

SOURCE(S):

Hitachi Magnetics Corp.

USA

YCM-8B
(Alloy or Grade)

Cast Alnico
(Type of Material)

5.2/1.7
(MMPA Brief Designation)

R1-1-5
(IEC Code Reference)

MAGNETIC PROPERTIES		PHYSICAL PROPERTIES	
Residual Induction B_r (G)	9000	Density (lbs./in.^{3})	.2680
Coercive Force (Oe) — Normal (H_c)	1465	General Mechanical Properties	HB
Coercive Force (Oe) — Intrinsic (H_{ci})	Approx. 1650	Tensile Strength (psi)	—
Maximum Energy Product BH_{max} (MGO)	5.2	Transverse Modulus of Rupture (psi)	—
B/H Ratio (Load Line) at BH_{max}	5.2	Coeff. of Therm. Expansion inches x 10^{-6}/°C	Orientation: Par. 11.4 Rt.Angl. —
Flux Density in Magnet B_m at BH_{max} (G)	5200	Resistivity (micro-ohms/cm/cm^2)	50 @ 25 °C
Coercive Force in Magnet H_m at BH_{max} (Oe)	1000	Hardness — Scale	50 Rockwell C
Energy Loss W_c (watt-sec/cycle/lb)	15.3	Basic Forming Method	CM
Importance of Operating Magnet at BH_{max}	C	Basic Finishing Method	WGR
Average Recoil Permeability	4.3	Relative Corrosion Resistance	G
Field Strength Required to Fully Saturate Magnet (Oe)	3000	Primary Chemical Composition	Al-Ni-Co-Cu-Fe
Ability to Withstand Demagnetizing Fields	E	Special Notes:	
Maximum Service Temperature (°F)	900	(1) Lowest temperature co-efficient of all materials. Little irreversible loss even at high temperatures.	
Approximate Temperature Permanently Affecting Magnet (°F)	1580	(2) Should be magnetized in final circuit.	
Inherent Orientation of Material	HT		

Note: All information based on available producers literature. In some cases of foreign materials interpretation or calculation required.

SOURCE(S):

Hitachi Magnetics Corp.

USA

YCM-8C	(Alloy or Grade)
Cast Alnico	(Type of Material)
6.0/1.8	(MMPA Brief Designation)
R1-1-5	(IEC Code Reference)

MAGNETIC PROPERTIES		PHYSICAL PROPERTIES	
Residual Induction B_r (G)	9000	Density (lbs./in.^{3})	.2680
Coercive Force (Oe) — Normal (H_c)	1600	General Mechanical Properties	HB
Coercive Force (Oe) — Intrinsic (H_{ci})	Approx. 1800	Tensile Strength (psi)	-
Maximum Energy Product BH_{max} (MGO)	6.0	Transverse Modulus of Rupture (psi)	-
B/H Ratio (Load Line) at BH_{max}	6.6	Coeff. of Therm. Expansion inches x 10^{-6}/°C	Orientation: Par. 11.4 Rt.Angl. -
Flux Density in Magnet B_m at BH_{max} (G)	6300	Resistivity (micro-ohms/cm/cm^2)	50 @ 25 °C
Coercive Force in Magnet H_m at BH_{max} (Oe)	950	Hardness Scale 50	Rockwell C
Energy Loss W_c (watt-sec/cycle/lb)	15.3	Basic Forming Method	CM
Importance of Operating Magnet at BH_{max}	C	Basic Finishing Method	WGR
Average Recoil Permeability	4.3	Relative Corrosion Resistance	G
Field Strength Required to Fully Saturate Magnet (Oe)	3000	Primary Chemical Composition	Al-Ni-Co-Cu-Fe
Ability to Withstand Demagnetizing Fields	E	Special Notes:	
Maximum Service Temperature (°F)	900	(1) Lowest temperature coefficient of all materials. Little irreversible loss even at high temperatures.	
Approximate Temperature Permanently Affecting Magnet (°F)	1580	(2) Should be magnetized in final circuit.	
Inherent Orientation of Material	HT		

Note: All information based on available producers literature. In some cases of foreign materials interpretation or calculation required.

SOURCE(S):	YCM-8D
Hitachi Magnetics Corp.	(Alloy or Grade)
	Cast Alnico
	(Type of Material)
USA	6.0/2.0
	(MMPA Brief Designation)
	R1-1-7
	(IEC Code Reference)

MAGNETIC PROPERTIES		PHYSICAL PROPERTIES	
Residual Induction B_r (G)	7900	Density (lbs./in.^{3})	.2680
Coercive Force (Oe) — Normal (H_c)	1800	General Mechanical Properties	HB
Coercive Force (Oe) — Intrinsic (H_{ci})	Approx. 2000	Tensile Strength (psi)	-
Maximum Energy Product BH_{max} (MGO)	6.0	Transverse Modulus of Rupture (psi)	-
B/H Ratio (Load Line) at BH_{max}	6.0	Coeff. of Therm. Expansion inches x 10^{-6}/°C	Orientation: Par. 11.4 Rt.Angl. -
Flux Density in Magnet B_m at BH_{max} (G)	1000	Resistivity (micro-ohms/cm/cm^2)	50 @ 25 °C
Coercive Force in Magnet H_m at BH_{max} (Oe)	6000	Hardness 50	Scale Rockwell C
Energy Loss W_c (watt-sec/cycle/lb)	15.3	Basic Forming Method	CM
Importance of Operating Magnet at BH_{max}	C	Basic Finishing Method	WGR
Average Recoil Permeability	4.3	Relative Corrosion Resistance	G
Field Strength Required to Fully Saturate Magnet (Oe)	3000	Primary Chemical Composition	Al-Ni-Co-Cu-Fe
Ability to Withstand Demagnetizing Fields	E	Special Notes: (1) Lowest temperature co-efficient of all materials. Little irreversible loss even at high temperatures. (2) Should be magnetized in final circuit.	
Maximum Service Temperature (°F)	900		
Approximate Temperature Permanently Affecting Magnet (°F)	1580		
Inherent Orientation of Material	HT		

Note: All information based on available producers literature. In some cases of foreign materials interpretation or calculation required.

SOURCE(S):

Hitachi Magnetics Corp.

USA

YCM-8E
(Alloy or Grade)

Cast Alnico
(Type of Material)

-
(MMPA Brief Designation)

-
(IEC Code Reference)

MAGNETIC PROPERTIES		PHYSICAL PROPERTIES	
Residual Induction B_r (G)	8000	Density (lbs./in.^{3})	.2680
Coercive Force (Oe) — Normal (H_c)	2000	General Mechanical Properties	HB
Coercive Force (Oe) — Intrinsic (H_{ci})	Approx. 2000	Tensile Strength (psi)	-
Maximum Energy Product BH_{max} (MGO)	6.0	Transverse Modulus of Rupture (psi)	-
B/H Ratio (Load Line) at BH_{max}	3.8	Coeff. of Therm. Expansion inches x 10^{-6}/°C	Orientation: Par. 11.4 Rt.Angl. -
Flux Density in Magnet B_m at BH_{max} (G)	4800	Resistivity (micro-ohms/cm/cm^2)	50 @ 25 °C
Coercive Force in Magnet H_m at BH_{max} (Oe)	1250	Hardness 50	Scale Rockwell C
Energy Loss W_c (watt-sec/cycle/lb)	15.3	Basic Forming Method	CM
Importance of Operating Magnet at BH_{max}	C	Basic Finishing Method	WGR
Average Recoil Permeability	4.3	Relative Corrosion Resistance	G
Field Strength Required to Fully Saturate Magnet (Oe)	3000	Primary Chemical Composition	Al-Ni-Co-Cu-Fe
Ability to Withstand Demagnetizing Fields	E	Special Notes: (1) Lowest temperature coefficient of all materials. Little irreversible loss even at high temperatures. (2) Should be magnetized in final circuit.	
Maximum Service Temperature (°F)	900		
Approximate Temperature Permanently Affecting Magnet (°F)	1580		
Inherent Orientation of Material	HT		

Note: All information based on available producers literature. In some cases of foreign materials interpretation or calculation required.

SOURCE(S):	YCM-9B
Hitachi Magnetics Corp.	(Alloy or Grade)
	Cast Alnico
	(Type of Material)
USA	10.0/1.5
	(MMPA Brief Designation)
	R1-1-6
	(IEC Code Reference)

MAGNETIC PROPERTIES		PHYSICAL PROPERTIES	
Residual Induction B_r (G)	10,500	Density (lbs./in.^{3})	.2680
Coercive Force (Oe) — Normal (H_c)	1525	General Mechanical Properties	HB
Coercive Force (Oe) — Intrinsic (H_{ci})	Approx. 1500	Tensile Strength (psi)	-
Maximum Energy Product BH_{max} (MGO)	10.0	Transverse Modulus of Rupture (psi)	-
B/H Ratio (Load Line) at BH_{max}	6.7	Coeff. of Therm. Expansion inches x 10^{-6}/°C	Orientation: Par. 11.4 Rt.Angl. -
Flux Density in Magnet B_m at BH_{max} (G)	8000	Resistivity (micro-ohms/cm/cm^2)	50 @ 25 °C
Coercive Force in Magnet H_m at BH_{max} (Oe)	1200	Hardness Scale	50 Rockwell C
Energy Loss W_c (watt-sec/cycle/lb)	15.3	Basic Forming Method	CM
Importance of Operating Magnet at BH_{max}	C	Basic Finishing Method	WGR
Average Recoil Permeability	4.3	Relative Corrosion Resistance	G
Field Strength Required to Fully Saturate Magnet (Oe)	3000	Primary Chemical Composition	Al-Ni-Co-Cu-Fe
Ability to Withstand Demagnetizing Fields	E	Special Notes: (1) Lowest temperature coefficient of all materials. Little irreversible loss even at high temperatures. (2) Should be magnetized in final circuit.	
Maximum Service Temperature (°F)	900		
Approximate Temperature Permanently Affecting Magnet (°F)	1580		
Inherent Orientation of Material	HT		

Note: All information based on available producers literature. In some cases of foreign materials interpretation or calculation required.

SOURCE(S):	YCM-11
	(Alloy or Grade)
Hitachi Magnetics Corp.	Cast Alnico
	(Type of Material)
USA	11.0/1.5
	(MMPA Brief Designation)
	R1-1-6
	(IEC Code Reference)

MAGNETIC PROPERTIES		PHYSICAL PROPERTIES	
Residual Induction B_r (G)	11,000	Density (lbs./in.^{3})	.2680
Coercive Force (Oe) — Normal (H_c)	1500	General Mechanical Properties	HB
Coercive Force (Oe) — Intrinsic (H_{ci})	Approx. 1500	Tensile Strength (psi)	-
Maximum Energy Product BH_{max} (MGO)	11.0	Transverse Modulus of Rupture (psi)	-
B/H Ratio (Load Line) at BH_{max}	7.0	Coeff. of Therm. Expansion inches x 10^{-6}/°C	Orientation: Par. 11.4 Rt.Angl. -
Flux Density in Magnet B_m at BH_{max} (G)	8800	Resistivity (micro-ohms/cm/cm^2)	50 @ 25 °C
Coercive Force in Magnet H_m at BH_{max} (Oe)	1250	Hardness Scale	50 Rockwell C
Energy Loss W_c (watt-sec/cycle/lb)	15.3	Basic Forming Method	CM
Importance of Operating Magnet at BH_{max}	C	Basic Finishing Method	WGR
Average Recoil Permeability	4.3	Relative Corrosion Resistance	G
Field Strength Required to Fully Saturate Magnet (Oe)	3000	Primary Chemical Composition	Al-Ni-Co-Cu-Fe
Ability to Withstand Demagnetizing Fields	E	Special Notes: (1) Lowest temperature co-efficient of all materials. Little irreversible loss even at high temperatures. (2) Should be magnetized in final circuit.	
Maximum Service Temperature (°F)	900		
Approximate Temperature Permanently Affecting Magnet (°F)	1580		
Inherent Orientation of Material	HT		

Note: All information based on available producers literature. In some cases of foreign materials interpretation or calculation required.

SOURCE(S):

Hitachi Magnetics Corp.

USA

YHJ-2
(Alloy or Grade)

Wrought
(Type of Material)

-
(MMPA Brief Designation)

-
(IEC Code Reference)

MAGNETIC PROPERTIES		PHYSICAL PROPERTIES	
Residual Induction B_r (G)	6000 to 12,500	Density (lbs./in.^{3})	.2782
Coercive Force (Oe) Normal (H_c)	50 to 150	General Mechanical Properties	HT
Coercive Force (Oe) Intrinsic (H_{ci})	-	Tensile Strength (psi)	-
Maximum Energy Product BH_{max} (MGO)	-	Transverse Modulus of Rupture (psi)	-
B/H Ratio (Load Line) at BH_{max}	-	Coeff. of Therm. Expansion inches x 10^{-6}/°C	Orientation: Par. - Rt.Angl. -
Flux Density in Magnet B_m at BH_{max} (G)	-	Resistivity (micro-ohms/cm/cm^2)	75 @ 25 °C
Coercive Force in Magnet H_m at BH_{max} (Oe)	-	Hardness Scale 270-480 (See Note 2)	Vickers
Energy Loss W_c (watt-sec/cycle/lb)	-	Basic Forming Method	W (See Note 3)
Importance of Operating Magnet at BH_{max}	A	Basic Finishing Method	WGR (See Note 3)
Average Recoil Permeability	-	Relative Corrosion Resistance	E
Field Strength Required to Fully Saturate Magnet (Oe)	2000	Primary Chemical Composition	Fe-Mn
Ability to Withstand Demagnetizing Fields	P	Special Notes: (1) Primary use: Hysteris material (2) Hardness range shown is before heat treatment. 580-600 (Vickers) after H.T. Work hardening type material (3) Suitable for sheet and wire configurations. May be punched, drawn or curled.	
Maximum Service Temperature (°F)	900		
Approximate Temperature Permanently Affecting Magnet (°F)	1200 min.		
Inherent Orientation of Material	none		

Note: All information based on available producers literature. In some cases of foreign materials interpretation or calculation required.

SOURCE(S):	**YHJ-3A**
	(Alloy or Grade)
Hitachi Magnetics Corp.	Wrought
	(Type of Material)
USA	-
	(MMPA Brief Designation)
	-
	(IEC Code Reference)

MAGNETIC PROPERTIES		PHYSICAL PROPERTIES	
Residual Induction B_r (G)	8000 to 11,500	Density (lbs./in.^{3})	.2782
Coercive Force (Oe) — Normal (H_c)	140 to 170	General Mechanical Properties	HT
Coercive Force (Oe) — Intrinsic (H_{ci})	-	Tensile Strength (psi)	-
Maximum Energy Product BH_{max} (MGO)	-	Transverse Modulus of Rupture (psi)	-
B/H Ratio (Load Line) at BH_{max}	-	Coeff. of Therm. Expansion inches x 10^{-6}/°C Orientation: Par. - Rt.Angl. -	
Flux Density in Magnet B_m at BH_{max} (G)	-	Resistivity (micro-ohms/cm/cm^2)	75 @ 25 °C
Coercive Force in Magnet H_m at BH_{max} (Oe)	-	Hardness Scale 270-480 (See Note 2)	Vickers
Energy Loss W_c (watt-sec/cycle/lb)	-	Basic Forming Method	W (See Note 3)
Importance of Operating Magnet at BH_{max}	A	Basic Finishing Method	WGR (See Note 3)
Average Recoil Permeability	-	Relative Corrosion Resistance	E
Field Strength Required to Fully Saturate Magnet (Oe)	2000	Primary Chemical Composition	Fe-Mn
Ability to Withstand Demagnetizing Fields	P	Special Notes: (1) Primary use: Hysteris material (2) Hardness range shown is before heat treatment. 580-600 (Vickers) after H.T. Work hardening type material. (3) Suitable for sheet and wire configurations. May be punched, drawn or curled.	
Maximum Service Temperature (°F)	900		
Approximate Temperature Permanently Affecting Magnet (°F)	1200 min.		
Inherent Orientation of Material	none		

Note: All information based on available producers literature. In some cases of foreign materials interpretation or calculation required.

SOURCE(S):	YHJ-3B
	(Alloy or Grade)
Hitachi Magnetics Corp.	Wrought
	(Type of Material)
USA	-
	(MMPA Brief Designation)
	-
	(IEC Code Reference)

MAGNETIC PROPERTIES		PHYSICAL PROPERTIES	
Residual Induction B_r (G)	8500	Density (lbs./in.^{3})	.2782
Coercive Force (Oe) — Normal (H_c)	145	General Mechanical Properties	HT
Coercive Force (Oe) — Intrinsic (H_{ci})	-	Tensile Strength (psi)	-
Maximum Energy Product BH_{max} (MGO)	-	Transverse Modulus of Rupture (psi)	-
B/H Ratio (Load Line) at BH_{max}	-	Coeff. of Therm. Expansion inches x 10^{-6}/°C	Orientation: Par. - Rt.Angl. -
Flux Density in Magnet B_m at BH_{max} (G)	-	Resistivity (micro-ohms/cm/cm^2)	75 @ 25 °C
Coercive Force in Magnet H_m at BH_{max} (Oe)	-	Hardness Scale 270-480 (See Note 2)	Vickers
Energy Loss W_c (watt-sec/cycle/lb)	-	Basic Forming Method	W (See Note 3)
Importance of Operating Magnet at BH_{max}	A	Basic Finishing Method	WGR (See Note 3)
Average Recoil Permeability	-	Relative Corrosion Resistance	E
Field Strength Required to Fully Saturate Magnet (Oe)	2000	Primary Chemical Composition	Fe-Mn
Ability to Withstand Demagnetizing Fields	P	Special Notes: (1) Primary use: Hysteris material (2) Hardness range shown is before heat treatment. 580-600 (Vickers) after H.T. Work and age hardening type material. (3) Suitable for sheet and wire configurations. May be punched, drawn or curled.	
Maximum Service Temperature (°F)	900		
Approximate Temperature Permanently Affecting Magnet (°F)	1200 min.		
Inherent Orientation of Material	none		

Note: All information based on available producers literature. In some cases of foreign materials interpretation or calculation required.

SOURCE(S):

Hitachi Magnetics Corp.

USA

YHJ-5
(Alloy or Grade)

Wrought
(Type of Material)

-
(MMPA Brief Designation)

-
(IEC Code Reference)

MAGNETIC PROPERTIES		PHYSICAL PROPERTIES	
Residual Induction B_r (G)	8000 to 10,000	Density (lbs./in.^{3})	.2782
Coercive Force (Oe) — Normal (H_c)	100 to 140	General Mechanical Properties	HT
Coercive Force (Oe) — Intrinsic (H_{ci})	-	Tensile Strength (psi)	-
Maximum Energy Product BH_{max} (MGO)	-	Transverse Modulus of Rupture (psi)	-
B/H Ratio (Load Line) at BH_{max}	-	Coeff. of Therm. Expansion inches x 10^{-6}/°C Orientation: Par. - Rt.Angl. -	
Flux Density in Magnet B_m at BH_{max} (G)	-	Resistivity (micro-ohms/cm/cm^2)	75 @ 25°C
Coercive Force in Magnet H_m at BH_{max} (Oe)	-	Hardness Scale 270-480 (See Note 2) Vickers	
Energy Loss W_c (watt-sec/cycle/lb)	-	Basic Forming Method	W (See Note 3)
Importance of Operating Magnet at BH_{max}	A	Basic Finishing Method	WGR (See Note 3)
Average Recoil Permeability	-	Relative Corrosion Resistance	E
Field Strength Required to Fully Saturate Magnet (Oe)	2000	Primary Chemical Composition	Fe-Mn
Ability to Withstand Demagnetizing Fields	P	Special Notes: (1) Primary use: Hysteris material (2) Hardness range shown is before heat treatment. 580-600 (Vickers) after H.T. Work and age hardening type material. (3) Suitable for sheet and wire configurations. May be punched, drawn or curled.	
Maximum Service Temperature (°F)	900		
Approximate Temperature Permanently Affecting Magnet (°F)	1200 min.		
Inherent Orientation of Material	none		

Note: All information based on available producers literature. In some cases of foreign materials interpretation or calculation required.

SOURCE(S):	**YHJ-25-8**
	(Alloy or Grade)
Hitachi Magnetics Corp.	Wrought
	(Type of Material)
USA	-
	(MMPA Brief Designation)
	-
	(IEC Code Reference)

MAGNETIC PROPERTIES		PHYSICAL PROPERTIES	
Residual Induction B_r (G)	8500 to 9500	Density (lbs./in.^{3})	.2782
Coercive Force (Oe) — Normal (H_c)	280 to 340	General Mechanical Properties	HT
Coercive Force (Oe) — Intrinsic (H_{ci})	-	Tensile Strength (psi)	-
Maximum Energy Product BH_{max} (MGO)	-	Transverse Modulus of Rupture (psi)	-
B/H Ratio (Load Line) at BH_{max}	-	Coeff. of Therm. Expansion inches x 10^{-6}/°C	Orientation: Par. - Rt.Angl. -
Flux Density in Magnet B_m at BH_{max} (G)	-	Resistivity (micro-ohms/cm/cm^2)	75 @ 25 °C
Coercive Force in Magnet H_m at BH_{max} (Oe)	-	Hardness Scale	230 (See Note 2) Vickers
Energy Loss W_c (watt-sec/cycle/lb)	-	Basic Forming Method	W (See Note 3)
Importance of Operating Magnet at BH_{max}	A	Basic Finishing Method	WGR (See Note 3)
Average Recoil Permeability	-	Relative Corrosion Resistance	E
Field Strength Required to Fully Saturate Magnet (Oe)	2000	Primary Chemical Composition	Fe-Cr-Co
Ability to Withstand Demagnetizing Fields	F	Special Notes:	
Maximum Service Temperature (°F)	900	(1) Primary use: Instruments (2) Hardness shown is before heat treatment. 300-400 (Vickers) after heat treatment. (3) Suitable for sheet and wire configurations. May be punched, drawn or curled.	
Approximate Temperature Permanently Affecting Magnet (°F)	1200 min.		
Inherent Orientation of Material	none		

Note: All information based on available producers literature. In some cases of foreign materials interpretation or calculation required.

SOURCE(S):

Hitachi Magnetics Corp.

USA

YHJ-30-10 (Anistropic)
(Alloy or Grade)

Wrought
(Type of Material)

-
(MMPA Brief Designation)

-
(IEC Code Reference)

MAGNETIC PROPERTIES		PHYSICAL PROPERTIES	
Residual Induction B_r (G)	11,500	Density (lbs./in.^{3})	.2782
Coercive Force (Oe) — Normal (H_c)	575	General Mechanical Properties	HT
Coercive Force (Oe) — Intrinsic (H_{ci})	-	Tensile Strength (psi)	-
Maximum Energy Product BH_{max} (MGO)	-	Transverse Modulus of Rupture (psi)	-
B/H Ratio (Load Line) at BH_{max}	-	Coeff. of Therm. Expansion inches x 10^{-6}/°C	Orientation: Par. - Rt.Angl. -
Flux Density in Magnet B_m at BH_{max} (G)	-	Resistivity (micro-ohms/cm/cm^2)	75 @ 25 °C
Coercive Force in Magnet H_m at BH_{max} (Oe)	-	Hardness Scale	230 (See Note 2) Vickers
Energy Loss W_c (watt-sec/cycle/lb)	-	Basic Forming Method	W (See Note 3)
Importance of Operating Magnet at BH_{max}	B	Basic Finishing Method	WGR (See Note 3)
Average Recoil Permeability	-	Relative Corrosion Resistance	E
Field Strength Required to Fully Saturate Magnet (Oe)	3000	Primary Chemical Composition	Fe-Cr-Co
Ability to Withstand Demagnetizing Fields	F	Special Notes:	
Maximum Service Temperature (°F)	900	(1) Primary use: Instruments (2) Hardness shown is before heat treatment. 300-400 (Vickers) after heat treatment. (3) Suitable for sheet and wire configurations. May be punched, drawn or curled.	
Approximate Temperature Permanently Affecting Magnet (°F)	1200 min.		
Inherent Orientation of Material	CR		

Note: All information based on available producers literature. In some cases of foreign materials interpretation or calculation required.

SOURCE(S):	YHJ-30-10 (Isotropic)
	(Alloy or Grade)
Hitachi Magnetics Corp.	Wrought
	(Type of Material)
USA	-
	(MMPA Brief Designation)
	-
	(IEC Code Reference)

MAGNETIC PROPERTIES		PHYSICAL PROPERTIES	
Residual Induction B_r (G)	9000	Density (lbs./in.^{3})	.2782
Coercive Force (Oe) — Normal (H_c)	400	General Mechanical Properties	HT
Coercive Force (Oe) — Intrinsic (H_{ci})	-	Tensile Strength (psi)	-
Maximum Energy Product BH_{max} (MGO)	-	Transverse Modulus of Rupture (psi)	-
B/H Ratio (Load Line) at BH_{max}	-	Coeff. of Therm. Expansion inches x 10^{-6}/°C	Orientation: Par. - Rt.Angl. -
Flux Density in Magnet B_m at BH_{max} (G)	-	Resistivity (micro-ohms/cm/cm^2)	75 @ 25 °C
Coercive Force in Magnet H_m at BH_{max} (Oe)	-	Hardness Scale	230 (See Note 2) Vickers
Energy Loss W_c (watt-sec/cycle/lb)	-	Basic Forming Method	W (See Note 3)
Importance of Operating Magnet at BH_{max}	B	Basic Finishing Method	WGR (See Note 3)
Average Recoil Permeability	-	Relative Corrosion Resistance	E
Field Strength Required to Fully Saturate Magnet (Oe)	3000	Primary Chemical Composition	Fe-Cr-Co
Ability to Withstand Demagnetizing Fields	F	Special Notes: (1) Primary use: Instruments (2) Hardness shown is before heat treatment. 300-400 (Vickers) after heat treatment. (3) Suitable for sheet and wire configurations. May be punched, drawn or curled.	
Maximum Service Temperature (°F)	900		
Approximate Temperature Permanently Affecting Magnet (°F)	1200 min.		
Inherent Orientation of Material	none		

Note: All information based on available producers literature. In some cases of foreign materials interpretation or calculation required.

SOURCE(S):	**YHJ-32-14**
	(Alloy or Grade)
Hitachi Magnetics Corp.	Wrought
	(Type of Material)
USA	-
	(MMPA Brief Designation)
	-
	(IEC Code Reference)

MAGNETIC PROPERTIES		PHYSICAL PROPERTIES	
Residual Induction B_r (G)	7000 to 10,000	Density (lbs./in.^{3})	.2782
Coercive Force (Oe) — Normal (H_c)	400 to 600	General Mechanical Properties	HT
Coercive Force (Oe) — Intrinsic (H_{ci})	-	Tensile Strength (psi)	-
Maximum Energy Product BH_{max} (MGO)	-	Transverse Modulus of Rupture (psi)	-
B/H Ratio (Load Line) at BH_{max}	-	Coeff. of Therm. Expansion inches x 10^{-6}/°C	Orientation: Par. _ Rt.Angl. _
Flux Density in Magnet B_m at BH_{max} (G)	-	Resistivity (micro-ohms/cm/cm^2)	75 @ 25 °C
Coercive Force in Magnet H_m at BH_{max} (Oe)	-	Hardness Scale	230 (See Note 2) Vickers
Energy Loss W_c (watt-sec/cycle/lb)	-	Basic Forming Method	W (See Note 3)
Importance of Operating Magnet at BH_{max}	B	Basic Finishing Method	WGR (See Note 3)
Average Recoil Permeability	-	Relative Corrosion Resistance	E
Field Strength Required to Fully Saturate Magnet (Oe)	3000	Primary Chemical Composition	Fe-Cr-Co
Ability to Withstand Demagnetizing Fields	F	Special Notes:	
Maximum Service Temperature (°F)	900	(1) Primary use: Instruments (2) Hardness shown is before heat treatment. 300-400 (Vickers) after heat treatment. (3) Suitable for sheet and wire configurations. May be punched, drawn or curled.	
Approximate Temperature Permanently Affecting Magnet (°F)	1200 min.		
Inherent Orientation of Material	CR		

Note: All information based on available producers literature. In some cases of foreign materials interpretation or calculation required.

SOURCE(S): **Hitachi Magnetics Corp.** USA	**YRM-2A** (Alloy or Grade)
	Bonded Ferrite (Type of Material)
	7.5/2.7 (MMPA Brief Designation)
	- (IEC Code Reference)

MAGNETIC PROPERTIES		PHYSICAL PROPERTIES	
Residual Induction B_r (G)	1800	Density (lbs./in.^{3})	-
Coercive Force (Oe) — Normal (H_c)	1600	General Mechanical Properties	FL
Coercive Force (Oe) — Intrinsic (H_{ci})	2700	Tensile Strength (psi)	-
Maximum Energy Product BH_{max} (MGO)	7.5	Transverse Modulus of Rupture (psi)	-
B/H Ratio (Load Line) at BH_{max}	1.3	Coeff. of Therm. Expansion inches x 10^{-6}/°C Orientation: Par. - Rt.Angl. -	
Flux Density in Magnet B_m at BH_{max} (G)	1000	Resistivity (micro-ohms/cm/cm^2)	- @ - °C
Coercive Force in Magnet H_m at BH_{max} (Oe)	750	Hardness Scale	-
Energy Loss W_c (watt-sec/cycle/lb)	-	Basic Forming Method	E
Importance of Operating Magnet at BH_{max}	E	Basic Finishing Method	SRD
Average Recoil Permeability	-	Relative Corrosion Resistance	E
Field Strength Required to Fully Saturate Magnet (Oe)	10,000	Primary Chemical Composition	Ba-Fe + binder
Ability to Withstand Demagnetizing Fields	Ex	Special Notes: (1) Polyethelene binder. Extruded or roller. (2) Maximum service temperature depends on binder used. (3) Excellent elasticity makes long shapes or thin sheets possible.	
Maximum Service Temperature (°F)	See Note		
Approximate Temperature Permanently Affecting Magnet (°F)	200		
Inherent Orientation of Material	none		

Note: All information based on available producers literature. In some cases of foreign materials interpretation or calculation required.

SOURCE(S):

Hitachi Magnetics Corp.

USA

YRM-2B
(Alloy or Grade)

Bonded Ferrite
(Type of Material)

1.1/3.0
(MMPA Brief Designation)

-
(IEC Code Reference)

MAGNETIC PROPERTIES	
Residual Induction B_r (G)	2300
Coercive Force (Oe) — Normal (H_c)	2000
Coercive Force (Oe) — Intrinsic (H_{ci})	2950
Maximum Energy Product BH_{max} (MGO)	1.1
B/H Ratio (Load Line) at BH_{max}	2.0
Flux Density in Magnet B_m at BH_{max} (G)	1500
Coercive Force in Magnet H_m at BH_{max} (Oe)	750
Energy Loss W_c (watt-sec/cycle/lb)	-
Importance of Operating Magnet at BH_{max}	E
Average Recoil Permeability	-
Field Strength Required to Fully Saturate Magnet (Oe)	10,000
Ability to Withstand Demagnetizing Fields	Ex
Maximum Service Temperature (°F)	See Note
Approximate Temperature Permanently Affecting Magnet (°F)	200
Inherent Orientation of Material	none

PHYSICAL PROPERTIES	
Density (lbs./in.^{3})	-
General Mechanical Properties	FL
Tensile Strength (psi)	-
Transverse Modulus of Rupture (psi)	-
Coeff. of Therm. Expansion inches x 10^{-6}/°C — Orientation: Par. - , Rt.Angl. -	
Resistivity (micro-ohms/cm/cm^2)	- @ - °C
Hardness - Scale -	
Basic Forming Method	E
Basic Finishing Method	SRD
Relative Corrosion Resistance	E
Primary Chemical Composition	Ba-Fe + binder

Special Notes:

(1) Polyethelene binder. Extruded or rolled.
(2) Maximum service temperature depends on binder used.
(3) Excellent elasticity makes long shapes or thin sheets possible.

Note: All information based on available producers literature. In some cases of foreign materials interpretation or calculation required.

SOURCE(S):

Magnetfabrik Schramberg GMBH & Co.

Germany

3/18P
(Alloy or Grade)

Bonded Ferrite
(Type of Material)

.4/2.3
(MMPA Brief Designation)

-
(IEC Code Reference)

MAGNETIC PROPERTIES		PHYSICAL PROPERTIES	
Residual Induction B_r (G)	1400	Density (lbs./in.^{3})	.1292
Coercive Force (Oe) — Normal (H_c)	1130	General Mechanical Properties	FL or HT See Note (1)
Coercive Force (Oe) — Intrinsic (H_{ci})	2260	Tensile Strength (psi)	-
Maximum Energy Product BH_{max} (MGO)	.4	Transverse Modulus of Rupture (psi)	-
B/H Ratio (Load Line) at BH_{max}	1.36	Coeff. of Therm. Expansion inches x 10^{-6}/°C Orientation: Par. / Rt.Angl.	- / -
Flux Density in Magnet B_m at BH_{max} (G)	750	Resistivity (micro-ohms/cm/cm^2 @ °C)	-
Coercive Force in Magnet H_m at BH_{max} (Oe)	553	Hardness Scale	-
Energy Loss W_c (watt-sec/cycle/lb)	-	Basic Forming Method	E or VM
Importance of Operating Magnet at BH_{max}	E	Basic Finishing Method	MCT/SCD/SRD
Average Recoil Permeability	1.05	Relative Corrosion Resistance	E
Field Strength Required to Fully Saturate Magnet (Oe)	10,000	Primary Chemical Composition	Ba or Sr-Fe + binder
Ability to Withstand Demagnetizing Fields	Ex	Special Notes:	
Maximum Service Temperature (°F)	212	(1) Injection molded or calendered depending on binder.	
Approximate Temperature Permanently Affecting Magnet (°F)	See Note (2)	(2) Maximum service temperature depends on binder used. Binder not identified by manufacturer.	
Inherent Orientation of Material	none		

Note: All information based on available producers literature. In some cases of foreign materials interpretation or calculation required.

SOURCE(S): **Stackpole Corp.** USA

7
(Alloy or Grade)

Sintered Ferrite
(Type of Material)

2.8/4.0
(MMPA Brief Designation)

S1-1-2
(IEC Code Reference)

MAGNETIC PROPERTIES	
Residual Induction B_r (G)	3400
Coercive Force (Oe) Normal (H_c)	3250
Coercive Force (Oe) Intrinsic (H_{ci})	4000
Maximum Energy Product BH_{max} (MGO)	2.8
B/H Ratio (Load Line) at BH_{max}	Est. .83
Flux Density in Magnet B_m at BH_{max} (G)	Est. 2500
Coercive Force in Magnet H_m at BH_{max} (Oe)	Est. 1800
Energy Loss W_c (watt-sec/cycle/lb)	-
Importance of Operating Magnet at BH_{max}	E
Average Recoil Permeability	Est. 1.1
Field Strength Required to Fully Saturate Magnet (Oe)	10,000
Ability to Withstand Demagnetizing Fields	Ex
Maximum Service Temperature (°F)	482
Approximate Temperature Permanently Affecting Magnet (°F)	842
Inherent Orientation of Material	WP

PHYSICAL PROPERTIES	
Density (lbs./in.^{3})	.1734
General Mechanical Properties	VHB
Tensile Strength (psi)	-
Transverse Modulus of Rupture (psi)	-
Coeff. of Therm. Expansion inches x 10^{-6}/°C	Orientation: Par. — Rt.Angl. —
Resistivity (micro-ohms/cm/cm^2)	1.0 @ 25 °C
Hardness	6.5 Scale Moh's
Basic Forming Method	PMF
Basic Finishing Method	WGD
Relative Corrosion Resistance	Ex
Primary Chemical Composition	Ba or Sr & Fe
Special Notes:	

Note: All information based on available producers literature. In some cases of foreign materials interpretation or calculation required.

SOURCE(S):	
Stackpole Corp.	**8** (Alloy or Grade)
	Sintered Ferrite (Type of Material)
USA	3.4/3.2 (MMPA Brief Designation)
	S1-1-5 (IEC Code Reference)

MAGNETIC PROPERTIES		PHYSICAL PROPERTIES	
Residual Induction B_r (G)	3850	Density (lbs./in.^{3})	.1734
Coercive Force (Oe) Normal (H_c)	3050	General Mechanical Properties	VHB
Coercive Force (Oe) Intrinsic (H_{ci})	3150	Tensile Strength (psi)	—
Maximum Energy Product BH_{max} (MGO)	3.4	Transverse Modulus of Rupture (psi)	—
B/H Ratio (Load Line) at BH_{max}	Est. 1.14	Coeff. of Therm. Expansion inches x 10^{-6}/°C	Orientation: Par. — Rt.Angl. —
Flux Density in Magnet B_m at BH_{max} (G)	Est. 1980	Resistivity (micro-ohms/cm/cm^2)	1.0 @ 25 °C
Coercive Force in Magnet H_m at BH_{max} (Oe)	Est. 1735	Hardness Scale	6.5 Moh's
Energy Loss W_c (watt-sec/cycle/lb)	—	Basic Forming Method	PMF
Importance of Operating Magnet at BH_{max}	E	Basic Finishing Method	WGD
Average Recoil Permeability	Est. 1.1	Relative Corrosion Resistance	Ex
Field Strength Required to Fully Saturate Magnet (Oe)	10,000	Primary Chemical Composition	Ba or Sr & Fe
Ability to Withstand Demagnetizing Fields	Ex	Special Notes:	
Maximum Service Temperature (°F)	482		
Approximate Temperature Permanently Affecting Magnet (°F)	842		
Inherent Orientation of Material	WP		

Note: All information based on available producers literature. In some cases of foreign materials interpretation or calculation required.

SOURCE(S):	8/22
	(Alloy or Grade)
Magnetfabrik Schramberg GMBH & Co.	Sintered Ferrite
	(Type of Material)
	1.1/2.9
Germany	(MMPA Brief Designation)
	-
	(IEC Code Reference)

MAGNETIC PROPERTIES		PHYSICAL PROPERTIES	
Residual Induction B_r (G)	2200	Density (lbs./in.^{3})	.1734
Coercive Force (Oe) — Normal (H_c)	1760	General Mechanical Properties	VHB
Coercive Force (Oe) — Intrinsic (H_{ci})	2890	Tensile Strength (psi)	-
Maximum Energy Product BH_{max} (MGO)	1.1	Transverse Modulus of Rupture (psi)	-
B/H Ratio (Load Line) at BH_{max}	Est. 1.1	Coeff. of Therm. Expansion inches x 10^{-6}/°C	Orientation: Par. - Rt.Angl. -
Flux Density in Magnet B_m at BH_{max} (G)	Est. 1125	Resistivity (micro-ohms/cm/cm^2)	1.0 @ 25 °C
Coercive Force in Magnet H_m at BH_{max} (Oe)	Est. 1000	Hardness Scale	6.5 Moh's
Energy Loss W_c (watt-sec/cycle/lb)	-	Basic Forming Method	PF
Importance of Operating Magnet at BH_{max}	E	Basic Finishing Method	WGD
Average Recoil Permeability	Est. 1.1	Relative Corrosion Resistance	Ex
Field Strength Required to Fully Saturate Magnet (Oe)	10,000	Primary Chemical Composition	Ba-Fe
Ability to Withstand Demagnetizing Fields	Ex	Special Notes:	
Maximum Service Temperature (°F)	482		
Approximate Temperature Permanently Affecting Magnet (°F)	842		
Inherent Orientation of Material	none		

Note: All information based on available producers literature. In some cases of foreign materials interpretation or calculation required.

SOURCE(S):	9/19P
	(Alloy or Grade)
Magnetfabrik Schramberg GMBH & Co.	Bonded Ferrite
	(Type of Material)
	1.2/2.4
Germany	(MMPA Brief Designation)
	-
	(IEC Code Reference)

MAGNETIC PROPERTIES		PHYSICAL PROPERTIES	
Residual Induction B_r (G)	2200	Density (lbs./in.^{3})	.1292
Coercive Force (Oe) Normal (H_c)	1880	General Mechanical Properties	FL or HT See Note (1)
Coercive Force (Oe) Intrinsic (H_{ci})	2390	Tensile Strength (psi)	-
Maximum Energy Product BH_{max} (MGO)	1.2	Transverse Modulus of Rupture (psi)	-
B/H Ratio (Load Line) at BH_{max}	1.2	Coeff. of Therm. Expansion inches x 10^{-6}/°C	Orientation: Par. - Rt.Angl. -
Flux Density in Magnet B_m at BH_{max} (G)	120	Resistivity (micro-ohms/cm/cm^2	- @ - °C
Coercive Force in Magnet H_m at BH_{max} (Oe)	1000	Hardness -	Scale -
Energy Loss W_c (watt-sec/cycle/lb)	-	Basic Forming Method	E or VM
Importance of Operating Magnet at BH_{max}	E	Basic Finishing Method	MCT/SCD/SRD
Average Recoil Permeability	1.05	Relative Corrosion Resistance	E
Field Strength Required to Fully Saturate Magnet (Oe)	10,000	Primary Chemical Composition	Ba or Sr-Fe + binder
Ability to Withstand Demagnetizing Fields	Ex	Special Notes: (1) Injection molded or calendered depending on binder (2) Maximum service temperature depends on binder used. Binder not identified by manufacturer.	
Maximum Service Temperature (°F)	212		
Approximate Temperature Permanently Affecting Magnet (°F)	See Note (2)		
Inherent Orientation of Material	none		

Note: All information based on available producers literature. In some cases of foreign materials interpretation or calculation required.

SOURCE(S):	**12/22P**
	(Alloy or Grade)
Magnetfabrik Schramberg GMBH & Co.	Bonded Ferrite
	(Type of Material)
	1.6/2.8
Germany	(MMPA Brief Designation)
	-
	(IEC Code Reference)

MAGNETIC PROPERTIES		PHYSICAL PROPERTIES	
Residual Induction B_r (G)	2550	Density (lbs./in.^{3})	.1292
Coercive Force (Oe) — Normal (H_c)	2260	General Mechanical Properties	FL or HT See Note (1)
Coercive Force (Oe) — Intrinsic (H_{ci})	2830	Tensile Strength (psi)	-
Maximum Energy Product BH_{max} (MGO)	1.6	Transverse Modulus of Rupture (psi)	-
B/H Ratio (Load Line) at BH_{max}	1.08	Coeff. of Therm. Expansion inches x 10^{-6}/°C	Orientation: Par. - Rt.Angl. -
Flux Density in Magnet B_m at BH_{max} (G)	1300	Resistivity (micro-ohms/cm/cm^2)	- @ - °C
Coercive Force in Magnet H_m at BH_{max} (Oe)	1200	Hardness - Scale -	
Energy Loss W_c (watt-sec/cycle/lb)	-	Basic Forming Method	E or VM
Importance of Operating Magnet at BH_{max}	E	Basic Finishing Method	MCT/SCD/SRD
Average Recoil Permeability	1.05	Relative Corrosion Resistance	E
Field Strength Required to Fully Saturate Magnet (Oe)	10,000	Primary Chemical Composition	Ba or Sr-Fe + binder
Ability to Withstand Demagnetizing Fields	Ex	Special Notes: (1) Injection molded or calendered depending on binder. (2) Maximum service temperature depends on binder used. Binder not identified by manufacturer.	
Maximum Service Temperature (°F)	212		
Approximate Temperature Permanently Affecting Magnet (°F)	See Note (2)		
Inherent Orientation of Material	none		

Note: All information based on available producers literature. In some cases of foreign materials interpretation or calculation required.

SOURCE(S):	24/16
Magnetfabrik Schramberg GMBH & Co. Germany	(Alloy or Grade)
	Sintered Ferrite
	(Type of Material)
	3.2/2.3
	(MMPA Brief Designation)
	-
	(IEC Code Reference)

MAGNETIC PROPERTIES		PHYSICAL PROPERTIES	
Residual Induction B_r (G)	3650	Density (lbs./in.^{3})	.1806
Coercive Force (Oe) Normal (H_c)	2200	General Mechanical Properties	VHB
Coercive Force (Oe) Intrinsic (H_{ci})	2260	Tensile Strength (psi)	-
Maximum Energy Product BH_{max} (MGO)	3.2	Transverse Modulus of Rupture (psi)	-
B/H Ratio (Load Line) at BH_{max}	1.12	Coeff. of Therm. Expansion inches x 10^{-6}/°C	Orientation: Par. - Rt.Angl. -
Flux Density in Magnet B_m at BH_{max} (G)	1900	Resistivity (micro-ohms/cm/cm^2)	1.0 @ 25 °C
Coercive Force in Magnet H_m at BH_{max} (Oe)	1700	Hardness Scale	6.5 Moh's
Energy Loss W_c (watt-sec/cycle/lb)	-	Basic Forming Method	PMF
Importance of Operating Magnet at BH_{max}	E	Basic Finishing Method	WGD
Average Recoil Permeability	Est. 1.1	Relative Corrosion Resistance	Ex
Field Strength Required to Fully Saturate Magnet (Oe)	10,000	Primary Chemical Composition	Ba-Fe
Ability to Withstand Demagnetizing Fields	Ex	Special Notes:	
Maximum Service Temperature (°F)	482		
Approximate Temperature Permanently Affecting Magnet (°F)	842		
Inherent Orientation of Material			

Note: All information based on available producers literature. In some cases of foreign materials interpretation or calculation required.

SOURCE(S):

Magnetfabrik Schramberg GMBH & Co.

Germany

24/23
(Alloy or Grade)

Sintered Ferrite
(Type of Material)

3.2/3.0
(MMPA Brief Designation)

-
(IEC Code Reference)

MAGNETIC PROPERTIES	
Residual Induction B_r (G)	3650
Coercive Force (Oe) Normal (H_c)	2890
Coercive Force (Oe) Intrinsic (H_{ci})	3010
Maximum Energy Product BH_{max} (MGO)	3.2
B/H Ratio (Load Line) at BH_{max}	1.05
Flux Density in Magnet B_m at BH_{max} (G)	1820
Coercive Force in Magnet H_m at BH_{max} (Oe)	1734
Energy Loss W_c (watt-sec/cycle/lb)	-
Importance of Operating Magnet at BH_{max}	E
Average Recoil Permeability	1.1
Field Strength Required to Fully Saturate Magnet (Oe)	10,000
Ability to Withstand Demagnetizing Fields	Ex
Maximum Service Temperature (°F)	482
Approximate Temperature Permanently Affecting Magnet (°F)	842
Inherent Orientation of Material	MP

PHYSICAL PROPERTIES	
Density (lbs./in.^{3})	.1734
General Mechanical Properties	VHB
Tensile Strength (psi)	-
Transverse Modulus of Rupture (psi)	-
Coeff. of Therm. Expansion inches x 10^{-6}/°C	Orientation: Par. - Rt.Angl. -
Resistivity (micro-ohms/cm/cm^2)	1.0 @ 25 °C
Hardness	6.5 Scale Moh's
Basic Forming Method	PMF
Basic Finishing Method	WGD
Relative Corrosion Resistance	Ex
Primary Chemical Composition	Sr-Fe
Special Notes:	

Note: All information based on available producers literature. In some cases of foreign materials interpretation or calculation required.

SOURCE(S):

Magnetfabrik Schramberg GMBH & Co.

Germany

26/16
(Alloy or Grade)

Sintered Ferrite
(Type of Material)

3.4/2.2
(MMPA Brief Designation)

-
(IEC Code Reference)

MAGNETIC PROPERTIES		PHYSICAL PROPERTIES	
Residual Induction B_r (G)	3800	Density (lbs./in.^{3})	.1806
Coercive Force (Oe) Normal (H_c)	2200	General Mechanical Properties	VHB
Coercive Force (Oe) Intrinsic (H_{ci})	2260	Tensile Strength (psi)	-
Maximum Energy Product BH_{max} (MGO)	3.4	Transverse Modulus of Rupture (psi)	-
B/H Ratio (Load Line) at BH_{max}	1.17	Coeff. of Therm. Expansion inches x 10^{-6}/°C	Orientation: Par. - Rt.Angl. -
Flux Density in Magnet B_m at BH_{max} (G)	2000	Resistivity (micro-ohms/cm/cm^2	1.0 @ 25 °C
Coercive Force in Magnet H_m at BH_{max} (Oe)	1710	Hardness Scale	6.5 Moh's
Energy Loss W_c (watt-sec/cycle/lb)	-	Basic Forming Method	PMF
Importance of Operating Magnet at BH_{max}	E	Basic Finishing Method	WGD
Average Recoil Permeability	1.1	Relative Corrosion Resistance	Ex
Field Strength Required to Fully Saturate Magnet (Oe)	10,000	Primary Chemical Composition	Ba-Fe
Ability to Withstand Demagnetizing Fields	Ex	Special Notes:	
Maximum Service Temperature (°F)	482		
Approximate Temperature Permanently Affecting Magnet (°F)	842		
Inherent Orientation of Material	MP		

Note: All information based on available producers literature. In some cases of foreign materials interpretation or calculation required.

SOURCE(S):	26/24
	(Alloy or Grade)
Magnetfabrik Schramberg GMBH & Co.	Sintered Ferrite
	(Type of Material)
Germany	3.4/3.1
	(MMPA Brief Designation)
	-
	(IEC Code Reference)

MAGNETIC PROPERTIES		PHYSICAL PROPERTIES	
Residual Induction B_r (G)	3800	Density (lbs./in.^{3})	.1734
Coercive Force (Oe) — Normal (H_c)	3010	General Mechanical Properties	VHB
Coercive Force (Oe) — Intrinsic (H_{ci})	3140	Tensile Strength (psi)	-
Maximum Energy Product BH_{max} (MGO)	3.4	Transverse Modulus of Rupture (psi)	-
B/H Ratio (Load Line) at BH_{max}	1.14	Coeff. of Therm. Expansion inches x 10^{-6}/°C Orientation: Par. - Rt.Angl. -	
Flux Density in Magnet B_m at BH_{max} (G)	1980	Resistivity (micro-ohms/cm/cm^2)	1.0 @ 25 °C
Coercive Force in Magnet H_m at BH_{max} (Oe)	1734	Hardness Scale	6.5 Moh's
Energy Loss W_c (watt-sec/cycle/lb)	-	Basic Forming Method	PMF
Importance of Operating Magnet at BH_{max}	E	Basic Finishing Method	WGD
Average Recoil Permeability	1.1	Relative Corrosion Resistance	Ex
Field Strength Required to Fully Saturate Magnet (Oe)	10,000	Primary Chemical Composition	Sr-Fe
Ability to Withstand Demagnetizing Fields	Ex	Special Notes:	
Maximum Service Temperature (°F)	482		
Approximate Temperature Permanently Affecting Magnet (°F)	842		
Inherent Orientation of Material	MP		

Note: All information based on available producers literature. In some cases of foreign materials interpretation or calculation required.

SOURCE(S):	28/16 (Alloy or Grade)
Magnetfabrik Schramberg GMBH & Co. Germany	Sintered Ferrite (Type of Material)
	3.8/2.1 (MMPA Brief Designation)
	- (IEC Code Reference)

MAGNETIC PROPERTIES		PHYSICAL PROPERTIES	
Residual Induction B_r (G)	4000	Density (lbs./in.^{3})	.1806
Coercive Force (Oe) — Normal (H_c)	2140	General Mechanical Properties	VHB
Coercive Force (Oe) — Intrinsic (H_{ci})	2140	Tensile Strength (psi)	-
Maximum Energy Product BH_{max} (MGO)	3.8	Transverse Modulus of Rupture (psi)	-
B/H Ratio (Load Line) at BH_{max}	1.31	Coeff. of Therm. Expansion inches x 10^{-6}/°C Orientation: Par. _ Rt.Angl. _	
Flux Density in Magnet B_m at BH_{max} (G)	2225	Resistivity (micro-ohms/cm/cm^2)	1.0 @ 25 °C
Coercive Force in Magnet H_m at BH_{max} (Oe)	1700	Hardness Scale	6.5 Moh's
Energy Loss W_c (watt-sec/cycle/lb)	-	Basic Forming Method	PMF
Importance of Operating Magnet at BH_{max}	E	Basic Finishing Method	WGD
Average Recoil Permeability	1.1	Relative Corrosion Resistance	Ex
Field Strength Required to Fully Saturate Magnet (Oe)	10,000	Primary Chemical Composition	Ba-Fe
Ability to Withstand Demagnetizing Fields	Ex	Special Notes:	
Maximum Service Temperature (°F)	482		
Approximate Temperature Permanently Affecting Magnet (°F)	842		
Inherent Orientation of Material	MP		

Note: All information based on available producers literature. In some cases of foreign materials interpretation or calculation required.

SOURCE(S):

Magnetfabrik Schramberg GMBH & Co.

Germany

28/26
(Alloy or Grade)

Sintered Ferrite
(Type of Material)

3.8/3.4
(MMPA Brief Designation)

-
(IEC Code Reference)

MAGNETIC PROPERTIES	
Residual Induction B_r (G)	3950
Coercive Force (Oe) — Normal (H_c)	3330
Coercive Force (Oe) — Intrinsic (H_{ci})	3450
Maximum Energy Product BH_{max} (MGO)	3.8
B/H Ratio (Load Line) at BH_{max}	1.11
Flux Density in Magnet B_m at BH_{max} (G)	2050
Coercive Force in Magnet H_m at BH_{max} (Oe)	1850
Energy Loss W_c (watt-sec/cycle/lb)	-
Importance of Operating Magnet at BH_{max}	E
Average Recoil Permeability	1.1
Field Strength Required to Fully Saturate Magnet (Oe)	10,000
Ability to Withstand Demagnetizing Fields	Ex
Maximum Service Temperature (°F)	482
Approximate Temperature Permanently Affecting Magnet (°F)	842
Inherent Orientation of Material	MP

PHYSICAL PROPERTIES	
Density (lbs./in.^{3})	.1734
General Mechanical Properties	VHB
Tensile Strength (psi)	-
Transverse Modulus of Rupture (psi)	-
Coeff. of Therm. Expansion inches x 10^{-6}/°C — Orientation: Par. / Rt.Angl.	- / -
Resistivity (micro-ohms/cm/cm^2)	1.0 @ 25 °C
Hardness Scale	6.5 Moh's
Basic Forming Method	PMF
Basic Finishing Method	WGD
Relative Corrosion Resistance	Ex
Primary Chemical Composition	Sr-Fe
Special Notes:	

Note: All information based on available producers literature. In some cases of foreign materials interpretation or calculation required.

SOURCE(S):	30/16
Magnetfabrik Schramberg GMBH & Co. Germany	(Alloy or Grade)
	Sintered Ferrite
	(Type of Material)
	4.1/2.1
	(MMPA Brief Designation)
	-
	(IEC Code Reference)

MAGNETIC PROPERTIES		PHYSICAL PROPERTIES	
Residual Induction B_r (G)	4100	Density (lbs./in.^{3})	.1806
Coercive Force (Oe) — Normal (H_c)	2140	General Mechanical Properties	VHB
Coercive Force (Oe) — Intrinsic (H_{ci})	2140	Tensile Strength (psi)	-
Maximum Energy Product BH_{max} (MGO)	3.9	Transverse Modulus of Rupture (psi)	-
B/H Ratio (Load Line) at BH_{max}	1.19	Coeff. of Therm. Expansion inches x 10^{-6}/°C	Orientation: Par. - Rt.Angl. -
Flux Density in Magnet B_m at BH_{max} (G)	2150	Resistivity (micro-ohms/cm/cm^2)	1.0 @ 25 °C
Coercive Force in Magnet H_m at BH_{max} (Oe)	1800	Hardness — Scale	6.5 Moh's
Energy Loss W_c (watt-sec/cycle/lb)	-	Basic Forming Method	PMF
Importance of Operating Magnet at BH_{max}	E	Basic Finishing Method	WGD
Average Recoil Permeability	1.1	Relative Corrosion Resistance	Ex
Field Strength Required to Fully Saturate Magnet (Oe)	10,000	Primary Chemical Composition	Ba-Fe
Ability to Withstand Demagnetizing Fields	Ex	Special Notes:	
Maximum Service Temperature (°F)	482		
Approximate Temperature Permanently Affecting Magnet (°F)	842		
Inherent Orientation of Material	MP		

Note: All information based on available producers literature. In some cases of foreign materials interpretation or calculation required.

SOURCE(S):

Magnetfabrik Schramberg GMBH & Co.

Germany

30/26
(Alloy or Grade)

Sintered Ferrite
(Type of Material)

3.9/3.4
(MMPA Brief Designation)

-
(IEC Code Reference)

MAGNETIC PROPERTIES		PHYSICAL PROPERTIES	
Residual Induction B_r (G)	4050	Density (lbs./in.^{3})	.1734
Coercive Force (Oe) Normal (H_c)	3330	General Mechanical Properties	VHB
Coercive Force (Oe) Intrinsic (H_{ci})	3390	Tensile Strength (psi)	-
Maximum Energy Product BH_{max} (MGO)	3.9	Transverse Modulus of Rupture (psi)	-
B/H Ratio (Load Line) at BH_{max}	1.14	Coeff. of Therm. Expansion inches x 10^{-6}/°C	Orientation: Par. _ Rt.Angl. _
Flux Density in Magnet B_m at BH_{max} (G)	2100	Resistivity (micro-ohms/cm/cm^2)	1.0 @ 25 °C
Coercive Force in Magnet H_m at BH_{max} (Oe)	1850	Hardness Scale	6.5 Moh's
Energy Loss W_c (watt-sec/cycle/lb)	-	Basic Forming Method	PMF
Importance of Operating Magnet at BH_{max}	E	Basic Finishing Method	WGD
Average Recoil Permeability	1.1	Relative Corrosion Resistance	Ex
Field Strength Required to Fully Saturate Magnet (Oe)	10,000	Primary Chemical Composition	Sr-Fe
Ability to Withstand Demagnetizing Fields	Ex	Special Notes:	
Maximum Service Temperature (°F)	482		
Approximate Temperature Permanently Affecting Magnet (°F)	842		
Inherent Orientation of Material	MP		

Note: All information based on available producers literature. In some cases of foreign materials interpretation or calculation required.

SOURCE(S):	90A
	(Alloy or Grade)
Hitachi Magnetics Corp.	Sintered Rare Earth
	(Type of Material)
USA	16/18
	(MMPA Brief Designation)
	R5-1
	(IEC Code Reference)

MAGNETIC PROPERTIES		PHYSICAL PROPERTIES	
Residual Induction B_r (G)	8200	Density (lbs./in.^{3})	.2999
Coercive Force (Oe) — Normal (H_c)	7500	General Mechanical Properties	HB
Coercive Force (Oe) — Intrinsic (H_{ci})	15,000	Tensile Strength (psi)	5000
Maximum Energy Product BH_{max} (MGO)	16.0	Transverse Modulus of Rupture (psi)	-
B/H Ratio (Load Line) at BH_{max}	1.2	Coeff. of Therm. Expansion inches x 10^{-6}/°C	Orientation: Par. .6 Rt.Angl. .12
Flux Density in Magnet B_m at BH_{max} (G)	4400	Resistivity (micro-ohms/cm/cm^2)	50 @ 25 °C
Coercive Force in Magnet H_m at BH_{max} (Oe)	3600	Hardness 500 Scale Vickers	
Energy Loss W_c (watt-sec/cycle/lb)	-	Basic Forming Method	DPM
Importance of Operating Magnet at BH_{max}	E	Basic Finishing Method	WGR
Average Recoil Permeability	1.05	Relative Corrosion Resistance	G
Field Strength Required to Fully Saturate Magnet (Oe)	20,000	Primary Chemical Composition	Sm-Co
Ability to Withstand Demagnetizing Fields	0	Special Notes: (1) Irreversible loss in magnetic properties occurs above maximum service temperature.	
Maximum Service Temperature (°F)	482		
Approximate Temperature Permanently Affecting Magnet (°F)	Est. 1472		
Inherent Orientation of Material	MP		

Note: All information based on available producers literature. In some cases of foreign materials interpretation or calculation required.

SOURCE(S):	**90B**
	(Alloy or Grade)
Hitachi Magnetics Corp.	Sintered Rare Earth
	(Type of Material)
USA	18/20
	(MMPA Brief Designation)
	R5-1
	(IEC Code Reference)

MAGNETIC PROPERTIES		PHYSICAL PROPERTIES	
Residual Induction B_r (G)	8700	Density (lbs./in.^{3})	.2994
Coercive Force (Oe) — Normal (H_c)	8200	General Mechanical Properties	HB
Coercive Force (Oe) — Intrinsic (H_{ci})	15,000	Tensile Strength (psi)	5000
Maximum Energy Product BH_{max} (MGO)	18.0	Transverse Modulus of Rupture (psi)	–
B/H Ratio (Load Line) at BH_{max}	1.4	Coeff. of Therm. Expansion inches x 10^{-6}/°C	Orientation: Par. .6 Rt.Angl. .12
Flux Density in Magnet B_m at BH_{max} (G)	5000	Resistivity (micro-ohms/cm/cm^2)	50 @ 25 °C
Coercive Force in Magnet H_m at BH_{max} (Oe)	3600	Hardness Scale	500 Vickers
Energy Loss W_c (watt-sec/cycle/lb)	–	Basic Forming Method	DPM
Importance of Operating Magnet at BH_{max}	E	Basic Finishing Method	WGR
Average Recoil Permeability	1.05	Relative Corrosion Resistance	G
Field Strength Required to Fully Saturate Magnet (Oe)	15,000	Primary Chemical Composition	Sm-Co
Ability to Withstand Demagnetizing Fields	0	Special Notes: (1) Irreversible loss in magnetic properties occurs above maximum service temperature.	
Maximum Service Temperature (°F)	482		
Approximate Temperature Permanently Affecting Magnet (°F)	Est. 1472		
Inherent Orientation of Material	MP		

Note: All information based on available producers literature. In some cases of foreign materials interpretation or calculation required.

SOURCE(S):	**96A**
	(Alloy or Grade)
Hitachi Magnetics Corp.	Sintered Rare Earth
	(Type of Material)
USA	18/15
	(MMPA Brief Designation)
	R-5-1
	(IEC Code Reference)

MAGNETIC PROPERTIES		PHYSICAL PROPERTIES	
Residual Induction B_r (G)	9000	Density (lbs./in.^{3})	.2994
Coercive Force (Oe) — Normal (H_c)	8300	General Mechanical Properties	HB
Coercive Force (Oe) — Intrinsic (H_{ci})	15,000	Tensile Strength (psi)	5000
Maximum Energy Product BH_{max} (MGO)	18.0	Transverse Modulus of Rupture (psi)	-
B/H Ratio (Load Line) at BH_{max}	1.3	Coeff. of Therm. Expansion inches x 10^{-6}/°C	Orientation: Par. Rt.Angl.
Flux Density in Magnet B_m at BH_{max} (G)	4750	Resistivity (micro-ohms/cm/cm^2)	50 @ 25 °C
Coercive Force in Magnet H_m at BH_{max} (Oe)	3750	Hardness 500 — Scale Vickers	
Energy Loss W_c (watt-sec/cycle/lb)	-	Basic Forming Method	DPM
Importance of Operating Magnet at BH_{max}	E	Basic Finishing Method	WGR
Average Recoil Permeability	1.05	Relative Corrosion Resistance	G
Field Strength Required to Fully Saturate Magnet (Oe)	20,000	Primary Chemical Composition	Sm-Co
Ability to Withstand Demagnetizing Fields	0	Special Notes: (1) Irreversible loss in magnetic properties occurs above maximum service temperature.	
Maximum Service Temperature (°F)	302		
Approximate Temperature Permanently Affecting Magnet (°F)	Est. 1472		
Inherent Orientation of Material	MP		

Note: All information based on available producers literature. In some cases of foreign materials interpretation or calculation required.

SOURCE(S):	**96B**
	(Alloy or Grade)
Hitachi Magnetics Corp.	Sintered Rare Earth
	(Type of Material)
USA	22/15
	(MMPA Brief Designation)
	R5-1
	(IEC Code Reference)

MAGNETIC PROPERTIES		PHYSICAL PROPERTIES	
Residual Induction B_r (G)	9400	Density (lbs./in.^{3})	.2994
Coercive Force (Oe) — Normal (H_c)	8800	General Mechanical Properties	HB
Coercive Force (Oe) — Intrinsic (H_{ci})	15,000	Tensile Strength (psi)	5000
Maximum Energy Product BH_{max} (MGO)	21.5	Transverse Modulus of Rupture (psi)	-
B/H Ratio (Load Line) at BH_{max}	1.8	Coeff. of Therm. Expansion inches x 10^{-6}/°C	Orientation: Par. .6 Rt.Angl. .12
Flux Density in Magnet B_m at BH_{max} (G)	6200	Resistivity (micro-ohms/cm/cm^2)	50 @ 25 °C
Coercive Force in Magnet H_m at BH_{max} (Oe)	3450	Hardness Scale	500 Vickers
Energy Loss W_c (watt-sec/cycle/lb)	-	Basic Forming Method	DPM
Importance of Operating Magnet at BH_{max}	E	Basic Finishing Method	WGR
Average Recoil Permeability	1.05	Relative Corrosion Resistance	G
Field Strength Required to Fully Saturate Magnet (Oe)	15,000	Primary Chemical Composition	Sm-Co
Ability to Withstand Demagnetizing Fields	0	Special Notes: (1) Irreversible loss in magnetic properties occurs above maximum srvice temperature.	
Maximum Service Temperature (°F)	482		
Approximate Temperature Permanently Affecting Magnet (°F)	Est. 1472		
Inherent Orientation of Material	MP		

Note: All information based on available producers literature. In some cases of foreign materials interpretation or calculation required.

SOURCE(S):

Hitachi Magnetics Corp.

USA

99A
(Alloy or Grade)

Sintered Rare Earth
(Type of Material)

22/6
(MMPA Brief Designation)

R5-1
(IEC Code Reference)

MAGNETIC PROPERTIES	
Residual Induction B_r (G)	9700
Coercive Force (Oe) — Normal (H_c)	6000
Coercive Force (Oe) — Intrinsic (H_{ci})	6500
Maximum Energy Product BH_{max} (MGO)	21.5
B/H Ratio (Load Line) at BH_{max}	1.4
Flux Density in Magnet B_m at BH_{max} (G)	5400
Coercive Force in Magnet H_m at BH_{max} (Oe)	4000
Energy Loss W_c (watt-sec/cycle/lb)	-
Importance of Operating Magnet at BH_{max}	E
Average Recoil Permeability	1.05
Field Strength Required to Fully Saturate Magnet (Oe)	15,000
Ability to Withstand Demagnetizing Fields	0
Maximum Service Temperature (°F)	482
Approximate Temperature Permanently Affecting Magnet (°F)	Est. 1472
Inherent Orientation of Material	MP

PHYSICAL PROPERTIES	
Density (lbs./in.^{3})	.2994
General Mechanical Properties	HB
Tensile Strength (psi)	5000
Transverse Modulus of Rupture (psi)	-
Coeff. of Therm. Expansion inches x 10^{-6}/°C — Orientation: Par. / Rt.Angl.	.6 / .12
Resistivity (micro-ohms/cm/cm^2)	50 @ 25 °C
Hardness / Scale	- / -
Basic Forming Method	DPM
Basic Finishing Method	WGR
Relative Corrosion Resistance	G
Primary Chemical Composition	Sm-Co

Special Notes:

(1) Irreversible loss in magnetic properties occurs above maximum service temperature.

Note: All information based on available producers literature. In some cases of foreign materials interpretation or calculation required.

SOURCE(S):

Hitachi Magnetics Corp.

USA

99B
(Alloy or Grade)

Sintered Rare Earth
(Type of Material)

24/6
(MMPA Brief Designation)

R5-1
(IEC Code Reference)

MAGNETIC PROPERTIES	
Residual Induction B_r (G)	10,000
Coercive Force (Oe) — Normal (H_c)	6000
Coercive Force (Oe) — Intrinsic (H_{ci})	6500
Maximum Energy Product BH_{max} (MGO)	24.0
B/H Ratio (Load Line) at BH_{max}	1.2
Flux Density in Magnet B_m at BH_{max} (G)	5400
Coercive Force in Magnet H_m at BH_{max} (Oe)	4500
Energy Loss W_c (watt-sec/cycle/lb)	-
Importance of Operating Magnet at BH_{max}	E
Average Recoil Permeability	1.05
Field Strength Required to Fully Saturate Magnet (Oe)	15,000
Ability to Withstand Demagnetizing Fields	0
Maximum Service Temperature (°F)	482
Approximate Temperature Permanently Affecting Magnet (°F)	Est. 1472
Inherent Orientation of Material	MP

PHYSICAL PROPERTIES	
Density (lbs./in.^{3})	.2994
General Mechanical Properties	HB
Tensile Strength (psi)	5000
Transverse Modulus of Rupture (psi)	-
Coeff. of Therm. Expansion inches x 10^{-6}/°C — Orientation: Par. / Rt.Angl.	
Resistivity (micro-ohms/cm/cm^2)	50 @ 25 °C
Hardness / Scale	- / -
Basic Forming Method	DPM
Basic Finishing Method	WGR
Relative Corrosion Resistance	G
Primary Chemical Composition	Sm-Co
Special Notes: (1) Irreversible loss in magnetic properties occurs above maximum service temperature.	

Note: All information based on available producers literature. In some cases of foreign materials interpretation or calculation required.

SOURCE(S):

Crucible Magnetics

USA

261
(Alloy or Grade)

Sintered Rare Earth
(Type of Material)

28/17
(MMPA Brief Designation)

R7-3
(IEC Code Reference)

MAGNETIC PROPERTIES		PHYSICAL PROPERTIES	
Residual Induction B_r (G)	10,800	Density (lbs./in.^{3})	.2730
Coercive Force (Oe) Normal (H_c)	10,100	General Mechanical Properties	HB
Coercive Force (Oe) Intrinsic (H_{ci})	17,000	Tensile Strength (psi)	-
Maximum Energy Product BH_{max} (MGO)	28.0	Transverse Modulus of Rupture (psi)	-
B/H Ratio (Load Line) at BH_{max}	Est. 1.35	Coeff. of Therm. Expansion inches x 10^{-6}/°C	Orientation: Par. - Rt.Angl. -
Flux Density in Magnet B_m at BH_{max} (G)	Est. 6400	Resistivity (micro-ohms/cm/cm^2)	- @ - °C
Coercive Force in Magnet H_m at BH_{max} (Oe)	Est. 4800	Hardness -	Scale -
Energy Loss W_c (watt-sec/cycle/lb)	-	Basic Forming Method	DPM
Importance of Operating Magnet at BH_{max}	E	Basic Finishing Method	WGR
Average Recoil Permeability	1.05	Relative Corrosion Resistance	P
Field Strength Required to Fully Saturate Magnet (Oe)	32,000	Primary Chemical Composition	Nd-Fe-B
Ability to Withstand Demagnetizing Fields	0	Special Notes:	
Maximum Service Temperature (°F)	302		
Approximate Temperature Permanently Affecting Magnet (°F)	590		
Inherent Orientation of Material	MP		

Note: All information based on available producers literature. In some cases of foreign materials interpretation or calculation required.

SOURCE(S):

Crucible Magnetics

USA

282
(Alloy or Grade)

Sintered Rare Earth
(Type of Material)

28/17
(MMPA Brief Designation)

R7-3
(IEC Code Reference)

MAGNETIC PROPERTIES			PHYSICAL PROPERTIES	
Residual Induction B_r (G)		10,800	Density (lbs./in.^{3})	.2730
Coercive Force (Oe)	Normal (H_c)	10,100	General Mechanical Properties	HB
	Intrinsic (H_{ci})	17,000	Tensile Strength (psi)	-
Maximum Energy Product BH_{max} (MGO)		Est. 28.0	Transverse Modulus of Rupture (psi)	-
B/H Ratio (Load Line) at BH_{max}		Est. 1.35	Coeff. of Therm. Expansion inches x 10^{-6}/°C	Orientation: Par. - Rt.Angl. -
Flux Density in Magnet B_m at BH_{max} (G)		Est. 6400	Resistivity (micro-ohms/cm/cm^2)	- @ - °C
Coercive Force in Magnet H_m at BH_{max} (Oe)		Est. 4800	Hardness Scale	- -
Energy Loss W_c (watt-sec/cycle/lb)		-	Basic Forming Method	DPM
Importance of Operating Magnet at BH_{max}		E	Basic Finishing Method	WGR
Average Recoil Permeability		1.05	Relative Corrosion Resistance	P
Field Strength Required to Fully Saturate Magnet (Oe)		32,000	Primary Chemical Composition	Nd-Fe-B
Ability to Withstand Demagnetizing Fields		0	Special Notes:	
Maximum Service Temperature (°F)		302		
Approximate Temperature Permanently Affecting Magnet (°F)		590		
Inherent Orientation of Material		MP		

Note: All information based on available producers literature. In some cases of foreign materials interpretation or calculation required.

SOURCE(S):

Crucible Magnetics

USA

301
(Alloy or Grade)

Sintered Rare Earth
(Type of Material)

30/20
(MMPA Brief Designation)

R7-3
(IEC Code Reference)

MAGNETIC PROPERTIES	
Residual Induction B_r (G)	11,000
Coercive Force (Oe) Normal (H_c)	10,600
Coercive Force (Oe) Intrinsic (H_{ci})	20,000
Maximum Energy Product BH_{max} (MGO)	30.0
B/H Ratio (Load Line) at BH_{max}	Est. 1.35
Flux Density in Magnet B_m at BH_{max} (G)	Est. 6400
Coercive Force in Magnet H_m at BH_{max} (Oe)	Est. 4750
Energy Loss W_c (watt-sec/cycle/lb)	-
Importance of Operating Magnet at BH_{max}	E
Average Recoil Permeability	1.05
Field Strength Required to Fully Saturate Magnet (Oe)	40,000
Ability to Withstand Demagnetizing Fields	O
Maximum Service Temperature (°F)	302
Approximate Temperature Permanently Affecting Magnet (°F)	590
Inherent Orientation of Material	MP

PHYSICAL PROPERTIES		
Density (lbs./in.^{3})		.2730
General Mechanical Properties		HB
Tensile Strength (psi)		-
Transverse Modulus of Rupture (psi)		-
Coeff. of Therm. Expansion inches x 10^{-6}/°C	Orientation: Par. Rt.Angl.	- -
Resistivity (micro-ohms/cm/cm^2)	- @	- °C
Hardness	Scale -	-
Basic Forming Method		DPM
Basic Finishing Method		WGR
Relative Corrosion Resistance		P
Primary Chemical Composition		Nd-Fe-B
Special Notes:		

Note: All information based on available producers literature. In some cases of foreign materials interpretation or calculation required.

SOURCE(S):

Crucible Magnetics

USA

___

315
(Alloy or Grade)

Sintered Rare Earth
(Type of Material)

31/14
(MMPA Brief Designation)

R7-3
(IEC Code Reference)

MAGNETIC PROPERTIES		PHYSICAL PROPERTIES	
Residual Induction B_r (G)	11,500	Density (lbs./in.^{3})	.270
Coercive Force (Oe) — Normal (H_c)	10,900	General Mechanical Properties	HB
Coercive Force (Oe) — Intrinsic (H_{ci})	14,000	Tensile Strength (psi)	-
Maximum Energy Product BH_{max} (MGO)	31.0	Transverse Modulus of Rupture (psi)	-
B/H Ratio (Load Line) at BH_{max}	Est. 1.0	Coeff. of Therm. Expansion inches x 10^{-6}/°C — Orientation: Par. / Rt.Angl.	- / -
Flux Density in Magnet B_m at BH_{max} (G)	Est. 5600	Resistivity (micro-ohms/cm/cm^2) @ °C	- @ -
Coercive Force in Magnet H_m at BH_{max} (Oe)	Est. 5600	Hardness / Scale	- / -
Energy Loss W_c (watt-sec/cycle/lb)	-	Basic Forming Method	DPM
Importance of Operating Magnet at BH_{max}	E	Basic Finishing Method	WGR
Average Recoil Permeability	1.05	Relative Corrosion Resistance	P
Field Strength Required to Fully Saturate Magnet (Oe)	32,000	Primary Chemical Composition	Nd-Fe-B
Ability to Withstand Demagnetizing Fields	O	Special Notes:	
Maximum Service Temperature (°F)	302		
Approximate Temperature Permanently Affecting Magnet (°F)	590		
Inherent Orientation of Material	MP		

Note: All information based on available producers literature. In some cases of foreign materials interpretation or calculation required.

SOURCE(S):

Crucible Magnetics

USA

322
(Alloy or Grade)

Sintered Rare Earth
(Type of Material)

32/17
(MMPA Brief Designation)

R7-3
(IEC Code Reference)

MAGNETIC PROPERTIES	
Residual Induction B_r (G)	11,600
Coercive Force (Oe) — Normal (H_c)	10,800
Coercive Force (Oe) — Intrinsic (H_{ci})	17,000
Maximum Energy Product BH_{max} (MGO)	32.0
B/H Ratio (Load Line) at BH_{max}	Est. 1.0
Flux Density in Magnet B_m at BH_{max} (G)	Est. 5700
Coercive Force in Magnet H_m at BH_{max} (Oe)	Est. 5700
Energy Loss W_c (watt-sec/cycle/lb)	-
Importance of Operating Magnet at BH_{max}	E
Average Recoil Permeability	1.05
Field Strength Required to Fully Saturate Magnet (Oe)	32,000
Ability to Withstand Demagnetizing Fields	0
Maximum Service Temperature (°F)	302
Approximate Temperature Permanently Affecting Magnet (°F)	590
Inherent Orientation of Material	MP

PHYSICAL PROPERTIES	
Density (lbs./in.^{3})	.2730
General Mechanical Properties	HB
Tensile Strength (psi)	-
Transverse Modulus of Rupture (psi)	-
Coeff. of Therm. Expansion inches x 10^{-6}/°C — Orientation: Par. / Rt.Angl.	- / -
Resistivity (micro-ohms/cm/cm^2)	- @ - °C
Hardness / Scale	- / -
Basic Forming Method	DPM
Basic Finishing Method	WGR
Relative Corrosion Resistance	P
Primary Chemical Composition	Nd-Fe-B
Special Notes:	

Note: All information based on available producers literature. In some cases of foreign materials interpretation or calculation required.

SOURCE(S):	355
	(Alloy or Grade)
Crucible Magnetics	Sintered Rare Earth
	(Type of Material)
USA	35/17
	(MMPA Brief Designation)
	R7-3
	(IEC Code Reference)

MAGNETIC PROPERTIES		PHYSICAL PROPERTIES	
Residual Induction B_r (G)	12,300	Density (lbs./in.^{3})	.2673
Coercive Force (Oe) — Normal (H_c)	11,300	General Mechanical Properties	HB
Coercive Force (Oe) — Intrinsic (H_{ci})	17,000	Tensile Strength (psi)	-
Maximum Energy Product BH_{max} (MGO)	35.0	Transverse Modulus of Rupture (psi)	-
B/H Ratio (Load Line) at BH_{max}	Est. 1.0	Coeff. of Therm. Expansion inches x 10^{-6}/°C	Orientation: Par. - Rt.Angl. -
Flux Density in Magnet B_m at BH_{max} (G)	Est. 5900	Resistivity (micro-ohms/cm/cm^2	- @ - °C
Coercive Force in Magnet H_m at BH_{max} (Oe)	Est. 5900	Hardness Scale	- -
Energy Loss W_c (watt-sec/cycle/lb)	-	Basic Forming Method	DPM
Importance of Operating Magnet at BH_{max}	E	Basic Finishing Method	WGR
Average Recoil Permeability	1.05	Relative Corrosion Resistance	P
Field Strength Required to Fully Saturate Magnet (Oe)	32,000	Primary Chemical Composition	Nd-Fe-B
Ability to Withstand Demagnetizing Fields	0	Special Notes:	
Maximum Service Temperature (°F)	302		
Approximate Temperature Permanently Affecting Magnet (°F)	590		
Inherent Orientation of Material	MP		

Note: All information based on available producers literature. In some cases of foreign materials interpretation or calculation required.

SOURCE(S):

Arnold Engineering Co.

USA

2002A
(Alloy or Grade)

Bonded Rare Earth
(Type of Material)

-
(MMPA Brief Designation)

-
(IEC Code Reference)

MAGNETIC PROPERTIES	
Residual Induction B_r (G)	4900
Coercive Force (Oe) Normal (H_c)	4100
Coercive Force (Oe) Intrinsic (H_{ci})	15,000
Maximum Energy Product BH_{max} (MGO)	5.0
B/H Ratio (Load Line) at BH_{max}	-
Flux Density in Magnet B_m at BH_{max} (G)	-
Coercive Force in Magnet H_m at BH_{max} (Oe)	-
Energy Loss W_c (watt-sec/cycle/lb)	-
Importance of Operating Magnet at BH_{max}	E
Average Recoil Permeability	-
Field Strength Required to Fully Saturate Magnet (Oe)	35,000
Ability to Withstand Demagnetizing Fields	0
Maximum Service Temperature (°F)	260
Approximate Temperature Permanently Affecting Magnet (°F)	594
Inherent Orientation of Material	none

PHYSICAL PROPERTIES	
Density (lbs./in.^{3})	.1800
General Mechanical Properties	HT
Tensile Strength (psi)	2500
Transverse Modulus of Rupture (psi)	-
Coeff. of Therm. Expansion inches x 10^{-6}/°C	Orientation: Par. - Rt.Angl. -
Resistivity (micro-ohms/cm/cm^2)	- @ - °C
Hardness 85	Scale Shore D
Basic Forming Method	E
Basic Finishing Method	MCT
Relative Corrosion Resistance	G
Primary Chemical Composition	Nd-Fe-B + binder (See Notes)
Special Notes: (1) Injection molded (2) Thermoplastic binder	

Note: All information based on available producers literature. In some cases of foreign materials interpretation or calculation required.

SOURCE(S):	**2002B**
	(Alloy or Grade)
Arnold Engineering Co.	Bonded Rare Earth
	(Type of Material)
USA	-
	(MMPA Brief Designation)
	-
	(IEC Code Reference)

MAGNETIC PROPERTIES		PHYSICAL PROPERTIES	
Residual Induction B_r (G)	5100	Density (lbs./in.^{3})	.1800
Coercive Force (Oe) — Normal (H_c)	3800	General Mechanical Properties	HT
Coercive Force (Oe) — Intrinsic (H_{ci})	7500	Tensile Strength (psi)	2500
Maximum Energy Product BH_{max} (MGO)	5.0	Transverse Modulus of Rupture (psi)	-
B/H Ratio (Load Line) at BH_{max}	-	Coeff. of Therm. Expansion inches x 10^{-6}/°C	Orientation: Par. - Rt.Angl. -
Flux Density in Magnet B_m at BH_{max} (G)	-	Resistivity (micro-ohms/cm/cm^2)	- @ - °C
Coercive Force in Magnet H_m at BH_{max} (Oe)	-	Hardness -	Scale -
Energy Loss W_c (watt-sec/cycle/lb)	-	Basic Forming Method	E
Importance of Operating Magnet at BH_{max}	E	Basic Finishing Method	MCT
Average Recoil Permeability	-	Relative Corrosion Resistance	G
Field Strength Required to Fully Saturate Magnet (Oe)	30,000	Primary Chemical Composition (See Notes)	Nd-Fe-B + binder
Ability to Withstand Demagnetizing Fields	0	Special Notes:	
Maximum Service Temperature (°F)	260	(1) Injection molded (2) Thermoplastic binder	
Approximate Temperature Permanently Affecting Magnet (°F)	594		
Inherent Orientation of Material	none		

Note: All information based on available producers literature. In some cases of foreign materials interpretation or calculation required.

SOURCE(S):	2004D
	(Alloy or Grade)
Arnold Engineering Co.	Bonded Rare Earth
	(Type of Material)
USA	-
	(MMPA Brief Designation)
	-
	(IEC Code Reference)

MAGNETIC PROPERTIES		PHYSICAL PROPERTIES	
Residual Induction B_r (G)	5000	Density (lbs./in.^{3})	.1800
Coercive Force (Oe) — Normal (H_c)	3900	General Mechanical Properties	FL
Coercive Force (Oe) — Intrinsic (H_{ci})	9000	Tensile Strength (psi)	450
Maximum Energy Product BH_{max} (MGO)	5.0	Transverse Modulus of Rupture (psi)	-
B/H Ratio (Load Line) at BH_{max}	-	Coeff. of Therm. Expansion inches x 10^{-6}/°C — Orientation: Par. - Rt.Angl. -	
Flux Density in Magnet B_m at BH_{max} (G)	-	Resistivity (micro-ohms/cm/cm^2)	- @ - °C
Coercive Force in Magnet H_m at BH_{max} (Oe)	-	Hardness 45	Scale Shore D
Energy Loss W_c (watt-sec/cycle/lb)	-	Basic Forming Method	E
Importance of Operating Magnet at BH_{max}	E	Basic Finishing Method	SCD/SRD
Average Recoil Permeability	-	Relative Corrosion Resistance	G
Field Strength Required to Fully Saturate Magnet (Oe)	35,000	Primary Chemical Composition	Nb-Fe-B + binder (See Notes)
Ability to Withstand Demagnetizing Fields	0	Special Notes: (1) Flexible material (2) Nitrile rubber binder	
Maximum Service Temperature (°F)	260		
Approximate Temperature Permanently Affecting Magnet (°F)	594		
Inherent Orientation of Material	none		

Note: All information based on available producers literature. In some cases of foreign materials interpretation or calculation required.

SOURCE(S):

Stackpole Corp.

USA

2236
(Alloy or Grade)

Sintered Ferrite
(Type of Material)

1.1/3.8
(MMPA Brief Designation)

S1-0-1
(IEC Code Reference)

MAGNETIC PROPERTIES	
Residual Induction B_r (G)	2200
Coercive Force (Oe) — Normal (H_c)	1800
Coercive Force (Oe) — Intrinsic (H_{ci})	3800
Maximum Energy Product BH_{max} (MGO)	1.0
B/H Ratio (Load Line) at BH_{max}	Est. 1.0
Flux Density in Magnet B_m at BH_{max} (G)	Est. 1000
Coercive Force in Magnet H_m at BH_{max} (Oe)	Est. 1000
Energy Loss W_c (watt-sec/cycle/lb)	-
Importance of Operating Magnet at BH_{max}	E
Average Recoil Permeability	Est. 1.1
Field Strength Required to Fully Saturate Magnet (Oe)	10,000
Ability to Withstand Demagnetizing Fields	Ex
Maximum Service Temperature (°F)	482
Approximate Temperature Permanently Affecting Magnet (°F)	842
Inherent Orientation of Material	none

PHYSICAL PROPERTIES	
Density (lbs./in.^{3})	.1734
General Mechanical Properties	VHB
Tensile Strength (psi)	-
Transverse Modulus of Rupture (psi)	-
Coeff. of Therm. Expansion inches x 10^{-6}/°C — Orientation: Par. / Rt.Angl.	- / -
Resistivity (micro-ohms/cm/cm^2)	1.0 @ 25 °C
Hardness	6.5 Moh's Scale
Basic Forming Method	PF
Basic Finishing Method	WGD
Relative Corrosion Resistance	Ex
Primary Chemical Composition	Ba or Sr & Fe
Special Notes:	

Note: All information based on available producers literature. In some cases of foreign materials interpretation or calculation required.

SOURCE(S):	
Stackpole Corp.	**2331** (Alloy or Grade)
	Sintered Ferrite (Type of Material)
USA	1.2/3.3 (MMPA Brief Designation)
	S1-0-1 (IEC Code Reference)

MAGNETIC PROPERTIES		PHYSICAL PROPERTIES	
Residual Induction B_r (G)	2200	Density (lbs./in.^{3})	.1734
Coercive Force (Oe) — Normal (H_c)	1800	General Mechanical Properties	VHB
Coercive Force (Oe) — Intrinsic (H_{ci})	3340	Tensile Strength (psi)	-
Maximum Energy Product BH_{max} (MGO)	.98	Transverse Modulus of Rupture (psi)	-
B/H Ratio (Load Line) at BH_{max}	Est. 1.6	Coeff. of Therm. Expansion inches x 10^{-6}/°C	Orientation: Par. — Rt.Angl. —
Flux Density in Magnet B_m at BH_{max} (G)	Est. 1300	Resistivity (micro-ohms/cm/cm^2)	1.0 @ 25 °C
Coercive Force in Magnet H_m at BH_{max} (Oe)	Est. 850	Hardness 6.5	Scale Moh's
Energy Loss W_c (watt-sec/cycle/lb)	-	Basic Forming Method	PF
Importance of Operating Magnet at BH_{max}	E	Basic Finishing Method	WGD
Average Recoil Permeability	Est. 1.1	Relative Corrosion Resistance	Ex
Field Strength Required to Fully Saturate Magnet (Oe)	10,000	Primary Chemical Composition	Ba or Sr & Fe
Ability to Withstand Demagnetizing Fields	Ex	Special Notes:	
Maximum Service Temperature (°F)	482		
Approximate Temperature Permanently Affecting Magnet (°F)	842		
Inherent Orientation of Material	none		

Note: All information based on available producers literature. In some cases of foreign materials interpretation or calculation required.

SOURCE(S):	
	2532 (sleeves)
	(Alloy or Grade)
Stackpole Corp.	Sintered Ferrite
	(Type of Material)
USA	1.4/3.4
	(MMPA Brief Designation)
	-
	(IEC Code Reference)

MAGNETIC PROPERTIES		PHYSICAL PROPERTIES	
Residual Induction B_r (G)	2575	Density (lbs./in.^{3})	.1734
Coercive Force (Oe) — Normal (H_c)	2130	General Mechanical Properties	VHB
Coercive Force (Oe) — Intrinsic (H_{ci})	3350	Tensile Strength (psi)	-
Maximum Energy Product BH_{max} (MGO)	1.35	Transverse Modulus of Rupture (psi)	-
B/H Ratio (Load Line) at BH_{max}	-	Coeff. of Therm. Expansion inches x 10^{-6}/°C	Orientation: Par. _ Rt.Angl. _
Flux Density in Magnet B_m at BH_{max} (G)	-	Resistivity (micro-ohms/cm/cm^2)	1.0 @ 25 °C
Coercive Force in Magnet H_m at BH_{max} (Oe)	-	Hardness Scale	6.5 Moh's
Energy Loss W_c (watt-sec/cycle/lb)	-	Basic Forming Method	PF
Importance of Operating Magnet at BH_{max}	E	Basic Finishing Method	WGD
Average Recoil Permeability	Est. 1.1	Relative Corrosion Resistance	Ex
Field Strength Required to Fully Saturate Magnet (Oe)	10,000	Primary Chemical Composition	Sr-Fe
Ability to Withstand Demagnetizing Fields	Ex	Special Notes: (1) Improved ceramagnet A20 sleeves	
Maximum Service Temperature (°F)	482		
Approximate Temperature Permanently Affecting Magnet (°F)	842		
Inherent Orientation of Material	none		

Note: All information based on available producers literature. In some cases of foreign materials interpretation or calculation required.

SOURCE(S):	**2732 (segments)**
	(Alloy or Grade)
Stackpole Corp.	Sintered Ferrite
	(Type of Material)
USA	1.8/3.4
	(MMPA Brief Designation)
	-
	(IEC Code Reference)

MAGNETIC PROPERTIES		PHYSICAL PROPERTIES	
Residual Induction B_r (G)	2850	Density (lbs./in.^{3})	.1734
Coercive Force (Oe) — Normal (H_c)	2400	General Mechanical Properties	VHB
Coercive Force (Oe) — Intrinsic (H_{ci})	3350	Tensile Strength (psi)	-
Maximum Energy Product BH_{max} (MGO)	1.8	Transverse Modulus of Rupture (psi)	-
B/H Ratio (Load Line) at BH_{max}	-	Coeff. of Therm. Expansion inches x 10^{-6}/°C Orientation: Par. _ Rt.Angl. _	
Flux Density in Magnet B_m at BH_{max} (G)	-	Resistivity (micro-ohms/cm/cm^2)	1.0 @ 25 °C
Coercive Force in Magnet H_m at BH_{max} (Oe)	-	Hardness Scale	6.5 Moh's
Energy Loss W_c (watt-sec/cycle/lb)	-	Basic Forming Method	PMF
Importance of Operating Magnet at BH_{max}	E	Basic Finishing Method	WGD
Average Recoil Permeability	Est. 1.1	Relative Corrosion Resistance	Ex
Field Strength Required to Fully Saturate Magnet (Oe)	10,000	Primary Chemical Composition	Sr-Fe
Ability to Withstand Demagnetizing Fields	Ex	Special Notes: (1) Radial orientation	
Maximum Service Temperature (°F)	482		
Approximate Temperature Permanently Affecting Magnet (°F)	842		
Inherent Orientation of Material	MP		

Note: All information based on available producers literature. In some cases of foreign materials interpretation or calculation required.

SOURCE(S): _____

Stackpole Corp.

USA

__3540__
(Alloy or Grade)

__Sintered Ferrite__
(Type of Material)

__3.1/4.2__
(MMPA Brief Designation)

__S1-1-2__
(IEC Code Reference)

MAGNETIC PROPERTIES		PHYSICAL PROPERTIES	
Residual Induction B_r (G)	3620	Density (lbs./in.^{3})	.1716
Coercive Force (Oe) — Normal (H_c)	3400	General Mechanical Properties	VHB
Coercive Force (Oe) — Intrinsic (H_{ci})	4200	Tensile Strength (psi)	-
Maximum Energy Product BH_{max} (MGO)	4.1	Transverse Modulus of Rupture (psi)	-
B/H Ratio (Load Line) at BH_{max}	.78	Coeff. of Therm. Expansion inches x $10^{-6}/°C$	Orientation: Par. _ Rt.Angl. _
Flux Density in Magnet B_m at BH_{max} (G)	2550	Resistivity (micro-ohms/cm/cm^2)	1.0 @ 25 °C
Coercive Force in Magnet H_m at BH_{max} (Oe)	2000	Hardness Scale	6.5 Moh's
Energy Loss W_c (watt-sec/cycle/lb)	-	Basic Forming Method	PMF
Importance of Operating Magnet at BH_{max}	E	Basic Finishing Method	WGD
Average Recoil Permeability	Est. 1.1	Relative Corrosion Resistance	Ex
Field Strength Required to Fully Saturate Magnet (Oe)	10,000	Primary Chemical Composition	Sr-Fe
Ability to Withstand Demagnetizing Fields	Ex	Special Notes: (1) Available as segments and in conventional shapes	
Maximum Service Temperature (°F)	482		
Approximate Temperature Permanently Affecting Magnet (°F)	842		
Inherent Orientation of Material	MP		

Note: All information based on available producers literature. In some cases of foreign materials interpretation or calculation required.

SOURCE(S):	3547
Stackpole Corp.	(Alloy or Grade)
	Sintered Ferrite
	(Type of Material)
USA	3.1/4.9
	(MMPA Brief Designation)
	-
	(IEC Code Reference)

MAGNETIC PROPERTIES		PHYSICAL PROPERTIES	
Residual Induction B_r (G)	3620	Density (lbs./in.^{3})	.1734
Coercive Force (Oe) — Normal (H_c)	3400	General Mechanical Properties	VHB
Coercive Force (Oe) — Intrinsic (H_{ci})	4850	Tensile Strength (psi)	-
Maximum Energy Product BH_{max} (MGO)	3.1	Transverse Modulus of Rupture (psi)	-
B/H Ratio (Load Line) at BH_{max}	.78	Coeff. of Therm. Expansion inches x 10^{-6}/°C	Orientation: Par. - Rt.Angl. -
Flux Density in Magnet B_m at BH_{max} (G)	1550	Resistivity (micro-ohms/cm/cm^2)	1.0 @ 25 °C
Coercive Force in Magnet H_m at BH_{max} (Oe)	2000	Hardness 6.5	Scale Moh's
Energy Loss W_c (watt-sec/cycle/lb)	-	Basic Forming Method	PMF
Importance of Operating Magnet at BH_{max}	E	Basic Finishing Method	WGD
Average Recoil Permeability	Est. 1.1	Relative Corrosion Resistance	Ex
Field Strength Required to Fully Saturate Magnet (Oe)	10,000	Primary Chemical Composition	Sr-Fe
Ability to Withstand Demagnetizing Fields	Ex	Special Notes: (1) Available as segments and in conventional shapes	
Maximum Service Temperature (°F)	482		
Approximate Temperature Permanently Affecting Magnet (°F)	842		
Inherent Orientation of Material	MP		

Note: All information based on available producers literature. In some cases of foreign materials interpretation or calculation required.

SOURCE(S):	3831
	(Alloy or Grade)
Stackpole Corp.	Sintered Ferrite
	(Type of Material)
USA	3.6/3.3
	(MMPA Brief Designation)
	S1-1-5
	(IEC Code Reference)

MAGNETIC PROPERTIES		PHYSICAL PROPERTIES	
Residual Induction B_r (G)	3920	Density (lbs./in.^{3})	.1770
Coercive Force (Oe) — Normal (H_c)	3175	General Mechanical Properties	VHB
Coercive Force (Oe) — Intrinsic (H_{ci})	3275	Tensile Strength (psi)	-
Maximum Energy Product BH_{max} (MGO)	3.6	Transverse Modulus of Rupture (psi)	-
B/H Ratio (Load Line) at BH_{max}	.9	Coeff. of Therm. Expansion inches x 10^{-6}/°C	Orientation: Par. - Rt.Angl. -
Flux Density in Magnet B_m at BH_{max} (G)	1800	Resistivity (micro-ohms/cm/cm^2)	1.0 @ 25 °C
Coercive Force in Magnet H_m at BH_{max} (Oe)	2000	Hardness Scale	6.5 Moh's
Energy Loss W_c (watt-sec/cycle/lb)	-	Basic Forming Method	PMF
Importance of Operating Magnet at BH_{max}	E	Basic Finishing Method	WGD
Average Recoil Permeability	Est. 1.1	Relative Corrosion Resistance	Ex
Field Strength Required to Fully Saturate Magnet (Oe)	10,000	Primary Chemical Composition	Ba or Sr & Fe
Ability to Withstand Demagnetizing Fields	Ex	Special Notes: (1) Available as segments and in conventional configurations	
Maximum Service Temperature (°F)	482		
Approximate Temperature Permanently Affecting Magnet (°F)	842		
Inherent Orientation of Material	MP		

Note: All information based on available producers literature. In some cases of foreign materials interpretation or calculation required.

SOURCE(S):	3838
	(Alloy or Grade)
Stackpole Corp.	Sintered Ferrite
	(Type of Material)
USA	3.6/4.0
	(MMPA Brief Designation)
	-
	(IEC Code Reference)

MAGNETIC PROPERTIES		PHYSICAL PROPERTIES	
Residual Induction B_r (G)	3920	Density (lbs./in.^{3})	.1752
Coercive Force (Oe) — Normal (H_c)	3520	General Mechanical Properties	VHB
Coercive Force (Oe) — Intrinsic (H_{ci})	4000	Tensile Strength (psi)	-
Maximum Energy Product BH_{max} (MGO)	3.6	Transverse Modulus of Rupture (psi)	-
B/H Ratio (Load Line) at BH_{max}	.9	Coeff. of Therm. Expansion inches x 10^{-6}/°C	Orientation: Par. - Rt.Angl. -
Flux Density in Magnet B_m at BH_{max} (G)	1800	Resistivity (micro-ohms/cm/cm^2)	1.0 @ 25 °C
Coercive Force in Magnet H_m at BH_{max} (Oe)	2000	Hardness Scale	6.5 Moh's
Energy Loss W_c (watt-sec/cycle/lb)	-	Basic Forming Method	PMF
Importance of Operating Magnet at BH_{max}	E	Basic Finishing Method	WGD
Average Recoil Permeability	Est. 1.1	Relative Corrosion Resistance	Ex
Field Strength Required to Fully Saturate Magnet (Oe)	10,000	Primary Chemical Composition	Sr-Fe
Ability to Withstand Demagnetizing Fields	Ex	Special Notes: (1) Available as segments and in conventional shapes	
Maximum Service Temperature (°F)	482		
Approximate Temperature Permanently Affecting Magnet (°F)	842		
Inherent Orientation of Material	MP		

Note: All information based on available producers literature. In some cases of foreign materials interpretation or calculation required.

SOURCE(S):	4130
	(Alloy or Grade)
Stackpole Corp.	Sintered Ferrite
	(Type of Material)
USA	4.0/3.1
	(MMPA Brief Designation)
	-
	(IEC Code Reference)

MAGNETIC PROPERTIES		PHYSICAL PROPERTIES	
Residual Induction B_r (G)	4150	Density (lbs./in.^{3})	.1788
Coercive Force (Oe) — Normal (H_c)	3000	General Mechanical Properties	VHB
Coercive Force (Oe) — Intrinsic (H_{ci})	3100	Tensile Strength (psi)	-
Maximum Energy Product BH_{max} (MGO)	4.0	Transverse Modulus of Rupture (psi)	-
B/H Ratio (Load Line) at BH_{max}	1.0	Coeff. of Therm. Expansion inches x 10^{-6}/°C — Orientation: Par. / Rt.Angl.	- / -
Flux Density in Magnet B_m at BH_{max} (G)	2000	Resistivity (micro-ohms/cm/cm^2)	1.0 @ 25 °C
Coercive Force in Magnet H_m at BH_{max} (Oe)	2000	Hardness Scale	6.5 Moh's
Energy Loss W_c (watt-sec/cycle/lb)	-	Basic Forming Method	PMF
Importance of Operating Magnet at BH_{max}	E	Basic Finishing Method	WGD
Average Recoil Permeability	Est. 1.1	Relative Corrosion Resistance	Ex
Field Strength Required to Fully Saturate Magnet (Oe)	10,000	Primary Chemical Composition	Sr-Fe
Ability to Withstand Demagnetizing Fields	Ex	Special Notes: (1) Available as segments and in conventional shapes	
Maximum Service Temperature (°F)	482		
Approximate Temperature Permanently Affecting Magnet (°F)	842		
Inherent Orientation of Material	MP		

Note: All information based on available producers literature. In some cases of foreign materials interpretation or calculation required.

Appendix 3

International Index of Permanent Magnet Materials

Note: In this edition, the Interchangeability/Replacement Guide has been eliminated. The great proliferation of new materials has made making even reasonable comparisons useless.

Notations Used in Data Sheets

Importance of Operating Magnet at BH_{max} point:
A-E: A = very important; E = least important.

Ability to Withstand Demagnetizing Fields:
P = Poor   G = Good   Ex = Exceptional
F = Fair   E = Excellent   O = Outstanding

Inherent Orientation of Material:
HT = In direction of heat treatment *after* casting or pressing.
CHT = In direction of crystal growth during casting process and heat treatment of finished magnet.
CR = In direction of mechanical deformation (rolling, swagging, etc.).
MP = In direction of magnetic field applied *during* pressing.

Magnetic Properties

Availability	GO	GO	GA	GS	GS	GO	GS	GA	GS	GS	GA	GS	GS
Alloy or Grade	Alnico 1, cast	Alnico 2, cast	Alnico 2, sintered	Alnico 3, cast	Alnico 4, cast	Alnico 4, sintered	Alnico 5, cast (un-oriented)	Alnico 5, cast	Alnico 5DG, cast	Alnico 5-7, cast	Alnico 5, sintered	Alnico 6, cast	Alnico 6 sintered
Residual Induction B_r (G)	7100	7200	6800	6700	5200	5200	7400	12,400	12,600	13,400	10,500	10,200	8800
Coercive Force (Oe) H_c Normal	400	540	520	450	700	700	520	640	670	730	590	770	800
Intrinsic H_{ci}	—	—	—	—	—	—	—	—	—	—	—	—	—
Maximum Energy Product BH_{max} (MGO)	1.3	1.6	1.5	1.4	1.2	1.2	1.4	5.5	6.3	7.5	3.5	3.8	2.8
B/H Ratio (Load Line) at BH_{max}	14.0	12.0	12.3	13.0	8.0	7.5	15.5	18.0	17.0	18.5	18.4	13.0	11.0
Flux Density in Magnet B_m at BH_{max} (G)	4200	4500	4300	4300	3000	3000	4600	10,000	10,400	11,150	8000	7000	5500
Coercive Force in Magnet H_m at BH_{max} (Oe)	300	365	350	320	380	400	300	550	600	650	435	535	500
Energy Loss W_c (watt-sec/cycle/lb)	7.4	9.5	—	7.1	10.8	—	—	15.3	—	—	—	17.8	—
Importance of Operating Magnet at BH_{max} Point	B	B	B	B	C	C	C	A	A	A	A	B	B
Average Recoil Permeability	6.8	6.4	6.4	6.5	4.1	4.1	—	4.3	4.0	3.8	4.0	5.3	5.0
Field Strength Required to Fully Saturate Magnet (Oe)	2000	2000	2000	2000	3000	3000	3000	3000	3000	3000	3000	3000	3000
Ability to Withstand Demagnetizing Fields	G	G	G	G	E	E	E	G	G	G	G	E	E
Approximate Temperature Permanently Affecting Magnet (°F)	900	900	900	900	900	900	900	900	900	900	900	900	900
Inherent Orientation of Material	none	none	none	none	none	none	none	none	CHT	CHT	HT	HT	HT

Magnetic Properties

Availability	GS	GS	GA	GA	LS	GO	GS	GS	GS	GS	GS	GS	GS
Alloy or Grade	Alnico 7, cast (un-oriented)	Alnico 7, cast	Alnico 8, cast	Alnico 8, sintered	Alnico 9, cast	Alnico 12, cast	Carbon Steel (.65% C)	Carbon Steel (1.0% C)	Chrome Steel (1.0% Cr)	Chrome Steel (2.0% Cr)	Chrome Steel (3.0% Cr)	Chrome Steel (3.5% Cr)	Chrome Steel (6.0% Cr)
Residual Induction B_r (G)	7500	8600	7600	7400	10,500	6100	10,000	9000	9500	9300	9800	9000	9500
Coercive Force (Oe) H_c Normal	890	1050	1620	1500	1450	1000	42	51	52	60	70	63	74
H_{ci} Intrinsic	—	—	—	—	—	—	—	—	—	—	—	—	—
Maximum Energy Product BH_{max} (MGO)	2.5	3.7	5.3	4.0	8.5	1.7	.18	.20	.23	.26	.29	.29	.30
B/H Ratio (Load Line) at BH_{max}	8.5	8.0	5.0	5.0	7.3	6.4	250	125	155	126	135	171	103
Flux Density in Magnet B_m at BH_{max} (G)	4500	5500	5150	4400	7700	3200	6500	5900	6500	6300	6200	6000	6200
Coercive Force in Magnet H_m at BH_{max} (Oe)	550	680	1025	880	1080	520	32	47	42	50	46	35	60
Energy Loss W_c (watt-sec/cycle/lb)	—	17.4	16.6	—	—	—	—	—	—	—	—	1.03	—
Importance of Operating Magnet at BH_{max} Point	B	B	D	D	B	C	A	A	A	A	A	A	A
Average Recoil Permeability	—	—	2.0	2.0	—	—	—	48.0	—	—	—	35.0	—
Field Strength Required to Fully Saturate Magnet (Oe)	3500	3500	5000	5000	5000	3000	300	300	300	300	300	300	300
Ability to Withstand De-magnetizing Fields	E	E	Ex	Ex	Ex	E	P	P	P	P	P	P	P
Approximate Temperature Permanently Affecting Magnet (°F)	900	900	900	900	900	900	900	900	900	900	900	900	900
Inherent Orientation of Material	none	HT	HT	HT	HT	none	CR	CR	CR	CR	CR	CR	CR

Magnetic Properties

Alloy or Grade	Availability	Cobalt-Chrome Steel (2.0% Cr)	Cobalt-Chrome Steel (16.0% Cr)	Cobalt Steel (3.0% Co)	Cobalt Steel (9% Co)	Cobalt Steel (15% Co)	Cobalt Steel (17% Co)	Cobalt Steel (35% Co)	Cobalt Steel (36% Co)	Cobalt Steel (40% Co)	Cunife (.030 in. diameter)	Cunife (.150 in. diameter)	Cunife 1	Cunife 2
		GS	GS	GS	GS	GS	GS	GS	GS	GS	GA	GA	GO	GO
Residual Induction B_r (G)		9800	8000	7200	7800	8200	9000	9000	9600	10,000	5400	4750	5700	7300
Coercive Force (Oe) Normal H_c		80	180	130	122	180	170	250	228	242	550	440	590	260
Intrinsic H_{ci}		—	—	—	—	—	—	—	—	—	—	—	—	—
Maximum Energy Product BH_{max} (MGO)		.32	.90	.35	.41	.62	.65	.95	.94	1.03	1.54	.99	1.85	.78
B/H Ratio (Load Line) at BH_{max}		123	50	50.5	50.1	44.5	53.6	37.0	45.0	41.0	9.4	11.0	9.8	30.0
Flux Density in Magnet B_m at BH_{max} (G)		6280	6700	4200	5100	5250	5900	5930	6300	6500	3800	3300	4200	4700
Coercive Force in Magnet H_m at BH_{max} (Oe)		51	135	83	100	118	110	160	140	159	405	300	440	152
Energy Loss W_c (watt-sec/cycle/lb)		—	—	—	—	—	—	—	4.5	—	—	—	—	—
Importance of Operating Magnet at BH_{max} Point		A	A	A	A	A	A	B	B	B	B	B	B	A
Average Recoil Permeability		30.0	—	18.5	16.5	—	—	12.0	12.0	12.0	1.7	2.2	—	—
Field Strength Required to Fully Saturate Magnet (Oe)		300	300	500	600	600	600	1000	1000	1000	2400	2400	2400	2400
Ability to Withstand Demagnetizing Fields		P	P	P	P	P	P	F	F	F	G	G	G	G
Approximate Temperature Permanently Affecting Magnet (°F)		900	900	900	900	900	900	900	900	900	900	900	900	900
Inherent Orientation of Material		CR	CR	CR	CR	CR	CR	CR	CR	CR	GR	GR	GR	GR

Magnetic Properties

Availability	GA	GA	GA	GA	GA	GS	GS	GS	GS	LS	LS	LS	LS
Alloy or Grade	Ferrite 1	Ferrite 2	Ferrite 3	Ferrite 4	Ferrite 5	Ferrite 6	Ferrite 7	Ferrite, rubber bonded	Ferrite, vinyl bonded	Lodex 30	Lodex 31	Lodex 32	Lodex 33
Residual Induction B_r (G)	2200	2700	3350	2550	3850	3300	3400	2100	1675	4000	6250	7300	8000
Coercive Force (Oe) Normal H_c	1825	2250	2400	2300	2200	3100	3200	1300	1220	1250	1140	940	860
Intrinsic H_{ci}	3135	—	—	—	2300	—	3800	—	—	—	—	—	—
Maximum Energy Product BH_{max} (MGO)	1.0	1.7	2.6	1.5	3.5	2.6	2.8	1.0	.575	1.68	3.4	3.4	3.2
B/H Ratio (Load Line) at BH_{max}	1.2	1.1	.9	1.1	1.1	1.06	1.05	1.2	1.4	3.4	5.3	8.2	10.5
Flux Density in Magnet B_m at BH_{max} (G)	1100	1350	1600	1300	1900	1650	1700	1050	899	2400	4250	5300	5800
Coercive Force in Magnet H_m at BH_{max} (Oe)	900	1225	1650	1150	1800	1600	1700	950	640	700	800	650	550
Energy Loss W_c (watt-sec/cycle/lb)	—	—	—	—	—	—	—	—	—	—	—	—	—
Importance of Operating Magnet at BH_{max} Point	E	E	E	E	E	E	E	E	E	D	C	B	B
Average Recoil Permeability	1.1	1.2	1.1	1.1	1.1	1.1	1.06	—	—	1.5	1.9	2.6	3.0
Field Strength Required to Fully Saturate Magnet (Oe)	10,000	10,000	10,000	10,000	10,000	10,000	10,000	10,000	10,000	6000	6000	5000	5000
Ability to Withstand Demagnetizing Fields	O	O	O	O	O	O	O	O	O	Ex	Ex	E	Ex
Approximate Temperature Permanently Affecting Magnet (°F)	1800	1800	1800	1800	1800	1800	1800	200	158	600	600	600	600
Inherent Orientation of Material	none	MP	MP	MP	MP	MP	MP	MP	none to MP	MP	MP	MP	MP

Magnetic Properties

Availability	LS	LS	LS	LS	LS	LS	LS	LS	LS	LS	LS	LO	LO
Alloy or Grade	Lodex 36	Lodex 37	Lodex 38	Lodex 40	Lodex 41	Lodex 42	Lodex 43	P-6 Alloy	Perma-flux C	Perma-flux Cl	Platinum Cobalt	Remalloy (Comol) Alloy 1	Remalloy (Comol) Alloy 2
Residual Induction B_r (G)	3400	5500	6200	2700	4300	5250	6000	14,000	5600	3000	6450	9700	10,000
Coercive Force (Oe) Normal H_c	1220	1000	840	1100	980	850	710	58	4800	2800	4300	210	230
Intrinsic H_{ci}	—	—	—	—	—	—	—	—	6200	4800	5300	—	—
Maximum Energy Product BH_{max} (MGO)	1.45	2.1	2.2	.85	1.4	1.4	1.3	.5	7.0	2.0	9.5	.95	1.1
B/H Ratio (Load Line) at BH_{max}	2.5	5.8	7.0	2.0	3.8	7.6	10.0	22.1	1.08	1.06	1.14	52.0	43.0
Flux Density in Magnet B_m at BH_{max} (G)	1700	3500	4000	1300	2300	3250	3600	10,500	2800	1600	3250	6800	6900
Coercive Force in Magnet H_m at BH_{max} (Oe)	850	600	550	650	600	430	360	48	2500	1250	2850	135	160
Energy Loss W_c (watt-sec/cycle/lb)	—	—	—	—	—	—	—	—	—	—	—	—	—
Importance of Operating Magnet at BH_{max} Point	E	D	C	E	D	D	D	A	E	E	E	C	C
Average Recoil Permeability	2.0	3.0	3.5	2.5	3.2	3.5	3.8	16.7	1.02	1.02	1.2	—	13.0
Field Strength Required to Fully Saturate Magnet (Oe)	5000	5000	5000	5000	5000	5000	5000	250	25,000	20,000	20,000	1000	1000
Ability to Withstand Demagnetizing Fields	E	E	E	E	E	E	E	P	O	O	O	G	G
Approximate Temperature Permanently Affecting Magnet (°F)	600	600	600	600	600	600	600	900	—	—	850	900	900
Inherent Orientation of Material	MP	MP	MP	none	none	none	none	CR	HT	none	CR	CR	CR

Magnetic Properties

Availability	LS	GS	GS	GO	GO
Alloy or Grade	Samarium-Cobalt	Vicalloy (HR bars, strips, and forgings)	Vicalloy (thin tape and wire)	Vicalloy 1	Vicalloy 2
Residual Induction B_r (G)	7500–9000	8000	3000	9000	10,000
Coercive Force (Oe) Normal H_c	7500–9000	250	225	300	450
Coercive Force (Oe) Intrinsic H_{ci}	25,000	—	—	—	—
Maximum Energy Product B/H_{max} (MGO)	16–20	.80	.325	1.0	3.0
B/H Ratio (Load Line) at BH_{max}	1.0	31.2	17.6	32.0	23.0
Flux Density in Magnet B_m at BH_{max} (G)	4500 at 20 MGO	5000	2200	5500	8200
Coercive Force in Magnet H_m at BH_{max} (Oe)	4500 at 20 MGO	160	125	180	300
Energy Loss W_c (watt-sec/cycle/lb)	—	—	—	—	—
Importance of Operating Magnet at BH_{max} Point	E	D	D	D	D
Average Recoil Permeability	—	—	—	—	—
Field Strength Required to Fully Saturate Magnet (Oe)	100,000	3000	3000	3000	3000
Ability to Withstand Demagnetizing Fields	O	F	F	F	F
Approximate Temperature Permanently Affecting Magnet (°F)	1562	600	600	600	600
Inherent Orientation of Material	MP	CR	CR	CR	CR

Notations Used in Data Sheets

General Mechanical Properties:
VHB = Very hard and brittle
HB   = Hard and brittle
HT   = Hard and tough
TC = Weak; tends to crumble
WB = Weak and brittle
D   = Ductile
FL = Soft and flexible

Basic Forming Method:
C   = Sand-, shell-, or investment-cast.
CM   = Sand-, shell-, or investment-cast and heat treated in a magnetic field of proper shape and intensity.
CCM = Shell molded with copper or steel "chill" in mold, followed by heat treatment in a magnetic field of proper shape and intensity.
DP   = Dry-pressed from powder.
DPM = Dry-pressed from powder in a magnetic field of proper shape and intensity.
P   = Pressed from a liquid slurry.
PM   = Pressed from a liquid slurry while in a magnetic field of proper shape and intensity.
PF   = Pressed from a liquid slurry and fired at a high temperature.
PMF = Pressed from a liquid slurry while in a magnetic field of proper shape and intensity and fired at a high temperature.
E   = Extruded or molded.
VM   = Calendered and vulcanized while in a magnetic field of proper shape and intensity.
W   = Mechanically deformed (reduced) from cast ingots.

Basic Finishing Method:
WGR = Wet-ground with rubber-bonded oxide wheels.
WGD = Wet-ground with diamond wheels.
MCT = Machined with conventional cutting tools.
SCD = Stamped with conventional dies.
SRD = Stamped with steel-rule dies or cut with a knife edge.

Availability:
GA   = Available from two or more mills in a variety of standard or special sizes and shapes in any quantity. Normally, stocked in standard shapes and/or fabricated by distributors. Considered a current, widely used material.
GS   = Available from two or more mills only on special order and usually only in sizable quantities. Not normally stocked by distributors. Considered a "seldom used" material; may become obsolete.
LS   = Available only from one mill and only on special order. Usually available only in sizable quantities. Not normally stocked by distributors.
GO   = Considered obsolete, but may be obtained from one or more mills on special order in sizable quantities only. Not normally available from distributors.
LO   = Considered obsolete; virtually unobtainable. Generally has been replaced by another material.

Relative Corrosion Resistance:
P = Poor
F = Fair
G = Good
E = Excellent
Ex = Exceptional

Physical Properties

Alloy or Grade	Alnico 1, cast	Alnico 2, cast	Alnico 2, sintered	Alnico 3, cast	Alnico 4, cast	Alnico 4, sintered	Alnico 5, cast (un-oriented)	Alnico 5, cast	Alnico 5-7, cast	Alnico 5DG, cast	Alnico 5, sintered	Alnico 6, cast	Alnico 6, sintered
Density (lbs./in.^{3})	.249	.256	.247	.249	.253	.250	.264	.264	.264	.264	.253	.268	.241
General Mechanical Properties	HB	HB	HT	HB	HB	HT	HB	HB	HB	HB	HT	HB	HB
Tensile Strength (psi)	4100	3000	65,000	12,000	9100	60,000	5450	5450	5000	5200	50,000	23,000	—
Transverse Modulus of Rupture (psi)	13,900	7200	70,000	22,500	24,000	85,000	10,500	10,500	8000	9000	55,000	45,000	—
Coefficient of Thermal Expansion (per °C × 10^{-6})	12.6	12.4	12.4	13.0	13.1	13.1	11.6	11.6	11.4	11.4	11.3	11.4	—
Resistivity at 25°C (micro-ohms/cm/cm^2)	75	65	68	60	75	68	47	47	47	47	50	50	—
Rockwell Hardness Scales	C45	C45	C43	C45	C45	C43	C50	C50	C50	C50	C44	C50	C43
Basic Forming Method	C	C	DP	C	C	DP	C	CM	CMM	CCM	DPM	CM	DPM
Basic Finishing Method	WRG	WRG	WRG	WRG	WRG	WRG	WRG	WRG	WRG	WRG	WRG	WRG	WRG
Relative Corrosion Resistance	F	F	G	F	F	G	F	F	F	F	G	F	G
Nominal Metallurgical Composition (percent)	12 Al, 21 Ni, 5 Co, 3 Cu; bal. Fe	10 Al, 19 Ni, 13 Co, 3 Cu; bal. Fe	10 Al, 19 Ni, 13 Co, 3 Cu; bal. Fe	12 Al, 25 Ni, 3 Co; bal. Fe	12 Al, 27 Ni, 5 Co; bal. Fe	12 Al, 27 Ni, 5 Co	8 Al, 14 Ni, 24 Co, 3 Cu; bal. Fe	8 Al, 14 Ni, 24 Co, 3 Cu; bal. Fe	8 Al, 14 Ni, 24 Co, 3 Cu; bal. Fe	8 Al, 1 Ni, 24 Co, 3 Cu; 1 Ti; bal. Fe	8 Al, 14 Ni, 24 Co, 3 Cu; bal. Fe	8 Al, 16 Ni, 24 Co, 3 Cu; bal. Fe	8 Al, 16 Ni, 24 Co, 1 Ti; bal. Fe

Physical Properties

Alloy or Grade	Alnico 7, cast (un-oriented)	Alnico 7, cast	Alnico 8, cast	Alnico 8, sintered	Alnico 9, cast	Alnico 12, cast	Carbon Steel (.65% C)	Carbon Steel (1.0% C)	Chrome Steel (1.0% Cr)	Chrome Steel (2.0% Cr)	Chrome Steel (3.0% Cr)	Chrome Steel (3.5% Cr)	Chrome Steel (6.0% Cr)
Density (lbs./in.^{3})	.265	.265	.265	.252	.265	.264	.283	.282	.282	.282	.282	.281	.281
General Mechanical Properties	HB	HB	HB	HB	HB	HB	HT	HT	HT	HT	HT	HT	HT
Tensile Strength (psi)	—	—	39,500	—	5000	39,500	300,000	300,000	300,000	300,000	300,000	300,000	300,000
Transverse Modulus of Rupture (psi)	2000	2000	30,000	55,000	8000	50,000	—	—	—	—	—	—	—
Coefficient of Thermal Expansion (per °C × 10^{-6})	11.4	11.4	11.0	—	11.0	11.0	13.3	12.4	13.4	13.1	—	12.6	12.2
Resistivity at 25°C (micro-ohms/cm/cm^2)	58	58	50	—	62	62	18	20	23	28	—	29	34
Rockwell Hardness Scales	C60	C60	C56	—	C58	C58	C65	C62	C65	C65	C62	C62	C65
Basic Forming Method	C	CM	CM	DPM	CCM	CM	W	W	W	W	W	W	W
Basic Finishing Method	WRG	WRG	WRG	WRG	WRG	WRG	WRG	WRG	WRG	WRG	WRG	WRG	WRG
Relative Corrosion Resistance	F	F	F	G	F	F	P	P	P	P	P	P	P
Nominal Metallurgical Composition (percent)	85 Al, 18 Ni, 24 Co, 3.25 Cu, 5 Ti; bal. Fe	8.5 Al, 18 Ni, 24 Co, 3.25 Cu, 5 Ti; bal. Fe	7 Al, 15 Ni, 35 Co, 4 Cu, 5 Ti; bal. Fe	7 Al, 15 Ni, 35 Co, 4 Cu, 5 Ti; bal. Fe	7 Al, 15 Ni, 35 Co, 4 Cu, 5 Ti; bal. Fe	6 Al, 18 Ni, 35 Co, 8 Ti; bal. Fe	.65 C, .85 Mn; bal. Fe	1 C, .5 Mn; bal. Fe	.9 Cr, .6 C, .45 Mn; bal. Fe	2.15 Cr, .9 C; bal. Fe	—	3.5 Cr, 1 C, .5 Mn; bal. Fe	6.0 Cr, 1.1 C, .4 Mn; bal. Fe

Physical Properties

Alloy or Grade	Cobalt-Chrome Steel (2.0% Cr)	Cobalt-Chrome Steel (16.0% Cr)	Cobalt Steel (3.0% Co)	Cobalt Steel (9.0% Co)	Cobalt Steel (15.0% Co)	Cobalt Steel (17.0% Co)	Cobalt Steel (35.0% Co)	Cobalt Steel (36.0% Co)	Cobalt Steel (40.0% Co)	Cunife (.030 in. diameter)	Cunife (.150 in. diameter)	Cunife 1	Cunife 2
Density (lbs/in.^{3})	.282	.281	.282	.286	.288	.302	.295	.296	.296	.311	.311	.311	.311
General Mechanical Properties	HT	HT	HT	HT	HT	HT	HT	HT	HT	HT	HT	HT	HT
Tensile Strength (psi)	300,000	300,000	300,000	300,000	300,000	300,000	300,000	300,000	300,000	100,000	100,000	100,000	100,000
Transverse Modulus of Rupture (psi)	—	—	—	—	—	—	—	—	—	—	—	—	—
Coefficient of Thermal Expansion (per °C × 10^{-6})	—	—	—	14.0	—	15.9	—	17.2	17.4	14.0	14.0	14.0	14.0
Resistivity at 25°C (micro-ohms/cm/cm^2)	—	—	—	30	—	28	—	27	25	18	18	18	18
Rockwell Hardness Scales	C60	C65	C60	C62	C63	C65	C61	C62	C63	B200	B200	B200	B200
Basic Forming Method	W	W	W	W	W	W	W	W	W	W	W	W	W
Basic Finishing Method	WRG	WRG	WRG	WRG	WRG	WRG	WRG	WRG	WRG	MCT	MCT	MCT	MCT
Relative Corrosion Resistance	P	P	P	P	P	P	F	F	F	G	G	G	G
Nominal Metallurgical Composition (percent)	2 Co, 4 Cr; bal. Fe	16 Co, 9 Cr, 1 C, .3 Mn; bal. Fe	1.05 C, 3 Co, 9 Cr, 1.5 Mo; bal. Fe	9 Co, .9 C, 1.25 W, 5 Cr; bal. Fe	—	17 Co, .7 C, 8.25 W, 2.5 Cr; bal. Fe	—	36 Co, .8 C, 3.75 W, 5.75 Cr; bal. Fe	40 Co, .7 C, 5 W, 4.25 Cr; bal. Fe	60 Cu, 20 Ni; bal. Fe	60 Cu, 20 Ni; bal. Fe	60 Cu, 20 Ni; bal. Fe	50 Cu, 20 Ni; 2.5 Co; bal. Fe

Physical Properties

Alloy or Grade	Ferrite 1	Ferrite 2	Ferrite 3	Ferrite 4	Ferrite 5	Ferrite 6	Ferrite 7	Ferrite, rubber bonded	Ferrite, vinyl bonded	Lodex 30	Lodex 31	Lodex 32	Lodex 33
Density (lbs./in.^{3})	.167	.162	.180	.160	.180	.162	.166	.133	.133	.364	.346	.337	.333
General Mechanical Properties	VHB	VHB	VHB	VHB	VHB	VHB	VHB	FL	FL	TC	TC	TC	TC
Tensile Strength (psi)	low	low	low	low	low	low	low	2000–7000	2000–7000	1000	1000	1000	1000
Transverse Modulus of Rupture (psi)	—	—	—	—	—	—	—	—	—	4500	4500	4500	4500
Coefficient of Thermal Expansion (per °C $\times 10^{-6}$)	10.0	10.0	10.3	10.3	10.3	10.3	—	2.61	—	18.0	18.0	18.0	18.0
Resistivity at 25 °C (micro-ohms/cm/cm^2)	10^{12}	10^{12}	10^{10}	10^{10}	10^{10}	10^{10}	10^{10}	3.5	—	120	120	120	120
Rockwell Hardness Scales	off C	off C	off C	off C	off C	off C	off C	50–100	Duro"	—	—	—	—
Basic Forming Method	PMF	PMF	PMF	PMF	PMF	PMF	PMF	VM	E	PM	PM	PM	PM
Basic Finishing Method	WGD	WGD	WDG	WDG	WDG	WDG	WDG	SRD	SRD	MCT	MCT	MCT	MCT
Relative Corrosion Resistance	O	O	O	O	O	O	O	O	O	E	E	E	E
Nominal Metallurgical Composition (percent)	Ba O.6 Fe$_2$O$_3$	Ba O.6 Fe$_2$O$_3$	Ba O.6 Fe$_2$O$_3$	Ba O.6 Fe$_2$O$_3$	Ba O.6 Fe$_2$O$_3$	Ba O.6 Fe$_2$O$_3$	—	Rubber + Ba O.6 Fe$_2$O$_3$	PVC + Ba O.6 Fe$_2$O$_3$	Fe-Co-Pb	Fe-Co-Pb	Fe-Co-Pb	Fe-Co-Pb

Physical Properties

Alloy or Grade	Lodex 36	Lodex 37	Lodex 38	Lodex 40	Lodex 41	Lodex 42	Lodex 43	P-6 Alloy	Perma-flux C	Perma-flux C1	Platinum-Cobalt	Remalloy (Comol) Alloy 1	Remalloy (Comol) Alloy 2
Density (lbs./in.^{3})	.368	.353	.346	.371	.365	.356	.340	.285	.300	.300	.565	.295	.295
General Mechanical Properties	TC	TC	TC	TC	TC	TC	TC	HT	TC	TC	D	HT	HT
Tensile Strength (psi)	9200	9200	9200	1000	1000	1000	1000	100,000	—	—	200,000	126,000	126,000
Transverse Modulus of Rupture (psi)	15,000	15,000	15,000	4000	4000	4000	4000	170,000	—	—	230,000	50,000	50,000
Coefficient of Thermal Expansion (per °C × 10^{-6})	18.0	18.0	18.0	18.0	18.0	18.0	18.0	13.0	—	—	11.4	9.3	9.3
Resistivity at 25°C (micro-ohms/cm/cm^2)	120	120	120	120	120	120	120	40	—	—	28	45	45
Rockwell Hardness Scales	—	—	—	—	—	—	—	C65	—	—	C26	C60	C60
Basic Forming Method	PM	PM	PM	P	P	P	P	W	CM	CM	W	W	W
Basic Finishing Method	MCT	MCT	MCT	MCT	MCT	MCT	MCT	MCT	MCT	MCT	MCT	WRG	WRG
Relative Corrosion Resistance	E	E	E	E	E	E	E	F	—	—	O	F	F
Nominal Metallurgical Composition (percent)	Fe-Co-Pb	Fe-Co-Pb	Fe-Co-Pb	Fe-Co-Pb	11.6 Co, 67.7 Pb; bal. Fe	10.3 Co, 72.7 Pb; bal. Fe	Fe-Co-Pb	45 Co, 6 Ni, 4 V; bal. Fe	Ce, Co, and Cu (% unknown)	Ce, Co, and Cu (% unknown)	76.7 Pt, 23.3 Co	15 Mo, 5 Co, 2.5 Cu; bal. Fe	17.0 Mo, 12.0 Co; bal. Fe

Physical Properties

Alloy or Grade	Samarium-Cobalt	Vicalloy (HR bars, strips, and forgings)	Vicalloy (thin tape and wire)	Vicalloy 1	Vicalloy 2
Density (lbs./in.^{3})	.290	.296	.296	.296	.293
General Mechanical Properties	HB	HT	HT	HT	HT
Tensile Strength (psi)	8000	153,000–209,000	153,000–209,000	—	—
Transverse Modules of Rupture (psi)	—	—	—	—	—
Coefficient of Thermal Expansion (per °C × 10^{-6})	—	—	—	11.2	11.2
Resistivity at 25°C (micro-ohms/cm/cm^2)	—	67	67	60	60
Rockwell Hardness Scales	—	C34–63	C34–63	C60	C60
Basic Forming Method	DPM	W	W	W	W
Basic Finishing Method	WGR	MCT/WGR	MCT/WGR	WRG	WRG
Relative Corrosion Resistance	—	E	E	E	E
Nominal Metallurgical Composition (percent)	SmCo$_5$	52 Co, 10V; bal. Fe	52 Co, 10 V; bal. Fe	9.5 V, 52 Co; bal. Fe	13 V, 52 Co; bal. Fe

Number, Grade or Alloy	Manufacturer or Country of Origin	Type of Material	B_r (kG)	H_c (kOe)	BH_{max} (MGO)
A	May refer to either Ceramagnet A or Ferrite A				
A	**See Ferram A, Ba or Str**				
A1	See Sermalloy A1				
A2	See Sermalloy A				
A9	See Ceramagnet A9				
A10	See Ceramagnet A10				
A15-7	UGIMag U.S.A.	Cast Alnico	13,400	730	7.5
A19	See Ceramagnet A19				
A20 Ceramag Segments	**Stackpole U.S.A.**	**Sintered Ferrite**	2850	2400	**1.8**
A20 Ceramag Sleeves	**Stackpole U.S.A.**	**Sintered Ferrite**	2575	2130	**1.35**
A20	See Ceramagnet A20				
A50	See Ceramagnet A50				
A70	See Ceramagnet A70				
A Ba	**See Ferram A Ba**				
AL1 [Cast]	UGIMag U.S.A.	Cast Alnico	7200	470	1.4
AL1 [Sintered]	UGIMag U.S.A.	Sintered Alnico	7100	550	1.5
AL2 [Cast]	Thomas & Skinner Inc., U.S.A.	Cast Alnico	7500	570	1.8
AL2 [Cast]	UGIMag U.S.A.	Cast Alnico	7500	560	1.7
AL2 [Sintered]	UGIMag U.S.A.	Sintered Alnico	7100	550	1.5
AL3	UGIMag U.S.A.	Cast Alnico	7000	480	1.35

Number, Grade or Alloy	Manufacturer or Country of Origin	Type of Material	B_r (kG)	H_c (kOe)	BH_{max} (MGO)
AL4	UGIMag U.S.A.	Cast Alnico	5600	720	1.35
AL5 [Cast]	Thomas & Skinner Inc., U.S.A.	Cast Alnico	12,700	660	5.5
AL5 [Cast]	UGIMag U.S.A.	Cast Alnico	12,800	640	5.5
AL5 [Sintered]	UGIMag U.S.A.	Sintered Alnico	10,900	620	3.9
AL5E	Thomas & Skinner Inc., U.S.A.	Cast Alnico	10,800	725	4.3
AL5-7	Thomas & Skinner Inc., U.S.A.	Cast Alnico	13,500	750	7.5
AL5-7	UGIMag U.S.A.	Cast Alnico	13,400	730	7.5
AL6 [Cast]	Thomas & Skinner Inc., U.S.A.	Cast Alnico	10,400	810	3.8
AL6 [Cast]	UGIMag U.S.A.	Cast Alnico	10,500	780	3.9
AL6 [Sintered]	UGIMag U.S.A.	Sintered Alnico	9400	800	2.9
AL7 [oriented]	UGIMag U.S.A.	Cast Alnico	8500	104	3.7
AL7 [Sintered]	UGIMag U.S.A.	Sintered Alnico	6200	1000	2.2
AL7 [unoriented]	UGIMag U.S.A.	Cast Alnico	7500	890	2.5
AL8H [Cast]	UGIMag U.S.A.	Cast Alnico	7500	1930	5.25
AL8H [Sintered]	UGIMag U.S.A.	Sintered Alnico	7250	1980	5.25
AL8HC	Thomas & Skinner Inc., U.S.A.	Cast Alnico	8100	2100	6.3
AL8HE [Cast]	Thomas & Skinner Inc., U.S.A.	Cast Alnico	8800	1800	6.8
AL8HE [Cast]	UGIMag U.S.A.	Cast Alnico	9100	1600	6.0

Number, Grade or Alloy	Manufacturer or Country of Origin	Type of Material	B_r (kG)	H_c (kOe)	BH_{max} (MGO)
AL8HE [Sintered]	UGIMag U.S.A.	Sintered Alnico	8500	1650	5.5
AL9	Thomas & Skinner Inc., U.S.A.	Cast Alnico	10,800	1520	10.5
AL9 [1985]	UGIMag U.S.A.	Cast Alnico	10,500	1600	10.0
AL9Nb	Thomas & Skinner Inc., U.S.A.	Cast Alnico	11,400	1640	11.8
Alcomax	**Trade name of S.G.Magnets, Ltd. for sintered alnico materials**				
Alcomax	Trade name used by several manufacturers in England and Italy for cast or sintered types of metallic magnet materials based on aluminum, nickel, cobalt and iron. (Refer to specific grade number)				
Alcomax	(Mfg. unknown) Italy	Cast Alnico	–	600	5.0
Alcomax (modified)	(Mfg. unknown) England	Cast Alnico	9500	475	2.25
Alcomax I	Most English magnet mfgs.	Cast Alnico	12,000	475	3.5
Alcomax II [Cast]	Most English magnet mfgs.	Cast Alnico	13,000	580	5.4
Alcomax II [Sintered]	(Mfg. unknown) England	Sintered Alnico	10,700	550	3.4
Alcomax II S.C.	Most English magnet mfgs.	Cast Alnico	13,700	600	5.9
Alcomax III [Cast]	Most English magnet mfg.	Cast Alnico	12,600	650	5.4
Alcomax III [Sintered]	Murex, Ltd. and Jessop-Saville, Ltd., England	Sintered Alnico	11,000–11,300	550–620	4.1–4.3
Alcomax III S.C.	Most English magnet mfg.	Cast Alnico	13,200	700	6.1
Alcomax IV [Cast]	Most English magnet mfg.	Cast Alnico	11,500	750	4.5
Alcomax IV [Sintered]	A few English magnet mfg.	Sintered Alnico	10,300	780	4.0

Number, Grade or Alloy	Manufacturer or Country of Origin	Type of Material	B_r (kG)	H_c (kOe)	BH_{max} (MGO)
Alcomax IV S.C.	Most English magnet mfgs.	Cast Alnico	12,200	780	5.2
Alcomax X	(Mfg. unknown) Italy	Cast Alnico	-	780	5.0
Alcomax 1	See Alcomax I (correct designation)				
Alcomax 2	See Alcomax II (correct designation)				
Alcomax 2 S.C.	See Alcomax II S.C. (correct designation)				
Alcomax 3	See Alcomax III (correct designation). Note there are two listings for this type.				
Alcomax 3 S.C.	See Alcomax III S.C. (correct designation)				
Alcomax 3A	**S. G. Magnets, Ltd., U.S.A.**	**Sintered Alnico**	**11,700**	**650**	**4.9**
Alcomax 4	See Alcomax IV (correct designation). Note there are two listings for this type.				
Alcomax 4	**S. G. Magnets, Ltd., U.S.A.**	**Sintered Alnico**	**10,600**	**720**	**4.1**
Alcomax 4 S.C.	See Alcomax IV S.C. (correct designation)				
Alcomax 10	See Alcomax X (correct designation)				
Alconit AN-O	(Mfg. unknown) Switzerland	Cast Alnico	6500	500	1.25
Alni	English, Italian & German magnet manufacturers	Cast Alnico	5000-6200	480-680	1.25
Alni (normal)	Several English magnet mfgs.	Cast Alnico	5600	580	1.25
Alni (High B_r)	Several English magnet mfgs.	Cast Alnico	6200	480	1.25
Alni (High H_c)	Several English magnet mfgs.	Cast Alnico	5000	680	1.25
Alni 090	Magnetfabrik-Bonn Germany	Cast Alnico	7600	280	.90
Alni 120 [Cast]	Magnetfabrik-Bonn Germany	Cast Alnico	5500	550	1.1
Alni 120 [Sintered]	**Magnetfabrik-Bonn Germany**	**Sintered Alnico**	**5500**	**550**	**1.10**

Number, Grade or Alloy	Manufacturer or Country of Origin	Type of Material	B_r (kG)	H_c (kOe)	BH_{max} (MGO)
Alni 120 Cu	Magnetfabrik-Bonn Germany	Cast Alnico	6500	500	1.14
Alni 120P	Magnetfabrik-Bonn Germany	Cast Alnico	3400	465	.50
Alni 120S	Magnetfabrik-Bonn Germany	Cast Alnico	5200	460	.98
Alni 1207	(Mfg. unknown) Germanmy	Bonded Alnico	-	500	.6
Alnic	(Mfg. unknown) England	Cast Alnico	-	-	-
Alnico	Most U. S., English & Italian magnet manufacturers	Cast or Sintered Alnico			
Alnico, Cast [no grade number]	Turton-Matthews Ltd. England	Cast Alnico	7250	560	1.7
Alnico, Sintered [no grade number]	Several English magnet mfgs.	Sintered Alnico	7200	550	1.6
Alnico (Normal)	Several English magnet mfgs.	Cast Alnico	7250	480	1.7
Alnico (High B_r)	Several English magnet mfgs.	Cast Alnico	8000	500	1.7
Alnico (High H_c)	Several English magnet mfgs.	Cast Alnico	6500	620	1.7
Alnico (Directional-Grain)	Refers to Alnico 5 manufactured by a method that produces crystal growth. Description applies to either Cast Alnico 5DG or Cast Alnico 5-7				
Alnico F	**S. G. Magnets, Ltd., U.S.A.**	**Sintered Alnico**	**6200**	**620**	**1.5**
Alnico H	(Mfg. unknown) Italy	Cast Alnico	-	700	5.0
Alnico HRF	**S. G. Magnets, Ltd., U.S.A.**	**Sintered Alnico**	**7600**	**470**	**1.6**
Alnico ISO.8	**Hitachi U.S.A.**	**Sintered Alnico**	**5800**	**1200**	**2.1**
Alnico I	Most U.S. magnet mfgs.	Cast Alnico	7100	400	1.3

Number, Grade or Alloy	Manufacturer or Country of Origin	Type of Material	B_r (kG)	H_c (kOe)	BH_{max} (MGO)
Alnico IA	Crucible Steel Co., U.S.A.	Cast Alnico	6600	540	1.4
Alnico IB	Crucible Steel Co., U.S.A.	Cast Alnico	7100	450	1.4
Alnico IC	Crucible Steel Co., U.S.A.	Cast Alnico	7600	400	1.4
Alnico II [Cast]	Most U.S. magnet mfgs.	Cast Alnico	7200	540	1.6
Alnico II [Sintered]	A few U.S. magnet mfgs.	Sintered Alnico	6800	520	1.5
Alnico IIA	Crucible Steel Co., U.S.A.	Cast Alnico	7000	650	1.7
Alnico IIB	Crucible Steel Co., U.S.A.	Cast Alnico	7500	580	1.7
Alnico IIC	Crucible Steel Co., U.S.A.	Cast Alnico	8000	425	1.6
Alnico IIH	Crucible Steel Co., U.S.A.	Cast Alnico	8400	600	2.0
Alnico III	Most U.S. magnet mfgs.	Cast Alnico	6700	450	1.4
Alnico IIIA	Crucible Steel Co., U.S.A.	Cast Alnico	6400	560	1.35
Alnico IIIB	Crucible Steel Co., U.S.A.	Cast Alnico	6800	490	1.35
Alnico IIIC	Crucible Steel Co., U.S.A.	Cast Alnico	7500	400	1.35
Alnico III-NF1	Westinghouse Corp., U.S.A.	Cast Alnico	6700	520	1.4
Alnico IV [Cast]	Most U.S. magnet mfgs.	Cast Alnico	5500	700	1.3
Alnico IV [Sintered]	A few U.S. magnet mfgs.	Sintered Alnico	5200	700	1.2
Alnico IVA	Crucible Steel Co., U.S.A.	Cast Alnico	5500	730	1.35
Alnico IVB	Crucible Steel Co., U.S.A.	Cast Alnico	6000	660	1.45

Number, Grade or Alloy	Manufacturer or Country of Origin	Type of Material	B_r (kG)	H_c (kOe)	BH_{max} (MGO)
Alnico IVH	Crucible Steel Co., U.S.A.	Cast Alnico	6000	790	1.65
Alnico V [Cast]	Most U.S. magnet manufacturers	Cast Alnico	12,400	640	5.5
Alnico V [Cast] (non-oriented on special order)	Most U.S. magnet manufacturers	Cast Alnico	7400	520	1.4
Alnico V [Sintered]	A few U.S. magnet mfgs.	Sintered Alnico	10,500	590	3.5
Alnico VA	Crucible Steel Co., U.S.A.	Cast Alnico	12,500	720	5.0
Alnico VAB	Crucible Steel Co., U.S.A.	Cast Alnico	12,500	685	5.5
Alnico VB	Crucible Steel Co., U.S.A.	Cast Alnico	12,700	650	5.5
Alnico VC	Crucible Steel Co., U.S.A.	Cast Alnico	13,200	580	5.5
Alnico VCC	Arnold Engr. U.S.A.	Cast Alnico	13,100	640	6.1
Alnico VE	Crucible Steel Co., U.S.A.	Cast Alnico	11,000	700	4.5
Alnico V-NC1	Westinghouse Corp., U.S.A.	Cast Alnico	12,100	740	5.0
Alnico V-NR1	Westinghouse Corp., U.S.A.	Cast Alnico	13,200	625	5.5
Alnico V-NW1	Westinghouse Corp., U.S.A.	Cast Alnico	12,600	650	5.5
Alnico VDG	Most U.S. magnet mfgs.	Cast Alnico	12,600	670	6.3
Alnico VAB DG	Crucible Steel Co., U.S.A..	Cast Alnico	13,100	700	6.5
Alnico VB DG	Crucible Steel Co., U.S.A.	Cast Alnico	13,300	685	6.5
Alnico VC DG	Crucible Steel Co., U.S.A.	Cast Alnico	14,200	610	6.5

Number, Grade or Alloy	Manufacturer or Country of Origin	Type of Material	B_r (kG)	H_c (kOe)	BH_{max} (MGO)
Alnico V-PR1	Westinghouse Corp., U.S.A.	Cast Alnico	13,500	675	6.5
Alnico V-OW1	Westinghouse Corp., U.S.A.	Cast Alnico	13,700	700	7.0
Alnico V-7	Most U.S. magnet mfgs.	Cast Alnico	13,400	730	7.5
Alnico V-OR1	Westinghouse Corp., U.S.A.	Cast Alnico	13,700	700	7.5
Alnico VI [Cast]	Most U.S. magnet mfgs.	Cast Alnico	10,200	770	3.8
Alnico VI [Sintered]	A few U.S. magnet mfgs.	Sintered Alnico	8800	800	2.8
Alnico VIA	Crucible Steel Co., U.S.A.	Cast Alnico	7500	975	2.75
Alnico VI AB	Crucible Steel Co., U.S.A.	Cast Alnico	8400	860	3.0
Alnico VIB	Crucible Steel Co., U.S.A.	Cast Alnico	10,500	760	3.65
Alnico VIC	Crucible Steel Co., U.S.A.	Cast Alnico	11,000	700	4.0
Alnico VI-NS1	Westinghouse Corp., U.S.A.	Cast Alnico	10,500	750	3.7
Alnico VI-NS2	Westinghouse Corp., U.S.A.	Cast Alnico	11,000	840	4.5
Alnico VII (oriented)	Most U.S. magnet mfgs.	Cast Alnico	8600	1050	3.7
Alnico VII (non-oriented)	Most U.S. magnet mfgs.	Cast Alnico	7500	890	2.5
Alnico VII	See sintered Alnico 7 (correct designation)				
Alnico VII-S (oriented)	UGIMag U.S.A.	Cast Alnico	8570	1040	3.7
Alnico VII-S (non-oriented)	UGIMag U.S.A.	Cast Alnico	7540	890	2.5
Alnico VII-C_i	See Alnico 7-C_i (correct designation)				
Alnico VIII [Cast]	Most U.S. magnet mfgs.	Cast Alnico	8500	1600	5.0

Number, Grade or Alloy	Manufacturer or Country of Origin	Type of Material	B_r (kG)	H_c (kOe)	BH_{max} (MGO)
Alnico VIII [Sintered]	A few U.S. magnet mfgs.	Sintered Alnico	7600	1550	4.5
Alnico VIII$_A$	See Alnico 8$_A$ (correct designation)				
Alnico VIII$_C$	See Alnico 8$_C$ (correct designation)				
Alnico VIIIA	See Alnico 8A (correct designation)				
Alnico VIIIB	See Alnico 8B (correct designation)				
Alnico VIII (Improved)	See Improved Alnico 8 (correct designation)				
Alnico VIII-ND1	Westinghouse Corp., U.S.A.	Cast Alnico	8700	1450	4.5
Alnico IX	A few U.S. magnet mfgs.	Cast Alnico	10,500	1450	8.5
Alnico X-900	Arnold Engr. U.S.A.	Cast Alnico	6500	850	1.8
Alnico XII	Most U.S. magnet mfgs.	Cast Alnico	6100	1000	1.7
Alnico XII-NM1	Westinghouse Corp., U.S.A.	Cast Alnico	6500	925	1.75
Alnico 1	Most U.S. magnet mfgs.	Cast Alnico	7100	400	1.3
Alnico 1A	See Alnico 1A (correct designation)				
Alnico 1B	See Alnico 1B (correct designation)				
Alnico 1C	See Alnico 1C (correct designation)				
Alnico 2 [Cast]	Most U.S. magnet mfgs.	Cast Alnico	7200	540	1.6
Alnico 2 [Sintered]	A few U.S. magnet mfgs.	Sintered Alnico	6800	520	1.5
Alnico 2 [Cast]	Hitachi U.S.A.	Cast Alnico	7200	540	1.6
Alnico 2 [Sintered]	Hitachi U.S.A.	Sintered Alnico	6800	520	1.5
Alnico 2A	See Alnico IIA (correct designation)				
Alnico 2B	See Alnico IIB (correct designation)				

Number, Grade or Alloy	Manufacturer or Country of Origin	Type of Material	B_r (kG)	H_c (kOe)	BH_{max} (MGO)
Alnico 2C	See Alnico IIC (correct designation)				
Alnico 2H	See Alnico IIH (correct designation)				
Alnico 3	Most U.S. magnet mfgs.	Cast Alnico	6700	450	1.4
Alnico 3	Hitachi U.S.A.	Cast Alnico	6700	450	1.4
Alnico 3A	See Alnico IIIA (correct designation)				
Alnico 3B	See Alnico IIIB (correct designation)				
Alnico 3C	See Alnico IIIC (correct designation)				
Alnico 3-NF1	See Alnico III-NF1 (correct designation)				
Alnico 3/0.6	**See Koerzit 3/0.6**				
Alnico 4 [Cast]	Most U.S. magnet mfgs.	Cast Alnico	5500	700	1.3
Alnico 4 [Sintered]	A few U.S. magnet mfgs.	Sintered Alnico	5200	700	1.2
Alnico 4 [Cast]	Hitachi U.S.A.	Cast Alnico	5200	700	1.2
Alnico 4A	See Alnico IVA (correct designation)				
Alnico 4B	See Alnico IVB (correct designation)				
Alnico 4/0.9	**See Koerzit 4/0.9**				
Alnico 4/1.0	**See Koerzit 4/1.0**				
Alnico 5 [Cast]	Most U.S. magnet manufacturers	Cast Alnico	12,400	640	5.5
Alnico 5 [Cast] (non-oriented on special order)	Most U.S. magnet manufacturers	Cast Alnico	7400	520	1.4
Alnico 5 [Sintered]	A few U.S. magnet mfgs.	Sintered Alnico	10,500	590	3.5
Alnico 5 [Cast]	Hitachi U.S.A.	Cast Alnico	12,400	640	5.5
Alnico 5A	See Alnico VA (correct designation)				

Number, Grade or Alloy	Manufacturer or Country of Origin	Type of Material	B_r (kG)	H_c (kOe)	BH_{max} (MGO)
Alnico 5AB	See Alnico VAB (correct designation)				
Alnico 5AB DG	See Alnico VAB DG (correct designation)				
Alnico 5B	See Alnico VB (correct designation)				
Alnico 5B DG	See Alnico VB DG (correct designation)				
Alnico 5C	See Alnico VC (correct designation)				
Alnico 5CC	See Alnico VCC (correct designation)				
Alnico 5cc	**Arnold Engr. U.S.A.**	**Cast Alnico**	**13,200**	**675**	**6.5**
Alnico 5C DG	See Alnico VC DG (correct designation)				
Alnico 5DG	Most U.S. magnet mfgs.	Cast Alnico	12,600	670	6.3
Alnico 5DG	Hitachi U.S.A.	Cast Alnico	12,600	670	6.25
Alnico 5E	See Alnico VE (correct designation)				
Alnico 5-NC1	See Alnico V-NC1 (correct designation)				
Alnico 5-NR1	See Alnico V-NR1 (correct designation)				
Alnico 5-NW1	See Alnico V-NW1 (correct designation)				
Alnico 5-OR1	See Alnico V-OR1 (correct designation)				
Alnico 5-OW1	See Alnico V-OW1 (correct designation)				
Alnico 5-PR1	See Alnico V-PR1 (correct designation)				
Alnico 5-7	Most U.S. magnet mfgs.	Cast Alnico	13,400	730	7.5
Alnico 5-7	Hitachi U.S.A.	Cast Alnico	13,000	730	7.25
Alnico 5/1.4	**See Koerzit 5/1.4**				
Alnico 6 [Cast]	Most U.S. magnet mfgs.	Cast Alnico	10,200	770	3.8
Alnico 6 [Sintered]	A few U.S. magnet mfgs.	Sintered Alnico	8800	800	2.8
Alnico 6 [Cast]	Hitachi U.S.A.	Cast Alnico	10,200	770	3.75

Number, Grade or Alloy	Manufacturer or Country of Origin	Type of Material	B_r (kG)	H_c (kOe)	BH_{max} (MGO)
Alnico 6A	See Alnico VIA (correct designation)				
Alnico 6B	See Alnico VIB (correct designation)				
Alnico 6C	See Alnico VIC (correct designation)				
Alnico 6-NS1	See Alnico VI-NS1 (correct designation)				
Alnico 6-NS2	See Alnico VI-NS2 (correct designation)				
Alnico 6-Sint.	Hitachi U.S.A.	Sintered Alnico	8600	790	3.0
Alnico 6/1.6	**See Koerzit 6/1.6**				
Alnico 7 [oriented]	Most U.S. magnet mfgs.	Cast Alnico	8600	1050	3.7
Alnico 7 [non-oriented]	Most U.S. magnet mfgs.	Cast Alnico	7500	890	2.5
Alnico 7 [Sintered]	Thomas & Skinner Inc., U.S.A.	Sintered Alnico	7800	775	2.4
Alnico 7-C_i	Thomas & Skinner Inc., U.S.A.	Cast Alnico	6850	960	2.5
Alnico 7-S	See Alnico VII-S (correct designation)				
Alnico 8 [Cast]	Most U.S. magnet mfgs.	Cast Alnico	8500	1600	5.0
Alnico 8 [Sintered]	A few U.S. magnet mfgs.	Sintered Alnico	7600	1550	4.5
Alnico 8 (Improved)	Arnold Engr. U.S.A.	Cast Alnico	8000	1800	5.76
Alnico 8A [Cast]	Hitachi U.S.A.	Cast Alnico	8500	1600	5.0
Alnico 8A [Cast]	Hitachi U.S.A.	Cast Alnico	8500	1600	5.2
Alnico 8A [Sintered]	Hitachi U.S.A.	Sintered Alnico	7600	1550	4.5
Alnico 8$_A$	Thomas & Skinner Inc., U.S.A.	Cast Alnico	8700	1600	5.3
Alnico 8B [Cast]	**Arnold Engr. U.S.A.**	**Cast Alnico**	**8300**	**1650**	**5.5**

Number, Grade or Alloy	Manufacturer or Country of Origin	Type of Material	B_r (kG)	H_c (kOe)	BH_{max} (MGO)
Alnico 8B [Cast]	Hitachi U.S.A.	Cast Alnico	7500	1850	5.3
Alnico 8B [Cast]	Hitachi U.S.A.	Cast Alnico	7800	1850	5.2
Alnico 8B [Sintered]	Hitachi U.S.A.	Sintered Alnico	6500	1800	4.5
Alnico 8C	Hitachi U.S.A.	Cast Alnico	9000	1480	5.2
Alnico 8C	Thomas & Skinner Inc., U.S.A.	Cast Alnico	7300	1900	5.5
Alnico 8H	**Arnold Engr. U.S.A.**	**Cast Alnico**	**7400**	**1900**	**5.5**
Alnico 8HE	**Arnold Engr. U.S.A.**	**Cast Alnico**	**9300**	**1550**	**6.0**
Alnico 8-ND1	See Alnico VIII-ND1 (correct designation)				
Alnico 9	A few U.S. magnet mfgs.	Cast Alnico	10,500	1450	8.5
Alnico 9	**Arnold Engr. U.S.A.**	**Cast Alnico**	**11,000**	**1500**	**10.5**
Alnico 11 [Cast]	Most U.S. magnet mfgs.	Cast Alnico	7200	540	1.6
Alnico 11 [Sintered]	A few U.S. magnet mfgs.	Sintered Alnico	6800	520	1.5
Alnico 11A	Crucible Steel Co., U.S.A.	Cast Alnico	7000	650	1.7
Alnico 11B	Crucible Steel Co., U.S.A.	Cast Alnico	7500	580	1.7
Alnico 11C	Crucible Steel Co., U.S.A.	Cast Alnico	8000	425	1.6
Alnico 11H	Crucible Steel Co., U.S.A.	Cast Alnico	8400	600	2.0
Alnico 12	Most U.S. magnet mfgs.	Cast Alnico	6100	1000	1.7
Alnico 12-NM1	See Alnico XII-NM1 (correct designation)				
Alnico .12/6	**See Oerstit 160**				

Number, Grade or Alloy	Manufacturer or Country of Origin	Type of Material	B_r (kG)	H_c (kOe)	BH_{max} (MGO)
Alnico 18/10	See Oerstit 260				
Alnico 29/6	See Oerstit 400				
Alnico 35/5	See Oerstit 500S				
Alnico 36/12	See Oerstit 450				
Alnico 38/5	See Oerstit 500G				
Alnico 40/12	Magnetfabrik-Bonn Germany	Sintered Alnico	8500	1500	5.0
Alnico 40/15	Magnetfabrik-Bonn Germany	Cast Alnico	7500	1800	5.0
Alnico 120	Magnetfabrik-Bonn Germany	Sintered Alnico	5500	550	1.1
Alnico 130	Magnetfabrik-Bonn Germany	Cast Alnico	5700	650	1.4
Alnico 130S	Magnetfabrik-Bonn Germany	Cast Alnico	5100	600	1.25
Alnico 160	Magnetfabrik-Bonn Germany	Sintered Alnico	6500	680	1.5
Alnico 160	Magnetfabrik-Bonn Germany	Cast Alnico	6400	680	1.5
Alnico 160P	(Mfg. unknown) Germany	Bonded Alnico	-	700	.8
Alnico 190 [Cast]	Magnetfabrik-Bonn Germany	Cast Alnico	7100	680	1.8
Alnico 190 [Sintered]	Magnetfabrik-Bonn Germany	Sintered Alnico	7100	680	1.8
Alnico 200	(Mfg. unknown) Germany	Cast Alnico	-	900	1.8
Alnico 220	Magnetfabrik-Bonn Germany	Cast Alnico	5600	900	1.9
Alnico 250	Magnetfabrik-Bonn Germany	Cast Alnico	6200	1000	2.3
Alnico 250P	(Mfg. unknown) Germany	Bonded Alnico	-	900	1.0
Alnico 250S	Magnetfabrik-Bonn Germany	Cast Alnico	5600	915	2.03

Number, Grade or Alloy	Manufacturer or Country of Origin	Type of Material	B_r (kG)	H_c (kOe)	BH_{max} (MGO)
Alnico 260	Magnetfabrik-Bonn Germany	Sintered Alnico	6000	1000	2.2
Alnico 300	Magnetfabrik-Bonn Germany	Cast Alnico	8800	710	2.7
Alnico 350 [Cast]	Magnetfabrik-Bonn Germany	Cast Alnico	8000	1050	3.3
Alnico 350 [Sintered]	Magnetfabrik-Bonn Germany	Sintered Alnico	8000	1050	3.3
Alnico 400	Magnetfabrik-Bonn Germany	Cast Alnico	11,000	580	4.0
Alnico 400S	Magnetfabrik-Bonn Germany	Cast Alnico	11,100	620	4.2
Alnico 450	Magnetfabrik-Bonn Germany	Sintered Alnico	8500	1300	4.5
Alnico 500	Magnetfabrik-Bonn Germany	Cast Alnico	11,200	590	4.4
Alnico 500	Magnetfabrik-Bonn Germany	Cast Alnico	11,400	600	4.5
Alnico 500S	Magnetfabrik-Bonn Germany	Cast Alnico	11,200	600	4.0
Alnico 550	UGIMag U.S.A.	Cast Alnico	13,500	590	5.6
Alnico 550 UGIMAX	UGIMag U.S.A.	Cast Alnico	13,500	620	6.0
Alnico 580	Magnetfabrik-Bonn Germany	Cast Alnico	12,500	680	5.5
Alnico 580	Magnetfabrik-Bonn Germany	Cast Alnico	12,000	680	5.2
Alnico 600	Magnet Appl. Ltd/Cookson England	Cast Alnico	12,600	630	5.3
Alnico 600	UGIMag U.S.A.	Cast Alnico	12,600	630	5.3
Alnico 600 Super UGIMAX	UGIMag U.S.A.	Cast Alnico	13,500	740	7.5
Alnico 600 UGIMAX	UGIMag U.S.A.	Cast Alnico	13,200	700	6.2

Number, Grade or Alloy	Manufacturer or Country of Origin	Type of Material	B_r (kG)	H_c (kOe)	BH_{max} (MGO)
Alnico 700	Magnetfabrik-Bonn Germany	Cast Alnico	13,000	680	6.5
Alnico 700	Magnetfabrik-Bonn Germany	Cast Alnico	13,000	680	6.2
Alnico 700	UGIMag U.S.A.	Cast Alnico	12,000	700	5.2
Alnico 800	UGIMag U.S.A.	Cast Alnico	11,000	800	4.3
Alnico 800 UGIMAX	UGIMag U.S.A.	Cast Alnico	12,000	850	6.0
Alnico 1300	UGIMag U.S.A.	Cast Alnico	8800	1400	5.0
Alnico 1500	UGIMag U.S.A.	Cast Alnico	9000	1570	5.7
Alnico 1500 Super UGIMAX	UGIMag U.S.A.	Cast Alnico	10,800	1550	10.3
Alnico 2000	UGIMag U.S.A.	Cast Alnico	8000	2000	6.5
Alnico 2200	UGIMag U.S.A.	Cast Alnico	7400	2200	5.5
Alnicus	U.S. Magnet & Alloy U.S.A.	Cast Alnico	-	-	-
Alpha ( )	See Westro Alpha (correct designation)				
AN-0	See Alconit AN-0 (correct designation)				
Arelec	Magnet Appl. Ltd./Cookson - Trade name for magnetic catch and holding magnet assemblies. Not covered in this handbook.				
ArKomax 800	Arnold Engr. U.S.A.	Cast Alnico	13,700	740	8.1
ArKomax 800 H_i-H_c	Arnold Engr. U.S.A.	Cast Alnico	13,200	810	8.1
Armax 18	Arnold Engr. U.S.A.	Sintered Rare Earth	8800	8500	18.0
Armax 20	Arnold Engr. U.S.A.	Sintered Rare Earth	9000	8900	20.0

Number, Grade or Alloy	Manufacturer or Country of Origin	Type of Material	B_r (kG)	H_c (kOe)	BH_{max} (MGO)
Armax 22H	Arnold Engr. U.S.A.	Sintered Rare Earth	9800	8700	22.0
Armax 26H	Arnold Engr. U.S.A.	Sintered Rare Earth	10,000	9700	26.0
Arnife 1	Arnold Engr. U.S.A.	Cast Alnico	10,000	170	.90
Arnife 2	Arnold Engr. U.S.A.	Cast Alnico	9300	230	1.1
Arnife 3	Arnold Engr. U.S.A.	Cast Alnico	9500	75	.40
Arnife 4	Arnold Engr. U.S.A.	Cast Alnico	9700	125	.70
Arnokrome III	Arnold Engr. U.S.A.	Ductile Iron-Chrome Cobalt	9000–12,000	50–300	.40–1.2
Arnox	Arnold Engr. USA trade name for Arnold Engineering's line of barium ferrites or iron-oxide/cobalt-oxide materials (refer to specific grade number)				
Arnox I	Arnold Engr. U.S.A.	Sintered Ferrite	2200	1600	1.9
Arnox III	Arnold Engr. U.S.A.	Bonded Ferrite	1500	1100	.40
Arnox IV	Arnold Engr. U.S.A.	Sintered Ferrite	3850	2100	3.5
Arnox 1	See Arnox I (correct designation)				
Arnox 3	See Arnox III (correct designation)				
Arnox 5	See Arnox V (correct designation)				
Arnox 7	Arnold Engr. U.S.A.	Sintered Ferrite	3450	3200	2.7
Arnox 8	Arnold Engr. U.S.A.	Sintered Ferrite	3850	2900	3.5
Arnox 8B	Arnold Engr. U.S.A.	Sintered Ferrite	4100	2900	4.0
Arnox 8H	Arnold Engr. U.S.A.	Sintered Ferrite	3800	3600	3.4

Number, Grade or Alloy	Manufacturer or Country of Origin	Type of Material	B_r (kG)	H_c (kOe)	BH_{max} (MGO)
Arnox 9	Arnold Engr. U.S.A.	Sintered Ferrite	3700	3500	3.25
Arnox 9H [High H_{ci}]	Arnold Engr. U.S.A.	Sintered Ferrite	3700	3500	3.25
Arnox 4000	Arnold Engr. U.S.A.	Sintered Ferrite	4000	3700	3.9
A Str	See Ferram A Str				
B	May refer to either Nial B or Spinalor B				
B-1013	Arnold Engr. U.S.A.	Bonded Ferrite	2450	2200	1.4
B-1030	Arnold Engr. U.S.A.	Bonded Ferrite	2450	2200	1.4
B-1033	Arnold Engr. U.S.A.	Bonded Ferrite	2500	2300	1.6
B-1035/B-1044/ B-1046	Arnold Engr. U.S.A.	Bonded Ferrite	-	-	1.4
B-1037	Arnold Engr. U.S.A.	Bonded Ferrite	2150	1650	1.0
B-1044	See B-1035				
B-1046	See B-1035				
B-1060	Arnold Engr. U.S.A.	Bonded Ferrite	2650	2425	1.7
B-1061	Arnold Engr. U.S.A.	Bonded Ferrite	2800	2250	1.9
B-1062	Arnold Engr. U.S.A.	Bonded Ferrite	2760	2650	1.9
B-1316/B-1317	Arnold Engr. U.S.A.	Thermoplastic Bonded Ferrite adhesive backed tape-contact manufacturer for magnetic properties			
BAB16	See BPA16				
BBB12	See BPB12				
BCB16	See BPC16				
BPA16	TDK Corp. U.S.A.	Bonded Ferrite	2550	2380	1.6

Number, Grade or Alloy	Manufacturer or Country of Origin	Type of Material	B_r (kG)	H_c (kOe)	BH_{max} (MGO)
BPB04	TDK Corp. U.S.A.	Bonded Ferrite	1300	1150	.4
BPB12	TDK Corp. U.S.A.	Bonded Ferrite	2250	2100	1.3
BPC05	TDK Corp. U.S.A.	Bonded Ferrite	1600	800	.5
BPC16	TDK Corp. U.S.A.	Bonded Ferrite	2600	2325	1.6
BQA14	TDK Corp. U.S.A.	Bonded Ferrite	2400	2225	1.4
BQB14 [High H_{ci}]	TDK Corp. U.S.A.	Bonded Ferrite	2400	2225	1.4
BQC14 [High H_{ci}]	TDK Corp. U.S.A.	Bonded Ferrite	2400	2225	1.4
BQK12	TDK Corp. U.S.A.	Bonded Ferrite	2300	2090	1.23
BRA70	TDK Corp. U.S.A.	Bonded Rare Earth	5600	4300	6.9
BRA90	TDK Corp. U.S.A.	Bonded Rare Earth	6100	4640	8.4
B1-00	See Ceramagnet B1-00				
BG	Stackpole U.S.A.	Barium Ferrite powder (only)	-	-	-
BG1	Stackpole U.S.A.	Barium Ferrite powder (only)	-	-	-
Barium Ferrite	Most ferrite magnet manufacturers	Generic name for $BaO.6\ Fe_3O_4$ magnet materials. Made under various trade names, such as Arnox, Genox, Indox. (Refer to specific name and grade number)			
Barium Ferrite 100	(Mfg. unknown) Germany	Sintered Ferrite	2000	1700	1.0
Barium Ferrite 300	(Mfg. unknown) Germany	Sintered Ferrite	3700	2000	2.8
Beta ( )	See Westro Beta (correct designation)				

Number, Grade or Alloy	Manufacturer or Country of Origin	Type of Material	B_r (kG)	H_c (kOe)	BH_{max} (MGO)
Bismanol	Crucible Steel Co., U.S.A.	Sintered Manganese Bismuth	4800	3700	5.3
Bremag	Magnet Appl. Ltd./Cookson England	Trade MAL/Cookson trade name for Rare Earth based materials			
Bremag 4NF	Properties not available at this time. Not released commercially				
Bremag 5N	Magnet Appl. Ltd./Cookson England	Bonded Rare Earth	4300	4000	3.8
Bremag 10N	Magnet Appl. Ltd./Cookson England	Bonded Rare Earth	6800	5780	10.0
Bremag 18	Magnet Appl. Ltd./Cookson England	Sintered Rare Earth	9000	8500	20.0
Bremag 26	Magnet Appl. Ltd./Cookson England	Sintered Rare Earth	10,300	8000	20.0
Bremag 27N	Magnet Appl. Ltd./Cookson England	Sintered Rare Earth	11,000	8750	27.0
Bremag 35N	Magnet Appl. Ltd./Cookson England	Sintered Rare Earth	12,100	11,600	35.0
C	See Ticonal C (correct designation)				
C	(Mfg. unknown) Germany	Wrought Alloy	9600	60	.29
CC	See Alnico VCC (correct designation)				
CM8B	TDK Corp. U.S.A.	Bonded Rare Earth	6400	8700	8.4
Carbon Steel [.65 percent C]	All commercial steel producers	Wrought Alloy	10,000	42	.18
Carbon Steel [1.0 percent C]	All commercial steel producers	Wrought Alloy	9000	51	.20
Carbon Steel No. 1095	All commercial steel producers	Wrought Alloy	9000	51	.20

Number, Grade or Alloy	Manufacturer or Country of Origin	Type of Material	B_r (kG)	H_c (kOe)	BH_{max} (MGO)
Caslox (no grade number)	(Mfg. unknown) England	Pressed Cobalt-Iron Oxide	1100	700	.2
Caslox I	See Caslox 1 (correct designation)				
Caslox II	See Caslox 2 (correct designation)				
Caslox III	Plessey Co. England	Bonded Ferrite	1400	1050	.4
Caslox VI	Plessey Co. England	Sintered Ferrite	2100	1600	.90
Caslox VII	Plessey Co. England	Sintered Ferrite	2700	2000	1.5
Caslox 1	Plessey Co. England	Pressed Cobalt-Iron Oxide	-	-	.5
Caslox 2	Plessey Co. England	Pressed Cobalt-Iron Oxide	2000	1700	1.0
Caslox 3	See Caslox III (correct designation)				
Caslox 6	See Caslox VI (correct designation)				
Caslox 7	See Caslox VII (correct designation)				
Ceramag A	**Stackpole U.S.A.**	**Sintered Ferrite**	**2300**	**1860**	**1.1**
Ceramagnet A	Stackpole U.S.A.	Sintered Ferrite	2300	1900	1.1
Ceramagnet A9	Stackpole U.S.A.	Sintered Ferrite	2550	2100	1.35
Ceramagnet A10	Stackpole U.S.A.	Sintered Ferrite	2550	2100	1.35
Ceramagnet A19	Stackpole U.S.A.	Sintered Ferrite	2700	2200	1.55
Ceramagnet A20	Stackpole U.S.A.	Sintered Ferrite	2850	2400	1.8
Ceramagnet A50	Stackpole U.S.A.	Sintered Ferrite	3850	2470	3.5
Ceramagnet A70	Stackpole U.S.A.	Sintered Ferrite	3400	2800	2.6

Number, Grade or Alloy	Manufacturer or Country of Origin	Type of Material	B_r (kG)	H_c (kOe)	BH_{max} (MGO)
Ceramagnet B1-00	Stackpole U.S.A.	Bonded Ferrite	1500	1200	3.5
Ceramic	Commonly used name for all ferrite magnet materials (Refer to specific grade number)				
Ceramic 1	The Electrodyne Co., Inc. U.S.A.	Bonded Ferrite	2200	1830	1.0
Chrome Steel (no other designation)	Several English magnet mfgs.	Wrought Alloy	9800	70	2.85
Chrome Steel (low chrome)	Probably refers to 1-percent chrome steel				
Chrome Steel (high chrome)	Probably refers to 3-1/2-percent chrome steel				
Chrome Steel (1 percent Cr)	Most wrought magnet mfgs.	Wrought Alloy	9500	52	.23
Chrome Steel (2 percent Cr)	Most wrought magnet mfgs.	Wrought Alloy	9300	62	.26
Chrome Steel (3 percent Cr)	Several English manufacturers	Wrought Alloy	9800	72	.285
Chrome Steel (3-1/2 percent Cr)	Most wrought magnet manufacturers	Wrought Alloy	9500	66	.29
Chrome Steel (4 percent Cr)	Jessop-Saville Ltd. England	Wrought Alloy	9800	60	.28
Chrome Steel (5 percent Cr)	(Mfg. unknown) Germany	Wrought Alloy	9400	65	.30
Chrome Steel (6 percent Cr)	Most wrought magnet mfgs.	Wrought Alloy	9500	74	.30
Chrome-Cobalt Steel	May refer to various chrome-cobalt-iron alloys. When no other specification is given, usually refers to 2-percent cobalt - 4-percent chrome steel (refer to specific grade number)				
Chrome-Manganese Steel	(Mfg. unknown) Germany	Wrought Alloy	9400	65	.30
Chrome-Platinum	No specific information available. Experimental material never commercially available on any scale				
Chrome-Silicon Steel	(Mfg. unknown) Germany	Wrought Alloy	9400	65	.30

Number, Grade or Alloy	Manufacturer or Country of Origin	Type of Material	B_r (kG)	H_c (kOe)	BH_{max} (MGO)
Chrome-Tungsten Steel	Metallic alloy of 6-percent tungsten-8-percent chrome (bal. Fe). Once sold under the trade name Cobaflex. Manufacturer unknown. No other information available.				
Co 040	(Mfg. unknown) Germany	Wrought Alloy	9800	80	.34
Co 045	(Mfg. unknown) Germany	Wrought Alloy	7200	130	.38
Co 050	(Mfg. unknown) Germany	Wrought Alloy	7800	145	.46
Co 060	(Mfg. unknown) Germany	Wrought Alloy	8400	165	.58
Co 070	(Mfg. unknown) Germany	Wrought Alloy	8500	180	.65
Co 090	(Mfg. unknown) Germany	Wrought Alloy	8600	230	.90
Co 100	(Mfg. unknown) Germany	Wrought Alloy	9000	250	.95
Cobaflex	See Chrome-Tungsten Steel				
Cobalt Steel (2-percent Co)	(Mfg. unknown) England	Wrought Alloy	9800	80	.34
Cobalt Steel (3-percent Co)	Most wrought magnet mfgs.	Wrought Alloy	7800	130	.35
Cobalt Steel (6-percent Co)	Most wrought magnet mfgs.	Wrought Alloy	7500	145	.44
Cobalt Steel (9-percent Co)	Most wrought magnet mfgs.	Wrought Alloy	7800	160	.50
Cobalt Steel (11-percent Co)	Manufacturer unknown	Wrought Alloy	8400	165	.58
Cobalt Steel (15-percent Co)	Most wrought magnet mfgs.	Wrought Alloy	8200	180	.62
Cobalt Steel (17-percent Co)	Most wrought magnet mfgs.	Wrought Alloy	9000	165	.62
Cobalt Steel (17-1/2 percent Co)	Hoskins Mfg. Co. U.S.A.	Wrought Alloy	9000	160	-
Cobalt Steel (20-percent Co)	(Mfg. unknown) England	Wrought Alloy	9000	210	.75

Number, Grade or Alloy	Manufacturer or Country of Origin	Type of Material	B_r (kG)	H_c (kOe)	BH_{max} (MGO)
Cobalt Steel (30-percent Co)	(Mfg. unknown) Germany	Wrought Alloy	8600	230	.90
Cobalt Steel (35-percent Co)	Most wrought magnet mfgs.	Wrought Alloy	9000	250	.95
Cobalt Steel (36-percent Co)	Most wrought magnet mfgs.	Wrought Alloy	9600	240	.95
Cobalt Steel (38-percent Co)	(Mfg. unknown) U.S.A.	Wrought Alloy	10,000	240	.98
Cobalt Steel (40-percent Co)	Most wrought magnet mfgs.	Wrought Alloy	10,000	242	1.03
Cobalt-Chrome Steel	May refer to various Cobalt-chrome-iron alloys. When no other designation is given, usually refers to 2-percent cobalt-4-percent chrome steel (refer to specific grade number)				
Cobalt-Chrome Steel (2-percent Cr)	Most wrought magnet manufacturers	Wrought Alloy	9800	80	.32
Cobalt-Chrome Steel (2-percent Co - 4 percent Cr)	Several English magnet manufacturers	Wrought Alloy	9800	80	.32
Cobalt-Chrome Steel (16-percent Co)	Most wrought magnet manufacturers	Wrought Alloy	8000	180	.9
Cobalt-Paladium	No specific information available. Experimental material never commercially available on any scale.				
Cobalt-Platinum	See Platinum-Cobalt (correct designation)				
Coercimax	(Mfg. unknown) Italy	Cast Alnico	–	700	5.0
Columax	Sev. English Cos. Also Thomas & Skinner	Cast Alnico	13,500	740	7.4
Columax V-7	See Columax 5-7 (correct designation)				
Columax IX	See Columax 9 (correct designation)				
Columax 5	Thomas & Skinner Inc., U.S.A.	Cast Alnico	13,500	740	7.5
Columax 5-7	Thomas & Skinner Inc., U.S.A.	Cast Alnico	Same as Columax 5		

Number, Grade or Alloy	Manufacturer or Country of Origin	Type of Material	B_r (kG)	H_c (kOe)	BH_{max} (MGO)
Columax 9	Thomas & Skinner Inc., U.S.A.	Cast Alnico	10,400	1580	10.0
COMO	Japan Steel Co. Japan	Designation for early grade of material now obsolete. No other information available.			
Comol (any grade)		See Remalloy Alloy 1 or Alloy 2.			
Comalloy	Murex Ltd. England	Sintered Metal Alloy	10,000	270	1.1
Conial 140	(Mfg. unknown) Austria	Cast Alnico	6300	650	1.4
Conial 160	(Mfg. unknown) Austria	Cast Alnico	7000	700	1.6
Cormax 2000	Sumitomo Spec. Metals Co. Ltd. Japan	Rare Earth	8900	8800	19.4
Cr	(Mfg. unknown) Germany	Wrought Alloy	9600	60	.29
Cr 030	(Mfg. unknown) Germany	Wrought Alloy	9600	65	.30
Cr 035	(Mfg. unknown) Germany	Wrought Alloy	9400	65	.30
Cromag	(Mfg. and origin unknown)	Sintered Ferrite	2000	1700	1.0
Crovac 10/130	Vacuumschmelze GmbH (VAC) Germany	Malleable Alnico	9000-10,000	410-480	1.3-1.8
Crovac 10/380	Vacuumschmelze GmbH (VAC) Germany	Malleable Alnico	12,000-13,000	540-620	3.8-4.3
Crovac 15/150	Vacuumschmelze GmbH (VAC) Germany	Malleable Alnico	8500-9500	450-530	1.4-1.9
Crovac 15/400	Vacuumschmelze GmbH (VAC) Germany	Malleable Alnico	11,500-12,500	580-650	4.0-4.5
Crovac 23/250	Vacuumschmelze GmbH (VAC) Germany	Malleable Alnico	8000-9000	600-680	2.0-2.5

Number, Grade or Alloy	Manufacturer or Country of Origin	Type of Material	B_r (kG)	H_c (kOe)	BH_{max} (MGO)
Crovac 23/500	Vacuumschmelze GmbH (VAC) Germany	Malleable Alnico	11,000-12,000	650-750	4.3-5.0
Crumax	Crucible Steel Co., U.S.A.	Trade name for Nd-Fe-B (rare earth) magnet materials			
Cunico	May refer to either one of two grades. Where no grade number is given, probably refers to Cunico 1.				
Cunico I	See Cunico 1 (correct designation)				
Cunico II	See Cunico 2 (correct designation)				
Cunico 1	Most wrought magnet mfgs.	Wrought Alloy	3400	710	.85
Cunico 2	Most wrought magnet mfgs.	Wrought Alloy	5300	450	.99
Cunife (no grade number)	UGIMag U.S.A.	Wrought Alloy	5400	500	1.3
Cunife (no grade number)	Hoskins Mfg. Co. U.S.A.	Wrought Alloy	5700	575	1.46
Cunife I	See Cunife 1 (correct designation)				
Cunife II	See Cunife 2 (correct designation)				
Cunife 1	Most wrought magnet mfgs.	Wrought Alloy	5700	590	1.85
Cunife 1	UGIMag U.S.A.	Wrought Alloy	5500	530	1.4
Cunife 2	Most wrought magnet mfgs.	Wrought Alloy	7300	260	.78
Cunife 4	UGIMag U.S.A.	Wrought Alloy	4600	750	1.6
D	May refer either to Oxalit D or Ticonal D				
Da	See Oxalit Da (correct designation)				
D55	See Ferroxdure D55 (correct designation)				
Directional Grain Alnico	See Alnico, Directional-Grain (correct designation)				
E	See Ticonal E (correct designation)				
EEC	Designation for various grades of rare earth materials made by Electron Energy Corp.				

Number, Grade or Alloy	Manufacturer or Country of Origin	Type of Material	B_r (kG)	H_c (kOe)	BH_{max} (MGO)
EEC 1:5-18	Electron Energy Corp. U.S.A.	Sintered Rare Earth	8600	8400	18.0
EEC 1:5TC-9	Electron Energy Corp. U.S.A.	Sintered Rare Earth	6100	6000	9.0
EEC 1:5TC-13	Electron Energy Corp. U.S.A.	Sintered Rare Earth	7300	7200	13.0
EEC 1:5TC-15	Electron Energy Corp. U.S.A.	Sintered Rare Earth	7800	7700	15.0
EEC 2:17-15	Electron Energy Corp. U.S.A.	Sintered Rare Earth	8000	7200	14.5
EEC 2:17-24	Electron Energy Corp. U.S.A.	Sintered Rare Earth	10,100	9300	24.0
EEC 2:17-27	Electron Energy Corp. U.S.A.	Sintered Rare Earth	10,800	10,100	27.5
EEC 2:17TC-15	Electron Energy Corp. U.S.A.	Sintered Rare Earth	8000	7200	14.5
EEC 2:17TC-18	Electron Energy Corp. U.S.A.	Sintered Rare Earth	9000	8200	18.5
EEC NEO 27	Electron Energy Corp. U.S.A.	Sintered Rare Earth	10,800	10,200	27.5
EEC NEO 33	Electron Energy Corp. U.S.A.	Sintered Rare Earth	12,200	10,600	33.5
Ergit Max 1	(Mfg. unknown) Hungary	Wrought Alloy	9000	250	.95
Ergit Max III	(Mfg. unknown) Hungary	Cast Alnico	6500	500	1.25
Eriem 2000	Used by Eriez Magnetics in material handling products				
Erium (all grades)	Refers to magnet materials of any type and grade which the Eriez Mfg. Co. uses as components in the manufacture of magnetic separation and handling equipment. Not a type of material.				
ESD (all grades)	Refers to elongated single domain magnet materials. Made in various sizes in the United States by the General Electric Co., only under the trade name Lodex.				
ETM	Japan Steel Co., Japan	Designation for early magnet material, now obsolete. No other information available.			
F	See Ticonal F (correct designation)				

Number, Grade or Alloy	Manufacturer or Country of Origin	Type of Material	B_r (kG)	H_c (kOe)	BH_{max} (MGO)
F-310	D.M.Steward Co. U.S.A.	Sintered Ferrite	2000	1700	1.0
F-330	D.M.Steward Co. U.S.A.	Sintered Ferrite	2385	1920	1.15
F-335	D.M.Steward Co. U.S.A.	Sintered Ferrite	2250	1750	1.03
F-340	D.M.Steward Co. U.S.A.	Sintered Ferrite	2400	2000	1.25
F-500	D.M.Steward Co. U.S.A.	Sintered Ferrite	4100	2350	3.75
F-620	D.M.Steward Co. U.S.A.	Sintered Ferrite	3445	3000	2.6
F-625	D.M.Steward Co. U.S.A.	Sintered Ferrite	3400	3000	2.65
F-730	D.M.Steward Co. U.S.A.	Sintered Ferrite	3300	3000	2.5
F-830	D.M.Steward Co. U.S.A.	Sintered Ferrite	4000	3000	3.65
FB	Designation for various grades of sintered ferrite materials made by TDK Corp.				
FB( )	TDK Corp. U.S.A.	Trade designation for various grades of sintered ferrite magnet materials. Grade number shown as subscript.			
FB_1	TDK Corp. U.S.A.	Sintered Ferrite	2000-2300	1800-1900	1.0-1.2
FB1	**TDK Corp. U.S.A.**	**Sintered Ferrite**	**2200**	**1900**	**1.1**
FB_2	TDK Corp. U.S.A.	Sintered Ferrite	3600-3800	1600-1900	2.7-3.1
FB2	**TDK Corp. U.S.A.**	**Sintered Ferrite**	**3850**	**2000**	**3.5**
FB_3	TDK Corp. U.S.A.	Sintered Ferrite	3600-3800	2600-2900	2.7-3.1
FB3B	**TDK Corp. U.S.A.**	**Sintered Ferrite**	**3550**	**3000**	**3.0**
FB3K	**TDK Corp. U.S.A.**	**Sintered Ferrite**	**3000**	**2750**	**2.1**

Number, Grade or Alloy	Manufacturer or Country of Origin	Type of Material	B_r (kG)	H_c (kOe)	BH_{max} (MGO)
FB3X	TDK Corp. U.S.A.	Sintered Ferrite	3750	2950	3.2
FB4	TDK Corp. U.S.A.	Sintered Ferrite	3800-4200	1700-2000	3.6-4.0
FB4A	TDK Corp. U.S.A.	Sintered Ferrite	4100	2200	4.0
FB4B	TDK Corp. U.S.A.	Sintered Ferrite	4000	3200	3.8
FB4H	TDK Corp. U.S.A.	Sintered Ferrite	3700	3450	3.25
FB4N	TDK Corp. U.S.A.	Sintered Ferrite	4300	2300	4.4
FB4X	TDK Corp. U.S.A.	Sintered Ferrite	4200	2950	4.2
FB5B	TDK Corp. U.S.A.	Sintered Ferrite	4200	3300	4.3
FB5E	TDK Corp. U.S.A.	Sintered Ferrite	3700	3550	3.2
FB5H	TDK Corp. U.S.A.	Sintered Ferrite	4050	3650	3.8
FB5N	TDK Corp. U.S.A.	Sintered Ferrite	4400	2850	4.5
FM5	Crucible Steel Co., U.S.A.	Sintered Ferrite	3950	2400	3.6
FM7A	Crucible Steel Co., U.S.A.	Sintered Ferrite	3500	3250	2.8
FM7B	Crucible Steel Co., U.S.A.	Sintered Ferrite	3800	3500	3.3
FM8A	Crucible Steel Co., U.S.A.	Sintered Ferrite	3900	3200	3.5
FM8B	Crucible Steel Co., U.S.A.	Sintered Ferrite	4200	2900	4.1
FM8C	Crucible Steel Co., U.S.A.	Sintered Ferrite	4300	2400	4.3
FM8D	Crucible Steel Co., U.S.A.	Sintered Ferrite	4000	3350	3.8
FPMG	See BQK12				

Number, Grade or Alloy	Manufacturer or Country of Origin	Type of Material	B_r (kG)	H_c (kOe)	BH_{max} (MGO)
FSD2	See Ferroxdure FXD2 (correct designation)				
FS10	Mfg. and origin unknown	Sintered Ferrite	3500	2500	2.6
FWM	Japan Steel Co. Japan	Designation for early magnet material now obsolete. No other information available.			
FXD1	See Ferroxdure FXD1 (correct designation)				
FXD1	Philips Co. Holland	Sintered Ferrite	2000	1700	.9
FXD2	See Ferroxdure FXD2 (correct designation)				
FXD300	**Philips Co. Holland**	**Sintered Ferrite**	**4000**	**2000**	**3.7**
FXD330	**Philips Co. Holland**	**Sintered Ferrite**	**3700**	**3100**	**3.2**
FXD380	**Philips Co. Holland**	**Sintered Ferrite**	**3900**	**3300**	**3.6**
FXD400	**Philips Co. Holland**	**Sintered Ferrite**	**4100**	**3300**	**3.9**
FXD 480	**Philips Co. Holland**	**Sintered Ferrite**	**3800**	**3500**	**3.4**
FXD500	**Philips Co. Holland**	**Sintered Ferrite**	**4000**	**3700**	**3.8**
FXD520	**Philips Co. Holland**	**Sintered Ferrite**	**4250**	**3100**	**4.2**
FXD580	**Philips Co. Holland**	**Sintered Ferrite**	**3850**	**3770**	**3.5**
Fama 100	(Mfg. unknown) Sweden	Cast Alnico	5500	1000	1.9
Fama 600	(Mfg. unknown) Sweden	Cast Alnico	6500	500	1.25
Fama 700	(Mfg. unknown) Sweden	Cast Alnico	6500	640	1.7
FeCrCo	See special grade number				
Feroba (bonded)	Darwins, Ltd. England	Darwins' trade name for Rubber PVC or Resin-Bonded Ferrites of various grades			

Number, Grade or Alloy	Manufacturer or Country of Origin	Type of Material	B_r (kG)	H_c (kOe)	BH_{max} (MGO)
Feroba I	Darwins, Ltd. England	Sintered Ferrite	2200	1700	1.0
Feroba II	Darwins, Ltd. England	Sintered Ferrite	3900	1700	3.2
Feroba III	Darwins, Ltd. England	Sintered Ferrite	3400	2500	2.7
Feroba 1	See Feroba I (correct designation)				
Feroba 2	See Feroba II (correct designation)				
Feroba 3	See Feroba III (correct designation)				
Feroba A	(Mfg. unknown) England	Bonded Ferrite	-	-	.4
Feroba B	(Mfg. unknown) England	Bonded Ferrite	-	-	.4
Feroba C	(Mfg. unknown) England	Bonded Ferrite	-	-	.5
Ferram	**Magnet Applications/Cookson trade name for sintered ferrites and bonded ferrites**				
Ferram I	**Magnet Appl. Ltd./Cookson England**	Sintered Ferrite	2200	2000	**1.1**
Ferram A-BA	**Magnet Appl. Ltd./Cookson England**	Sintered Ferrite	3900	1800	**3.5**
Ferram A-Str	**Magnet Appl. Ltd./Cookson England**	Sintered Ferrite	3600	2900	**3.2**
Ferriflex	See Plastoferroxdure (correct designation)				
Ferrimag	**Crucible Materials Corporation's trade name for ferrite (ceramic) magnet materials**				
Ferrimag I	Crucible Steel Co., U.S.A.	Sintered Ferrite	2200	1800	1.0
Ferrimag II	Crucible Steel Co., U.S.A.	Sintered Ferrite	3800	2200	3.5
Ferrimag 1	See Ferrimag I (correct designation)				
Ferrimag 2	See Ferrimag II (correct designation)				
Ferrit 100	See Barium Ferrite 100 (correct designation)				

Number, Grade or Alloy	Manufacturer or Country of Origin	Type of Material	B_r (kG)	H_c (kOe)	BH_{max} (MGO)
Ferrit 300	See Barium Ferrite 300 (correct designation)				
Ferrite A	(Mfg. unknown) France	Bonded Ferrite	-	-	.5
Ferrite I	See Ferrite 1 (correct designation)				
Ferrite II	See Ferrite 2 (correct designation)				
Ferrite III	See Ferrite 3 (correct designation)				
Ferrite IV	See Ferrite 4 (correct designation)				
Ferrite V	See Ferrite 5 (correct designation)				
Ferrite VI	See Ferrite 6 (correct designation)				
Ferrite VII	See Ferrite 7 (correct designation)				
Ferrite 1	Most U.S. magnet mfgs.	Sintered Ferrite	2200	1825	1.0
Ferrite 2	Most U.S. magnet mfgs.	Sintered Ferrite	2700	2250	1.7
Ferrite 3	Most U.S. magnet mfgs.	Sintered Ferrite	3350	2400	2.6
Ferrite 4	Most U.S. magnet mfgs.	Sintered Ferrite	2550	2300	1.5
Ferrite 5	Most U.S. magnet mfgs.	Sintered Ferrite	3850	2200	3.5
Ferrite 6	Most U.S. magnet mfgs.	Sintered Ferrite	3300	3100	2.6
Ferrite 7	Most U.S. magnet mfgs.	Sintered Ferrite	3400	3200	2.8
Ferrites, PVC (bonded)	See Bonded Ferrites				
Ferrites, plastic-bonded	See Bonded Ferrites				
Ferrites, resin-bonded (rigid)	Made in the United States by Arnold Engineering as Arnox III and by the Stackpole Carbon Co. as Ceramagnet B1-00.				
Ferrites, rubber-bonded	Made in the United States by the 3 M Company in several grades, under the trade name Plastiform (refer to specific grade number)				

Number, Grade or Alloy	Manufacturer or Country of Origin	Type of Material	B_r (kG)	H_c (kOe)	BH_{max} (MGO)
Ferrites, thermoplastic-bonded	Made in the United States by the B. F. Goodrich Co. under the trade name Koroseal and by H.O. Canfield as "Magnetized Plastic."				
Ferrites, vinyl-bonded	See Bonded Ferrites				
Ferrogum (all grades)	Japan Spec. Steel Japan	Trade name for various grades of bonded ferrites.			
Ferrogum X3	Japan Spec. Steel Japan	Bonded Ferrite	7000	620	.11
Ferrogum X5	Japan Spec. Steel Japan	Bonded Ferrite	1000	850	.23
Ferrogum X7	Japan Spec. Steel Japan	Bonded Ferrite	1250	1000	.35
Ferrogum X10	Japan Spec. Steel Japan	Bonded Ferrite	1500	1100	.48
Ferrogum X12	Japan Spec. Steel Japan	Bonded Ferrite	1600	1150	.52
Ferrogum Y3	Japan Spec. Steel Japan	Bonded Ferrite	900	800	.18
Ferrogum Y5	Japan Spec. Steel Japan	Bonded Ferrite	1350	1000	.40
Ferrogum Y7	Japan Spec. Steel Japan	Bonded Ferrite	1600	1150	.56
Ferrogum Y10	Japan Spec. Steel Japan	Bonded Ferrite	1900	1350	.75
Ferromax	Japan Spec. Steel Japan	Sintered Ferrite	2050-2200	1500-1600	.8-.9
Ferroxdure (all grades)	Trade name used by Philips Co., Holland for Dutch, Italian and French sintered ferrites and for bonded ferrites				
Ferroxdure FXD1	Philips Co. Holland	Sintered Ferrite	2000	1700	.9
Ferroxdure FXD2	Philips Co. Holland	Sintered Ferrite	3600	2100	2.6
Ferroxdure I	See Ferroxdure 1 (correct designation)				
Ferroxdure II	See Ferroxdure 2 (correct designation)				
Ferroxdure III	See Ferroxdure 3 (correct designation)				

Number, Grade or Alloy	Manufacturer or Country of Origin	Type of Material	B_r (kG)	H_c (kOe)	BH_{max} (MGO)
Ferroxdure 1	Holland-France Italy (Mfg. unknown)	Sintered Ferrite	2000	1700	1.0
Ferroxdure 2	Holland-France Italy (Mfg. unknown)	Sintered Ferrite	4000	1600	3.2
Ferroxdure 3	Holland-France Italy (Mfg. unknown)	Sintered Ferrite	3300	3000	2.7
Ferroxdure GP	France (Mfg. unknown)	Sintered Ferrite	3700	2700	2.8
Ferroxdure D55	Philips Co. Holland	Bonded Ferrite	1700	1400	.6
Ferroxdure P30	Philips Co. Holland	Bonded Ferrite	1250	1100	.35
Ferroxdure P40	Philips Co. Holland	Bonded Ferrite	1450	1200	.45
Ferroxdure SP50	Philips Co. Holland	Bonded Ferrite	1600	1275	.55
Ferroxdure 100	Philips Co. Holland	Sintered Ferrite	2200	1650	.95
Ferroxdure 280K	Philips Co. Holland	Sintered Ferrite	3550	3000	2.8
Ferroxdure 3ooR	Philips Co. Holland	Sintered Ferrite	3900	1900	3.3
Ferroxdure 330K	Philips Co. Holland	Sintered Ferrite	3700	3000	3.4
Ferroxdure 330 (RAD)	Philips Co. Holland	Sintered Ferrite	3500	3000	2.9
Ferroxdure 360R 360R	Philips Co. Holland	Sintered Ferrite	3900	2200	3.6
Flexam	Magnet Appl. Ltd./Cookson England	Trade name for bonded ferrite material			
Flexam	Magnet Appl. Ltd./Cookson England	Bonded Ferrite	2000	1600	.8

Number, Grade or Alloy	Manufacturer or Country of Origin	Type of Material	B_r (kG)	H_c (kOe)	BH_{max} (MGO)
Flexam P5	Magnet Appl. Ltd./Cookson England	Bonded Ferrite	1400	1200	.5
Flexam P8	Magnet Appl. Ltd./Cookson England	Bonded Ferrite	2000	1600	.8
Flexor	Magnet Appl. Ltd./Cookson England	Trade name for rubber bonded ferrite material			
Flexor 15	Magnet Appl. Ltd./Cookson England	Bonded Ferrite	2200	1950	1.15
Flexor 45	Magnet Appl. Ltd./Cookson England	Bonded Ferrite	2500	2000	1.45
G	May refer to Koerox G, Oxalit G or Ticonal G.				
GA	See Oxalit Ga (correct designation)				
GG	See Ticonal GG (correct designation)				
GP	See Ferroxdure GP (correct designation)				
GPC	See Maxpowr GPC				
GX	See Ticonal GX (correct designation)				
G1-1a	Magnetfabrik Schramberg Germany	Sintered Rare Earth	8800	8670	19.4
G1-1i	Magnetfabrik Schramberg Germany	Sintered Rare Earth	9400	8920	21.2
G2-1a	Magnetfabrik Schramberg Germany	Sintered Rare Earth	10,000	9000	24.0
G2-1i	Magnetfabrik Schramberg Germany	Sintered Rare Earth	10,500	9500	27.0
G3-1a	Magnetfabrik Schramberg Germany	Sintered Rare Earth	11,500	10,500	31.0
G3-1i	Magnetfabrik Schramberg Germany	Sintered Rare Earth	12,200	11,000	34.0

Number, Grade or Alloy	Manufacturer or Country of Origin	Type of Material	B_r (kG)	H_c (kOe)	BH_{max} (MGO)
G3-2a	Magnetfabrik Schramberg Germany	Sintered Rare Earth	10,800	10,100	26.0
G3-2i	Magnetfabrik Schramberg Germany	Sintered Rare Earth	11,600	10,800	32.0
G3-3a	Magnetfabrik Schramberg Germany	Sintered Rare Earth	10,500	10,000	25.0
G3-3i	Magnetfabrik Schramberg Germany	Sintered Rare Earth	11,200	10,500	28.0
G4-1	Magnetfabrik Schramberg Germany	Bonded Rare Earth	6000	5000	9.3
G4-1a	Magnetfabrik Schramberg Germany	Bonded Rare Earth	6000	5000	9.4
G4-2	Magnetfabrik Schramberg Germany	Bonded Rare Earth	6400	5300	10.0
G4-2a	Magnetfabrik Schramberg Germany	Bonded Rare Earth	6300	4800	9.4
G4-3	Magnetfabrik Schramberg Germany	Bonded Rare Earth	6000	5000	9.4
G4-4	Magnetfabrik Schramberg Germany	Bonded Rare Earth	6400	5000	10.0
Gamma ( )	See Westro Gamma (correct designation)				
Gaussit 180	(Mfg. unknown) Germany	Wrought Aloy	9000	300	1.0
Gaussit 1600	(Mfg. unknown) Germany	Sintered Ferrite	2000	1700	1.0
Genox	General Magnetics Co., U.S.A. trade name for Sintered Ferrite materials				
Genox I	See Genox 1 (correct designation)				
Genox V	See Genox 5 (correct designation)				

Number, Grade or Alloy	Manufacturer or Country of Origin	Type of Material	B_r (kG)	H_c (kOe)	BH_{max} (MGO)
Genox VI	See Genox 6 (correct designation)				
Genox 1	General Magnetics Co. U.S.A.	Sintered Ferrite	-	-	-
Genox 5	General Magnetics Co. U.S.A.	Sintered Ferrite	3900	2380	3.4
Genox 6	General Magnetics Co. U.S.A.	Sintered Ferrite	-	-	-
Genox 8	General Magnetics Co. U.S.A.	Sintered Ferrite	3850	2940	3.4
Genox 8H	General Magnetics Co. U.S.A.	Sintered Ferrite	4100	2850	3.9
Genox 39x25	See Genox 5				
Genox 41x29	See Genox 8H				
Genox 39x31	See Genox 8				
Grade 5	General Magnetics Co. U.S.A.	Sintered Ferrite	3900	2380	2.44
Grade 7	General Magnetics Co. U.S.A.	Sintered Ferrite	3450	3080	4.0
Grade 8	General Magnetics Co. U.S.A.	Sintered Ferrite	3850	2940	3.08
Gumox (no grade number)	Magnetfabrik-Bonn Germany	Bonded Ferrite	1900	1400	.9
Gumox I	See Gumox 1 (correct designation)				
Gumox II	See Gumox 2 (correct designation)				
Gumox 1	Magnetfabrik-Bonn Germany	Bonded Ferrite	1300	800	.25
Gumox 2	Magnetfabrik-Bonn Germany	Bonded Ferrite	1900	1400	.8
H	May refer to Alnico H, Koerzit H, Spinalor H or Ticonal H				
H-13S	Hitachi U.S.A.	Sintered Rare Earth	6500-7500	6000 minimum	10.0-14.0
H-18B	Hitachi U.S.A.	Sintered Rare Earth	8000-9000	7000-9000	16.0-19.0

Number, Grade or Alloy	Manufacturer or Country of Origin	Type of Material	B_r (kG)	H_c (kOe)	BH_{max} (MGO)
H-20SV	Hitachi U.S.A.	Sintered Rare Earth	9300	8750	20.0
H-21EV	Hitachi U.S.A.	Sintered Rare Earth	8700-9700	6000 minimum	17.0-23.0
H-22A	Hitachi U.S.A.	Sintered Rare Earth	8500-9500	8000-9500	18.0-22.0
H-23B	Hitachi U.S.A.	Sintered Rare Earth	9400-10,400	5000-8500	19.0-24.0
H-23CV	Hitachi U.S.A.	Sintered Rare Earth	9500-10,500	7500-10,000	20.0-26.0
H-23EH	Hitachi U.S.A.	Sintered Rare Earth	9300-10,300	7000 minimum	19.0-25.0
H-23EV	Hitachi U.S.A.	Sintered Rare Earth	9300-10,300	6500 minimum	19.0-25.0
H-25B	Hitachi U.S.A.	Sintered Rare Earth	10,200-11,000	5000-8500	22.0-27.0
H-25EH	Hitachi U.S.A.	Sintered Rare Earth	9700-10,700	7000 minimum	2.0-27.0
H-30CH	Hitachi U.S.A.	Sintered Rare Earth	10,400-11,400	8000-20,000	25.0-31.0
H-90A	See 90A				
H-90B	See 90B				
H-96A	See 96A				
H-96B	See 96B				
H-99A	See 99A				
H-99B	See 99B				
HA 0	(Mfg. unknown) Belgium	Wrought Alloy	9600	60	.29
HA 1	(Mfg. unknown) Belgium	Wrought Alloy	7200	130	.38
HA 2	(Mfg. unknown) Belgium	Wrought Alloy	7800	145	.46
HA 3	(Mfg. unknown) Belgium	Wrought Alloy	8500	180	.65

Number, Grade or Alloy	Manufacturer or Country of Origin	Type of Material	B_r (kG)	H_c (kOe)	BH_{max} (MGO)
HA 4	(Mfg. unknown) Belgium	Wrought Alloy	9000	250	.9
HB-061	Hitachi U.S.A.	Bonded Rare Earth	5700	4600	6.4
HB-081	Hitachi U.S.A.	Bonded Rare Earth	6400	5000	8.3
H.C.A.	Maurex, Ltd. England	Sintered Ferrite	6400	980	2.0
HEC	See Maxpowr HEC				
HH	See Spinalor HH				
HIMAG	Hitachi U.S.A.	Cast Alnico	13,000-14,000	700-800	6.8 8.2
HS-20BR [Ring Only]	Hitachi U.S.A.	Sintered Rare Earth	9600	9000	21.0
HS-25BR [Ring Only]	Hitachi U.S.A.	Sintered Rare Earth	10,600	10,000	26.0
HS-25CR [Ring Only]	Hitachi U.S.A.	Sintered Rare Earth	10,400	9800	24.0
HS-27CV	Hitachi U.S.A.	Sintered Rare Earth	10,700	9500	27.0
HS-30BV	Hitachi U.S.A.	Sintered Rare Earth	11,500	10,500	28.0
HS-30CH	Hitachi U.S.A.	Sintered Rare Earth	11,300	10,800	30.0
HS-30CR [Ring Only]	Hitachi U.S.A.	Sintered Rare Earth	11,200	10,700	30.0
HS-30CV	Hitachi U.S.A.	Sintered Rare Earth	11,200	10,700	30.0
HS-32BV	Hitachi U.S.A.	Sintered Rare Earth	11,600	11,000	32.0
HS-35BH	Hitachi U.S.A.	Sintered Rare Earth	12,300	11,500	35.0
HS-35CH	Hitachi U.S.A.	Sintered Rare Earth	12,100	11,500	35.0
HS-37BH	Hitachi U.S.A.	Sintered Rare Earth	12,500	11,800	37.0

Number, Grade or Alloy	Manufacturer or Country of Origin	Type of Material	B_r (kG)	H_c (kOe)	BH_{max} (MGO)
HT-3/19P	See Sprox HT-3/19P				
HT-11/21P	See Sprox 11/21P				
HT-14/21P	See Sprox 14/21P				
HT-16/19	See Sprox 16/19				
Hardyne (non-oriented)	(Mfg. and origin unknown)	Pressed Oxide	1400	800	.297
Hardyne (oriented)	(Mfg. and origin unknown)	Pressed Oxide	1900	1340	.725
Hartferrit	Trade name for group of ferrite magnet materials made by Magnetfabrik Schramberg-Germany				
Hartferrit 7/21	See Oxit 100				
Hartferrit 24/23	See Oxit 360				
Hartferrit 27/26	See Oxit 380				
Hartferrit 27/30	See Oxit 360				
Hartferrit 28/34	See Oxit 400HC				
Hartferrit 30/26	See Oxit 400				
Hartferrit 30/30	See Oxit 400C				
Hartferrit 33/26	See Oxit 420				
Hicorex	Hitachi U.S.A.	Hitachi brand name for sintered rare earth materials			
High-Chrome Steel	See Chrome Steel (3-1/2 percent)				
Honda Alloy	(Mfg. unknown) Japan	Cast Alnico	6500	900	2.0
Honda Steel	(Mfg. and origin unknown)	Cast Alnico	7100	780	2.0
Hycomax (no grade number)	Generic name used in England for various grades of cast and sintered Alnico alloys. When no grade is stated, material is probably Hycomax. Also may have been an early Italian material				
Hycomax	Trade name of S.G. Magnets Ltd. for sintered alnico materials				
Hycomax (no grade number)	(Mfg. unknown) Italian	Cast Alnico	-	700	5.0

843

Number, Grade or Alloy	Manufacturer or Country of Origin	Type of Material	B_r (kG)	H_c (kOe)	BH_{max} (MGO)
Hycomax I	Several English magnet mfg.	Cast Alnico	9000	825	3.2
Hycomax I	Most English magnet mfg.	Sintered Alnico	8000	800	2.6
Hycomax II	Several English magnet mfg.	Cast Alnico	8500	1200	4.0
Hycomax III	Several English magnet mfg.	Cast Alnico	8800	1450	5.0
Hycomax III	Murex, Ltd. England	Sintered Alnico	8200	1450	4.5
Hycomax 1	**S.G.Magnets, Ltd. U.S.A.**	**Sintered Alnico**	**7900**	**910**	**3.0**
Hycomax 3	**S.G.Magnets, Ltd. U.S.A.**	**Sintered Alnico**	**7800**	**1600**	**4.8**
Hycomax 4	**S.G.Magnets, Ltd. U.S.A.**	**Sintered Alnico**	**6800**	**1940**	**5.0**
Hycorex	See specific grade number				
Hycorex 90A	Hitachi U.S.A.	Rare Earth Cobalt	8200	7500	16.0
Hycorex 90B	Hitachi U.S.A.	Rare Earth Cobalt	8700	8200	18.0
Hycorex 96A	Hitachi U.S.A.	Rare Earth Cobalt	9000	8300	19.5
Hycorex 96B	Hitachi U.S.A.	Rare Earth Cobalt	9400	8800	21.5
Hycorex 99A	Hitachi U.S.A.	Rare Earth Cobalt	9700	6000	21.5
Hycorex 99B	Hitachi U.S.A.	Rare Earth Cobalt	10,000	6000	24.0
Hycorex 99C	Hitachi U.S.A.	Rare Earth Cobalt	10,600	6000	27.0
Hyflux	Obsolete trade name used by Indiana General Corp. (now UGIMag) for cast alnico materials. At one time also used to designate pressed, powdered-iron type materials similar to PF grades				
Hyflux	See specific grade number				

Number, Grade or Alloy	Manufacturer or Country of Origin	Type of Material	B_r (kG)	H_c (kOe)	BH_{max} (MGO)
Hynical	(Mfg. unknown) England	Cast Alnico	5250	675	1.15
Hynico (no grade number)	(Mfg. unknown) England	Cast Alnico	7250	628	1.63
Hynico	**Trade name of S.G. Magnets, Ltd. for sintered alnico materials**				
Hynico	**S.G. Magnets, Ltd. U.S.A.**	**Sintered Alnico**	**6200**	**1100**	**2.5**
Hynico II	Morrison & Catherall, Ltd. England	Cast Alnico	6000	900	1.80
IGH 250	See Indalloy IGH 250				
IM	See Maxpowr IM				
Incor	**UGIMAG U.S.A.**	**Trade name for Samarium Cobalt materials**			
Incor 4PB	UGIMag U.S.A.	Bonded Rare Earth	4200	3600	3.8
Incor 6	UGIMag U.S.A.	Sintered Rare Earth Cobalt	5000	4950	6.0
Incor 9PB	UGIMag U.S.A.	Bonded Rare Earth	6400	5600	9.0
Incor 11PB	UGIMag U.S.A.	Bonded Rare Earth	6950	6100	11.0
Incor 16	UGIMag U.S.A.	Sintered Rare Earth Cobalt	8100	7900	16.0
Incor 18	UGIMag U.S.A.	Sintered Rare Earth Cobalt	8800	8200	18.0
Incor 18	**UGIMag U.S.A.**	**Sintered Rare Earth**	**8750**	**8600**	**19.0**
Incor 21	UGIMag U.S.A.	Sintered Rare Earth Cobalt	9400	8300	21.0
Incor 21	**UGIMag U.S.A.**	**Sintered Rare Earth**	**9300**	**9100**	**21.0**

Number, Grade or Alloy	Manufacturer or Country of Origin	Type of Material	B_r (kG)	H_c (kOe)	BH_{max} (MGO)
Incor 21HE	UGIMag U.S.A.	Sintered Rare Earth Cobalt	9600	8500	22.0
Incor 24	**UGIMag U.S.A.**	**Sintered Rare Earth**	**10,100**	**9500**	**24.0**
Incor 24HE	UGIMag U.S.A.	Sintered Rare Earth Cobalt	10,200	9200	24.0
Incor 26	**UGIMag U.S.A.**	**Sintered Rare Earth**	**10,500**	**10,100**	**26.0**
Incor 26HE	UGIMag U.S.A.	Sintered Rare Earth Cobalt	10,600	9400	26.0
Indalloy	See specific grade number				
Indalloy	(Mfg. and origin unknown)	Sintered Metal Alloy	-	-	-
Indalloy 2	UGIMag U.S.A.	Iron-Chrome Cobalt	8800	460	1.6
Indalloy 5	UGIMag U.S.A.	Iron-Chrome Cobalt	13,500	600	5.25
Indalloy IGH 250	UGIMag U.S.A.	Iron-Chrome Cobalt	14,000	250	2.0
Indox	Trade name used by Indiana General Corp. (now UGIMag) for various grades of sintered ferrite magnet materials				
Indox I	UGIMag, U.S.A.	Same as Ferrite 1			
Indox II	UGIMag, U.S.A.	Same as Ferrite 2			
Indox III	UGIMag, U.S.A.	Same as Ferrite 3			
Indox IV	UGIMag, U.S.A.	Same as Ferrite 4			
Indox V	UGIMag, U.S.A.	Same as Ferrite 5			
Indox VI-A	UGIMag, U.S.A.	Same as Ferrite 6			
Indox VII	UGIMAg, U.S.A.	Same as Ferrite 7			
Indox 1	See Indox I (correct designation)				
Indox 2	See Indox II (correct designation)				
Indox 3	See Indox III (correct designation)				

Number, Grade or Alloy	Manufacturer or Country of Origin	Type of Material	B_r (kG)	H_c (kOe)	BH_{max} (MGO)
Indox 4	See Indox IV (correct designation)				
Indox 5	See Indox V (correct designation)				
Indox 6-A	See Indox VI-A (correct designation)				
Indox 7	See Indox VII (correct designation)				
Indox 20	UGIMag U.S.A.	Sintered Ferrite	1530	1300	.5
Iron-Molybdenum-Chrome	(Mfg. and origin unknown)	Wrought Alloy	7000	250	.80
Iron-Molybdenum-Cobalt	(Mfg. and origin unknown)	Wrought Alloy	10,000	230	1.1
Iron-Platinum	See Platinum-Iron (correct designation)				
Iron-Titanium	(Mfg. and origin unknown)	Wrought Alloy	10,100	48	.22
ISO .8 [Cast]	Hitachi U.S.A.	Cast Alnico	5800	1300	2.20
ISO .8 [Sintered]	Hitachi U.S.A.	Sintered Alnico	5800	1200	2.10
Isotropic Plastic	The Electrodyne Co., Inc. U.S.A	Bonded Ferrite	1900	1460	.77
Jobmax 18	Jobmaster U.S.A.	Samarium Cobalt	8500	8300	20.0
Jobmax 27	Jobmaster U.S.A.	Rare Earth-Neodymium Iron Boron	10,800	9500	27.0
K	May refer to either Nialco K or Ticonal K				
K2	(Mfg. unknown) Germany	Wrought Alloy	9800	80	.34
K6	(Mfg. unknown) Germany	Wrought Alloy	7800	145	.46
K11	(Mfg. unknown) Germany	Wrought Alloy	8400	165	.58
K16	(Mfg. unknown) Germany	Wrought Alloy	8500	180	.65
K30	(Mfg. unknown)	Wrought Alloy	8600	230	.90

Number, Grade or Alloy	Manufacturer or Country of Origin	Type of Material	B_r (kG)	H_c (kOe)	BH_{max} (MGO)
KHJ-1	Hitachi U.S.A.	Cast Fe-Cr-Co	13,000	600	5.7
KHJ-1	Hitachi U.S.A.	Fe-Cr-Co	12,900	600	5.6
KHJ-1D	Hitachi U.S.A.	Fe-Cr-Co	12,000	550	4.5
KHJ-2	Hitachi U.S.A.	Fe-Cr-Co	11,500	750	4.5
KHJ-3A	Hitachi U.S.A.	Cast Fe-Cr-Co	13,500	650	6.5
KHJ-3B	Hitachi U.S.A.	Fe-Cr-Co	13,500	700	7.0
KHJ-4DA	Hitachi U.S.A.	Fe-Cr-Co	9000	50	.30
KHJ-4DB	Hitachi U.S.A.	Fe-Cr-Co	9000	250	.90
KHJ-4DC	Hitachi U.S.A.	Fe-Cr-Co	9000	400	1.5
KHJ-4DD	Hitachi U.S.A.	Fe-Cr-Co	8000	550	2.0
KHJ-7DA	Hitachi U.S.A.	Fe-Cr-Co	13,000	120	1.0
KPM-2A	Hitachi U.S.A.	Bonded Ferrite	2600	2200	1.45
KPM-3	Hitachi U.S.A.	Bonded Ferrite	1400	1200	.4
KS Steel	(Mfg. and origin unknown)	Cast Alnico	7100	780	2.0
KS 1	(Mfg. unknown) Japan	Wrought Alloy	9000	250	.95
KS 2	(Mfg. unknown) Japan	Wrought Alloy	8500	180	.65
KS 3	(Mfg. unknown) Japan	Wrought Alloy	8000	160	.52
KS 4	(Mfg. unknown) Japan	Wrought Alloy	7200	130	.38

Number, Grade or Alloy	Manufacturer or Country of Origin	Type of Material	B_r (kG)	H_c (kOe)	BH_{max} (MGO)
Kato's Oxide	(Mfg. and origin unknown)	Sintered Oxide	1600	900	.5
KENYMAX	Trade name of TMC Magnetics Inc. for some materials				
Kobalt 100	(Mfg. unknown) Germany	Wrought Alloy	7200	130	.38
Kobalt 125	(Mfg. unknown) Germany	Wrought Alloy	7800	145	.46
Kobalt 160	(Mfg. unknown) Germany	Wrought Alloy	8000	160	.52
Kobalt 200	(Mfg. unknown) Germany	Wrought Alloy	8500	180	.65
Kobalt 300	(Mfg. unknown) Germany	Wrought Alloy	9000	250	.95
Koerdym	Krupp trade name for Nd-Fe-B materials				
Koerdym P	Krupp trade name for bonded rare earth (Nd-Fe-B) materials				
Koerdym 32P	Krupp Widia Germany	Bonded Rare Earth	4500	3770	4.0
Koerdym 35P	Krupp Widia Germany	Bonded Rare Earth	4800	3770	4.4
Koerdym 38P	Krupp Widia Germany	Bonded Rare Earth	4900	4020	4.8
Koerdym 42P	Krupp Widia Germany	Bonded Rare Earth	5200	4020	5.3
Koerdym 55/100P	Krupp Widia Germany	Bonded Rare Earth	6000	5100	7.5
Koerdym 60P	Krupp Widia Germany	Bonded Rare Earth	5700	4780	6.9
Koerdym 75P	Krupp Widia Germany	Bonded Rare Earth	6900	5030	9.4
Koerdym 190	Krupp Widia Germany	Sintered Rare Earth	10,000	9420	23.9
Koerdym 210	Krupp Widia Germany	Sintered Rare Earth	10,500	9800	26.4
Koerdym 230	Krupp Widia Germany	Sintered Rare Earth	10,800	10,420	28.9

Number, Grade or Alloy	Manufacturer or Country of Origin	Type of Material	B_r (kG)	H_c (kOe)	BH_{max} (MGO)
Koerdym 240	Krupp Widia Germany	Sintered Rare Earth	11,000	10,550	30.1
Koerdym 260	Krupp Widia Germany	Sintered Rare Earth	11,500	11,050	32.6
Koerdym 280	Krupp Widia Germany	Sintered Rare Earth	12,000	11,560	35.1
Koerflex (all grades)	Trade name used by Krupp Widia (Germany) for various grades of wrought metallic alloys, based on iron, cobalt and vanadium				
Koerflex	Krupp trade name for malleable magnet materials				
Koerflex 30	Krupp Widia Germany	Wrought Alloy	17,000-18,000	20-40	.4-.6
Koerflex 160 [8/2]	Krupp Widia Germany	Wrought FeCrCo Alloy	9400	300	1.3
Koerflex 160 [10/4]	Krupp Widia Germany	Wrought FeCrCo Alloy	8400	500	1.5
Koerflex 200	Krupp Widia Germany	Wrought Alloy	13,000-16,000	60-230	.5-1.5
Koerflex 300 (strip)	Krupp Widia Germany	Wrought FeCrCo Alloy	9500	330	1.8
Koerflex 300 (tape)	Krupp Widia Germany	Wrought Alloy	9500-11,000	280-330	1.7
Koerflex 300 (wire)	Krupp Widia Germany	Wrought FeCrCo Alloy	11,000	380	2.5
Koerflex 300 (wire)	Krupp Widia Germany	Wrought Alloy	19,800-12,200	340-410	2.8-3.4
Koermax	Krupp Widia trade name for rare earth-cobalt materials				
Koermax 150	Krupp Widia Germany	Sintered Rare Earth	8900	5400	17.6
Koermax 160	Krupp Widia Germany	Sintered Rare Earth	9400	8540	20.0
Koermax 170	Krupp Widia Germany	Sintered Rare Earth	9600	5650	21.3
Koermax 200	Krupp Widia Germany	Sintered Rare Earth	10,200	9290	25.1
Koerox	Krupp Widia trade name for sintered ferrite (ceramic) materials				

Number, Grade or Alloy	Manufacturer or Country of Origin	Type of Material	B_r (kG)	H_c (kOe)	BH_{max} (MGO)
Koerox (all grades)	Trade name used by Krupp Widia for various grades of sintered or bonded ferrites.				
Koerox G	Krupp Widia Germany	Bonded-Ferrite	1500-2000	1300-1500	.5-1.0
Koerox P	**Krupp Widia trade name for bonded ferrites**				
Koerox P	Krupp Widia Germany	Bonded Ferrite	1400-1600	1100-1300	.4-.6
Koerox VS31	Krupp Widia Germany	Bonded Ferrite	1400-1700	1300-1500	.6-.8
Koerox 1/18P	**Krupp Widia Germany**	**Bonded Ferrite**	**830**	**820**	**.16**
Koerox 2/20P	**Krupp Widia Germany**	**Bonded Ferrite**	**1180**	**1010**	**.30**
Koerox 4/22P	**Krupp Widia Germany**	**Bonded Ferrite**	**1500**	**1190**	**.46**
Koerox 8/19P	**Krupp Widia Germany**	**Bonded Ferrite**	**2100**	**1570**	**1.0**
Koerox 10/22P	**Krupp Widia Germany**	**Bonded Ferrite**	**2300**	**1950**	**1.31**
Koerox 12/22P	**Krupp Widia Germany**	**Bonded Ferrite**	**2600**	**2260**	**1.63**
Koerox 100 [Sintered]	Krupp Widia Germany	Sintered Ferrite	1800-2200	1600-2000	.8-1.1
Koerox 100 [Sintered]	**Krupp Widia Germany**	**Sintered Ferrite**	**2300**	**1950**	**1.13**
Koerox 150	**Krupp Widia Germany**	**Sintered Ferrite**	**2700**	**2200**	**1.5**
Koerox 300 [Sintered]	Krupp Widia Germany	Sintered Ferrite	3500-3900	1900-2200	2.7-3.2
Koerox 300 [Sintered]	**Krupp Widia Germany**	**Sintered Ferrite**	**3650**	**2640**	**3.0**
Koerox 300K	Krupp Widia Germany	Sintered Ferrite	2900-3500	2300-2800	2.0-2.7
Koerox 300R	Krupp Widia Germany	Sintered Ferrite	3700-4100	1500-1800	2.8-3.5
Koerox 330 [Sintered]	Krupp Widia Germany	Sintered Ferrite	3400-3800	2600-3200	3.0-3.4

Number, Grade or Alloy	Manufacturer or Country of Origin	Type of Material	B_r (kG)	H_c (kOe)	BH_{max} (MGO)
Koerox 330 [Sintered]	**Krupp Widia Germany**	**Sintered Ferrite**	3700	3020	3.3
Koerox 350	**Krupp Widia Germany**	**Sintered Ferrite**	3700	3140-	3.3
Koerox 360 [Sintered]	Krupp Widia Germany	Sintered Ferrite	3700-4100	2000-2900	3.3-3.8
Koerox 360 [Sintered]	**Krupp Widia Germany**	**Sintered Ferrite**	4000	2140	3.8
Koerox 400	**Krupp Widia Germany**	**Sintered Ferrite**	4000	3200	3.9
Koerox 420	**Krupp Widia Germany**	**Sintered Ferrite**	4200	3330	4.3
Koerstit	Krupp Widia Germany	Cast Alnico	-	1000	3.5
Koerzit	Trade name used by Krupp Widia for alnico type materials				
Koerzit (no grade number)	**Trade name used by Krupp Widia for an early metallic material similar to KS steel. Later types are wrought steels or alnicos with grade numbers**				
Koerzit H	Krupp Widia Germany	Wrought Alloy	14,000-16,000	60-200	1.0
Koerzit T (strip)	Krupp Widia Germany	Wrought Alloy	9500-11,000	280-320	1.7-2.2
Koerzit T (wire)	Krupp Widia Germany	Wrought Alloy	10,800-12,200	340-410	2.8-3.4
Koerzit 3/0.6	**Krupp Widia Germany**	**Sintered Alnico**	9900	80	.4
Koerzit 4/0.9	**Krupp Widia Germany**	**Sintered Alnico**	9000	110	.5
Koerzit 4/1.0	**Krupp Widia Germany**	**Sintered Alnico**	8800	125	.55
Koerzit 5/1.4	**Krupp Widia Germany**	**Sintered Alnico**	8400	180	.65
Koerzit 6/1.6	**Krupp Widia Germany**	**Sintered Alnico**	8200	200	.8
Koerzit 50	Krupp Widia Germany	Cast or Sintered Alnico	8500	090-110	.5

Number, Grade or Alloy	Manufacturer or Country of Origin	Type of Material	B_r (kG)	H_c (kOe)	BH_{max} (MGO)
Koerzit 55	Krupp Widia Germany	Cast or Sintered Alnico	-	2000	5.0
Koerzit 120 [Cast or Sintered]	Krupp Widia Germany	Cast or Sintered Alnico	5300-6500	550-700	1.1-1.4
Koerzit 120 [Sintered]	**Krupp Widia Germany**	**Sintered Alnico**	**5300**	**650**	**1.2**
Koerzit 120K	Krupp Widia Germany	Cast or Sintered Alnico	5300-6500	500-700	1.1-1.5
Koerzit 120R	Krupp Widia Germany	Cast or Sintered Alnico	6500-7500	350-500	1.3-1.5
Koerzit 130	Krupp Widia Germany	Cast or Sintered Alnico	-	600	1.3
Koerzit 130K	**Krupp Widia Germany**	**Sintered Alnico**	**5500**	**680**	**1.3**
Koerzit 160 [Cast or Sintered]	Krupp Widia Germany	Cast or Sintered Alnico	5800-7200	600-780	1.5-1.9
Koerzit 160 [Sintered]	**Krupp Widia Germany**	**Sintered Alnico**	**6500**	**720**	**1.7**
Koerzit 190 [Cast or Sintered]	Krupp Widia Germany	Cast or Sintered Alnico	7400-8600	650-800	1.8-2.4
Koerzit 190 [Sintered]	**Krupp Widia Germany**	**Sintered Alnico**	**7500**	**750**	**2.1**
Koerzit 220	Krupp Widia Germany	Cast or Sintered Alnico	5800-6800	1150-1300	2.0-2.6
Koerzit 260 [Cast or Sintered]	Krupp Widia Germany	Cast or Sintered Alnico	5800-6800	1100-1300	2.0-2.8
Koerzit 260 [Sintered]	**Krupp Widia Germany**	**Sintered Alnico**	**6400**	**1320**	**2.8**
Koerzit 300	Krupp Widia Germany	Cast or Sintered Alnico	8500-9800	700-800	2.6-3.6

Number, Grade or Alloy	Manufacturer or Country of Origin	Type of Material	B_r (kG)	H_c (kOe)	BH_{max} (MGO)
Koerzit 350	Krupp Widia Germany	Cast or Sintered Alnico	7500-9000	1050-1250	3.2-4.2
Koerzit 400	Krupp Widia Germany	Cast or Sintered Alnico	11,000-12,000	570-650	3.8-4.5
Koerzit 400K [Cast or Sintered]	Krupp Widia Germany	Cast or Sintered Alnico	9000-10,500	780-850	3.5-4.0
Koerzit 400K [Sintered]	**Krupp Widia Germany**	**Sintered Alnico**	**11,000**	**800**	**4.0**
Koerzit 450 [Cast or Sintered]	Krupp Widia Germany	Cast or Sintered Alnico	7500-9200	1100-1300	4.0-4.8
Koerzit 450 [Sintered]	**Krupp Widia Germany**	**Sintered Alnico**	**8800**	**1450**	**5.5**
Koerzit 500 [Cast or Sintered]	Krupp Widia Germany	Cast or Sintered Alnico	11,500-12,500	600-700	4.5-5.5
Koerzit 500 [Sintered]	**Krupp Widia Germany**	**Sintered Alnico**	**12,400**	**640**	**5.2**
Koerzit 600	Krupp Widia Germany	Cast or Sintered Alnico	12,000-13,000	680-760	5.2-6.5
Koerzit 700	**Krupp Widia Germany**	**Cast Alnico**	**13,500**	**730**	**7.8**
Koerzit 1800	**Krupp Widia Germany**	**Sintered Alnico**	**7400**	**1880**	**5.5**
Koroseal	Trade name used by B. F. Goodrich Co. for its line of bonded ferrites				
Koroseal	**Trade name of RJF International Corp. for bonded materials**				
Koroseal 1-M-2208	B.F.Goodrich Co. U.S.A.	Bonded Ferrite	1675	1110	.575
Koroseal 1-M-2320	B.F.Goodrich Co. U.S.A.	Bonded Ferrite	1675	1220	.575
Koroseal 22-024	B.F.Goodrich Co. U.S.A.	Bonded Ferrite	1675	1220	.575

Number, Grade or Alloy	Manufacturer or Country of Origin	Type of Material	B_r (kG)	H_c (kOe)	BH_{max} (MGO)
L	See Ticonal L (correct designation)				
LP	See Oxalit LP (correct designation)				
Lodex	Trade name used by the General Electric Co. (now Hitachi) for elongated single domain materials. No other ESD source in the United States. Equivalents shown are the best substitutes of other types of materials.				
Lodex	See specific grade number				
Lodex 30	Hitachi U.S.A.	ESD	4000	1250	1.68
Lodex 31	Hitachi U.S.A.	ESD	6250	1140	3.40
Lodex 31 [1985]	Hitachi U.S.A.	ESD	6500	940	2.9
Lodex 32	Hitachi U.S.A.	ESD	7300	940	3.4
Lodex 32 [1985]	Hitachi U.S.A.	ESD	6770	885	2.9
Lodex 33	Hitachi U.S.A.	ESD	8000	860	3.2
Lodex 33 [1985]	Hitachi U.S.A.	ESD	7050	860	3.0
Lodex 36	Hitachi U.S.A.	ESD	3400	1220	1.45
Lodex 37	Hitachi U.S.A.	ESD	5500	1000	2.1
Lodex 37 [1985]	Hitachi U.S.A.	ESD	5100	920	1.7
Lodex 38	Hitachi U.S.A.	ESD	6200	840	2.2
Lodex 38 [1985]	Hitachi U.S.A.	ESD	5600	780	1.7
Lodex 40	Hitachi U.S.A.	ESD	2700	1100	.85
Lodex 41	Hitachi U.S.A.	ESD	4300	980	1.4
Lodex 41 [1985]	Hitachi U.S.A.	ESD	4200	950	1.4
Lodex 42	Hitachi U.S.A.	ESD	5250	850	1.4
Lodex 42 [1985]	Hitachi U.S.A.	ESD	5200	840	1.5
Lodex 43	Hitachi U.S.A.	ESD	6000	710	1.3
Low-Chrome Steel	See Chrome Steel (1 percent) (correct designation)				
M	See Ticonal M (correct designation)				
MA1	Tokushu Seiko Co. Japan	Wrought Alloy	9500-11,000	55-60	.25

Number, Grade or Alloy	Manufacturer or Country of Origin	Type of Material	B_r (kG)	H_c (kOe)	BH_{max} (MGO)
MA2	Tokushu Seiko Co. Japan	Wrought Alloy	10,000-11,000	60-70	.28
MA3	Tokushu Seiko Co. Japan	Wrought Alloy	8500-9500	70-80	.30
MA4	Tokushu Seiko Co. Japan	Wrought Alloy	8000-9000	80-90	.33
MA5	Tokushu Seiko Co. Japan	Wrought Alloy	8000-9000	95-110	.36
MA6	Tokushu Seiko Co. Japan	Wrought Alloy	8000-9000	120-130	.45
MGO-1016	**Magnetic strip and sheeting (refer to Manufacturer for properties)**				
MK Alloy	(Mfg. unknown) Japan	Cast Metal Alloy	-	-	-
Mo 1	(Mfg. unknown) Japan	Wrought Alloy	7000	250	.80
M-01	Allen Bradley Co. U.S.A.	Sintered Ferrite	2200	1800	1.0
M01C	Allen Bradley Co. U.S.A.	Sintered Ferrite	2350	2000	1.2
M05A	Allen Bradley Co. U.S.A.	Sintered Ferrite	3800	1050	3.4
M05B	Allen Bradley Co. U.S.A.	Sintered Ferrite	3800	2050	3.3
M05C	Allen Bradley Co. U.S.A.	Sintered Ferrite	3300	2300	2.6
M06C	Allen Bradley Co. U.S.A.	Sintered Ferrite	3300	2800	2.5
MS6	(Mfg. unknown) Switzerland	Wrought Alloy	7800	145	.46
MS15	(Mfg. unknown) Switzerland	Wrought Alloy	8500	180	.65
MS35	(Mfg. unknown) Switzerland	Wrought Alloy	9000	250	.95
M7	Allen Bradley Co. U.S.A.	Sintered Ferrite	3400	3250	2.7

Number, Grade or Alloy	Manufacturer or Country of Origin	Type of Material	B_r (kG)	H_c (kOe)	BH_{max} (MGO)
M8	Allen Bradley Co. U.S.A.	Sintered Ferrite	3850	2950	3.5
Magloy 1	(Mfg. unknown) England	Cast Alnico	-	600	5.0
Magloy 2	(Mfg. unknown) England	Cast Alnico	-	800	4.5
Magnadur 1	Mullard, Ltd. England	Sintered Ferrite	2000	1750	.95
Magnadur 2	Mullard, Ltd. England	Sintered Ferrite	3900	2200	3.6
Magnadur 3	Mullard, Ltd. England	Sintered Ferrite	3700	3000	3.3
Magnadur II	(Mfg. unknown) Holland	Sintered Ferrite	3500	1350	2.3
Magnadur III	(Mfg. unknown) Holland	Sintered Ferrite	3000	1950	2.2
Magnalox 7680	Xolox Corp. U.S.A.	Bonded Ferrite	2600	2000	1.7
Magnetoflex 12	(Mfg. unknown) Germany	Wrought Alloy	3600	680	1.0
Magnetoflex 20	(Mfg. unknown) Germany	Wrought Alloy	5500	550	1.7
Magnetoflex 35 (strip)	Vacuumschmelze GmbH (VAC) Germany	Malleable Iron-Cobalt	8000-9500	300-380	1.4-2.0
Magnetoflex 35 (wire)	Vacuumschmelze GmbH (VAC) Germany	Malleable Iron-Cobalt	9000-10,500	250-340	2.0-3.0
Magnetoflex 40 (aniso)	Vacuumschmelze GmbH (VAC) Germany	Malleable Iron-Cobalt	10,000-12,000	110-140	.5-1.0
Magnetoflex 40 (iso)	Vacuumschmelze GmbH (VAC) Germany	Malleable Iron-Cobalt	8000	150	.8
Magnico	(Mfg. unknown) USSR	Cast Alnico	11,100	620	4.2
Magnite I	Magno-Ceram Co. U.S.A.	Sintered Ferrite	2200	1800	1.0

Number, Grade or Alloy	Manufacturer or Country of Origin	Type of Material	B_r (kG)	H_c (kOe)	BH_{max} (MGO)
Magnite V	Magno-Ceram Co. U.S.A.	Sintered Ferrite	3900	2000	3.5
Magnite 1	See Magnite I (correct designation)				
Magnite 5	See Magnite V (correct designation)				
Magnyl	(Mfg. and origin unknown)	Bonded Ferrite	-	-	-
Manganese-Aluminum	(Mfg. and origin unknown)	Sintered Metal Alloy	4300	2800	3.5
Manganese-Bismuth	See Bismanol (correct designation)				
Manganese-Steel	(Mfg. and origin unknown)	Wrought Alloy	10,000	43	.18
Manganese-Chrome Steel	See Chrome-Manganese Steel (correct designation)				
Marathon Grades	Not actual material grades or types. Marathon Co. is [was] the U.S. sales agent for some Germany magnet manufacturers.				
Max I	See Ergit Max 1 (correct designation)				
Max III	See Ergit Max 111 (correct designation)				
Maxalco	(Mfg. unknown) Italy	Cast Alnico	-	600	5.0
Maxpowr GPC	Panasote, Inc. U.S.A.	Bonded Ferrite	1650-1800	1300-1400	.55-.70
Maxpowr HEC	Panasote, Inc. U.S.A.	Bonded Ferrite	1900-2050	1700-1900	.90-1.00
Maxpowr IM	Panasote, Inc. U.S.A.	Bonded Ferrite	2530	2130	1.5
Mishima Alloy	(Mfg. and origin unknown)	Cast Alnico	-	-	-
Mi-T-Mag	Trade name used by the National Moldite Co. for various grades of sintered ferrites				
Modified Alcomax	See Alcomax modified (correct designation)				
Moldite	Trade name used by National Moldite Co. (refer to specific grade number)				

Number, Grade or Alloy	Manufacturer or Country of Origin	Type of Material	B_r (kG)	H_c (kOe)	BH_{max} (MGO)
Moldite XL	National Moldite Co., U.S.A.	Sintered Ferrite	2100	1800	1.0
Molybdenum Steels (10-19 percent Mo; 0-12 percent Co)	Most wrought magnet manufacturers	Wrought Alloy	8400-13,000	20-300	-
Murex	Trade name used by Murex, Ltd., England for sintered Alnico magnet materials				
Nd94EA	Hitachi U.S.A.	Sintered Rare Earth	10,800	10,000	28.0
Nd94EB	Hitachi U.S.A.	Sintered Rare Earth	11,500	10,700	31.0
Nd94FA	Hitachi U.S.A.	Sintered Rare Earth	10,300	9500	24.5
Nd97EA	Hitachi U.S.A.	Sintered Rare Earth	11,500	10,000	30.0
Nd97EB	Hitach U.S.A.	Sintered Rare Earth	12,400	10,100	34.5
NFW-3C	Japan Spec. Steel Japan	Cast Alnico	6000-6500	500-600	1.3
NFW-3D	Japan Spec. Steel Japan	Cast Alnico	5300-5700	650-700	1.3
NFW-5	Japan Spec. Steel Japan	Cast Alnico	10,500-11,500	670-760	3.9
NFW-7	Japan Spec. Steel Japan	Cast Alnico	12,000-12,500	600-650	4.8
NFW-7SS	Japan Spec. Steel Japan	Cast Alnico	12,200-13,200	650-720	5.8
NFW-U9	Japan Spec. Steel Japan	Cast Alnico	13,200-14,000	680-750	7.0
N.K.S.3	(Mfg. unknown) Japan	Cast Alnico	12,500	670	5.1
NQ1C	See Neolit NQ1C				
NQ1D	See Neolit NQ1D				
NQ2E	See Neolit NQ2E				
NQ2F	See Neolit NQ2F				

Number, Grade or Alloy	Manufacturer or Country of Origin	Type of Material	B_r (kG)	H_c (kOe)	BH_{max} (MGO)
NQ3E	See Neolit NQ3E				
NQ3F	See Neolit NQ3F				
NQ3G	See Neolit NQ3G				
NeIGT	See specific grade number				
NeIGT	UGIMag U.S.A.	Neodymium Iron-Boron	10,800	9600	27.0
NeIGT 35	UGIMag U.S.A.	Neodymium Iron-Boron	12,300	9800	35.0
NEO	Stackpole U.S.A.	Trade name for bonded rare earth materials			
NEO 27	See EEC NEO 27				
NEO 27	Electron Energy Corp. U.S.A.	Sintered Rare Earth	12,300	9100	35.0
NEO 31	Electron Energy Corp. U.S.A.	Sintered Rare Earth	11,300	10,100	31.0
NEO 33	See EEC NEO 33				
NEO 33	Electron Energy Corp. U.S.A.	Sintered Rare Earth	12,300	9100	35.0
NEO 2001A	Stackpole U.S.A.	Bonded Rare Earth	4600	4100	4.5
NEO 2001B	Stackpole U.S.A.	Bonded Rare Earth	5200	4300	5.5
Neodure	Trade name of Philips for rare earth group of magnet materials				
Neofer 31/100p	Magnetfabrik-Bonn Germany	Bonded Rare Earth	4100	3640	3.8
Neofer 35/100p	Magnetfabrik-Bonn Germany	Bonded Rare Earth	4500	3760	4.3
Neofer 37/60p	Magnetfabrik-Bonn Germany	Bonded Rare Earth	4800	3760	4.6
Neofer 44/60p	Magnetfabrik-Bonn Germany	Bonded Rare Earth	5400	4000	5.5
Neofer 55/100p	Magnetfabrik-Bonn Germany	Bonded Rare Earth	5800	5000	6.9

Number, Grade or Alloy	Manufacturer or Country of Origin	Type of Material	B_r (kG)	H_c (kOe)	BH_{max} (MGO)
Neofer 62/60p	Magnetfabrik-Bonn Germany	Bonded Rare Earth	6500	4770	7.7
Neofer 230/80	Magnetfabrik-Bonn Germany	Sintered Rare Earth	11,000	9500	29.0
Neofer 230/120	Magnetfabrik-Bonn Germany	Sintered Rare Earth	11,000	9800	29.0
Neolit	Thyssen Germany	Trade name for rare earth (Nd-Fe-B) materials			
Neolit NQ1C	Thyssen Germany	Sintered Rare Earth	6400	5700	8.8
Neolit NQ1D	Thyssen Germany	Sintered Rare Earth	7000	5700	9.5
Neolit NQ2E	Thyssen Germany	Sintered Rare Earth	8000	7000	14.0
Neolit NQ2F	Thyssen Germany	Sintered Rare Earth	8000	7500	14.0
Neolit NQ3E	Thyssen Germany	Sintered Rare Earth	12,500	11,200	36.0
Neolit NQ3F	Thyssen Germany	Sintered Rare Earth	12,300	11,500	35.0
Neolit NQ3G	Thyssen Germany	Sintered Rare Earth	11,400	10,700	31.0
NEOREC	Trade name for Nd-Fe-B materials made by TDK Corp.				
NEOREC-27SH	TDK Corp. U.S.A.	Sintered Rare Earth	10,600	10,000	27.0
NEOREC-30SH	TDK Corp. U.S.A.	Sintered Rare Earth	11,200	10,700	30.0
NEOREC-32H	TDK Corp. U.S.A.	Sintered Rare Earth	11,500	10,700	31.0
NEOREC-33	TDK Corp. U.S.A.	Sintered Rare Earth	11,700	10,500	32.0
NEOREC-35H	TDK Corp. U.S.A.	Sintered Rare Earth	12,100	11,500	37.0
NEOREC-38	TDK Corp. U.S.A.	Sintered Rare Earth	12,600	11,500	37.0
New Honda Steel	(Mfg. and origin unknown)	Cast Alnico	7150	785	2.03

Number, Grade or Alloy	Manufacturer or Country of Origin	Type of Material	B_r (kG)	H_c (kOe)	BH_{max} (MGO)
New KS Steel	(Mfg. unknown) Japan	Cast Alnico	7150	785	2.03
Nial	(Mfg. unknown) England	Cast Alnico	6500	500	1.25
Nial	(Mfg. unknown) Belgium	Cast Alnico	6000	560	1.25
Nial 110	(Mfg. unknown) Austria	Cast Alnico	7800	260	1.1
Nial 120	(Mfg. unknown) Austria	Cast Alnico	5800	500	1.1
Nial B	French standard designation; same as Nialco I				
Nialco	**UGIMag U.S.A.**	Trade name for non-oriented cast alnico and/or bonded alnico materials			
Nialco	(Mfg. unknown) Belgium	Cast Alnico	7250	580	1.7
Nialco 200	(Mfg. unknown) Austria	Cast Alnico	9500	650	2.4
Nialco K	French standard designation; same as Nialco II				
Nialco W	French standard designation; same as Nialco III				
Nialco I	French Std. designation	Cast Alnico	–	600	1.36
Nialco IA	French Std. designation	Bonded Alnico	–	500	.6
Nialco II	French Std. designation	Cast Alnico	–	650	1.5
Nialco III	French Std. designation	Cast Alnico	–	650	1.7
Nialco IIIA	French Std. designation	Bonded Alnico	–	700	.8
Nialco IV	French Std. designation	Cast Alnico	–	900	1.8
Nialco IVA	French Std. designation	Bonded Alnico	–	900	1.2
Nialco V	French Std. designation	Cast Alnico	–	800	2.0

Number, Grade or Alloy	Manufacturer or Country of Origin	Type of Material	B_r (kG)	H_c (kOe)	BH_{max} (MGO)
Nialco 1	UGIMag U.S.A.	Cast Alnico	7000	550	1.5
Nialco 1	UGIMag U.S.A.	Bonded Alnico	3200	500	.60
Nialco 3	UGIMag U.S.A.	Cast Alnico	6900	700	1.75
Nialco 4	UGIMag U.S.A.	Bonded Alnico	3000	1000	1.0
Nialco 5	UGIMag U.S.A.	Cast Alnico	8500	800	2.9
Nialco 6	UGIMag U.S.A.	Bonded Alnico	3600	1100	1.1
Nialco 7	UGIMag U.S.A.	Cast Alnico	6400	1150	2.3
Nialco 8	UGIMag U.S.A.	Cast Alnico	5800	1650	2.8
Nifal	(Mfg. unknown) England	Cast Alnico	6500	500	1.25
Nipermag	(Mfg. and origin unknown)	Cast Alnico	5600	660	1.34
OP	See Vectolite (correct designation)				
Oerstit	Trade name used by Magnetfabrik-Dortmund, Germany for various types and grades of metallic magnet materials (refer to specific grade number)				
Oerstit	Thyssen Germany	Trade name for alnico materials			
Oerstit 90	Magnetfabrik-Dortmund Germany	Cast or Sintered Alnico	-	400	1.2
Oerstit 120	Magnetfabrik-Dortmund Germany	Cast or Sintered Alnico	5300-6400	580-700	1.1-1.4
Oerstit 130	Magnetfabrik-Dortmund Germany	Cast or Sintered Alnico	-	600	1.36
Oerstit 160	Magnetfabrik-Dortmund Germany	Cast or Sintered Alnico	6000-7000	640-770	1.5-1.9

Number, Grade or Alloy	Manufacturer or Country of Origin	Type of Material	B_r (kG)	H_c (kOe)	BH_{max} (MGO)
Oerstit 160	**Thyssen Germany**	**Cast Alnico**	**7000**	**800**	**1.8**
Oerstit 190	Magnetfabrik-Dortmund Germany	Cast or Sintered Alnico	7000-8000	760-880	1.8-2.4
Oerstit 200	Magnetfabrik-Dortmund Germany	Pressed Iron	3400	710	.85
Oerstit 220	Magnetfabrik-Dortmund Germany	Cast or Sintered Alnico	5600-6600	1100-1250	2.0-2.6
Oerstit 250	Magnetfabrik-Dortmund Germany	Cast or Sintered Alnico	-	1000	3.5
Oerstit 260	Magnetfabrik-Dortmund Germany	Cast or Sintered Alnico	5600-6600	1100-1300	2.0-2.8
Oerstit 260	**Thyssen Germany**	**Cast Alnico**	**6100**	**1200**	**2.6**
Oerstit 350	Magnetfabrik-Dortmund Germany	Cast or Sintered Alnico	7700-8800	1050-1200	3.2-4.2
Oerstit 400	Magnetfabrik-Dortmund Germany	Cast or Sintered Alnico	-	800	4.5
Oerstit 400	**Thyssen Germany**	**Cast Alnico**	**10,000**	**700**	**3.8**
Oerstit 400K	Magnetfabrik-Dortmund Germany	Cast or Sintered Alnico	9200-10,400	700-800	3.6-4.2
Oerstit 400R	Magnetfabrik-Dortmund Germany	Cast or Sintered Alnico	11,000-12,000	570-650	3.6-4.2
Oerstit 450	Magnetfabrik-Dortmund Germany	Cast or Sintered Alnico	7800-9000	1100-1400	3.5-4.8
Oerstit 450	**Thyssen Germany**	**Cast Alnico**	**8300**	**1500**	**5.2**
Oerstit 500	Magnetfabrik-Dortmund Germany	Cast or Sintered Alnico	11,500-12,800	590-680	4.5-5.5

Number, Grade or Alloy	Manufacturer or Country of Origin	Type of Material	B_r (kG)	H_c (kOe)	BH_{max} (MGO)
Oerstit 500G	Thyssen Germany	Cast Alnico	12,400	600	5.0
Oerstit 500S	Thyssen Germany	Cast Alnico	11,600	600	4.5
Oerstit 600	Magnetfabrik-Dortmund Germany	Cast or Sintered Alnico	12,500-13,500	670-750	5.5-6.5
Oerstit 700	Magnetfabrik-Dortmund Germany	Cast or Sintered Alnico	12,500	700	6.5
Oerstit 800	Magnetfabrik-Dortmund Germany	Cast Alnico	6500	750	1.9
Oerstit 900CP	Magnetfabrik-Dortmund Germany	Platinum Cobalt	5500-6000	4200-4800	8.8-9.3
Oerstit 1000	Magnetfabrik-Dortmund Germany	Sintered Alnico	-	-	-
Oerstit 1000	(Mfg. unknown) Germany	Cast Alnico	6000	950	1.8
Orthodur	Not a permanent-magnet material. Trade name for soft magnetic materials made by Thomas & Skinner Co., USA.				
OX	Trade name used by Magnetfabrik-Dortmund, Germany for various types and grades of sintered ferrites (refer to specific grade number)				
OX 100	Magnetfabrik-Bonn Germany	Sintered Ferrite	2100	1700	1.0
OX 100	Magnetfabrik-Dortmund Germany	Sintered Ferrite	2100	1700	1.0
OX 300	Magnetfabrik Bonn Germany	Sintered Ferrite	3600	1800	2.8
OX 300	Magnetfabrik-Dortmund Germany	Sintered Ferrite	3500	1800	2.7
OX 330	Magnetfabrik-Bonn Germany	Sintered Ferrite	3500	2700	3.0

Number, Grade or Alloy	Manufacturer or Country of Origin	Type of Material	B_r (kG)	H_c (kOe)	BH_{max} (MGO)
OX 330	Magnetfabrik-Dortmund Germany	Sintered Ferrite	3700	3000	3.3
OX 360	Magnetfabrik-Dortmund Germany	Sintered Ferrite	3900	2200	3.6
OX 380	**Magnetfabrik-Bonn Germany**	**Sintered Ferrite**	**3800**	2200	**3.4**
OX 400	**Magnetfabrik-Bonn Germany**	**Sintered Ferrite**	**3800**	3050	**3.8**
OX 400	Magnetfabrik-Dortmund Germany	Sintered Ferrite	3800	2800	3.6
Oxalit	Trade name used by Magnetfabrik-Dortmund, Germany for various grades of bonded ferrites (refer to specific grade number)				
Oxalit D	Magnetfabrik-Dortmund Germany	Bonded Ferrite	1500-1800	1300-1500	.5-.65
Oxalit Da	Magnetfabrik-Dortmund Germany	Bonded Ferrite	1800-2000	1300-1350	.7-.8
Oxalit G	Magnetfabrik-Dortmund Germany	Bonded Ferrite	1500-1800	1300-1500	.5-.65
Oxalit Ga	Magnetfabrik-Dortmund Germany	Bonded Ferrite	1900-2100	1200-1300	.9-1.1
Oxalit LP	Mangetfabrik-Dortmund Germany	Bonded Ferrite	1500-1800	1300-1500	.5-.65
Oxalit P	Magnetfabrik-Dortmund Germany	Bonded Ferrite	1500-1800	1300-1500	.5-.65
Oximall	Not a permanent-magnet material. Soft magnetic material made by Magnetfabrik-Dortmund and used for temperature compensation in permanent magnet circuits.				
Oxit	**Thyssen Germany**	**Trade name for ferrite (ceramic) materials**			

Number, Grade or Alloy	Manufacturer or Country of Origin	Type of Material	B_r (kG)	H_c (kOe)	BH_{max} (MGO)
Oxit	Trade name used by Magnetfabrik-Dortmund for various grades of sintered ferrites				
Oxit 100	Magnetfabrik-Dortmund Germany	Sintered Ferrite	1900-2000	1600-1900	.8-1.0
Oxit 100	Thyssen Germany	Sintered Ferrite	2100	1700	.9 Max.
Oxit 300	Magnetfabrik-Dortmund Germany	Sintered Ferrite	4000	1600	3.2
Oxit 300K	Magnetfabrik-Dortmund Germany	Sintered Ferrite	3600-3900	1900-2300	2.7-3.2
Oxit 300KK	Magnetfabrik-Dortmund Germany	Sintered Ferrite	2800-3500	2300-2800	2.0-2.7
Oxit 300R	Magnetfabrik-Dortmund Germany	Sintered Ferrite	3750-4100	1600-2000	2.9-3.4
Oxit 330	Magnetfabrik-Dortmund Germany	Sintered Ferrite	3600-3800	2600-3100	3.0-3.4
Oxit 360	Magnetfabrik-Dortmund Germany	Sintered Ferrite	3750-4000	1900-2300	3.3-3.7
Oxit 360	Thyssen Germany	Sintered Ferrite	3800	2900	3.2 Max.
Oxit 380	Thyssen Germany	Sintered Ferrite	3900	3600	3.6 Max.
Oxit 380C	Thyssen Germany	Sintered Ferrite	3900	3600	3.2 Max.
Oxit 400	Thyssen Germany	Sintered Ferrite	4100	3300	4.0 Max.
Oxit 400C	Thyssen Germany	Sintered Ferrite	4100	3600	4.0 Max.
Oxit 400HC	Thyssen Germany	Sintered Ferrite	4000	3800	3.8 Max.
Oxit 420	Thyssen Germany	Sintered Ferrite	4300	3300	4.3 Max.

Number, Grade or Alloy	Manufacturer or Country of Origin	Type of Material	B_r (kG)	H_c (kOe)	BH_{max} (MGO)
P	May refer to either Koerox P or Oxalit P				
P5	**See Flexam P5**				
P6	Hitachi U.S.A.	Wrought Alloy	14,000	58	.5
P8	**See Flexam P8**				
P30	**See Ferroxdure P30 (correct designation)**				
P30	**Philips Co. U.S.A.**	**Bonded Ferrite**	1250	1110	.35
P40	**See Ferroxdure P40 (correct designation)**				
P40B	**Philips Co. U.S.A.**	**Bonded Ferrite**	1450	1210	.45
PF1	See Permet PF1 (correct designation)				
PF2	See Permet PF2 (correct designation)				
PM 8	Poly-Mag Inc., U.S.A.	Bonded Ferrite	1700	1200	.80
PM 10	Poly-Mag Inc., U.S.A.	Bonded Ferrite	2000	1800	.90
PM 12	Poly-Mag Inc., U.S.A.	Bonded Ferrite	2200	2000	1.20
PM 14	Poly-Mag Inc., U.S.A.	Bonded Ferrite	2400	2200	1.40
PM 60	Poly-Mag Inc., U.S.A.	Bonded Samarium Cobalt	5500	4500	6.0
PM 80	Poly-Mag Inc., U.S.A.	Bonded Samarium Cobalt	6200	5600	8.0
PM 160	Poly-Mag Inc., U.S.A.	Sintered Samarium Cobalt	8000	7500	16.0
PM 200	Poly-Mag Inc., U.A.S.	Sintered Samarium Cobalt	9000	8500	20.0
PVC	May refer to bonded ferrite materials				
Palladium-Cobalt	Experimental material. Never commercially available on any scale.				

Number, Grade or Alloy	Manufacturer or Country of Origin	Type of Material	B_r (kG)	H_c (kOe)	BH_{max} (MGO)
Permaflux C	Sel-Rex Corp. U.S.A.	Rare Earth	5600	4800	7.0
Permaflux C1	Sel-Rex Corp. U.S.A.	Rare Earth	3000	2800	2.0
Permanit	(Mfg. and origin unknown)	Same as Cobaflex or chrome-tungsten steel.			
Permet	(Mfg. unknown) U.S.A.	Pressed Iron	3400	710	.85
Permet PF1	(Mfg. and origin unknown)	Pressed Iron Particles	5700	470	1.1
Permet PF1	(Mfg. and origin unknown)	Pressed Iron Particles	6050	570	1.08
Permet PF2	(Mfg. and origin unknown)	Pressed Particles	6000	625	1.52
Permet PF2	(Mfg. and origin unknown)	Pressed Iron Particles	9500	405	1.72
Placam/Placor	**Magnet Appl. Ltd./Cookson England**	**Trade name for steel based magnetic rubber. Not covered in this handbook.**			
Placo	Magnetfabrik-Bonn Germany	Platinum Cobalt	5500	4000	7.0
Placovar	Hamilton Watch Co. U.S.A.	Platinum Cobalt	6450	4300	9.5
Placovar	Lancaster Metal Science Corp. (LMS Corp.) USA	Platinum Cobalt	6400	5200	9.5
Plastalloy	**Trade name of Electrodyne Co. for bonded ferrite materials**				
Plastalloy 1A	**The Electrodyne Co., Inc. U.S.A.**	**Bonded Ferrite**	2200	1900	**1.1**
Plastalloy 3	**The Electrodyne Co., Inc. U.S.A.**	**Bonded Ferrite**	2450	2200	**1.4**
Plastic (magnetized)	Term used by H.O. Canfield to describe their bonded ferrite materials				
Plastiform	**Arnold Engr. U.S.A.**	**Trade name for bonded materials**			

Number, Grade or Alloy	Manufacturer or Country of Origin	Type of Material	B_r (kG)	H_c (kOe)	BH_{max} (MGO)
Plastiform 1	3M Co. U.S.A.	Bonded Ferrite	2200	1480	1.06
Plastiform 1H	3M Co. U.S.A.	Bonded Ferrite	2140	1940	1.04
Plastoferrite	See Ferrite A				
Plasto-ferroxdure	(Mfg. unknown) Holland	Bonded Ferrite	-	-	.4
Plasto-ferroxdure	(Mfg. unknown) France	Bonded Ferrite	-	-	.4
Platinox II	J. Bishop Co. USA Johnson-Matthey (England)	Platinum Cobalt	6400	4800	9.2
Platinum-Chrome	Experimental material. Never commercially available on any scale.				
Platinum-Cobalt	Several U.S. manufacturers	Platinum Cobalt	6450	4300	9.5
Platinum-Cobalt 77/23	Vacuumschmelze GmbH (VAC) Germany	Platinum Cobalt	6000-6400	4398-6283	8.0-9.5
Platinum-Iron	(Mfg. and origin unknown)	Platinum Iron	5830	1570	3.07
Polycore	See specific grade number				
Polymag	See specific grade number				
Prac 120	Magnetfabrik-Bonn Germany	Bonded Alni	2800	470	.38
Prac 120	Magnetfabrik-Bonn Germany	Bonded Alnico	2900	460	.4
Prac 160	Magnetfabrik-Bonn Germany	Bonded Alnico	3200	580	.65
Prac 160	Magnetfabrik-Bonn Germany	Bonded Alnico	3400	580	.65
Prac 250	Magnetfabrik-Bonn Germany	Bonded Alnico	4000	850	1.0
Prac 260K	Magnetfabrik-Bonn Germany	Bonded Alnico	3800	1000	1.1

Number, Grade or Alloy	Manufacturer or Country of Origin	Type of Material	B_r (kG)	H_c (kOe)	BH_{max} (MGO)
Prac 260u.260T	**Magnetfabrik-Bonn Germany**	**Bonded Alnico**	3600	900	.9
Pro-Mag	Trade name of Magnetic Specialty Inc. for bonded magnet products				
Prox	Magnetfabrik-Bonn Germany	Bonded Oxide	-	-	.5
Prox 200 (Parallel to orientation)	Magnetfabrik-Bonn-Germany	Bonded Oxide	2800	2100	1.8
Prox 200 (at right angles to orientation)	Magnetfabrik-Bonn Germany	Bonded Oxide	2600	1900	1.6
Rafcore 60	R. Audemars, SA Switzerland	Bonded Samarium Cobalt	5100	5100	3.6
Rafcore 65	R. Audemars, SA Switzerland	Bonded Samarium Cobalt	5700	5700	11.2
Rafcore 80	R. Audemars, SA Switzerland	Bonded Samarium Cobalt	6600	6600	15.1
Rafcore 125	R. Audemars, SA Switzerland	Sintered Samarium Cobalt	8000	6900	13.7
Rafcore 130	R. Audemars, SA Switzerland	Sintered Samarium Cobalt	8500	7500	16.2
Rafcore 145	R. Audemars, SA Switzerland	Sintered Samarium Cobalt	9000	8100	18.6
Rafcore 155	R. Audemars, SA Switzerland	Sintered Samarium Cobalt	9200	8800	20.1
Rafcore 175	R. Audemars, SA Switzerland	Sintered Samarium Cobalt	9600	9100	19.4
Rafcore 190	R. Audemars, SA Switzerland	Sintered Samarium Cobalt	10,000	9500	18.8

Number, Grade or Alloy	Manufacturer or Country of Origin	Type of Material	B_r (kG)	H_c (kOe)	BH_{max} (MGO)
Rafcore 200	R. Audemars, SA Switzerland	Sintered Samarium Cobalt	10,100	9900	16.2
Reance	Trade name of Electrodyne Co. for bonded rare earth materials				
REC	Designation for various grades of Sm-Co materials made by TDK Corp.				
REC-18	TDK Corp. U.S.A.	Sintered Rare Earth	8500	8000	18.0
REC-18B	TDK Corp. U.S.A.	Sintered Rare Earth	8600	7200	18.0
REC-20	TDK Corp. U.S.A.	Sintered Rare Earth	9000	8750	20.0
REC-22	TDK Corp. U.S.A.	Sintered Rare Earth	9550	8600	22.0
REC-22B	TDK Corp. U.S.A.	Sintered Rare Earth	9600	7800	22.0
REC-24	TDK Corp. U.S.A.	Sintered Rare Earth	10,000	6400	23.0
REC-26	TDK Corp. U.S.A.	Sintered Rare Earth	10,500	9200	26.0
REC-26A	TDK Corp. U.S.A.	Sintered Rare Earth	10,400	8000	25.0
REC-30	TDK Corp. U.S.A.	Sintered Rare Earth	11,000	6400	30.0
REC-32A	TDK Corp. U.S.A.	Sintered Rare Earth	11,300	8000	30.0
Reco 1	Philips Co. Holland	Cast Alnico	–	600	1.36
Reco 2A	Mullard, Ltd. England	Cast Alnico	5500	1000	1.92
Reco 3A	Mullard, Ltd. England	Cast Alnico	7200	645	1.7
Reco 100	Philips Co. Holland	Cast Alnico	6200	480	1.1
Reco 120	Philips Co. Holland	Cast Alnico	5900	600	1.3

Number, Grade or Alloy	Manufacturer or Country of Origin	Type of Material	B_r (kG)	H_c (kOe)	BH_{max} (MGO)
Reco 140	Philips Co. Holland	Cast Alnico	6500	565	1.4
Reco 160	Philips Co. Holland	Cast Alnico	6600	680	1.65
Reco 170	Philips Co. Holland	Cast Alnico	5600	890	1.65
Reco 220	Philips Co. Holland	Cast Alnico	6300	1200	2.3
Recoma	**UGIMag, Inc. trade name for Samarium Cobalt materials**				
Recoma 2.5	Aimants/UGIMag France	Sintered Rare Earth	3250	3100	2.5
Recoma 5	Aimants/UGIMag France	Sintered Rare Earth	4600	4200	5.0
Recoma 10	Aimants/UGIMag France	Sintered Rare Earth	6400	6000	10.0
Recoma 20	Aimants/UGIMag France	Sintered Rare Earth	9000	8800	20.0
Recoma 22	**UGIMag U.S.A.**	**Sintered Rare Earth**	**9400**	**9300**	**22.0.**
Recoma 25	**Aimants/UGIMag France**	**Sintered Rare Earth**	**10,000**	**9500**	**25.0**
Recoma 25	**UGIMag U.S.A.**	**Sintered Rare Earth**	**10,000**	**9500**	**25.0**
Recoma 28	**Aimants/UGIMag France**	**Sisntered Rare Earth**	**10,700**	**9000**	**28.0**
Recoma 28	**UGIMag U.S.A.**	**Sintered Rare Earth**	**10,700**	**10,300**	**28.0**
Recoma Stabo-0	Aimants/UGIMag France	Sintered Rare Earth	6400	5900	10.0
Recoma Stabo-0.02	Aimants/UGIMag France	Sintered Rare Earth	7700	7200	15.0
Remalloy Alloy 1	(Mfg. and origin unknown)	Wrought Alloy	9700	210	.95
Remalloy Alloy 2	(Mfg. and origin unknown)	Wrought Alloy	10,300	230	1.1
Remalloy 17	Simmonds Steel U.S.A.	Wrought Alloy	10,000	250	1.1

Number, Grade or Alloy	Manufacturer or Country of Origin	Type of Material	B_r (kG)	H_c (kOe)	BH_{max} (MGO)
Remalloy 20	Simmonds Steel U.S.A.	Wrought Alloy	8550	355	1.25
Remco 16	Electron Energy Corp. U.S.A.	Sintered Rare Earth	8000	7500	16.0
Remco 18	Electron Energy Corp. U.S.A.	Sintered Rare Earth	8500	8300	18.0
RES 190	Philips Co. Holland	Sintered Rare Earth	8900	8420	19.4
RES 195	Philips Co. Holland	Sintered Rare Earth	8900	8480	19.4
RES 230	Philips Co. Holland	Sintered Rare Earth	10,000	8170	23.2
RES 239	Philips Co. Holland	Sintered Rare Earth	9600	9400	23.0
RES 255	Philips Co. Holland	Sintered Rare Earth	10,500	9425	25.1
RES 257	Philips Co. Holland	Sintered Rare Earth	10,200	9400	25.0
RES 270	Philips Co. Holland	Sintered Rare Earth	11,200	10,700	30.0
RES 275	Philips Co. Holland	Sintered Rare Earth	11,500	10,700	31.0
RES 300	Philips Co. Holland	Sintered Rare Earth	12,000	11,300	34.0
RES 302	Philips Co. Holland	Sintered Rare Earth	11,900	11,300	33.0
RES 303	Philips Co. Holland	Sintered Rare Earth	11,600	11,000	32.0
RES 305	Philips Co. Holland	Sintered Rare Earth	11,500	10,680	30.2
RES 350	Philips Co. Holland	Sintered Rare Earth	12,000	10,680	35.2
RES 421	Philips Co. Holland	Sintered Rare Earth	13,000	11,550	40.0
Rubber Ferrite	See Bonded Ferrites				
RUMAX	**Trade name of TMC Magnetics Inc. for some materials**				

Number, Grade or Alloy	Manufacturer or Country of Origin	Type of Material	B_r (kG)	H_c (kOe)	BH_{max} (MGO)
S	May refer to Spinalor S or Ticonal S				
SP02	Philips Co. Holland	Bonded Ferrite	1,000	1010	.25
SP10	Philips Co. Holland	Bonded Ferrite	800	729	.11
SP50	See Ferroxdure SP50				
SP160	Philips Co. Holland	Bonded Ferrite	2450	2260	1.5
SP170	Philips Co. Holland	Bonded Ferrite	2700	2460	1.75
Samarium-Cobalt	Raytheon Co. U.S.A.	Sintered Rare Earth	7500-9000	7500-9000	20.0
SECO	See Koermax 200				
SECO	See Secolit materials				
SECO 50/60p	Magnetfabrik-Bonn Germany	Bonded Rare Earth	5000	4000	6.0
SECO 130/46	See Koermax 150				
SECO 140/120	See Koermax 160				
SECO 140/120	Magnetfabrik-Bonn Germany	Sintered Rare Earth	8500	8000	18.0
SECO 150/46	See Koermax 170				
SECO 170/120	Magnetfabrik-Bonn Germany	Sintered Rare Earth	9500	8000	21.0
Secolit	Thyssen Germany trade name for rare earth (Sm-Co) materials				
Secolit 100TK	Thyssen Germany	Sintered Rare Earth	7300	7200	13.0 Max.
Secolit 170	Thyssen Germany	Sintered Rare Earth	9500	9200	23.0 Max.
Secolit 215	Thyssen Germany	Sintered Rare Earth	11,000	10,200	26.0 Max.
Secolit 215N	Thyssen	Sintered Rare Earth	11,000	7500	26.0 Max.
Series 1-5	Designation for TDK Corp. group of Sm-Co materials				

Number, Grade or Alloy	Manufacturer or Country of Origin	Type of Material	B_r (kG)	H_c (kOe)	BH_{max} (MGO)
Series 2-17	Designation for TDK Corp. group of Sm-Co materials				
Sermalloy A1	(Mfg. unknown) France	Cast Alnico	7400	2120	6.0
Sermalloy A2	(Mfg. unknown) France	Cast Alnico	14,300	530	5.0
Silicon-Chrome Steel	See Chrome-Silicon Steel (correct designation)				
Silmanol	Hitachi U.S.A.	Wrought Alloy	595	570	.085
Silver-Manganese Aluminum Alloy	Same as Silmanol				
Spinalor	UGIMag trade name for ceramic ferrite segments				
Spinalor B	(Mfg. unknown) France	Sintered Ferrite	4000	1600	3.2
Spinalor H	(Mfg. unknown) France	Sintered Ferrite	3500	2500	2.6
Spinalor HH	(Mfg. unknown) France	Sintered Ferrite	3300	3000	2.7
Spinalor S	(Mfg. unknown) France	Sintered Ferrite	3700	2000	2.8
Sprox	Trade name used by Magnetfabrik-Bonn, Germany for their bonded ferrites made in various grades				
Sprox HT3/19p	Magnetfabrik-Bonn Germany	Bonded Ferrite	1350	1000	.35
Sprox HT11/21p	Magnetfabrik-Bonn Germany	Bonded Ferrite	2300	2000	1.3
Sprox HT14/21p	Magnetfabrik-Bonn Germany	Bonded Ferrite	2600	2200	1.75
Sprox HT16/19p	Magnetfabrik-Bonn Germany	Bonded Ferrite	2800	2200	1.9
Sprox 1	Magnetfabrik-Bonn Germany	Bonded Ferrite	1000	800	.25
Sprox 1F	Magnetfabrik-Bonn Germany	Bonded Ferrite	700	600	.11
Sprox 1H	Magnetfabrik-Bonn Germany	Bonded Ferrite	750	650	.13

Number, Grade or Alloy	Manufacturer or Country of Origin	Type of Material	B_r (kG)	H_c (kOe)	BH_{max} (MGO)
Sprox 2	Magnetfabrik-Bonn Germany	Bonded Ferrite	1500	1100	.5
Sprox 2F	Magnetfabrik-Bonn Germany	Bonded Ferrite	1100	900	.28
Sprox 2FE	**Magnetfabrik-Bonn Germany**	**Bonded Ferrite**	**1500**	**1300**	**.5**
Sprox 2FE	Magnetfabrik-Bonn Germany	Bonded Ferrite	1500	1150	.50
Sprox 2H	Magnetfabrik-Bonn Germany	Bonded Ferrite	1600	1200	.55
Sprox 4Fu.4FE	**Magnetfabrik-Bonn Germany**	**Bonded Ferrite**	**2200**	**1950**	**1.15**
Sprox 5F	**Magnetfabrik-Bonn Germany**	**Bonded Ferrite**	**2400**	**2100**	**1.4**
STAB 0	**UGIMag U.S.A.**	**Sintered Rare Earth**	**6400**	**5900**	**10.0**
STAB 0.02	**UGIMag U.S.A.**	**Sintered Rare Earth**	**7700**	**7200**	**15.0**
Stabon	Stackpole U.S.A. trade name for bonded ferrites				
Stabon E15N	Stackpole U.S.A.	Bonded Ferrite	2600	2100	1.55
Stabon E120	Stackpole U.S.A.	Bonded Ferrite	2250	2000	1.2
Stabon E140	Stackpole U.S.A.	Bonded Ferrite	2450	2200	1.4
Stabon IM140	Stackpole U.S.A.	Bonded Ferrite	2450	2200	1.4
Stabon IM160	Stackpole U.S.A.	Bonded Ferrite	2550	2300	1.6
Stabon IM180	Stackpole U.S.A.	Bonded Ferrite	2700	2280	1.8
Strontium Ferrite	Class of sintered ferrites similar to barium ferrites				
T	See Koerzit T (correct designation)				
T500	See Tromalit T500 (correct designation)				
T700	See Tromalit 700 (correct designation)				

Number, Grade or Alloy	Manufacturer or Country of Origin	Type of Material	B_r (kG)	H_c (kOe)	BH_{max} (MGO)
T1000	See Tromalit 1000 (correct designation)				
T1100	See Tromalit 1100 (correct designation)				
T1400	See Taomalit 1400 (correct designation)				
TSK1	Tokushu Seiko Co. Japan	Cast Alnico	6500	450	1.2
TSK2	Tokushu Seiko Co. Japan	Cast Alnico	7500	550	1.6
TSK5	Tokushu Seiko Co. Japan	Cast Alnico	12,000	650	5.5
Ticonal	Trade name used by English, French and Dutch (Netherlands) producers for various grades of Alnico-type materials				
Ticonal-Columnar	Refers to partially or fully crystal-oriented Alnico materials made by various manufacturers under the trade name Ticonal and with various grade designations.				
Ticonal C	Mullard, Ltd. England	Cast Alnico	12,500	680	5.0
Ticonal D	Mullard, Ltd. England	Cast Alnico	12,000	600	3.8
Ticonal E	Mullard, Ltd. England	Cast Alnico	11,070	740	4.1
Ticonal F	Mullard, Ltd. England	Cast Alnico	12,400	600	3.8
Ticonal G	Mullard, Ltd. England	Cast Alnico	13,480	583	5.7
Ticonal GG	(Mfg. unknown) France	Cast Alnico	-	880	7.7
Ticonal GX	Mullard, Ltd. England	Cast Alnico	13,500	720	7.5
Ticonal H	Mullard, Ltd. England	Cast Alnico	11,800	770	4.5
Ticonal K	Mullard, Ltd. England	Cast Alnico	8500	1150	3.6
Ticonal L	(Mfg. unknown) Holland	Cast Alnico	-	800	2.2

Number, Grade or Alloy	Manufacturer or Country of Origin	Type of Material	B_r (kG)	H_c (kOe)	BH_{max} (MGO)
Ticonal M	(Mfg. unknown) Holland	Cast Alnico	-	800	4.5
Ticonal S	Mullard, Ltd. England	Sintered Alnico	11,070	620	4.2
Ticonal X	Mullard, Ltd. England	Cast Alnico	9000	1300	4.0
Ticonal 2A	Mullard, Ltd. England	Cast Alnico	-	-	-
Ticonal 3A	Mullard, Ltd. England	Cast Alnico	-	650	1.7
Ticonal 190	Philips Co. Holland	Cast Alnico	8000	730	2.1
Ticonal 360	Philips Co. Holland	Cast Alnico	10,700	710	3.6
Ticonal 400	Philips Co. Holland	Cast Alnico	11,600	640	4.0
Ticonal 450	Philips Co. Holland	Cast Alnico	8500	1335	4.25
Ticonal 500	Philips Co. Holland	Cast Alnico	12,500	630	4.8
Ticonal 600	Philips Co. Holland	Cast Alnico	13,500	645	5.7
Ticonal 650	Philips Co. Holland	Cast Alnico	12,800	630	4.8
Ticonal 750	Philips Co. Holland	Cast Alnico	13,400	760	7.5
Ticonal 800	Philips Co. Holland	Cast Alnico	-	800	4.5
Ticonal 900	Philips Co. Holland	Cast Alnico	10,600	1400	9.0
Ticonal 1500	(Mfg. unknown) France	Cast Alnico	-	1500	5.0
Ticonal 2000	(Mfg. unknown) France	Cast Alnico	-	2000	5.0
Transcor 5	Thomas & Skinner Inc., U.S.A.	Sintered Rare Earth	4800	4500	5.4

Number, Grade or Alloy	Manufacturer or Country of Origin	Type of Material	B_r (kG)	H_c (kOe)	BH_{max} (MGO)
Transcor 18	Thomas & Skinner Inc., U.S.A.	Sintered Rare Earth	8600	8400	18.0
Transcor 19	Thomas & Skinner Inc., U.S.A.	Sintered Rare Earth	10,700	4000	19.0
Transcor 21	Thomas & Skinner Inc., U.S.A.	Sintered Rare Earth	9200	8800	21.0
Transcor 22	Thomas & Skinner Inc., U.S.A.	Sintered Rare Earth	9700	8000	22.0
Transcor 22H	Thomas & Skinner Inc., U.S.A.	Sintered Rare Earth	9900	8800	22.0
Transcor 27	Thomas & Skinner Inc., U.S.A.	Sintered Rare Earth	10,700	8800	27.0
Transcor 27H	Thomas & Skinner Inc., U.S.A.	Sintered Rare Earth	10,700	14,000	27.0
Tri-Neo	Trade name used by Tridus International for rare earth (Nd-Fe-B) materials				
Tromadur	Baermann Germany trade name for bonded ferrite materials				
Tromadur 3/16p	Baermann Germany	Bonded Ferrite	1350	1131	.38
Tromadur 9/21p	Baermann Germany	Bonded Ferrite	2200	1847	1.1
Tromadur 13/22p	Baermann Germany	Bonded Ferrite	2600	2325	1.6
Tromadur 16/25p	Baermann Germany	Bonded Ferrite	2950	2639	2.0
Tromadym	Baermann Germany trade name for bonded rare earth types				
Tromadym 35/70p	Baermann Germany	Bonded Rare Earth	4700	3770	4.4
Tromadym 50/70p	Baermann Germany	Bonded Rare Earth	5500	4524	6.3
Tromadym 60/70p	Baermann Germany	Bonded Rare Earth	6000	5027	7.5
Tromadym 60/110p	Baermann Germany	Bonded Rare Earth	6000	5027	7.5

Number, Grade or Alloy	Manufacturer or Country of Origin	Type of Material	B_r (kG)	H_c (kOe)	BH_{max} (MGO)
Tromaflex 3/24p	Baermann Germany	Bonded Ferrite	1270	1144	.38
Tromaflex 3/25p	Baermann Germany	Bonded Ferrite	1420	1232	.43
Tromaflex 4/24p	Baermann Germany	Bonded Ferrite	1630	1382	.57
Tromaflex 9/28p	Baermann Germany	Bonded Ferrite	2200	2136	1.1
Tromaflex 11/20p	Baermann Germany	Bonded Ferrite	2450	2086	1.4
Tromaflex 13/23p	Baermann Germany	Bonded Ferrite	2700	2287	1.6
Tromaflex 324	See Tromaflex 3/24p				
Tromaflex 325	See Tromaflex 3/25p				
Tromaflex 424	See Tromaflex 4/24p				
Tromaflex 928	See Tromaflex 9/28p				
Tromaflex 1120	See Tromaflex 11/20p				
Tromaflex 1323	See Tromaflex 13/23p				
Tromalit	Trade name used by Baermann for bonded alnico magnet materials of various grades				
Tromalit	Baermann trade name for bonded ferrite and Alnico materials				
Tromalit Alni 090	Bearmann Germany	Bonded Alnico	4400	250	.40
Tromalit 6/6p	Baermann Germany	Bonded Alnico	4350	628	.75
Tromalit 11/9p	Baermann Germany	Bonded Alnico	4350	1018	1.38
Tromalit 500	Baermann Germany	Bonded Alnico	3700	525	.57
Tromalit 600	Baermann Germany	Bonded Alnico	3500	600	.63
Tromalit 700	Baermann Germany	Bonded Alnico	3800	700	.76

Number, Grade or Alloy	Manufacturer or Country of Origin	Type of Material	B_r (kG)	H_c (kOe)	BH_{max} (MGO)
Tromalit 800	Baermann Germany	Bonded Alnico	4200	800	.97
Tromalit 800s	Baermann Germany	Bonded Alnico	5000	800	1.23
Tromalit 1000	Baermann Germany	Bonded Alnico	4300	955	1.3
Tromalit T500	Baermann Germany	Bonded Alnico	3730	523	.56
Tromalit T700	Baermann Germany	Bonded Alnico	4350	635	.85
Tromalit T1000	Baermann Germany	Bonded Alnico	4350	955	1.28
Tromalit T1100	Baermann Germany	Bonded Alnico	4500	1020	1.35
Tromalit T1400	Baermann Germany	Bonded Alnico	1800	1370	.61
Truecore	See specific grade number				
Tungsten Steel (5 percent W)	(Mfg. unknown) U.S.A.	Wrought Alloy	10,300	70	.32
Tungsten Steel (6 percent W)	(Mfg. unknown) U.S.A.	Wrought Alloy	10,500	66	.30
Tungsten-Chrome Steel	See Chrome-Tungsten Steel (correct designation)				
USM55-16	U.S. Magnet & Alloy U.S.A.	Cast Alnico	11,700	735	5.5
USM55-20	U.S. Magnet & Alloy U.S.A.	Cast Alnico	12,700	650	5.5
USM55-24	U.S. Magnet & Alloy U.S.A.	Cast Alnico	13,200	635	5.5
USM65-16	U.S. Magnet & Alloy U.S.A.	Cast Alnico	12,800	790	6.5
USM65-20	U.S. Magnet & Alloy U.S.A.	Cast Alnico	13,500	690	6.5
USM75-16	U.S. Magnet & Alloy U.S.A.	Caast Alnico	13,100	820	7.5
USM75-20	U.S. Magnet & Alloy U.S.A.	Cast Alnico	13,900	740	7.5

Number, Grade or Alloy	Manufacturer or Country of Origin	Type of Material	B_r (kG)	H_c (kOe)	BH_{max} (MGO)
USM80H-12	U.S. Magnet & Alloy U.S.A.	Cast Alnico	9600	800	3.75
USM90H-12	U.S. Magnet & Alloy U.S.A.	Cast Alnico	10,600	900	5.0
USM95H-8	U.S. Magnet & Alloy U.S.A.	Cast Alnico	7000	950	3.0
USM105H-8	U.S. Magnet & Alloy U.S.A.	Cast Alnico	8400	1050	4.0
UGIMAX	**UGIMag U.S.A.**	**Trade name for Nb-Fe-B materials**			
UGIMAX 30K	**UGIMag U.S.A.**	**Sintered Rare Earth**	**11,400**	**10,800**	**30.0**
UGIMAX 31H	**UGIMag U.S.A.**	**Sintered Rare Earth**	**11,500**	**11,000**	**31.0**
UGIMAX 34B	**UGIMag U.S.A.**	**Sintered Rare Earth**	**12,000**	**11,000**	**34.0**
UGIMAX 34K	**UGIMag U.S.A.**	**Sintered Rare Earth**	**12,000**	**11,500**	**34.0**
UGIMAX 35H	**UGIMag U.S.A.**	**Sintered Rare Earth**	**12,100**	**11,700**	**35.0**
UGIMAX 37B	**UGIMag U.S.A.**	**Sintered Rare Earth**	**12,500**	**11,700**	**37.0**
UGIMAX 600	Ugine France	Cast Alnico	-	700	6.3
UGIMAX 600 Super	Ugine France	Cast Alnico	-	800	7.7
UGISTAB	**UGIMag U.S.A.**	**Trade name for Nd-Fe-B materials**			
UGISTAB 26x4	**UGIMag U.S.A.**	**Sintered Rare Earth**	**10,400**	**10,000**	**26.0**
UGISTAB 30x4	**UGIMag**	**Sintered Rare Earth**	**11,100**	**10,800**	**30.0**
Ultra-Mag	Trade name of Flexmag Industries, Inc. for bonded materials				
Ultra-Mag	Magnets, Inc. U.S.A.	Bonded Ferrite	1890	1460	.77
Ultramag	P.R. Mallory U.S.A.	Platnumi-Cobalt	6400	4800	9.2

Number, Grade or Alloy	Manufacturer or Country of Origin	Type of Material	B_r (kG)	H_c (kOe)	BH_{max} (MGO)
Unox 1	(Mfg. unknown) Germany	Sintered Ferrite	2000	1700	1.0
Unox 5	(Mfg. unknown) Germany	Sintered Ferrite	3700	2000	2.8
Unox 6	(Mfg. unknown) Germany	Sintered Ferrite	3500	2500	2.6
Unox 9	(Mfg. unknown) Germany	Sintered Ferrite	3300	3000	2.7
Unox 10	(Mfg. unknown) Germany	Sintered Ferrite	3500	2500	2.6
VS 31	See Koerox VS 31 (correct designation)				
VS 35	(Mfg. unknown) Germany	Cast Alnico	-	1000	2.0
VS 55	(Mfg. unknown) Germany	Cast Alnico	-	2000	5.0
Vacodym 335HR	Vacuumschmelze GmbH (VAC) Germany	Sintered Rare Earth	12,500	11,300	3.7
Vacodym 335WZ	Vacuumschmelze GmbH (VAC) Germany	Sintered Rare Earth	11,600	10,700	32.0
Vacodym 351HR	Vacuumschmelze GmbH (VAC) Germany	Sintered Rare Earth	12,400	11,800	36.5
Vacodym 351WZ	Vacuumschmelze GmbH (VAC) Germany	Sintered Rare Earth	11,500	11,000	31.5
Vacodym 362HR	Vacuumschmelze GmbH (VAC) Germany	Sintered Rare Earth	13,300	12,800	42.5
Vacodym 362WZ	Vacuumschmelze GmbH (VAC) Germany	Sintered Rare Earth	12,300	11,700	36.0
Vacodym 370HR	Vacuumschmelze GmbH (VAC) Germany	Sintered Rare Earth	12,000	11,400	34.5
Vacodym 370WZ	Vacuumschmelze GmbH (VAC) Germany	Sintered Rare Earth	11,200	10,700	30.0

Number, Grade or Alloy	Manufacturer or Country of Origin	Type of Material	B_r (kG)	H_c (kOe)	BH_{max} (MGO)
Vacodym 383HR	Vacuumschmelze GmbH (VAC) Germany	Sintered Rare Earth	12,700	12,200	39.0
Vacodym 383WZ	Vacuumschmelze GmbH (VAC) Germany	Sintered Rare Earth	11,800	11,200	33.0
Vacodym 400HR	Vacuumschmelze GmbH (VAC) Germany	Sintered Rare Earth	11,500	11,000	31.5
Vacodym 400WZ	Vacuumschmelze GmbH (VAC) Germany	Sintered Rare Earth	10,600	10,100	27.0
VACOMAX	See Rafcore Series Materials				
Vacomax 65K	Vacuumschmelze GmbH (VAC) Germany	Sintered Rare Earth	5500-6400	5278-6283	7.5-10.0
Vacomax 80T	Vacuumschmelze GmbH (VAC) Germany	Sintered Rare Earth	6000-6600	5718-6535	8.8-10.7
Vacomax 95T	Vacuumschmelze GmbH (VAC) Germany	Sintered Rare Earth	6800-7400	6472-7357	11.3-13.8
Vacomax 145 Sonderqualitat	Vacuumschmelze GmbH (VAC) Germany	Sintered Rare Earth	9000	8300	20.0
Vacomax 145	Vacuumschmelze GmbH (VAC) Germany	Sintered Rare Earth	9000	8300	20.0
Vacomax 170	Vacuumschmelze GmbH (VAC) Germany	Sintered Rare Earth	9500	9000	23.0
Vacomax 200	Vacuumschmelze GmbH (VAC) Germany	Sintered Rare Earth	10,100	9500	25.0
Vacomax 225HR	Vaccumschmelze GmbH (VAC) Germany	Sintered Rare Earth	11,000	9400	27.0
Vacomax 225WZ	Vacuumschmelze GmbH (VAC) Germany	Sintered Rare Earth	10,000	8800	22.0

Number, Grade or Alloy	Manufacturer or Country of Origin	Type of Material	B_r (kG)	H_c (kOe)	BH_{max} (MGO)
Vacozet 200	**Vacuumschmelze GmbH (VAC) Germany**	**Malleable Iron Cobalt**	11,000–15,000	13–21	.12–.25
Vacozet 258	**Vacuumschmelze GmbH (VAC) Germany**	**Malleable Iron-Cobalt**	13,000–14,500	25–44	.25–.38
Vacozet 655	**Vacuumschmelze GmbH (VAC) Germany**	**Malleable Iron Cobalt**	13,000–14,500	38–63	.38–.63
Vectolite	(Mfg. and origin unknown)	Iron & Cobalt Oxides	1600	900	.5
Vicalloy	General name for vanadium-iron wrought alloys. Made under this name by various U.S. producers in various grades				
Vicalloy (no grade number)	Westinghouse Corp. U.S.A.	Wrought Alloy	8000	200–250	.9
Vicalloy (no grade number)	Thomas & Skinner Inc., U.S.A.	Wrought Alloy	9500	265	1.4
Vicalloy I	Several U.S. wrought magnet manufacturers	Wrought Alloy	9000	300	1.0
Vicalloy II	Several U.S. wrought magnet manufacturers	Wrought Alloy	10,000	450	3.0
Vicalloy 1	See Vicalloy I (correct designation)				
Vicalloy 2	See Vicalloy II (correct designation)				
Vinyl Ferrite	See Ferrites, Plastic-Bonded				
W	See Nialco W (correct designation)				
W	(Mfg. unknown) Germany	Wrought Alloy	10,200	70	.30
WH	(Mfg. unknown) Germany	Wrought Alloy	9800	80	.34
Westro Alpha	Westinghouse Corp. U.S.A.	Sintered Ferrite	4000	2200	3.66
Westro Beta	Westinghouse Corp. U.S.A.	Sintered Ferrite	3550	3150	3.0

Number, Grade or Alloy	Manufacturer or Country of Origin	Type of Material	B_r (kG)	H_c (kOe)	BH_{max} (MGO)
Westro Gamma	Westinghouse Corp. U.S.A.	Sintered Ferrite	No information available		
X	See Ticonal X (correct designation)				
X3	See Ferrogum X3 (correct designation)				
X5	See Ferrogum X5 (correct designation)				
X7	See Ferrogum X7 (correct designation)				
X10	See Ferrogum X10 (correct designation)				
X12	See Ferrogum X12 (correct designation)				
X900	See Alnico X900 (correct designation)				
XL	See Moldite XL (correct designation)				
Xolox 7000	Xolox Corp. U.S.A.	Bonded Rare Earth	5700	4200	7.1
Xolox 7480	Xolox Corp. U.S.A.	Bonded Ferrite	1370	1180	.40
Xolox 7540	Xolox Corp. U.S.A.	Bonded Ferrite	2200	2000	1.1
Y3	See Ferrogum Y3 (correct designation)				
Y5	See Ferrogum Y5 (correct designation)				
Y7	See Ferrogum Y7 (correct designation)				
Y10	See Ferrogum Y10 (correct designation)				
YBM 1A	**Hitachi U.S.A.**	**Sintered Ferrite**	**3900**	**1900**	**3.5**
YBM 1B	**Hitachi U.S.A.**	**Sintered Ferrite**	**4000**	**2000**	**3.0**
YBM 1BB	**Hitachi U.S.A.**	**Sintered Ferrite**	**3950**	**2400**	**3.7**
YBM 2B	Hitachi U.S.A.	Sintered Ferrite	3950	3100	3.70
YBM 2BA	Hitach U.S.A.	Sintered Ferrite	4280	2400	4.35
YBM 2BB	Hitachi U.S.A.	Sintered Ferrite	4230	2950	4.20

Number, Grade or Alloy	Manufacturer or Country of Origin	Type of Material	B_r (kG)	H_c (kOe)	BH_{max} (MGO)
YBM 2BC	Hitachi U.S.A.	Sintered Ferrite	3600	3200	3.05
YBM 2BD	Hitachi U.S.A.	Sintered Ferrite	3850	3100	3.50
YBM 2BE	Hitachi U.S.A.	Sintered Ferrite	3800	3550	3.4
YBM 2BF	Hitachi U.S.A.	Sintered Ferrite	3750	3550	3.3
YBM 2C	Hitachi U.S.A.	Sintered Ferrite	3700	3000	3.2
YBM 2CS	Hitachi U.S.A.	Sintered Ferrite	3850	2750	3.45
YBM 2D	Hitachi U.S.A.	Sintered Ferrite	3500	2950	2.9
YBM 3 Rad. orient.	Hitachi U.S.A.	Sintered Ferrite	3500	2950	2.85
YBM 4A	Hitachi U.S.A.	Sintered Ferrite	2500+	1900+	1.3+
YBM 4B	Hitachi U.S.A.	Sintered Ferrite	2700+	2200+	1.5+
YBM 4D	Hitachi U.S.A.	Sintered Ferrite	2500+	1900+	1.3+
YBM 4E	Hitachi U.S.A.	Sintered Ferrite	[Not readily measurable]		
YBM 4F	Hitachi U.S.A.	Sintered Ferrite	[Not readily measurable]		
YBM 5BB	Hitachi U.S.A.	Sintered Ferrite	4300	2850	4.4
YBM 5BD	Hitachi U.S.A.	Sintered Ferrite	4100	3450	4.0
YBM 5BE	Hitachi U.S.A.	Sintered Ferrite	4000	3750	3.8
YBM 5BF	Hitachi U.S.A.	Sintered Ferrite	3900	3650	3.6
YBM 7BE	Hitachi U.S.A.	Sintered Ferrite	3800	3300+	3.4

Number, Grade or Alloy	Manufacturer or Country of Origin	Type of Material	B_r (kG)	H_c (kOe)	BH_{max} (MGO)
YCM 1B	Hitachi U.S.A.	Cast Alnico	13,000	630	5.2
YCM 1D	Hitachi U.S.A.	Cast Alnico	13,000	690	5.8
YCM 2B	Hitachi U.S.A.	Cast Alnico	10,750	725	4.0
YCM 2C	Hitachi U.S.A.	Cast Alnico	10,000	800	3.5
YCM 4A	Hitachi U.S.A.	Cast Alnico	6000	750	1.5
YCM 4B	Hitachi U.S.A.	Cast Alnico	6500	600	1.7
YCM 4C	Hitachi U.S.A.	Cast Alnico	7500	500	1.6
YCM 4D	Hitachi U.S.A.	Cast Alnico	6500	1075	2.4
YCM 5AB	Hitachi U.S.A.	Cast Alnico	8000	150	–
YCM 5CD	Hitachi U.S.A.	Cast Alnico	8500	195	–
YCM 8B	Hitachi U.S.A.	Cast Alnico	9000	1465	5.2
YCM 8C	Hitachi U.S.A.	Cast Alnico	9000	1600	6.0
YCM 8D	Hitachi U.S.A.	Cast Alnico	7900	1800	6.0
YCM 8E	Hitachi U.S.A.	Cast Alnico	8000	2000	6.0
YCM 9B	Hitachi U.S.A.	Cast Alnico	10,500	1525	10.0
YCM 11	Hitachi U.S.A.	Cast Alnico	11,000	1500	11.0
YHJ 2	Hitachi U.S.A.	Wrought Alloy	6000–12,500	50–150	–
YHJ 3A	Hitachi U.S.A.	Wrought Alloy	8000–11,500	140–170	–

Number, Grade or Alloy	Manufacturer or Country of Origin	Type of Material	B_r (kG)	H_c (kOe)	BH_{max} (MGO)
YHJ 3B	Hitachi U.S.A.	Wrought Alloy	8500	145	-
YHJ 5	Hitachi U.S.A.	Wrought Alloy	8000-10,000	800-1000	-
YHJ 25-8	Hitachi U.S.A.	Wrought Alloy	8500-9500	280-340	-
YHJ 30-10	Hitachi U.S.A.	Wrought [Isotropic]	8000-10,000	350-460	1.5
YHJ 30-10	Hitachi U.S.A.	Wrought [Anisotropic]	11,500	575	-
YHJ 32-14	Hitachi U.S.A.	Wrought Alloy	7000-10,000	400-600	-
YMB 1BB	Hitachi U.S.A.	Sintered Ferrite	3950	2400	3.7
YRM 2A	Hitachi U.S.A.	Bonded Ferrite	1800	1600	.75
YRM 2B	Hitachi U.S.A.	Bonded Ferrite	2300	2000	1.1
ZKANTU 24-15	French Std. Designation	Cast Alnico	-	1000	3.5
ZKANTU 35-15	French Std. Designation	Same as Ticonal X			
ZKANTU 40-15	French Std. Designation	Same as Ticonal 2000			
ZKNAU 24-14	French Std. Designation	Same as Ticonal 600, Ticonal 700, Ugimax 600 or Ugimax 600 Super			
ZKNAUT 24-14	French Std. Designation	Same as Ticonal 800			
ZKNAUT 30-18	French Std. Designation	Cast Alnico	-	1000	2.0
ZNA 20-12	French Std. Designation	Cast Alnico	-	400	1.2
ZNA 25-12	French Std. Designation	Same as Nialco I			
ZNAKU 25-12	French Std. Designation	Same as Nialco II			

Number, Grade or Alloy	Manufacturer or Country of Origin	Type of Material	B_r (kG)	H_c (kOe)	BH_{max} (MGO)
ZNKAU 20-12	French Std. Designation	Same as Nialco III			
ZNKAU 20-12	French Std. Designation	Same as Nialco IV			
ZNKAUT 20-16	French Std. Designation	Sames as Nialco V			
.65 percent Carbon Steel	See Carbon Steel, 65 percent C (correct designation)				
1 percent Carbon Steel	See Carbon Steel, 1.0 percent C (correct designation)				
1 percent Chrome Steel	See Chrome Steel 1 percent Cr (correct designation)				
2 percent Chrome Steel	See Chrome Steel 2 percent Cr (correct designation)				
2 percent Cobalt 4 percent Chrome Steel	See Cobalt-Chrome Steel, 2 percent Co, 4 percent Cr (correct designation)				
3 percent Chrome Steel	See Chrome Steel, 3 percent Cr (correct designation)				
3 percent Cobalt Steel	See Cobalt Steel 3 percent Co (correct designation)				
3 percent Fe-Co	Thomas & Skinner Inc., U.S.A.	Wrought Alloy	10,000	65	.5
3-1/2 percent Chrome Steel	Same as Chrome Steel (3-1/2 percent Cr)				
4 percent Chrome Steel	Same as Chrome Steel (4 percent Cr)				
5 percent Tungsten Steel	Same as Tungsten Steel (5 percent W)				
6 percent Chrome Steel	Same as Chrome Steel (6 percent Cr)				
6 percent Cobalt Steel	Same as Cobalt Steel (6 percent Co)				
6 percent Tungsten Steel	Same as Tungsten Steel (6 percent W)				
9 percent Cobalt Steel	Same as Cobalt Steel (9 percent Co)				

Number, Grade or Alloy	Manufacturer or Country of Origin	Type of Material	B_r (kG)	H_c (kOe)	BH_{max} (MGO)
15 percent Cobalt Steel	Same as Cobalt Steel (15 percent Co)				
17 percent Cobalt Steel	Same as Cobalt Steel (17 percent Co)				
17 percent Fe-Co	Thomas & Skinner Inc., U.S.A.	Wrought Alloy	10,400	175	.75
17-1/2 percent Cobalt Steel	Same as Cobalt Steel (17-1/2 percent Co)				
35 percent Cobalt Steel	Same as Cobalt Steel (35 percent Co)				
35 percent Fe-Co	Thomas & Skinner Inc., U.S.A.	Wrought Alloy	10,400	260	.9
36 percent Cobalt Steel	Same as Cobalt Steel (36 percent Co)				
40 percent Cobalt Steel	Same as Cobalt Steel (40 percent Co)				
0	**See Recoma, Stabo-0 or STAB 0**				
.02	**See Recoma or Stabo-0 .02**				
.8	See ISO .8				
030	See Cr 030 (correct designation)				
035	See Cr 035 (correct designation)				
040	See Co 040 (correct designation)				
045	See Co 045 (correct designation)				
050	See Co 050 (correct designation)				
060	See Co 060 (correct designation)				
061	**See HB 061**				
070	See Co 070 (correct designation)				
081	**See HS 081**				
090	May refer to Alni 090, Co 090 or Tromalit Alni 090				
1	May refer to Cast Alnico 1, Sintered Alnico 1, Ferrite 1, Nialco 1, Alcomax 1, Caslox 1, Cunife 1 or Cunico 1				
1	See Ceramic 1 (this appendix)				

Number, Grade or Alloy	Manufacturer or Country of Origin	Type of Material	B_r (kG)	H_c (kOe)	BH_{max} (MGO)
1	See Ferram I or Arnife 1				
1A	See Alnico IA (correct designation)				
1A	**See YBM 1A**				
1B	See Alnico IB (correct designation)				
1B	**See YCM 1B or YBM 1B**				
1BB	**See YMB 1BB**				
1C	See Alnico IC (correct designation)				
1D	**See YCM 1D**				
1H	See Plastiform IH (correct designation)				
1P	See Oxalit IP (correct designation)				
1-M-2208	See Koroseal 1-M-2208 (correct designation)				
1-M-2320	See Koroseal 1-M-2320 (correct designation)				
1-5	**Designation for TDK Corp. group of Sm-Co materials**				
1:5TC-9	See EEC 1:5TC-9				
1:5TC-13	See EEC 1:5TC-13				
1:5TC-15	See EEC 1:5TC-15				
1:5-18	See EEC 1:5-18				
1/18p	See Koerox 1/18p				
2	May refer to Cast Alnico 2, Sintered Alnico 2, Alcomax 2, Ferrite 2, Cunife 2, Cunico 2 or Nialco 2				
2	May refer also to Indalloy 2 or KHJ-2				
2	**See Arnife 2**				
2A	May refer to Alnico IIA, Reco 2A or Ticonal 2A				
2A	**See KPM 2A or YRM 2A**				
2B	See Alnico IIB (correct designation)				
2B	**See YBM 2B**				
2B	**See YCM 2B or YBM 2B or YRM 2B**				
2BA	**See YBM 2BA**				

Number, Grade or Alloy	Manufacturer or Country of Origin	Type of Material	B_r (kG)	H_c (kOe)	BH_{max} (MGO)
2BB	**See YBM 2BB**				
2BC	**See YBM 2BC**				
2BD	**See YBM 2BD**				
2BE	**See YBM 2BE**				
2BF	**See YBM 2BF**				
2C	See Alnico IIC (correct designation)				
2C	**See YCM 2C or YBM 2C**				
2CS	See YBM 2CS				
2D	See YBM 2D				
2FE	**See Sprox 2FE**				
2H	See Alnico IIH (correct designation)				
2 S.C.	See Alcomax II S.C. (correct designation)				
2.5	See Recoma 2.5				
2-17	**Designation for TDK Corp. group of Sm-Co+ materials**				
2:17TC-15	**See EEC 2:17TC-15**				
2:17TC-18	**See EEC 2:17TC-18**				
2:17-24	**See EEC 2:17-24**				
2:17-27	**See EEC 2:17-27**				
2/20p	**See Koerox 2/20p**				
3	May refer to Cast Alnico 3, Sintered Alnico 3, Alcomax III, Ferrite 3 or Caslox III				
3	**See Arnife 3 or YBM 3 or KPM 3**				
3	See YBM 3 and Plastaloy 3				
3A	May refer to Alnico IIIA, Reco 3A or Ticonal 3A				
3B	See Alnico IIIB (correct designation)				
3C	See Alnico IIIC (correct designation)				
3-NF1	See Alnico III-NF1 (correct designation)				
3 S.C.	See Alcomax III S.C. (correct designation)				

Number, Grade or Alloy	Manufacturer or Country of Origin	Type of Material	B_r (kG)	H_c (kOe)	BH_{max} (MGO)
/0.6	See Koerzit 3/0.6				
/4p	See Prac 120				
/18p	Magnetfabrik Schramberg Germany	Bonded Ferrite	1400	1130	.4
/19	See Sprox HT 3/19p				
	May refer to Cast Alnico 4, Sintered Alnico 4, Ferrite 4 or Alcomax IV				
	See Arnife 4				
A	See Alnico IVA (correct designation)				
A	See YBM 4A				
A	See YCM 4A				
B	See Alnico IVB (correct designation)				
B	See YBM 4B				
B	See YCM 4B				
C	See YCM 4C				
D	See YCM 4D				
Fu.4FE	See Sprox 4Fu.4Fe				
PB	See Incor 4PB				
S.C.	See Alcomax IV S.C. (correct designation)				
/0.9	See Koerzit 4/0.9				
/1.0	See Koerzit 4/1.0				
/19p	See Sprox 2FE				
/22p	See Koerox 4/22p				
	May refer to Alnico 5 or Ferrite 5				
	May refer also to Indalloy 2 or Transcor 5 or Recoma 5				
A	See Alnico VA (correct designation)				
AB	See Alnico VAB (correct designation)				
AB	See YCM 5AB				

Number, Grade or Alloy	Manufacturer or Country of Origin	Type of Material	B_r (kG)	H_c (kOe)	BH_{max} (MGO)
5B	See Alnico VB (correct designation)				
5BB	**See YBM 5BF**				
5BD	**See YBM 5BD**				
5BE	**See YBM 5BE**				
5BF	**See YBM 5F**				
5C	See Alnico VC (correct designation)				
5CD	**See YCM 5CD**				
5E	See Alnico VE (correct designation)				
5F	**See Sprox 5F**				
5N	**See Bremag 5N**				
5-NC1	See Alnico V-NC1 (correct designation)				
5-NR1	See Alnico V-NR1 (correct designation)				
5-NW1	See Alnico V-NW1 (correct designation)				
5DG	See Alnico 5DG (correct designation)				
5AB-DG	See Alnico VAB DG (correct designation)				
5B DG	See Alnico VB DG (correct designation)				
5C DG	See Alnico VC DG (correct designation)				
5-PR1	See Alnico V-PR1 (correct designation)				
5-OW1	See Alnico V-OW1 (correct designation)				
5-7	See Alnico 5-7 (correct designation)				
5-OR1	See Alnico V-OR1 (correct designation)				
5/1.4	**See Koerzit 5/1.4**				
5/5p	**See Prac 160**				
6	May refer to Cast Alnico 6, Sintered Alnico 6 or Ferrite 6				
6	See Incor 6				
6B	See Alnico VIB (correct designation)				
6C	See Alnico VIC (correct designation)				

Number, Grade or Alloy	Manufacturer or Country of Origin	Type of Material	B_r (kG)	H_c (kOe)	BH_{max} (MGO)
6-NS1	See Alnico VI-NS1 (correct designation)				
6-NS2	See Alnico VI-NS2 (correct designation)				
6/1.6	See Koerzit 6/1.6				
6/6p	See Tromalit 6/6p				
7	May refer to Cast Alnico 7, Oriented or Non-oriented Sintered Alnico 7 or Ferrite 7				
7	Stackpole U.S.A.	Sintered Ferrite	3400	3250	2.8
7BE	See YBM 7BE				
7S	See Alnico 7S				
7C_i	See Alnico 7C_i				
7/8p	See Prac 260u.260T				
7/21	See Oxit 100				
8	See Cast Alnico 8 or Sintered Alnico 8				
8	See PM 8 and Alnico 8C				
8	Stackpole U.S.A.	Sintered Ferrite	3850	3050	3.4
8$_A$	See Alnico 8$_A$ (correct designation)				
8$_C$	See Alnico 8$_C$ (correct designation)				
8A	See Alnico 8A (correct designation)				
8B	See Alnico 8B or YCM 8B				
8C	See YCM 8C				
8D	See YCM 8D				
8E	See YCM 8E				
8H	See Alnico 8H or AL 8H				
8HE	See Alnico 8HE or AL 8HE				
8 (Improved)	See Alnico 8, Improved				
8-ND1	See Alnico VIII-ND1 (correct designation)				
8/2	See Koerflex 160				

Number, Grade or Alloy	Manufacturer or Country of Origin	Type of Material	B_r (kG)	H_c (kOe)	BH_{max} (MGO)
8/5	See Koerzit 120				
8.8/16	See Neolit NQ1C				
8/19p	See Koerox 8/19p				
8/22	See Ox 100 or Hartferite 8/22				
8/22	Magnetfabrik Schramberg Germany	Sintered Ferrite	2200	1760	1.18
8/27	See Koerox 100				
9	See Alnico 9				
9B	See YCM 9				
9PB	See Incor 9PB				
9/5	See Alni 120 or Koerzit 130K				
9.5/11	See Neolit NQ1D				
9/9p	See Prac 260K				
9/19p	Magnetfabrik Schramberg Germany	Bonded Ferrite	2200	1880	1.2
9/21p	See Sprox 4Fu.4FE				
10	May refer to Alcomax X, Unox 10, PM 10 or Recoma 10				
10N	See Bremag 10N				
10/4	See Koerflex 160				
10/22p	See Koerox 10/22P				
10/27	See Koerox 150, YCM 11PB or Incor 11PB				
11	See YCM 11				
11PB	See Incor 11PB				
11/2	See Koerflex 300 (strip)				
11/9	See Tromalit 11/9p				
11/21p	See Sprox HT 11/21p				
11/23p	See Sprox 5F				
12	See Alnico 12 or PM 12				

Number, Grade or Alloy	Manufacturer or Country of Origin	Type of Material	B_r (kG)	H_c (kOe)	BH_{max} (MGO)
12-NM1	See Alnico XII-NM1 (correct designation)				
12/6	See Alnico 160, Koerzit 160 or Oerstit 160				
12/12p	Magnetfabrik Schramberg Germany	Bonded Ferrite	2550	2260	1.6
12/22p	See Koerox 12/22p				
12/45/46	(Mfg. unknown) Germany	Wrought Alloy	9600	060	.29
13S	See H 13S				
14	See PM 14				
14/6	See Alnico 190				
14/18	See Neolit NQ2E				
14/20	See Neolit NQ2F				
14/21p	See Sprox HT 14/21p				
14/135/46	(Mfg. unknown) Germany	Wrought Alloy	9400	65	.30
15	See Flexor 15				
15/6	See Koerzit 190				
16	See Incor 16 or Remco 16				
16/2	See Koerflex 300 (wire)				
16/19p	See Sprox 16/19p				
16/120/48	(Mfg. unknown) Germany	Wrought Alloy	9800	80	.34
17	See Remalloy 17				
18	See Bremag 18, Incor 18 or REC 18				
18	See Incor 18, Transcor 18, Remco 18 or Jobmax 18				
18	See REC 18				
18B	See H 18B or REC 18B				
18/9	See Alnico 260				
18/10	See Oerstit 260				

Number, Grade or Alloy	Manufacturer or Country of Origin	Type of Material	B_r (kG)	H_c (kOe)	BH_{max} (MGO)
18/97/47	(Mfg. unknown)	Wrought Alloy	7200	130	.38
19	See Transcor 19				
19/11	See Koerzit 260				
20	See Armax 20 or REC 20				
20	See Recoma 20 or Remalloy 20				
20/68/44	(Mfg. unknown) Germany	Wrought Alloy	7800	145	.46
20BR	See HS 20BR				
20SV	See H 20SV				
20-12	May refer to ZNKAU 20-12 or ZNA 20-12				
20-16	See ZNKAUT 20-16				
20-20	See ZNKAU 20-20				
21	See Incor 21				
21	See Incor 21 or Transcor 21				
21EV	See H 21EV				
21HE	See Incor 21HE				
22	See RACOMA 22 or REC 22				
22	See Transcor 22				
22A	See H 22A				
22B	See REC 22B				
22H	See Armax 22H				
22H	See Transcor 22A				
22/15	See OX 300				
22-924	See Koroseal 22-924				
23B	See H 23B				
23CV	See H 23CV				
23EH	See H 23EH				
23EV	See H 23EV				

Number, Grade or Alloy	Manufacturer or Country of Origin	Type of Material	B_r (kG)	H_c (kOe)	BH_{max} (MGO)
23/21	See Koerox 300				
24	See Incor 24 REC 24				
24HE	See Incor 24HE				
24-14	May refer to ZKNAU 24-14 or ZKNAUT 24-14				
24-15	See ZKANTU 24-15				
24/16	Magnetfabrik Schramberg Germany	Sintered Ferrite	3650	2200	3.2
24/23	Magnetfabrik Schramberg Germany	Sintered Ferrite	3650	2890	3.2
24/23	See OX 330 or Oxit 360				
24/24	See Koerox 330				
25	See RACOMA 25				
25B	See HG 25B				
25BR	See HS 25BR				
25CR	See HS 25CR				
25EH	See H 25EH				
25-12	May refer to ZNA 25-12 or ZNAKU 25-12				
25/51/43	(Mfg. unknown) Germany	Wrought Alloy	8000	160	.52
26	See Bremag 26, Incor 26 or REC 26				
26A	See REC 26A				
26H	See Armax 26H				
26HE	See Incor 26HE				
26/9	See Alnico 350				
26x4	See UGISTAB 26x4				
26/16	Magnetfabrik Schramberg Germany	Sintered Ferrite	3800	2200	3.4

Number, Grade or Alloy	Manufacturer or Country of Origin	Type of Material	B_r (kG)	H_c (kOe)	BH_{max} (MGO)
26/24	Magnetfabrik Schramberg Germany	Sintered Ferrite	3800	3010	3.4
26/29	See Koerox 350				
27	See NeIGT 27				
27	See NEO 27				
27	See NeIGT 27, Transcor 27, NEO 27 or Jobmax 27				
27CV	See HS 27CV				
27H	See Transcor 27H				
27N	See Bremag 27N				
27SH	See NeoREC 27SH				
27/18	See Ox 380				
27/26	See Oxit 380				
27/30	See Oxit 380C				
28	See RACOMA 28				
28/6	See Koerzit 400K				
28/15	See Koerox 360				
28/16	Magnetfabrik Schramberg Germany	Sintered Ferrite	4000	2100	3.8
28/26	Magnetfabrik Schramberg Germany	Sintered Ferrite	3950	3330	3.8
28/34	See Oxit 400HC				
28/46/43	(Mfg. unknown) Germany	Wrought Alloy	8500	180	.65
28/90p	See Koerdym 32p				
29/6	See Oerstit 400				
29/24	See Koerox 400				
30	May refer to Lodex 30 or Koerflex 30				
30	See REC 30				

Number, Grade or Alloy	Manufacturer or Country of Origin	Type of Material	B_r (kG)	H_c (kOe)	BH_{max} (MGO)
30BV	See HS 30BV				
30CH	See HS 30CH or H 30CH				
30CR	See HS 30CR				
30CV	See HS 30CV				
30K	See UGIMAX 30K				
30SH	See NEOREC 30SH				
30/16	Magnetfabrik Schramberg Germany	Sintered Ferrite	4100	2140	3.9
30-18	See ZKNAUT 30-18				
30/26	Magnetfabrik Schramberg Germany	Sintered Ferrite	4050	3330	3.9
30/26	See OX 400, Oxit 400 or Oxit 420				
30/30	See Oxit 400C				
30/54p	See Koerdym 35p				
30x4	See UGISTAB 30x4				
31	See Lodex 31 or NEO 31				
31H	See UGIMAX 31H				
31/23	See Neolit NQ3G				
31/100p	See Neofer 31/100p				
32	See Lodex 32				
32A	See REC 32A				
32BV	See HS 32BV				
32H	See NEOREC 32H				
32p	See Koerdym 32p				
32/5	See Alnico 500				
32/26	See Koerox 420				
33	See Lodex 33				
33	See Neo 33 or NEOREC 33				

Number, Grade or Alloy	Manufacturer or Country of Origin	Type of Material	B_r (kG)	H_c (kOe)	BH_{max} (MGO)
34	See Lodex 34				
34B	See UGIMAX 34B				
34K	See UGIMAX 34K				
34/90p	See Koerdym 38p				
35	See Magnetoflex 35				
35	See NeIGT 35 or NEO 35				
35BH	See SH 35BH				
35CH	See SH 35CH				
35H	See NEOREC 35H or UGIMAX 35H				
35N	See Bremag 35N				
35p	See Koerdym 35p				
35/5	See Alnico 500 or Oerstit 500S				
35-15	See ZKANTU 35-15				
35/17	See Neolit NQ3F				
35/54p	See Koerdym 42p				
35/70p	See Tromadym 35/70p				
35/100p	See Neofer 35/100p				
36	See Lodex 36				
36/10	See Alnico 450				
36/12	See Oerstit 450				
36/13	See Neolit NQ3E				
37	See Lodex 37				
37B	See UGIMAX 37B				
37BH	See SH 37BH				
37/5	See Koerzit 500				
37/60p	See Neofer 37/60p				
38	See Lodex 38				
38	See NEOREC 38				

Number, Grade or Alloy	Manufacturer or Country of Origin	Type of Material	B_r (kG)	H_c (kOe)	BH_{max} (MGO)
38p	See Koerdym 38p				
38/5	See Oerstit 500G				
39/12	See Koerzit 450				
39/15	See Koerzit 1800				
39x25	See Genox 5				
39x31	See Genox 8				
40	See Lodex 40				
40	See Magnetoflex 40				
40/12	See Alnico 40/12				
40/15	See Alnico 40/15				
40-15	See ZKANTU 40-15				
40/35/42	(Mfg. unknown) Germany	Wrought Alloy	9000	250	.95
41	See Lodex 41				
41x29	See Genox 8H				
42	See Lodex 42				
42p	See Koerdym 42p				
43	See Lodex 43				
44/6	See Alnico 580				
44/60p	See Neofer 44/60p				
45	See Flexor 45				
45/28/47	(Mfg. unknown) Germany	Cast Alnico	7600	280	.90
50	See Koerzit 50				
50/60p	See Seco 50/60p				
50/70p	See Tromadym 50/70p				
50/12/38	(Mfg. unknown) Germany	Cast Alnico	5800	500	1.1
52/6	See Alnico 700 or Koerzit 700				

Number, Grade or Alloy	Manufacturer or Country of Origin	Type of Material	B_r (kG)	H_c (kOe)	BH_{max} (MGO)
55	See Koerzit 55				
55-16	See USM 55-16 (correct designation)				
55-20	See USM 55-20 (correct designation)				
55-24	See USM 55-24 (correct designation)				
55/95p	**See Koerdym 60p**				
55/100p	**See Koerdym 55/100p or Neofer 55/00p**				
60	See PM 60 or Rafcore 60				
60p	**See Koerdym 60p**				
60/70p	**See Tromadym 60/70p, G4-2 or G4-4**				
60/100	**See Vacomax 65K**				
60/100p	**See G4-1 or G4-3**				
60/110p	**See Tromadym 60/100p**				
60/160	**See Vacomax 80T**				
62/60p	**See Neofer 62/60p**				
65	See Rafcore 65				
65K	**See Vacomax 65K**				
65-16	See USM 65-16 (correct designation)				
65/16/37	(Mfg. unknown) Germany	Cast Alnico	6400	630	1.45
68/62p	**See Koerdym 75p**				
73-3.5 percent Co	Simmonds Steel U.S.A.	Wrought Alloy	10,300	60	.302
75p	**See Koerdym 75p**				
75-16	See USM 75-16 (correct designation)				
75-20	See USM 75-20 (correct designation)				
77/23	**See Platinum-Cobalt 77/23**				
80	See PM 80 or Rafcore 80				
80H-12	See USM 80H-12 (correct designation)				
80T	**See Vacomax 80T**				

Number, Grade or Alloy	Manufacturer or Country of Origin	Type of Material	B_r (kG)	H_c (kOe)	BH_{max} (MGO)
80/9/36	(Mfg. unknown) Germany	Cast Alnico	6500	750	1.9
80/160	See Vacomax 95T				
81-18.5 percent Co	Simmonds Steel U.S.A.	Wrought Alloy	10,700	160	.690
83-3 percent Co	Simmonds Steel U.S.A.	Wrought Alloy	9700	81	.382
85/7/35	(Mfg. unknown) Germany	Cast Alnico	6000	950	1.8
85/120	See Secolit 100TK				
90	See Oerstit 90				
90A	See Hycorex 90A				
90A	Hitachi U.S.A.	Sintered Rare Earth	8200	7500	16.0
90B	See Hycorex 90B				
90B	Hitachi U.S.A.	Sintered Rare Earth	8700	8200	18.0
90H-8	See USM 90H-8 (correct designation)				
90H-12	See USM 90H-12 (correct designation)				
95T	See Vacomax 95T				
96A	See Hycorex 96A				
96A	Hitachi U.S.A.	Sintered Rare Earth	9000	8800	19.5
96B	See Hycorex 96B				
96B	Hitachi U.S.A.	Sintered Rare Earth	9400	8800	21.5
99A	See Hycorex 99A				
99A	Hitachi U.S.A.	Sintered Rare Earth	9700	6000	21.5
99B	See Hycorex 99B				
99B	Hitachi U.S.A.	Sintered Rare Earth	10,000	6000	21.5
99C	See Hycorex 99C				

Number, Grade or Alloy	Manufacturer or Country of Origin	Type of Material	B_r (kG)	H_c (kOe)	BH_{max} (MGO)
100	**May refer to Barium Ferrite 100, Ferroxdure 100, Koerox 100, Oxalit 100, Reco 100, Co 100, Fama 100 or Kobalt 100**				
100	**See Koerox 100, Oxit 100 or OX 100**				
100TK	**See Secolit 100TK**				
105H-B	See USM 105H-B (correct designation)				
110	See Nial 110 (correct designation)				
120	May refer to Alnico 120, Koerzit 120, Prac 102, Reco 120 or Nial 120.				
120	**See Koerzit 120, Prac 120 or Alnico 120**				
120Cu	See Alni 120Cu (correct designation)				
120K	See Koerzit 120K				
120P	See Alni P (correct designation)				
120R	See Koerzit 120R				
120S	See Alni S (correct designation)				
125	See Kobalt 125 or Rafcore 125				
130	May refer to Alnico 130, Koerzit 130, Oerstit 130 or Rafcore				
130K	**See Koerzit 130K**				
130S	See Alnico 130S (correct designation)				
130/46	**See Koermax 150**				
140	See Conial 140 (correct designation)				
140/120	**See Koermax 160, Vacomax 145 or Seco 140/120**				
145	See Rafcore 145				
145	**See Vacomax 145**				
145/200	**See G1-1a**				
150	See Koermax 150 or Koerox 150				
150/46	**See Koermax 170**				
150/80	**See Vacomax 225WZ**				
155	See Rafcore 155				

Number, Grade or Alloy	Manufacturer or Country of Origin	Type of Material	B_r (kG)	H_c (kOe)	BH_{max} (MGO)
160	May refer to Alnico 160, Koerzit 160, Oerstit 160, Prac 160, Reco 160, Conial 160 or Kobalt 160				
160	See Koermax 160, Koerzit 160, Oerstit 160, Prac 160 or Alnico 160				
160	See PM 160				
160P	See Alnico 160P				
160/18/55	(Mfg. unknown) Germany	Cast Alnico	12,500	670	5.1
160/20	See Vacomax 170				
160/120	See Secolit 170				
160/200	See G1-1i				
170	See Koermax 170, Secolit 170 or Vacomax 170				
170P	See Alnico 170P				
170/120	See Seco 170/120				
170/150	See Koerdym 190				
170/170	See Vacodym 400WZ				
175	See Rafcore 175				
180	See Gaussit 180				
180/100	See Vacomax 200				
185/200	See G2-1a				
190	May refer to Alnico 190, Koerzit 190, Oerstit 190 or Ticonal 190				
190	See Alnico 190, Koerdym 190 or Koerzit 190				
190/80	See Vacomax 225HR				
190/120	See Koerdym 210				
190/135	See Vacodym 370WZ				
190/Var	See G2-2a				
200	May refer to Alnico 200, Koerflex 200, Oerstit 200, Kobalt 200 or Nialco 200, PM, Rafcore 200, Vacomax 200 or Vacozet 200				
200	See Koermax 200				

Number, Grade or Alloy	Manufacturer or Country of Origin	Type of Material	B_r (kG)	H_c (kOe)	BH_{max} (MGO)
200/80	See Secolit 215N				
200/120	See Secolit 215				
200/140	See G3-3a				
200/Var	See G2-2i				
205/120	See G3-2a				
210/8	See Koerdym 230				
210/80	See Vacodym 335WZ				
210/105	See Vacodym 351WZ				
210/150	See Koerdym 240				
210/170	See Vacodym 400HR				
210/200	See G2-1i				
215	See Secolit 215				
215N	See Secolit 215N				
220	May refer to Alnico 220, Koerzit 220, Oerstit 220 or Reco 220				
220/140	See G3-3i				
225HR	See Vacomax 225HE				
225WZ	See Vacomax 225WZ				
230	See Koerdym 230				
230/80	See Neofer 230/80				
230/90	See G3-1a				
230/120	See Koerdym 260 or Neofer 230/120				
240	See Koerdym 240				
240/120	See G3-2i				
240/135	See Vacodym 370HR				
250	May refer to Alnico 250, Oerstit 250, Prac 250, or Indalloy IGH 250				
250P	See Alnico 250P				
250S	See Alnico 250S (correct designation)				

Number, Grade or Alloy	Manufacturer or Country of Origin	Type of Material	B_r (kG)	H_c (kOe)	BH_{max} (MGO)
258	See Vacozet 258				
260	May refer to Koerzit 260 or Oerstit 260				
260	See Alnico 260, Koerdym 260, Koerzit 260 or Oerstit 260				
260K	See Prac 260K				
260/80	See Koerdym 280 or Vacodym 335HR				
260/90	See G3-1i				
260/105	See Vacodym 351HR				
260u.260T	See Prac 260u.260T				
261	Crucible Steel Co., U.S.A.	Sintered Rare Earth	10,400	10,000	26.0
280	See Koerdym 280				
280K	See Ferroxdure 280K				
282	Crucible Steel Co., U.S.A.	Sintered Rare Earth	10,800	10,100	28.0
300	May refer to Barium Ferrite 300, Koerflex 300 (wire or tape), Koerzit 300, Koerox 300, Oxalit 300, Kobalt 300 or Nialco 300				
300	See Koerflex 300, Koerox 300 or OX 300				
300K	May refer to Koerox 300K or Oxalit 300K				
300KK	See Ferroxdure 300 (RAD)				
300R	May refer to Ferroxdure 300R, Koerox 300R or Oxalit 300R				
301	Crucible Steel Co., U.S.A.	Sintered Rare Earth	11,000	10,600	30.0
315	Crucible Steel Co., U.S.A.	Sintered Rare Earth	11,500	10,900	31.0
322	Crucible Steel Co., U.S.A.	Sintered Rare Earth	11,600	10,800	32.0
324	See Tromaflex 3/24p				
325	See Tromaflex 3/25p				
330	May refer to Koerox 330 or Oxalit 330				

Number, Grade or Alloy	Manufacturer or Country of Origin	Type of Material	B_r (kG)	H_c (kOe)	BH_{max} (MGO)
330	See Koerox 330 or OX 330				
330K	See Ferroxdure 330K				
330 (RAD)	See Ferroxdure 330 (RAD)				
335HR	See Vacodym 335HR				
335WZ	See Vacodym 335WZ				
340	See F 340				
350	May refer to Alnico 350, Koerzit 350 or Oerstit 350				
350	See Alnico 350 or Koerox 350				
351HR	See Vacodym 351HR				
351WZ	See Vacodym 351WZ				
355	Crucible Steel Co., U.S.A.	Sintered Rare Earth	12,300	11,300	35.0
360	May refer to Koerox 360, Oxalit 360 or Ticonal 360				
360	See Koerox 360 or Oxit 360				
360R	See Ferroxdure 360R				
362HR	See Vacodym 362HR				
362WZ	See Vacodym 362WZ				
370HR	See Vacodym 370HR				
370WZ	See Vacodym 370WZ				
380	See Oxit 380				
380C	See Oxit 380C				
383HR	See Vacodym 383HR				
383WZ	See Vacodym 383WZ				
400	May refer to Alnico 400, Koerzit 400, Oerstit 400 or Ticonal 400				
400	See Koerox 400, Oerstit 400 or Oxit 400				
400C	See Oxit 400C				
400HC	See Oxit 400HC				
400HR	See Vacodym 400HR				

Number, Grade or Alloy	Manufacturer or Country of Origin	Type of Material	B_r (kG)	H_c (kOe)	BH_{max} (MGO)
400K	May refer to Koerzit 400K or Oerstit 400K				
400K	See Koerzit 400K				
400R	See Oerstit 400R				
400S	See Alnico 400S (correct designation)				
400WZ	See Vacodym 400WZ				
420	See Koerox 420 or Oxit 420				
424	See Tromaflex 4/24p				
450	May refer to Koerzit 450, Oerstit 450 or Ticonal 450				
450	See Alnico 450, Koerzit 450 or Oerstit 450				
500	May refer to Tromalit 500, Koerzit 500, Oerstit 500 or Ticonal 500				
500	See Alnico 500 or Koerzit 500				
500G	See Oerstit 500G				
500S	See Alnico 500S or Oerstit 500S				
550	See Alnico 550				
550 UGIMAX	See Alnico 550 UGIMAX				
580	See Alnico 580				
600	May refer to Tromalit 600, Koerzit 600, Oerstit 600, Ticonal 600 or Fama 600				
600	See Alnico 550 or Alnico 600 [Magnet Appl. Inc.]				
600 Super UGIMAX	See Alnico 600 Super UGIMAX				
600 UGIMAX	See Alnico 600 Super UGIMAX				
625	See F 625				
650	See Ticonal 650				
655	See Vacozet 655				
700	May refer to Tromalit 700, Oerstit 700, Koerzit 700 or Fama 700				
700	See Alnico 700 or Koerzit 700				
730	See F 730				

Number, Grade or Alloy	Manufacturer or Country of Origin	Type of Material	B_r (kG)	H_c (kOe)	BH_{max} (MGO)
750	See Ticonal 750				
800	**May refer to Tromalit 800, Ticonal 800 or Oerstit 800**				
800	**See Alnico 800 or ArKrome 800**				
800 UGIMAX	**See Alnico 800 UGIMAX**				
800HiHc	**See ArKrome 800HiHc**				
800s	See Tromalit 800s				
830	See F 830				
900	See Ticonal 900				
900CP	See Oerstit 900CP				
928	**See Tromaflex 9/28p**				
1000	May refer to Tromalit 2000 or Oerstit 1000				
1095	See Carbon Steel No. 1095				
1120	**See Tromaflex 11/20p**				
1207	See Alni 1207				
1300	**See Alnico 1300**				
1323	See Tromaflex 13/23p				
1500 Super UGIMAX	**See Alnico 1500 Super UGIMAX**				
1500	See Ticonal 1500				
1600	See Gaussit 1600				
1800	**See Koerzit 1800**				
2000	**See Alnico 2000**				
2000	See Cormax 2000				
2000	See Ticonal 2000				
2002A	Arnold Engr. U.S.A.	Bonded Rare Earth	4900	4100	5.0
2002B	Arnold Engr. U.S.A.	Bonded Rare Earth	5100	3800	5.0
2004D	Arnold Engr. U.S.A.	Bonded Rare Earth	5000	3900	5.0

Number, Grade or Alloy	Manufacturer or Country of Origin	Type of Material	B_r (kG)	H_c (kOe)	BH_{max} (MGO)
2200	See Alnico 2200				
2236	Stackpole U.S.A.	Sintered Ferrite	2260	1940	1.1
2331	Stackpole U.S.A.	Sintered Ferrite	2360	1930	1.1
2532 (Sleeves)	Stackpole U.S.A.	Sintered Ferrite	2575	2130	1.35
2732 (Segments)	Stackpole U.S.A.	Sintered Ferrite	2850	2400	1.8
3500-37 percent Co	Simmonds Steel U.S.A.	Wrought Alloy	10,400	230	.982
3540	Stackpole U.S.A.	Sintered Ferrite	3620	3400	3.1
3547	Stackpole U.S.A.	Sintered Ferrite	3620	3400	3.1
3831	Stackpole U.S.A.	Sintered Ferrite	3920	3175	3.6
3838	Stackpole U.S.A.	Sintered Ferrite	3920	3520	3.6
4130	Stackpole U.S.A.	Sintered Ferrite	4150	3000	4.0
7000	See Xolox 7000				
7480	See Xolox 7480				
7540	See Xolox 7540				
7680	See Magnalox 7680				
I	May refer to Cast Alnico I, Sintered Alnico I, Ferrite I, Caslox I, Cunife I, Cunico I, Nialco I or Alcomax I				
IA	May refer to Alnico IA or Nialco IA				
IB	See Alnico IB				
IC	See Alnico IC				
II	May refer to Cast Alnico II, Sintered Alnico II, Ferrite II, Caslox II, Cunife II, Cunico II, Nialco II or Alcomax II				
IIA	See Alnico IIA				

Number, Grade or Alloy	Manufacturer or Country of Origin	Type of Material	B_r (kG)	H_c (kOe)	BH_{max} (MGO)
IIB	See Alnico IIB				
IIC	See Alnico IIC				
IIH	See Alnico IIH				
II S.C.	See Alcomax II S.C.				
III	May refer to Cast Alnico III, Sintered Alnico III, Alcomax III, Ferrite III or Caslox III				
III	**See Archrome III**				
IIIA	May refer to Alnico IIIA or Nialco IIIA				
IIIB	See Alnico IIIB				
IIIC	See Alnico IIIC				
III-NF1	See Alnico III-NF1				
III S.C.	See Alcomax III S.C.				
IV	May refer to Alnico IV, Alcomax IV or Ferrite IV				
IVA	May refer to Alnico IVA or Nialco IVA				
IVB	See Alnico IVB				
IVH	See Alnico IVH (correct designation)				
IV S.C.	See Alcomax IV S.C.				
V	May refer to Alnico V, Alxomax V or Ferrite V				
VA	See Alnico VA				
VAB	See Alnico VAB				
VB	See Alnico VB				
VC	See Alnico VC				
VCC	See Alnico VCC				
VE	See Alnico VE				
V-NC1	See Alnico V-NC1				
V-NR1	See Alnico V-NR1				
V-NW1	See Alnico V-NW1				
VDG	See Alnico VDG				

Number, Grade or Alloy	Manufacturer or Country of Origin	Type of Material	B_r (kG)	H_c (kOe)	BH_{max} (MGO)
VAB DG	See Alnico VAB DG				
VB DG	See Alnico VB DG				
VC DG	See Alnico VC DG				
V-PR1	See Alnico V-PR1				
V-OW1	See Alnico V-OW1				
V-7	See Alnico V-7				
V-OR1	See Alnico V-OR1				
VI	May refer to Cast Alnico VI, Sintered Alnico VI, Caslox VI or Ferrite VI				
VI-AB	See Alnico VI-AB (correct designation)				
VIB	See Alnico VIB				
VIC	See Alnico VIC				
VI-NS1	See Alnico VI-NS1				
VI-NS2	See Alnico VI-NS2				
VII	May refer to Alnico 7, Ferrite 7 or Caslox VII				
VIIS	See Alnico VIIS				
VII-C_i	See Alnico 7-C_i (correct designation)				
VIII	May refer to Cast Alnico 8 or Sintered Alnico 8				
VIII$_A$	See Alnico 8$_A$				
VIII$_C$	See Alnico 8$_C$ (correct designation)				
VIIIA	See Alnico 8A (correct designation)				
VIIIB	See Alnico 8B (correct designation)				
VIII (Improved)	See Alnico VIII Improved				
VIII-ND1	See Alnico VIII-ND1				
IX	See Alnico 9 (correct designation)				
X	See Alcomax X				
XII	See Cast Alnico 12 (correct designation)				
XII-NM1	See Alnico XII-NM1				

Appendix 4

Conversion Factors

CGS to SI Units Conversions

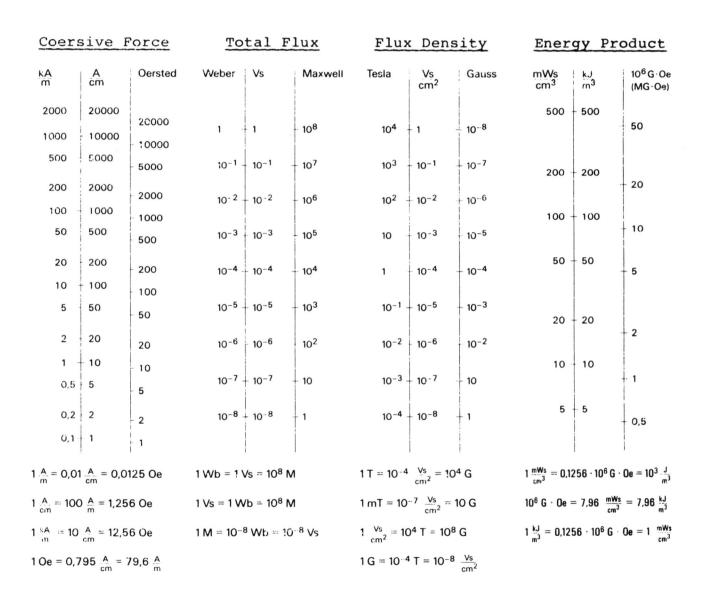

Magnetic Units: Conversion Factors

Magnetic Parameter	Unit in cgs System	Multiply cgs Unit by Factor Shown to give...		Unit in mks System	Multiply mks Unit by Factor Shown to give...		Unit in English System	Multiply English Unit by Factor Shown to give...	
		mks Units	English Units		cgs Units	English Units		cgs Units	mks Units
Total flux ϕ	Maxwells (1 Maxwell = 1 line of force)	0.1	1.0	Webers (1 Weber = 10^8 lines)	10	10	Maxwells (1 Maxwell = 1 line of force)	0.1	1.0
Flux density B	Gauss	10^{-4}	6.45	Webers/m^2	10^4	6.45×10^4	lines/in.^{2}	15.5×10^{-2}	15.5×10^{-6}
Magnetizing force H	Oersteds or Gilberts/centimeters	79.6	2.02	Ampere-turns/m	125.7	2.54×10^2	Ampere-turns/in.	.4940	39.4
Magnetomotive force F	Gilberts	.796	.796	Ampere-turns	1.257	1.257	Ampere-turns	1.257	1.257
Mechanical force F_p	dynes	1.02×10^{-6}	2.25×10^{-6}	Kilograms	$.98 \times 10^6$	2.21	Pounds	$.445 \times 10^6$	.454
Area A	Square centimeter	10^{-4}	1550×10^{-4}	Square meters	10^4	1550	Square Inch	6.45	6.45×10^{-4}
Length L	centimeters	.01	.3937	meters	100	39.37	inches	2.54	2.54×10^{-2}
Permeability of a vacuum or of air μ_0	1.0	—	—	$.4\pi \times 10^{-6}$	—	—	3.192	—	—
Work or energy W_0	dyne-cm (ergs)	1.0197×10^{-8}	7.736×10^{-8}	kg/m	9.820×10^7	7.233	ft/lbs	1.355×10^7	.1383

English Metric-Length Conversions

Measurement		Conversions						
in.	=	millimeters						
mm	=	----------	inches					
ft	=	----------	---------	meters				
m	=	----------	---------	---------	feet	---------	yards	
yd	=	----------	---------	---------	---------	meters		
miles	=	----------	---------	---------	---------	---------	---------	kilometers
km	=	----------	---------	---------	---------	---------	---------	---------	miles
1		25.400 1	0.039 370	0.304 801	3.280 83	0.914 402	1.093 61	1.609 35	0.621 37
2		50.800 1	0.078 740	0.609 601	6.561 67	1.828 80	2.187 22	3.218 69	1.242 74
3		76.200 2	0.118 110	0.914 402	9.842 50	2.743 21	3.280 83	4.828 04	1.864 11
4		101.600	0.157 480	1.219 20	13.123 3	3.657 61	4.374 44	6.437 39	2.485 48
5		127.000	0.196 850	1.524 00	16.404 2	4.572 01	5.468 06	8.046 74	3.106 85
6		152.400	0.236 220	1.828 80	19.685 0	5.486 41	6.561 67	9.656 08	3.728 22
7		177.800	0.275 590	2.133 60	22.965 8	6.400 81	7.655 28	11.265 4	4.349 59
8		203.200	0.314 960	2.438 40	26.246 7	7.315 21	8.748 89	12.874 8	4.970 96
9		228.600	0.354 330	2.743 21	29.527 5	8.229 62	9.842 50	14.484 1	5.592 33
10		254.001	0.393 700	3.048 01	32.808 3	9.144 02	10.936 1	16.093 5	6.213 70
11		279.401	0.433 070	3.352 81	36.089 2	10.058 4	12.029 7	17.702 8	6.835 07
12		304.801	0.472 440	3.657 61	39.370 0	10.972 8	13.123 3	19.312 2	7.456 44
13		330.201	0.511 810	3.962 41	42.650 8	11.887 2	14.216 9	20.921 5	8.077 81
14		355.601	0.551 180	4.267 21	45.931 7	12.801 6	15.310 6	22.530 9	8.699 18
15		381.001	0.590 550	4.572 01	49.212 5	13.716 0	16.404 2	24.140 2	9.320 55
16		406.401	0.629 920	4.876 81	52.943 3	14.630 4	17.497 8	25.749 6	9.941 92
17		431.801	0.669 290	5.181 61	55.774 2	15.544 8	18.591 4	27.358 9	10.563 3
18		457.201	0.708 660	5.486 41	59.005 0	16.459 2	19.685 0	28.968 2	11.184 7
19		482.601	0.748 030	5.791 21	62.335 8	17.373 6	20.778 6	30.577 6	11.806 0
20		508.001	0.787 400	6.096 01	65.616 7	18.288 0	21.872 2	32.186 9	12.427 4
21		533.401	0.826 770	6.400 81	68.896 5	19.202 4	22.965 8	33.196 3	13.048 8
22		558.801	0.866 140	6.705 61	72.178 3	20.116 8	24.059 4	35.405 6	13.670 1
23		584.201	0.905 510	7.010 41	75.459 2	21.031 2	25.153 1	37.015 0	14.291 5
24		609.601	0.944 880	7.315 21	78.740 0	21.945 6	26.246 7	38.624 3	14.912 9
25		635.001	0.984 250	7.620 02	82.020 8	22.860 0	27.340 3	40.233 7	15.534 2
26		660.401	1.023 62	7.924 82	85.301 7	23.774 4	28.433 9	41.843 0	16.155 6
27		685.801	1.062 99	8.229 62	88.582 5	24.688 9	29.527 5	43.452 4	16.777 0
28		711.201	1.102 36	8.534 42	91.863 3	25.603 3	30.621 1	45.061 7	17.398 4
29		736.601	1.141 73	8.939 22	95.144 2	26.517 7	31.714 7	46.671 1	18.019 7
30		762.002	1.181 10	9.144 02	98.425 0	27.432 1	32.808 3	48.280 4	18.641 1
31		787.402	1.220 47	9.448 82	101.706	28.346 5	33.901 9	49.889 8	19.262 5
32		812.802	1.259 84	9.753 62	104.987	29.260 9	34.995 6	51.499 1	19.883 8
33		838.202	1.299 21	10.058 4	108.268	30.175 3	36.089 2	53.108 5	20.505 2
34		863.602	1.338 58	10.363 2	111.548	31.089 7	37.182 8	54.717 8	21.126 6
35		889.002	1.377 95	10.668 0	114.829	32.004 1	38.276 4	56.327 2	21.747 9
36		914.402	1.417 32	10.972 8	118.110	32.918 5	39.370 0	57.936 5	22.369 3
37		939.802	1.456 69	11.277 6	121.391	33.832 9	40.463 6	59.545 8	22.990 7
38		965.202	1.496 06	11.582 4	124.672	34.747 3	41.557 2	61.155 2	23.612 1
39		990.602	1.535 43	11.887 2	127.953	35.661 7	42.650 8	62.764 5	24.233 4
40		1 016.00	1.574 80	12.192 0	131.233	36.576 1	43.744 4	64.373 9	24.854 8
41		1 041.40	1.614 17	12.496 8	134.514	37.490 5	44.838 1	65.983 2	25.476 2
42		1 066.80	1.653 54	12.801 6	137.795	38.404 9	45.931 7	67.592 6	26.097 5
43		1 092.20	1.692 91	13.106 4	141.076	39.319 3	47.025 3	69.201 9	26.718 9
44		1 117.60	1.732 28	13.411 2	144.357	40.233 7	48.118 9	70.811 3	27.340 3
45		1 143.00	1.771 65	12.716 0	147.638	41.148 1	49.212 5	72.420 6	27.961 6
46		1 168.40	1.811 02	14.020 8	150.918	42.062 5	50.306 1	74.030 0	28.583 0
47		1 193.80	1.850 39	14.325 6	154.199	42.976 9	51.399 7	75.639 3	29.204 4
48		1 219.20	1.889 76	14.630 4	157.480	43.891 3	52.493 3	77.248 7	29.825 8
49		1 244.60	1.929 13	14.935 2	160.761	44.805 7	53.586 9	78.858 0	30.447 1
50		1 270.00	1.968 50	15.240 0	164.042	45.720 1	54.680 6	80.467 4	31.068 5

English Metric-Length Conversions

Measurement				Conversions					
in. =	millimeters								
mm =	----------	inches							
ft =	----------	--------	meters						
m =	----------	--------	--------	feet	--------	yards			
yd =	----------	--------	--------	--------	meters				
miles =	----------	--------	--------	--------	--------	--------	kilometers		
km =	----------	--------	--------	--------	--------	--------	--------	miles	
51	1 295.40	2.007 87	15.544 8	167.323	46.634 5	55.774 2	82.076 7	31.689 9
52	1 320.80	2.047 24	15.849 6	170.603	47.548 9	56.867 8	83.686 1	32.311 2
53	1 346.20	2.086 61	16.154 4	173.884	48.463 3	57.961 4	85.295 4	32.932 6
54	1 371.60	2.125 98	16.549 2	177.165	49.377 7	59.055 0	86.904 7	33.554 0
55	1 397.00	2.165 35	16.764 0	180.446	50.292 1	60.148 6	88.514 1	34.175 3
56	1 422.40	2.204 72	17.068 8	183.727	51.206 5	61.242 2	90.123 4	34.796 7
57	1 447.80	2.244 09	17.373 6	187.008	52.120 9	62.335 8	91.732 8	35.418 1
58	1 473.20	2.283 46	17.678 4	190.288	53.035 3	63.429 4	93.342 1	36.039 5
59	1 498.60	2.322 83	17.983 2	193.569	53.949 7	64.523 1	94.951 5	36.660 8
60	1 524.00	2.362 20	18.288 0	196.850	54.864 1	65.616 7	96.560 8	37.282 2
61	1 549.40	2.401 57	18.592 8	200.131	55.778 5	66.710 3	98.170 2	37.903 6
62	1 574.80	2.440 94	18.897 6	203.412	56.692 9	67.803 9	99.779 5	38.524 9
63	1 600.20	2.480 31	19.202 4	206.693	57.607 3	68.897 5	101.389	39.146 3
64	1 625.60	2.519 68	19.507 2	209.973	58.521 7	69.991 1	102.998	39.767 7
65	1 651.00	2.559 05	19.812 0	213.254	59.436 1	71.084 7	104.608	40.389 0
66	1 676.40	2.598 42	20.116 8	216.535	60.350 5	72.178 3	106.217	41.010 4
67	1 701.80	2.637 79	20.421 6	219.816	61.264 9	73.271 9	107.826	41.631 8
68	1 727.20	2.677 16	20.726 4	223.097	62.179 3	74.365 6	109.436	42.253 2
69	1 752.60	2.716 53	21.031 2	226.378	63.093 7	75.459 2	111.045	42.874 5
70	1 778.00	2.755 90	21.336 0	229.658	64.008 1	76.552 8	112.654	43.495 9
71	1 803.40	2.795 27	21.640 8	232.939	64.922 5	77.646 4	114.264	44.117 3
72	1 828.80	2.834 64	21.945 6	236.220	65.836 9	78.740 0	115.873	44.738 6
73	1 854.20	2.874 01	22.250 4	239.501	66.751 3	79.833 6	117.482	45.360 0
74	1 879.60	2.913 38	22.555 2	242.782	67.665 7	80.927 2	119.092	45.981 4
75	1 905.00	2.952 75	22.860 0	246.063	68.580 1	82.020 8	120.701	46.602 7
76	1 930.40	2.992 12	23.164 8	249.343	69.494 5	83.114 4	122.310	47.224 1
77	1 955.80	3.031 49	23.469 6	252.624	70.408 9	84.208 1	123.920	47.845 5
78	1 981.20	3.070 86	23.774 4	255.905	71.323 3	85.301 7	125.529	48.466 9
79	2 006.60	3.110 23	24.079 2	259.186	72.237 7	86.395 3	127.138	49.088 2
80	2 032.00	3.149 60	24.384 0	262.467	73.152 1	87.488 9	128.748	49.709 6
81	2 057.40	3.188 97	24.688 8	265.748	74.066 5	88.582 5	130.357	50.331 0
82	2 082.80	3.228 34	24.993 6	269.028	74.981 0	89.676 1	131.966	50.952 3
83	2 108.20	3.267 71	25.298 4	272.309	75.895 4	90.769 7	133.576	51.573 7
84	2 133.60	3.307 08	25.603 3	275.590	76.809 8	91.863 3	135.185	52.195 1
85	2 159.00	3.346 45	25.908 1	278.871	77.724 2	92.956 9	136.795	52.816 4
86	2 184.40	3.385 82	26.212 9	282.152	78.638 6	94.050 6	138.404	53.437 8
87	2 209.80	3.425 19	26.517 7	285.433	79.553 0	95.144 2	140.013	54.059 2
88	2 235.20	3.464 56	26.822 5	288.713	80.467 4	96.237 8	141.623	54.680 6
89	2 260.60	3.503 93	27.127 3	291.994	81.381 8	97.331 4	143.232	55.301 9
90	2 286.00	3.543 30	27.432 1	295.275	82.296 2	98.425 0	144.841	55.923 3
91	2 311.40	3.582 67	27.736 9	298.556	83.210 6	99.518 6	146.451	56.544 7
92	2 336.80	3.622 04	28.041 7	301.837	84.125 0	100.612	148.060	57.160 0
93	2 362.20	3.661 41	28.346 5	305.118	85.039 4	101.706	149.669	57.787 4
94	2 387.60	3.700 78	28.651 3	308.398	85.953 8	102.799	151.279	58.408 8
95	2 413.00	3.740 15	28.956 1	311.679	86.868 2	103.893	152.888	59.030 1
96	2 438.40	3.779 52	29.260 9	314.960	87.782 6	104.987	154.497	59.651 5
97	2 463.80	3.818 89	29.565 7	318.241	88.697 0	106.080	156.107	60.272 9
98	2 489.20	3.858 26	29.870 5	321.522	89.611 4	107.174	157.716	60.894 3
99	2 514.60	3.897 63	30.175 3	324.803	90.525 8	108.268	159.325	61.515 6
100	2 540.01	3.937 00	30.480 1	328.083	91.440 2	109.361	160.935	62.137 0

Weight Conversions—Look up reading in center column: If it is in grams, read ounces in right column; if it is in ounces, read grams in left column.

Grams		Ounces	Grams		Ounces	Grams		Ounces
28.35	1	.0353	1304.10	46	1.6238	2579.85	91	3.2123
56.70	2	.0706	1332.45	47	1.6501	2608.20	92	3.2476
85.05	3	.1059	1360.80	48	1.6944	2636.55	93	3.2829
113.40	4	.1412	1389.15	49	1.7197	2664.90	94	3.3182
141.75	5	.1765	1417.50	50	1.7650	2693.25	95	3.3535
170.10	6	.2118	1445.85	51	1.8003	2721.60	96	3.3888
198.45	7	.2471	1474.10	52	1.8356	2749.95	97	3.4241
226.80	8	.2824	1502.55	53	1.8709	2778.30	98	3.4594
255.15	9	.3177	1530.90	54	1.9062	2806.65	99	3.4947
283.50	10	.3530	1559.25	55	1.9415	2835.00	100	3.5300
311.85	11	.3883	1587.60	56	1.9768	2863.35	101	3.5653
340.20	12	.4236	1615.95	57	2.0121	2891.70	102	3.6006
368.55	13	.4589	1644.30	58	2.0474	2920.05	103	3.6357
396.90	14	.4942	1672.65	59	2.0827	2948.40	104	3.6712
425.25	15	.5295	1701.00	60	2.1180	2976.75	105	3.7065
453.60	16	.5648	1729.35	61	2.1533	3005.10	106	3.7418
481.95	17	.6001	1757.70	62	2.1886	3033.45	107	3.7771
510.30	18	.6354	1786.05	63	2.2239	3061.80	108	3.8124
538.65	19	.6707	1814.40	64	2.2592	3090.15	109	3.8477
567.00	20	.7060	1842.75	65	2.2945	3118.50	110	3.8830
595.35	21	.7413	1871.10	66	2.3298	3146.85	111	3.9183
623.70	22	.7766	1899.45	67	2.3651	3175.20	112	3.9536
652.05	23	.8119	1927.80	68	2.4004	3203.55	113	3.9889
680.40	24	.8472	1956.15	69	2.4357	3231.90	114	4.0242
708.75	25	.8825	1984.50	70	2.4710	3260.25	115	4.0595
737.10	26	.9178	2012.85	71	2.5063	3288.60	116	4.0948
765.45	27	.9531	2041.20	72	2.5416	3316.95	117	4.1301
793.80	28	.9884	2069.55	73	2.5769	3345.30	118	4.1654
822.15	29	1.0237	2097.90	74	2.6122	3373.65	119	4.2007
850.50	30	1.0590	2126.25	75	2.6475	3402.00	120	4.2360
878.85	31	1.0943	2154.60	76	2.6828	3430.35	121	4.2713
907.20	32	1.1296	2182.95	77	2.7181	3458.70	122	4.3066
935.55	33	1.1649	2211.30	78	2.7534	3487.05	123	4.3419
963.90	34	1.2002	2239.65	79	2.7887	3515.40	124	4.3772
992.25	35	1.2355	2268.00	80	2.8240	3543.75	125	4.4125
1020.60	36	1.2708	2296.35	81	2.8593	3572.10	126	4.4478
1048.45	37	1.3061	2324.70	82	2.8946	3600.45	127	4.4831
1077.30	38	1.3414	2353.05	83	2.9249	3638.80	128	4.5184
1105.65	39	1.3767	2381.40	84	2.9652	3652.15	129	4.5537
1134.00	40	1.4120	2409.75	85	3.0005	3685.50	130	4.5890
1162.35	41	1.4473	2438.10	86	3.0358	3713.85	131	4.6243
1190.70	42	1.4826	2466.45	87	3.0711	3742.20	132	4.6596
1219.05	43	1.5179	2494.80	88	3.1064	3770.55	133	4.6949
1247.40	44	1.5532	2523.15	89	3.1417	3798.90	134	4.7302
1275.75	45	1.5885	2551.50	90	3.1770	3822.25	135	4.7655

Temperature Conversions: +1000 to +3000. Look up reading in center column: If it is in °C, read °F in right column; if it is in °F, read °C in left column.

°C		°F	°C		°F	°C		°F	°C		°F
538	1000	1832	816	1500	2732	1093	2000	3632	1371	2500	4532
543	1010	1850	821	1510	2750	1099	2010	3650	1377	2510	4550
549	1020	1868	827	1520	2768	1104	2020	3668	1382	2520	4568
554	1030	1886	832	1530	2786	1110	2030	3686	1388	2530	4586
560	1040	1904	838	1540	2804	1116	2040	3704	1393	2540	4604
566	1050	1922	843	1550	2822	1121	2050	3722	1399	2550	4622
571	1060	1940	849	1560	2840	1127	2060	3740	1404	2560	4640
577	1070	1958	854	1570	2858	1132	2070	3758	1410	2570	4658
582	1080	1976	860	1580	2876	1138	2080	3776	1416	2580	4676
588	1090	1994	866	1590	2894	1143	2090	3794	1421	2590	4694
593	1100	2012	871	1600	2912	1149	2100	3812	1427	2600	4712
599	1110	2030	877	1610	2930	1154	2110	3830	1432	2610	4730
604	1120	2048	882	1620	2948	1160	2120	3848	1438	2620	4748
610	1130	2066	888	1630	2966	1166	2130	3866	1443	2630	4766
616	1140	2084	893	1640	2984	1171	2140	3884	1449	2640	4784
621	1150	2102	899	1650	3002	1177	2150	3902	1454	2650	4802
627	1160	2120	904	1660	3020	1182	2160	3920	1460	2660	4820
632	1170	2138	910	1670	3038	1188	2170	3938	1466	2670	4838
638	1180	2156	916	1680	3056	1193	2180	3956	1471	2680	4856
643	1190	2174	921	1690	3074	1199	2190	3974	1477	2690	4874
649	1200	2192	927	1700	3092	1204	2200	3992	1482	2700	4892
654	1210	2210	932	1710	3110	1210	2210	4010	1488	2710	4910
660	1220	2228	938	1720	3128	1216	2220	4028	1493	2720	4928
666	1230	2246	943	1730	3146	1221	2230	4046	1499	2730	4946
671	1240	2264	949	1740	3164	1227	2240	4064	1504	2740	4964
677	1250	2282	954	1750	3182	1232	2250	4082	1510	2750	4982
682	1260	2300	960	1760	3200	1238	2260	4100	1516	2760	5000
688	1270	2318	966	1770	3218	1243	2270	4118	1521	2770	5018
693	1280	2336	971	1780	3236	1249	2280	4136	1527	2780	5036
699	1290	2354	977	1790	3254	1254	2290	4154	1532	2790	5054
704	1300	2372	982	1800	3272	1260	2300	4172	1538	2800	5072
710	1310	2390	988	1810	3290	1266	2310	4190	1543	2810	5090
716	1320	2408	993	1820	3308	1271	2320	4208	1549	2820	5108
721	1330	2426	999	1830	3326	1277	2330	4226	1554	2830	5126
727	1340	2444	1004	1840	3344	1282	2340	4244	1560	2840	5144
732	1350	2462	1010	1850	3362	1288	2350	4262	1566	2850	5162
738	1360	2480	1016	1860	3380	1293	2360	4280	1571	2860	5180
743	1370	2498	1021	1870	3398	1299	2370	4298	1577	2870	5198
749	1380	2516	1027	1880	3416	1304	2380	4316	1582	2880	5216
754	1390	2534	1032	1890	3434	1310	2390	4334	1588	2890	5234
760	1400	2552	1038	1900	3452	1316	2400	4352	1593	2900	5252
766	1410	2570	1043	1910	3470	1321	2410	4370	1599	2910	5270
771	1420	2588	1049	1920	3488	1327	2420	4388	1604	2920	5288
777	1430	2606	1054	1930	3506	1332	2430	4406	1610	2930	5306
782	1440	2624	1060	1940	3524	1338	2440	4424	1616	2940	5324
788	1450	2642	1066	1950	3542	1343	2450	4442	1621	2950	5342
793	1460	2660	1071	1960	3560	1349	2460	4460	1627	2960	5360
799	1470	2678	1077	1970	3578	1354	2470	4478	1632	2970	5378
804	1480	2696	1082	1980	3596	1360	2480	4496	1638	2980	5396
810	1490	2714	1088	1990	3614	1366	2490	4514	1643	2990	5414
			1093	2000	3632				1649	3000	5432

Temperature Conversions: −459.4 to +1000. Look up reading in center column: If it is in °C, read °F in right column; if it is in °F, read °C in left column.

°C	°F	°C		°F	°C		°F	°C		°F	°C		°F
−273	−459.4	−17.8	0	32	10.0	50	122.0	38	100	212	260	500	932
−268	−450	−17.2	1	33.8	10.6	51	123.8	43	110	230	266	510	950
−262	−440	−16.7	2	35.6	11.1	52	125.6	49	120	248	271	520	968
−257	−430	−16.1	3	37.4	11.7	53	127.4	54	130	266	277	530	986
−251	−420	−15.6	4	39.2	12.2	54	129.2	60	140	284	282	540	1004
−246	−410	−15.0	5	41.0	12.8	55	131.0	66	150	302	288	550	1022
−240	−400	−14.4	6	42.8	13.3	56	132.8	71	160	320	293	560	1040
−234	−390	−13.9	7	44.6	13.9	57	134.6	77	170	338	299	570	1058
−229	−380	−13.3	8	46.4	14.4	58	136.4	82	180	356	304	580	1076
−223	−370	−12.8	9	48.2	15.0	59	138.2	88	190	374	310	590	1094
−218	−360	−12.2	10	50.0	15.6	60	140.0	93	200	392	316	600	1112
−212	−350	−11.7	11	51.8	16.1	61	141.8	99	210	410	321	610	1130
−207	−340	−11.1	12	53.6	16.7	62	143.6	100	212	413.6	327	620	1148
−201	−330	−10.6	13	55.4	17.2	63	145.4	104	220	428	332	630	1166
−196	−320	−10.0	14	57.2	17.8	64	147.2	110	230	446	338	640	1184
−190	−310	− 9.4	15	59.0	18.3	65	149.0	116	240	464	343	650	1202
−184	−300	− 8.9	16	60.8	18.9	66	150.8	121	250	482	349	660	1220
−179	−290	− 8.3	17	62.6	19.4	67	152.6	127	260	500	354	670	1238
−173	−280	− 7.8	18	64.4	20.0	68	154.4	132	270	518	360	680	1256
−169	−273	−459.4	− 7.2	19	66.2	20.6	69	156.2	138	280	536	366	690	1274
−168	−270	−454	− 6.7	20	68.0	21.1	70	158.0	143	290	554	371	700	1292
−162	−260	−436	− 6.1	21	69.8	21.7	71	159.8	149	300	572	377	710	1310
−157	−250	−418	− 5.6	22	71.6	22.2	72	161.6	154	310	590	382	720	1328
−151	−240	−400	− 5.0	23	73.4	22.8	73	163.4	160	320	608	388	730	1346
−146	−230	−382	− 4.4	24	75.2	23.3	74	165.2	166	330	626	393	740	1364
−140	−220	−364	− 3.9	25	77.0	23.9	75	167.0	171	340	644	399	750	1382
−134	−210	−346	− 3.3	26	78.8	24.4	76	168.8	177	350	662	404	760	1400
−129	−200	−328	− 2.8	27	80.6	25.0	77	170.6	182	360	680	410	770	1418
−123	−190	−310	− 2.2	28	82.4	25.6	78	172.4	188	370	698	416	780	1436
−118	−180	−292	− 1.7	29	84.2	26.1	79	174.2	193	380	716	421	790	1454
−112	−170	−274	− 1.1	30	86.0	26.7	80	176.0	199	390	734	427	800	1472
−107	−160	−256	− 0.6	31	87.8	27.2	81	177.8	204	400	752	432	810	1490
−101	−150	−238	0.0	32	89.6	27.8	82	179.6	210	410	770	438	820	1508
− 96	−140	−220	0.6	33	91.4	28.3	83	181.4	216	420	788	443	830	1526
− 90	−130	−202	1.1	34	93.2	28.9	84	183.2	221	430	806	449	840	1544
− 84	−120	−184	1.7	35	95.0	29.4	85	185.0	227	440	824	454	850	1562
− 79	−110	−166	2.2	36	96.8	30.0	86	186.8	232	450	842	460	860	1580
− 73	−100	−148	2.8	37	98.6	30.6	87	188.6	238	460	860	466	870	1598
− 68	− 90	−130	3.3	38	100.4	31.1	88	190.4	243	470	878	471	880	1616
− 62	− 80	−112	3.9	39	102.2	31.7	89	192.2	249	480	896	477	890	1634
− 57	− 70	− 94	4.4	40	104.0	32.2	90	194.0	254	490	914	482	900	1652
− 51	− 60	− 76	5.0	41	105.8	32.8	91	195.8				488	910	1670
− 46	− 50	− 58	5.6	42	107.6	33.3	92	197.6				493	920	1688
− 40	− 40	− 40	6.1	43	109.4	33.9	93	199.4				499	930	1706
− 34	− 30	− 22	6.7	44	111.2	34.4	94	201.2				504	940	1724
− 29	− 20	− 4	7.2	45	113.0	35.0	95	203.0				510	950	1742
− 23	− 10	14	7.8	46	114.8	35.6	96	204.8				516	960	1760
− 17.8	0	32	8.3	47	116.6	36.1	97	206.6				521	970	1778
			8.9	48	118.4	36.7	98	208.4				527	980	1796
			9.4	49	120.2	37.2	99	210.2				532	990	1814
						37.8	100	212.0				538	1000	1832

Fractions of an Inch: Decimal Equivalents

		1/64	.015625			33/64	.51562
	1/32		.03125		17/32		.53125
		3/64	.046875			35/64	.54687
1/16			.0625	9/16			.5625
		5/64	.078125			37/64	.578125
	3/32		.09375		19/32		.59375
		7/64	.109375			39/64	.609375
1/8			.125	5/8			.625
		9/64	.140625			41/64	.640625
	5/32		.15625		21/32		.65625
		11/64	.171875			43/64	.671875
3/16			.1875	11/16			.6875
		13/64	.203125			45/64	.703125
	7/32		.21875		23/32		.71875
		15/64	.234375			47/64	.734375
1/4			.250	3/4			.750
		17/64	.265625			49/64	.765625
	9/32		.28125		25/32		.78125
		19/64	.296875			51/64	.796875
5/16			.3125	13/16			.8125
		21/64	.328125			53/64	.828125
	11/32		.34375		27/32		.84375
		23/64	.359375			55/64	.859375
3/8			.375	7/8			.875
		25/64	.390625			57/64	.890625
	13/32		.40625		29/32		.90625
		27/64	.421875			59/64	.921875
7/16			.4375	15/16			.9375
		29/64	.453125			61/64	.953125
	15/32		.46875		31/32		.96875
		31/64	.484375			63/64	.984375
1/2			.500	1			1.0000

Standard Gauges (in.)

Number of Wire Gauge	American or Brown & Sharpe	Birmingham or Stubs' Iron Wire	U.S. Standard Gauge for Sheet and Plate Iron and Steel (revised)	Washburn & Moen, or Steel Wire Gauge	Music Wire Gauge	Stubs' Steel Wire
00000000	—	—	—	—	—	—
0000000	—	—	.50	.4900	—	—
000000	.580	—	.46875	.4615	.004	—
00000	.5165	—	.4375	.4305	.005	—
0000	.460	.454	.40625	.3938	.006	—
000	.409642	.425	.375	.3625	.007	—
00	.364796	.380	.34375	.3310	.008	—
0	.324861	.340	.3125	.3065	.009	—
1	.289297	.300	.28125	.2830	.010	.227
2	.257627	.284	.26563	.2625	.011	.219
3	.229423	.259	.250	.2437	.012	.212
4	.204307	.238	.23438	.2253	.013	.207
5	.18194	.220	.21875	.2070	.014	.204
6	.162023	.203	.20313	.1920	.016	.201
7	.144285	.180	.1875	.1770	.018	.199
8	.12849	.165	.17188	.1620	.020	.197
9	.114423	.148	.15625	.1483	.022	.194
10	.101897	.134	.14063	.1350	.024	.191
11	.090742	.120	.125	.1205	.026	.188
12	.080808	.109	.10938	.1055	.029	.185
13	.071962	.095	.09375	.0915	.031	.182
14	.064084	.083	.07813	.0800	.033	.180
15	.057068	.072	.07031	.0720	.035	.178
16	.050821	.065	.0625	.0625	.037	.175
17	.045257	.058	.05625	.0540	.039	.172
18	.040303	.049	.050	.0475	.041	.168
19	.03589	.042	.04375	.0410	.043	.164
20	.031961	.035	.0375	.0348	.045	.161
21	.028462	.032	.03438	.03175	.047	.157
22	.025346	.028	.03125	.0286	.049	.155
23	.022572	.025	.02813	.0258	.051	.153
24	.020101	.022	.025	.0230	.055	.151
25	.0179	.020	.02188	.0204	.059	.148
26	.015941	.018	.01875	.0181	.063	.146
27	.014195	.016	.01719	.0173	.067	.143
28	.012641	.014	.01563	.0162	.071	.139
29	.011257	.013	.01406	.0150	.075	.134
30	.010025	.012	.0125	.0140	.080	.127
31	.008928	.010	.01094	.0132	.085	.120
32	.00795	.009	.01016	.0128	.090	.115
33	.00708	.008	.00938	.0118	.095	.112
34	.006305	.007	.00859	.0104	—	.110
35	.005615	.005	.00781	.0095	—	.108
36	.005	.004	.00703	.0090	—	.106
37	.004453	—	.00664	.0085	—	.103
38	.003965	—	.00625	.0080	—	.101
39	.003531	—	—	.0075	—	.099
40	.003144	—	—	.0070	—	.097

Hardness Conversions: Rockwell to Brinell

Rockwell C Scale Hardness No.	Brinell Hardness No. (10-mm tungsten-carbide ball; 3000-kg load)	Rockwell Hardness No. B Scale, 100-kg Load (1/16 in. diameter ball)	Rockwell Hardness No. A Scale, 60-kg Load (Brale Penetrator)	Diamond Pyramid Hardness No. (Vickers)	Shore Sceleroscope Hardness No.	Tensile Strength (approximate) in 1000 psi
68	—	—	85.6	940	97	—
67	—	—	85.0	900	95	—
66	—	—	84.5	865	92	—
65	739	—	83.9	832	91	—
64	722	—	83.4	800	88	—
63	705	—	82.8	772	87	—
62	688	—	82.3	746	85	—
61	670	—	81.8	720	83	—
60	654	—	81.2	697	81	—
59	634	—	80.7	674	80	326
58	615	—	80.1	653	78	315
57	595	—	79.6	633	76	305
56	577	—	79.0	613	75	295
55	560	—	78.5	595	74	287
54	543	—	78.0	577	72	278
53	525	—	77.4	560	71	269
52	512	—	76.8	544	69	262
51	496	—	76.3	528	68	253
50	481	—	75.9	513	67	245
49	469	—	75.2	498	66	239
48	455	—	74.7	484	64	232
47	443	—	74.1	471	63	225
46	432	—	73.6	458	62	219
45	421	—	73.1	446	60	212
44	409	—	72.5	434	58	206
43	400	—	72.0	423	57	201
42	390	—	71.5	412	56	196
41	381	—	70.9	402	55	191
40	371	—	70.4	392	54	186
39	362	—	69.9	382	52	181
38	353	—	69.4	372	51	176
37	344	—	68.9	363	50	172
36	336	(109.0)	68.4	354	49	168
35	327	(108.5)	67.9	345	48	163
34	319	(108.0)	67.4	336	47	159
33	311	(107.5)	66.8	327	46	154
32	301	(107.0)	66.3	318	44	150
31	294	(106.0)	65.8	310	43	146
30	286	(105.5)	65.3	302	42	142
29	279	(104.5)	64.7	294	41	138
28	271	(104.0)	64.3	286	41	134
27	264	(103.0)	63.8	279	40	131
26	258	(102.5)	63.3	272	38	127
25	253	(101.5)	62.8	266	38	124
24	247	(101.0)	62.4	260	37	121
23	243	100.0	62.0	254	36	118
22	237	99.0	61.5	248	35	115
21	231	98.5	61.0	243	35	113
20	226	97.8	60.5	238	34	110

Hardness Conversions: Brinell to Rockwell

| Brinell Indentation Diameter mm | Brinell Hardness No. (10-mm tungsten-carbide ball; 8000-kg load) | Rockwell Hardness No. | | | Diamond Pyramid Hardness No. (Vickers) | Shore Scleroscope Hardness No. | Tensile Strength (approximate) in 1000 psi |
		C Scale, 150-kg Load (brale Penetrator)	B Scale, 100-kg Load (1/16-in. diameter ball)	A Scale 60-kg Load (Brale Penetrator)			
2.25	745	65.3	—	84.1	840	91	—
—	733	64.7	—	83.8	820	90	—
—	722	64.0	—	83.4	800	88	—
2.30	712	—	—	—	—	—	—
—	710	63.3	—	83.0	780	87	—
—	698	62.5	—	82.6	760	86	—
—	684	61.8	—	82.2	740	—	—
2.35	682	61.7	—	82.2	737	84	—
—	670	61.0	—	81.8	720	83	—
—	656	60.1	—	81.3	700	—	—
2.40	653	60.0	—	81.2	697	81	—
—	647	59.7	—	81.1	690	—	—
—	638	59.2	—	80.8	680	80	329
—	630	58.8	—	80.6	670	—	324
2.45	627	58.7	—	80.5	667	79	323
—	—	59.1	—	80.7	677	—	328
2.50	601	57.3	—	79.8	640	77	309
—	—	57.3	—	79.8	640	—	309
2.55	578	56.0	—	79.1	615	75	297
—	—	55.6	—	78.8	607	—	293
2.60	555	54.7	—	78.4	591	73	285
—	—	54.0	—	78.0	579	—	279
2.65	534	53.5	—	77.8	569	71	274
—	—	52.5	—	77.1	553	—	266
2.70	514	52.1	—	76.9	547	70	263
—	—	51.6	—	76.7	539	—	259
2.75	—	51.1	—	76.4	530	—	254
—	495	51.0	—	76.3	528	68	253
—	—	50.3	—	75.9	516	—	247
2.80	—	49.6	—	75.6	508	—	243
—	477	49.6	—	75.6	508	66	243
—	—	48.8	—	75.1	495	—	237
2.85	—	48.5	—	74.9	491	—	235
—	461	48.5	—	74.9	491	65	235
—	—	47.2	—	74.3	474	—	226
2.90	—	47.1	—	74.2	472	—	225
—	444	47.1	—	74.2	472	63	225
2.95	429	45.7	—	73.4	455	61	217
3.00	415	44.5	—	72.8	440	59	210
3.05	401	43.1	—	72.0	425	58	202
3.10	388	41.8	—	71.4	410	56	195
3.15	375	40.4	—	70.6	396	54	188
3.20	363	39.1	—	70.0	383	52	182
3.25	352	37.9	(110.0)	69.3	372	51	176
3.30	341	36.6	(109.0)	68.7	360	50	170
3.35	331	35.5	(108.5)	68.1	350	48	166
3.40	321	34.3	(108.0)	67.5	339	47	160
3.45	311	33.1	(107.5)	66.9	328	46	155
3.50	302	32.1	(107.0)	66.3	319	45	150
3.55	293	30.9	(106.0)	65.7	309	43	145
3.60	285	29.9	(105.5)	65.3	301	—	141
3.65	277	28.8	(104.5)	64.6	292	41	137
3.70	269	27.6	(104.0)	64.1	284	40	133
3.75	262	26.6	(103.0)	63.6	276	39	129
3.80	255	25.4	(102.0)	63.0	269	38	126
3.85	248	24.2	(101.0)	62.5	261	37	122
3.90	241	22.8	100.0	61.8	253	36	118
3.95	235	21.7	99.0	61.4	247	35	115
4.00	229	20.5	98.2	60.8	241	34	111
4.05	223	(18.8)	97.3	—	234	—	—
4.10	217	(17.5)	96.4	—	228	33	105

Brinell Indentation Diameter mm	Brinell Hardness No. (10-mm tungsten-carbide ball; 8000-kg load)	Rockwell Hardness No.			Diamond Pyramid Hardness No. (Vickers)	Shore Sleroscope Hardness No.	Tensile Strength (approximate) in 1000 psi
		C Scale, 150-kg Load (brale Penetrator)	B Scale, 100-kg Load (1/16-in. diameter ball)	A Scale, 60-kg Load (brale Penetrator)			
4.15	212	(16.0)	95.5	—	222	—	102
4.20	207	(15.2)	94.6	—	218	32	100
4.25	201	(13.8)	93.8	—	212	31	98
4.30	197	(12.7)	92.8	—	207	30	95
4.35	192	(11.5)	91.9	—	202	29	93
4.40	187	(10.0)	90.7	—	196	—	90
4.45	183	(9.0)	90.0	—	192	28	89
4.50	179	(8.0)	89.0	—	188	27	87
4.55	174	(6.4)	87.8	—	182	—	85
4.60	170	(5.4)	86.8	—	178	26	83
4.65	167	(4.4)	86.0	—	175	—	81
4.70	163	(3.3)	85.0	—	171	25	79
4.80	156	(0.9)	82.9	—	163	—	76
4.90	149	—	80.8	—	156	23	73
5.00	143	—	78.7	—	150	22	71
5.10	137	—	76.4	—	143	21	67
5.20	131	—	74.0	—	137	—	65

Area Conversions

Multiply by → to Obtain ↓	Circular Mils	Square Centimeters	Square Feet	Square Inches	Square Meters	Square Millimeters
Circular Mils	1	1.97×10^{-5}	1.83×10^{8}	1.27×10^{6}	1.97×10^{9}	1973
Square Centimeters	5.07×10^{-6}	1	929	6.45	10^{4}	.01
Square Feet	—	1.08×10^{-3}	1	6.94×10^{-3}	10.76	1.076×10^{-5}
Square Inches	7.85×10^{-7}	.1550	144	1	1550	1.55×10^{-3}
Square Meters	—	.0001	.0929	6.45×10^{-4}	1	10^{-6}
Square Millimeters	5.07×10^{-4}	100	9.29×10^{4}	645.2	10^{6}	1

Note: 1 circular mil = area of a circle .001 inch in diameter.
1 circular inch = area of a circle 1 inch in diameter.
1 square inch = 1.27 circular inches.
1 square mil = 1.27 circular mils
1 circular inch = 10^{6} circular mils = $.7854 \times 10^{6}$ square mils.
1 square inch = 1.27×10^{6} circular mils = 10^{6} square mils.

Volume Conversions

Multiply by → to Obtain ↓	Cubic Centimeters	Cubic Feet	Cubic Inches	Cubic Meters	Cubic Yards	Gallons (Liquid)	Liters	Pints (Liquid)	Quarts (Liquid)
Cubic Centimeters	1	2.832×10^{4}	16.39	10^{6}	7.65×10^{5}	3785	1000	473.2	946.4
Cubic Feet	3.531×10^{-5}	1	5.787×10^{-4}	35.31	27	.1337	3.53×10^{-2}	1.67×10^{-2}	3.34×10^{-2}
Cubic Inches	6.102×10^{-2}	1728	1	6.1×10^{4}	46,656	231	61.02	28.87	57.75
Cubic Meters	10^{-6}	2.832×10^{-2}	1.639×10^{-5}	1	.7646	3.79×10^{-3}	.001	4.732×10^{-4}	9.46×10^{-4}
Cubic Yards	1.31×10^{-6}	3.704×10^{-2}	2.143×10^{-5}	1.31	1	4.95×10^{-3}	1.31×10^{-3}	6.189×10^{-4}	1.24×10^{-3}
Gallons (Liquid)	2.642×10^{-4}	7.481	4.33×10^{-3}	264.2	202	1	.2642	.125	.25
Liters	.001	28.32	1.64×10^{-2}	1000	764.6	3.79	1	.4732	.9464
Pints (Liquid)	2.113×10^{-3}	59.84	3.463×10^{-2}	2113	1616	8	2.113	1	2
Quarts (Liquid)	1.057×10^{-3}	29.92	1.732×10^{-2}	1057	807.9	4	1.057	.5	1

Note: 10 milliliters = 1 centiliter = .338 fluid ounces.
10 centiliters = 1 deciliter = .845 fluid gills.
10 deciliters = 1 liter = 1.0567 fluid quarts.
10 liters = 1 dekaliter = 2.6417 fluid gallons.

Force Conversions

Multiply → by ↘ to Obtain ↓	Dynes	Grams	Joules/ Centimeter	Joules/ Meter	Kilograms	Pounds	Poundals
Dynes	1	980.7	10^7	10^5	9.807×10^5	4.448×10^5	1.383×10^4
Grams	1.02×10^{-3}	1	1.02×10^4	102.0	1000	453.6	141.1
Joules/ Centimeter	10^{-7}	9.81×10^{-5}	1	.01	9.807×10^{-2}	4.448×10^{-2}	1.383×10^{-3}
Joules/ Meter	10^{-5}	9.81×10^{-3}	100	1	9.807	4.448	.1383
Kilograms	1.02×10^{-6}	.001	10.2	.102	1	.4536	1.410×10^{-2}
Pounds	2.248×10^{-6}	2.205×10^{-3}	22.48	.225	2.205	1	3.108×10^{-2}
Poundals	7.233×10^{-5}	7.093×10^{-2}	723.3	7.233	7.093	32.17	1

Energy–Work–Heat Conversions

Multiply → by ↘ to Obtain ↓	BTU	Centimeter- Grams	Ergs (or cm dynes)	Foot- Pounds	Horse- power- Hours	Joules (or watt-sec)	Kilo- gram- Calories	Kilo- watt- Hours	Meter- Kilo- grams	Watt- Hours
BTU	1	9.297×10^{-8}	9.48×10^{-11}	1.285×10^{-3}	2545	9.48×10^{-4}	3.97	34.13	9.297×10^{-3}	3.413
Centimeter- Grams	1.076×10^7	1	1.02×10^{-3}	1.383×10^4	2.737×10^{10}	1.02×10^4	42.69×10^6	367.1×10^8	10^5	36.7×10^6
Ergs (or cm dynes)	1.055×10^{10}	980.7	1	1.356×10^7	2.684×10^{13}	10^7	418.6×10^8	36×10^{12}	98.1×10^6	360×10^8
Foot- Pounds	778.3	7.233×10^{-5}	7.367×10^{-8}	1	198×10^4	.7376	3087	2.655×10^6	7.233	2655
Horsepower- Hours	3.929×10^{-4}	3.654×10^{-11}	3.72×10^{-14}	5.05×10^{-7}	1	$.3722 \times 10^{-6}$	1.56×10^{-3}	1.341	3.653×10^{-6}	$.134 \times 10^{-2}$
Joules (or watt-sec)	1054.8	9.807×10^{-5}	10^{-7}	1.356	2.684×10^6	1	4186	3.6×10^6	9.81	3600
Kilogram- Calories	.2520	2.343×10^{-8}	2.389×10^{-11}	3.239×10^{-4}	641.3	2.39×10^{-4}	1	860	$.2343 \times 10^{-2}$	.86
Kilowatt- Hours	2.93×10^{-4}	2.729×10^{-11}	2.778×10^{-14}	3.766×10^{-7}	.7457	$.2778 \times 10^{-6}$	11.63×10^{-4}	1	2.724×10^{-6}	.001
Meter- Kilograms	107.6	10^5	1.02×10^{-8}	.1383	27.37×10^4	.102	426.9	36.71×10^4	1	367.1
Watt- Hours	.293	2.724×10^{-8}	2.778×10^{-11}	3.766×10^{-4}	745.7	2.778×10^{-4}	1.163	1000	$.2724 \times 10^{-2}$	1

Angular Velocity Conversions

Multiply by to Obtain ↓	Degrees/Second	Radians/Second	RPM	RPS
Degrees/Second	1	57.3	6	360
Radians/Second	1.745×10^{-2}	1	.1047	6.283
RPM	.1667	9.549	1	60
RPS	2.778×10^{-3}	.1592	1.667×10^{-2}	1

Torque Conversions

Multiply by to Obtain ↓	Dyne Centimeters	Gram-Centimeters	Kilogram-Meters	Foot-Pounds
Dyne-Centimeters	1	980.7	9.807×10^7	1.356×10^7
Gram-Centimeters	1.020×10^{-3}	1	10^5	1.383×10^4
Kilogram-Meters	1.020×10^{-8}	10^{-5}	1	.1383
Foot-Pounds	7.376×10^{-8}	7.233×10^{-5}	7.233	1

Power Conversions

Multiply by to Obtain ↓	BTU/Minute	Ergs/Second	Foot-Pounds/Second	Foot-Pounds/Minute	Horsepower	Kilogram-Calories/Minute	Kilowatts	Metric Horsepower	Watts
BTU/Minute	1	$.569 \times 10^{-8}$	7.71×10^{-2}	12.85×10^{-4}	42.41	3.97	56.9	41.83	5.69×10^{-2}
Ergs/Second	175.8×10^6	1	1.356×10^7	2.26×10^5	7.46×10^9	697.7×10^6	10^{10}	73.55×10^8	10^7
Foot-Pounds/Second	12.97	7.376×10^{-8}	1	1.667×10^{-2}	550	51.44	737.6	542.5	.7376
Foot-Pounds/Minute	778	4.426×10^{-6}	60	1	3.3×10^4	3087	4.426×10^4	325.5×10^2	44.26
Horsepower	2.357×10^{-2}	1.34×10^{-10}	18.2×10^{-4}	$.303 \times 10^{-4}$	1	9.355×10^{-2}	.7457	.9863	1.341×10^{-3}
Kilogram-Calories/Minute	.2520	1.433×10^{-9}	1.943×10^{-2}	3.24×10^{-4}	10.69	1	14.33	10.54	1.433×10^{-2}
Kilowatts	1.758×10^{-2}	10^{-10}	1.356×10^{-3}	2.26×10^{-5}	.7457	69.8×10^{-3}	1	.7355	10^{-3}
Metric Horsepower	2.39×10^{-2}	136×10^{-12}	1.843×10^{-3}	3.07×10^{-5}	1.014	9.485×10^{-2}	1.36	1	$.136 \times 10^{-2}$
Watts	17.58	10^{-7}	1.356	2.26×10^{-2}	745.7	69.77	1000	735.5	1

Bibliography

"A Bright Future for Magnetic Bubble Circuits." *Automation* (April 1973): 18.

"A Disk-Shaped DC Motor with Adjustable Ferroxdure Field Magnets." *Electronic Applications, Components, and Materials* 26 (3) (1965/1966): 107-22. 107-22.

"A Scientist Comes Up with a Perfect Car: Detroit Just Yawns." *The Wall Street Journal* (March 15, 1971): 1.

Adams, W.J. "Linear Induction Motors." *Machine Design* (March 1970).

"An Introduction to Burgler/Intruder Alarms VII. Microswitches, Button Switches, and Magnetic Switches." *Locksmith Ledger* (March 1970).

"Anaerobic Adhesives." *American Ceramic Society Bulletin* (April 1971).

Angus, H.C. "Platinum-Based, Permanent-Magnet Alloys." *IEE (London) Conference Publication* 33 (September 1967): 37-39.

Appels, J.T., and Geels, B.H. *Handbook of Relay Switching Techniques.* New York: Springer-Verlag, 1966.

"Applications of Magnetic Reed Switches." *Electronic Components* (July 1963): 733.

"Applying Indox Permanent Magnets to DC Motors." *Applied Magnetics* II (1) (1st and 2nd quarter, 1963) Valparaiso, Ind.: Indiana General Corp.

Arendt, C.H., Jr. *Designer's Handbook: The When, Why and How of Magnetic Shielding.* Blairsville, Pa.: Westinghouse Metals Division.

Arrott, W. "New Developments in Magnetic Materials and Applications." *Electrical Manufacturing* (February 1959).

Asbell, W.E. "Sealed Contact Relays on Printed Wiring Boards." *Bell Laboratories Record* (February 1971).

ASTM Standards, Part 8. Philadelphia: American Society for Testing Materials, 1967.

"Automation in Action-Conveying Through a Sound Barrier." *Automation* (July 1969).

Baasch, T.L. "Understanding Digital Stepping Motors." *Electronic Products* (August 1970).

Baeker, F.T. "A Magnetic Journal Bearing." *Philips Technical Review* 22 (1961).

Balmer, T.R. "Transient Load Weighed in Motion: Watthour Meter and Photocell Weigh Bulk Material in Transit." *Design News* (July 7, 1969).

Bancel, D.A. "Talking Turkey About a Metal of Many Uses [for fastening magnets]." *American Metal Market* (November 21, 1969).

Banks, B.A., and Bechtel, R.T. *1000-Hour Endurance Test of Glass-Coated Accelerator Grid on a 15-Centimeter Diameter Kaufman Thruster.* NASA TND-5891 (July 1970).

Baran, W.K.A. "Influence of Different Magnetic Field Profiles on Eddy-Current Braking." *IEEE Trans. on Magnetics* MAG-6 (2) (June 1970): 260-63.

Bardell, P.R. "Magnetic Materials in the Electrical Industry," Second Edition. London: MacDonald & Company, 1960.

Barnothy, M.F. (ed.). *Biological Effects of Magnetic Fields.* New York: Plenum Press, 1964.

Barnwell, F.H., and Brown, F.A., Jr. "Response of Planarians and Snails." *Biological Effects of Magnetic Fields.* M.F. Barnothy (ed.) New York: Plenum Press, 1964, p. 277.

Barone, D., and DeLigios, S. "Temperature Compensation in Eddy-Current Instruments." *IEEE Trans. on Magnetics* MAG-6 (June 1970).

Barrett, W.T., et al. "Rapid Method of Evaluating Magnetic Separator Force Patterns." *Transactions* of the *Society of Mining Engineers,* (September 1970).

Barta, G.T. "Factors Affecting Magnet Stability." *IEEE Intermag Conference,* Washington, D.C., April 1970.

———. *Permanent-Magnet Processing.* Valparaiso, Ind.: Indiana General Corp., 1967.

———. "Permanent-Magnet Processing." *Electro-Technology* (May 1966).

———. "Testing Magnets Automatically." *Instrumentation* 14 (1) (1961): 8-10.

——— and Rosine, L.L. "Permanent Magnets." *Electro-Technology* (May 1969).

Bartnik, J.A. "Magnetic Flocculation." *Minerals Processing* (August 1970).

Becker, J.J., et al. "Permanent-Magnet Materials." *IEEE Trans. on Magnetics* MAG-4 (2) (June 1968).

Bell, A. "Permanent Magnets vs. Electromagnets." *Proceedings of the IEE* (London), 112 (9) (September 1965): 1707-11.

Berkowitz, A.E., and Kneller, E. (eds.). *Magnetism and Metallurgy.* New York: Academic Press, 1969.

Beroset, J.E., Griesemer, H.A., Large, D.M., and Whitefield, K.C. "Magnetic Suspension-Parts Handling." *Western Electric Engineer* 11 (July 1967): 36-41.

Betz, W.K., and McDade, E.J. "Magnetic Pick-Up Tool [for fragile ferrous parts]." *RCA Technical Notes* (April 15, 1970).

Bianculli, A.J. "Stepper Motors: Application and Selection." *IEEE Spectrum* (December 1970): 25-29.

"Big Savings Claimed for Permanent-Magnet Motors." *Steel* (June 25, 1962).

Binns, K.J., and Barnard, W.R. "Novel Design of Self-Starting Motor." *Proceedings of the IEE* (London) (February 1971).

Bischoff, A.F. "Designing a High-Speed Relay." *Tele-Tech & Electronic Industries* (October 1954): 84-86.

Bitter, F. "Strong Magnets." *International Science & Technology* (April 1962).

Blacklock, D. "Putting Reed Switches to Work." *Electronics World* (August 1970).

Bolger, J.C., and Lysaght, M.J. "New Heating Methods and Cures Expand Uses for Epoxy Bounding [fastening magnets]." *Assembly Engineering* (March 1971).

Bolwell, A.J. "Tuned-Reed Relays Permit Centralization of Railroad Signaling." *Design News* (August 1970).

Boning, P. "A New Homopolar Motor." *Journal of The Franklin Institute* (Philadelphia) (July 1954).

"Boom in Magnetic Technology: Boon to Innovators." *Steel* (January 10, 1966).

Borcherts, R.H. "Mathematical Analysis of Permanent-Magnet Suspension Systems." *Journal of Applied Physics* (March 1971).

"Bouncing Magnet Snaps Keyboard Switch." *Machine Design* (June 1970).

Bowes, D.H. "Magnets Control Many Problems." *Pollution Engineering* (January/February 1972).

Bowman, C.G., Skelton, R.G., and Walsh, L. "Flexible Anisotropic Magnets." *IEE (London) Conference Publication* 33 (September 1967): 55-60.

Bozorth, R.M. "Ferromagnetism." Princeton, N.J.: D. Van Nostrand, 1951.

———. *Magnetic Properties of Metals and Alloys.* Cleveland, Ohio: American Society for Metals, 1959.

Bradley, F.N. *Materials for Magnetic Functions.* New York: Hayden Book Company, 1971.

Brailsford, F. *Magnetic Materials.* New York: John Wiley & Sons, 1960.

———. *Physical Principles of Magnetism.* Princeton, N.J.: D. Van Nostrand, 1964.

Brainard, M.W., and Strauss, F. *Synchronous Motors with Rotating Permanent Magnet Fields: Part I, Characteristics and Mechanical Constrution; Part II, Magnetic and Electrical Design Considerations.* AIEE Summer General Meeting. Minneapolis, Minn., June 1952.

Breazeale, J.B., McIlwraith, C.G., and Dacus, E.N. "Factors Limiting a Magnetic Suspension System." *Journal of Applied Physics* 29 (3) (March 1958): 414-15.

Brockman, F.G., et al. "A New Automatic Hysteresis Curve Recorder." *Philips Technical Review* 16 (3) (September 1954): 79-87.

Brod, R.C., and Schlang, H.A. "Removal of Metallic Foreign Bodies by a Magnetic Force." *Journal of American Medical Association* 17 (January 13, 1962): 164-65.

Bronkala, W.J. "Magnetic Removal of Tramp Iron." *Minerals Processing* (August 1970).

Brown, G.L. "Magnetism: Theory and Design Practices." *Electromechanical Design* (November 1966).

Bulman, W.E. "Applications of the Hall Effect." *Solid State Electronics* 9 (1966): 361.

Bungay, H.R. "Hanging Magnetic Stirrer Which Minimizes Call Disruption." *Applied Microbiology* (October 1970).

Burnett, J.R., and Koestle, F.L. "Acylic Generator: A New Power Generating Tool for Industry." *Direct Current* (England) (July 1963): 196-203.

Buschow, K.H. "Magnet Material with a BH_{max} of 18.5 MGO." *Philips Technical Review* 29 (11) (1968): 336-37.

Byrnes, W.S., and Crawford, R.G. *Improved Torque Magnetometer.* Wright Air Development Center, Dayton, Ohio, WADC Tech. Note 58-307, ASTIA Doc. No. AD 204219, September 1958.

Cameron, J., and Godfrey, E.I. "The Design and Operation of High-Power Magnetic Drives." *Biotechnology and Bioengineering* (September 1969).

Carey, R., and Isaac, E.D. *Magnetic Domains and Techniques for Their Observation.* New York: Academic Press, 1966.

Carpenter Alloys for Electronic, Magnetic, and Electrical Applications. Reading, Pa. Carpenter Steel Company, 1972.

Carpenter Alloys for Electronic Tubes. Reading, Pa.: Carpenter Steel Company, 1965, pp. 9-12.

Casarella, W.J., et al. "The Magnetically Guided Bronchial Catheter of Modified POD Design." *Radiology* 93 (1969).

"Catheters to the Brain." *Government Executive* (July 1969): 26.

Cavanaugh, R.J. "Timing and Stepper Motors." *Machine Design* (April 1969).

Ceramic Permanent Magnets and Materials. Bulletin No. 61C. Stackpole Carbon Company, St. Mary's, Pa., 1966.

Chang, K.K.N. "Beam Focusing by Periodic and Complementary Fields." *Proceedings of the IEEE* (June 1955): 62-71.

"Chart Recorders." *Systems Designer's Handbook, Electromechanical Design.* (September 1969).

Chiarella, L.J. "Rotation by the Digits with Permanent-Magnet Stopper Motors." *Machine Design* (November 1970).

Chikazumi, S. *Physics of Magnetism.* New York: John Wiley & Sons, 1964.

Chironis, N.P. "Magnetically Suspended Bearings." *Product Engineering* (July 24, 1961): 46-48.

Cioffi, P.P. "Relation Between Permanent-Magnet Configuration and Performance." *AIEE Paper 59-1120* (October 1959).

Clawson, A.R., and Wieder, H.H. "Bibliography of the Hall Effect, Theory and Application." *Solid State Electronics* 7 (1964): 387.

Clerk, R.C. "The Utilization of Flywheel Energy." *SAE Paper No. 711A* (June 1963).

Clogston, A.M., and Heffner, H. "Focusing of an Electron Beam by Periodic Fields." *Journal of Applied Physics* 25 (4) (April 1954): 436-447.

Clurman, S.P. *On Hunting in Hysteresis Motors and New Damping Techniques.* IEEE Intermag Conference, Denver, Colorado, April 1971.

Cochardt, A. "Loudspeaker Structures with Strontium Ferrite Magnets." *IRE Trans. on Audio* AU-10 (6) (November/December 1962): 164-70.

———. "Recent Ferrite Magnet Developments." *Journal of Applied Physics* 37 (3) (March 1966): 1113-16.

Cochrane, J.H. "Surveying the Field of Permanent-Magnet Materials." *Machine Design* (September 15, 1966).

Coggshall, J., et al. "Control of Biomedical Implants by DC Magnetic Fields." *Proceedings of the Nineteenth Annual Conference on Engineering in Medicine and Biology* 8 (November 1966): 121.

Cohen, D. "Magnetic Fields Around the Torso: Production of Electrical Activity of the Human Heart." *Science* 156 (1967): 652-64.

Cohn, D. "Magnetoencephalography: Evidence of Magnetic Fields Produced by Alpha-Rhythm Currents." *Science* 161 (1968): 784-86.

———. "DC Motors: Wound Field or Permanent Magnet?" *Machine Design* (February 1971).

Collins, Frank. "Permanent-Magnet Motors." *Machine Design* (July 23, 1970).

Conley, C.C. *A Review of the Biological Effects of Very Low Magnetic Fields.* NASA TND-5902 (August 1970).

Cordwell, J.F. "Electromagnets and [Permanent-Magnet] Separators in the Foundry." *Foundry Trade Journal* (November 1970).

Cornell, A.W., and Parker, R.J. "Stabilization Interactions in Small Permanent-Magnet DC Motors." *IEEE Trans. on Magnetics* MAG-6 (2) (June 1970).

"Corrosives, Abrasives, Viscous Polymers Handled by Flowmeter." *Chemical Processing* (February 1969).

Craik, D.J. "Domain Theory and Observations." *Journal of Applied Physics* 38 (3) (March 1967): 931-38.

Crandall, W.B. "Modern Ceramics—Magnetic Ferrites." *Ordinance* (March/April 1971).

Crawford, W.E. "Stepper Motor Drive Electronics for the Solar Electric Thrust Vector Control Subsystem." *Cal Tech JPL Space Programs Summary* III (1970): 108-110.

Cronk, E.R. "Recent Developments in High-Energy Alnico Alloys." *Journal of Applied Physics* 37 (3) (March 1966): 1097-100.

Cross, B. "New Developments in Magnetic Bearings." *Product Engineering* (April 1, 1963): 79.

Davidson, R.S., and Groulay, R.D. "Apply the Hall Effect to Angular Transducers." *Solid State Electronics* 9 (1966): 471.

Davis, J.F. "Integral Horsepower DC Motors." *Machine Design* (April 1970).

Davis, S.A. "Component Selection and Application—Fractional Horsepower Motors." *Electromechanical Design* (December 1970).

———. "DC Servomotors in Commercial Applications." *Electromechanical Design* (September 1970).

———. "Servomotors: AC and DC." *Electromechanical Design* (January 1969).

DeMott, E.G. "Integrating Fluxmeter with Digital Readout." *IEEE Trans. on Magnetics* MAG-6 (2) (June 1970).

Dennis, W.H. "Ferrite Magnets Offer a Cheap Source of Magnetomotive Force." *Electrical Review* (April 3, 1970).

Derr, A.J., and DeCleene, D.F. "A Study of Permanent Magnetic Devices for Improvement of Adhesion in Railway Vehicles." *Proceedings of the Permanent Magnet Users Association* (Philadelphia) (1969).

Design and Application of Permanent Magnets. Manual 7. Valparaiso, Ind.: Indiana General Corp., 1964.

Designing DC Motors with Indox Permanent Magnets. Bulletin 40. Valparaiso, Ind.: Indiana General Corp., 1968.

Desmond, D.J. "The Economic Utilization of Modern Permanent Magnets." *Proceedings of the IEE* (London) 92 (2) (1945): 229-52.

Dietrich H. "Contribution to the Temperature Dependence of the Magnetic Properties of Permanent Magnets." *Cobalt* 35 (June 1967): 78-97.

"Difficult-to-Handle Scrap Being Removed by Conveyor." *American Metal Market* (March 3, 1971).

"Digital Readout: Accuracy, Readability, Low Error." *Electromechanical Design* (May 1970).

Dorman, F., et al. "Progress in the Design of a Centrifugal Cardiac-Assist Pump with Transcutaneous Energy Transmission by Magnetic Coupling." *Trans. of the American Society of Artificial Internal Organs* XV (1969).

Dranetz, A.I. "Electromechanical Transducers." *Machine Design* (January 9, 1958).

Driller, J. "Kinetics of Magnetically Guided Catheters," *IEEE Trans. on Magnetics* MAG-6 (September 1970).

———. et al. "The POD Bronchial Catheter." *IEEE Trans. on Magnetics* MAG-6 (June 1970).

"Drive and Transmission Components: Clutches and Brakes." *Electromechanical Design* (January 1969).

Edson, A.P., and Peters, D.T. *Calculation of Electromechanical Permeance of Alnico-Magnet Materials in Electrical Motors and Generators.* IEEE Intermag Conference, Denver, Colorado, April 1971.

Edwards, A. "Rubber Magnets," *IEE (London) Conference Publication* 33 (September 1967): 51-54.

Eggert, A.A., et al. "ELLA [The Experimental Link-Laboratory Analytical System] Applied to Experimental Control." *Analytic Chemistry* (May 1971).

Eklund, A. "A Novel Design for a Lightweight Ka-Bank Magnetron." *Microwave Journal* (November 1969).

Elarde, P.F. "All-Electronic Magnetic Hysteresisgraph." *Western Electric Engineer* (January 1965): 8-14.

Electrical Materials Handbook. Pittsburgh: Allegheny Ludlum Steel Company, 1961, pp. V/1-V/39.

Electromagnetic Shielding Design Manual. Camden, N.J.: Magnetic Metals Company, 1972.

"Electromagnetic Transducers: Synchros, Resolvers, LVDTs." *Systems Designer's Handbook, Electromechanical Design* (April 1968): 99-107.

"Electromechanical Relays." *Electromechanical Design* (January 1969).

"Electromechanical Switches." *Electromechanical Design* (January 1969).

Ellis, J.N., and Collins, F.A. *Design and Evaluation of Brushless Electrical Generators.* NASA Tech. Brief 70-10554 (October 1970).

Ellis, L.F. "Electromagnetic Force Transducer." *Advances in Instrumentation.* Paper 744-70.

Elzinga, W.E., and Kaufman, W.M. "Major Coronary Thrombi Formation in the Unanesthetized Animal." *American Journal of Surgery* (December 1970).

Engelsted, J.W. "Demagnetize Before Assembly." *Product Engineering* (July 4, 1960): 28-32.

Enz, U., et al. "Determining Magnetic Quantities by Displacement Measurements." *Philips Technical Review* 25 (8) (1963/1964): 207-13.

Fahlenbrach, H. "Properties and Fields of Applications of Vicalloy-Type, Permanent-Magnet Alloys." *Cobalt* 25 (December 1964): 1-8.

Falce, L.R. *Traveling-Wave Tubes Using Rare-Earth/Cobalt Periodic Permanent-Magnet Focusing.* IEEE Intermag Conference, Denver, Colorado, April 1971.

Falk, R.B. "A Current Review of Lodex Permanent-Magnet Technology." *Journal of Applied Physics* 27 (3) (1966): 1108-12.

Fay, L.E., III. "The Hall-Effect Applications in Electrical Measurements." *Semiconductor Products* (May 1960): 39.

Fehely, F.C. "Angle-Plate Edge Finder Held on by Magnets." *American Machinist* (May 31, 1971).

Fink, R.A. "The Brushless Motor: Types and Sources." *Control Engineering* (August 1970).

"Floating Magnet Triggers Flow Switches." *Machine Design* (March 1970).

Floros, J. "Fail-Safe Lifting Magnet Protects Against Power Failure." *Plant Engineering* (March 1971).

———. "MH Equipment, Space Tight? Try Magnetic Conveying." *Plant Engineering* (March 6, 1969).

"Flying Magnet Motion Improves Keyboard Feel." *EDN* (June 1, 1970).

Foner, S. "Versatile and Sensitive Vibrating Sample Magnetometer." *Review of Scientific Instruments* 30 (7) (July 1959): 548-57.

Forster, G.A. "Performance of Permanent-Magnet, Flow-Through Type Sodium Flowmeters. . . ." *IEEE Trans. on Nuclear Science* NS-18 (February 1971).

Fow, P.B. "Couple Your Servo Load Through a Magnetic Particle Clutch." *Electromechanical Design* (April 1970): 56-57.

Foy, J.E., and Parker, R.J. "Permanent-Magnet Leakage/Permeance Evaluation Based on Polar Radiation Analogy." *Journal of Applied Physics* V (31) (May 1960).

"Free-Flying Magnets Control Switch-Pulse Time." *Electromechanical Design* (March 1971).

Frei, E.H., et al. "The POD and Its Applications." *Medical Research Engineering* (4th quarter, 1966).

Freiser, M.J. "A Survey of Magneto-Optic Effects." *IEEE Trans. on Magnetics* MAG-4 (2) (June 1968): 152-61.

French, D.P. "Slip Rings." *Electromechanical Design* (October 1969).

"Frictionless Transducer Uses Hall Effect." *Electronics Design* (July 5, 1970).

Fujishima, M., et al. "Effects of Experimental Occlusion by Magnetic Localization of Iron Filings on Cerebral Blood Flow." *Neurology* (September 1970).

Gaster, G.R. "Quick Method for Designing Radial Magnetic Couplings." *Machine Design* (September 1970).

General Electric Permanent-Magnet Manual. Edmore, Mich.: General Electric Company, Magnetic Materials Business Section.

Gibson, R.J. "A Monograph on Magnetic Fields for Life Scientists." *Journal of The Franklin Institute* (Philadelphia) (1969).

Gillot, J.H., et al. "Eddy-Current Loss in Saturated Solid Magnetic Plates, Rods, and Conductors." *IEEE Trans. on Magnetics* MAG-1 (2) (June 1965): 126-37.

Gilmore, J.P., and Feldman, J. "Gyroscope in Torque to Balance Strap-down Application." *Journal of Spacecraft and Rockets* (September 1970).

Ginsberg, D., and Misenheimer, L.J. *Design Calculations for Permanent-Magnet Generators.* AIEE Tech. Paper No. 53-9 (January 1953).

Glass, M. "Straight-Field Permanent Magnets of Minimum Weight for TWT-Focusing Design and Graphic Aids in Design," *Proceedings of the IRE* 45 (8) (August 1957): 1100-05.

Goeddecke, H. "Flux Measurements with Additional Magnets for Adjustment." *IEEE Trans. on Magnetics* MAG-6 (June 1970).

Gollhardt, J.B. *Description of Basic Types of Permanent-Magnet Applications.* Milwaukee, Wis.: Allen-Bradley Company, 1970.

———. *Principles of Permanent-Magnet Field, DC-Motor Design.* IEEE Intermag Conference, Denver, Colorado, April 1971.

———, and Beaudoin, L.W. "Ideas for Designers: New Magnet Materials Spur DC Motor Design." *Electromechanical Design* (August 1970).

Gordon, D.I., et al. "Nuclear Radiation Effects in Magnetic Core Materials and Permanent Magnets." *Materials Research in the Navy,* ONR-5 Symposium (1959).

Gould, J.E. "Permanent-Magnet Applications." *IEEE Trans. on Magnetics* MAG-5 (December 1969).

———. "Permanent-Magnet, Eddy-Current Braking at High Speeds." *Cobalt* 39 (June 1968): 75-80.

———. "Permanent Magnets, Their Materials and Applications." *Electrical Review* (England) 6 (12) (1963): 680-84.

———. "Some Aspects of Permanent-Magnet, DC Motors." *IEEE Trans. on Magnetics* MAG-6 (June 1970).

———. Stability of Permanent Magnets. *Journal of the Permanent Magnet Association* (England) (2) (June 1966).

———. Summary of Useful Relationships in the Design and Testing of Permanent Magnets." *Journal of the Permanent Magnet Association* (England) 9 (June 1967).

Greenwald, F.S. *An Engineering Approach to Material [Permanent Magnet] Selection.* IEEE Intermag Conference, Denver, Colorado, April 1971.

———. "Permanent-Magnet Rotors for Alternators." *Applied Magnetics* 4 (2) (March/April 1956). Valparaiso, Ind.: Indiana General Corp.

Grob, D., and Stein, P. "Samarium-Cobalt Magnetic Palpebral Prostheses." *Journal of Applied Physics* (March 15, 1971).

Guerdan, D.A. "Hermetic Magnetic Couplings." *Product Engineering* (April 28, 1958): 98-100.

Guthrie, G.L. "Sensitive AC Hall-Effect Circuit." *Review of Scientific Instruments* 36 (1965): 1177.

Hadfield, D. "Magnets in the Dounreay Reactor." *Iron & Steel* (December 1959).

———. *Permanent Magnets and Magnetism.* New York: John Wiley & Sons, 1962.

———. "The Relative Fields of Utility of Alloy and Ceramic Ferrite Magnets." *Journal of the Permanent Magnet Association* (England), 6 (February 1968).

Hagedorn, G. "Hetropolar Synchronous Machine with Unwound Rotor." *The Engineers Digest* (April 1958): 56.

"Hall-Effect Compass Technical Note." *Journal of The Franklin Institute* (Philadelphia) (May 1968).

Hammond, E.F., et al. "Feasibility Study and Preliminary Design for a 100-Kilowatt Permanent Magnet Generator." *U.S. Government R&D Report on Contract #DAAKO2-69-C-9763* (June 1970).

Hanson, K.L. "Synchronous Motors." *Machine Design* (April 1970).

Harding, J.T. *Progress in Magnetic Suspension Applied to High-Speed Ground Transportation.* American Institute of Physics, 17th Annual Conference on Magnetism and Magnetic Materials, Chicago, Ill., November 1971.

Harrold, W.J. *Permanent-Magnet Design by Computer.* IEEE Intermag Conference, Denver, Colorado, April 1971.

———. *Permanent Magnets in Microwave Devices.* American Institute of Physics, 17th Annual Conference on Magnetism and Magnetic Materials, Chicago, Ill., November 1971.

———, and Reid, W.R. "Permanent Magnets for Microwave Devices," *IEEE Trans. on Magnetics* MAG-4 (3) (September 1968): 229-39.

Hellwege, K.H., and Hellwege, A.M. *Magnetic Properties* Vols. 1 and 2. New York: Springer-Verlag, 1962, 1967.

Helmer, J.C. "Magnetic Circuits Employing Ceramic Magnets." *Proceedings of the IRE* (October 1963): 1528-37.

Helmick, C.G., and Chapman, J.H. "The Modern Hysteresis Motor." *Westinghouse Engineer* (July 1961): 127-28.

Hender, B.S. "The Future of the Battery Electric Car." *Journal of the IEE (England)* 8 (1964): 250-54.

Hennig, G.R. "Applying the Hall Effect to Practical Magnetic Testing." *Electrical Manufacturing* (April 1958).

Henschke, W.O. "Flywheel Alternators with Ferromagnetic Wheels." *SAE Transactions.* SAE International Congress, Detroit, Mich., January 1961.

Hilsum, C. "Galvanomagnetic Effects and Their Application." *British Journal of Applied Physics* 12 (1961): 85.

Hinegardner, R. "Magnetic Stirring Device." *U.S. Government R&D Report* 70(22) (November 25, 1970): 142.

Hoffman, G.A. "Future Electric Car." *SAE Paper No. 690073* (January 1969).

Hoffman, W. "Stray Field Neutralization." *IEEE Trans. on Magnetics* MAG-6 (June 1970).

Hohenemser, K., and McCaull, J. "The Wind-Up Car." *Environment* 12 (5) (June 1970): 14-21.

"How To Specify Permanent Magnets." *Journal of the Permanent Magnet Association* (England) 3 (June 1966).

Iden, D.J., et al. *Present and Future Applications of High-Coersive Force Magnets.* American Institute of Physics, 17th Annual Conference on Magnetism and Magnetic Materials, Chicago, Ill., November 1971.

"Improved Magnetic Materials." *Microwave Journal* (June 1966).

"Indox-V Speaker Magnets: A Guide to Reduce Costs and Improve Designs." Engineering Data Form 381A. Valparaiso, Ind.: Indiana General Corp., October 1, 1964.

"Instruments." *Electromechanical Design* (January 1969).

Ireland, J.R. *Ceramic Permanent-Magnet Motors: The Key to Cordless Appliances.* Valparaiso, Ind.: Indiana General Corp., 1968.

———. "New Figure of Merit for Ceramic Permanent Magnets Intended for DC Motor Applications." *Journal of Applied Physics* 38 (3) (March 1968): 1011-12.

Jacobs, I.S. "Role of Magnetism in Technology." *Journal of Applied Physics* 40 (3) (March 1969): 917-27.

Janicke, J.M. *How to Magnetize, Measure, and Stabilize Permanent Magnets.* Boonton, N.J.: RFL Industries, 1970.

———, "New Developments in Magnetic Instrumentation." *Measurements and Data* (November/December 1971).

Jarrett, G.D.R. "Production and Properties of Magnetic Coatings." *Electroplating and Metal Finishing* (November 1969).

Jarrett, J.H. "Potentially Useful Properties of Magnetic Materials." *Journal of Applied Physics* 40 (3) (March 1969): 938-44.

———. "Traveling-Wave Tube and Its Manufacture." *Western Electric Engineer* (January 1961): 2-11.

Jauch, W.L., et al. *Magnetics—History, Application, Measurement.* Boonton, N.J.: RFL Industries, 1970.

Johnson, H.L. "Designing Loudspeaker Magnets with Indox-V." *Applied Magnetics* 9 (1) (1st quarter, 1961): 2-5. Valparaiso, Ind.: Indiana General Corp.

———. "Permanent Magnetic Design Guidelines for Linear Actuators." *Electromechanical Design* (December 1970).

Jones, F.G., and Parker, R.J. *Demagnetizing Coefficients (B/H) vs. Geometry Relationships for High-Coersive Force, Cobalt Rare-Earth Magnets.* IEEE Intermag Conference, Denver, Colorado, April 1971.

Kappius, F. "DC Motor with Electronic Commution, Permanent-Magnet Excitation, and Galvanomagnetic Signal Transmitter." *IEEE Trans. on Magnetics* MAG-6 (June 1970).

Katz, I. "Motors and Controls." *Electromechanical Design* (January 1969).

———. "Permanent Magnets: Materials and Applications." *Electromechanical Design* (July 1966).

———. "Permanent Magnets: Materials and Applications." *Electromechanical Design* (July 1967).

Keller, E.A. "The Hall Compass." *Proceedings of the National Electronics Conference*, (Chicago) 18 (1962): 753.

Kerr, C., and Lynn, C. "The Roller Road." *Westinghouse Engineer* (March 1961).

Kessler, J.N. "Introducing the Electronic Car: Anti-skid, Anti-polluting, Pro-Driver." *Electronic Design* (January 1971).

Kholdov, Yu A. *The Effects of Electromagnetic and Magnetic Fields on the Central Nervous System.* NASA Technical Translation F-465 (June 1967).

King, George. "Magnetic Powder Clutches/Brakes." *Electromechanical Design* (September 1969).

Kinne, H.M. "Monitoring Machine Vibrations." *Automation* (July 1969).

Kittel, C. "Physical Theory of Ferromagnetic Domains." *Reviews of Modern Physics* (October 1949): 521.

Kituchi, Y. "Magnetostrictive Materials and Applications." *IEEE Trans. on Magnetics* MAG-4 (2) (June 1968): 107-17.

Kloeffler, Royce G., Kerchner, Russell M., and Brenneman, Jesse L., *Direct-Current Machinery, Second Edition.* New York: Macmillan, 1958.

Kneller, E. *Ferromagnetismus.* New York: Springer-Verlag, 1962.

Knight, F. "Magnetizing of Permanent Magnets." *Journal of the Permanent Magnet Association* (England) 7 (November 1966).

Koch, J. "Calculation of Center-Pole Loudspeaker Magnet." *Electronic Applications, Components, and Materials* (1970).

———. "Energetic Problems of Permanent Magnetic Attraction Systems with Variable Leakage Permeance Depending on Air Gap." *IEEE Trans. on Magnetics* MAG-6 (June 1970).

Kolm, H., et al. *Magnetic Filtration.* American Institute of Physics, 17th Annual Conference on Magnetism and Magnetic Materials, Chicago, Ill., November 1971.

Koroseal Flexible Magnetic Strip. No. IPC-1061-2. Akron, Ohio: B.F. Goodrich Industrial Company, 1958.

Koslow, H. "Two Set-Up Ideas for Lathe Chuck-Control, Work-Mounting Alignment." *American Machinist* (February 23, 1970).

Kronenberg, K.J., and Bohlmann, M.A. *Long-Term Magnetic Stability of Alnico V and Other Permanent-Magnet Materials.* WADC Technical Report 58-535, U.S. Office of Technical Services (1958).

Kueser, P.E., et al. *Properties of Magnetic Materials for Use in High-Temperature, Space Power Applications.* NASA-SP-3043 (1967).

Kusko, A. "Analysis of Forces in Permanent-Magnet Transducers." *IEEE Trans. on Magnetics* MAG-6 (September 1970).

Kusserow, B.K. "The Use of a Magnetic Field to Remotely Power an Implantable Blood Pump." *Trans. of the American Society of Artificial Internal Organs* (1960).

Lavie, A.M. "The Swimming of the POD: Theoretical Analysis and Experimental Results." *IEEE Trans. on Magnetics* MAG-6 (June 1970).

Lavoie, F.J. "Magnetic Fluids." *Machine Design* (June 1, 1972): 78-82.

Lawrence, L.G. "Geomagnetic Observations." *Electronics World* (February 1969).

Lax, B., and Button, K.J. *Microwave Ferrites and Ferrimagnetics.* New York: McGraw-Hill, 1962.

Lee, E.W. *Magnetism.* London: Pelican Original, 1963.

Leow, P., and Gollhardt, J. "Ceramic Magnets for a New Generation of PM Motors." *Insulation/Circuits* (July 1970).

"Let's Look at Permanent Magnets." *Electronic Capabilities* (1966).

Lock, J.M. "Magnetism and the Rare-Earth Metals." *IRE Trans. on Component Parts* CP5-7 (June 1959): 93-105.

"Long-Term Stability of Alnico and Barium-Ferrite Magnets." *Journal of Applied Physics* 31 (May 1960): 825-45.

Loucks, J. "Mechanical Design of Permanent Magnets." *Machine Design* (July 24, 1969): 125-27.

"Low-Mass Linear Motor Drives Floating Plotter Head." *Electromechanical Design* (April 1970).

"Low Melting-Point Alloy Holds Magnets in Rings." *Machine Tool Blue Book* (July 1969).

Luborsky, F.E. "Permanent Magnets." *Electrotechnology* (July/August 1962).

———. "Permanent Magnets in Use Today." *Journal of Applied Physics* 37 (3) (March 1966): 1091-94.

Ludwig, J.T. "Design of Optimum Inductors Using Magnetically Hard Ferrites in Combination with Magnetically Soft Materials." *Journal of Applied Physics* 29 (3) (March 1958): 409-10.

———. "Inductors Biased with Permanent Magnets. Part I: Theory and Analysis; Part II: Design and Synthesis." *AIEE Communications and Electronics* (July 1960).

Lujic, A. "Controlling Brushless DC Motors." *Machine Design* (October 1969).

Lunas, L.J. "Switchboard and Panel Instruments." *Instruments and Control Systems* (January 1969).

Mackay, R.S. *Biomedical Telemetry.* New York: John Wiley & Sons, 1969.

MacLean, K.S. "High Magnetic Fields in Advanced Malignancy and Other Disorders." *Abstracts from the Conference on the Effects of Diffuse Electrical Currents on Physiological Mechanisms* 4 (1967): 58.

"Magnet Diet Prevents Hardware Disease." *Inco Nickel Topics* 8 (8) (1955): 5. New York: The International Nickel Company.

"Magnet Makers Turn Up the Heat." *Chemical Week* (September 10, 1965).

"Magnet Measures Quality of Capsules at Large Pharmaceutical Firm." *Magnetic Attractions* 24 (1970): 5. Erie, Pa.: Eriez Magnetics.

"Magnet Replaces Torque-Wrench Spring." *Machine Design* (June 11, 1970).

"Magnetic Boost Lowers Coil Power Requirements." *Machine Design* (January 1970).

"Magnetic Detaction of Moving Metal." *Journal of The Franklin Institute* (Philadelphia) (1968).

Magnetic Devices. Engineering Data Form 382. Valparaiso, Ind.: Indiana General Corp., April 22, 1968.

Magnetic Drives. Engineering Data Form 382. Valparaiso, Ind.: Indiana General Corp., June 15, 1966.

"Magnetic Flocculation Speeds Water Treating." *American Metal Market* (July 14, 1969).

"Magnetic Force Gives Aid for Cleanest Steel Strip." *American Metal Market* (March 18, 1971).

Magnetic Handling Application. Handbook 3500. Valparaiso, Ind.: Indiana General Corp., 1965.

Magnetic Materials. EM-18. Pittsburgh: Allegheny Ludlum Steel Corp., 1963.

"Magnetic Measurements: Home Study Course No. 7," *Measurements and Data* (January/February 1968).

Magnetic Modules. Handbook 3600. Valparaiso, Ind.: Indiana General Corp., 1965.

"Magnetic Printing Cylinder." *Industrial Equipment News* (April 1970).

"Magnetic Properties and Design Data for Standard P.M.A. Materials." *Journal of the Permanent Magnet Association* (England) 1A (September 1965).

"Magnetic Properties of Metals and Alloys," *Journal of the American Society for Metals* (1959).

"Magnetic Rolls Convey Rolled Sections." *Iron and Steel Engineer* (February 1970).

"Magnetic Sorting Doubles Scrap Value." *Modern Materials Handling* (June 1969).

"Magnetic Speed Changer." *Electromechanical Design* (December 1970).

Magnetic Testing: Theory and Nomenclature STP371. Philadelphia: American Society for Testing and Materials, 1965.

Magnetism: A Treatise on Modern Theory and Materials, 3 vols. George T. Rado and H. Suhl (eds.). New York: Academic Press, 1963-1966.

Magnetism and Magnetic Materials. R.L. White and K.A. Wickersheim (eds.). New York: Academic Press, 1965.

Magnetostriction. New York: The International Nickel Company, R&D Division, 1952.

"Magnets Support Sled," *Electrical Review* (January 30, 1970).

"Major Improvement in Traveling-Wave Tubes: Samarium-Cobalt." *Aviation Week & Space Technology* (November 16, 1970).

Mandel, J.T., et al. "Electron-Beam Focusing with Periodic Permanent-Magnet Fields." *Proceedings of the IRE* (May 1954): 800-10.

Mann, P.J. "Magnetism: Parts, Pieces, and Containers Are in Its Power." *Material Handling Engineering* (August 1967): 88-92.

———. "Magnetism: Power Is There for Heavy Lifts." *Material Handling Engineering* (October 1967).

———. "Magnetism Keeps Intruders Out of Bulk." *Material Handling Engineering* (July/August and September/October 1967).

———. "Magnetism's Magic Can Help Your Handling." *Material Handling Engineering* (June 1965).

Manual Techniques Des Aimants Permanents. Paris: Chambre Syndicale Des Producteurs D'Aciers Fins et Speciaux, January 1967.

Marshall, R. "Solid Brushes on the Canberra Homopolar Generator." *Nature* (England) (1964): 1079.

"Materials for Permanent Magnets." *Insulation/Circuits* (July 1970).

Matthews, R.W. "Instrument Motors." *Machine Design* (April 1970).

Maynard, C.A. "Analysis and Design of Permanent-Magnet Assemblies." *Machine Design* (April 18, 1957).

———. *Applied Magnetics* 3 (5) Valparaiso, Ind.: Indiana General Corp. (1955): 3-7.

McCaig, M. "Permanent-Magnet Performance at Temperatures Above 550° C." *Cobalt* 41 (December 1968): 196-98.

———. "Permanent Magnets for Repulsion Devices." *Journal of the Permanent Magnet Association* (England) 4 (January 1965).

———. "Permanent Magnets for Repulsion Systems," *Electrical Review* (England) (September 1961).

———. "Present and Future Technological Applications of Permanent Magnets." *IEEE Trans. on Magnetics* MAG-4 (3) (September 1968): 221-28.

———. "Recent Developments in Permanent Magnetism." *Journal of Applied Physics* 35 (3) (March 1964).

McConnell, D.R. "Microwave Ovens: Revolution in Cooking. I. Operating Principles and Design." *Electronics World* (August 1970).

McGann, D.W. "Development of a Low-Loss, High-Precision Permeameter for the Evaluation of High-

Energy Permanent Magnets." *Audio Engineer Society Journal* (October 1961).

Merriam, E.R. *The Paradox of Flat Block Magnets in Small-Motor Design.* Preprint 710096. New York: Society of Automotive Engineers, 1971.

Merrill, F.W. *The Permanent-Field Synchronous Motor.* New York: C.M. Technical Publications Corp., 1961.

Mesch, F. "Magnetic Components for the Attitude Control of Space Vehicles." *IEEE Trans. on Magnetics* MAG-5 (3) (September 1969): 586-92.

Milligan, N.P., and Burgess, J.P. "Hall-Effect Devices for Low-Level Magnetic Detection." *Solid State Electronics* 7 (1964): 323.

Montgomery, D.B. "A Magnetically Guided Catheter System for Intercranial Use in Man." *IEEE Trans. on Magnetics* MAG-6 (June 1970).

———. *Solenoid Magnet Design.* New York: John Wiley & Sons, 1969.

———, et al. "Superconducting Magnet System for Intravascular Navigation." *Journal of Applied Physics* (5) (April 1969).

———, and Weggel, R.J. "Magnetic Forces for Medical Applications." *Journal of Applied Physics* 40 (3) (March 1969).

Moreau, L. "The Place of Magnets in Modern Industry." *IEEE Trans. on Magnetics* MAG-6 (June 1970).

Moskowitz, L.R. "Handling Materials with Magnetic Devices." *Automation* (April 1965).

———. "Hard Magnetic Materials," *Trans. of the Golden Gate Metals Conference* (San Francisco) (February 1964).

———. *Magnetic Handling for Automatic Assembly* M S66-142. Dearborn, Mich.: American Society of Tool and Manufacturing Engineers, 1965.

———. "Magnets: Silent Helpers for Welding Operations." *Plant Engineering* (November 1965).

———. "Selecting Magnets for Reed-Switch Actuation." *Automation.* (October 1968).

———. (ed.) "First-User Seminar on Permanent-Magnet Engineering and Application," *Proceedings of the Permanent Magnet Users Association* (Philadelphia) (1969).

———, and Israelson, A.F. "Magnetic Handling Devices." *Automation* (January 1958).

"Moving Chips Magnetically." *American Machinist* (May 1971).

Murray, R.W. "Permanent-Magnet Motors for Mills . . . ." *The Iron Age* (February 23, 1967): 63-65.

Nakamura, T., et al. "Magneto-Medicine: Biological Aspects of Ferromagnetic Fine Particles." *Journal of Applied Physics* (March 15, 1971).

Nave, P.M.W. "Tactile Rifle Orientation Indicator," *Proceedings of the Permanent Magnet Users Association* (Philadelphia) (1970).

Nesbitt, E.A. "New Permanent-Magnet Materials Containing Rare-Earth Metals." *Journal of Applied Physics* 40 (3) (March 1969): 1455-57.

"New Material Produces Superior Permanent Magnets." *Ceramic Industry* (May 1973).

"New Transistor Responds to Magnet." *EDN* (February 1, 1969).

Niklas, W.F. "An Improved Ion-Trap Magnet." *Philips Technical Review* 15 (8/9) (February 1952): 258-62.

Nishihara, H., and Terada, M. "Measurements on Misalignment of the Axis of a Magnetic Field for Electron-Beam Focusing." *Journal of Applied Physics* (July 1970).

Norfolk, B.A., and Towndraw, P.E. "Design and Analysis of Permanent-Magnet, Eddy-Current Systems Typically Used for Automobile Speedometers and Tachometers." *IEEE Trans. on Magnetics* MAG-6 (June 1970).

Okamoto, T. "Dynamic Analysis of Polar Relays." *Fujitsu Science & Technology Journal* (December 1970).

Oliver, F.J. *Practical Instrument Transducers.* New York: Hayden Book Company, 1971.

Olsen, E. *Applied Magnetism.* New York: Springer-Verlag, 1966.

Owen, T.D., et al. "The Effect of Reactor Radiation on the Magnetic Moments of Colomax Magnets," D.E.G. Memo 194 (W). U.K. Atomic Energy Authority, 1959.

Packard, H. "Permanent-Magnet Torquers." *Electromechanical Design* (September 1967): 28-29.

Paramentier, J.C. "Integrated Rate Gyros." *Electromechanical Design* (May 1970).

Parker, R.J. *Analytical Methods for Permanent-Magnet Design: Part 1, Electrical Manufacturing; Part 2, Electro-Technology.* September/October 1960.

———. "Magnetizing and Demagnetizing Permanent Magnets." *Electrical Manufacturing* (September 1956).

———. *Properties and Potential Uses of Rare-Earth Permanent Magnets.* IEEE Intermag Conference, Denver, Colorado, April 1971.

———. "Trends in Permanent-Magnet Technology." *Insulation/Circuits* (July 1970).

———. "Understanding and Predicting Permanent-Magnet Performance by Electrical Analog Methods." *Journal of Applied Physics* 29 (3) (March 1958): 409-10.

———, and Studders, R.V. *Permanent Magnets and Their Applications.* New York: John Wiley & Sons, 1962.

Patzer, W.A. "Coil-Operated Controls Check for Spurious Coins." *Machine Design* (February 1951): 141-45.

"Permanent Electromagnetic Chuck for Tool-Post Milling." *Mass Production* (January 1970).

Permanent-Magnet Design. Bulletin No. M303. Indianapolis: Thomas and Skinner, Inc., 1962.

Permanent-Magnet Guidelines. Evanston, Ill.: Magnetic Materials Producers Association, November 1969.

"Permanent-Magnet Hoist." *Machine Design* (September 18, 1969).

"Permanent Magnet Key Element in Idaho Power Research." *American Metal Market* (December 28, 1970).

Permanent-Magnet Manual. Edmore, Mich.: General Electric Company, Magnetic Materials Section.

"Permanent Magnetic Separators Yield High-Grade Fe Concentrate at BC Mine." *Engineering and Mining Journal* (April 1963).

"Permanent Magnets Attract Industry." *Electrical Review* (October 10, 1969).

"Permanent Magnets Can Be Switched." *Steel* (February 8, 1965).

"Picking Right Motor Gets Tougher." *Steel* (October 13, 1969): 37-41.

"Pipe Processing is Speeded by Magnetic Roll System." *Steel* (June 26, 1961).

Pittman, U.J. "Growth Reaction and Magnetatropism in Roots of Winter Wheat (Kharkov 22 M.E.)." *Canadian Journal of Plant Science* 42 (1962): 430-36.

Platts, S. *Magnetic Amplifiers: Theory and Applications.* Englewood Cliffs, N.J.: Prentice-Hall, 1958.

"PM Armature Speeds Direct Magnetic Tape Drive." *EDN* (February 1, 1969).

"PM Panels Simplify Display Design." *Design Engineering* (March 1971).

Polgreen, G.R. "High-Capacity Lifting Magnets Based on Ferrites." *Electrical Review* (June 12, 1970).

———. *New Applications of Modern Magnets.* London: MacDonald & Company, 1966.

———. "Transport Possibilities with Magnetic Suspension." *Electrical Times* (England) (August 1965).

Porter, E. "Fractional Horsepower DC Motors." *Machine Design* (April 1970).

"Probing the Body's Canals by Magnet." *The New Scientist* (January 1970).

"Proximity Switches Control High-Speed Terminal Staking." *Assembly Engineering* (May 1970).

Puchstein, A.F. *The Design of Small Direct-Current Motors.* New York: John Wiley & Sons, 1968.

"Pulse-Actuated Solenoid Latches at Both Ends." *Electromechanical Design* (April 1970).

Rabenhorst, D.W. *Primary Energy Storage and the Super Flywheel.* APL/JHU Report No. TG-1081, September 1969.

Rahman, E.P. *Minor-Loop Hysteresis Losses in Hysteresis Torque Devices Using Permanent-Magnet Materials.* IEEE Intermag Conference, Washington, D.C., April 1970.

"Reed Switch Controls Electric-Watch Jogging." *Machine Design* (February 1970).

"Reeds—The Miniature Mechanical Relays." *Design Engineering* (January 1971).

"Reeds: Switches and Relays; Systems Designer Handbook." *Electromechanical Design* (April 1968): 87-97.

Reichart, K. "The Calculation of Magnetic Circuits with Permanent Magnets by Digital Computer." *IEEE Trans. on Magnetics* MAG-6 (June 1970).

Resler, E.L., Jr., and Rosensweig, R.E. "Magnetocaloric Power." *AIAA Journal* 1418, 2 (8) (May 1964).

"Resonant and Transient Theory Applied to Control Power in Demagnetization." *General Motors Engineering Journal* (October 1954): 34-37.

Rex, Harold B. "The Transductor." *Instruments* 20 (December 1947): 1102-09, and 21 (April 1948): 332, and 352-62.

Reynst, M.F. "Design of Small DC Motors Embodying Ferroxdure Field Magnets." *Electronic Applications Components and Materials,* Philips Technical Review, 26 (1) (1965/1966): 26-42.

———, and Langendam, W.T. "Design of Ferroxdure Loudspeaker Magnets." *Philips Technical Review* 24 (4/5) (1962/1963): 150-56.

Richards, P.L. *Magnetic Levitation for High-Speed Ground Transportation.* American Institute of Physics, 17th Annual Conference on Magnetism and Magnetic Materials, Chicago, Ill., November 1971.

Richardson, K.I.T. *The Gyroscope Applied.* Hutchinson's Scientific and Technical Publications, 1954.

Riches, E.E. "Magnetic Ceramics for Microwave Devices." *Ceramics* (January 1970).

Richter, E. *The Implication of New Permanent-Magnet Materials for Rotating Electrical Machinery.* American Institute of Physics, 17th Annual Conference on Magnetism and Magnetic Materials, Chicago, Ill., November 1971.

Riviere, M.R., et al. *Effects de Champs Electromagetiques sur un Lymphosarcome Lymphobastique Transplantable du Rat.* Paris: C.R. Acad. Se., (1965): 15 ferrier t. 260, pp. 2099-102.

Roberts, W.H., and Mitchell, S.L. *Effects of High Temperature on the Performance of Alnico V and Alnico VI Permanent Magnets.* Report APEX-384. U.S. Office of Technical Services, 1958.

Robinson, D.J., and Taft, C.K. "Dynamic Analysis of Magnetic Stepping Motors." *IEEE Trans. on Industrial Electronics and Control Instrumentation* (September 1969).

Rosener, A.A., and Hanger, A.W. "The Use of Permanent Magnets in Zero-Gravity Mobility and Restraint Footwear Concept." *IEEE Trans. on Magnetics* (September 1970).

Rosensweig, R.E. "Magnetic Fluids," *International Science & Technology* 55 (July 1966): 48.

———. *Preparation of Magnetic Ferrofluids in Alternative Carrier Liquids.* NASA Tech. Brief 70-10011 (July 1970).

Ross, I.M., et al. "The Hall-Effect Compass." *Journal of Scientific Instruments* 34 (1957): 479.

Roth, D., and Alksne, J. "Repair of Aneurysms in the Brain Using Colloidal Iron." *Boston Herald* (May 1, 1966).

Rotors, H.C. *Electromagnetic Devices.* New York: John Wiley & Sons, 1941.

"Rubber-Bonded Magnets." *Electrotechnology* (August 1964).

Rubin, M. *Practical Electricity and Magnetism.* Chemical Rubber Publishing Company (Tudor), 1951.

Rush, F. "Versatile Tool Holder for Diamond Dresser." *Modern Machine Shop* (July 1969).

Schindler, M.J. "Design of High-Coercivity Permanent Magnets Exposed to External Fields." *Communications and Electronics* (September 1961).

———. "Improved Procedure for Design of Periodic Permanent-Magnet Assemblies." *IEEE Trans. on Electron Devices* ED-13 (12) (December 1966).

———. "The Magnetic Field and Flux Distribution in a Periodic Focusing Stack for Traveling-Wave Tubes." *RCA Review* XXI (3) (September 1960).

Scholes, R. "Application of Operational Amplifiers to Magnetic Measurements." *IEEE Trans. on Magnetics* MAG-6 (June 1970).

Schoten, R.A. "Shortcut for Holding Magnet Design." *Product Engineering* (September 1957): 94-100.

Schuder, J.C., et al. "Response of Dogs and Mice to Long-Term Exposure to Electromagnetic Field Required to Power an Artificial Heart." *Trans. of the American Society of Artificial Internal Organs* 14 (1968): 291-95.

Schuringa, T.M. "Reed Switches for Telephony Switching." *Philips Telecommunication Review* 27 (3) (May 1968): 105-23.

Schwabe, E. "New Permanent-Magnet Couplings Using Sintered Oxide Materials." *The Engineers Digest* (April 1958): 137-55.

Scott, M. "Reeds: Switches, Relays, and Associated Devices." *Electromechanical Design, Systems Designers Handbook* (July 1967): 185-205.

Seely, E. *The Magnetization of Rare-Earth Magnet Materials.* Boonton, N.J.: RFL Industries, 1972.

Selection, Engineering, and Fabrication of Carpenter Alloys for Electronic, Magnetic, and Electrical Applications. Reading, Pa.: Carpenter Steel Company, 1964.

Seliger, R.L. "Analysis of the Expected Thrust Misalignment of Kaufman Thrusters." *Journal of Spacecraft and Rockets* (April 1970).

Sery, R.S., et al. *Radiation Damage Thresholds for Permanent Magnets* NOLTR61-45, White Oak, Md.: U.S. Naval Ordnance Laboratory, May 18, 1961.

"Shortcuts to Production Efficiency: Attracting the Right Position." *Electronic Packaging and Production* (August 1970).

Sisson, E.D. "Hall-Pak Magnetic Circuit Applications." *Electrical Design News* (February 1963).

Skinner, J.C. "Permanent Magnets, Materials, and Devices." *Industrial Research* (August 1965).

"Small Motors Attract Ceramic Magnets." *Iron Age* (November 13, 1969).

Smart, J.S. *Effective Field Theories of Magnetism.* Philadelphia: W.B. Saunders, 1965.

Smith, S.S. *Permanent-Magnet Design Equations.* IEEE Intermag Conference, Washington, D.C., April 1970.

Soderholm, L.G. "Flux Control Increases Force of Permanent Magnet." *Design News* (November 9, 1966).

———. "Ring-Magnet Drive Eliminates Thrust on Pump Impeller." *Design News* (March 30, 1966).

———. "Tape-Drive Armature Accelerates at 370 Gs," *Design News* (February 16, 1970).

Solov'eva, G.R. "Present State and Future Prospects of Using a Constant Magnetic Field in Medicine." *Biomedical Engineering* (May/June 1970).

"Special Report on Magnetic Materials." *Electronic Products* (June 1968).

Spreadbury, F.G. *Permanent Magnets.* London: Pitman & Sons, Ltd., 1949.

Staats, G.W. "Improvements in Performance of Bulk Separation Equipment Using Permanent Ceramic Magnets." *Journal of Applied Physics* 37 (3) (March 1966): 1154-56.

"Stability of Permanent Magnets." *Applied Magnetics* 15 (1) (1968): 1-12, Valparaiso, Ind.: Indiana General Corp.

Stablein, H. *A Review of Permanent-Magnet Materials.* American Institute of Physics, 17th Annual Conference on Magnetism and Magnetic Materials, Chicago, Ill., November 1971.

Standard Specifications for Permanent-Magnet Materials, MMPA Standard #0100-72. Magnetic Materials Producers Association, 3525 W. Peterson Avenue, Chicago, Ill. 60645.

"Standard Specifications for Permanent Magnets." *Journal of the Permanent Magnet Association* (England) 10 (April 1967).

"State-of-the-Art Keyboards." *Electromechanical Design* (July 1970).

Stefanides, E.J. "Flux is Carried Around Corners to Couple Intersecting Shafts." *Design News* (October 11, 1968).

———. "Permanent Magnets: Your Silent Servants." *Design News* (February 15, 1967).

"Stepper Motors." *Electromechanical Design, Systems Designers Handbook* (July 1967): 77-89.

"Stepper Motors." *Electromechanical Design, Systems Designers Handbook* (April 1968): 23-33.

"Stepper Motors and Controls." *Electromechanical Design* (October 1969).

Stern, R.A. *A Fast 3-Millimeter Ferrite Switch.* U.S. Army Electronics Command Report ECOM-3264 (April 1970).

Strnat, K.J. *Current Problems and Trends in Permanent-Magnet Materials.* American Institute of Physics, 17th Annual Conference on Magnetism and Magnetic Materials, Chicago, Ill., November 1971.

Studer, P.A. *Magnetic Bearings for Spacecraft.* IEEE Intermag Conference, Denver, Colorado, April 1971.

"Study of a System for Permanent Implantation of Cannulae." *Proceedings of the Permanent Magnet Users Association* (Philadelphia) (1969).

Stuijts, A.L., et al. "Ferroxdure II and III: Anisotropic Permanent-Magnet Materials." *Philips Technical Review* 16 (5/6) (1954): 141-47.

Sugatani, S., et al. "Magneto-Optical Display Element Using Permalloy Films." *IEEE Trans. on Magnetics* MAG-5 (3) (1969): 464-67.

Summers, G.D. "An Alternative Technique for Powering an Artificial Heart Through the Intact Skin." *Journal of Medical Instrumentation* 3 (1969): 129-34.

———. "Controlled Transport of Fluids Within the Body by a Passively Powered Implanted Pump." *Proceedings of the 19th Annual Conference on Engineering in Medicine and Biology* (Denver: November 1966).

———. "Electromechanical Control of an Artificial Leg." *IEEE International Convention Record* (Pittsburgh: March 1967).

———. *Magnetics for Power and Control of Body Implants.* 5th National Instrument Society of America, Biomedical Instrumentation Symposium, 1967.

———. "A Model for the Effects of Magnetic Fields on the Interaction Between Lymphocytes and Malignant Neoplasms." *Abstr. of the Neurolectric Conference,* San Francisco, 1969.

———. "Toward an Artificial Implant for Control of a Neurogenic Bladder." *Proceedings of 8th International Conference on Medicine and Biology,* Chicago, Ill., 1969.

———, and Mathews, E.S. "A New Miniature Pump for the Treatment of Hydrocephalus." *Journal of the Association for the Advancement of Medical Instrumentation* (May/June 1967).

Swords, D.L. "Magnetic Clutches for Continuous Slip." *Product Engineering* (August 28, 1961): 23-28.

Symposium on Magnetic Testing. American Society for Testing Materials, Philadelphia, Pa., 1948.

Taeler, D.H. "The Future of Hard Ferrites." *Ceramic Age* (August 1968): 28.

Taimuty, S.I. *Effects of Nuclear Radiation on Magnetic and Ferroelectric Materials and Quartz: A Literature Survey.* U.S. Department of Commerce, Office of Technical Services, PB161115, July 28, 1959.

Tape-Wound Core Design Manual. Camden, N.J.: Magnetic Metals Company.

Taren, J.A., and Gabrielson, T.O. "Radio Frequency Coagulation of Vascular Malformations with a Transvascular Magnetic Catheter." *IEEE Trans. on Magnetics* MAG-6 (June 1970).

Tebble, R.S., and Craik, D.J. *Magnetic Materials.* New York: John Wiley & Sons, 1969.

"Technology Abroad: A Lifting Magnet That Releases Its Load When Power Is Applied." *Electronic Design* (March 15, 1971).

Temperature and Radiation Effects on Permanent Magnets. Detroit, Mich.: General Magnetics Corp., 1966.

"Temperature Effects on the Remanence of Permanent Magnets." *Applied Magnetics* 9 (2) (1961): 3-9. Valparaiso, Ind.: Indiana General Corp.

Tenzer, R.K. "Magnet Design: Pt. 1—Estimating Leakage Factors for Permanent Magnets from Geometry of Magnetic Circuits; Pt. 2—Figuring Air Gaps for Maximum Pull of Opposing Electromagnets." *Electrical Manufacturing* (February 1957).

———. *A Simple Method to Calculate Leakage Factors for Magnetic Circuits with Permanent Magnets.* AIEE Paper T-91. Conference on Magnetism and Magnetic Materials, Boston, Mass. October 1956.

———. *Temperature Effects on the Remanence of Permanent Magnets.* Tech. Doc. Report ASD-TDR-63-500. U.S. Department of Commerce, Office of Technical Services, AD-420235.

"The Inverted Stator: Better Synchronous Motors." *Electromechanical Design* (August 1966): 40-44.

Thees, R. "Small Electric Motors." *Philips Technical Review* 26 (4/5/6) (1965): 143-47.

"Thickness Gauge." *Product Finishing* (January 1970).

Thornton, R.D. "Flying Low with Maglev." *IEEE Spectrum* (April 1973).

———, and Navon, D.H. *Application of Solid-State Electronics to Electric Propulsion.* No. PB173-620. Cambridge, Mass.: Massachusetts Institute of Technology, 1967.

Tillander, H. "Selective Aniography with a Catheter Guided by a Magnet." *IEEE Trans. on Magnetics* MAG-6 (June 1970).

Tippins, W.C., Jr. "Removal of Foreign Objects with

a Magnetic Field." *Journal of the Georgia Medical Association* (April 1963).

Trigg, T. "Customer Engineering Clinic: Don't Forget Magnetic Blowout." *EDN* (January 1, 1971).

Tustin, W. "Vibration-Detection and Protection Systems." *Engineers Digest* (December 1970).

"Twisting Jet Tube Forms a Low-Inertia Recorder Pen." *Machine Design* (December 1969).

Underhill, E.M., et al. *Permanent-Magnet Handbook.* Pittsburgh, Pa.: Crucible Steel Company, 1957.

VanderBurgt, C.M. "Ferroxcube Material for Piezomagnetic Vibrator." *Philips Technical Review* 18 (10) (March 1957): 285-97.

Verber, F., et al. "Development of an In-Core Permanent-Magnet, Probe-Type Sodium Flow-Sensor . . . ." *IEEE Trans. on Nuclear Science* NS-18 (February 1971).

Verhoef, J.A. "The Focusing of Television Picture Tubes with Ferroxdure Magnets." *Philips Technical Review* 15 (7) (1954): 214-20.

Von Aulock, W.H. *Handbook of Microwave Ferrite Materials.* New York: Academic Press, 1965.

Wachob, J.C., and Erlandson, J.C. "The Modern Permanent-Magnet DC Motor." *Westinghouse Engineer* 29 (2) (March 1969).

Waddington, C.J. "Paleomagnetic Field Reversals and Cosmic Radiation." *Science* 156 (1967): 913-14.

Wagner, I.F. "Oriented Barium-Ferrite, Straight-Field Focusing Structure." *IEEE Trans. on Magnetics* MAG-6 (June 1972).

Walker, J.H. "High-Frequency Alternators." *Journal of the IEE* (England) 93 (1946): 67-80.

Warbasse, J.R., et al. "Physiologic Evaluation of a Catheter-Tip Electromagnetic Velocity Probe." *American Journal of Cardiology* 23 (1969).

Watanabe, T., et al. "Transistor Motor: A Brushless DC Motr." *Toshiba Review* (January/February 1970).

Weak Magnetic Fields. Philadelphia: Sperry Rand Corp., 1969.

Weisman, P. *Magnetic Focusing of Beta Rays.* Memorandum No. 9091-6-6884-1. Pittsburgh, Pa.: Westinghouse Electric Corp., New Products Engineering Department, August 25, 1958.

"Welding at Work: Built in Magnets Reduce Hazards, Up Production." *Welding Design and Fabrication* (January 1970).

Went, J.J., et al. "Ferroxdure: A Class of New Permanent-Magnet Materials." *Philips Technical Review* 13 (1952): 195.

Wentworth, B.W., and Ellis, E.L. *Stabilizing Prediction for Permanent-Magnet Field Motor and Generators.* Chicago, Ill.: Magnet Materials Producers Association, November 11, 1965, conference papers.

Westendorp, F.F. *Recent Experiments on $SmCo_5$ Magnets.* IEEE Intermag Conference, Washington, D.C., April 1970.

White, B. "New Ideas for the Linear Induction Motor." *Power Transmission Design* (May 1969).

White, D.C., et al. *Some Problems Related to Electric Propulsion.* No. PB173-639. Cambridge, Mass.: Massachusetts Institute of Technology, 1966.

White, H.A., and Gigliotti, O.V. *Magnetization, Demagnetization, and Measurement of Permanent Magnets.* Kane, Pa.: Stackpole Carbon Company, 1970.

———. "Magnetization of Permanent Magnets." *Machine Design* (July 24, 1969): 128-31.

Wijn, H.P.J. (ed.), *Magnetism: Pt. 1—Ferromagnetism, Pt. 2—Paramagnetism.* New York: Springer-Verlag, 1966/1968.

Williamson, W.J. "Magnetoresistance Displacement Transducers." *IEEE Trans. on Magnetics* MAG-4 (2) (June 1968).

Wilson, S.E. "Remember the Magnet in Magnetic Reed Switches." *EDN* (1970).

Wojtowicz, P.J. "Semiconducting Ferromagnetic Compounds." *IEEE Trans. on Magnetics* MAG-5 (4) (December 1961): 840-47.

Woodring, E.D. "Magnetic Turbine Flowmeters." *Instruments and Control Systems* (June 1969).

Young, R.W. *Ceramagnet Design Guide.* St. Marys, Pa.: Stackpole Carbon Company, 1964.

Zinder, D.A. "Constant-Speed Motor Control Using Tachometer Feedback." *Control and Instrumentation* (March 1971).

Zingery, W.L., et al. "Evaluation of Long-Term Magnet Stability." *Journal of Applied Physics* (1) (March 1966).

Glossary

Symbols Used in Permanent-Magnet Design and Application

σ	–	Leakage flux
φ	–	Total flux or magnetic flux (general)
φ_g	–	Total flux at some defined area of the air gap in a magnetic circuit
φ_m	–	Total flux at some defined area in the magnet itself
μ	–	Permeability (absolute or general)
μ_0	–	Permeability (initial)
μ_d	–	Permeability (differential)
μ_i	–	Permeability (intrinsic)
μ_{max}	–	Permeability (maximum)
μ_r	–	Permeability (recoil)
μ_v	–	Permeability (relative or space)
A	–	Area (general)
A_g	–	Area of an air gap at some specific point
A_m	–	Cross-sectional area of a magnet at some specific point
AT	–	Ampere-Turns. Also written NI
B	–	Flux density or induction (general)
B_d	–	Remanence
B_g	–	Flux density in air gap
B_h	–	Hysteresis loss (also written as P_h)
B_i	–	Intrinsic induction
B_{is}	–	Saturation induction (intrinsic)
B_m	–	Flux density in the magnet at any operating point on the normal demagnetization curve
B_r	–	Residual induction
B_s	–	Saturation induction
BH_{max}	–	Maximum energy product (also written as $(B_d H_d)_{max}$)
B/H ratio	–	The relationship between the induction and the magnetizing force for a magnet under a particular set of operating conditions (same as permeance coefficient, load line, slope, or shear line)
EDM	–	Electrical discharge machining
emf	–	Electromotive force
ESD	–	Elongated single-domain (magnet materials)
f	–	Reluctance factor or coefficient
F	–	Mechanical force, leakage factor, or magnetomotive force (all general)
F_p	–	Mechanical force (pull or tractive)
F_r	–	Mechanical force (repulsion)
H	–	Magnetizing force (general)
H_c	–	Coercive force (normal)
H_{ci}	–	Coercive force (intrinsic)
H_{cr}	–	Coercive Force (relaxation)
H_{cs}	–	Coercivity
H_d	–	Demagnetizing force
H_g	–	Magnetizing force in air gap
H_m	–	Magnetizing force in magnet (also written as H_d)
H_s	–	Magnetizing force required to saturate magnet
J	–	Intensity of magnetization
K	–	Magnetic susceptibility
kg	–	Kilogram
L	–	Length (general)
L_g	–	Length of air gap

L_m	–	Mean length of magnet
L/d ratio	–	Dimensional ratio (also called length/diameter ratio and written as L/D ratio)
←M→	–	Preferred direction of magnetization to obtain best properties (anisotropy)
MGO	–	Million Gauss-Oersteds (G-Oe × 10^6)
M	–	Unit magnetic pole, north (also written as m)
M'	–	Unit magnetic pole, south (also written as m')
mmf	–	Magnetomotive force
$n\phi$	–	Flux interlinkages
N	–	North pole of magnet
NI	–	Ampere-turns (also written as At.)
P	–	Permeance, pull, or power (all general)
P_c	–	Core loss or iron loss
P_e	–	Eddy-current loss
P_g	–	Permeance of air gap
P_T	–	Permeance (total)
r_f	–	Reluctance coefficient
R	–	Reluctance (general)
R_g	–	Reluctance of air gap
R_T	–	Reluctance (total)
S	–	South pole of magnet
T	–	Torque (general) or temperature
t	–	Temperature (general)
T_c		Curie temperature (also written as t_c)
ν	–	Reluctivity
V	–	Volume (general)
V_m	–	Volume of magnet
W	–	Weight or work (general)
W_0	–	Work (general)
W_m	–	Weight of magnet

Terms and Definitions Used in Permanent–Magnet Design and Application

Air gap: A nonmagnetic discontinuity in a ferromagnetic circuit.

Ampere-turn (NI or At): *A unit of magnetomotive force.*

Anisotropic: When magnetic properties are not the same in all directions. The preferred direction of magnetization to obtain best properties is indicated by ←M→.

Anisotropy: Relates to the importance of direction in determining properties of a magnetic material.

Area (A): The cross-sectional area of a particular volume of space.

Cast process: The forming by means of molten metal poured into some type of mold.

Ceramic magnet or materials: Magnets or materials based on barium, strontium, or lead ferrite.

Closed circuit, magnetic: A magnetic path without air gaps of any significant size.

Coercimeter: An instrument for measuring the normal and intrinsic coercive properties of a magnetic material or of a sample taken from a basic magnet. (See also Permeameter.)

Coercive force, intrinsic (H_{ci}): The demagnetizing force required to reduce the intrinsic induction to zero.

Coercive force, normal (H_c): The demagnetizing force required to reduce the normal induction to zero. Usually designates the demagnetizing force required for a fully saturated magnet. The term "normal" is frequently omitted in normal usage.

Coercive force, relaxation (H_{cr}): The reversed magnetizing force which, when reduced to zero, results in zero induction.

Coercivity (H_{cs}): The property of a magnetic material measured by the maximum value of its coercive force.

Core loss (P_c): The power expended in a magnetic material when it is subjected to a changing magnetizing force.

Curie temperature (T_c): The temperature at which a material changes from a ferromagnetic to a paramagnetic state.

Cyclically magnetized state: A magnet or magnetic material in such a state of operation that it is under a magnetizing or demagnetizing force which changes it from one level or direction of magnetization to another. The successive hysteresis loops for the material are identical.

Demagnetization curve: The second quadrant portion of the saturated condition hysteresis loop of a permanent-magnet material (frequently called the B/H curve).

Demagnetizer: A device for reducing or eliminating the remanent induction in a magnet.

Demagnetizing force (H_d): A force applied to a previously magnetized material in such a way that it reduces the remanent induction to zero.

Diamagnetic material: Any material that is repelled by a magnetic field. The permeability of such materials is less than 1. Diamagnetic materials include bismuth and several of the rare earths.

Dimensional ratio (L/d ratio): The ratio of the mean length of a magnet to its diameter. For an open-circuited magnet, the dimensional ratio can be related to the B/H ratio.

Direction of magnetization: The direction, in relation to the configuration of the magnet, in which the magnetizing force is applied. Ideally, the direction of magnetization coincides with the direction of orientation of the material (where the material is oriented).

Dry press: The process by which some types of magnets are fabricated by compaction from a dry powder or a granular form of the basic material. (*See also* Sinter.)

Dyne: The force producing an acceleration of 1 cm/sec when applied to a mass of 1 g.

Eddy current: An electron flow produced in a conducting material by an induced electromotive force (emf) that is commonly induced by the relative motion between the conductor and a magnetic field.

Eddy–current loss (P_e): The power loss ($I^2 R$ loss) produced by the flow of eddy currents in a conducting material.

Electrical discharge machining (EDM): A process which utilizes electrical sparks to form cavities of desired shapes in hard materials. (Used to perforate very hard, metallic magnet materials.)

Electromagnet (general): A device consisting of a current-carrying winding that produces a specific magnetic field. (Usually used in reference to an iron-core type unit.)

Electromagnetizer: *See* Magnetizer, electromagnetic.

Elongated single-domain (EDS) *magnets or materials:* Permanent magnets made of domain-sized particles of pure or alloyed iron in a suitable matrix.

Energy (W): *See* work.

Energy product curve: A curve obtained by plotting the product of B_d and H_d against B_d.

Erg: The work done by a force of 1 dyne moving through a distance of 1 cm.

Ferrite material: May refer to either hard (permanent) or soft magnetic materials. When used in reference to permanent-magnet materials, the material may be barium, strontium, or lead ferrite—all oxides of iron. Barium ferrite is most common. (Frequently called "ceramic" or "oxide" magnets.)

Ferromagnetic material: Any material that has a permeability substantially greater than 1 and that exhibits hysteresis properties. Strongly attracted by a magnetic field.

Flexible magnets or materials: A general class of permanent-magnet materials based on a ferrite dispersed in a rubber, vinyl, or other plastic matrix.

Flux density (B): Lines of flux per unit area.

Flux, total (ϕ): *See* Magnetic flux.

Fluxmeter: An instrument, instrument system, or method for measuring the total flux produced in a specific area. Most often, a device consisting of a highly damped, taut-band galvanometer combined with the necessary calibrating controls and individually made "search coils."

Function: The end purpose that a magnet or magnet assembly is intended to serve.

Galvanometer, ballistic: A highly damped, highly sensitive instrument used in many types of magnetic measurements.

Gauss: The cgs unit of magnetic induction.

Gaussmeter: An instrument for measuring the flux density of a magnetic field. Most commonly refers to Hall-effect instruments.

Gilbert: The cgs unit of magnetomotive force.

Half-cycle magnetizer: *See* Magnetizer, half-cycle.

Hard magnetic material: Any material that exhibits ferromagnetic properties and that has a substantial remanence after exposure to a magnetizing force.

Hysteresis: A material property where the condition at any instant depends upon the preceding condition.

Hysteresis loop (major): A curve showing the full cyclic relationship between magnetizing force and induction in a magnetic material. (The designation "major" is usually omitted.)

Hysteresis loop (minor): A graph of the relationship between magnetizing force and induction in a magnetic material when the air gap of the magnet or the circuit in which it operates is cyclically changed.

Hysteresis loss (E_h or P_h): The energy expended in a material when it is cycled over the complete range of its possible conditions.

Impulse magnetizer: *See* Magnetizer, impulse.

Induced poles: *See* Pole, induced.

Induction curve, intrinsic: The curve showing the relationship between intrinsic induction and magnetizing force.

Induction curve, normal: The curve showing the relationship between normal induction and magnetizing force in a magnetic material; the first quadrant of the hysteresis loop. (The designation "normal" is usually omitted.)

Induction, intrinsic (B_i): The excess induction in a magnetic material to the induction in a vacuum for a specific magnetizing force.

Induction, magnetic (general) (B): The flux per unit area measured at right angles to the direction of the flux.

Induction, residual (B_r): The induction remaining in a

closed ring of magnetic material when the magnetizing force adequate to saturate the material is reduced to zero.

Induction, saturation, intrinsic (B_{is}): The largest induction possible in a magnetic material.

Induction, saturation, normal (B_s): The maximum induction in a magnetic material.

Intensity of magnetization (J): The number of "unit poles" per unit of area.

Interlinkages, flux ($n\phi$): The product of the number of turns in a circuit and the average value of the flux linking the circuit.

Intrinsic coercive force (H_{ci}): See Coercive force, intrinsic.

Intrinsic induction curve: See Induction curve, intrinsic.

Investment casting (lost-wax process): The process by which wax replicas of the final part are used to form molds, and these molds are, in turn, used to cast the final part in metal after the wax has been melted out. This process is best for dimensional control and for intricate parts. It is the most costly magnet-making process.

Iron loss: See Core loss.

Isotropic: When magnetic properties are substantially the same in all directions. Commonly used in referring to unoriented magnets or materials.

Keeper: A ferromagnetic part temporarily added to a magnetic circuit for the purpose of forming a closed-circuit. (Used to protect magnets in handling, shipping, etc.)

Leakage factor (σ or F): The ratio of the total flux present at the neutral section of a magnet to the useful flux produced by that magnet.

Leakage flux: That portion of the magnetic flux that exists in regions other than the working air gap.

Line of flux: A term used to describe magnetic flux. (1 line of flux = 1 Maxwell)

Line of force, magnetic: See Magnetic line of force

Load Line: The line on the demagnetization curve of a magnetic material that shows the relationship between residual induction and coercive force under operating conditions. (Also called the B/H ratio or the shear line.)

Magnet assembly: A physical entity made up of one or more basic magnets and their auxiliary parts to direct, contain, shunt, or control the flux, as well as the necessary hardware to secure these components.

Magnet, basic: A single physical entity made up entirely of a permanent-magnet material.

Magnet, standard: A basic magnet or a magnet assembly whose performance is precisely known and can therefore be used as a reference in magnetic testing. Also, a basic magnet that is commonly available commercially from several sources without tooling costs and in any quantities. (*See also* Magnet, stock.)

Magnet, stock: A geometric shape of a given permanent-magnet material that is readily available from both basic producers and distributors without tooling changes and in any quantity.

Magnetic circuit: The path followed by magnetic lines of flux within or emanating from a source of such flux.

Magnetic field: A region in which magnetic lines of flux or force occur.

Magnetic field strength (H): The magnitude of the force that produces a magnetic field.

Magnetic flux (ϕ): The total quantity of lines of flux that exist in a given area.

Magnetic induction (B): See Flux density or Induction, magnetic (general).

Magnetic line of force: An imaginary line representing the points in a magnetic field that produce the same force on an infinitely small ferrous object.

Magnetic pole: A region where lines of flux appear to be concentrated.

Magnetic potential difference: The line integral to the magnetizing force between two points in a magnetic field.

Magnetizer, electromagnetic: A device for applying a magnetizing force to a permanent magnet based on a continuous flow of direct current in a winding that surrounds an iron core. (Available in a wide variety of configurations.)

Magnetizer (general): Any device for applying a magnetizing force to a permanent magnet—another permanent magnet, an electromagnet, or a capacitor-charge/discharge system.

Magnetizer, half-cycle: A device for applying a magnetizing force to a permanent magnet based on an electrical coil (with or without an iron core), where one-half of a single cycle of alternating current (a sinusoidal pulse of DC) is applied to the winding.

Magnetizer, impulse: A power source used in conjunction with single conductors, air-core windings, or iron-core windings to magnetize permanent magnets. The power source is essentially a small rectifier, capacitor storage bank, and high-energy switching means for discharging the stored energy into the conductors or winding. (Also called a capacitor/discharge magnetizer.)

Magnetization curve (general): The first quadrant of the hysteresis loop of a magnetic material.

Magnetization curve (virgin or initial): The portion of the first quadrant of a hysteresis loop that shows the relationship between magnetizing force and induction *for a magnetic material magnetized from an initially *completely* demagnetized state.

Magnetizing force (H): The magnetomotive force per unit of magnet length.

Magnetometer: Usually, a highly sensitive instrument used to measure very small magnetic fields. Also a wide variety of laboratory or production testing devices used when two successive tests are made on a magnet at different operating levels.

Magnetomotive force (mmf or *F*): That phenomenon which tends to produce a magnetic field.

Match plate: Tooling used to manufacture magnets using the shell-molding process.

Maximum energy product $(BH_{max}$ or $(B_d H_d)_{max})$: The highest product of *B* and *H* from the demagnetization curve of a magnetic material.

Maxwell: The cgs unit of total magnetic flux.

Mechanical force (F or P): The attraction or repulsion exerted by a magnetic field on a ferrous object or by the interaction of two or more magnetic fields.

Metallic magnets or materials: Those magnets or materials based on alloys of pure metal (like Alnico, Cunife, and Cunico).

Minor hysteresis loop: See Hysteresis loop (minor).

Nonmagnetic material: Any material that is totally unaffected by a magnetic field. For practical purposes, the permeability of such materials is substantially the same as that of a vacuum.

Normal induction curve: See Induction curve, normal.

Normal permeability: See Permeability, absolute.

North Pole (N): The portion of a magnetized object that, if free to move, will point toward the portion of the Earth geographically designated as North. By accepted magnetic convention, lines of flux emanate from the north pole and enter the south pole.

Oersted: The cgs unit of magnetizing force.

Open-circuit, magnetic: Usually, a magnet without the necessary supplemental ferromagnetic parts to form a complete path which does not, in itself, form a complete circuit. Also, any magnetic circuit that is not fully continuous (that contains an air gap).

Operating point: The *B/H* ratio, the characteristic curve of the material, and the state of magnetization at the instant under consideration.

Orientation: The alignment of all or a substantial portion of the available ferromagnetic domains in the magnet material. May be accomplished by heat treatment, cold-working, pressing in a magnetic field, or by some combination of methods.

Paramagnetic material: Any material whose permeability is slightly greater than 1. (Weakly attracted by a magnetic field.)

Permeability, absolute (μ): The ratio of the flux density in a material to the magnetizing force producing it.

Permeability, differential (μ_d): The slop of the normal induction curve.

Permeability, incremental $(\mu\Delta)$: The ratio of cyclic change in induction to change in magnetizing force when the induction is greater than zero.

Permeability, initial (μ_0): The slope of the induction curve when induction and magnetizing force approach zero.

Permeability, intrinsic (μ_i): The ratio of intrinsic induction to the corresponding magnetizing force.

Permeability, maximum (μ_{max}): The greatest value of the slope of the induction curve.

Permeability, normal: See Permeability, absolute.

Permeability, recoil (μ_r): The average slope of the minor hysteresis loop.

Permeability, space or relative (μ_v): The ratio of the flux produced in a material to the flux produced by the same magnetizing force applied to a vacuum.

Permeameter: An instrument for measuring the induction and coercive properties of a permanent-magnet material. Normally applicable to limited sizes, regular sections, or samples of straight-line oriented materials. (*See also* Coercimeter.)

Permeance (P): A term describing the relative ease with which flux passes through a given material or space. The reciprocal of reluctance.

Permeance coefficient: Same as the *B/H* ratio.

Pole, induced: A region of apparent flux concentration produced by a magnetizing force from a separate source than the magnetic part in which the induced pole occurs.

Pole piece: A soft magnetic member added to a magnet to direct, distribute, or otherwise control the magnetic flux in the manner necessary for the device to function. A pole piece may also be added for protective or environmental reasons.

Potential difference, magnetic: See Magnetic potential difference.

Power (P): Work per unit of time.

Pull (P or F): The mechanical attractive force exerted on a ferromagnetic object by a magnetic field.

Relaxation coercive force: See Coercive force, relaxation.

Reluctance (R): The relative resistance of a material or space to the passage of flux. The reciprocal of permeance.

Reluctance coefficient $(r_f$ or *f*): The ratio of the total magnetomotive force produced by a magnet to the magnetomotive force in its air gap.

Reluctivity (v): The reciprocal of permeability.

Remanence (B_d): The magnetic induction retained by a magnet (with an air gap) after the initial magnetizing force is removed.

Remanent induction: See Remanence.

Residual induction: See Induction, residual.

Retentivity: Same as residual induction.

Rubber magnet or material: Commonly refers to barium ferrite materials dispersed in a rubber matrix, but may also refer to such materials in vinyl or other plastics. (*See also* Flexible magnets or materials.)

Saturation: A condition where all of the available elementary magnetic moments in a ferromagnetic material are aligned in substantially the same direction.

Saturation induction: See Induction, saturation, normal.

Search coil: A winding of fine wire used with suitable instruments to measure total flux or flux density in a particular region of space.

Shear line: Same as load line or *B/H* ratio.

Shell molding: The process by which a prepared sand-resin mix is used with a pattern (match plate) to form a thin-walled mold from which the final part is cast in metal. This is the most widely used process for making metallic magnets.

Shield: A soft magnetic material used to prevent the passage of magnetic flux between regions.

Shunt: A soft magnetic material used to by-pass, divert, or redirect the magnetic flux from the primary air gap.

Sinter: The process by which a finely divided solid is compacted under heat and pressure until it is re-formed into a desired shape. Usually, the method of making small Alnico permanent magnets, but may also be used in making ferrites. (*See also* Dry press.)

Slope: Same as *B/H* ratio.

Soft magnetic material: Any material exhibiting ferromagnetic properties but having a remanence induction that is substantially zero after exposure to a magnetizing force.

South pole (S): The portion of a magnetized object that, if free to move, will point toward the portion of the Earth geographically designated as South. By accepted magnetic convention, lines of flux emanate from the north pole and enter the south pole.

Stabilization: Treatment of a magnetic material or a magnet by thermal, electrical, physical, magnetic, nuclear, or other means to preclude or minimize future changes in performance.

Standard magnets: See Magnet, standard.

Susceptibility (K): The ratio of the intensity of magnetization to the magnetizing force.

Tolerance ($\pm$): The allowable limits of variation in the properties of a material or object.

Torque (T): A force acting through a distance from the center of rotation of an object and producing an angular displacement of that object.

Total flux (ϕ): See Magnetic flux.

Tractive force: See Pull.

Unit poles (m *and* m' *or* M *and* M'): A fictitious concept used by early magnetic investigators to assign values to the intensity of force exerted between two magnetic bodies in free space and divorced of any association with a pole of opposite polarity in the same body.

Weber: The mks unit of total flux.

Wet press: The process by which most ferrite magnets are made. The ferrite powder is dispersed in a liquid (water) slurry to facilitate forming it into a final shape and to develop its optimum properties.

Work (W_0): A force acting through a distance.

Index

Absolute permeability, definition of, 377; equation for, 49
AC demagnetizers, 161, 167
AC motors, 229, 230. *See also* Stator magnets; Rotors
Actuation methods for reed switches, 214-219
Advantages and disadvantages, general, 183; of magnetically operated products, 210; of PM vs. EM operated products, 211; of various types of material, 183-184
Aging, 193
Air gap: in couplings, 87, 104; definition of, 10, 39; effect of objects in, 77; maximum, 61, 68, 77, 94; requirement for, 39; variable, 78, 94
Alni (Alnico): origin of, 5
Alnico: classification, 15; comparison with other materials, 183-184; demagnetization curves and work sheets of all common materials, 269; directional grain, 23, 25; environmental effects, 110-127; manufacturing processes, 22-30, 172; origin of, 5; properties of various types and grades, 269, 286, 296, 297
Alnicos, cast, oriented: crystal method, 23, 25; heat treat method, 23, 24; manufacturing process, 22-26
Alnicos, cast, unoriented: properties of, 286
Alnicos, sintered: heat treat oriented, 26; manufacturing process, 23-26; unoriented, 26
Alternators: typical rotor configurations, 72, 226
Alvord, C.: coating thickness gauge, 226
Ambient, effect of. *See* Environmental effects
Analog (electrical) of magnetic circuit, 56, 59, 80
Aneurysm-repair (in humans), 261, 265
Angular velocity unit conversions, 358

Animal protection, 258
Applications of permanent magnets: by industry, 11; by products, 12
Arc quenching, 211
Arc shaped magnets. *See* Horseshoe shaped magnets
Arc welding, 259
Area of magnet required for specific application, 49
Area unit conversions, 356
Artificial: eye, 260; heart, 260
Assemblies, magnet: definition of, 9; simple holding/reaching types of, 65-69
Assembly, magnetization after, 161
ASTM null method solenoid tester, 129. *See also* Permeameters
Automatic magnetizers, production of, 163, 164
Automation equipment, 234
Automobile: other applications, 229; speedometers, 232
Availability of commercial materials: classification, 19; types from each U.S. mill, 184
Axial forces in coupling, 104
Axial pole tube magnets. *See* Tube magnets

Babbit permeameter, 130
Baked sand molding, 15
Baking pan separation-recycling system, 249
Bar/rod shaped magnets: fixtures for heat treat orientation of Alnicos, 27, 28; fixtures for magnetizing, 156, 157, 164, 165, 168, 169; flux densities, 47, 48; L/D to B/H ratio conversions, 45; L/t to B/H ratio conversions, 45
Barium ferrite materials: bonded (composite), 14; dry pressed, unoriented, 13; machining, 173; wet pressed, oriented, 26, 31, 32. *See also* Ferrites; Magnets

Barrel, wood: fork truck handling attachment, 249
Basic permanent magnet: definition of, 9
Beam balance indicators, 212
Beam switching tubes, 221
Becker magnetizer/demagnetizer, 161
BH_{max}. *See* Maximum energy product
B/H ratio: conversions form L/D ratios, 45, 46; definition of, 44; optimum for various materials, 286, 291
Biasing of reed switches, 220, 221
Biomedical applications, 213
Bismonol: origin of, 5; properties of, 313
Bismuth coil-resistance bridge gaussmeter, 135
Blocking poles, 54. *See also* Leakage and fringing fluxes
Bobbin controls for textile machines, 233
Bonded materials, 28. *See also* Rubber bonded materials; Vinyl bonded materials
Bottle cap control system, 251
Boundaries, domain, 37
Bowlshaped magnets: fixture for heat treat orientation of Alnicos, 27
Br. *See* Residual induction
Braking system for bottle caps, 251
Brinell hardness to Rockwell conversions, 353-355
Burglar alarms for windows and doors, 212
Burrows permeameter, 130
Buus, H.W.: drum separator element, 238

Calculations: using computer methods, 56; energy/work, 101; exact, 94; external fields using demagnetization curve, 96; permeance, 55; pull vs. long distance, 107; simplified design, 49; torque coupling, 87, 90; unit conversions, 345

Calibration, 10, 170
Cam timer, 212
Can: conveying and special handling, 247, 248; routing system, 251; testing system, 248; washing/inverting system, 248
Capability: general, 39; as related to recoil permeability, 103; restoring, 103
Capacitor discharge type magnetizers, 151
Capacity requirements for magnetizers, 150, 160
Cardiac pacers, 213
Catheter guidance system, 260
Cathode ray tubes (focusing), 223
Ceiling tile holding system, 84
Cementing, 177
Ceramic materials. See Ferrites
Cerium-copper materials: origin of, 6; process, 29; properties of, 314. See also Rare earth magnet materials
CGS to English or MKS unit conversions, 345
Changes in performance due to environmental effects, 108
Chattock potentiometer, 140
Chemical: composition of various materials, 195, 201, 296; properties, 178
Chipping, 171
Chrome steels: process, 32; properties of, 124, 125, 297
Chute type separators, 240, 241, 244, 246
Circuit effects, 61
Circuit performance testing, 144
Circuits, magnetic: definition of, 9; multiple magnet types, 85
Clamping, foundry flask, 250
Classification of materials: categories, 11; by commercial availability, 19; by corrosion resistance, 17; by direction of magnetization, 17; by electrical properties, 18; by energy per unit volume, 19; by energy per unit weight, 20; by machinability, 17; by magnetic properties, 15, 21; by manufacturing process, 15; by material type, 13; by physical properties, 17; by relative cost, 18; by shape, 13
Clegg, A.G., 1
Clock motors, 229
Closed rings (no air gap), 39
Coating/painting, 175
Coating racks, 257
Coating thickness gauges, 226
Cobalt-platinum. See Platinum-cobalt
Cobalt steels: process, 32; properties, 124, 125, 288
Coefficients of leakage, 50, 55
Coefficient of thermal expansion of various materials, 296
Coersive force (normal): comparison table, 21; materials classified by, 16, 21; relationship to intrinsic coersive, 127; standard tolerances, 181; tables, 285. See also Intrinsic coersive force
Collandering (process), 15
Commercial materials classified by availability, 19
Communications, electronics applications of, 220
Comparison of various types of materials, 183
Comparison magnetometer, 145
Compensators, temperature. See Temperature compensation
Composite materials, 14. See also Bonded materials
Composition of various materials, 293
Computer calculation of magnetic circuits, 56
Concentration of flux: occurrence and effects of, 74-76
Concrete form holders, 256
Configuration/dimension guide for sintered ferrites, 173, 174
Construction industry and equipment, 256
Contact demagnetization. See Spurious contacts, effects on performance
Contacts, spurious. See Spurious contacts, effects on performance
Continuous: belt separators, 239, 241, 247, 248; demagnetizers, 168; foundry flask clamping, 250
Contraceptive loops, 258
Control circuits for magnetizers, 151, 158, 160, 161
Control system for bottle caps, 251
Conversion curves and tables: L/D ratio to B/H ratio, 45, 49; various magnetic and physical units of measure, 345
Conveyor, 234; rolls, 255, 256; slider beds (rails), 247; vibratory, 231
Correlation of tests, methods for, 146
Corrosion resistance: classified by, 17; general factors of, 178; of pumps, 235; of various materials, 295
Costs, relative: of demagnetizers, stabilizers, and calibrators, 170; effect of size, shape, and orientation, 171; of machining, 174; of magnetizers, 166; of materials, 182; test equipment, 147; of tooling, 171
Couplings: calculation of, 90, 104; circuits for, 233, 234; design of, 86, 87, 104; force characteristics of, for three standard side pole types, 90; synchronization of, 87; tote box, 254
Courses in permanent magnet design, 1
Cow magnets, 258
Cracks, chips, burrs, voids, inclusions, 171
Critical temperature for various materials, 111
Cross-belt separators, 239
Cross-reference of all U.S. and foreign materials, 303
Crushing rolls. See Rolls, magnetic
Cunico: origin of, 5; process, 33; properties of, 111, 316
Cunife: origin of, 5; process, 33; properties of, 111, 119, 124, 125, 288, 298, 316
Cup-shaped magnets: fixture for heat treat orientation of Alnicos, 27
Cutting and machining. See Machining
Curie alloys of temperature compensation, 178

D'Arsonval: magnetometer for testing, 146; meter movements, 225
DC bias/AC inrush type magnetizers, 161
DC electromagnet type magnetizers, 154, 156-160
DC magnetization curves, 79
DC motors, 228
Decimal to fraction conversions, 351
DeCleen, D.: clamping method for foundry flasks, 250
Defects, physical, 171, 181, 190, 207
Definitions (and terms): limited, 8-10; complete, 373
Degreasing racks, 257
Demagnetization: in couplings, 87; definition of, 10; due to environmental effects, 108; due to spurious contacts, 115, 126; methods (deliberate), 167; relative resistance of various materials, 283; self-demagnetization, 39, 43, 97
Demagnetization curve, intrinsic: comparison with normal curve, 127; use in exact design method for effects of external demagnetizing fields, 96
Demagnetization curve, normal: determining potential energy from, 101; relationship to hysteresis loop, 35-39, 43; use in design, 39, 94; use of for variable G and P calculations, 94; worksheets for common materials, 269
Demagnetization effects of external fields: calculation of, 96
Demagnetizer(s): costs of, 170; definition of, 167; sources of standard units for, 186; types of, 167
Demagnetizing force (-Hc): in relation to hysteresis loop, 38, 39
Density (weight) of various materials, 293
Density, flux. See Flux density
Dental plate retainer, 264
Design calculations: computer methods in, 56; equations used in, 57; exact method of, 94; simplified method of, 44, 49; unit conversions for, 345
Design considerations, general, 40, 171
Design problems: calculating pull vs. distance, 107; calculating volume, configuration, and material to produce a specific flux in an air gap, 49; flux density that will be produced in

a specific circuit, 56, 69; torque coupling, 104
Diametrical pole tube magnets. See Tube magnets
Die casting of magnets, 176
Die sets (holding assemblies), 258
Dies, wet pressed-sintered ferrites: construction and operation, 32
Dimension conversion tables, 345
Dimensions and tolerances, MMPA standard, 202-205
Direction of magnetization, classification, and standard methods, 17, 18
Disadvantages and advantages, general: of all magnetically operated products, 210; of PM vs. EM operated products, 211; of various types of materials, 183
Disk-type separators, 240. See also Couplings
Distance, effects of: on pull, 107
Distributor-fabricators: industry structure and function, 8, 172; list of U.S. distributor-fabricators, 185
Domains: boundaries, 36-37; rotation, 36-37, 108; saturation induction, 37, 38; theory, 35, 36
Door/window alarms, 212
Drag system: for bottle caps, 251; for rotating shafts, 233; for steel tote boxes, 254
Drilling, 172
Drives. See Couplings
Drum separators: general construction, 236; magnetic circuits, 237, 238
Dry pressing (process), 15
Drying system for cans, 248
Dutch (Holland): list of producers, 185
Dynamic operation (of a magnet). See Air gap, variable

Eddy current: in couplings, 88; speedometers, 232; synchronous couplings, 233. See also Couplings
Effect of magnet location in circuit, 53
Efficiency: length, 47, 48; volumetric, 19
Eh. See Hysteresis loss
Electrical analog of magnetic circuits, 56, 59, 80
Electrical methods of controlling the strength of a magnet, 87
Electrical properties, 18, 178; tables of, 293
Electrical resistance classified by, 18
Electrodeposition process, 15
Electromagnet type (DC) magnetizers, 154, 186
Electromagnets: general advantages and disadvantages, 210; for magnetizing, 154, 186; used with reed switches, 218
Electron beam focusing, 91, 92
Electronics/communications applications, 220
Electroplating, standard method for Alnico magnets, 181

Elevating conveyors, 247, 249, 251, 253
Elongated single domain materials, 5, 14. See also Lodex or ESD materials
Encapsulating, 176
Energy capability: general, 39; as related to recoil permeability, 103; restoring, 103
Energy loss (Wc) for various materials, 283
Energy per unit volume: comparison of common materials, 19
Energy per unit weight: comparison of common materials, 20
Energy, potential, 101, 104
Energy product: classified by, 16, 21; maximum, 283; relative, 21
Energy-work-heat: calculations, 101; unit conversions, 345
English producers, 185
English to CGS or MKS unit conversions, 345
Environmental effects, 108, 190: B/H ratio for ferrites to avoid temperature effects, 123; changes with elevated temperature for various metallic materials, 113-119; changes with reduced temperatures for various Alnicos, 120; changes with temperatures of various ferrites, 121, 122; critical temperatures for various materials, 111; external fields on various Alnicos, 127; general curves for temperature effects on Alnicos and ferrites, 112, 119; impact on various metallic materials, 124; mechanical stress of cunife, 124; nuclear effects on various materials, 125; remagnetizable time change for various materials, 110, 111; shock on various grades of Alnicos, 124; spurious contacts on cast Alnicos, 126; vibration, shock, and mechanical stress, 113; vibration on various metallic materials, 124
Equations: permeance, 55; used in all magnet design, 57
ESD materials, 5, 14. See also Lodex and Elongated single domain materials
Ewing isthmus permeameter, 129
Ewing permeameter, modified, 129
External fields: calculation of effects by using demagnetization curves, 96; effect on various Alnicos, 127; in environment, 108
Extrusion process, 15

Fabricating of magnets for stock materials: general, 171, 172; list of U.S. distributor-fabricators, 185
Factors: that affect performance, 40, 108; leakage, in simplified design calculation method, 41; in material selection, 108, 183; reluctance, in simplified design calculation method, 40

Fahy simplex permeameter, 130
Fahy super H permeameter, 129
Fanners, sheet steel: construction, 252; illustrations, 253, 263; principle of operation, 251
Fastening, 176
Federal specifications, 180
Feeders and conveyors, vibratory, 231
Ferrite materials, 5, 13, 26; properties, 289, 299
Ferrites, hard: bonded, 14, 28; classification, 13; demagnetization methods, 167; die cost, 27; die structure, wet pressed ferrites, 32; origin, 4; pressed and sintered, 26; process and process chart, wet, 26, 31, 32, 173
Ferrites, oriented and unoriented: dry pressed, 13; wet pressed, 26, 31, 32
Ferromagnetic materials: definition of, 8
Fiber cording crushing rolls, 64
Field reversals in magnetizers, 161
Fields, magnetic: definition of, 9; external fields, calculation of effects of, 96; required to saturate various materials, 283; shape and magnitude for magnetizers, 150; testing for pattern of, 137
File cards, self-separating, 263
Filters: open channel, 246; pipe line, 246
Finishes, surface: for Alnico magnets, 181; imperfections in, 171; QQ-M-60, 197; standard method for electroplating, 181; supplemental, 175, 182; types of, 197, 202-204
Finishing methods for various materials, 293. See also Manufacturing processes
Fixtures and methods: for heat treat orientation of Alnico magnets, 26-29; for magnetizing, 152-156; for measurement testing, 138, 141, 142
Flags, magnetically held, 262, 263
Flasks, foundry clamping, 250
Flat grate separators, 244. See also Grate separators
Floaters for sheet steel, 251-253. See also Fanners, sheet steel
Floor sweepers, 243, 244, 246
Flow gages, 212
Fluid pumps, 235. See also Couplings
Flux concentrations, 51, 74-76
Flux density (B): in an air gap, equation for, 48, 49; of bar magnets of ends and at various air gaps for end, 47, 48; testing for, 132
Flux, magnetic: concentrations, 51-55, 74; definition of, 9; leakage and fringing, 51; minimizing leakage, 52, 54; testing for total, 132; total, equations for, 47, 48, 57
Flux measurement, absolute: fluxmeters, 132; gaussmeters, 134; magnetometers, 136; types of instruments, 128

Fluxmeters: circuit, 133; principles, 132
Focusing arrays (tubes), 223
Focusing magnets (electron beam), 91, 92
Force, control of, 84
Force, demagnetizing (-Hc): as related to hysteresis loop, 38
Force/distance curves, development of, 107
Force, magnetizing (+Hc): applied, 35; required to saturate, 38, 165, 166
Force, magnetomotive (F or mmf): definition of, 377; equations for, 48, 57
Force, mechanical: axial for couplings, 104; force distance curves, development of, 107; latching, 97; pull-in torque, 106; repulsion force testers, 139, 140; testing for traction or repulsion, 137; torque for three standard side pole rotors, 90; tractive force testers, 137, 138; unit conversions, 345
Foreign material removal from human body, 258
Foreign materials: all materials, 303; list of producers, 185
Fork truck: sweepers for, 243, 244; wood barrel handling attachment, 246
Form holders, concrete, 256
Forming method for various materials, 293. See also Manufacturing processes
Foundry flask clamping, 250
Fraction to decimal conversions, 351
French: list of producers, 185
Fringing (flux): methods for minimizing, 52, 54
Function of mills and distributor/fabricators, 8
Functions of a magnet, 1
Fundamentals of magnetism, 35

Gages. See Gauges
Galvanometers: typical PM core configurations, 72
Gap, air: definition of, 10, 39; flux density at various air gaps (on end) of bar magnets, 47, 48; variable, design calculations, 94
Gas welding, 260, 261
Gate, sorting and routing for cans, 251
Gauge, standard wire, 352
Gauges, coating thickness and heat treat temperature, 226
Gaussmeters, 134-136. See also Hall effect gaussmeters
General design considerations, 40, 171
Generators and alternators: designs, 222; typical rotor configurations, 72
German (West): list of producers, 185
Gilbert, Sir William, 4
Glossary of symbols and terms, 373
Gradient, magnetic: definition of and effects of, 63; examples of, 65-67

Grain-oriented Alnico, 23, 25
Grate separators, 244, 245
Gravity chute separators, 240-242, 244, 245, 246
Greenwald, F.S.: drum separator element, 238
Ground (surface), definition of, 9

Half-cycle type magnetizers, 155, 160
Hall effect gaussmeters: fixtures for performance tests, 142-147; principles, circuits, and uses, 134
Hard magnetic material, definition of, 9
Hardness conversions, Rockwell-Brinell-Rockwell, 353-355
Hardness (Rockwell) for various materials, 293
Hardyne: origin, 5; properties, 321
Hc. See Coersive force (normal)
Hci. See Intrinsic coersive force
Heart pacemaker control, 213
Heat demagnetization of ferrites, 167
Heat-energy-work: calculations, 101; unit conversions, 357
Heat-treat temperature gage, 226
Heat treatment: Hensler's alloys, 5; method for orienting Alnico magnets, 22, 25-29; QQ-M-60, 193
History of permanent magnets: brief, 4; further reading, 7; projection for future, 6
Hoff, R.C.: vibratory motor, 232
Holding magnets: designing for maximum holding effects, 63; examples (pull curves), 65-69; simple rectangular types with performance characteristics, 74-76; simple circular types with performance characteristics, 77, 78; small standard (die set) assemblies, 258
Holes and inserts, 171
Holland (Dutch): list of producers, 185
Honda, K., 5
Hopkinson permeameter, 129
Hopper separators: disk type, 244; grate type, 244, 245
Horseshoe shaped magnets: calculating pull vs. distance, 107; fixtures for heat treat orientation, 26, 29; fixtures for magnetizing, 153-156, 163; L/W to B/H ratio conversions, 46; pull curves for family of standard sizes, 62
Hump-type separators, 241-242
Hydrocephalic pump system, 260, 264
Hypersyn motors, 230
Hysteresis: couplings, 233, 234; motors, 104, 229; tension controls, 233
Hysteresis loop, major: analysis of, 37-39
Hysteresis loop, minor. See Minor hysteresis loop
Hysteresis loss: in couplings, 88; comparison tables, 21; materials classified by, 16, 21

Ignition systems, magneto, 227
Immersion system for floating containers, 248
Impact, effects on performance: general, 113; on various metallic materials, 113, 124. See also Shock
Imperfections, physical (crack, chips, burrs, voids, and inclusions), 171
Implantable pumps (biomedical), 260
Impulse type magnetizers, 151
Inclined conveyors, 247; for tote boxes, 254, 255
Indicators, 220
Induction: applied, 35; motors, 229, 230; saturation, 38
Inherent orientation characteristics of various materials, 283
Inserts and holes, 172
Inspection-sorting system, 248
Instrument applications, 220
Interchangeability/replacement guide, U.S. and foreign materials, 303
Interfloor vertical conveyor (for tote boxes), 253
International index of magnetic materials, 303
Intrinsic coersive force: for common materials, 283; comparison table of, 21; materials classified by, 16, 21; relationship of, to normal coersive, 127
Intrinsic demagnetization curve. See Demagnetization curve, intrinsic
Inverting system for cans, 248
Investment casting, 15
Iron filings tests, 137
Irreversible changes. See Non-remagnetizable changes, environmentally caused
Irreversible temperature stability: materials classified by, 17, 21. See also Environmental effects
Isolators, load (microwave), 223
Isotropic materials. See Unoriented magnets
Israelson, A.F.: domain theory, 35, 36

Japanese producers, 185
Joint reluctance, 49, 75, 80
Jonas, B., 5

Keeper magnetic: definition of, 10
Kirkhoff's laws applied to a magnetic circuit, 56, 80
Knight, Gowin, 4

Large magnets, definition of, 9
Large quantities, definition of, 9
Latching: force, 97; design, 97
L/D ratio to B/H ratio conversions for various shapes, 45, 46
Lead ferrites. See Ferrite materials
Leakage and fringing fluxes: factors, 50, 51, 53; graphic description, 51, 52; kinds, 51; methods for minimizing, 52-55

Leakage factors (coefficients): accurate calculation method, 51; coefficients, 50; in simplified design calculation method, 50
Length conversions, English-Metric-English, 345-347
Length of magnet required: to compensate for losses (reluctance) in pole pieces, 49, 68; for specific application, 49
Length, optimum (for maximum flux density) of bar magnets, 47, 48
Levitation, magnetic, 88, 93
Lights, magnetically attached, 263
Linear motors, 231
Liquid line separators, 247
Load isolators, 223
Load line. *See* B/H ratio
Locating magnet in circuit: effect on leakage, 53
Lock, door, 256
Lodestone, 4
Lodex: classification, 14; origin, 5; process, 29; properties, 289, 290, 299, 300. *See also* Elongated single domain materials or ESD materials
Loop, hysteresis. *See* Hysteresis loop
Losses: leakage and fringing flux, 49-51; in pole pieces (reluctance), 73. *See* Hysteresis loss; Stability, temperature
Loudspeakers: construction details, 224; interaction of fields, 96; magnetic circuits, 70
L/t to B/H ratio conversions for rectangular bar magnets, 45
L/W to B/H ratio conversions for horseshoe magnets, 46

Machinability: classification by, 17
Machining: Alnicos, 172; cost guide for, 174; ferrites, 173; general, 34
Magnetic: circuit calculations by computer methods, 56; circuits for loudspeakers, 70; circuits performance testing, 144; design methods, 40, 94; fields, external, calculations for, 96, 127; fundamentals, 35; gradient, 63; leakage fluxes, 49, 51; levitation, 85, 88, 93; measurement and testing, 128, 187; permanent magnet, 9; springs, 93; strength (force) control methods, 84; symbols, terms and definitions, 8-10, 373; units conversions, 345
  definitions: air gap, 10, 374; circuit, 9; field, 9, 376; flux, 9, 375; keeper, 10, 376; materials, hard, 9, 375; materials, soft, 8, 378; reluctance, 10, 377; saturation, 10, 378
  properties: classified by magnetic, 15, 19-21; classified by physical, 17; relative, 9, 21
Magnetic Materials Producers Association, 180; MMPA Standard No. 0100-72, 198

Magnetization: after assembly, choice of materials for, 150, 158, 161-163; classification by direction of, 17, 18; curves for various hard and soft materials, 79; curve, virgin, 36, 37; definition of, 10; patterns for rectangular and ring shapes with performance characteristics, 74-79; standard methods for, 18
Magnetizers: comparative costs, 166; definition of, 150; improving and determining capacity requirements, 160; requirements, 150; sources, 186; types, methods, and fixtures, 150
Magnetizing force (H): applied, 35; equation for, 48, 49; of magnetizers, 165; required to saturate, 38, 165
Magneto ignition system, 227
Magnetometers: principle, 136; types, 136, 145, 146
Magnetomotive force (F or mmf): definition of, 377; equation for, 48
Magnetron-type tubes, 221
Magnets: advantages and disadvantages of PM's and EM's, 183; designed for "holding power," 63; designed for "reaching power," 63; effect of location in circuit, 53; examples of pull curves, 65-69; ferrite configuration/dimension guide, 173, 174; first commercially available material, 5; fixtures for heat treat orientation of Alnicos, 26-29; machining, 172; magnetization, demagnetization, stabilization, and calibration, 150; magnets in series or parallel, 86; measurement and testing, 128; methods of controlling strength, 85-87; motor rotors, typical, 72; motor stators, typical, 71; plate type structures, 88, 89; reed switches and relays, 215; reference, 135, 143; simple rectangular and circular types with performance characteristics, 74-78; standard methods of magnetization, 18; tooling costs, production of special magnets, 27, 171
  classified by: application, 11-13; direction of magnetization, 17; industry, 11; manufacturing process, 15; products, 12, 13; shape, 13, 14
  definition of: assemblies, 9; basic, 9; calibration, 10; demagnetization, 10; magnetization, 10; oriented and unoriented, 9; quantity, 9 (*See also* QQ-M-60; MMPA Standard No. 0100-72); size, 9; special, 9; stabilization, 10; standard, 9; surface finish, 197
Manufacturing processes: Alnico, cast, 22-25; Alnico, sintered, 23, 30; classification, 15; die construction and operation wet pressed, sintered ferrites, 32, 171; fabricating of magnets from stock materials, 34, 172; ferrites, bonded, 28; ferrites, wet pressed and sintered, process chart,

27, 31; fixtures for heat treat orientation of Alnicos, 26-29; forming and finishing methods for various materials, 293; platinum cobalt, 34; process factors that affect properties, 22; rare earths process, 28, 29; standard magnetization methods, 18; wrought alloys (*See* Wrought materials)
Market sectors (suppliers): definitions of, 8
Material handling equipment, 234
Materials: chemical, 178; demagnetizing curves and work sheets for common materials, 269; demagnetizing, stabilizing, and calibration, 167; environmental effect on, 108; measurement and testing, 128; mechanical properties, 177; properties standards, nominal, 181; standards and specifications, 180; thermal properties, 108, 177; tolerances on normal properties, 181
  classified by: coersive force (normal), 16; commercial availability, 19; electrical properties, 18; hysteresis loss, 16; intrinsic coersive force, 16; irreversible temperature stability, 17; machinability, 17; magnetic properties, 15, 19-21; maximum energy product, 16; physical properties, 17; relative cost, 18; residual induction, 16; reversible temperature stability, 16; type, 13
Maximum energy product: for common materials, 283; comparison table, 21, 183; materials classified by, 16, 21; relationship to demagnetization curve, 39, 43; standard tolerances, 181
Measurement and testing: absolute flux measuring instruments types, 131; circuit performance testing, 136; correlation of tests, 146; costs of commercial test equipment, 147; fluxmeters, 132; gaussmeters, 134; magnet performance testing, 136; materials properties, 128; permeameters, 129, 139; physical (mechanical, 147, 179; (*See also* Stabilization; Calibration); production testing (for Q.C.), 140, 147; purposes, 128; standard sampling plan for statistical Q.C., 148, 149; unit conversions, 345
Mechanical considerations: encapsulation, 176; fastening, 176; imperfections, 171; inserts, holes, and configurations, 172; machining, 34, 172, 174; properties, 176, 293; size, shape, and orientation, 171; supplemental finishing, 175; tooling costs, 27, 171
Mechanical force. *See* Pull; Force; Repulsion force
Mechanical methods of controlling the strength of a magnet, 85-87
Mechanical properties, 17, 177: for common materials, 293; testing, 147

INDEX 957

Mechanical stress, effects on performance. *See* Stress, mechanical
Mechanical type repulsion force testers, 139, 140
Mechanical type tractive force testers, 138, 139
Metallic materials, classification, 13
Metallurgical: changes due to environmental effects, (*See* Environmental effects); nominal composition of various materials, 293
Meters: applications, 220; panel, D'Arsonvol type, 225; watt hour, 177
Methods: of actuating reed switches, 217-219; of design, 49, 94; of forming and finishing various materials, 293; magnetizing, demagnetizing, and calibrating, 150; standards for magnetization, 18; testing, 128
Microphones: designs, 224
Microwave: tubes, 91, (*See also* specific type); load isolators, 223
MIL-STD-IOA, surface: finish specifications, 181
Mills: industry structure and function, 8, 172; list of foreign mills, 185; list of U.S. mills, 184
Minimizing: fringing fluxes, 54, 55; leakage fluxes, 52, 54
Minor hysteresis loop: relation to hysteresis loop (major), 37-39
Mishma, T., 5
MKS to English or CGS unit conversions, 345
MMPA Standard No. 0100-72, 198
Molding, plastic (process), 15
Molding, shell or baked sand (process), 15
Moskowitz, L.R.: bases for vehicular flags, 262, 263; clamping methods for foundry flasks, 250; coating thickness gauge, 226; vehicular sign, 262; vibratory motor, 232
Motors: effect on varying air gap, 94; rotary, 228-230; stepping, 231; types, 226; typical rotor configurations, 72; typical stator configurations, 71; vibratory, 232
Moving magnet-moving coil magnetometers, 136
Multiple magnet circuits, 85
Multipolar magnetizing: fixtures, 153, 154, 156, 162; requirements for, 153

National Bureau of Standards polarity identification, 179
Neurological applications, 260
Nickel-iron alloys for temperature compensation, 178
Non-magnetic material, 8
Non-remagnetizable changes, environmentally caused: corrective action for, 109; definition of, 8; effects causing, 108. *See also* Environmental effects

Normal coercive force. *See* Coersive force (normal)
Nuclear radiation, effects on performance: general, 120; on various materials, 125

Oil spill cleanup, 256
Ophthalmological applications, 260
Optimum (optimization): length of bar magnets for maximum flux density, 47, 48; of orientation to minimize leakage flux, 54; of shape to minimize leakage flux, 54
Orientation: definition of, 9; of domains, theory, 36; fixtures for heat treatment of Alnico, 26-29; methods for cast Alnico, 23, 25-29; optimum to reduce leakage flux, 54; properties of various materials, 283
Oscillators (tubes), 221
Output, control of magnet, 84

P-6 alloy: origin, 5; process, 33; properties, 330
Pacemaker control, 213
Packaging, 182, 193
Painting/coating, 175
Painting racks, 257
Panel meters. *See* Meters
Part being affected by magnet: effect of saturation, 81, 82; effect of thickness on pull for sheet steel or plate, 81, 82; factors to consider, 77
Parts racks for painting, coating, plating, and degreasing of small parts and sheets, 257
Performance: definition of, 9; factors that affect (list), 40; multiple paint tests, 143, 144; single point tests, 142, 143; testing of magnets, 136
Permanent magnet materials, definition of, 9
Permanent magnet type magnetizers, 151
Permasyn motors, 230
Permeability, absolute: definition of, 377; equation for, 49; low permeability type testers, 129. *See also* Permeameters
Permeability curves for temperature conpensating materials, 178
Permeability, recoil, 103, 377
Permeameters: general, 128; kinds (circuits), 129, 130, 146
Permeance (P): definition of, 51, 377; equations, 55; methods for calculating, manual or computer, 55
Permeance coefficient. *See* B/H ratio
Personnel safety, 166
Phonograph pickups, 224
Physical imperfections (cracks, chips, burrs, voids, inclusions), 171
Physical properties: classification by, 17; testing, 147

Pickups, phone, 224
Pinch rolls for conveying non-magnetic materials, 89
Pipe handling rolls, 256
Pipeline traps, 64, 246
Plastic bonded materials, 14; Alnicos, 14; ferrites, 14. *See also* Composite materials; Bonded materials
Plastics machine hopper separators. *See* Hopper separators
Plate magnets: basic structures, 88, 89
Plate type separators, 241
Plating, 175: standard method, 181
Plating racks, 257
Platinum-cobalt: origin, 5; process, 34; properties, 111, 125
Polarity: identification standards, 179; reversal to determine magnetization effectiveness, 163; reversal in material as related to hysteresis loop, 38
Pole pieces: effect on leakage and fringing, 53; effect of shape and location on performance, 53, 61, 65-69; design, 68; losses in, 68; for magnetizers, 161; purposes of, 66; requirements for, 66
Pole spacing, effects, 61
Polishing, 175
Potential energy (stored and restored), 104
Potentiometer tests, 137
Power generation, 222
Power supplies for magnetizers, 161
Power unit conversions, 358
Precious metal alloys. *See* Platinum-cobalt; Silmanol
Pressing, wet and dry (process), 26, 27, 31, 32
Preventing sliding, 81
Printed circuit motors, 228
Printing industry, 256
Process charts: for cast Alnicos, 24; for sintered Alnicos, 30; for wet-pressed sintered ferrites, 31
Production: automatic magnetizers-calibrators, 163, 164; demagnetizers, 168; magnetizers, 150-165; stabilizers-sorters, 163, 164; testing, 147. *See also* Measurements and testing
Properties: chemical, 178, 201; definition of, 9; electrical, 178; mechanical, 177, 190, 201; thermal, 177; tolerances on normal materials properties, 181
Prosthetic devices, 258
Pull: axial in couplings, 104; effect of part thickness (steel sheet and plate), 81, 82; methods of controlling, 84; testing for, 139, 140; vs. distance calculations, 107

characteristics: of Alnico vs. ferrite horseshoes, 69; for family of standard horseshoe shapes, 62; of two grades of ferrites in assemblies, 68;

for rectangular magnets and assemblies, 65, 66, 68, 69, 74-77; of rectangular vs. circular assemblies, 68; for ring magnets and assemblies, 68
Pull-in torque, 104
Pulleys, 255
Pumps, 232, 235. *See also* Couplings
Purchasing, 182: cost comparisons, 182; sources of magnets, 184, 185; sources of magnetizers, demagnetizers, test equipment, 186, 187
Pyromagnetic motor, 230

Quality control: MIL-STD statistical sampling plan, 148, 149; production testing for, 147; standards, 180
Quantities: definition of, large and small, 9
Quick remove type ceiling, 84
QQ-M-60 federal specifications, 189

Racks for painting, coating, plating, and degreasing: small parts and sheets, 257
Radial pole magnetizing fixtures, 154, 156
Radial pole rotor magnets. *See* Rotors
Radiation, nuclear. *See* Nuclear radiation, effects on performance
Radus, R.: flux transfer principle, 87
Rails (conveyor), 247
Rankin, H.: clamping method for foundry flasks, 250
Rare earth magnet materials: cerium copper (process), 29; classification, 13; origin, 4, 6; samarium cobalt (process), 28; types, 13
Reaching magnets: designing for maximum reaching effects, 63, 107; examples, 65-69
Reciprocating motors, 231
Recoil: line, 96; permeability, 103; average values for various materials, 283
Rectangular bar magnets, 74-78. *See also* Bar/rod shaped magnets
Reducing leakage and fringing flux losses. *See* Leakage and fringing fluxes
Reed switches, principles, 211, 213
Reference magnets, 135, 143
Relative costs (of materials and equipment), 182. *See* Costs, relative
Relays: general, 211; reed, 212, 218, 219
Relays, stepping: typical rotor configurations, 72
Reluctance factors: exact determination, 68, 73, 75, 80; joint reluctance, 49; in simplified design calculation method, 49
Reluctance, magnetic; definition of, 10, 377; equation for, 73; exact determination, 73; factors for simplified design calculation method, 49; joint reluctance, 49, 73, 75, 80; losses in pole pieces, 73; motors, 226

Remagnetizable changes, environmentally caused: definition of, 108; effects that cause change, 108; summary of corrective action, 109. *See also* Environmental effects
Remalloy, properties, 111
Repulsion balance type tester, 140
Repulsion force: testing for, 137
Residual induction: for common materials, 283; comparison table, 21; materials classified by, 16, 21; standard tolerances, 181
Resistance, electrical, classified by, 18
Retentivity: definition of, 378; values for various materials, 283
Reversible changes, environmentally caused: definition of, 108; effects that cause change, 108; summary of corrective action, 109. *See also* Environmental effects
Reversible temperature stability: materials classified by, 16, 21; comparison table, 21. *See also* Environmental effects
Ring-shaped magnets: closed (no air gap), 39; L/D to B/H ratio conversions, 45; radial fixture for heat treat orientation of Alnicos, 26, 27; simple holding types with performance characteristics, 77, 78; standard methods for magnetization, 18
Road sweepers, 243
Rockwell hardness: conversions to brinell, 353; of various materials, 293
Rod/bar shaped magnets. *See* Bar/rod shaped magnets
Roller conveyors: slippage control of steel tote boxes, 254, 255; speed control of steel tote boxes on, 254
Rolling (process), 15
Rolls, magnetic: crushing used in textile fiber cording, 64; induced for roller conveyors, 255; internally powered, flat, 255, 256; "pinch" type for conveying non-magnetic materials, 89; pipe handling, 255, 256
Rotary grate separators, 245
Rotation, domain, 38
Rotors: coupling force for three standard side pole types, 90; fixtures for heat treating Alnico, 27, 28; for motors, tachometers, generators, and alternators, typical configurations, 72
Rotors, Herbert C., 55
Round bar magnets. *See* Bar/rod shaped magnets
Routing system for cans, 251
Rubber bonded materials, 14. *See* Bonded materials

Safety: contraceptive loops, 258; door lock, 256, of personnel handling magnets, 166; skiing, 256
Salient pole rotors. *See* Rotors

Samarium-cobalt: origin, 6; process, 28; properties, 291, 301. *See also* Rare earth materials
Sampling plan for statistical quality control, 148, 149, 192, 208
Sand casting (process), 15
Sanford-Bennett permeameter, 130; high H type, 129
Sanford-Winter M-H permeameter, 129
Saturation: effects of, 74; field required for various materials, 165, 283; levels for various hard and soft magnetic materials, 79, 165; magnetizing force required to saturate, 36, 165
Saturation (magnetic): definition of, 10, 378; field required for various materials, 165, 283; induction, 37
Search coil and galvanometer type fluxmeter, 133
Sectional variations, effect of, 172
Segments, motor stator. *See* Stator magnets
Self-cleaning separators, 236, 239, 240
Self-demagnetization, 39, 43, 97, 125
Semi-continuous DC (electromagnet) type magnetizers, 154
Separators, 234; continuous belt, 239; drum, 236-238; effect of changing air gap, 63; grate, 244, 245; hopper, 240, 244, 245; liquid line, 64, 246; plate, 88, 89, 241
Servo motors, 288
Shape: classification of magnets, 13, 14; of objects in air gap (effect), 77; optimum to reduce leakage flux, 54; to reduce costs, 171
Sheet steel floaters. *See* Fanners, sheet steel
Shell molding (process), 15
Shock absorbers for tote boxes, 254
Shock, effects on performance: general, 113; on various grades of Alnicos, 124
Shunt-type DC motors, 228
Shunting materials (soft magnetic), 178
Shunting methods (for control), 85
Shut-off methods for a magnet, 84-87
Side-pole rotor magnets. *See* Rotors
Signs, 262, 263
Silmanol: origin, 5; properties, 332
Simplified design method, 44
Single conductor: calculation of magnetizing force produced in, 153
Sintering: for Alnicos, 23, 30; general process, 15
Slider beds (belt conveyor), 247
Sliding-pull ratio: control of sliding, 83; types and principles, 81; typical S/P value, 83
Slippage control of: any magnet on steel surface, 81; steel tote boxes on conveyors, 254, 255
Sloan, C.D.: classifications of magnets by shape, 13, 14; drum separator element, 238

Slug-shaped magnets. *See* Bar/rod shaped magnets
Small parts racks, 257
Small quantities, definition of, 9
Snap action mechanism, 211
Soft magnetic materials: definition of, 8; magnetization curves (initial), 79; radiation effects, 125
Soldering, 176
Solenoid type low permeability tester, 129. *See also* Permeameters
Solenoids, 129, 156, 211
Sorting inspection system: for aerosol cans, 248; for magnets (production), 163; for steel ended fiber containers, 248
Spacing of poles (effects), 61
Special magnets, definition of, 9
Special purpose magnetizers, 158
Specifications, 180, 181: MMPA Standard No. 0100-72, 198; QQ-M-60 federal specifications, 189; standard method for electroplating Alnico magnets, 181
Speed control for steel tote boxes, 254
Speedometers, 232
Springs, magnetic, 93
Spurious contacts, effects on performance: on cast Alnico 5 horseshoes and rods, 126; general, 115
Squirrel cage motors, 230
Stability, temperature: irreversible, 17, 21; reversible, 16, 21. *See also* Environmental effects
Stabilization: calculation of effects, 96; choice of material when stabilization required, 170; definition of, 10; methods, 170
Stainless steel, magnetic effects, 80
Standard magnets, definition of, 9
Standard methods for: electroplating Alnico magnets, 181; magnetization, 18
Standards for material specifications, 180: MMPA Standard No. 0100-72, 198; QQ-M-60 federal specifications, 189
Standards for polarity: identification, 179
Standard wire gauges, 352
Statistical quality control: sampling plan (MIL-STD), 148, 149
Stator magnets: fixture for heat treat orientation of stator segments and arcs, 27, 29; for motors, typical configurations, 71
Steels: stainless, 80. *See* Chrome steels; Cobalt steels
Stepping motors, 231
Strength, control of magnet, 84–87
Stress, mechanical, effects on performance: on cunife, 124; general, 113
Strontium ferrites, 5. *See also* Ferrite materials

Structure, permanent magnet industry, 8
Supplemental finishing (plating, painting, etc.), 175
Surface finishes, 9; federal specifications standards QQ-M-60 and MIL-STD-10A, 180; supplemental (plating, painting, etc.), 175
Suspended separators, 239
Sweepers, road, yard, and floor, 243, 246
Switches, 211
Synchronization of coupling elements, 87, 88
Synchronous: couplings, 87, 88, 233, 234; motors, 229, 230

Table, hold-down for welding, 261
Tables: of all magnetic and physical properties, 283, 293, 303; of relative magnetic properties, 21; unit conversions, 345
Tachometers: typical rotor configurations, 72
Telemetry, 260
Temperature compensation: alloys (curves), 178; general, 177; in watt hour meters, 177
Temperature, critical for various materials, 111
Temperature, effect on performance: B/H ratio for ferrites, 127; changes with elevated temperatures in demagnetization curve, 113–119; critical temperatures for various materials, 111; general curves for Alnicos and ferrites, 112, 119; materials classified by degree of irreversible stability, 17, 21; materials classified by degree of reversible stability, 16, 21; various Alnicos and ferrites at reduced temperatures, 120–122
Temperature stability, comparison tables, 21
Temperature unit conversions, 349, 350
Temporary wall holding system, 83
Tensile strength of various materials, 293
Tension controls, 233
Terms and definitions: complete, 373; limited, 8–10
Test equipment, sources, 187
Testing, 192, 208. *See also* Measurement and testing
Textile machinery: crushing rolls for carding, 64; others, 256; tension control, 233
Thermal demagnetization of ferrites, 119; 121, 122
Thermal properties: coefficients of expansion, 293; general, 108
Thermostats, 211
Thickness gauges, 226
Time duration for magnetizing, 150

Time effects on performance: general, 110; remagnetizable chance for various materials, 110, 111
Timer, cam, 212
Toggle mechanism, 211
Tolerances: on materials properties, normal, 181
Tooling costs of magnets: production, 27, 171; sample, 171
Torque couplings. *See* Couplings
Torque, motor, 228
Torque, pull-in, 104
Torque unit conversions, 358
Total flux (ϕ): equations for, 39; tests, 132
Tote boxes, steel: couplings shock absorbers for, 254; slippage control on inclined conveyor, 254, 255; speed control for, 254; vertical conveyor for, 253
Traction, railroad car wheels, 250
Tractive force testing, 138, 139
Transducers, electromechanical: loudspeakers, microphones, and phono pickups, 70, 71, 224; motors and generators, 222; panel meters, 225; vibratory motors, 231, 232
Transportation equipment, 250
Transverse modulus of rupture of various materials, 293
Traps, pipeline, 64, 246
Traveling wave tube focusing magnets, 92
Tube magnets: internal structure, 63; performance characteristics, 63; special magnetizer for, 162
Tubes, electronic: focusing arrays, 91, 92, 223; klystrons, 222; magnetrons, 221; oscillators, 221
TWT focusing magnets, 92
Types of materials: classification, 13

U-shaped magnets. *See* Horseshoe-shaped magnets
Unit (of measurement) conversions, 345
Unit poles, 52
United States: list of distributor/fabricators, 185; list of mills, 184
Unoriented magnets, definition of, 9
Usage of magnets, world wide, 1
Users, industry structure and function of, 8

Variable gap calculations, use of demagnetized curve for, 94
Variations, sectional (physical), 172
Vehicle signs, flags, and lights, 261, 262
Velocity, angular unit conversions, 358
Vertical conveyor for steel tote boxes, 253
Vibration, effects on performance: general, 113; various metallic materials, 124
Vibrator, high frequency, 212

Vibratory motors, 226, 231, 232
Vicalloy: origin, 5; process, 33; properties, 111
Vinyl bonded materials. *See also* Bonded materials, 14
Virgin magnetization curves: magnetic materials, 79; principles, 35, 37; for various hard and soft
Volume: comparison of energy per unit for common materials, 19; unit conversions, 356

Wall holding system, movable, 83
Washing systems for cans, 248
Watt hour meters, 177
Weight: comparison of energy per unit for common materials, 20; unit conversions, 348; of various materials, 293
Welding: of magnets, 176; releasable table design, 261; techniques using magnets, 242, 259, 260
Went, J.J. *See* Ferrites, hard, origin
Wet pressing (process), 15

Weyant, C.: vibratory motor, 232
Window and door alarms, 212
Wire gauges, standard, 352
Work-energy-heat: calculations, 101; unit conversions, 357
Wrought materials: processes, 32-34; properties, 111, 119, 124

Yard sweepers, 243, 246
Yield from wet pressed-sintered ferrite process, 27

About the Author

Lester R. Moskowitz is the former Dean of the Technology Division at Spring Garden College in Philadelphia and currently President of his own consulting firm. Mr. Moskowitz holds a B.S. degree from Virginia Polytechnic Institute and State University and has done graduate work at the University of Pittsburgh, Carnegie-Mellon University and Temple University.

In the industrial field, the author has served as Chief Development Engineer for the Eriez Manufacturing Company and as President of Zeta-Northern Company all of Erie, Pennsylvania. He has also been Engineering Manager for the Metals Division of Westinghouse Electric Corporation (Blairsville, Pa.), Manager of Development Engineering for the Indiana General Corporation (Milwaukee, Wisc.), and Engineering Manager for Permag Corporation (Jamaica, N.Y.).

Mr. Moskowitz was also a special staff member in the Electrical, Mechanical and Industrial Engineering Departments at Gannon College, in Erie, Pennsylvania, and a technical instructor for the U.S. Army Signal Corps Officer Candidate School, in Fort Monmouth, New Jersey. He is an Executive Director of the Permanent Magnet Users Association and has written numerous articles for such publications as *Modern Materials Handling* and *Material Handling Engineering*.